Birkhäuser

Progress in Mathematics
Volume 322

More information about this series at http://www.springer.com/series/4848

Vladimir Turaev • Alexis Virelizier

Monoidal Categories and Topological Field Theory

Birkhäuser

Vladimir Turaev
Department of Mathematics
Indiana University
Bloomington, IN, USA

Alexis Virelizier
Laboratoire Paul Painlevé
Université de Lille
Villeneuve d'Ascq, France

ISSN 0743-1643 ISSN 2296-505X (electronic)
Progress in Mathematics
ISBN 978-3-319-49833-1 ISBN 978-3-319-49834-8 (eBook)
DOI 10.1007/978-3-319-49834-8

Library of Congress Control Number: 2017945221

Mathematics Subject Classification (2010): 18D10, 18C20, 57M27, 57R56

Printed on acid-free paper

This book is published under the trade name Birkhäuser, www.birkhauser-science.com
The registered company is Springer International Publishing AG
The registered company address is: Gewerbestrasse 11, 6330 Cham, Switzerland

Ferran Sunyer i Balaguer (1912–1967) was a self-taught Catalan mathematician who, in spite of a serious physical disability, was very active in research in classical mathematical analysis, an area in which he acquired international recognition. His heirs created the Fundació Ferran Sunyer i Balaguer inside the Institut d'Estudis Catalans to honor the memory of Ferran Sunyer i Balaguer and to promote mathematical research.

Each year, the Fundació Ferran Sunyer i Balaguer and the Institut d'Estudis Catalans award an international research prize for a mathematical monograph of expository nature. The prize-winning monographs are published in this series. Details about the prize and the Fundació Ferran Sunyer i Balaguer can be found at

`http://ffsb.espais.iec.cat/en`

This book has been awarded the Ferran Sunyer i Balaguer 2016 prize.

The members of the scientific commitee of the 2016 prize were:

Lluís Alsedà
Universitat Autònoma de Barcelona

Jiang-Hua Lu
The University of Hong Kong

Gábor Lugosi
Universitat Pompeu Fabra

Eero Saksman
University of Helsinki

Yuri Tschinkel
Courant Institute of Mathematical Sciences, New York University

Ferran Sunyer i Balaguer Prize winners since 2005:

2005 Antonio Ambrosetti and Andrea Malchiodi
Perturbation Methods and Semilinear Elliptic Problems on R^n, PM 240

José Seade
On the Topology of Isolated Singularities in Analytic Spaces, PM 241

2006 Xiaonan Ma and George Marinescu
Holomorphic Morse Inequalities and Bergman Kernels, PM 254

2007 Rosa Miró-Roig
Determinantal Ideals, PM 264

2008 Luis Barreira
Dimension and Recurrence in Hyperbolic Dynamics, PM 272

2009 Timothy D. Browning
Quantitative Arithmetic of Projective Varieties, PM 277

2010 Carlo Mantegazza
Lecture Notes on Mean Curvature Flow, PM 290

2011 Jayce Getz and Mark Goresky
Hilbert Modular Forms with Coefficients in Intersection Homology and Quadratic Base Change, PM 298

2012 Angel Cano, Juan Pablo Navarrete and José Seade
Complex Kleinian Groups, PM 303

2013 Xavier Tolsa
Analytic capacity, the Cauchy transform, and non-homogeneous Calderón–Zygmund theory, PM 307

2014 Veronique Fischer and Michael Ruzhansky
Quantization on Nilpotent Lie Groups, Open Access, PM 314

Contents

Introduction

The theory of monoidal categories is intimately related to topology. Morphisms in monoidal categories and operations on these morphisms may be represented by pictures, as was pointed out by Roger Penrose in 1971. This leads to a beautiful graphical calculus which allows to replace lengthy algebraic computations by elementary topological arguments. Monoidal categories also play a crucial role in quantum topology, an area of mathematics and theoretical physics founded by Vaughan Jones and Edward Witten in the 1980s. Quantum topology studies Topological Quantum Field Theories (TQFTs) and related invariants of topological objects such as knots, links, manifolds, homeomorphisms, etc. This monograph is concerned with connections between monoidal categories and 3-dimensional TQFTs.

Two fundamental constructions of 3-dimensional TQFTs were introduced in the late 1980s by the first named author jointly with Nikolai Reshetikhin and Oleg Viro. The Reshetikhin–Turaev construction uses surgery on 3-manifolds to derive a TQFT from a modular category. The Turaev–Viro construction uses state sums on skeletons of 3-manifolds to derive a TQFT from a spherical fusion category. In its original form, the latter construction involved representations of quantum groups; the categorical formulation was given by John Barrett and Bruce Westbury in 1994. In 1995, the first named author conjectured that the state sum TQFT derived from a spherical fusion category is isomorphic to the surgery TQFT derived from the Drinfeld–Joyal–Street center of that category. This provides, in particular, a surgery computation of the state sum invariants of closed oriented 3-manifolds.

The goal of the present monograph is to prove the above mentioned conjecture. The monograph consists of four parts. Part I is an introduction to monoidal categories. We start with basic definitions of the theory of categories and proceed to a discussion of monoidal categories and of important classes of monoidal categories: rigid, pivotal, spherical, fusion, braided, and modular categories. We also describe the Penrose graphical calculus and define the center of monoidal categories. Part I culminates with a statement of two fundamental theorems of Michael Müger on the center of pivotal fusion categories (Theorems 5.3 and 5.4).

Part II of the book is devoted to an algebraic description of the center of monoidal categories based on the theory of Hopf monads due to Alain Bruguières and the second named author. We start by reviewing algebras, bialgebras, and

Hopf algebras in braided categories. Then we discuss monads, bimonads, Hopf monads on monoidal categories and study their connections with the Drinfeld–Joyal–Street center. This leads us to an algebraic description of the center that we use to prove the aforementioned theorems of Müger.

In Part III we give a formal definition of a TQFT and construct the state sum TQFT associated with a spherical fusion category. In generalization of the previous constructions, we work with arbitrary (possibly, non-generic) skeletons of 3-manifolds. We first introduce geometric transformations relating any two skeletons of a 3-manifold. Then we associate certain tensors with finite graphs in the 2-sphere and use these tensors to define a state sum on a skeleton of a 3-manifold. This turns out to be a topological invariant of the manifold which leads us to the state sum TQFT.

In Part IV we extend the state sum TQFT to a so-called graph TQFT which applies to 3-manifolds with colored ribbon graphs. We show how to present ribbon graphs by diagrams on skeletons and introduce local transformations relating any diagrams of a ribbon graph. Then we define a graph TQFT using the state sum method. The main theorem of Part IV (Theorem 17.1) interprets this state sum graph TQFT as a surgery TQFT. The book ends with several appendices.

Parts III and IV (with exception of the proof of Theorem 16.2) are independent from Part II and may be read directly after Part I. The chapters devoted to skeletons and ribbon graphs (Chapters 11 and 14) are purely geometric and may be read independently from the rest of the book.

The reader will find a number of exercises in the book. In Appendix I at the end of the book we give hints and solutions to the exercises marked by the asterisk in the text.

Throughout the book, the symbol $\Bbbk$ denotes a commutative ring with $1_{\Bbbk} \neq 0$.

Part I

Monoidal Categories

Chapter 1

Monoidal categories and functors

The study of monoidal categories originated in the work of Jean Bénabou [Ben] and Saunders Mac Lane [ML1]. In this chapter, we review the basics of the theory of monoidal categories.

1.1 Categories and functors

1.1.1 Categories

A *category* $\mathcal{C}$ consists of the following data:

- a class $\mathrm{Ob}(\mathcal{C})$, whose elements are called *objects* of $\mathcal{C}$;
- for any $X, Y \in \mathrm{Ob}(\mathcal{C})$, a set $\mathrm{Hom}_{\mathcal{C}}(X, Y)$, whose elements are called *morphisms* from X to Y and represented by arrows $X \to Y$;
- for any $X, Y, Z \in \mathrm{Ob}(\mathcal{C})$, a map

  $$\circ\colon \mathrm{Hom}_{\mathcal{C}}(Y, Z) \times \mathrm{Hom}_{\mathcal{C}}(X, Y) \to \mathrm{Hom}_{\mathcal{C}}(X, Z)$$

  called *composition*. The image of a pair (g, f) under this map is denoted $g \circ f$ or just gf;
- for every $X \in \mathrm{Ob}(\mathcal{C})$, a morphism $\mathrm{id}_X \in \mathrm{Hom}_{\mathcal{C}}(X, X)$, called the *identity* of X.

It is required that the composition is associative and unitary in the following sense:

$$(h \circ g) \circ f = h \circ (g \circ f) \quad \text{and} \quad f \circ \mathrm{id}_X = f = \mathrm{id}_Y \circ f$$

for all morphisms $f\colon X \to Y$, $g\colon Y \to Z$, $h\colon Z \to T$ with $X, Y, Z, T \in \mathrm{Ob}(\mathcal{C})$.

Given a morphism $f: X \to Y$ in a category $\mathcal{C}$, the object X is called the *source* and the object Y the *target* of f. Two morphisms g, f in $\mathcal{C}$ are *composable* if the source of g coincides with the target of f. For $X \in \mathrm{Ob}(\mathcal{C})$, the set $\mathrm{Hom}_{\mathcal{C}}(X,X)$ is denoted by $\mathrm{End}_{\mathcal{C}}(X)$, and its elements are called *endomorphisms* of X. The set $\mathrm{End}_{\mathcal{C}}(X)$ is a monoid with product $gf = g \circ f$ for any $f, g \in \mathrm{End}_{\mathcal{C}}(X)$ and unit id_X.

A morphism $f: X \to Y$ in $\mathcal{C}$ is an *isomorphism* if there exists a morphism $g: Y \to X$ in $\mathcal{C}$ such that $gf = \mathrm{id}_X$ and $fg = \mathrm{id}_Y$. Such a g is uniquely determined by f, is called the *inverse* of f and denoted f^{-1}. Two objects X, Y of $\mathcal{C}$ are *isomorphic* if there exists an isomorphism $X \to Y$. Isomorphism of objects is an equivalence relation on $\mathrm{Ob}(\mathcal{C})$ denoted $\simeq$.

The *opposite* of a category $\mathcal{C}$ is the category $\mathcal{C}^{\mathrm{op}}$ defined by $\mathrm{Ob}(\mathcal{C}^{\mathrm{op}}) = \mathrm{Ob}(\mathcal{C})$ and $\mathrm{Hom}_{\mathcal{C}^{\mathrm{op}}}(X,Y) = \mathrm{Hom}_{\mathcal{C}}(Y,X)$ for all $X, Y \in \mathrm{Ob}(\mathcal{C})$ with composition $\circ^{\mathrm{op}}$ defined by $g \circ^{\mathrm{op}} f = fg$.

A *subcategory* of a category $\mathcal{C}$ is a category $\mathcal{D}$ such that every object of $\mathcal{D}$ is an object of $\mathcal{C}$, for any $X, Y \in \mathrm{Ob}(\mathcal{D})$, the set $\mathrm{Hom}_{\mathcal{D}}(X,Y)$ is a subset of $\mathrm{Hom}_{\mathcal{C}}(X,Y)$, the composition in $\mathcal{D}$ is the restriction of that in $\mathcal{C}$, and the identity morphisms in $\mathcal{D}$ are the same as in $\mathcal{C}$. A subcategory $\mathcal{D}$ of $\mathcal{C}$ is *full* if $\mathrm{Hom}_{\mathcal{D}}(X,Y) = \mathrm{Hom}_{\mathcal{C}}(X,Y)$ for all $X, Y \in \mathrm{Ob}(\mathcal{D})$.

1.1.2 Example

Sets and maps between them form a category denoted Set. Finite sets and maps between them form a full subcategory of Set.

1.1.3 Example

Left modules over the ring $\Bbbk$ and $\Bbbk$-linear homomorphisms (with the usual composition) form a category denoted $\mathrm{Mod}_{\Bbbk}$.

1.1.4 Example

Given a set G, we define a category $G_{\Bbbk}$ as follows. The objects of $G_{\Bbbk}$ are elements of G. By definition, $\mathrm{Hom}_{G_{\Bbbk}}(g,g) = \Bbbk$ for all $g \in G$ and $\mathrm{Hom}_{G_{\Bbbk}}(g,h) = \{0\} \subset \Bbbk$ for any distinct $g, h \in G$. The composition of morphisms in $G_{\Bbbk}$ is induced by multiplication in $\Bbbk$. The identity of an object $g \in G$ is $\mathrm{id}_g = 1_{\Bbbk}$.

1.1.5 Functors and natural transformations

Functors are morphisms of categories and natural transformations are morphisms of functors. More precisely, a *functor* $F: \mathcal{C} \to \mathcal{D}$ from a category $\mathcal{C}$ to a category $\mathcal{D}$ assigns to each object X of $\mathcal{C}$ an object $F(X)$ of $\mathcal{D}$ and to each morphism $f: X \to Y$ in $\mathcal{C}$ a morphism $F(f): F(X) \to F(Y)$ in $\mathcal{D}$ so that

$$F(gf) = F(g)F(f) \quad \text{and} \quad F(\mathrm{id}_X) = \mathrm{id}_{F(X)}$$

for all composable morphisms g, f in $\mathcal{C}$ and all $X \in \mathrm{Ob}(\mathcal{C})$. For example, the identity functor $1_\mathcal{C}\colon \mathcal{C} \to \mathcal{C}$ carries every object/morphism in $\mathcal{C}$ to itself. The composition of two functors $F\colon \mathcal{C} \to \mathcal{D}$ and $G\colon \mathcal{D} \to \mathcal{E}$ is defined in the obvious way and yields a functor $GF\colon \mathcal{C} \to \mathcal{E}$. Every functor $\mathcal{C} \to \mathcal{D}$ determines a functor $\mathcal{C}^{\mathrm{op}} \to \mathcal{D}^{\mathrm{op}}$ in the obvious way.

A *natural transformation* $F \to G$ between two functors $F, G\colon \mathcal{C} \to \mathcal{D}$ is a family

$$\varphi = \{\varphi_X\colon F(X) \to G(X)\}_{X \in \mathrm{Ob}(\mathcal{C})}$$

of morphisms in $\mathcal{D}$ such that

$$\varphi_Y F(f) = G(f)\varphi_X$$

for all morphisms $f\colon X \to Y$ in $\mathcal{C}$. A natural transformation $\varphi\colon F \to G$ is *invertible* if φ_X is an isomorphism for all $X \in \mathrm{Ob}(\mathcal{C})$. Then the family of morphisms

$$\{\varphi_X^{-1}\colon G(X) \to F(X)\}_{X \in \mathrm{Ob}(\mathcal{C})}$$

is a natural transformation $G \to F$ called the *inverse* of φ and denoted φ^{-1}. Invertible natural transformations of functors are called *natural isomorphisms*. Clearly, the inverse of a natural isomorphism is a natural isomorphism. Two functors $\mathcal{C} \to \mathcal{D}$ are *isomorphic* if there is a natural isomorphism between them.

1.1.6 Products of categories and functors

The *product* of two categories $\mathcal{C}$ and $\mathcal{C}'$ is the category $\mathcal{C} \times \mathcal{C}'$ defined by $\mathrm{Ob}(\mathcal{C} \times \mathcal{C}') = \mathrm{Ob}(\mathcal{C}) \times \mathrm{Ob}(\mathcal{C}')$ and

$$\mathrm{Hom}_{\mathcal{C} \times \mathcal{C}'}((X, X'), (Y, Y')) = \mathrm{Hom}_\mathcal{C}(X, Y) \times \mathrm{Hom}_\mathcal{C}(X', Y')$$

for all $X, Y \in \mathrm{Ob}(\mathcal{C})$, $X', Y' \in \mathrm{Ob}(\mathcal{C}')$ with composition $(g, g') \circ (f, f') = (gf, g'f')$ and identity morphisms $\mathrm{id}_{(X,X')} = (\mathrm{id}_X, \mathrm{id}_{X'})$.

The *product* of two functors $F\colon \mathcal{C} \to \mathcal{D}$ and $F'\colon \mathcal{C}' \to \mathcal{D}'$ is the functor

$$F \times F'\colon \mathcal{C} \times \mathcal{C}' \to \mathcal{D} \times \mathcal{D}'$$

carrying (X, X') to $(F(X), F'(X'))$ for any $X \in \mathrm{Ob}(\mathcal{C})$, $X' \in \mathrm{Ob}(\mathcal{C}')$ and carrying (f, f') to $(F(f), F'(f'))$ for any morphisms f in $\mathcal{C}$ and f' in $\mathcal{C}'$.

1.1.7 Isomorphisms and equivalences of categories

A functor $F\colon \mathcal{C} \to \mathcal{D}$ between categories $\mathcal{C}$ and $\mathcal{D}$ is an *isomorphism* if there is a functor $G\colon \mathcal{D} \to \mathcal{C}$ such that $GF = 1_\mathcal{C}$ and $FG = 1_\mathcal{D}$. Such a functor G is uniquely determined by F, is an isomorphism, and is called the *inverse* of F. Two categories are *isomorphic* if there is an isomorphism between them.

A *quasi-inverse* of a functor $F\colon \mathcal{C} \to \mathcal{D}$ is a functor $G\colon \mathcal{D} \to \mathcal{C}$ such that there are natural isomorphisms $GF \simeq 1_\mathcal{C}$ and $FG \simeq 1_\mathcal{D}$. A functor is an *equivalence* if it has a quasi-inverse. Note that any quasi-inverse of an equivalence is an equivalence and the composition of two composable equivalences is an equivalence. Two categories are *equivalent* if there is an equivalence between them. It is clear from the definitions that isomorphisms of categories are equivalences and isomorphic categories are equivalent.

Any equivalence of categories $F\colon \mathcal{C} \to \mathcal{D}$ is *essentially surjective* in the sense that each object of $\mathcal{D}$ is isomorphic to $F(X)$ for some $X \in \mathrm{Ob}(\mathcal{C})$ and *fully faithful* in the sense that for all $X, Y \in \mathrm{Ob}(\mathcal{C})$, the map

$$\mathrm{Hom}_{\mathcal{C}}(X,Y) \to \mathrm{Hom}_{\mathcal{D}}(F(X),F(Y)), \quad f \mapsto F(f)$$

is bijective. If one assumes the axiom of choice, then all essentially surjective and fully faithful functors are equivalences.

1.1.8 Exercise

Verify that the isomorphism and the equivalence of categories are equivalence relations (i.e., are reflexive, symmetric, and transitive).

1.1.9 Exercise*

Let $F\colon \mathcal{C} \to \mathcal{D}$ be an equivalence of categories and $G\colon \mathcal{D} \to \mathcal{C}$ be a quasi-inverse of F. Prove that for any natural isomorphism $\alpha\colon 1_{\mathcal{C}} \to GF$, there is a unique natural isomorphism $\beta\colon FG \to 1_{\mathcal{D}}$ such that for all $X \in \mathrm{Ob}(\mathcal{C})$ and $Y \in \mathrm{Ob}(\mathcal{D})$,

$$\beta_{F(X)}F(\alpha_X) = \mathrm{id}_{F(X)} \quad \text{and} \quad G(\beta_Y)\,\alpha_{G(Y)} = \mathrm{id}_{G(Y)}. \tag{1.1}$$

1.1.10 Exercise

Let $\mathcal{S}$ be a full subcategory of a category $\mathcal{C}$ such that each object of $\mathcal{C}$ is isomorphic to exactly one object of $\mathcal{S}$. Prove that $\mathcal{C}$ is equivalent to $\mathcal{S}$.

1.1.11 Exercise*

Let $f\colon X \to Y$ be a morphism in a category $\mathcal{C}$. For any $Z \in \mathrm{Ob}(\mathcal{C})$, consider the map

$$h_Z\colon \mathrm{Hom}_{\mathcal{C}}(Y,Z) \to \mathrm{Hom}_{\mathcal{C}}(X,Z), \quad \alpha \mapsto h_Z(\alpha) = \alpha f.$$

Prove that the following conditions are equivalent:

(i) f is an isomorphism;

(ii) h_Z is a bijection for all $Z \in \mathrm{Ob}(\mathcal{C})$;

(iii) h_X is surjective and h_Y is injective.

1.2 Generalities on monoidal categories

A monoid is a set with associative multiplication having a 2-sided unit. Categorifying the notion of a monoid one obtains the notion of a monoidal category. We review monoidal categories and discuss several fundamental examples.

1.2.1 Definition

A *monoidal category* is a category $\mathcal{C}$ endowed with

- a functor $\otimes\colon \mathcal{C} \times \mathcal{C} \to \mathcal{C}$, called the *monoidal product*;
- an object $\mathbb{1} \in \mathrm{Ob}(\mathcal{C})$, called the *unit object*;
- a family of isomorphisms

  $$a = \{a_{X,Y,Z}\colon (X \otimes Y) \otimes Z \to X \otimes (Y \otimes Z)\}_{X,Y,Z \in \mathrm{Ob}(\mathcal{C})},$$

  called the *associativity constraint*;
- a family of isomorphisms $l = \{l_X\colon \mathbb{1} \otimes X \to X\}_{X \in \mathrm{Ob}(\mathcal{C})}$, called the *left unitality constraint*;
- a family of isomorphisms $r = \{r_X\colon X \otimes \mathbb{1} \to X\}_{X \in \mathrm{Ob}(\mathcal{C})}$, called the *right unitality constraint*.

It is required that:

(i) for all objects X, Y, Z, W of $\mathcal{C}$, the following diagram commutes:

$$\begin{array}{ccccc} & & (X \otimes Y) \otimes (Z \otimes W) & & \\ & \nearrow^{a_{X\otimes Y,Z,W}} & & \searrow^{a_{X,Y,Z\otimes W}} & \\ ((X \otimes Y) \otimes Z) \otimes W & & & & X \otimes (Y \otimes (Z \otimes W)) \\ \downarrow{\scriptstyle a_{X,Y,Z}\otimes \mathrm{id}_W} & & & & \uparrow{\scriptstyle \mathrm{id}_X \otimes a_{Y,Z,W}} \\ (X \otimes (Y \otimes Z)) \otimes W & & \xrightarrow{\quad a_{X,Y\otimes Z,W} \quad} & & X \otimes ((Y \otimes Z) \otimes W), \end{array}$$

(ii) for all objects X, Y of $\mathcal{C}$, the following diagram commutes:

$$\begin{array}{ccccc} & & X \otimes Y & & \\ & \nearrow^{r_X \otimes \mathrm{id}_Y} & & \nwarrow^{\mathrm{id}_X \otimes l_Y} & \\ (X \otimes \mathbb{1}) \otimes Y & & \xrightarrow{\quad a_{X,\mathbb{1},Y} \quad} & & X \otimes (\mathbb{1} \otimes Y), \end{array}$$

(iii) for all morphisms $f\colon X \to X'$, $g\colon Y \to Y'$, $h\colon Z \to Z'$ in $\mathcal{C}$, the following diagram commutes:

$$\begin{array}{ccc} (X \otimes Y) \otimes Z & \xrightarrow{(f\otimes g)\otimes h} & (X' \otimes Y') \otimes Z' \\ {\scriptstyle a_{X,Y,Z}}\downarrow & & \downarrow{\scriptstyle a_{X',Y',Z'}} \\ X \otimes (Y \otimes Z) & \xrightarrow[f\otimes(g\otimes h)]{} & X' \otimes (Y' \otimes Z'), \end{array}$$

(iv) for every morphism $f\colon X \to X'$ in $\mathcal{C}$, the following diagram commutes:

$$\begin{array}{ccc} \mathbb{1} \otimes X & \xrightarrow{\mathrm{id}_{\mathbb{1}}\otimes f} & \mathbb{1} \otimes X' \\ {\scriptstyle l_X}\downarrow & & \downarrow{\scriptstyle l_{X'}} \\ X & \xrightarrow[f]{} & X', \end{array}$$

(v) for every morphism $f\colon X \to X'$ in $\mathcal{C}$, the following diagram commutes:

$$\begin{array}{ccc} X \otimes \mathbb{1} & \xrightarrow{f\otimes\mathrm{id}_{\mathbb{1}}} & X' \otimes \mathbb{1} \\ {\scriptstyle r_X}\downarrow & & \downarrow{\scriptstyle r_{X'}} \\ X & \xrightarrow[f]{} & X'. \end{array}$$

That $\otimes$ is a functor means that $\mathrm{id}_X \otimes \mathrm{id}_Y = \mathrm{id}_{X\otimes Y}$ for all $X, Y \in \mathrm{Ob}(\mathcal{C})$ and

$$(g \otimes g')(f \otimes f') = gf \otimes g'f' \tag{1.2}$$

for all pairs of composable morphisms g, f and g', f' in $\mathcal{C}$.

The commutativity condition (i) is called the *pentagon coherence* and (ii) the *triangle coherence*. The condition (iii) may be rephrased by saying that the associativity constraint $a = \{a_{X,Y,Z}\}_{X,Y,Z\in\mathrm{Ob}(\mathcal{C})}$ is a natural isomorphism from the functor $\otimes(\otimes \times 1_{\mathcal{C}})$ to the functor $\otimes(1_{\mathcal{C}} \times \otimes)$. The condition (iv) means that the left unitality constraint $l = \{l_X\}_{X\in\mathrm{Ob}(\mathcal{C})}$ is a natural isomorphism from the functor $\mathbb{1}\otimes ?\colon \mathcal{C} \to \mathcal{C}$ to the functor $1_{\mathcal{C}}\colon \mathcal{C} \to \mathcal{C}$. The condition (v) means that the right unitality constraint $r = \{r_X\}_{X\in\mathrm{Ob}(\mathcal{C})}$ is a natural isomorphism from $? \otimes \mathbb{1}\colon \mathcal{C} \to \mathcal{C}$ to $1_{\mathcal{C}}$. Here, the functors $\mathbb{1}\otimes ?$ et $? \otimes \mathbb{1}$ are defined by

$$\begin{aligned} (\mathbb{1}\otimes ?)(X) &= \mathbb{1} \otimes X, & \qquad (? \otimes \mathbb{1})(X) &= X \otimes \mathbb{1}, \\ (\mathbb{1}\otimes ?)(f) &= \mathrm{id}_{\mathbb{1}} \otimes f, & \qquad (? \otimes \mathbb{1})(f) &= f \otimes \mathrm{id}_{\mathbb{1}}, \end{aligned}$$

for any $X \in \mathrm{Ob}(\mathcal{C})$ and any morphism f in $\mathcal{C}$.

A monoidal category is said to be *strict* if for any objects X, Y, Z of this category, we have $(X \otimes Y) \otimes Z = X \otimes (Y \otimes Z)$ and $\mathbb{1} \otimes X = X = X \otimes \mathbb{1}$, and the associativity and unitality constraints are the identity morphisms.

1.2.2 Related notions

A *monoidal subcategory* of a monoidal category $\mathcal{C}$ is a subcategory $\mathcal{D}$ of $\mathcal{C}$ which contains the unit object $\mathbb{1}$ of $\mathcal{C}$ and is closed under the monoidal product $\otimes$ of $\mathcal{C}$ in the sense that for any $X, Y \in \mathrm{Ob}(\mathcal{D}) \subset \mathrm{Ob}(\mathcal{C})$, we have $X \otimes Y \in \mathrm{Ob}(\mathcal{D})$. Such a subcategory $\mathcal{D}$ is monoidal with unit object $\mathbb{1}$ and with monoidal product and monoidal constraints inherited from $\mathcal{C}$ by restriction.

Each monoidal category $\mathcal{C} = (\mathcal{C}, \otimes, \mathbb{1}, a, l, r)$ gives rise to three *opposite* monoidal categories:

$$\begin{aligned} \mathcal{C}^{\mathrm{op}} &= (\mathcal{C}^{\mathrm{op}}, \otimes, \mathbb{1}, a^{\mathrm{op}}, l^{\mathrm{op}}, r^{\mathrm{op}}), \\ \mathcal{C}^{\otimes\mathrm{op}} &= (\mathcal{C}, \otimes^{\mathrm{op}}, \mathbb{1}, a^{\otimes\mathrm{op}}, l^{\otimes\mathrm{op}}, r^{\otimes\mathrm{op}}), \\ \mathcal{C}^{\mathrm{rev}} &= (\mathcal{C}^{\mathrm{op}}, \otimes^{\mathrm{op}}, \mathbb{1}, a^{\mathrm{rev}}, l^{\mathrm{rev}}, r^{\mathrm{rev}}). \end{aligned}$$

Here, the functor $\otimes^{\mathrm{op}} \colon \mathcal{C} \times \mathcal{C} \to \mathcal{C}$ is the *opposite monoidal product* of $\mathcal{C}$ defined by $X \otimes^{\mathrm{op}} Y = Y \otimes X$ for all $X, Y \in \mathrm{Ob}(\mathcal{C})$ and similarly for morphisms. The above associativity and unitality constraints are given for all $X, Y, Z \in \mathrm{Ob}(\mathcal{C})$ by

$$\begin{aligned} &(a^{\mathrm{op}})_{X,Y,Z} = (a_{X,Y,Z})^{-1}, && (l^{\mathrm{op}})_X = (l_X)^{-1}, && (r^{\mathrm{op}})_X = (r_X)^{-1}, \\ &(a^{\otimes\mathrm{op}})_{X,Y,Z} = (a_{Z,Y,X})^{-1}, && (l^{\otimes\mathrm{op}})_X = r_X, && (r^{\otimes\mathrm{op}})_X = l_X, \\ &(a^{\mathrm{rev}})_{X,Y,Z} = a_{Z,Y,X}, && (l^{\mathrm{rev}})_X = (r_X)^{-1}, && (r^{\mathrm{rev}})_X = (l_X)^{-1}. \end{aligned}$$

The transformations $\mathcal{C} \mapsto \mathcal{C}^{\mathrm{op}}$, $\mathcal{C} \mapsto \mathcal{C}^{\otimes\mathrm{op}}$, and $\mathcal{C} \mapsto \mathcal{C}^{\mathrm{rev}}$ are involutive, commute with each other, and each of them is the composition of the other two. In particular, $\mathcal{C}^{\mathrm{rev}} = (\mathcal{C}^{\otimes\mathrm{op}})^{\mathrm{op}} = (\mathcal{C}^{\mathrm{op}})^{\otimes\mathrm{op}}$.

1.2.3 Example

The category of sets (see Example 1.1.2) is a monoidal category with monoidal product given by the cartesian product of sets, unit object given by any singleton, and the obvious associativity and unitality constraints.

1.2.4 Example

The category $\mathrm{Mod}_{\Bbbk}$ from Example 1.1.3 is a monoidal category with monoidal product given by the standard tensor product $\otimes_{\Bbbk}$ over $\Bbbk$, unit object $\Bbbk$, and the obvious associativity and unitality constraints.

1.2.5 Example

A *super* $\Bbbk$*-module* is a (left) $\Bbbk$-module X endowed with a decomposition as a direct sum of two $\Bbbk$-modules $X = X_0 \oplus X_1$. A $\Bbbk$-linear homomorphism between super $\Bbbk$-modules $f \colon X \to Y$ is *grading-preserving* if $f(X_0) \subset Y_0$ and $f(X_1) \subset Y_1$. Super $\Bbbk$-modules and grading-preserving $\Bbbk$-linear homomorphisms (with usual composition) form a category denoted $\mathrm{SMod}_{\Bbbk}$. The monoidal product of two super $\Bbbk$-modules X and Y is the super $\Bbbk$-module

$$X \otimes Y = (X \otimes Y)_0 \oplus (X \otimes Y)_1$$

where

$$(X \otimes Y)_0 = (X_0 \otimes_{\Bbbk} Y_0) \oplus (X_1 \otimes_{\Bbbk} Y_1),$$
$$(X \otimes Y)_1 = (X_0 \otimes_{\Bbbk} Y_1) \oplus (X_1 \otimes_{\Bbbk} Y_0).$$

The monoidal product of two grading-preserving $\Bbbk$-linear homomorphisms between super $\Bbbk$-modules is the usual tensor product over $\Bbbk$ of these homomorphisms. This monoidal product turns $\mathrm{SMod}_{\Bbbk}$ into a monoidal category with monoidal unit $\Bbbk = \Bbbk \oplus 0$ and the obvious associativity and unitality constraints.

1.2.6 Example

Given a monoid G, the category $G_{\Bbbk}$ from Example 1.1.4 is strict monoidal with monoidal product defined by $g \otimes h = gh$ and $k \otimes l = kl$ for all $g, h \in \mathrm{Ob}(G_{\Bbbk}) = G$ and all morphisms k, l in $G_{\Bbbk}$ (which are elements of $\Bbbk$). The unit object of $G_{\Bbbk}$ is the unit element $1 \in G$.

A more general construction starts with a 3-cocycle α of G with values in the group $\Bbbk^*$ of invertible elements of $\Bbbk$. By definition, α is a map $G \times G \times G \to \Bbbk^*$ such that for all $g_1, g_2, g_3, g_4 \in G$,

$$\alpha(g_1g_2, g_3, g_4)\, \alpha(g_1, g_2, g_3g_4) = \alpha(g_1, g_2, g_3)\, \alpha(g_1, g_2g_3, g_4)\, \alpha(g_2, g_3, g_4). \tag{1.3}$$

We define a monoidal category $G_{\Bbbk}^{\alpha}$ in the same way as $G_{\Bbbk}$ but use the associativity and unitality constraints

$$a^{\alpha}_{g,h,j} \colon (g \otimes h) \otimes j \to g \otimes (h \otimes j), \quad l^{\alpha}_{g} \colon 1 \otimes g \to g, \quad r^{\alpha}_{g} \colon g \otimes 1 \to g$$

given by

$$a^{\alpha}_{g,h,j} = \alpha(g, h, j) \in \Bbbk, \quad l^{\alpha}_{g} = \alpha(1, 1, g)^{-1} \in \Bbbk, \quad \text{and} \quad r^{\alpha}_{g} = \alpha(g, 1, 1) \in \Bbbk$$

for any $g, h, j \in G$. For the trivial cocycle $\alpha = 1$, we have $G_{\Bbbk}^{\alpha} = G_{\Bbbk}$.

1.2.7 Exercise

Prove that the monoidal unit of a monoidal category is unique up to isomorphism. More precisely, let $\mathcal{C}$ be a category, $\otimes \colon \mathcal{C} \times \mathcal{C} \to \mathcal{C}$ be a functor, and $a = \{a_{X,Y,Z} \colon (X \otimes Y) \otimes Z \to X \otimes (Y \otimes Z)\}_{X,Y,Z \in \mathrm{Ob}(\mathcal{C})}$ be a natural isomorphism satisfying the pentagon coherence. Consider triples $(\mathbb{1}, l, r)$, where $\mathbb{1} \in \mathrm{Ob}(\mathcal{C})$,

$$l = \{l_X \colon \mathbb{1} \otimes X \to X\}_{X \in \mathrm{Ob}(\mathcal{C})} \quad \text{and} \quad r = \{r_X \colon X \otimes \mathbb{1} \to X\}_{X \in \mathrm{Ob}(\mathcal{C})}$$

are natural isomorphisms, such that $(\mathcal{C}, \otimes, \mathbb{1}, a, l, r)$ is a monoidal category. Prove that for any two such triples $(\mathbb{1}, l, r)$ and $(\mathbb{1}', l', r')$, there is a unique isomorphism $\phi \colon \mathbb{1}' \to \mathbb{1}$ such that $l'_X = l_X(\phi \otimes \mathrm{id}_X)$ and $r'_X = r_X(\mathrm{id}_X \otimes \phi)$ for all $X \in \mathrm{Ob}(\mathcal{C})$.

1.3 The ground monoid and coherence

We introduce the ground monoid of a monoidal category and define pure monoidal categories. Then we briefly discuss Mac Lane's coherence theorem and formulate our conventions concerning the associativity and unitality constraints.

1.3.1 The ground monoid

A monoidal category $\mathcal{C} = (\mathcal{C}, \otimes, \mathbb{1}, a, l, r)$ determines a monoid $\mathrm{End}_{\mathcal{C}}(\mathbb{1})$, called the *ground monoid* of $\mathcal{C}$. Its product is the composition of morphisms and its unit is $\mathrm{id}_{\mathbb{1}}$.

Lemma 1.1. *The monoid* $\mathrm{End}_{\mathcal{C}}(\mathbb{1})$ *is commutative.*

Proof. We first prove that for all $X, Y \in \mathrm{Ob}(\mathcal{C})$,

$$l_{X \otimes Y} = (l_X \otimes \mathrm{id}_Y)\, a^{-1}_{\mathbb{1},X,Y}. \tag{1.4}$$

Indeed, consider the following diagram:

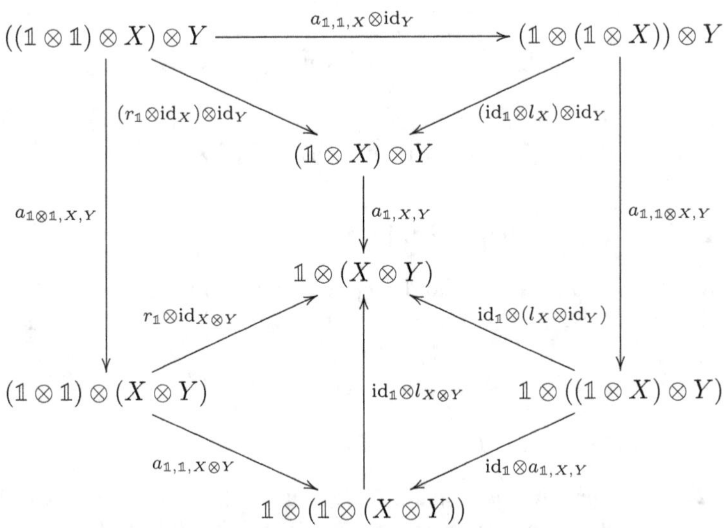

The pentagon coherence implies that the outside pentagon commutes. The naturality of a implies that the two quadrangles commute. The triangle coherence implies that the upper triangle and the lower left triangle commute. Consequently, the lower right triangle commutes. Thus,

$$\mathrm{id}_{\mathbb{1}} \otimes (l_{X \otimes Y} a_{\mathbb{1},X,Y}) = \mathrm{id}_{\mathbb{1}} \otimes (l_X \otimes \mathrm{id}_Y).$$

Note that the invertibility and naturality of l imply that two morphisms f, g in $\mathcal{C}$ with the same source and the same target are equal if and only if $\mathrm{id}_{\mathbb{1}} \otimes f = \mathrm{id}_{\mathbb{1}} \otimes g$. Therefore $l_{X \otimes Y} a_{\mathbb{1},X,Y} = l_X \otimes \mathrm{id}_Y$, which implies (1.4).

We prove next that for any $Y \in \mathrm{Ob}(\mathcal{C})$,

$$\mathrm{id}_{\mathbb{1}} \otimes l_Y = l_{\mathbb{1}\otimes Y}. \tag{1.5}$$

The naturality of l says that $l_Y(\mathrm{id}_{\mathbb{1}} \otimes f) = f l_X$ for any morphism $f\colon X \to Y$ in $\mathcal{C}$. In particular, $l_Y(\mathrm{id}_{\mathbb{1}} \otimes l_Y) = l_Y\, l_{\mathbb{1}\otimes Y}$. Since l_Y is an isomorphism, we deduce (1.5).

We show now that

$$l_{\mathbb{1}} = r_{\mathbb{1}} \colon \mathbb{1} \otimes \mathbb{1} \to \mathbb{1}. \tag{1.6}$$

We have:

$$l_{\mathbb{1}} \otimes \mathrm{id}_{\mathbb{1}} \overset{\text{(i)}}{=} l_{\mathbb{1}\otimes\mathbb{1}} a_{\mathbb{1},\mathbb{1},\mathbb{1}} \overset{\text{(ii)}}{=} (\mathrm{id}_{\mathbb{1}} \otimes l_{\mathbb{1}}) a_{\mathbb{1},\mathbb{1},\mathbb{1}} \overset{\text{(iii)}}{=} r_{\mathbb{1}} \otimes \mathrm{id}_{\mathbb{1}}.$$

Here (i) follows from (1.4), (ii) from (1.5), and (iii) from the triangle coherence. The equality $l_{\mathbb{1}} \otimes \mathrm{id}_{\mathbb{1}} = r_{\mathbb{1}} \otimes \mathrm{id}_{\mathbb{1}}$ implies (1.6) using the invertibility and naturality of r.

Finally, we prove the commutativity of $\mathrm{End}_{\mathcal{C}}(\mathbb{1})$. For any $\alpha, \beta \in \mathrm{End}_{\mathcal{C}}(\mathbb{1})$,

$$\begin{aligned}\alpha\beta &\overset{\text{(i)}}{=} r_{\mathbb{1}}(\alpha \otimes \mathrm{id}_{\mathbb{1}}) r_{\mathbb{1}}^{-1} l_{\mathbb{1}} (\mathrm{id}_{\mathbb{1}} \otimes \beta) l_{\mathbb{1}}^{-1} \overset{\text{(ii)}}{=} r_{\mathbb{1}}(\alpha \otimes \mathrm{id}_{\mathbb{1}})(\mathrm{id}_{\mathbb{1}} \otimes \beta) l_{\mathbb{1}}^{-1} \overset{\text{(iii)}}{=} r_{\mathbb{1}}(\alpha \otimes \beta) l_{\mathbb{1}}^{-1}\\ &\overset{\text{(iv)}}{=} r_{\mathbb{1}}(\mathrm{id}_{\mathbb{1}} \otimes \beta)(\alpha \otimes \mathrm{id}_{\mathbb{1}}) l_{\mathbb{1}}^{-1} \overset{\text{(v)}}{=} l_{\mathbb{1}}(\mathrm{id}_{\mathbb{1}} \otimes \beta) l_{\mathbb{1}}^{-1} r_{\mathbb{1}}(\alpha \otimes \mathrm{id}_{\mathbb{1}}) r_{\mathbb{1}}^{-1} \overset{\text{(vi)}}{=} \beta\alpha.\end{aligned}$$

Here, (i) and (vi) follow from the invertibility and naturality of l and r, (ii) and (v) from (1.6), and (iii) and (iv) from the functoriality of $\otimes$. □

1.3.2 Actions of the ground monoid

Let $\mathcal{C} = (\mathcal{C}, \otimes, \mathbb{1}, a, l, r)$ be a monoidal category. For any $X, Y \in \mathrm{Ob}(\mathcal{C})$, the set $\mathrm{Hom}_{\mathcal{C}}(X,Y)$ carries left and right actions of the monoid $\mathrm{End}_{\mathcal{C}}(\mathbb{1})$ defined by

$$\alpha \cdot f = l_Y(\alpha \otimes f)\, l_X^{-1} \quad \text{and} \quad f \cdot \alpha = r_Y(f \otimes \alpha)\, r_X^{-1}$$

for any $\alpha \in \mathrm{End}_{\mathcal{C}}(\mathbb{1})$ and $f \in \mathrm{Hom}_{\mathcal{C}}(X,Y)$.

Lemma 1.2. *The formulas above define left and right monoid actions of* $\mathrm{End}_{\mathcal{C}}(\mathbb{1})$ *on* $\mathrm{Hom}_{\mathcal{C}}(X,Y)$ *which commute with each other. The left and right actions of* $\mathrm{End}_{\mathcal{C}}(\mathbb{1})$ *on itself are given by the monoid product in* $\mathrm{End}_{\mathcal{C}}(\mathbb{1})$.

Proof. For $f \in \mathrm{Hom}_{\mathcal{C}}(X,Y)$, the naturality of l implies that

$$\mathrm{id}_{\mathbb{1}} \cdot f = l_Y(\mathrm{id}_{\mathbb{1}} \otimes f) l_X^{-1} = l_Y l_Y^{-1} f = f.$$

Let us check that $\alpha\cdot(\beta\cdot f) = (\alpha\beta)\cdot f$ for any $\alpha, \beta \in \mathrm{End}_{\mathcal{C}}(\mathbb{1})$. The last computation in the proof of Lemma 1.1 gives $\alpha\beta = r_{\mathbb{1}}(\alpha \otimes \beta) l_{\mathbb{1}}^{-1}$. Using (1.6), we deduce that

$$l_{\mathbb{1}}(\alpha \otimes \beta) l_{\mathbb{1}}^{-1} = \alpha\beta = r_{\mathbb{1}}(\alpha \otimes \beta) r_{\mathbb{1}}^{-1}. \tag{1.7}$$

Therefore,

$$\begin{aligned}\alpha\cdot(\beta\cdot f) &\overset{\text{(i)}}{=} l_Y(\alpha\otimes(l_Y(\beta\otimes f)l_X^{-1}))l_X^{-1}\\ &\overset{\text{(ii)}}{=} l_Y(\mathrm{id}_{\mathbb{1}}\otimes l_Y)(\alpha\otimes(\beta\otimes f))(\mathrm{id}_{\mathbb{1}}\otimes l_X^{-1})l_X^{-1}\\ &\overset{\text{(iii)}}{=} l_Y l_{\mathbb{1}\otimes Y}(\alpha\otimes(\beta\otimes f))l_{\mathbb{1}\otimes X}^{-1}l_X^{-1}\\ &\overset{\text{(iv)}}{=} l_Y\, l_{\mathbb{1}\otimes Y}\, a_{\mathbb{1},\mathbb{1},Y}\,((\alpha\otimes\beta)\otimes f)\, a_{\mathbb{1},\mathbb{1},X}^{-1}\, l_{\mathbb{1}\otimes X}^{-1}\, l_X^{-1}\\ &\overset{\text{(v)}}{=} l_Y(l_{\mathbb{1}}(\alpha\otimes\beta)l_{\mathbb{1}}^{-1}\otimes f)l_X^{-1} \overset{\text{(vi)}}{=} l_Y((\alpha\beta)\otimes f)l_X^{-1} \overset{\text{(vii)}}{=} (\alpha\beta)\cdot f.\end{aligned}$$

Here, (i) and (vii) follow from the definitions, (ii) from the functoriality of $\otimes$, (iii) from (1.5), (iv) from the invertibility and naturality of a, (v) from (1.4), and (vi) from (1.7). Thus the left action of $\mathrm{End}_{\mathcal{C}}(\mathbb{1})$ on $\mathrm{Hom}_{\mathcal{C}}(X,Y)$ defined above is a monoid action. One proves similarly that the right action of $\mathrm{End}_{\mathcal{C}}(\mathbb{1})$ on $\mathrm{Hom}_{\mathcal{C}}(X,Y)$ defined above is a monoid action. These actions commute because

$$\begin{aligned}\alpha\cdot(f\cdot\beta) &\overset{\text{(i)}}{=} l_Y(\alpha\otimes(r_Y(f\otimes\beta)r_X^{-1}))l_X^{-1}\\ &\overset{\text{(ii)}}{=} r_Y l_{Y\otimes\mathbb{1}}(\alpha\otimes(f\otimes\beta))l_{X\otimes\mathbb{1}}^{-1}r_X^{-1}\\ &\overset{\text{(iii)}}{=} r_Y l_{Y\otimes\mathbb{1}}a_{\mathbb{1},Y,\mathbb{1}}((\alpha\otimes f)\otimes\beta)a_{\mathbb{1},X,\mathbb{1}}^{-1}l_{X\otimes\mathbb{1}}^{-1}r_X^{-1}\\ &\overset{\text{(iv)}}{=} r_Y(l_Y(\alpha\otimes f)l_X^{-1}\otimes\beta)r_X^{-1} \overset{\text{(v)}}{=} (\alpha\cdot f)\cdot\beta.\end{aligned}$$

Here, (i) and (v) follow from the definitions, (ii) from the functoriality of $\otimes$ and the naturality of l (which yields that $l_Y(\mathrm{id}_{\mathbb{1}}\otimes r_Y)=r_Y l_{Y\otimes\mathbb{1}}$), (iii) from the invertibility and naturality of a, and (iv) from (1.4). Finally, Formula (1.7) shows that both left and right actions of $\mathrm{End}_{\mathcal{C}}(\mathbb{1})$ on itself are given by the monoid product in $\mathrm{End}_{\mathcal{C}}(\mathbb{1})$. □

Lemma 1.3. *The actions of* $\mathrm{End}_{\mathcal{C}}(\mathbb{1})$ *on the sets of morphisms are compatible with composition of morphisms in the following sense: for any* $\alpha\in\mathrm{End}_{\mathcal{C}}(\mathbb{1})$ *and any composable morphisms* g,f *in* $\mathcal{C}$, *we have*

$$\alpha\cdot(g\circ f)=(\alpha\cdot g)\circ f=g\circ(\alpha\cdot f)$$

and

$$(g\circ f)\cdot\alpha=(g\cdot\alpha)\circ f=g\circ(f\cdot\alpha).$$

Proof. Let $f\colon X\to Y$ and $g\colon Y\to Z$ be morphisms in $\mathcal{C}$. The functoriality of $\otimes$ and the naturality of l imply that for any $\alpha\in\mathrm{End}_{\mathcal{C}}(\mathbb{1})$,

$$\begin{aligned}\alpha\cdot(g\circ f)&=l_Z(\alpha\otimes gf)l_X^{-1}=l_Z(\alpha\otimes g)(\mathrm{id}_{\mathbb{1}}\otimes f)l_X^{-1}\\ &=l_Z(\alpha\otimes g)l_Y^{-1}f=(\alpha\cdot g)\circ f.\end{aligned}$$

Likewise $\alpha\cdot(g\circ f)=g\circ(\alpha\cdot f)$. The compatibility of the right action with composition is proved similarly using the naturality of r. □

Lemma 1.4. *The actions of* $\mathrm{End}_{\mathcal{C}}(\mathbb{1})$ *on the sets of morphisms are compatible with monoidal product of morphisms in the following sense: for any* $\alpha \in \mathrm{End}_{\mathcal{C}}(\mathbb{1})$ *and any morphisms* f, g *in* $\mathcal{C}$, *we have*

$$\alpha \cdot (f \otimes g) = (\alpha \cdot f) \otimes g \quad \textit{and} \quad (f \otimes g) \cdot \alpha = f \otimes (g \cdot \alpha).$$

Proof. We prove the first equality for morphisms $f\colon X \to Y$ and $g\colon A \to B$:

$$\begin{aligned}
\alpha \cdot (f \otimes g) &\overset{\text{(i)}}{=} l_{Y \otimes B}\, (\alpha \otimes (f \otimes g))\, l^{-1}_{X \otimes A} \\
&\overset{\text{(ii)}}{=} (l_Y \otimes \mathrm{id}_B)\, a^{-1}_{\mathbb{1},Y,B}\, (\alpha \otimes (f \otimes g))\, a_{\mathbb{1},X,A}\, (l_X \otimes \mathrm{id}_A)^{-1} \\
&\overset{\text{(iii)}}{=} (l_Y\, (\alpha \otimes f)\, l_X^{-1}) \otimes g \overset{\text{(iv)}}{=} (\alpha \cdot f) \otimes g.
\end{aligned}$$

Here, (i) and (iv) follow from the definitions, (ii) from (1.4), and (iii) from the naturality and invertibility of a. The second equality is proved similarly. □

1.3.3 Pure categories

A monoidal category $\mathcal{C}$ is *pure* if the left and right actions of $\mathrm{End}_{\mathcal{C}}(\mathbb{1})$ on the sets of morphisms in $\mathcal{C}$ coincide. Thus, $\mathcal{C}$ is pure if $\alpha \cdot f = f \cdot \alpha$ for any $\alpha \in \mathrm{End}_{\mathcal{C}}(\mathbb{1})$ and any morphism f in $\mathcal{C}$. In fact, it suffices to require that $\alpha \cdot \mathrm{id}_X = \mathrm{id}_X \cdot \alpha$ for any $\alpha \in \mathrm{End}_{\mathcal{C}}(\mathbb{1})$ and any $X \in \mathrm{Ob}(\mathcal{C})$. Indeed, this condition implies that for any morphism $f\colon X \to Y$ in $\mathcal{C}$, we have

$$\alpha \cdot f = \alpha \cdot (f \circ \mathrm{id}_X) = f \circ (\alpha \cdot \mathrm{id}_X) = f \circ (\mathrm{id}_X \cdot \alpha) = (f \circ \mathrm{id}_X) \cdot \alpha = f \cdot \alpha$$

where the second and fourth equalities follow from Lemma 1.3. For a pure monoidal category $\mathcal{C}$, Lemma 1.4 implies the identities

$$\alpha \cdot (f \otimes g) = (\alpha \cdot f) \otimes g = f \otimes (\alpha \cdot g)$$

for all $\alpha \in \mathrm{End}_{\mathcal{C}}(\mathbb{1})$ and all morphisms f, g in $\mathcal{C}$.

1.3.4 Mac Lane's coherence

In the proof of Lemma 1.1, we showed that for any objects X, Y of a monoidal category $\mathcal{C} = (\mathcal{C}, \otimes, \mathbb{1}, a, l, r)$, the diagram

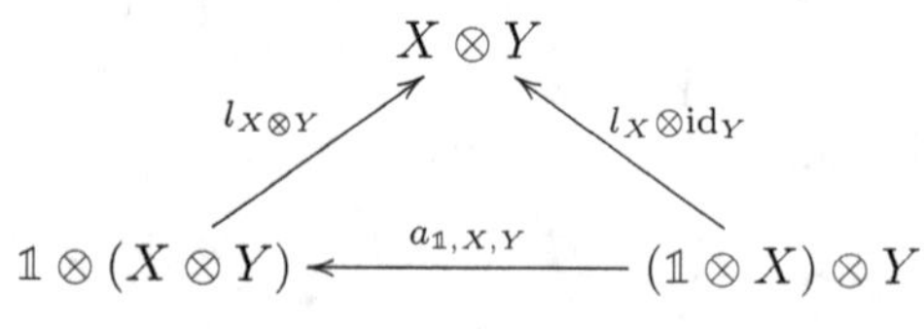

commutes. More generally, Mac Lane's coherence theorem asserts that every diagram in a monoidal category made up of the associativity and unitality constraints commutes, see [ML1, ML2].

1.4 Monoidal functors and natural transformations

Categorifying the notion of a monoid homomorphism between monoids one obtains a notion of a monoidal functor between monoidal categories. We review monoidal functors and discuss important classes of such functors.

1.4.1 Monoidal functors

Let $\mathcal{C} = (\mathcal{C}, \otimes, \mathbb{1}, a, l, r)$ and $\mathcal{D} = (\mathcal{D}, \otimes', \mathbb{1}', a', l', r')$ be monoidal categories. A *monoidal functor* from $\mathcal{C}$ to $\mathcal{D}$ is a functor $F\colon \mathcal{C} \to \mathcal{D}$ endowed with a morphism $F_0\colon \mathbb{1}' \to F(\mathbb{1})$ in $\mathcal{D}$ and with a natural transformation

$$F_2 = \{F_2(X,Y)\colon F(X) \otimes' F(Y) \to F(X \otimes Y)\}_{X,Y \in \mathrm{Ob}(\mathcal{C})}$$

between the functors $F \otimes' F = \otimes'(F \times F)\colon \mathcal{C} \times \mathcal{C} \to \mathcal{D}$ and $F\otimes\colon \mathcal{C} \times \mathcal{C} \to \mathcal{D}$ such that for all $X, Y, Z \in \mathrm{Ob}(\mathcal{C})$, the following three diagrams commute:

$$\begin{CD}
(F(X) \otimes' F(Y)) \otimes' F(Z) @>{a'_{F(X),F(Y),F(Z)}}>> F(X) \otimes' (F(Y) \otimes' F(Z))\\
@V{F_2(X,Y) \otimes' \mathrm{id}_{F(Z)}}VV @VV{\mathrm{id}_{F(X)} \otimes' F_2(Y,Z)}V\\
F(X \otimes Y) \otimes' F(Z) @. F(X) \otimes' F(Y \otimes Z)\\
@V{F_2(X \otimes Y, Z)}VV @VV{F_2(X, Y \otimes Z)}V\\
F((X \otimes Y) \otimes Z) @>>{F(a_{X,Y,Z})}> F(X \otimes (Y \otimes Z)),
\end{CD}$$

$$\begin{CD}
\mathbb{1}' \otimes' F(X) @>{l'_{F(X)}}>> F(X)\\
@V{F_0 \otimes \mathrm{id}_{F(X)}}VV @AA{F(l_X)}A\\
F(\mathbb{1}) \otimes' F(X) @>>{F_2(\mathbb{1},X)}> F(\mathbb{1} \otimes X),
\end{CD}$$

$$\begin{CD}
F(X) \otimes' \mathbb{1}' @>{r'_{F(X)}}>> F(X)\\
@V{\mathrm{id}_{F(X)} \otimes F_0}VV @AA{F(r_X)}A\\
F(X) \otimes' F(\mathbb{1}) @>>{F_2(X,\mathbb{1})}> F(X \otimes \mathbb{1}).
\end{CD}$$

The morphisms F_0 and F_2 are called the *monoidal constraints* associated with F. Recall that the naturality of F_2 means that for arbitrary morphisms $f\colon X \to X'$

and $g\colon Y \to Y'$ in $\mathcal{C}$, the following diagram commutes:

$$\begin{array}{ccc} F(X) \otimes' F(Y) & \xrightarrow{F_2(X,Y)} & F(X \otimes Y) \\ {\scriptstyle F(f)\otimes F(g)}\downarrow & & \downarrow{\scriptstyle F(f\otimes g)} \\ F(X') \otimes' F(Y') & \xrightarrow[F_2(X',Y')]{} & F(X' \otimes Y'). \end{array}$$

For example, the identity functor $1_{\mathcal{C}} : \mathcal{C} \to \mathcal{C}$ endowed with the monoidal constraints $(1_{\mathcal{C}})_0 = \mathrm{id}_{\mathbb{1}}\colon \mathbb{1} \to \mathbb{1}$ and $(1_{\mathcal{C}})_2(X,Y) = \mathrm{id}_{X\otimes Y}\colon X \otimes Y \to X \otimes Y$ for all $X, Y \in \mathrm{Ob}(\mathcal{C})$ is a monoidal functor. It is again denoted $1_{\mathcal{C}}$.

The *composition* of two monoidal functors $F\colon \mathcal{C} \to \mathcal{D}$ and $G\colon \mathcal{D} \to \mathcal{E}$ is the monoidal functor $GF\colon \mathcal{C} \to \mathcal{E}$ with

$$(GF)_0 = G(F_0)\, G_0 \quad \text{and} \quad (GF)_2(X,Y) = G(F_2(X,Y))\, G_2(F(X), F(Y))$$

for all $X, Y \in \mathrm{Ob}(\mathcal{C})$. The composition of monoidal functors is associative with identity functors being the units.

A monoidal functor (F, F_2, F_0) from a monoidal category $\mathcal{C}$ to a monoidal category $\mathcal{D}$ is *strict* if F_0 and $F_2(X,Y)$ are identity morphisms for all $X, Y \in \mathrm{Ob}(\mathcal{C})$. For example, the identity functor $1_{\mathcal{C}}\colon \mathcal{C} \to \mathcal{C}$ is strict.

A monoidal functor (F, F_2, F_0) is *strong* if F_0 and $F_2(X,Y)$ are isomorphisms for all $X, Y \in \mathrm{Ob}(\mathcal{C})$. Clearly, all strict monoidal functors are strong. The composition of two strict (respectively, strong) monoidal functors is strict (respectively, strong). A strong monoidal functor (F, F_2, F_0) from $\mathcal{C}$ to $\mathcal{D}$ induces a morphism of monoids $\mathrm{End}_{\mathcal{C}}(\mathbb{1}) \to \mathrm{End}_{\mathcal{D}}(\mathbb{1}')$ by $\alpha \mapsto F_0^{-1} F(\alpha) F_0$ for all $\alpha \in \mathrm{End}_{\mathcal{C}}(\mathbb{1})$.

Each monoidal functor $F\colon \mathcal{C} \to \mathcal{D}$ induces a monoidal functor $F^{\otimes\mathrm{op}}\colon \mathcal{C}^{\otimes\mathrm{op}} \to \mathcal{D}^{\otimes\mathrm{op}}$, which is the same functor F with monoidal constraints

$$(F^{\otimes\mathrm{op}})_0 = F_0 \quad \text{and} \quad (F^{\otimes\mathrm{op}})_2(X,Y) = F_2(Y,X)$$

for all $X, Y \in \mathrm{Ob}(\mathcal{C})$. A strong monoidal functor $F\colon \mathcal{C} \to \mathcal{D}$ induces strong monoidal functors $F^{\mathrm{op}}\colon \mathcal{C}^{\mathrm{op}} \to \mathcal{D}^{\mathrm{op}}$ and $F^{\mathrm{rev}}\colon \mathcal{C}^{\mathrm{rev}} \to \mathcal{D}^{\mathrm{rev}}$. Both are equal to F as functors and have the following monoidal constraints:

$$(F^{\mathrm{op}})_0 = (F^{\mathrm{rev}})_0 = F_0^{-1}$$

and for all $X, Y \in \mathrm{Ob}(\mathcal{C})$,

$$(F^{\mathrm{op}})_2(X,Y) = F_2(X,Y)^{-1} \quad \text{and} \quad (F^{\mathrm{rev}})_2(X,Y) = F_2(Y,X)^{-1}.$$

Note that $F^{\mathrm{rev}} = (F^{\otimes\mathrm{op}})^{\mathrm{op}} = (F^{\mathrm{op}})^{\otimes\mathrm{op}}$.

1.4.2 Monoidal natural transformations

A natural transformation φ from a monoidal functor $F\colon \mathcal{C} \to \mathcal{D}$ to a monoidal functor $G\colon \mathcal{C} \to \mathcal{D}$ is *monoidal* if

$$\varphi_{\mathbb{1}} F_0 = G_0 \quad \text{and} \quad \varphi_{X\otimes Y} F_2(X,Y) = G_2(X,Y)(\varphi_X \otimes \varphi_Y)$$

for all $X, Y \in \mathrm{Ob}(\mathcal{C})$. If $\varphi_X : F(X) \to G(X)$ is an isomorphism for all $X \in \mathrm{Ob}(\mathcal{C})$, then such a φ is a *monoidal natural isomorphism*. The functors F and G are *monoidally isomorphic* if there is a monoidal natural isomorphism $F \to G$.

1.4.3 Isomorphisms and equivalences of monoidal categories

Two monoidal categories $\mathcal{C}$ and $\mathcal{D}$ are *isomorphic* if there is a strong monoidal isomorphism between them, that is, a strong monoidal functor $\mathcal{C} \to \mathcal{D}$ which is an isomorphism of the underlying categories.

The monoidal categories $\mathcal{C}$ and $\mathcal{D}$ are *equivalent* if there is a strong monoidal equivalence between them, that is, a strong monoidal functor $\mathcal{C} \to \mathcal{D}$ which is an equivalence of the underlying categories.

Note that if F is a strong monoidal equivalence between monoidal categories, then so are the induced functors $F^{\otimes\mathrm{op}}$, F^{op}, and F^{rev}. As a consequence, if $\mathcal{C}$ and $\mathcal{D}$ are equivalent monoidal categories, then so are $\mathcal{C}^{\otimes\mathrm{op}}$ and $\mathcal{D}^{\otimes\mathrm{op}}$, $\mathcal{C}^{\mathrm{op}}$ and $\mathcal{D}^{\mathrm{op}}$, $\mathcal{C}^{\mathrm{rev}}$ and $\mathcal{D}^{\mathrm{rev}}$.

1.4.4 Example

Let G, H be monoids and let $\alpha\colon G^3 \to \mathbb{k}^*$, $\beta\colon H^3 \to \mathbb{k}^*$ be 3-cocycles. Consider the monoidal categories $\mathcal{C} = G_{\mathbb{k}}^{\alpha}$ and $\mathcal{D} = H_{\mathbb{k}}^{\beta}$, see Example 1.2.6. Any map $f\colon G \to H$ induces a functor $F\colon \mathcal{C} \to \mathcal{D}$ as follows: $F(g) = f(g)$ for any $g \in \mathrm{Ob}(\mathcal{C}) = G$ and $F(k) = k$ for any morphism $k \in \mathbb{k}$ in $\mathcal{C}$. The functor F lifts to a strong monoidal functor $\mathcal{C} \to \mathcal{D}$ if and only if f is a monoid homomorphism and the 3-cocycles α and $f^*(\beta)$ determine the same cohomology class, i.e., there is a map (a 2-cochain) $\gamma\colon G \times G \to \mathbb{k}^*$ such that for all $g, h, j \in G$,

$$\beta(f(g), f(h), f(j)) = \alpha(g,h,j)\,\gamma(h,j)^{-1}\,\gamma(gh,j)\,\gamma(g,hj)^{-1}\,\gamma(g,h) \in \mathbb{k}^*. \tag{1.8}$$

In particular, if f is a monoid homomorphism, then each such γ determines a strong monoidal functor

$$F_{(f,\gamma)} = (F, F_0, F_2)\colon \mathcal{C} \to \mathcal{D}$$

as follows:

$$F_0 = \gamma(1,1)^{-1} \in \mathbb{k} = \mathrm{Hom}_{\mathcal{D}}(1, f(1))$$

and for all $g, h \in G$,

$$F_2(g,h) = \gamma(g,h) \in \mathbb{k} = \mathrm{Hom}_{\mathcal{D}}\big(f(g)f(h), f(gh)\big).$$

Given two such pairs (f,γ) and (f',γ'), the strong monoidal functors $F_{(f,\gamma)}$ and $F_{(f',\gamma')}$ are monoidally isomorphic if and only if $f = f'$ and the 2-cochains γ, γ' determine the same cohomology class, i.e., there is a map (a 1-cochain) $\chi\colon G \to \Bbbk^*$ such that for all $g,h \in G$,

$$\gamma'(g,h) = \gamma(g,h)\,\chi(h)\,\chi(gh)^{-1}\chi(g) \in \Bbbk^*.$$

In particular, if $f = f'$, then each such χ determines the monoidal isomorphism

$$\{\chi(g) \in \Bbbk = \mathrm{Hom}_{\mathcal{D}}(f(g), f(g))\}_{g\in G}\colon F_{(f,\gamma)} \to F_{(f,\gamma')}.$$

1.4.5 Conventions

Unless explicitly stated to the contrary, we will suppress in our formulas the associativity and unitality constraints of monoidal categories. This does not lead to ambiguity because by Mac Lane's coherence theorem (see Section 1.3.4), all legitimate ways of inserting these constraints give the same results. Unless indicated otherwise, we will denote by $\otimes$ and $\mathbb{1}$ the monoidal product and unit object of a monoidal category. For any objects $X_1,\ldots,X_n$ of a monoidal category with $n \geq 2$, we set

$$X_1 \otimes X_2 \otimes \cdots \otimes X_n = (\cdots((X_1 \otimes X_2) \otimes X_3) \otimes \cdots \otimes X_{n-1}) \otimes X_n$$

and similarly for morphisms. Under these conventions, a monoidal functor from a monoidal category $\mathcal{C}$ to a monoidal category $\mathcal{D}$ is nothing but a functor $F\colon \mathcal{C} \to \mathcal{D}$ endowed with a morphism $F_0\colon \mathbb{1} \to F(\mathbb{1})$ in $\mathcal{D}$ and a natural transformation

$$F_2 = \{F_2(X,Y)\colon F(X) \otimes F(Y) \to F(X \otimes Y)\}_{X,Y\in \mathrm{Ob}(\mathcal{C})}$$

such that for all $X,Y,Z \in \mathrm{Ob}(\mathcal{C})$, the following three diagrams commute:

$$\begin{array}{ccc} F(X)\otimes F(Y)\otimes F(Z) & \xrightarrow{\mathrm{id}_{F(X)}\otimes F_2(Y,Z)} & F(X)\otimes F(Y\otimes Z) \\ {\scriptstyle F_2(X,Y)\otimes\mathrm{id}_{F(Z)}}\big\downarrow & & \big\downarrow{\scriptstyle F_2(X,Y\otimes Z)} \\ F(X\otimes Y)\otimes F(Z) & \xrightarrow[F_2(X\otimes Y,Z)]{} & F(X\otimes Y\otimes Z), \end{array}$$

$$\begin{array}{ccc} & F(\mathbb{1})\otimes F(X) & \\ {\scriptstyle F_0\otimes\mathrm{id}_{F(X)}}\nearrow & & \searrow{\scriptstyle F_2(\mathbb{1},X)} \\ F(X) & \xrightarrow[\mathrm{id}_{F(X)}]{} & F(X), \end{array} \qquad \begin{array}{ccc} & F(X)\otimes F(\mathbb{1}) & \\ {\scriptstyle \mathrm{id}_{F(X)}\otimes F_0}\nearrow & & \searrow{\scriptstyle F_2(X,\mathbb{1})} \\ F(X) & \xrightarrow[\mathrm{id}_{F(X)}]{} & F(X). \end{array}$$

1.4.6 Exercise*

Let $F\colon \mathcal{C} \to \mathcal{D}$ be a strong monoidal equivalence between monoidal categories. Let $G\colon \mathcal{D} \to \mathcal{C}$ be a quasi-inverse of F. Prove that for any natural isomorphism $\alpha\colon 1_{\mathcal{C}} \to GF$, the functor G has a unique structure of a strong monoidal functor so that α is monoidal and then prove that the natural isomorphism $\beta\colon FG \to 1_{\mathcal{D}}$ associated with α in Exercise 1.1.9 is monoidal.

1.4.7 Exercise

Prove that the isomorphism and the equivalence of monoidal categories are equivalence relations.

1.5 Pairings in monoidal categories

We define non-degenerate pairings in monoidal categories and then discuss pairings in the category of $\Bbbk$-modules.

1.5.1 Pairings

A *pairing* between two objects X, Y of a monoidal category $\mathcal{C}$ is a morphism $X \otimes Y \to \mathbb{1}$ in $\mathcal{C}$. A pairing $\omega\colon X \otimes Y \to \mathbb{1}$ is *non-degenerate* if there is a morphism $\Omega\colon \mathbb{1} \to Y \otimes X$ in $\mathcal{C}$ such that

$$(\mathrm{id}_Y \otimes \omega)(\Omega \otimes \mathrm{id}_Y) = \mathrm{id}_Y \quad \text{and} \quad (\omega \otimes \mathrm{id}_X)(\mathrm{id}_X \otimes \Omega) = \mathrm{id}_X. \tag{1.9}$$

Such an Ω is called the *inverse* of ω and is determined by ω uniquely. Indeed, if $\Omega'\colon \mathbb{1} \to Y \otimes X$ is another morphism with the same properties, then

$$\begin{aligned}\Omega' &= \mathrm{id}_{Y\otimes X}\, \Omega' = (\mathrm{id}_Y \otimes \mathrm{id}_X)\, \Omega' = (\mathrm{id}_Y \otimes (\omega \otimes \mathrm{id}_X)(\mathrm{id}_X \otimes \Omega))\, \Omega' \\ &= (\mathrm{id}_Y \otimes \omega \otimes \mathrm{id}_X)(\Omega' \otimes \Omega) = ((\mathrm{id}_Y \otimes \omega)(\Omega' \otimes \mathrm{id}_Y) \otimes \mathrm{id}_X)\, \Omega \\ &= (\mathrm{id}_Y \otimes \mathrm{id}_X)\, \Omega = \mathrm{id}_{Y\otimes X}\, \Omega = \Omega.\end{aligned}$$

Note the following useful property of a non-degenerate pairing $\omega\colon X \otimes Y \to \mathbb{1}$: for any $Z, T \in \mathrm{Ob}(\mathcal{C})$, the map

$$\mathrm{Hom}_{\mathcal{C}}(Z, Y \otimes T) \to \mathrm{Hom}_{\mathcal{C}}(X \otimes Z, T), \quad \alpha \mapsto (\omega \otimes \mathrm{id}_T)(\mathrm{id}_X \otimes \alpha) \tag{1.10}$$

is a bijection. The inverse map carries any $\beta \in \mathrm{Hom}_{\mathcal{C}}(X \otimes Z, T)$ to the morphism $(\mathrm{id}_Y \otimes \beta)(\Omega \otimes \mathrm{id}_Z)$, where Ω is the inverse of ω. Similarly, the map

$$\mathrm{Hom}_{\mathcal{C}}(Z, T \otimes X) \to \mathrm{Hom}_{\mathcal{C}}(Z \otimes Y, T), \ \alpha \mapsto (\mathrm{id}_T \otimes \omega)(\alpha \otimes \mathrm{id}_Y) \tag{1.11}$$

is a bijection whose inverse carries any $\beta \in \mathrm{Hom}_{\mathcal{C}}(Z \otimes Y, T)$ to $(\beta \otimes \mathrm{id}_X)(\mathrm{id}_Z \otimes \Omega)$. The proof of these properties is best formulated in terms of the graphical calculus discussed in the next chapter.

Strong monoidal functors transform non-degenerate pairings into non-degenerate pairings as explained in the following lemma.

Lemma 1.5. *If $F\colon \mathcal{C} \to \mathcal{D}$ is a strong monoidal functor between monoidal categories $\mathcal{C}, \mathcal{D}$ and $\omega\colon X \otimes Y \to \mathbb{1}$ is a non-degenerate pairing in $\mathcal{C}$, then the pairing*

$$\omega^F = F_0^{-1} F(\omega) F_2(X,Y)\colon F(X) \otimes F(Y) \to \mathbb{1}$$

is non-degenerate with inverse

$$\Omega^F = F_2(Y,X)^{-1} F(\Omega) F_0\colon \mathbb{1} \to F(Y) \otimes F(X),$$

where $\Omega\colon \mathbb{1} \to Y \otimes X$ is the inverse of ω.

Proof. Using the definition of a monoidal functor, we obtain that

$$\begin{aligned}
&(\mathrm{id}_{F(Y)} \otimes \omega^F)(\Omega^F \otimes \mathrm{id}_{F(Y)}) \\
&= (\mathrm{id}_{F(Y)} \otimes F_0^{-1} F(\omega) F_2(X,Y))(F_2(Y,X)^{-1} F(\Omega) F_0 \otimes \mathrm{id}_{F(Y)}) \\
&= (\mathrm{id}_{F(Y)} \otimes F_0^{-1} F(\omega))(\mathrm{id}_{F(Y)} \otimes F_2(X,Y)) \\
&\quad \circ (F_2(Y,X)^{-1} \otimes \mathrm{id}_{F(Y)})(F(\Omega) F_0 \otimes \mathrm{id}_{F(Y)}) \\
&= (\mathrm{id}_{F(Y)} \otimes F_0^{-1} F(\omega)) F_2(Y, X \otimes Y)^{-1} F_2(Y \otimes X, Y)(F(\Omega) F_0 \otimes \mathrm{id}_{F(Y)}) \\
&= (\mathrm{id}_{F(Y)} \otimes F_0^{-1}) F_2(Y, \mathbb{1})^{-1} F(\mathrm{id}_Y \otimes \omega) F(\Omega \otimes \mathrm{id}_Y) F_2(\mathbb{1}, Y)(F_0 \otimes \mathrm{id}_{F(Y)}) \\
&= F((\mathrm{id}_Y \otimes \omega)(\Omega \otimes \mathrm{id}_Y)).
\end{aligned}$$

By the assumptions on ω and Ω, the latter expression is equal to $F(\mathrm{id}_Y) = \mathrm{id}_{F(Y)}$. The formula

$$(\omega^F \otimes \mathrm{id}_{F(X)})(\mathrm{id}_{F(X)} \otimes \Omega^F) = \mathrm{id}_{F(X)}$$

is proved similarly. Note that it may be easier to check the computations in this proof using the graphical calculus of the next chapter. □

1.5.2 Pairings between $\Bbbk$-modules

Recall the monoidal category $\mathrm{Mod}_\Bbbk$ of $\Bbbk$-modules from Example 1.2.4. By definition, a pairing in $\mathrm{Mod}_\Bbbk$ between $\Bbbk$-modules X, Y is nothing but a $\Bbbk$-linear homomorphism $\omega\colon X \otimes_\Bbbk Y \to \Bbbk$. The pairing ω is non-degenerate in the sense of Section 1.5.1 if there exists a $\Bbbk$-linear homomorphism $\Omega\colon \Bbbk \to Y \otimes_\Bbbk X$ satisfying (1.9). The vector

$$*_\omega = \Omega(1_\Bbbk) \in Y \otimes_\Bbbk X$$

is called the *contraction vector* of ω. The following two lemmas reformulate the non-degeneracy of pairings in terms of dual modules and matrices. Recall that the *dual* of a $\Bbbk$-module X is the $\Bbbk$-module $X^\star = \mathrm{Hom}_\Bbbk(X, \Bbbk)$ consisting of all $\Bbbk$-linear homomorphisms $X \to \Bbbk$ with the $\Bbbk$-module structure given by $(kf)(x) = k\,f(x)$ for all $k \in \Bbbk$, $f \in X^\star$, $x \in X$. A $\Bbbk$-module is *projective of finite type* if it is a direct summand of a free $\Bbbk$-module of finite rank.

Lemma 1.6. *Let* $\omega\colon X\otimes_{\Bbbk}Y\to\Bbbk$ *be a pairing in* $\mathrm{Mod}_{\Bbbk}$ *between* $\Bbbk$*-modules* X *and* Y*. The following three conditions are equivalent:*

(i) ω *is non-degenerate in the sense of Section* 1.5.1*;*

(ii) X *is a projective* $\Bbbk$*-module of finite type and the homomorphism* $Y\to X^\star$ *adjoint to* ω *is an isomorphism;*

(iii) Y *is a projective* $\Bbbk$*-module of finite type and the homomorphism* $X\to Y^\star$ *adjoint to* ω *is an isomorphism.*

Proof. Let $\rho\colon X\otimes_{\Bbbk}X^\star\to\Bbbk$ be the evaluation pairing defined by $\rho(x\otimes_{\Bbbk}f)=f(x)$ for all $x\in X$ and $f\in X^\star$. We prove first that if the $\Bbbk$-module X is projective of finite type, then ρ is non-degenerate (in the sense of Section 1.5.1). The $\Bbbk$-module X is projective of finite type if and only if for some integer $n\geq 1$ there are linear maps $e\colon\Bbbk^n\to X$ and $\ell\colon X\to\Bbbk^n$ such that $e\circ\ell=\mathrm{id}_X$. Denoting by e_i the image under e of the ith standard basis vector of $\Bbbk^n$ and denoting by ℓ_i the ith coordinate function of ℓ, we obtain that X is projective of finite type if and only if there are finite families $\{e_i\}_{i=1}^n\subset X$ and $\{\ell_i\}_{i=1}^n\subset X^\star$ with $n\geq 1$ such that $x=\sum_{i=1}^n\ell_i(x)e_i$ for all $x\in X$. Having such sets, we define a $\Bbbk$-linear map $R\colon\Bbbk\to X^\star\otimes_{\Bbbk}X$ by $R(1_{\Bbbk})=\sum_{i=1}^n\ell_i\otimes_{\Bbbk}e_i$. For $x\in X$ and $f\in X^\star$,

$$\big((\mathrm{id}_{X^\star}\otimes_{\Bbbk}\rho)(R\otimes_{\Bbbk}\mathrm{id}_{X^\star})(f)\big)(x)=\sum_{i=1}^n f(e_i)\ell_i(x)=f\left(\sum_{i=1}^n\ell_i(x)e_i\right)=f(x),$$

$$(\rho\otimes_{\Bbbk}\mathrm{id}_X)(\mathrm{id}_X\otimes_{\Bbbk}R)(x)=\sum_{i=1}^n\ell_i(x)e_i=x.$$

Thus, the pairing ρ is non-degenerate with inverse R.

Let us prove the equivalence (i) $\Leftrightarrow$ (ii) of the lemma. The homomorphism $a\colon Y\to X^\star$ adjoint to ω carries any $y\in Y$ to $\omega(\,\cdot\otimes_{\Bbbk}y)\in X^\star$. For any $x\in X$,

$$\omega(x\otimes_{\Bbbk}y)=a(y)(x)=\rho(x\otimes_{\Bbbk}a(y))=\rho(\mathrm{id}_X\otimes_{\Bbbk}a)(x\otimes_{\Bbbk}y).$$

Hence $\omega=\rho(\mathrm{id}_X\otimes_{\Bbbk}a)$. This yields the implication (ii) $\Rightarrow$ (i): if X is projective of finite type, then the pairing ρ is non-degenerate, and if, additionally, a is an isomorphism, then the pairing $\omega=\rho(\mathrm{id}_X\otimes_{\Bbbk}a)$ is non-degenerate as well. Conversely, suppose that ω is non-degenerate with inverse $\Omega\colon\Bbbk\to Y\otimes_{\Bbbk}X$. Pick an expansion $\Omega(1_{\Bbbk})=\sum_i e'_i\otimes_{\Bbbk}e_i$ where i runs over a finite set of indices and $e_i\in X$, $e'_i\in Y$. For each i, set

$$\ell_i=\omega(\,\cdot\otimes_{\Bbbk}e'_i)\in X^\star\quad\text{and}\quad\ell'_i=\omega(e_i\otimes_{\Bbbk}\cdot\,)\in Y^\star.$$

Formulas (1.9) imply that $\sum_i\ell_i(x)e_i=x$ and $\sum_i\ell'_i(y)e'_i=y$ for all $x\in X$ and $y\in Y$. Thus X and Y are projective of finite type. Let $a'\colon X^\star\to Y$ be the map carrying any $f\in X^\star$ to $\sum_i f(e_i)e'_i\in Y$. For any $y\in Y$,

$$a'a(y)=\sum_i a(y)(e_i)e'_i=\sum_i\omega(e_i\otimes_{\Bbbk}y)e'_i=\sum_i\ell'_i(y)e'_i=y.$$

For any $f \in X^*$ and $x \in X$,

$$\begin{aligned}\big(aa'(f)\big)(x) &= \sum_i f(e_i)\, a(e_i')(x) = \sum_i f(e_i)\, \omega(x \otimes e_i') \\ &= \sum_i f(e_i)\, \ell_i(x) = f\left(\sum_i \ell_i(x) e_i\right) = f(x).\end{aligned}$$

Thus $a'a = \mathrm{id}_Y$ and $aa' = \mathrm{id}_{X^*}$, and so a is an isomorphism. Hence (i) $\Rightarrow$ (ii).

The equivalence v (i) $\Leftrightarrow$ (iii) follows from the equivalence (i) $\Leftrightarrow$ (ii) by noticing that a pairing $X \otimes_{\Bbbk} Y \to \Bbbk$ is non-degenerate if and only if its composition with the flip $Y \otimes_{\Bbbk} X \to X \otimes_{\Bbbk} Y$ is a non-degenerate pairing. □

Lemma 1.7. *A pairing $\omega\colon X \otimes_{\Bbbk} Y \to \Bbbk$ in $\mathrm{Mod}_{\Bbbk}$ between free $\Bbbk$-modules X, Y is non-degenerate if and only if X and Y have the same finite rank n and for some bases $(x_i)_{i=1}^n$ of X and $(y_j)_{j=1}^n$ of Y, the matrix $[\omega(x_i \otimes_{\Bbbk} y_j)]_{i,j=1}^n$ is invertible. The contraction vector of ω is then computed by*

$$*_\omega = \sum_{i,j=1}^n \Omega_{i,j}\, y_i \otimes_{\Bbbk} x_j \in Y \otimes_{\Bbbk} X,$$

where $[\Omega_{i,j}]_{i,j=1}^n$ is the inverse matrix of $[\omega(x_i \otimes_{\Bbbk} y_j)]_{i,j=1}^n$.

Proof. Assume first that ω is invertible. By Lemma 1.6, the free modules X and Y have finite rank and X is isomorphic to $Y^\star$. If Y has rank n, then so do $Y^\star$ and X. Pick bases $(x_i)_{i=1}^n$ of X and $(y_j)_{j=1}^n$ of Y, and consider the dual bases $(x_i^\star)_{i=1}^n$ of $X^\star$ and $(y_j^\star)_{j=1}^n$ of $Y^\star$. Set $\Omega_{i,j} = (y_i^\star \otimes_{\Bbbk} x_j^\star)\Omega(1_{\Bbbk}) \in \Bbbk$ for any $1 \le i, j \le n$, where $\Omega\colon \Bbbk \to Y \otimes_{\Bbbk} X$ is the inverse of ω. Using the identity $\mathrm{id}_Y = \sum_{1 \le k \le n} y_k y_k^\star$, we obtain for all $1 \le i, j \le n$,

$$\begin{aligned}&\sum_{1\le k\le n} \omega(x_i \otimes_{\Bbbk} y_k)\, \Omega_{k,j} = \sum_{1\le k\le n} \omega(x_i \otimes_{\Bbbk} y_k)(y_k^\star \otimes_{\Bbbk} x_j^\star)\Omega(1_{\Bbbk}) \\ &= \omega(x_i \otimes_{\Bbbk} \mathrm{id}_Y)(\mathrm{id}_Y \otimes_{\Bbbk} x_j^\star)\Omega(1_{\Bbbk}) = x_j^\star\big((\omega \otimes_{\Bbbk} \mathrm{id}_X)(\mathrm{id}_X \otimes_{\Bbbk} \Omega)(x_i)\big) \\ &= x_j^\star(\mathrm{id}_X(x_i)) = x_j^\star(x_i) = \delta_{i,j},\end{aligned}$$

where $\delta_{i,j} = 1_{\Bbbk}$ if $i = j$ and $\delta_{i,j} = 0$ otherwise. Thus the matrix $[\omega(x_i \otimes_{\Bbbk} y_j)]_{i,j=1}^n$ is invertible with inverse $[\Omega_{i,j}]_{i,j=1}^n$.

Conversely, assume that the modules X and Y have the same finite rank n and bases $(x_i)_{i=1}^n$ and $(y_j)_{j=1}^n$ such that the matrix $[\omega(x_i \otimes_{\Bbbk} y_j)]_{i,j=1}^n$ is invertible with inverse $[\Omega_{i,j}]_{i,j=1}^n$. Define the $\Bbbk$-linear homomorphism $\Omega\colon \Bbbk \to Y \otimes_{\Bbbk} X$ by

$$\Omega(1_{\Bbbk}) = \sum_{i,j=1}^n \Omega_{i,j}\, y_i \otimes_{\Bbbk} x_j \in Y \otimes_{\Bbbk} X.$$

We claim that $(\omega \otimes_{\Bbbk} \mathrm{id}_X)(\mathrm{id}_X \otimes_{\Bbbk} \Omega) = \mathrm{id}_X$. Indeed, for any $1 \le a \le n$,

$$
\begin{aligned}
(\omega \otimes_{\Bbbk} \mathrm{id}_X)(\mathrm{id}_X \otimes_{\Bbbk} \Omega)(x_a) &= \sum_{1\le k\le n} x_k^\star((\omega \otimes_{\Bbbk} \mathrm{id}_X)(\mathrm{id}_X \otimes_{\Bbbk} \Omega)(x_a))\, x_k \\
&= \sum_{1\le i,j,k\le n} \Omega_{i,j}\, x_k^\star((\omega \otimes_{\Bbbk} \mathrm{id}_X)(x_a \otimes_{\Bbbk} y_i \otimes_{\Bbbk} x_j))\, x_k \\
&= \sum_{1\le i,j,k\le n} \omega(x_a \otimes_{\Bbbk} y_i)\, \Omega_{i,j}\, x_i^\star(x_j)\, x_k = \sum_{1\le i,k\le n} \omega(x_a \otimes_{\Bbbk} y_i)\, \Omega_{i,k}\, x_k \\
&= \sum_{1\le k\le n} \delta_{a,k}\, x_k = x_a.
\end{aligned}
$$

Likewise, $(\mathrm{id}_Y \otimes_{\Bbbk} \omega)(\Omega \otimes_{\Bbbk} \mathrm{id}_Y) = \mathrm{id}_Y$. Therefore the pairing ω is non-degenerate with inverse Ω. We conclude using that $*_\omega = \Omega(1_{\Bbbk})$. □

1.6 Rigid categories and dual functors

We discuss duality in monoidal categories and introduce rigid categories.

1.6.1 Rigid categories

A *left dual* of an object X of a monoidal category $\mathcal{C}$ is a pair $({}^\vee X, \mathrm{ev}_X)$, where ${}^\vee X$ is an object of $\mathcal{C}$ and $\mathrm{ev}_X\colon {}^\vee X \otimes X \to \mathbb{1}$ is a non-degenerate pairing. The pairing ev_X is called the *left evaluation* and its inverse $\mathrm{coev}_X\colon \mathbb{1} \to X \otimes {}^\vee X$ the *left coevaluation*. A left dual of the object X, if it exists, is unique up to a unique isomorphism preserving the evaluation pairing. More precisely, if $(Y, e\colon Y \otimes X \to \mathbb{1})$ is another left dual of X, then

$$(e \otimes \mathrm{id}_{{}^\vee X})(\mathrm{id}_Y \otimes \mathrm{coev}_X)\colon Y \to {}^\vee X$$

is the unique isomorphism $a\colon Y \to {}^\vee X$ such that $e = \mathrm{ev}_X(a \otimes \mathrm{id}_X)$.

A *left duality* in a monoidal category $\mathcal{C}$ is a family $\{({}^\vee X, \mathrm{ev}_X)\}_{X\in \mathrm{Ob}(\mathcal{C})}$ where, for every $X \in \mathrm{Ob}(\mathcal{C})$, the pair $({}^\vee X, \mathrm{ev}_X)$ is a left dual of X. A *left rigid* category is a monoidal category admitting a left duality. A *left rigid category with distinguished left duality* is a left rigid category endowed with a left duality.

Similarly, a *right dual* of $X \in \mathrm{Ob}(\mathcal{C})$ is a pair $(X^\vee, \widetilde{\mathrm{ev}}_X)$ where $X^\vee \in \mathrm{Ob}(\mathcal{C})$ and $\widetilde{\mathrm{ev}}_X\colon X \otimes X^\vee \to \mathbb{1}$ is a non-degenerate pairing. The pairing $\widetilde{\mathrm{ev}}_X$ is called the *right evaluation* and its inverse $\widetilde{\mathrm{coev}}_X\colon \mathbb{1} \to X^\vee \otimes X$ the *right coevaluation.* A right dual of an object of $\mathcal{C}$, if it exists, is unique up to a unique isomorphism preserving the evaluation pairing.

A *right duality* in a monoidal category $\mathcal{C}$ is a family $\{(X^\vee, \widetilde{\mathrm{ev}}_X)\}_{X\in \mathrm{Ob}(\mathcal{C})}$ where, for every $X \in \mathrm{Ob}(\mathcal{C})$, the pair $(X^\vee, \widetilde{\mathrm{ev}}_X)$ is a right dual of X. A *right rigid* category is a monoidal category admitting a right duality. A *right rigid category with distinguished right duality* is a right rigid category endowed with a right duality.

A *rigid* (or *autonomous*) category is a monoidal category which is both left rigid and right rigid, that is, which admits both a left duality and a right duality. A *rigid category with distinguished duality* is a rigid category endowed with a left duality and a right duality.

1.6.2 Example

The monoidal category $G^\alpha_\Bbbk$ from Example 1.2.6 is left (respectively, right) rigid if and only if every element of the monoid G has a left (respectively, a right) inverse. The category $G^\alpha_\Bbbk$ is rigid if and only if G is a group.

1.6.3 Dual functors

A left duality in a left rigid category $\mathcal{C}$ determines a functor

$$^\vee ?\colon \mathcal{C}^{\mathrm{rev}} = (\mathcal{C}^{\mathrm{op}}, \otimes^{\mathrm{op}}, \mathbb{1}) \to \mathcal{C}$$

which carries each $X \in \mathrm{Ob}(\mathcal{C}) = \mathrm{Ob}(\mathcal{C}^{\mathrm{rev}})$ to $^\vee X$ and carries each morphism $f\colon X \to Y$ in $\mathcal{C}$ (that is a morphism $Y \to X$ in $\mathcal{C}^{\mathrm{rev}}$) to its *left dual*

$$^\vee f = (\mathrm{ev}_Y \otimes \mathrm{id}_{^\vee X})(\mathrm{id}_{^\vee Y} \otimes f \otimes \mathrm{id}_{^\vee X})(\mathrm{id}_{^\vee Y} \otimes \mathrm{coev}_X)\colon {}^\vee Y \to {}^\vee X.$$

The functor $^\vee ?$ is strong monoidal with monoidal constraints $^\vee ?_0 = \mathrm{coev}_{\mathbb{1}} \colon \mathbb{1} \to {}^\vee\mathbb{1}$ and $^\vee ?_2(X,Y)\colon {}^\vee X \otimes {}^\vee Y \to {}^\vee(Y \otimes X)$ defined by

$$^\vee ?_2(X,Y) = (\mathrm{ev}_X \otimes \mathrm{id}_{^\vee(Y\otimes X)})(\mathrm{id}_{^\vee X} \otimes \mathrm{ev}_Y \otimes \mathrm{id}_{X \otimes {}^\vee(Y\otimes X)})(\mathrm{id}_{^\vee X \otimes {}^\vee Y} \otimes \mathrm{coev}_{Y\otimes X}).$$

The functor $^\vee ?$ is called the *left dual functor* associated with the given left duality. The uniqueness of the left duals of objects implies that the left dual functors associated with different left dualities are monoidally isomorphic in a canonical way.

A right duality in a right rigid category $\mathcal{C}$ determines a functor $?^\vee \colon \mathcal{C}^{\mathrm{rev}} \to \mathcal{C}$ carrying each object X of $\mathcal{C}$ to $X^\vee$ and each morphism $f\colon X \to Y$ in $\mathcal{C}$ to its *right dual*

$$f^\vee = (\mathrm{id}_{X^\vee} \otimes \widetilde{\mathrm{ev}}_Y)(\mathrm{id}_{X^\vee} \otimes f \otimes \mathrm{id}_{Y^\vee})(\widetilde{\mathrm{coev}}_X \otimes \mathrm{id}_{Y^\vee})\colon Y^\vee \to X^\vee.$$

The functor $?^\vee$ is strong monoidal, with monoidal constraints $?^\vee_0 = \widetilde{\mathrm{coev}}_{\mathbb{1}} \colon \mathbb{1} \to \mathbb{1}^\vee$ and $?^\vee_2(X,Y)\colon X^\vee \otimes Y^\vee \to (Y \otimes X)^\vee$ defined by

$$?^\vee_2(X,Y) = (\mathrm{id}_{(Y\otimes X)^\vee} \otimes \widetilde{\mathrm{ev}}_Y)(\mathrm{id}_{(Y\otimes X)^\vee \otimes Y} \otimes \widetilde{\mathrm{ev}}_X \otimes \mathrm{id}_{Y^\vee})(\widetilde{\mathrm{coev}}_{Y\otimes X} \otimes \mathrm{id}_{X^\vee \otimes Y^\vee}).$$

The functor $?^\vee$ is called the *right dual functor* associated with the given right duality. The right dual functors associated with different right dualities are monoidally isomorphic in a canonical way.

For a rigid category $\mathcal{C}$ with distinguished duality, the left and right dual functors ${}^\vee?\colon \mathcal{C}^{\mathrm{rev}} \to \mathcal{C}$ and $?^\vee\colon \mathcal{C}^{\mathrm{rev}} \to \mathcal{C}$ are strong monoidal equivalences with respective quasi-inverses $(?^\vee)^{\mathrm{rev}}\colon \mathcal{C} \to \mathcal{C}^{\mathrm{rev}}$ and $({}^\vee?)^{\mathrm{rev}}\colon \mathcal{C} \to \mathcal{C}^{\mathrm{rev}}$. For $X \in \mathrm{Ob}(\mathcal{C})$, the corresponding monoidal natural isomorphisms ${}^\vee(X^\vee) \simeq X \simeq ({}^\vee X)^\vee$ are

$$(\widetilde{\mathrm{ev}}_X \otimes \mathrm{id}_{{}^\vee(X^\vee)})(\mathrm{id}_X \otimes \mathrm{coev}_{X^\vee})\colon X \to {}^\vee(X^\vee),$$
$$(\mathrm{id}_{({}^\vee X)^\vee} \otimes \mathrm{ev}_X)(\widetilde{\mathrm{coev}}_{{}^\vee X} \otimes \mathrm{id}_X)\colon X \to ({}^\vee X)^\vee.$$

1.6.4 Duality and monoidal functors

A strong monoidal functor $F\colon \mathcal{C} \to \mathcal{D}$ between monoidal categories carries any object having a left (respectively, right) dual to an object having a left (respectively, right) dual. Indeed, consider an object X of $\mathcal{C}$ with left dual $({}^\vee X, \mathrm{ev}_X)$. By Lemma 1.5, the non-degeneracy of ev_X implies the non-degeneracy of the pairing

$$(\mathrm{ev}_X)^F = F_0^{-1} F(\mathrm{ev}_X) F_2({}^\vee X, X)\colon F({}^\vee X) \otimes F(X) \to \mathbb{1}.$$

Thus $(F({}^\vee X), (\mathrm{ev}_X)^F)$ is a left dual of $F(X)$. Similarly, if $X \in \mathrm{Ob}(\mathcal{C})$ has a right dual $(X^\vee, \widetilde{\mathrm{ev}}_X)$, then $(F(X^\vee), (\widetilde{\mathrm{ev}}_X)^F)$ is a right dual of $F(X)$.

A *lift* of a left duality $\{({}^\vee Y, \mathrm{ev}_Y)\}_{Y \in \mathrm{Ob}(\mathcal{D})}$ in $\mathcal{D}$ along a strong monoidal functor $F\colon \mathcal{C} \to \mathcal{D}$ is a left duality $\{({}^\vee X, \mathrm{ev}_X)\}_{X \in \mathrm{Ob}(\mathcal{C})}$ in $\mathcal{C}$ such that

$$(F({}^\vee X), (\mathrm{ev}_X)^F) = ({}^\vee F(X), \mathrm{ev}_{F(X)})$$

for all $X \in \mathrm{Ob}(\mathcal{C})$. Similarly, one defines a lift of a right duality in $\mathcal{D}$ along F.

A strong monoidal functor $F\colon \mathcal{C} \to \mathcal{D}$ between left rigid categories with distinguished left duality determines a monoidal natural isomorphism

$$F^l = \{F^l(X)\colon F({}^\vee X) \to {}^\vee F(X)\}_{X \in \mathrm{Ob}(\mathcal{C})}$$

from the functor $F\,{}^\vee?\colon \mathcal{C}^{\mathrm{rev}} \to \mathcal{D}$ to the functor ${}^\vee?\,F^{\mathrm{rev}}\colon \mathcal{C}^{\mathrm{rev}} \to \mathcal{D}$. It is defined as follows. For each $X \in \mathrm{Ob}(\mathcal{C})$, both $(F({}^\vee X), (\mathrm{ev}_X)^F)$ and $({}^\vee F(X), \mathrm{ev}_{F(X)})$ are left duals of $F(X)$. By the uniqueness of a left dual, there is a unique isomorphism

$$F^l(X)\colon F({}^\vee X) \to {}^\vee F(X)$$

preserving the evaluation pairing, i.e., such that

$$(\mathrm{ev}_X)^F = \mathrm{ev}_{F(X)}(F^l(X) \otimes \mathrm{id}_{F(X)}).$$

The isomorphism $F^l(X)$ is computed by

$$F^l(X) = ((\mathrm{ev}_X)^F \otimes \mathrm{id}_{{}^\vee F(X)})(\mathrm{id}_{F({}^\vee X)} \otimes \mathrm{coev}_{F(X)}).$$

We leave it as an exercise to verify that $F^l = \{F^l(X)\}_{X \in \mathrm{Ob}(\mathcal{C})}$ is a monoidal natural isomorphism. Note that the distinguished left duality in $\mathcal{C}$ is a lift along F

of the distinguished left duality in $\mathcal{D}$ if and only if $F^l(X) = \mathrm{id}_{^\vee F(X)}$ for all $X \in \mathrm{Ob}(\mathcal{C})$.

Likewise, a strong monoidal functor $F\colon \mathcal{C} \to \mathcal{D}$ between right rigid categories with distinguished right duality determines a monoidal natural isomorphism

$$F^r = \{F^r(X)\colon F(X^\vee) \to F(X)^\vee\}_{X \in \mathrm{Ob}(\mathcal{C})}$$

from $F?^\vee\colon \mathcal{C}^{\mathrm{rev}} \to \mathcal{D}$ to $?^\vee F^{\mathrm{rev}}\colon \mathcal{C}^{\mathrm{rev}} \to \mathcal{D}$. It is computed by

$$F^r(X) = (\mathrm{id}_{F(X)^\vee} \otimes (\widetilde{\mathrm{ev}}_X)^F)(\widetilde{\mathrm{coev}}_{F(X)} \otimes \mathrm{id}_{F(X^\vee)})$$

for any $X \in \mathrm{Ob}(\mathcal{C})$. The distinguished right duality in $\mathcal{C}$ is a lift along F of the distinguished right duality in $\mathcal{D}$ if and only if $F^r(X) = \mathrm{id}_{F(X)^\vee}$ for all $X \in \mathrm{Ob}(\mathcal{C})$.

1.6.5 Exercise*

Let $F, G\colon \mathcal{C} \to \mathcal{D}$ be strong monoidal functors between monoidal categories and let $\alpha\colon F \to G$ be a monoidal natural transformation. Prove that if $\mathcal{C}$ is left rigid or right rigid, then α is an isomorphism.

1.7 Pivotal categories and functors

We review pivotal categories and pivotal functors.

1.7.1 Pivotal categories

A *pivotal* (or *sovereign*) category is a rigid category with distinguished duality such that the induced left and right dual functors coincide as monoidal functors. In other words, a pivotal category is a monoidal category $\mathcal{C}$ endowed with a *pivotal duality*, that is, a family of triples

$$\{(X^*, \mathrm{ev}_X, \widetilde{\mathrm{ev}}_X)\}_{X \in \mathrm{Ob}(\mathcal{C})},$$

where

- X^* is an object of $\mathcal{C}$ called the *dual* of X,
- $\mathrm{ev}_X\colon X^* \otimes X \to \mathbb{1}$ is a non-degenerate pairing in $\mathcal{C}$,
- $\widetilde{\mathrm{ev}}_X\colon X \otimes X^* \to \mathbb{1}$ is a non-degenerate pairing in $\mathcal{C}$,

such that the left dual functor associated with the left duality $\{(X^*, \mathrm{ev}_X)\}_{X \in \mathrm{Ob}(\mathcal{C})}$ and the right dual functor associated with the right duality $\{(X^*, \widetilde{\mathrm{ev}}_X)\}_{X \in \mathrm{Ob}(\mathcal{C})}$ coincide as monoidal functors. The pairings ev_X and $\widetilde{\mathrm{ev}}_X$ are called the *left evaluation* and the *right evaluation*, respectively. We let $\mathrm{coev}_X\colon \mathbb{1} \to X \otimes X^*$ and $\widetilde{\mathrm{coev}}_X\colon \mathbb{1} \to X^* \otimes X$ denote the inverses of these pairings. These two morphisms are called respectively the *left coevaluation* and the *right coevaluation*. The equality of the left and right dual functors means that:

(i) for any morphism $f\colon X \to Y$ in $\mathcal{C}$,

$$(\mathrm{ev}_Y \otimes \mathrm{id}_{X^*})(\mathrm{id}_{Y^*} \otimes f \otimes \mathrm{id}_{X^*})(\mathrm{id}_{Y^*} \otimes \mathrm{coev}_X) \\ = (\mathrm{id}_{X^*} \otimes \widetilde{\mathrm{ev}}_Y)(\mathrm{id}_{X^*} \otimes f \otimes \mathrm{id}_{Y^*})(\widetilde{\mathrm{coev}}_X \otimes \mathrm{id}_{Y^*})\colon Y^* \to X^*;$$

(ii) $\mathrm{coev}_{\mathbb{1}} = \widetilde{\mathrm{coev}}_{\mathbb{1}}\colon \mathbb{1} \to \mathbb{1}^*$;

(iii) for all $X, Y \in \mathrm{Ob}(\mathcal{C})$, we have the following equality of morphisms from $X^* \otimes Y^*$ to $(Y \otimes X)^*$:

$$(\mathrm{ev}_X \otimes \mathrm{id}_{(Y\otimes X)^*})(\mathrm{id}_{X^*} \otimes \mathrm{ev}_Y \otimes \mathrm{id}_{X\otimes(Y\otimes X)^*})(\mathrm{id}_{X^*\otimes Y^*} \otimes \mathrm{coev}_{Y\otimes X}) \\ = (\mathrm{id}_{(Y\otimes X)^*} \otimes \widetilde{\mathrm{ev}}_Y)(\mathrm{id}_{(Y\otimes X)^*\otimes Y} \otimes \widetilde{\mathrm{ev}}_X \otimes \mathrm{id}_{Y^*})(\widetilde{\mathrm{coev}}_{Y\otimes X} \otimes \mathrm{id}_{X^*\otimes Y^*}).$$

Clearly, a pivotal category is, in particular, a rigid category with distinguished duality. The left and right dual functors form a single functor $?^*\colon \mathcal{C}^{\mathrm{rev}} \to \mathcal{C}$ called the *dual functor* of $\mathcal{C}$. It carries any object $X \in \mathrm{Ob}(\mathcal{C}^{\mathrm{rev}}) = \mathrm{Ob}(\mathcal{C})$ to X^* and any morphism $f\colon X \to Y$ in $\mathcal{C}$ (that is a morphism $Y \to X$ in $\mathcal{C}^{\mathrm{rev}}$) to its *dual* $f^*\colon Y^* \to X^*$ defined as the left-hand side (or the right-hand side) of the equality in (i) above. The monoidal constraints

$$?^*_0\colon \mathbb{1} \to \mathbb{1}^* \quad \text{and} \quad ?^*_2(X,Y)\colon X^* \otimes Y^* \to (Y \otimes X)^*$$

of the dual functor $?^*$ are the morphisms defined by (ii) and (iii), respectively.

The duality identities

$$(\mathrm{id}_X \otimes \mathrm{ev}_X)(\mathrm{coev}_X \otimes \mathrm{id}_X) = \mathrm{id}_X = (\widetilde{\mathrm{ev}}_X \otimes \mathrm{id}_X)(\mathrm{id}_X \otimes \widetilde{\mathrm{coev}}_X), \tag{1.12}$$

$$(\mathrm{ev}_X \otimes \mathrm{id}_{X^*})(\mathrm{id}_{X^*} \otimes \mathrm{coev}_X) = \mathrm{id}_{X^*} = (\mathrm{id}_{X^*} \otimes \widetilde{\mathrm{ev}}_X)(\widetilde{\mathrm{coev}}_X \otimes \mathrm{id}_{X^*}) \tag{1.13}$$

imply that $(?^*_0)^{-1} = \mathrm{ev}_{\mathbb{1}} = \widetilde{\mathrm{ev}}_{\mathbb{1}}\colon \mathbb{1}^* \to \mathbb{1}$ and

$$\big(?^*_2(X,Y)\big)^{-1} \\ = (\mathrm{ev}_{Y\otimes X} \otimes \mathrm{id}_{X^*\otimes Y^*})(\mathrm{id}_{(Y\otimes X)^*\otimes Y} \otimes \mathrm{coev}_X \otimes \mathrm{id}_{Y^*})(\mathrm{id}_{(X\otimes Y)^*} \otimes \mathrm{coev}_Y) \\ = (\mathrm{id}_{X^*\otimes Y^*} \otimes \widetilde{\mathrm{ev}}_{Y\otimes X})(\mathrm{id}_{X^*} \otimes \widetilde{\mathrm{coev}}_Y \otimes \mathrm{id}_{X\otimes(Y\otimes X)^*})(\widetilde{\mathrm{coev}}_X \otimes \mathrm{id}_{(Y\otimes X)^*}).$$

These formulas are best checked using the graphical calculus of the next chapter.

If $\mathcal{C}$ is a pivotal category, then the opposite monoidal categories

$$\mathcal{C}^{\mathrm{op}} = (\mathcal{C}^{\mathrm{op}}, \otimes, \mathbb{1}), \quad \mathcal{C}^{\otimes\mathrm{op}} = (\mathcal{C}, \otimes^{\mathrm{op}}, \mathbb{1}), \quad \mathcal{C}^{\mathrm{rev}} = (\mathcal{C}^{\mathrm{op}}, \otimes^{\mathrm{op}}, \mathbb{1})$$

from Section 1.2.2 are pivotal in a canonical way. The dual objects in them are the same as in $\mathcal{C}$ and the evaluation morphisms are

$$\mathrm{ev}^{\mathrm{op}}_X = \widetilde{\mathrm{coev}}_X, \qquad \mathrm{ev}^{\otimes\mathrm{op}}_X = \widetilde{\mathrm{ev}}_X, \qquad \mathrm{ev}^{\mathrm{rev}}_X = \mathrm{coev}_X,$$

$$\widetilde{\mathrm{ev}}^{\mathrm{op}}_X = \mathrm{coev}_X, \qquad \widetilde{\mathrm{ev}}^{\otimes\mathrm{op}}_X = \mathrm{ev}_X, \qquad \widetilde{\mathrm{ev}}^{\mathrm{rev}}_X = \widetilde{\mathrm{coev}}_X.$$

1.7.2 Example

By Lemma 1.6, a $\Bbbk$-module has a left (or a right) dual in the monoidal category $\mathrm{Mod}_\Bbbk$ if and only if this module is projective of finite type. Let $\mathrm{proj}_\Bbbk$ be the full subcategory of $\mathrm{Mod}_\Bbbk$ consisting of projective $\Bbbk$-modules of finite type. It is clear that this subcategory is closed under $\otimes_\Bbbk$ and $\Bbbk \in \mathrm{Ob}(\mathrm{proj}_\Bbbk)$. Therefore $\mathrm{proj}_\Bbbk$ is a monoidal subcategory of $\mathrm{Mod}_\Bbbk$. The category $\mathrm{proj}_\Bbbk$ has a canonical structure of a pivotal category with pivotal duality

$$\{(X^\star, \mathrm{ev}_X, \widetilde{\mathrm{ev}}_X)\}_{X\in \mathrm{Ob}(\mathrm{proj}_\Bbbk)}$$

defined as follows. The dual of any projective $\Bbbk$-module of finite type X is the $\Bbbk$-module $X^\star = \mathrm{Hom}_\Bbbk(X, \Bbbk) \in \mathrm{Ob}(\mathrm{proj}_\Bbbk)$. The pairings $\mathrm{ev}_X \colon X^\star \otimes_\Bbbk X \to \Bbbk$ and $\widetilde{\mathrm{ev}}_X \colon X \otimes_\Bbbk X^\star \to \Bbbk$ are defined by

$$\mathrm{ev}_X(f \otimes_\Bbbk x) = f(x) = \widetilde{\mathrm{ev}}_X(x \otimes_\Bbbk f)$$

for all $x \in X$ and $f \in X^\star$.

1.7.3 Example

Let α be a $\Bbbk^*$-valued 3-cocycle of a group G. The monoidal category $G_\Bbbk^\alpha$ from Example 1.2.6 has a canonical structure of a pivotal category with pivotal duality $\{(g^* = g^{-1}, \mathrm{ev}_g, \widetilde{\mathrm{ev}}_g)\}_{g\in G}$ given by

$$\mathrm{ev}_g = 1_\Bbbk \in \Bbbk = \mathrm{Hom}_\mathcal{C}(g^{-1} \otimes g, 1) \quad \text{and} \quad \widetilde{\mathrm{ev}}_g = \lambda_g \in \Bbbk = \mathrm{Hom}_\mathcal{C}(g \otimes g^{-1}, 1),$$

where

$$\lambda_g = \alpha(1,1,g)\,\alpha(g,g^{-1},g)\,\alpha(g,1,1) \in \Bbbk^*.$$

The corresponding coevaluation morphisms are computed by

$$\mathrm{coev}_g = \lambda_g^{-1} \in \Bbbk = \mathrm{Hom}_\mathcal{C}(1, g \otimes g^{-1}) \quad \text{and} \quad \widetilde{\mathrm{coev}}_g = 1_\Bbbk \in \Bbbk = \mathrm{Hom}_\mathcal{C}(1, g^{-1} \otimes g).$$

More generally, any group homomorphism $d \colon G \to \Bbbk^*$ induces a pivotal duality

$$\big\{\big(g^* = g^{-1}, \mathrm{ev}_g = 1_\Bbbk, \widetilde{\mathrm{ev}}_g = d(g)\lambda_g\big)\big\}_{g\in G}$$

in $G_\Bbbk^\alpha$ with coevaluation morphisms $\mathrm{coev}_g = \lambda_g^{-1}$ and $\widetilde{\mathrm{coev}}_g = d(g)^{-1}$. The resulting pivotal category is denoted $G_\Bbbk^{\alpha,d}$. Observe that the dual functor of $G_\Bbbk^{\alpha,d}$ acts as the inversion on objects and the identity on morphisms.

1.7.4 Remark

A pivotal category may be equivalently defined as a left rigid category $\mathcal{C}$ with distinguished left duality $\{(X^\vee, \widetilde{\mathrm{ev}}_X)\}_{X\in\mathrm{Ob}(\mathcal{C})}$ and distinguished monoidal natural isomorphism $\psi \colon 1_\mathcal{C} \to {}^{\vee\vee}?$ where ${}^{\vee\vee}? \colon \mathcal{C} \to \mathcal{C}$ is the strong monoidal functor defined by ${}^{\vee\vee}? = {}^{\vee}? \circ ({}^{\vee}?)^{\mathrm{rev}}$. Indeed, this data turns $\mathcal{C}$ into a pivotal category (in the sense of Section 1.7.1) with pivotal duality

$$\big\{\big(X^* = {}^{\vee}X,\ \mathrm{ev}_X,\ \widetilde{\mathrm{ev}}_X = \mathrm{ev}_{{}^{\vee}X}(\psi_X \otimes \mathrm{id}_{{}^{\vee}X})\colon X \otimes X^* \to \mathbb{1}\big)\big\}_{X\in\mathrm{Ob}(\mathcal{C})}.$$

Conversely, a pivotal category in the sense of Section 1.7.1 determines in a canonical way a monoidal natural isomorphism ψ as above, see Section 2.5.

1.7.5 Pivotal functors

A *pivotal functor* is a strong monoidal functor F between pivotal categories such that the associated monoidal natural isomorphisms F^l and F^r defined in Section 1.6.4 are equal. Set then $F^1 = F^l = F^r$. The composition of two pivotal functors is pivotal. If F is a pivotal functor, then so are $F^{\otimes\mathrm{op}}, F^{\mathrm{op}}, F^{\mathrm{rev}}$, see Section 1.4.1.

A *strictly pivotal functor* is a pivotal functor $F\colon \mathcal{C} \to \mathcal{D}$ such that F^1 is the identity, that is, $F(X^*) = F(X)^*$ and $F^1(X) = \mathrm{id}_{F(X)^*}$ for all $X \in \mathrm{Ob}(\mathcal{C})$. For example, given a pivotal category $\mathcal{C}$, the identity functor $1_{\mathcal{C}}\colon \mathcal{C} \to \mathcal{C}$ and the dual functor $?^*\colon \mathcal{C}^{\mathrm{rev}} \to \mathcal{C}$ are strictly pivotal. Note that a strict monoidal functor $F\colon \mathcal{C} \to \mathcal{D}$ between pivotal categories is strictly pivotal if and only if $F(X^*) = F(X)^*$, $F(\mathrm{ev}_X) = \mathrm{ev}_{F(X)}$, and $F(\widetilde{\mathrm{ev}}_X) = \widetilde{\mathrm{ev}}_{F(X)}$ for all $X \in \mathrm{Ob}(\mathcal{C})$. This means that the distinguished left (respectively) right duality in $\mathcal{C}$ is a lift along F of the left (respectively) right duality in $\mathcal{D}$ (see Section 1.6.4). We say then that the pivotal duality in $\mathcal{C}$ is a *lift along* F of the pivotal duality in $\mathcal{D}$.

Pivotal categories $\mathcal{C}$ and $\mathcal{D}$ are *isomorphic* (respectively, *equivalent*) if there is a pivotal isomorphism (respectively, equivalence) $\mathcal{C} \to \mathcal{D}$, that is, a pivotal functor $\mathcal{C} \to \mathcal{D}$ which is an isomorphism (respectively, equivalence) of the underlying categories. For example, for any pivotal category $\mathcal{C}$, the dual functor $?^*\colon \mathcal{C}^{\mathrm{rev}} \to \mathcal{C}$ is a pivotal equivalence. Consequently, the pivotal categories $\mathcal{C}$ and $\mathcal{C}^{\mathrm{rev}}$ are pivotal equivalent and so are the pivotal categories $\mathcal{C}^{\otimes\mathrm{op}} = (\mathcal{C}^{\mathrm{op}})^{\mathrm{rev}}$ and $\mathcal{C}^{\mathrm{op}}$.

1.7.6 Example

Let G and H be groups, $\alpha\colon G^3 \to \Bbbk^*$ and $\beta\colon H^3 \to \Bbbk^*$ be 3-cocycles, $d\colon G \to \Bbbk^*$ and $d'\colon H \to \Bbbk^*$ be group homomorphisms. Consider the pivotal categories $G_{\Bbbk}^{\alpha,d}$ and $H_{\Bbbk}^{\beta,d'}$ from Example 1.7.3. By Example 1.4.4, any pair (f,γ), where $f\colon G \to H$ is a group homomorphism and $\gamma\colon G\times G \to \Bbbk^*$ is a 2-cocycle verifying (1.8), induces a strong monoidal functor $G_{\Bbbk}^{\alpha,d} \to H_{\Bbbk}^{\beta,d'}$. It is pivotal if and only if $d'f = d$. Consequently, if there is an isomorphism $f\colon G \to H$ such that the two 3-cocycles $\alpha, f^*(\beta)\colon G^3 \to \Bbbk^*$ represent the same class in $H^3(G;\Bbbk^*)$ and $d'f = d$, then $G_{\Bbbk}^{\alpha,d}$ and $H_{\Bbbk}^{\beta,d'}$ are equivalent pivotal categories.

1.7.7 Exercise

Let $F\colon \mathcal{C} \to \mathcal{D}$ be a pivotal equivalence between pivotal categories. Let $G\colon \mathcal{D} \to \mathcal{C}$ be a quasi-inverse of F and let $\alpha\colon 1_{\mathcal{C}} \to GF$ be a natural isomorphism. Endow G with the unique structure of a strong monoidal functor such that α is monoidal (see Exercise 1.4.6). Prove that G is pivotal.

1.7.8 Exercise

Prove that isomorphism and equivalence of pivotal categories are equivalence relations.

1.7.9 Exercise

Let $\mathcal{C}$ be a pivotal category and let $1_{\mathcal{C}}$ be its identity endofunctor. Prove that for any strong monoidal natural automorphisms φ, μ of $1_{\mathcal{C}}$, the family of triples

$$\big\{\big(X^*,\, \mathrm{ev}_X^{\varphi} = \mathrm{ev}_X(\mathrm{id}_{X^*} \otimes \varphi_X),\, \widetilde{\mathrm{ev}}_X^{\mu} = \widetilde{\mathrm{ev}}_X(\mu_X \otimes \mathrm{id}_{X^*})\big)\big\}_{X \in \mathrm{Ob}(\mathcal{C})}$$

is a pivotal duality in the underlying monoidal category of $\mathcal{C}$. The resulting pivotal category is denoted by $\mathcal{C}^{\varphi,\mu}$. Prove that the left and right dual functors of $\mathcal{C}^{\varphi,\mu}$ coincide with those of $\mathcal{C}$. Prove that for any strong monoidal natural automorphisms $\varphi, \mu, \varphi', \mu'$ of $1_{\mathcal{C}}$, the (strict monoidal) endofunctor $1_{\mathcal{C}}$ yields a pivotal isomorphism $\mathcal{C}^{\varphi,\mu} \to \mathcal{C}^{\varphi',\mu'}$ if and only if $\varphi^{-1}\mu = (\varphi')^{-1}\mu'$. Consequently, the pivotal categories $\mathcal{C}^{\varphi,\mu}$ and $\mathcal{C}^{1,\varphi^{-1}\mu}$ are isomorphic, where $1 = \{\mathrm{id}_X\}_{X \in \mathrm{Ob}(\mathcal{C})}$.

1.7.10 Exercise

Let G, α, d be as in Example 1.7.3 and let $G_{\Bbbk}^{\alpha}$ and $G_{\Bbbk}^{\alpha,d}$ be the pivotal categories defined there. By Example 1.4.4 (where $H = G$, $\beta = \alpha$, $f = f' = \mathrm{id}_G$, $\gamma = \gamma' = 1$, and $\chi = d$), the homomorphism $d\colon G \to \Bbbk^*$ determines a strong monoidal automorphism μ of the identity endofunctor of $G_{\Bbbk}^{\alpha}$. Prove that in the notation of Exercise 1.7.9, we have $G_{\Bbbk}^{\alpha,d} = (G_{\Bbbk}^{\alpha})^{1,\mu}$.

Chapter 2

The graphical calculus

A beautiful method, first suggested by Roger Penrose [Pe] in a special case, allows to represent morphisms in categories by planar diagrams. We discuss this method and its applications focusing on the case of pivotal categories.

2.1 Pictorial representation of morphisms

We discuss the basics of the Penrose graphical calculus.

2.1.1 The case of categories

Given a category $\mathcal{C}$, the identity id_X of an object X of $\mathcal{C}$, a morphism $f\colon X \to Y$ in $\mathcal{C}$, and the composition of two morphisms $f\colon X \to Y$ and $g\colon Y \to Z$ may be graphically represented as follows:

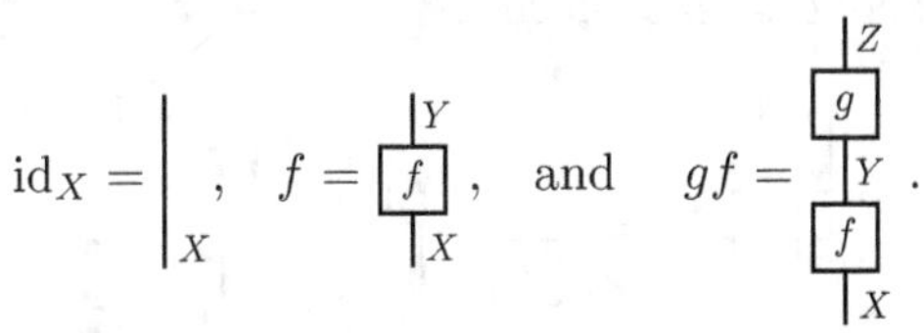

We call X, Y, Z the *colors* of the corresponding arcs of the diagram and call f, g the *colors* of the boxes.

2.1.2 The case of monoidal categories

Suppose that $\mathcal{C}$ is a monoidal category. The monoidal product of two morphisms $f\colon X \to Y$ and $g\colon U \to V$ in $\mathcal{C}$ is represented by juxtaposition:

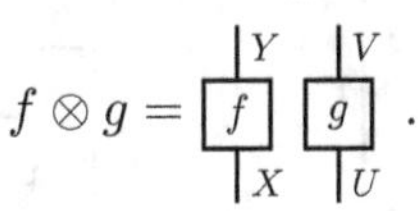

We can also use boxes with several strands attached to their horizontal sides. For example, a morphism $f\colon X \otimes Y \to A \otimes B \otimes C$ with $X, Y, A, B, C \in \mathrm{Ob}(\mathcal{C})$ may be represented in various ways, such as

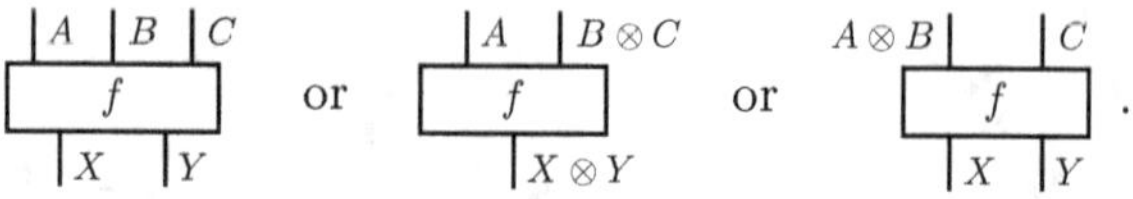

In accordance with conventions of Section 1.4.5, we ignore here the associativity constraint between the objects $A \otimes B \otimes C = (A \otimes B) \otimes C$ and $A \otimes (B \otimes C)$.

A box whose lower/upper side has no attached strands represents a morphism with source/target $\mathbb{1}$. For example, morphisms $\alpha\colon \mathbb{1} \to \mathbb{1}$, $\beta\colon \mathbb{1} \to X$, $\gamma\colon X \to \mathbb{1}$ with $X \in \mathrm{Ob}(\mathcal{C})$ may be represented by the diagrams

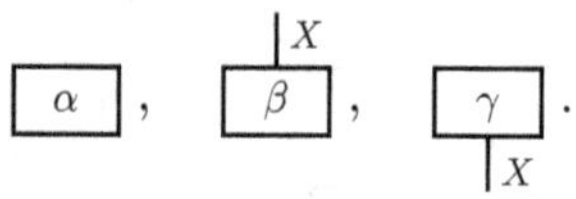

In accordance with conventions of Section 1.4.5, we ignore the unitality constraints. Therefore, deleting a strand colored by $\mathbb{1}$ in a diagram does not change the morphism represented by the diagram.

The *level-exchange property* allows us to push boxes lying on the same horizontal level up or down so that they move to different horizontal levels. The functoriality of $\otimes$ ensures that this operation keeps the morphism represented by the diagram. The simplest case of the level-exchange property is shown in Figure 2.1, where $f\colon X \to Y$ and $g\colon U \to V$ are any morphisms in $\mathcal{C}$. Figure 2.1 graphically expresses the formulas

$$f \otimes g = (\mathrm{id}_Y \otimes g)(f \otimes \mathrm{id}_U) = (f \otimes \mathrm{id}_V)(\mathrm{id}_X \otimes g).$$

In Figure 2.1 and in the sequel, the equality sign between the diagrams means the equality of the corresponding morphisms.

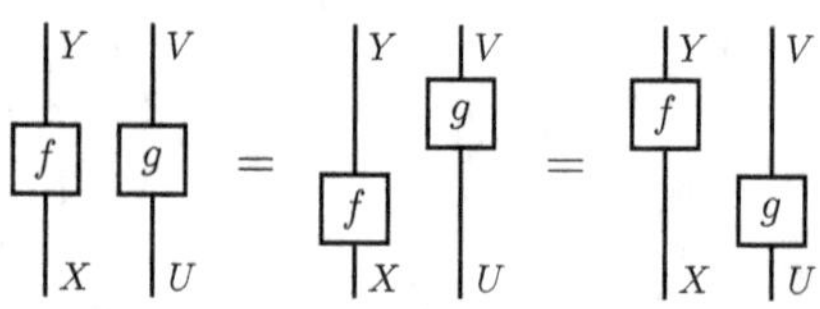

Figure 2.1: The level-exchange property

For example, consider a non-degenerate pairing $\omega\colon X \otimes Y \to \mathbb{1}$ in $\mathcal{C}$ with inverse Ω. Conditions (1.9) are graphically represented as

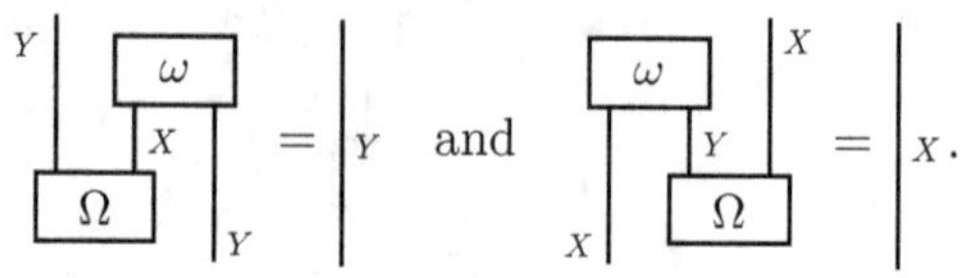

(2.1)

The uniqueness of the inverse may be graphically proved as follows:

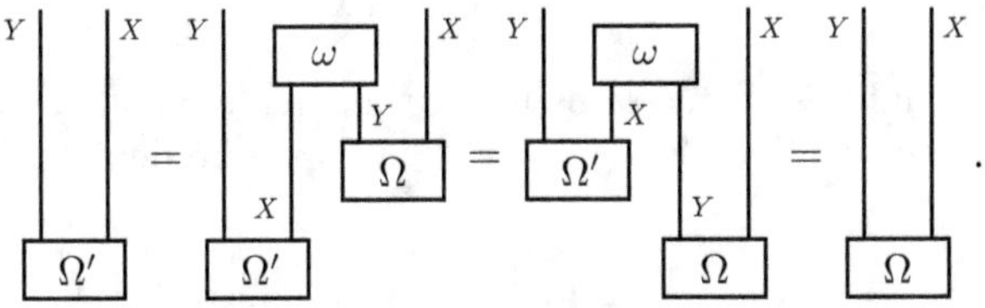

The bijection (1.10) and its inverse are graphically represented as

$\alpha \mapsto$ 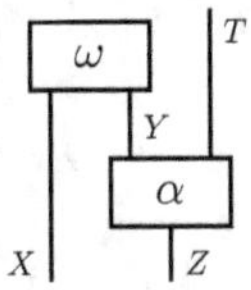 and $\beta \mapsto$ 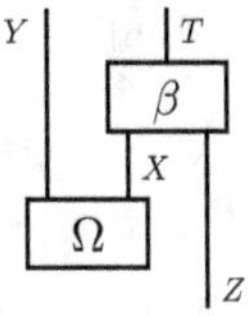 .

Suppose now that $\mathcal{C}$ is a left rigid category with distinguished left duality $\{({}^\vee X, \mathrm{ev}_X)\}_{X \in \mathrm{Ob}(\mathcal{C})}$. By Section 1.6.3, the left dual of a morphism $f\colon X \to Y$ in $\mathcal{C}$ is the morphism ${}^\vee f\colon {}^\vee Y \to {}^\vee X$ defined by

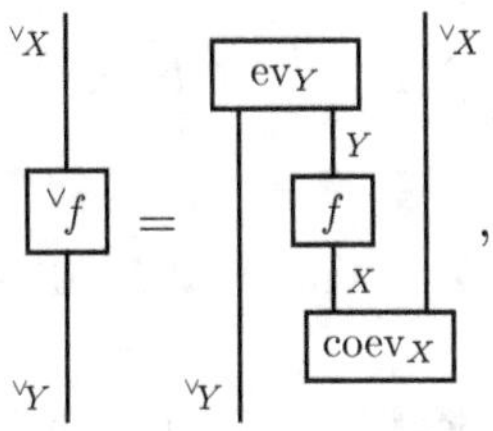

where coev_X is the inverse of the pairing ev_X. It follows from (2.1) that

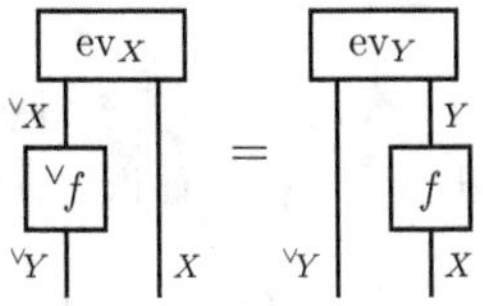 and 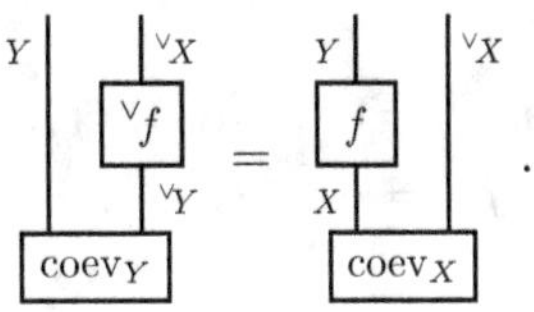 . (2.2)

Similar identities hold for the right duals of morphisms in a right rigid category with distinguished right duality.

2.1.3 The case of pivotal categories

Suppose now that $\mathcal{C}$ is a pivotal category. In this case, the graphical calculus is modified as follows. First, all arcs in the diagrams must be oriented. An arc colored with $X \in \mathrm{Ob}(\mathcal{C})$ and oriented downward contributes X to the source/target of morphisms (as in the pictures above where all arcs are secretly oriented downward). An arc colored with $X \in \mathrm{Ob}(\mathcal{C})$ and oriented upward contributes X^* to the source/target of morphisms. For example,

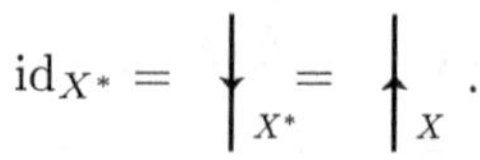

Here, as above, equalities of diagrams mean the equalities of the associated morphisms. A morphism $f\colon X^* \otimes Y \to U \otimes V^* \otimes W$ may be depicted as:

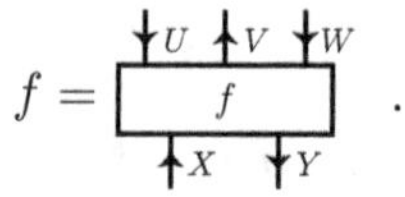

Second, we picture the duality morphisms in $\mathcal{C}$ as follows:

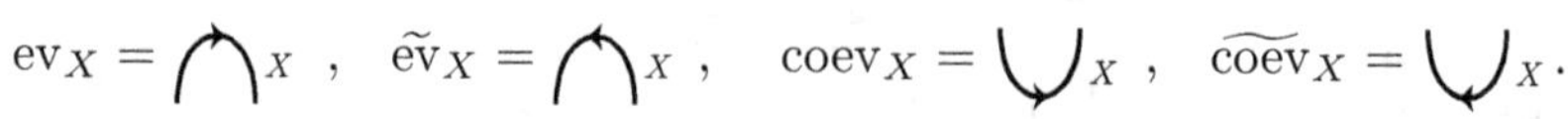

The two leftmost pictures are called *caps*, and the two rightmost pictures are called *cups*. The duality identities (1.12) and (1.13) are graphically expressed in Figure 2.2.

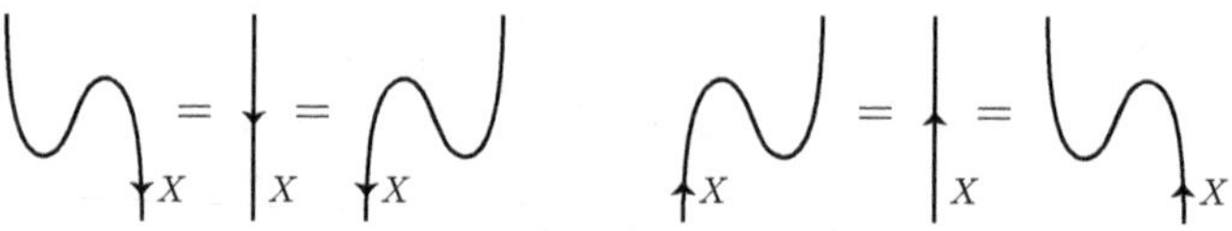

Figure 2.2: The duality identities

The expressions given in Section 1.7.1 for the dual $f^*\colon Y^* \to X^*$ of a morphism $f\colon X \to Y$ in $\mathcal{C}$ are graphically presented in Figure 2.3. These expressions and the duality identities imply the *dual morphism identities*

$$\mathrm{ev}_X(f^* \otimes \mathrm{id}_X) = \mathrm{ev}_Y(\mathrm{id}_{Y^*} \otimes f), \qquad (\mathrm{id}_Y \otimes f^*)\mathrm{coev}_Y = (f \otimes \mathrm{id}_{X^*})\mathrm{coev}_X,$$
$$\widetilde{\mathrm{ev}}_X(\mathrm{id}_X \otimes f^*) = \widetilde{\mathrm{ev}}_Y(f \otimes \mathrm{id}_{Y^*}), \qquad (f^* \otimes \mathrm{id}_Y)\widetilde{\mathrm{coev}}_Y = (\mathrm{id}_{X^*} \otimes f)\widetilde{\mathrm{coev}}_X,$$

see Figure 2.4. Note that the equalities in the first line are special cases of (2.2).

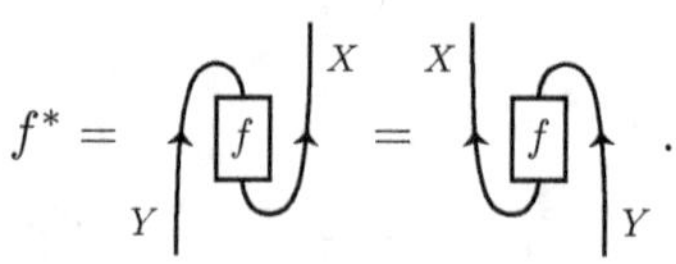

Figure 2.3: Two expressions for the dual morphism

The expressions for the monoidal constraint $?_2^*(X,Y)\colon X^* \otimes Y^* \to (Y \otimes X)^*$

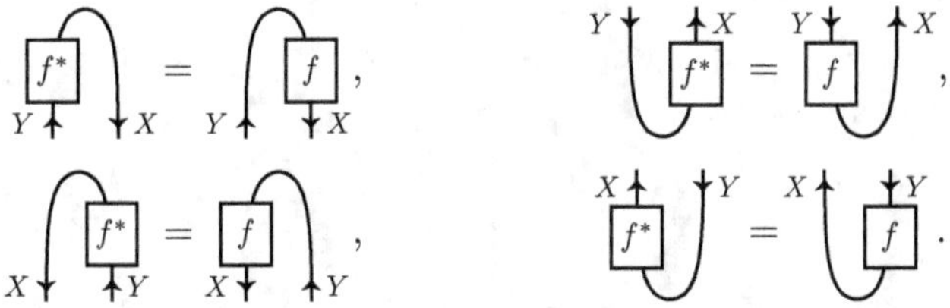

Figure 2.4: The dual morphism identities

given in Section 1.7.1 are depicted as

$?_2^*(X,Y) =$

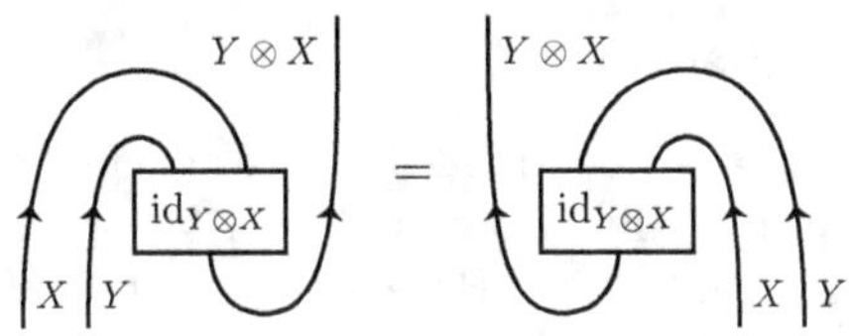

and those for its inverse as

$(?_2^*(X,Y))^{-1} =$

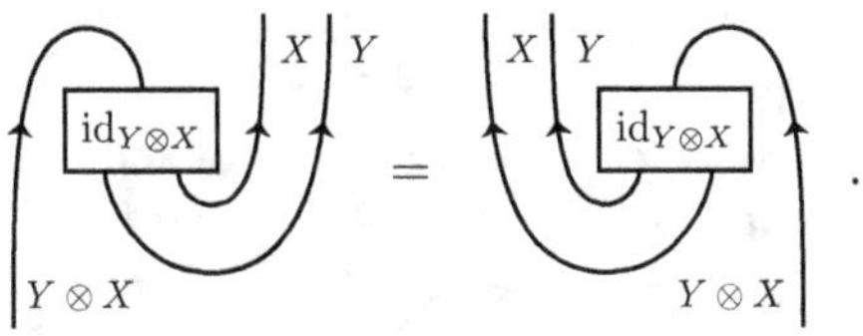

2.1.4 Exercise

Give a graphical proof of the bijectivity of the maps (1.10) and (1.11).

2.1.5 Exercise

Let $\omega\colon X \otimes Y \to \mathbb{1}$ be a non-degenerate pairing in a monoidal category $\mathcal{C}$ with inverse $\Omega\colon \mathbb{1} \to Y \otimes X$. Prove that the map

$$\mathrm{End}_{\mathcal{C}}(X) \to \mathrm{End}_{\mathcal{C}}(Y), \quad f \mapsto (\mathrm{id}_Y \otimes \omega)(\mathrm{id}_Y \otimes f \otimes \mathrm{id}_Y)(\Omega \otimes \mathrm{id}_Y)$$

is an anti-isomorphism of monoids with inverse

$$\mathrm{End}_{\mathcal{C}}(Y) \to \mathrm{End}_{\mathcal{C}}(X), \quad g \mapsto (\omega \otimes \mathrm{id}_X)(\mathrm{id}_X \otimes g \otimes \mathrm{id}_X)(\mathrm{id}_X \otimes \Omega).$$

2.2 Penrose diagrams and associated morphisms

We give a formal description of the graphical calculus in pivotal categories. Throughout this section, we fix a pivotal category $\mathcal{C}$.

2.2.1 Signed objects

A *signed object* of $\mathcal{C}$ is a pair (X, ε) where $X \in \mathrm{Ob}(\mathcal{C})$ and $\varepsilon \in \{+, -\}$. Such a pair (X, ε) determines an object X^ε of $\mathcal{C}$ by

$$X^\varepsilon = \begin{cases} X & \text{if } \varepsilon = +, \\ X^* & \text{if } \varepsilon = -. \end{cases}$$

More generally, any tuple $S = ((X_1, \varepsilon_1), \ldots, (X_n, \varepsilon_n))$ of $n \geq 1$ signed objects of $\mathcal{C}$ determines an object

$$X_S = X_1^{\varepsilon_1} \otimes \cdots \otimes X_n^{\varepsilon_n} \in \mathrm{Ob}(\mathcal{C}).$$

For an empty tuple of signed objects $S = \emptyset$, we set $X_\emptyset = \mathbb{1}$.

2.2.2 Penrose diagrams

We call the pictures as in Section 2.1 Penrose diagrams. More precisely, a *Penrose diagram* $\mathcal{P}$ is a finite family of boxes, directed loops, and directed arcs lying in the horizontal strip $\mathbb{R} \times [0,1] \subset \mathbb{R}^2$. Here a *box* is a (filled) rectangle in $\mathbb{R} \times (0,1)$ with sides parallel to the vertical and horizontal axes, a *loop* is the image of a smooth embedding $S^1 \hookrightarrow \mathbb{R} \times (0,1)$, and an *arc* is the image of a smooth embedding $[0,1] \hookrightarrow \mathbb{R} \times [0,1]$. The interiors of the arcs of $\mathcal{P}$ must lie in $\mathbb{R} \times (0,1)$. The boxes, loops, and interiors of the arcs of $\mathcal{P}$ must be pairwise disjoint. The endpoints of the arcs must lie on the horizontal sides of the boxes or on the lines $\mathbb{R} \times \{0\}$ and $\mathbb{R} \times \{1\}$. The endpoints of the arcs lying on $\mathbb{R} \times \{0\}$ and $\mathbb{R} \times \{1\}$ are called the *free ends* of $\mathcal{P}$. We always assume that the arcs are vertical near their endpoints. The arcs and the loops of a Penrose diagram $\mathcal{P}$ are called the *strands* of $\mathcal{P}$.

A Penrose diagram $\mathcal{P}$ is $\mathcal{C}$*-pre-colored* if every strand of $\mathcal{P}$ is labeled with an object of $\mathcal{C}$ called the *color*. A $\mathcal{C}$-pre-colored Penrose diagram $\mathcal{P}$ gives rise to two tuples $S_\bullet$ and $S^\bullet$ of signed objects of $\mathcal{C}$. The tuple $S_\bullet$ is obtained by moving from the left to the right along $\mathbb{R} \times \{0\}$ and recording the colors and the orientations of the free ends. Here an arc directed downward contributes the sign $+$, and an arc directed upward contributes the sign $-$. The tuple $S^\bullet$ is obtained similarly by moving from the left to the right along $\mathbb{R} \times \{1\}$. Set

$$\mathcal{P}_\bullet = X_{S_\bullet} \in \mathrm{Ob}(\mathcal{C}) \quad \text{and} \quad \mathcal{P}^\bullet = X_{S^\bullet} \in \mathrm{Ob}(\mathcal{C}).$$

In a similar way, the bottom and top horizontal sides of each box Q of $\mathcal{P}$ give rise to two tuples of signed objects of $\mathcal{C}$ and to two objects $Q_\bullet$ and $Q^\bullet$ of $\mathcal{C}$.

A Penrose diagram $\mathcal{P}$ is $\mathcal{C}$*-colored* if it is $\mathcal{C}$-pre-colored and every box Q of $\mathcal{P}$ is labeled with a morphism $Q_\bullet \to Q^\bullet$ in $\mathcal{C}$. The pictures of Section 2.1.3 are examples of $\mathcal{C}$-colored Penrose diagrams. Figure 2.5 presents one more $\mathcal{C}$-colored Penrose diagram $\mathcal{P}$, where A, B, U, V, X, Y, Z are objects of $\mathcal{C}$ and

$$g\colon X \otimes Z^* \to X \otimes V, \quad h\colon A \to Y^* \otimes B, \quad k\colon \mathbb{1} \to Z^* \otimes Y, \quad l\colon \mathbb{1} \to \mathbb{1}$$

are morphisms in $\mathcal{C}$. We have $\mathcal{P}_\bullet = A^* \otimes A$ and $\mathcal{P}^\bullet = U \otimes U^* \otimes A^* \otimes V \otimes B$.

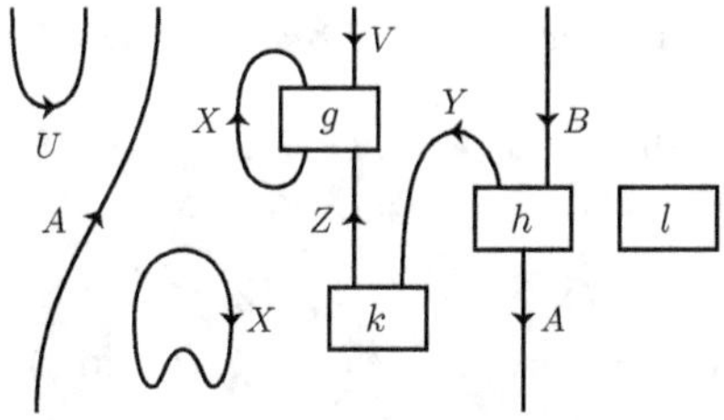

Figure 2.5: A $\mathcal{C}$-colored Penrose diagram

2.2.3 The associated morphism

A Penrose diagram $\mathcal{P}$ is *generic* if one can draw a finite number of horizontal lines in $\mathbb{R} \times (0,1)$ which do not meet the boxes of $\mathcal{P}$, meet the strands of $\mathcal{P}$ transversely, and which, together with the lines $\mathbb{R} \times \{0\}$ and $\mathbb{R} \times \{1\}$, split $\mathbb{R} \times [0,1]$ into several horizontal strips each meeting $\mathcal{P}$ at a juxtaposition of caps, cups, vertical segments, and boxes with attached vertical segments. Here a *vertical segment* is a subarc of a strand of $\mathcal{P}$ which projects injectively to the vertical axis. For example, the Penrose diagram in Figure 2.5 is generic and may be sliced in this way into five horizontal strips:

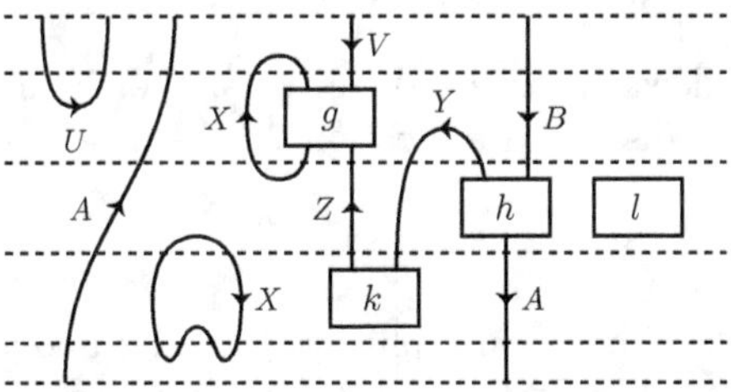

A generic $\mathcal{C}$-colored Penrose diagram $\mathcal{P}$ determines a morphism $\mathcal{P}_\bullet \to \mathcal{P}^\bullet$ in $\mathcal{C}$ as follows. First, draw horizontal lines as above. The resulting caps, cups and vertical segments inherit directions and colors from $\mathcal{P}$. To a vertical segment colored with $X \in \mathrm{Ob}(\mathcal{C})$ we associate id_X if the segment is directed downward and id_{X^*} if it is directed upward. To a box we associate its color. To caps and cups we associate the duality morphisms of $\mathcal{C}$:

X  $= \mathrm{ev}_X$, X  $= \widetilde{\mathrm{ev}}_X$, X  $= \mathrm{coev}_X$, X  $= \widetilde{\mathrm{coev}}_X$.

For each strip, we take the monoidal product of the morphisms associated in this way with the vertical segments, caps, cups, and boxes forming the intersection of the strip with $\mathcal{P}$ and enumerated from the left to the right. Finally, we compose the resulting morphisms proceeding from the bottom to the top. This gives the morphism $\mathcal{P}_\bullet \to \mathcal{P}^\bullet$ *represented by* $\mathcal{P}$. The functoriality of the monoidal product implies that this morphism is independent of the choice of horizontal lines. For example, the Penrose diagram in Figure 2.5 gives rise to the morphism

$$\begin{aligned}
&(\mathrm{id}_U \otimes \mathrm{id}_{U^*} \otimes \mathrm{id}_{A^*} \otimes \mathrm{ev}_X \otimes \mathrm{id}_V \otimes \mathrm{id}_B)\\
&\circ (\mathrm{coev}_U \otimes \mathrm{id}_{A^*} \otimes \mathrm{id}_{X^*} \otimes g \otimes \widetilde{\mathrm{ev}}_Y \otimes \mathrm{id}_B)\\
&\circ (\mathrm{id}_{A^*} \otimes \mathrm{ev}_X \otimes \widetilde{\mathrm{coev}}_X \otimes \mathrm{id}_{Z^*} \otimes \mathrm{id}_Y \otimes h \otimes l)\\
&\circ (\mathrm{id}_{A^*} \otimes \mathrm{id}_{X^*} \otimes \widetilde{\mathrm{ev}}_X \otimes \mathrm{id}_X \otimes k \otimes \mathrm{id}_A)\\
&\circ (\mathrm{id}_{A^*} \otimes \widetilde{\mathrm{coev}}_X \otimes \widetilde{\mathrm{coev}}_X \otimes \mathrm{id}_A)
\end{aligned}$$

from $A^* \otimes A$ to $U \otimes U^* \otimes A^* \otimes V \otimes B$.

2.2.4 Transformations of diagrams

Below we often apply to generic $\mathcal{C}$-colored Penrose diagrams the following color-preserving transformations:

(i) an isotopy in the class of generic $\mathcal{C}$-colored Penrose diagrams;

(ii) a deformation of a vertical segment as in Figure 2.2;

(iii) a transformation pushing a cap/cup/box higher up or lower down.

Lemma 2.1. *The morphism in $\mathcal{C}$ associated to a generic $\mathcal{C}$-colored Penrose diagram is invariant under the transformations* (i)–(iii).

Proof. This lemma follows from the definitions, the duality identities, and the level-exchange property discussed in Section 2.1.2. □

2.2.5 Exercise

Verify using Lemma 2.1 that the morphisms $?_2^*(X,Y)$ and $?_2^*(X,Y)^{-1}$ defined in Section 1.7.1 and depicted in Section 2.1.3 are inverse to each other.

2.3 Two constructions in pivotal categories

We discuss two useful constructions in a pivotal category: canonical isomorphisms between objects and their double duals and generalized evaluations.

Throughout this section, we fix a pivotal category $\mathcal{C}$.

2.3.1 The double duals

For each $X \in \mathrm{Ob}(\mathcal{C})$, we set $X^{**} = (X^*)^*$ and define a morphism $\psi_X \colon X \to X^{**}$ by

$\psi_X =$ 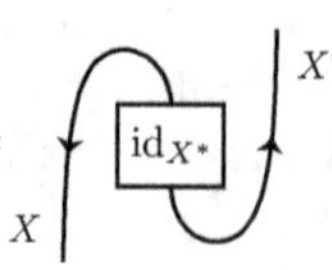

$= (\widetilde{\mathrm{ev}}_X \otimes \mathrm{id}_{X^{**}})(\mathrm{id}_X \otimes \mathrm{coev}_{X^*}).$

Lemma 2.2. *For any $X \in \mathrm{Ob}(\mathcal{C})$, the morphism ψ_X is invertible. Also,*

$\psi_X =$ 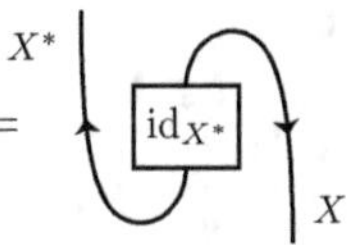 *and* $\psi_X^{-1} =$ 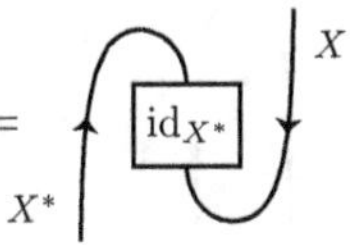 $=$ 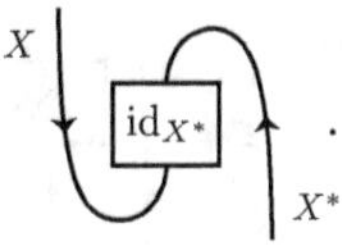 .

Proof. Consider the morphisms

$\phi_X =$ 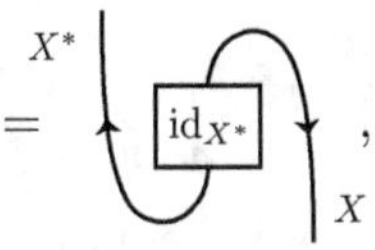 , $\vartheta_X =$ 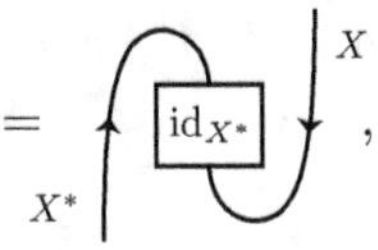 , and $\nu_X =$ 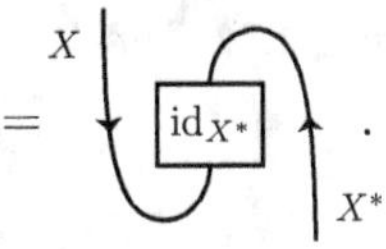 .

The duality identities imply that

$$\vartheta_X \psi_X = \mathrm{id}_X = \nu_X \phi_X \quad \text{and} \quad \psi_X \vartheta_X = \mathrm{id}_{X^{**}} = \phi_X \nu_X.$$

Thus ψ_X and ϕ_X are invertible, $\psi_X^{-1} = \vartheta_X$, and $\phi_X^{-1} = \nu_X$. We must prove that $\psi_X = \phi_X$ or, equivalently, that $\vartheta_X = \nu_X$. For this, consider the pairing:

$$e = (\widetilde{\mathrm{ev}}_X^{\mathrm{rev}})^{?^*} = (?^*)_0^{-1} \circ (\widetilde{\mathrm{coev}}_X)^* \circ (?^*)_2(X, X^*) \colon X^* \otimes X^{**} \to \mathbb{1}.$$

The three morphisms on the right-hand side can be computed using either the left duality morphisms ev, coev or the right duality morphisms $\widetilde{\mathrm{ev}}, \widetilde{\mathrm{coev}}$, see Section 1.7.1. Using the left duality morphisms we obtain

$e =$ 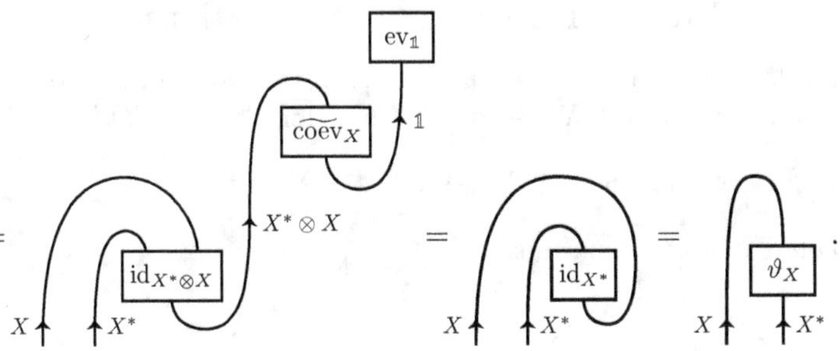

Using the right duality morphisms we similarly obtain

$e =$ 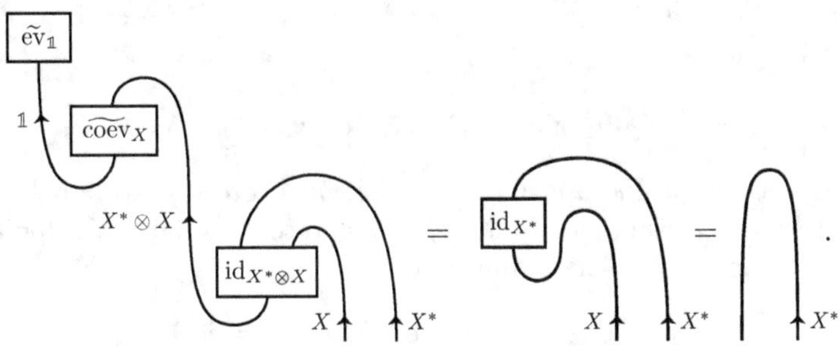

Therefore

$e =$ 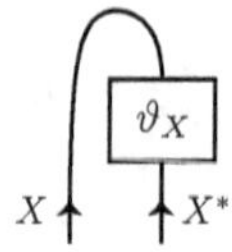 $=$ 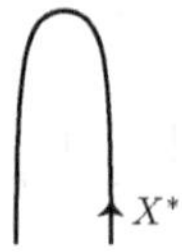 and so $\vartheta_X =$ 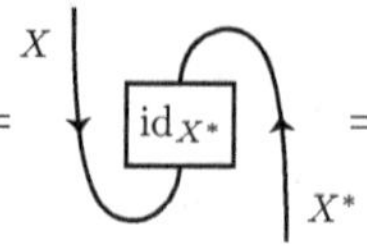 $= \nu_X.$

This concludes the proof of the lemma. □

Lemma 2.3. *The left and right (co)evaluations are related as follows:*

$$\widetilde{\mathrm{ev}}_X = \mathrm{ev}_{X^*}(\psi_X \otimes \mathrm{id}_{X^*}), \qquad \widetilde{\mathrm{coev}}_X = (\mathrm{id}_{X^*} \otimes \psi_X^{-1})\, \mathrm{coev}_{X^*},$$
$$\mathrm{ev}_X = \widetilde{\mathrm{ev}}_{X^*}(\mathrm{id}_{X^*} \otimes \psi_X), \qquad \mathrm{coev}_X = (\psi_X^{-1} \otimes \mathrm{id}_{X^*})\, \widetilde{\mathrm{coev}}_{X^*},$$

for any $X \in \mathrm{Ob}(\mathcal{C})$.

Proof. From the definition of ψ_X, we obtain that

$\mathrm{ev}_{X^*}(\psi_X \otimes \mathrm{id}_{X^*}) =$ 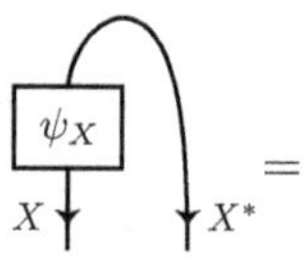 $=$ 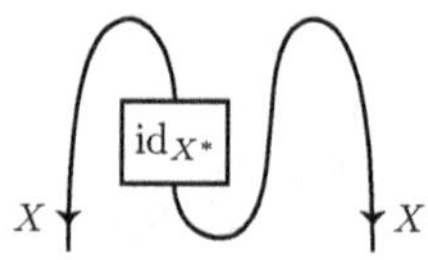 $=$ 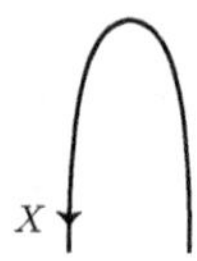 $= \widetilde{\mathrm{ev}}_X.$

The other three formulas are proved similarly using the computations of $\psi_X^{\pm 1}$ in Lemma 2.2. □

2.3.2 Generalized evaluations and coevaluations

By Section 2.2.1, each tuple $S = ((X_1, \varepsilon_1), \dots, (X_n, \varepsilon_n))$ of $n \geq 0$ signed objects of $\mathcal{C}$, determines an object $X_S \in \mathrm{Ob}(\mathcal{C})$. Consider also the *dual tuple*

$$S^* = ((X_n, -\varepsilon_n), \dots, (X_1, -\varepsilon_1)).$$

The $\mathcal{C}$-colored Penrose diagrams

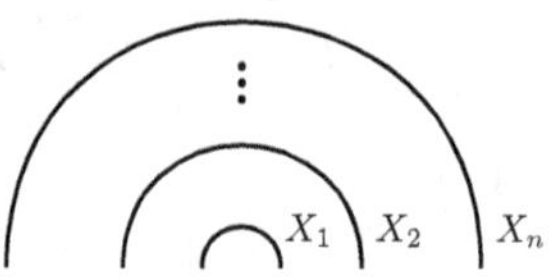 and 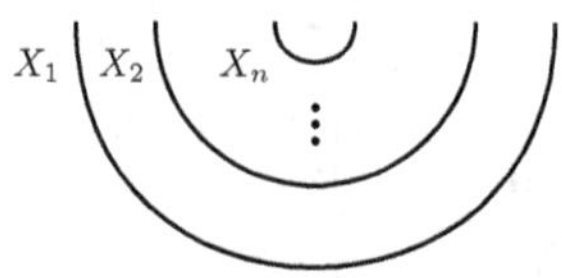

represent morphisms

$$\mathrm{ev}_S \colon X_{S^*} \otimes X_S \to \mathbb{1} \quad \text{and} \quad \mathrm{coev}_S \colon \mathbb{1} \to X_S \otimes X_{S^*}$$

called respectively *evaluation* and *coevaluation*. Here the arc labeled with X_i is oriented toward the right endpoint if $\varepsilon_i = +$ and toward the left endpoint if $\varepsilon_i = -$. It is easy to check using the graphical calculus that

$$(\mathrm{id}_{X_S} \otimes \mathrm{ev}_S)(\mathrm{coev}_S \otimes \mathrm{id}_{X_S}) = \mathrm{id}_{X_S} \tag{2.3}$$

and

$$(\mathrm{ev}_S \otimes \mathrm{id}_{X_{S^*}})(\mathrm{id}_{X_{S^*}} \otimes \mathrm{coev}_S) = \mathrm{id}_{X_{S^*}}. \tag{2.4}$$

Thus, the pairing ev_S is non-degenerate with inverse coev_S. For $n = 0$, by definition, $\emptyset^* = \emptyset$ and $\mathrm{ev}_\emptyset = \mathrm{coev}_\emptyset = \mathrm{id}_{\mathbb{1}}$. For $n = 1$,

$$\mathrm{ev}_{(X,+)} = \mathrm{ev}_X, \quad \mathrm{coev}_{(X,+)} = \mathrm{coev}_X, \quad \mathrm{ev}_{(X,-)} = \widetilde{\mathrm{ev}}_X, \quad \mathrm{coev}_{(X,-)} = \widetilde{\mathrm{coev}}_X.$$

The following example illustrates the case $n = 2$:

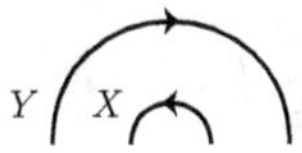

$$= \mathrm{ev}_{((X,-),(Y,+))} \colon (Y^* \otimes X) \otimes (X^* \otimes Y) \to \mathbb{1}.$$

The tuple $S = ((X_1, \varepsilon_1), \ldots, (X_n, \varepsilon_n))$ also determines an isomorphism

$$\Psi_S \colon X_S \to (X_{S^*})^*.$$

For $n = 0$, we set $\Psi_\emptyset = \mathrm{coev}_{\mathbb{1}} = \widetilde{\mathrm{coev}}_{\mathbb{1}} \colon \mathbb{1} \to \mathbb{1}^*$. For $n = 1$ and $X \in \mathrm{Ob}(\mathcal{C})$, set

$$\Psi_{(X,-)} = \mathrm{id}_{X^*} \colon X^* \to X^* \quad \text{and} \quad \Psi_{(X,+)} = \psi_X \colon X \to X^{**},$$

where ψ_X is defined in Section 2.3.1. For $n \geq 2$, we define Ψ_S as the composition of the isomorphism

$$\Psi_{(X_1,\varepsilon_1)} \otimes \cdots \otimes \Psi_{(X_n,\varepsilon_n)} \colon X_S = X_1^{\varepsilon_1} \otimes \cdots \otimes X_n^{\varepsilon_n} \to (X_1^{-\varepsilon_1})^* \otimes \cdots \otimes (X_n^{-\varepsilon_n})^*$$

with the isomorphism

$$(X_1^{-\varepsilon_1})^* \otimes \cdots \otimes (X_n^{-\varepsilon_n})^* \simeq (X_n^{-\varepsilon_n} \otimes \cdots \otimes X_1^{-\varepsilon_1})^* = (X_{S^*})^*$$

induced by the monoidal constraints of the dual functor. The following lemma generalizes Lemma 2.3.

Lemma 2.4. *For any tuple S of signed objects of $\mathcal{C}$,*

$$\mathrm{ev}_{X_S}(\Psi_{S^*} \otimes \mathrm{id}_{X_S}) = \mathrm{ev}_S = \widetilde{\mathrm{ev}}_{X_{S^*}}(\mathrm{id}_{X_{S^*}} \otimes \Psi_S),$$
$$(\mathrm{id}_{X_S} \otimes \Psi_{S^*}^{-1})\, \mathrm{coev}_{X_S} = \mathrm{coev}_S = (\Psi_S^{-1} \otimes \mathrm{id}_{X_{S^*}})\, \widetilde{\mathrm{coev}}_{X_{S^*}}.$$

Proof. The equality $\mathrm{ev}_{X_S}(\Psi_{S^*} \otimes \mathrm{id}_{X_S}) = \mathrm{ev}_S$ is proved by induction on the length of S. The case $S = \emptyset$ follows from the formula $\mathrm{id}_{\mathbb{1}} = \mathrm{ev}_{\mathbb{1}} \widetilde{\mathrm{coev}}_{\mathbb{1}}$. For $S = (X, +)$ with $X \in \mathrm{Ob}(\mathcal{C})$, the required equality follows from the formula $\mathrm{id}_{X^*} \otimes \mathrm{id}_X = \mathrm{id}_{X^* \otimes X}$. For $S = (X, -)$ with $X \in \mathrm{Ob}(\mathcal{C})$, the required equality $\mathrm{ev}_{X^*}(\psi_X \otimes \mathrm{id}_{X^*}) = \widetilde{\mathrm{ev}}_X$ follows from Lemma 2.3. If $\mathrm{ev}_S = \mathrm{ev}_{X_S}(\Psi_{S^*} \otimes \mathrm{id}_{X_S})$ for two tuples of signed objects $S = U$ and $S = V$, then for the concatenated tuple UV,

$$\begin{aligned}
\mathrm{ev}_{UV} &\overset{\text{(i)}}{=} \mathrm{ev}_V(\mathrm{id}_{X_{V^*}} \otimes \mathrm{ev}_U \otimes \mathrm{id}_{X_V}) \\
&\overset{\text{(ii)}}{=} \mathrm{ev}_{X_V}(\mathrm{id}_{X_V^*} \otimes \mathrm{ev}_{X_U} \otimes \mathrm{id}_{X_V})(\Psi_{V^*} \otimes \Psi_{U^*} \otimes \mathrm{id}_{X_U \otimes X_V}) \\
&\overset{\text{(iii)}}{=} \mathrm{ev}_{X_U \otimes X_V}\big(((?^*)_2(X_{V^*}, X_{U^*})(\Psi_{V^*} \otimes \Psi_{U^*}) \otimes \mathrm{id}_{X_U \otimes X_V}\big) \\
&\overset{\text{(iv)}}{=} \mathrm{ev}_{X_U \otimes X_V}(\Psi_{V^*U^*} \otimes \mathrm{id}_{X_{UV}}) \overset{\text{(v)}}{=} \mathrm{ev}_{X_{UV}}(\Psi_{(UV)^*} \otimes \mathrm{id}_{X_{UV}}).
\end{aligned}$$

Here (i) follows from the definitions, (ii) from the assumptions on U and V, (iii) from the definition of the monoidal constraint $(?^*)_2$, (iv) from the definition of Ψ, and (v) from the obvious equalities $X_U \otimes X_V = X_{UV}$ and $V^*U^* = (UV)^*$. Therefore $\mathrm{ev}_S = \mathrm{ev}_{X_S}(\Psi_{S^*} \otimes \mathrm{id}_{X_S})$ for $S = UV$.

The equality $\widetilde{\mathrm{ev}}_{X_{S^*}}(\mathrm{id}_{X_{S^*}} \otimes \Psi_S) = \mathrm{ev}_S$ is proved similarly. Here, the case $S = (X, -)$ follows from the formula $\mathrm{id}_X \otimes \mathrm{id}_{X^*} = \mathrm{id}_{X \otimes X^*}$, and the case $S = (X, +)$ amounts to the equality $\widetilde{\mathrm{ev}}_{X^*}(\mathrm{id}_{X^*} \otimes \psi_X) = \mathrm{ev}_X$ which follows from Lemma 2.3.

Using the formula $\mathrm{ev}_S = \mathrm{ev}_{X_S}(\Psi_{S^*} \otimes \mathrm{id}_{X_S})$ and (2.3), we obtain

$$\begin{aligned}(\mathrm{id}_{X_S} \otimes \mathrm{ev}_S)((\mathrm{id}_{X_S} \otimes \Psi_{S^*}^{-1})\mathrm{coev}_{X_S} \otimes \mathrm{id}_{X_S}) &= (\mathrm{id}_{X_S} \otimes \mathrm{ev}_{X_S})(\mathrm{coev}_{X_S} \otimes \mathrm{id}_{X_S}) \\ &= \mathrm{id}_{X_S} = (\mathrm{id}_{X_S} \otimes \mathrm{ev}_S)(\mathrm{coev}_S \otimes \mathrm{id}_{X_S}).\end{aligned}$$

The uniqueness of the inverse of a non-degenerate pairing implies the formula $(\mathrm{id}_{X_S} \otimes \Psi_{S^*}^{-1})\,\mathrm{coev}_{X_S} = \mathrm{coev}_S$. The formula $(\Psi_S^{-1} \otimes \mathrm{id}_{X_{S^*}})\,\widetilde{\mathrm{coev}}_{X_{S^*}} = \mathrm{coev}_S$ is proved similarly. □

2.3.3 Exercise

Prove that $\psi_{X^*} = (\psi_X^{-1})^* = (\psi_X^*)^{-1}$ for any $X \in \mathrm{Ob}(\mathcal{C})$.

2.4 Isotopy invariance

In this section, $\mathcal{C}$ is a pivotal category. We extend the graphical calculus to arbitrary $\mathcal{C}$-colored Penrose diagrams and show that the morphism in $\mathcal{C}$ represented by such a diagram is preserved under ambient isotopies of the diagram.

2.4.1 Isotopy of Penrose diagrams

Two $\mathcal{C}$-colored Penrose diagrams $\mathcal{P}$ and $\mathcal{Q}$ are *isotopic* if there is an ambient isotopy of the band $\mathbb{R} \times [0,1]$ in itself that carries the strands and boxes of $\mathcal{P}$ onto the strands and boxes of $\mathcal{Q}$ preserving the orientations and the colors. Clearly, $\mathcal{P}_\bullet = \mathcal{Q}_\bullet$ and $\mathcal{P}^\bullet = \mathcal{Q}^\bullet$.

Lemma 2.5. *Isotopic generic $\mathcal{C}$-colored Penrose diagrams represent the same morphism in $\mathcal{C}$.*

Proof. Recall the transformations (i)–(iii) of generic $\mathcal{C}$-colored Penrose diagrams introduced in Section 2.2.4. We need another transformation (iv) rotating a box of a Penrose diagram clockwise to the angle of 360° and letting the attached strands follow the movement. It is understood that the diagram must be generic before and after each transformation. A well-known folklore result says that any two isotopic generic $\mathcal{C}$-colored Penrose diagrams can be related by a generic isotopy which splits as a composition of a finite number of transformations (i)–(iv) and the inverse transformations. By Lemma 2.1, the transformations (i)–(iii) preserve the associated morphism in $\mathcal{C}$. It remains to prove the invariance of this morphism under rotation of a box. Let $S = ((X_1, \varepsilon_1), \dots, (X_m, \varepsilon_m))$ and $T = ((Y_1, \nu_1), \dots, (Y_n, \nu_n))$ be the tuples of signed objects of $\mathcal{C}$ associated with the bottom and top sides of the box, respectively, and let $f\colon X_S \to X_T$ be the color of the box. It is enough to prove that the diagrams

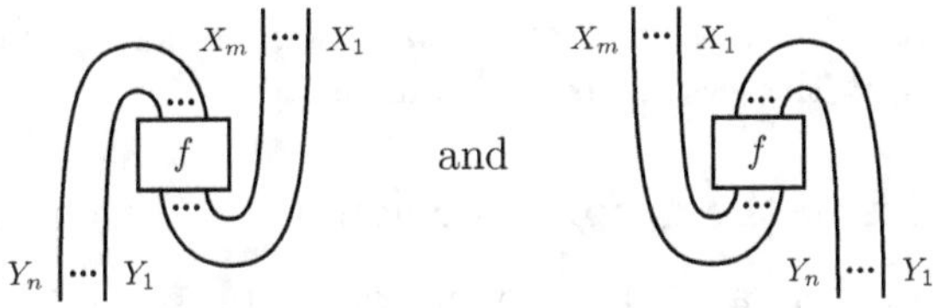

represent the same morphism in $\mathcal{C}$ (the directions of the arcs determined by ε_i, ν_j are not shown in the picture.) We give a graphical proof as follows:

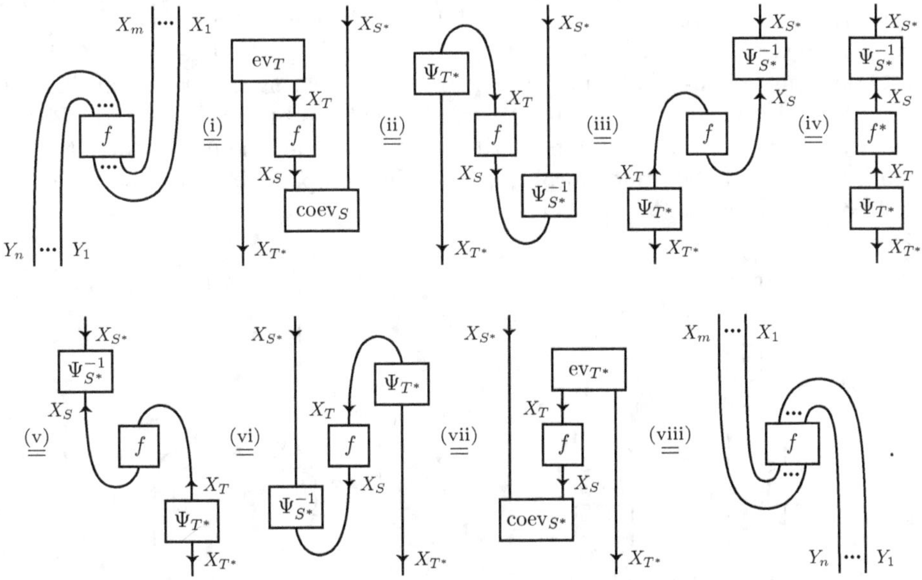

Here, the equalities (i) and (viii) follow from the definition of the generalized (co)evaluations (see Section 2.3.2), (ii) and (vii) from Lemma 2.4, (iii) and (vi) from the level-exchange property (see Figure 2.1), and (iv) and (v) from the equalities in Figure 2.3. □

2.4.2 Extension of the graphical calculus

We now extend the graphical calculus to arbitrary (not necessarily generic) $\mathcal{C}$-colored Penrose diagrams. Observe that any such diagram $\mathcal{P}$ is isotopic to a generic $\mathcal{C}$-colored Penrose diagram. The latter represents a morphism $\langle\mathcal{P}\rangle\colon \mathcal{P}_\bullet \to \mathcal{P}^\bullet$ in $\mathcal{C}$. Any two generic diagrams isotopic to $\mathcal{P}$ are isotopic to each other and so represent the same morphism in $\mathcal{C}$ by Lemma 2.5. Therefore the morphism $\langle\mathcal{P}\rangle$ depends only on $\mathcal{P}$. The following theorem, due to Joyal and Street [JS3, JS4], directly follows from Lemma 2.5.

Theorem 2.6. *Isotopic $\mathcal{C}$-colored Penrose diagrams represent the same morphism in $\mathcal{C}$.*

In our computations, we will often make no difference between $\mathcal{C}$-colored Penrose diagrams and the associated morphisms in $\mathcal{C}$.

2.4.3 Properties of the graphical calculus

Besides the isotopy, there are other modifications of $\mathcal{C}$-colored Penrose diagrams which keep the associated morphisms. One such modification trades a rectangular piece of a $\mathcal{C}$-colored Penrose diagram $\mathcal{P} \subset \mathbb{R} \times [0,1]$ for a single box. More precisely, suppose that $Q \subset \mathbb{R} \times (0,1)$ is a rectangle with sides parallel to the axes such that the vertical sides of Q do not meet $\mathcal{P}$ while the horizontal sides of Q are disjoint from the boxes of $\mathcal{P}$ and meet the strands of $\mathcal{P}$ transversely. The part of $\mathcal{P}$ lying in Q determines a $\mathcal{C}$-colored Penrose diagram representing a certain morphism f in $\mathcal{C}$. Replacing this part of $\mathcal{P}$ with a single box colored by f, we transform $\mathcal{P}$ into a new $\mathcal{C}$-colored Penrose diagram $\mathcal{P}_Q$. It follows from the functoriality of $\otimes$ that $\mathcal{P}$ and $\mathcal{P}_Q$ represent the same morphism in $\mathcal{C}$. This property includes the following replacements discussed in Section 2.1:

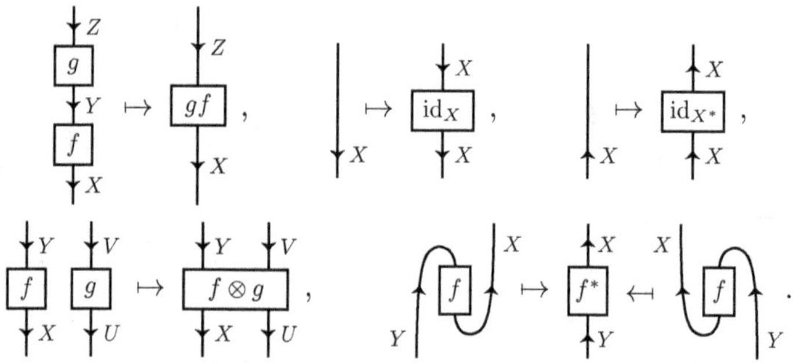

This property also implies the dual morphism identities in Figure 2.4.

We state one more useful property of Penrose diagrams. Suppose that some strands and boxes of a $\mathcal{C}$-colored Penrose diagram $\mathcal{P}$ form a $\mathcal{C}$-colored Penrose diagram $\mathcal{P}_0$ which has no free ends and is not connected to $\mathcal{P}' = \mathcal{P} \setminus \mathcal{P}_0$ by any arcs. Note that $\mathcal{P}'$ is itself a $\mathcal{C}$-colored Penrose diagram. If $\mathcal{C}$ is pure, then the morphisms $\alpha \in \mathrm{End}_{\mathcal{C}}(\mathbb{1})$, f, f' represented respectively by $\mathcal{P}_0$, $\mathcal{P}$, $\mathcal{P}'$ satisfy $f = \alpha \cdot f' = f' \cdot \alpha$ where $\cdot$ is the action of $\mathrm{End}_{\mathcal{C}}(\mathbb{1})$ on the Hom-sets of $\mathcal{C}$, see Sections 1.3.2 and 1.3.3. The equality $f = \alpha \cdot f'$ is obtained by pushing $\mathcal{P}_0$ across vertical arcs of $\mathcal{P}'$ into a position to the left of $\mathcal{P}'$.

The definition of a $\mathcal{C}$-colored Penrose diagram may be generalized replacing the band $\mathbb{R} \times [0,1]$ by $\mathbb{R} \times [a,b]$ for any real numbers $a < b$. For diagrams without free ends, one may replace $\mathbb{R} \times [0,1]$ with the whole plane $\mathbb{R}^2$. All the definitions and results above extend to such generalized diagrams with the obvious changes.

2.4.4 Exercise*

Let $\mathcal{P}$ be a $\mathcal{C}$-colored Penrose diagram having a loop c. Let $\mathcal{P}'$ be the $\mathcal{C}$-colored Penrose diagram obtained from $\mathcal{P}$ by replacing the color $X \in \mathrm{Ob}(\mathcal{C})$ of c by an object Y of $\mathcal{C}$ isomorphic to X. Let $\mathcal{P}''$ be the $\mathcal{C}$-colored Penrose diagram obtained from $\mathcal{P}$ by reversing the orientation of c and replacing its color X by X^*. Prove that $\langle \mathcal{P} \rangle = \langle \mathcal{P}' \rangle = \langle \mathcal{P}'' \rangle$.

2.5 Pivotal structures

We discuss double dual functors and pivotal structures of pivotal categories.

2.5.1 The double dual functor

Let $\mathcal{C}$ be a pivotal category. The *double dual functor* $?^{**}\colon \mathcal{C} \to \mathcal{C}$ carries any object X of $\mathcal{C}$ to $X^{**} = (X^*)^*$ and any morphism f in $\mathcal{C}$ to $f^{**} = (f^*)^*$. Recall from Section 2.3.1 the isomorphism

$\psi_X =$ 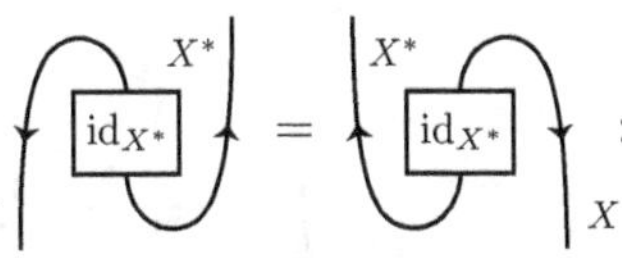 $: X \to X^{**}.$

The dual morphism identities (Figure 2.4) imply that the family $\psi = \{\psi_X\}_{X \in \mathrm{Ob}(\mathcal{C})}$ is a natural transformation from the identity functor $1_\mathcal{C}\colon \mathcal{C} \to \mathcal{C}$ to the double dual functor $?^{**}\colon \mathcal{C} \to \mathcal{C}$. Lemma 2.2 implies that ψ is a natural isomorphism. It is called the *pivotal structure* of $\mathcal{C}$.

To state the monoidality properties of $?^{**}$, note that the identity functor $1_\mathcal{C}$ is strict monoidal and the dual functor $?^*\colon \mathcal{C}^{\mathrm{rev}} \to \mathcal{C}$ is strong monoidal, see Section 1.7.1. Consider the strong monoidal functor $(?^*)^{\mathrm{rev}}\colon \mathcal{C} \to \mathcal{C}^{\mathrm{rev}}$ induced by $?^*$, see Section 1.4.1. Then $?^{**} = ?^* \circ (?^*)^{\mathrm{rev}}\colon \mathcal{C} \to \mathcal{C}$ is a composition of strong monoidal functors and hence is a strong monoidal functor, see Section 1.4.1.

Theorem 2.7. *The pivotal structure $\psi\colon 1_\mathcal{C} \to ?^{**}$ is a monoidal natural isomorphism.*

Proof. To prove that ψ is monoidal, we must show that $\psi_\mathbb{1} = (?^{**})_0$ and

$$\psi_{X \otimes Y} = (?^{**})_2(X,Y)(\psi_X \otimes \psi_Y)$$

for all $X, Y \in \mathrm{Ob}(\mathcal{C})$. By Section 1.4.1,

$$(?^{**})_0 = \big((?^*)_0^{-1}\big)^*(?^*)_0 = \big((?^*)_0^*\big)^{-1}(?^*)_0$$

and

$$(?^{**})_2(X,Y) = \big((?^*)_2(Y,X)^{-1}\big)^*(?^*)_2(X^*,Y^*) = \big((?^*)_2(Y,X)^*\big)^{-1}(?^*)_2(X^*,Y^*).$$

So, we need to prove that $(?^*)_0^* \psi_{\mathbb{1}} = (?^*)_0$ and

$$(?^*)_2(Y,X)^* \psi_{X\otimes Y} = (?^*)_2(X^*,Y^*)(\psi_X \otimes \psi_Y).$$

By the dual morphism identities in Figure 2.4,

$$\psi_{\mathbb{1}} = (\widetilde{\mathrm{ev}}_{\mathbb{1}} \otimes \mathrm{id}_{\mathbb{1}^{**}})\,\mathrm{coev}_{\mathbb{1}^*} = (\widetilde{\mathrm{ev}}_{\mathbb{1}})^* \mathrm{coev}_{\mathbb{1}}.$$

Thus

$$(?^*)_0^* \psi_{\mathbb{1}} = (\widetilde{\mathrm{coev}}_{\mathbb{1}})^* (\widetilde{\mathrm{ev}}_{\mathbb{1}})^* \mathrm{coev}_{\mathbb{1}} = (\widetilde{\mathrm{ev}}_{\mathbb{1}} \widetilde{\mathrm{coev}}_{\mathbb{1}})^* \mathrm{coev}_{\mathbb{1}} = (\mathrm{id}_{\mathbb{1}})^* \mathrm{coev}_{\mathbb{1}} = (?^*)_0.$$

The isotopy invariance of the graphical calculus (Theorem 2.6), the dual morphism identities, and the formulas $\mathrm{id}_{X\otimes Y} = \mathrm{id}_X \otimes \mathrm{id}_Y$, $(\mathrm{id}_X)^* = \mathrm{id}_{X^*}$, and $(\mathrm{id}_Y)^* = \mathrm{id}_{Y^*}$ imply that

$(?^*)_2(Y,X)^* \psi_{X\otimes Y} =$ 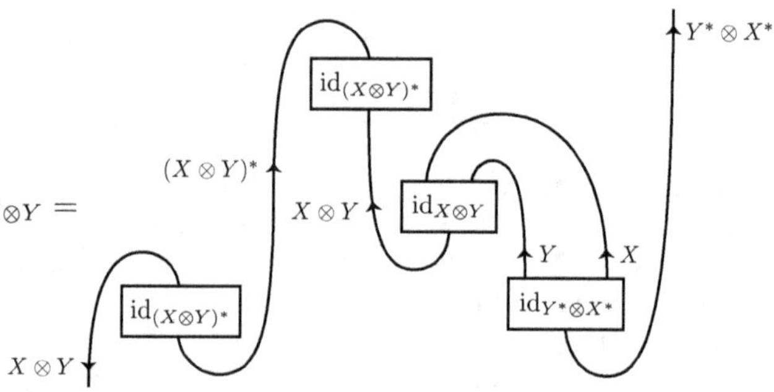

$=$ 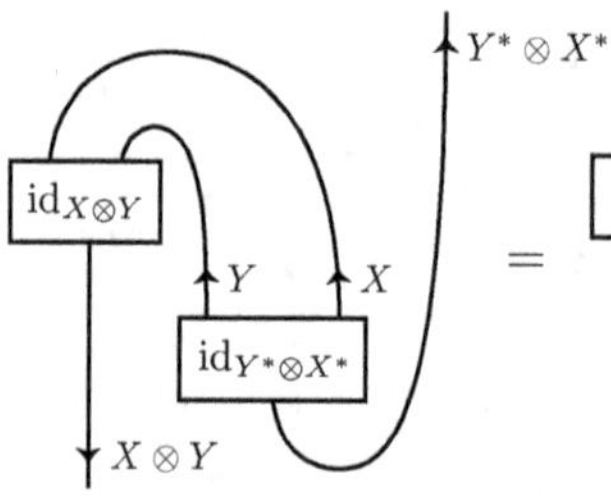

$=$ 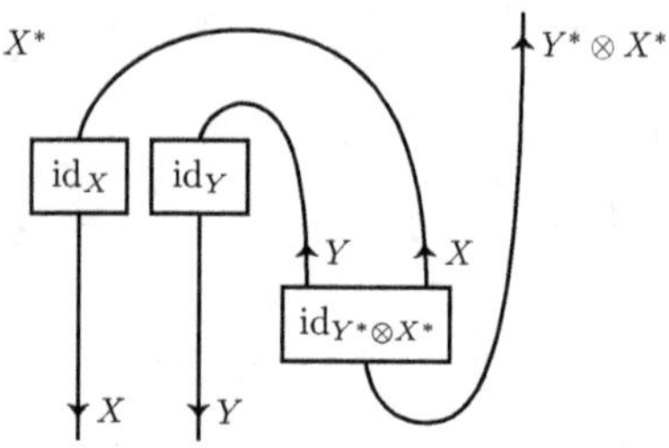

$=$ 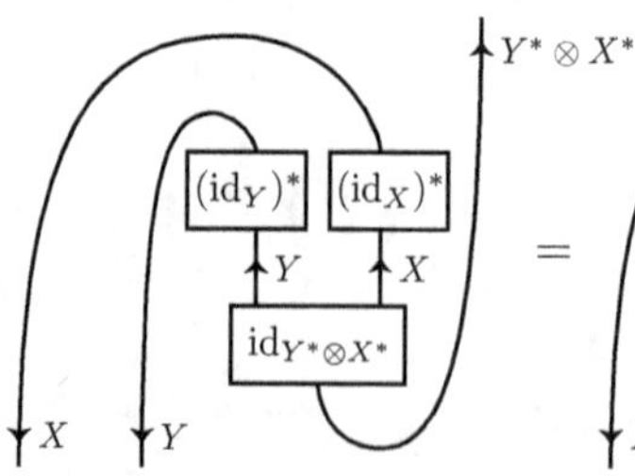

$=$ 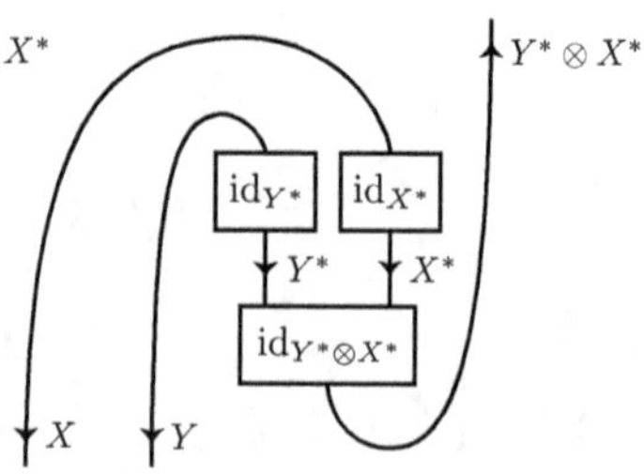

$=$ 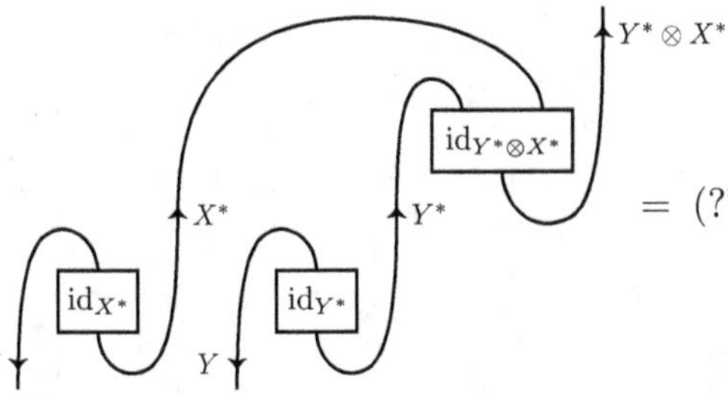 $= (?^*)_2(X^*, Y^*)(\psi_X \otimes \psi_Y).$

Hence, ψ is monoidal. □

2.5.2 Exercise

Prove that any pivotal functor $F: \mathcal{C} \to \mathcal{D}$ between pivotal categories preserves the pivotal structure in the sense that for all $X \in \mathrm{Ob}(\mathcal{C})$,

$$F^1(X^*)F(\psi^{\mathcal{C}}_X) = (F^1(X))^* \, \psi^{\mathcal{D}}_{F(X)} : F(X) \to F(X^*)^*,$$

where $\psi^{\mathcal{C}}$ and $\psi^{\mathcal{D}}$ are the pivotal structures of $\mathcal{C}$ and $\mathcal{D}$, respectively.

2.6 Traces and dimensions

We discuss traces of endomorphisms and dimensions of objects in pivotal categories.

2.6.1 Traces

Let $\mathcal{C}$ be a pivotal category. The *left trace* of an endomorphism f of an object X of $\mathcal{C}$ is defined by

$\mathrm{tr}_l(f) =$ 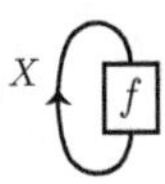 $= \mathrm{ev}_X(\mathrm{id}_{X^*} \otimes f)\,\widetilde{\mathrm{coev}}_X \in \mathrm{End}_{\mathcal{C}}(\mathbb{1}).$

The *right trace* of f is defined by

$\mathrm{tr}_r(f) =$ 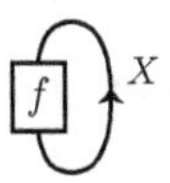 $= \widetilde{\mathrm{ev}}_X(f \otimes \mathrm{id}_{X^*})\,\mathrm{coev}_X \in \mathrm{End}_{\mathcal{C}}(\mathbb{1}).$

Both tr_l and tr_r are symmetric: for any morphisms $p: X \to Y$ and $q: Y \to X$ in $\mathcal{C}$,

$$\mathrm{tr}_l(pq) = \mathrm{tr}_l(qp) \quad \text{and} \quad \mathrm{tr}_r(pq) = \mathrm{tr}_r(qp).$$

Indeed, Theorem 2.6 and Section 2.4.3 imply that

$\mathrm{tr}_l(pq) =$ 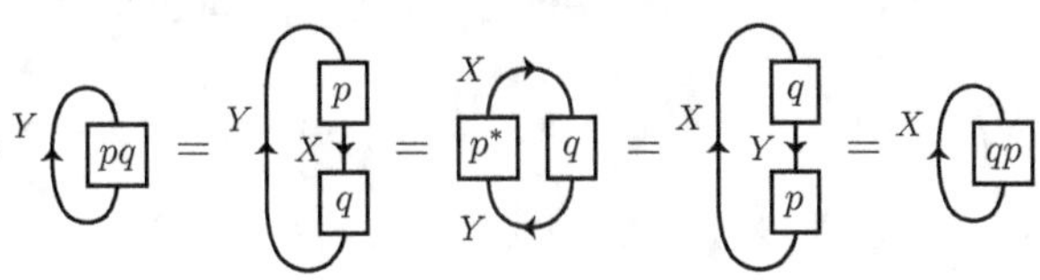 $= \mathrm{tr}_l(qp),$

and similarly for the right trace. Likewise, one proves using the graphical calculus that for any $\alpha \in \mathrm{End}_{\mathcal{C}}(\mathbb{1})$ and any endomorphisms f, g in $\mathcal{C}$,

$$\begin{gathered}\mathrm{tr}_l(\alpha) = \mathrm{tr}_r(\alpha) = \alpha, \quad \mathrm{tr}_l(f \cdot \alpha) = \alpha\, \mathrm{tr}_l(f), \quad \mathrm{tr}_r(\alpha \cdot f) = \alpha\, \mathrm{tr}_r(f),\\ \mathrm{tr}_l(f \otimes g) = \mathrm{tr}_l(\mathrm{tr}_l(f) \cdot g), \quad \mathrm{tr}_l(f) = \mathrm{tr}_r(f^*),\\ \mathrm{tr}_r(f \otimes g) = \mathrm{tr}_r(f \cdot \mathrm{tr}_r(g)), \quad \mathrm{tr}_r(f) = \mathrm{tr}_l(f^*).\end{gathered}$$

Here the dot stands for the left/right actions of the ground monoid $\mathrm{End}_{\mathcal{C}}(\mathbb{1})$, see Section 1.3.2. These formulas imply the identities

$$\mathrm{tr}_l(f) = \mathrm{tr}_l(f^{**}) \quad \text{and} \quad \mathrm{tr}_r(f) = \mathrm{tr}_r(f^{**}).$$

Lemma 2.3 implies that for any $X \in \mathrm{Ob}(\mathcal{C})$, the left and right traces of an endomorphism $f\colon X^* \to X^*$ are computed by

$\mathrm{tr}_l(f) =$ 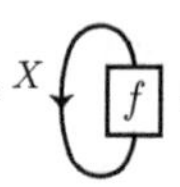 $= \widetilde{\mathrm{ev}}_X(\mathrm{id}_X \otimes f)\, \mathrm{coev}_X$

and

$\mathrm{tr}_r(f) =$ 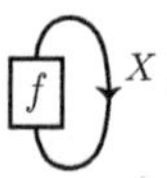 $= \mathrm{ev}_X(f \otimes \mathrm{id}_X)\, \widetilde{\mathrm{coev}}_X.$

If $\mathcal{C}$ is pure (see Section 1.3.3), then the traces are $\otimes$-*multiplicative*: for any endomorphisms f, g in $\mathcal{C}$,

$$\mathrm{tr}_l(f \otimes g) = \mathrm{tr}_l(f)\, \mathrm{tr}_l(g) \quad \text{and} \quad \mathrm{tr}_r(f \otimes g) = \mathrm{tr}_r(f)\, \mathrm{tr}_r(g).$$

Indeed,

$$\mathrm{tr}_l(f \otimes g) = \mathrm{tr}_l(\mathrm{tr}_l(f) \cdot g) = \mathrm{tr}_l(g \cdot \mathrm{tr}_l(f)) = \mathrm{tr}_l(f)\, \mathrm{tr}_l(g),$$

and similarly with l replaced by r.

2.6.2 Dimensions

The *left dimension* of an object X of a pivotal category $\mathcal{C}$ is defined by

$\dim_l(X) =$  $= \mathrm{ev}_X\, \widetilde{\mathrm{coev}}_X \in \mathrm{End}_{\mathcal{C}}(\mathbb{1}).$

The *right dimension* of X is defined by

$\dim_r(X) =$  $= \widetilde{\mathrm{ev}}_X\, \mathrm{coev}_X \in \mathrm{End}_{\mathcal{C}}(\mathbb{1}).$

Note that $\dim_l(X) = \mathrm{tr}_l(\mathrm{id}_X)$ and $\dim_r(X) = \mathrm{tr}_r(\mathrm{id}_X)$. Therefore

$$\dim_l(X) = \dim_r(X^*) = \dim_l(X^{**})$$

and

$$\dim_r(X) = \dim_l(X^*) = \dim_r(X^{**}).$$

The symmetry of the traces implies that isomorphic objects have the same dimensions. Clearly, $\dim_l(\mathbb{1}) = \dim_r(\mathbb{1}) = \mathrm{id}_{\mathbb{1}}$. If $\mathcal{C}$ is pure, then the dimensions are $\otimes$-*multiplicative*: for any $X, Y \in \mathrm{Ob}(\mathcal{C})$,

$$\dim_l(X \otimes Y) = \dim_l(X)\dim_l(Y) \quad \text{and} \quad \dim_r(X \otimes Y) = \dim_r(X)\dim_r(Y).$$

2.6.3 Behavior under pivotal functors

The following lemma shows that the traces of endomorphisms and dimensions of objects are preserved under pivotal functors.

Lemma 2.8. *Let $F\colon \mathcal{C} \to \mathcal{D}$ be a pivotal functor between pivotal categories. Then for any $X \in \mathrm{Ob}(\mathcal{C})$ and any endomorphism $f\colon X \to X$ in $\mathcal{C}$,*

$$\mathrm{tr}_l(F(f)) = F_0^{-1}F(\mathrm{tr}_l(f))F_0 \quad \textit{and} \quad \dim_l(F(X)) = F_0^{-1}F(\dim_l(X))F_0,$$

and similarly with l replaced by r.

Proof. The functoriality of F implies that

$$F_0^{-1}F(\mathrm{tr}_l(f))F_0 = F_0^{-1}F(\mathrm{ev}_X)F(\mathrm{id}_{X^*} \otimes f)F(\widetilde{\mathrm{coev}}_X)F_0.$$

It follows from the pivotality of F that

$F_0^{-1}F(\mathrm{ev}_X) =$ 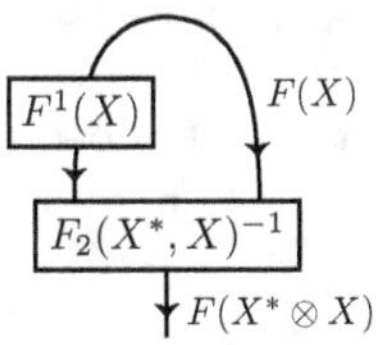 and $F(\widetilde{\mathrm{coev}}_X)F_0 =$ 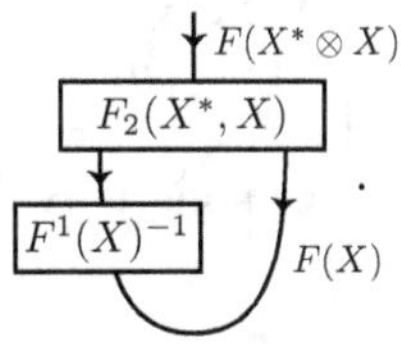 .

Using the naturality of F_2, we obtain

$F_0^{-1}F(\mathrm{tr}_l(f))F_0 =$ 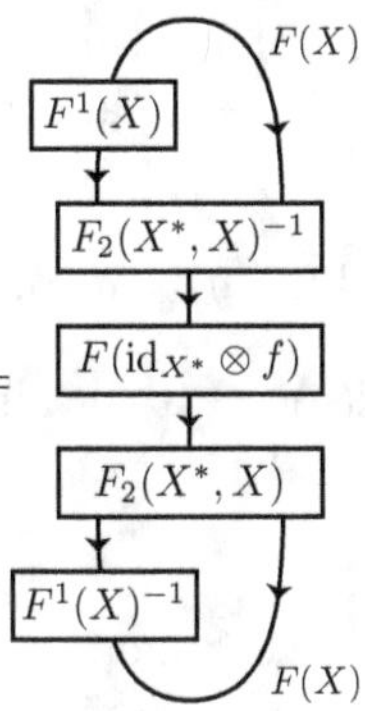 $=$ 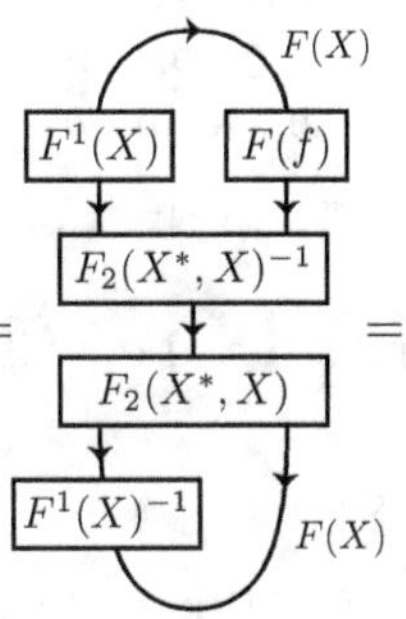 $=$ 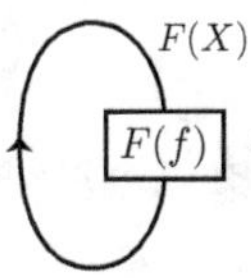 $= \mathrm{tr}_l(F(f))$.

The second equality in the claim of the lemma is obtained by taking $f = \mathrm{id}_X$. Right traces and right dimensions are treated similarly. □

2.7 Spherical categories

We introduce and study spherical categories.

2.7.1 Sphericity

A *spherical category* is a pivotal category whose left and right traces coincide, that is, $\mathrm{tr}_l(f) = \mathrm{tr}_r(f)$ for every endomorphism f in the category. Then $\mathrm{tr}(f) = \mathrm{tr}_l(f) = \mathrm{tr}_r(f)$ is the *trace of* f.

In a spherical category, the left and right dimensions of any object X are equal. Then $\dim(X) = \dim_l(X) = \dim_r(X)$ is the *dimension of* X.

The properties of the traces imply that in any spherical category, tr and dim are $\otimes$-multiplicative. Indeed, for any endomorphisms f, g,

$$\mathrm{tr}(f \otimes g) = \mathrm{tr}_l(f \otimes g) = \mathrm{tr}_l(\mathrm{tr}_l(f) \cdot g) = \mathrm{tr}_r(\mathrm{tr}_l(f) \cdot g) = \mathrm{tr}_l(f)\,\mathrm{tr}_r(g) = \mathrm{tr}(f)\,\mathrm{tr}(g).$$

Lemma 2.9. *If $\mathcal{C}$ is a spherical category, then the morphism in $\mathcal{C}$ represented by a $\mathcal{C}$-colored Penrose diagram $\mathcal{P}$ without free ends is invariant under isotopies of $\mathcal{P}$ in the 2-dimensional sphere $S^2 = \mathbb{R}^2 \cup \{\infty\}$.*

Proof. In view of Theorem 2.6, it suffices to prove the invariance of the associated morphism under an isotopy pushing a strand of $\mathcal{P}$ across the point ∞. Removing from $\mathcal{P}$ a small piece of this strand near ∞, we obtain a $\mathcal{C}$-colored Penrose diagram with one bottom free end and one top free end. The latter diagram represents a morphism $f\colon X^\varepsilon \to X^\varepsilon$ where $X \in \mathrm{Ob}(\mathcal{C})$ is the color of our strand and the sign $\varepsilon = \pm$ is determined by the direction of this strand near ∞. Before (respectively, after) hitting ∞, the diagram $\mathcal{P}$ represents $\mathrm{tr}_l(f)$ (respectively, $\mathrm{tr}_r(f)$) or vice versa. The sphericity of $\mathcal{C}$ implies that $\mathrm{tr}_l(f) = \mathrm{tr}_r(f)$. □

2.7.2 Example

Let α be a $\Bbbk^*$-valued 3-cocycle on a group G. Then the pivotal category $G_\Bbbk^\alpha$ from Example 1.7.3 is spherical and $\dim(g) = 1_\Bbbk \in \Bbbk$ for all $g \in \mathrm{Ob}(G_\Bbbk^\alpha) = G$. More generally, consider a group homomorphism $d\colon G \to \Bbbk^*$ and the pivotal category $G_\Bbbk^{\alpha,d}$ from Example 1.7.3. Then for any $g \in \mathrm{Ob}(G_\Bbbk^{\alpha,d}) = G$,

$$\dim_l(g) = d(g)^{-1} \quad \text{and} \quad \dim_r(g) = d(g).$$

The category $G_\Bbbk^{\alpha,d}$ is spherical if and only if $d(g)^2 = 1_\Bbbk$ for all $g \in G$.

2.7.3 Sphericity of $\mathrm{proj}_\Bbbk$

Consider the pivotal category $\mathrm{proj}_\Bbbk$ of projective $\Bbbk$-modules of finite type (see Example 1.7.2). We identify $\mathrm{End}_{\mathrm{proj}_\Bbbk}(\Bbbk)$ with $\Bbbk$ by assigning to each $\Bbbk$-linear homomorphism $\Bbbk \to \Bbbk$ its value at $1_\Bbbk \in \Bbbk$. The following lemma implies that the category $\mathrm{proj}_\Bbbk$ is spherical.

Lemma 2.10. *Left and right traces of endomorphisms in* $\mathrm{proj}_{\Bbbk}$ *coincide with the standard (Hattori–Stallings) trace. Left and right dimensions of objects of* $\mathrm{proj}_{\Bbbk}$ *coincide with the standard (Hattori–Stallings) dimension of projective* $\Bbbk$*-modules.*

Proof. Let f be an endomorphism of a projective $\Bbbk$-module X of finite type. Pick a $\Bbbk$-module R such that $Y = X \oplus R$ is a free $\Bbbk$-module of finite rank. Let $p\colon Y \to X$ be the projection and $j\colon X \to Y$ be the inclusion. The Hattori–Stallings trace of f is the usual trace $\mathrm{Tr}(jfp)$ of the matrix of jfp with respect to a basis of Y. We prove that this trace is equal to $\mathrm{tr}_l(f)$. Pick a basis $(e_i)_i$ of Y and the dual basis $(e_i^*)_i$ of $Y^\star$. Then $\widetilde{\mathrm{coev}}_Y(1_{\Bbbk}) = \sum_i e_i^* \otimes_{\Bbbk} e_i$ and, under the identification $\mathrm{End}_{\mathrm{proj}_{\Bbbk}}(\Bbbk) = \Bbbk$, we obtain that

$$\begin{aligned}\mathrm{Tr}(jfp) &= \sum_i e_i^*\big(jfp(e_i)\big) = \mathrm{ev}_Y(\mathrm{id}_{Y^\star} \otimes_{\Bbbk} jfp)\,\widetilde{\mathrm{coev}}_Y(1_{\Bbbk})\\ &= \mathrm{tr}_l(jfp) = \mathrm{tr}_l(fpj) = \mathrm{tr}_l(f\mathrm{id}_X) = \mathrm{tr}_l(f).\end{aligned}$$

The Hattori–Stallings dimension of X, defined as the Hattori–Stallings trace of id_X, is equal to $\mathrm{tr}_l(\mathrm{id}_X) = \dim_l(X)$. The right traces and dimensions in $\mathrm{proj}_{\Bbbk}$ are treated similarly (or, see Lemma 3.5 below). □

2.7.4 Exercise

Prove that if two pivotal categories are equivalent and one of them is spherical, then so is the other category.

Chapter 3

Braided categories

We review braidings and symmetries in monoidal categories. Then we discuss the graphical calculus for braided pivotal categories and define ribbon categories.

3.1 Braidings and symmetries

3.1.1 Braidings

A braiding in a monoidal category $\mathcal{C}$ is a commutativity constraint for the monoidal product in $\mathcal{C}$. More precisely, a *braiding* in $\mathcal{C}$ is a natural family of isomorphisms

$$\tau = \{\tau_{X,Y} \colon X \otimes Y \to Y \otimes X\}_{X,Y \in \mathrm{Ob}(\mathcal{C})}$$

which is $\otimes$-*multiplicative* in X and Y in the sense that for any $X, Y, Z \in \mathrm{Ob}(\mathcal{C})$,

$$\tau_{X,Y\otimes Z} = (\mathrm{id}_Y \otimes \tau_{X,Z})(\tau_{X,Y} \otimes \mathrm{id}_Z), \tag{3.1}$$

$$\tau_{X\otimes Y,Z} = (\tau_{X,Z} \otimes \mathrm{id}_Y)(\mathrm{id}_X \otimes \tau_{Y,Z}). \tag{3.2}$$

The naturality of τ means that τ is a natural isomorphism from $\otimes$ to $\otimes^{\mathrm{op}}$, that is,

$$\tau_{X',Y'}(f \otimes g) = (g \otimes f)\tau_{X,Y}$$

for any morphisms $f \colon X \to X'$ and $g \colon Y \to Y'$ in $\mathcal{C}$.

A *braided category* is a monoidal category endowed with a braiding. Any braided category is pure. This follows from the fact that for any braiding τ in a monoidal category $\mathcal{C}$ and any $X \in \mathrm{Ob}(\mathcal{C})$, we have:

$$\tau_{X,\mathbb{1}} = \tau_{\mathbb{1},X} = \mathrm{id}_X. \tag{3.3}$$

A *braided subcategory* of a braided category $\mathcal{C}$ is a monoidal subcategory of $\mathcal{C}$ endowed with the braiding obtained by restricting the one in $\mathcal{C}$.

A *braided functor* between braided categories $(\mathcal{C}, \tau)$ and $(\mathcal{C}', \tau')$ is a monoidal functor $F\colon \mathcal{C} \to \mathcal{C}'$ such that for all $X, Y \in \mathrm{Ob}(\mathcal{C})$,

$$F_2(Y, X)\, \tau'_{F(X),F(Y)} = F(\tau_{X,Y})\, F_2(X, Y).$$

Two braided categories are *isomorphic* (respectively, *equivalent*) if there is a braided strong monoidal isomorphism (respectively, equivalence) between them.

3.1.2 Symmetries

A braiding τ in a monoidal category $\mathcal{C}$ is *symmetric* if for all $X, Y \in \mathrm{Ob}(\mathcal{C})$,

$$\tau_{Y,X} \circ \tau_{X,Y} = \mathrm{id}_{X\otimes Y}\colon X \otimes Y \to X \otimes Y.$$

For shortness, symmetric braidings are called *symmetries*.

A *symmetric category* is a monoidal category endowed with a symmetry. A *symmetric subcategory* of a symmetric category $\mathcal{C}$ is a braided subcategory of $\mathcal{C}$. A *symmetric functor* is a braided monoidal functor between symmetric categories.

Two symmetric categories are *isomorphic* (respectively, *equivalent*) if they are isomorphic (respectively, equivalent) as braided categories.

3.1.3 Example

The monoidal category of sets from Example 1.2.3 carries a symmetry formed by the flips

$$X \times Y \to Y \times X, \quad (x, y) \mapsto (y, x)$$

for all sets X, Y and all $x \in X$, $y \in Y$.

3.1.4 Example

The monoidal category $\mathrm{Mod}_{\Bbbk}$ from Example 1.2.4 carries a symmetry formed by the flips

$$X \otimes_{\Bbbk} Y \to Y \otimes_{\Bbbk} X, \quad x \otimes_{\Bbbk} y \mapsto y \otimes_{\Bbbk} x$$

for all $X, Y \in \mathrm{Ob}(\mathrm{Mod}_{\Bbbk})$ and all $x \in X$, $y \in Y$.

3.1.5 Example

The monoidal category $\mathrm{SMod}_{\Bbbk}$ from Example 1.2.5 carries a symmetry formed by the $\Bbbk$-linear isomorphisms

$$X \otimes Y \to Y \otimes X, \quad x \otimes_{\Bbbk} y \mapsto (-1)^{ij} y \otimes_{\Bbbk} x$$

for all super $\Bbbk$-modules $X = X_0 \oplus X_1$, $Y = Y_0 \oplus Y_1$ and all $x \in X_i$, $y \in Y_j$ with $i, j \in \{0, 1\}$.

3.1.6 Example

Let G be a monoid and $\alpha\colon G^3 \to \Bbbk^*$ be a 3-cocycle. The monoidal category $G_{\Bbbk}^{\alpha}$ from Example 1.2.6 is braided if and only if G is abelian. If G is abelian, then the braidings in $G_{\Bbbk}^{\alpha}$ bijectively correspond to maps $c\colon G \times G \to \Bbbk^*$ such that, for all $g, h, j \in G$,

$$\begin{aligned} c(gh, j) &= c(g, j)\, c(h, j)\, \alpha(g, h, j)^{-1}\, \alpha(h, g, j)\, \alpha(h, j, g)^{-1}, \\ c(g, hj) &= c(g, h)\, c(g, j)\, \alpha(g, h, j)\, \alpha(g, j, h)^{-1}\, \alpha(j, g, h). \end{aligned}$$

The braiding determined by c is given by $c(g, h) \in \Bbbk = \mathrm{Hom}_{\mathcal{C}}(gh, hg)$ for $g, h \in G$. Note that this braiding may be symmetric or not depending on c. For example, if $\alpha = 1$, then $c\colon G \times G \to \Bbbk^*$ is just a bilinear form and the associated braiding is symmetric if and only if $c(h, g) = c(g, h)^{-1}$ for all $g, h \in G$.

3.1.7 Example

The *mirror* of a braiding τ in a monoidal category $\mathcal{C}$ is the braiding in $\mathcal{C}$

$$\overline{\tau} = \{\overline{\tau}_{X,Y} = \tau_{Y,X}^{-1}\}_{X,Y \in \mathrm{Ob}(\mathcal{C})}.$$

The braided category $(\mathcal{C}, \overline{\tau})$ is called the *mirror* of the braided category $(\mathcal{C}, \tau)$.

3.1.8 Example

Given a braided category $\mathcal{C}$ with braiding τ, the opposite monoidal categories $\mathcal{C}^{\mathrm{op}}$, $\mathcal{C}^{\otimes\mathrm{op}}$, $\mathcal{C}^{\mathrm{rev}}$ (see Section 1.2.2) are braided with braidings τ^{op}, $\tau^{\otimes\mathrm{op}}$, τ^{rev} defined respectively by

$$\tau_{X,Y}^{\mathrm{op}} = \tau_{X,Y}^{-1}, \quad \tau_{X,Y}^{\otimes\mathrm{op}} = \tau_{Y,X}, \quad \tau_{X,Y}^{\mathrm{rev}} = \tau_{Y,X}^{-1}$$

for all $X, Y \in \mathrm{Ob}(\mathcal{C})$. If $\mathcal{C}$ is symmetric, then so are $\mathcal{C}^{\mathrm{op}}$, $\mathcal{C}^{\otimes\mathrm{op}}$, and $\mathcal{C}^{\mathrm{rev}}$.

3.1.9 Exercise

Write the axioms of a braiding of a monoidal category using explicitly the associativity constraint of the category.

3.1.10 Exercise

Let $F\colon \mathcal{C} \to \mathcal{D}$ be a braided strong monoidal equivalence between braided categories. Let $G\colon \mathcal{D} \to \mathcal{C}$ be a quasi-inverse of F and $\alpha\colon 1_{\mathcal{C}} \to GF$ be a natural isomorphism. Endow G with the unique structure of a strong monoidal functor such that α is monoidal (see Exercise 1.4.6). Prove that G is braided.

3.1.11 Exercise

Prove that the isomorphism and equivalence of braided categories satisfy the usual requirements on equivalence relations (reflexivity, symmetry, transitivity).

3.1.12 Exercise*

A *weak braiding* in a monoidal category $\mathcal{C}$ is a family of morphisms $\tau = \{\tau_{X,Y} : X \otimes Y \to Y \otimes X\}_{X,Y \in \mathrm{Ob}(\mathcal{C})}$ which is natural in the sense of Section 3.1.1 and satisfies (3.1), (3.2), (3.3) for all $X, Y, Z \in \mathrm{Ob}(\mathcal{C})$. For example, all braidings in $\mathcal{C}$ are weak braidings. Prove that any weak braiding τ in $\mathcal{C}$ satisfies the following Yang–Baxter equation: for all $X, Y, Z \in \mathrm{Ob}(\mathcal{C})$,

$$(\tau_{Y,Z} \otimes \mathrm{id}_X)(\mathrm{id}_Y \otimes \tau_{X,Z})(\tau_{X,Y} \otimes \mathrm{id}_Z) = (\mathrm{id}_Z \otimes \tau_{X,Y})(\tau_{X,Z} \otimes \mathrm{id}_Y)(\mathrm{id}_X \otimes \tau_{Y,Z}).$$

Prove that if $\mathcal{C}$ is left (or right) rigid, then any weak braiding in $\mathcal{C}$ is a braiding.

3.2 The Reidemeister calculus

We extend the graphical calculus in a braided pivotal category from Penrose diagrams to more general Reidemeister diagrams.

3.2.1 Reidemeister diagrams

Reidemeister diagrams generalize Penrose diagrams by allowing the strands to have double transversal crossings such that at each crossing, one of the two meeting passes is distinguished. The distinguished pass at a crossing is called an *over-pass* and the other one is called an *under-pass*. The under-passes are pictorially represented by broken curves like

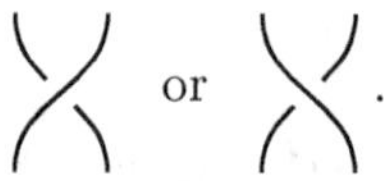

Given a pivotal category $\mathcal{C}$, the definition of a $\mathcal{C}$-colored Penrose diagram extends to Reidemeister diagrams word for word. As in the theory of Penrose diagrams, a $\mathcal{C}$-colored Reidemeister diagram $\mathcal{R}$ may have free ends lying on the lines $\mathbb{R} \times \{0\}$ and $\mathbb{R} \times \{1\}$; the orientations and the colors of the strands attached to the free ends yield two tuples of signed objects of $\mathcal{C}$ and determine two objects $\mathcal{R}_\bullet$, $\mathcal{R}^\bullet$ of $\mathcal{C}$. If $\mathcal{C}$ carries a braiding τ, then we define a morphism $\langle \mathcal{R} \rangle : \mathcal{R}_\bullet \to \mathcal{R}^\bullet$ as follows. First, we isotop $\mathcal{R}$ in $\mathbb{R} \times [0, 1]$ to ensure that the strands of $\mathcal{R}$ are directed downward at all crossings. For example,

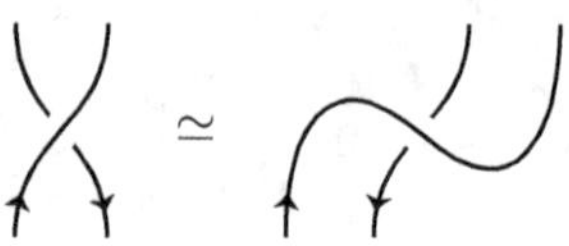

is such an isotopy for one choice of orientations. Then we replace each crossing by a box colored with the appropriate braiding isomorphism or its inverse, see Figure 3.1. This gives a $\mathcal{C}$-colored Penrose diagram with the same free ends as $\mathcal{R}$, and we let $\langle\mathcal{R}\rangle\colon \mathcal{R}_\bullet \to \mathcal{R}^\bullet$ be the associated morphism. Theorem 2.6 implies that $\langle\mathcal{R}\rangle$ is preserved under ambient isotopies of $\mathcal{R}$ in $\mathbb{R}\times[0,1]$.

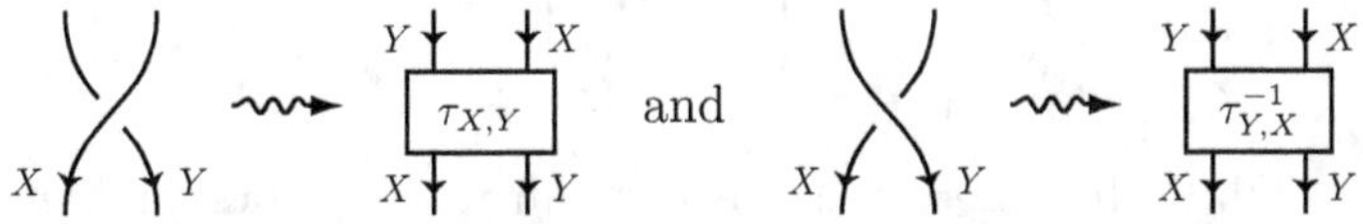

Figure 3.1: Transformation near a crossing

3.2.2 Reidemeister moves

Figure 3.2 defines five local moves $R_1, \dots, R_5$ on colored Reidemeister diagrams. Each move modifies a diagram in a small disk preserving the orientations and the colors of all strands and coupons. The inverse moves $R_1^{-1}, \dots, R_5^{-1}$ are defined by reversing the arrows in Figure 3.2. Observe that $R_i^{-1} = R_i$ for $i = 3, 4, 5$. The following lemma describes the behavior of the associated morphism under the moves $R_1, \dots, R_5$.

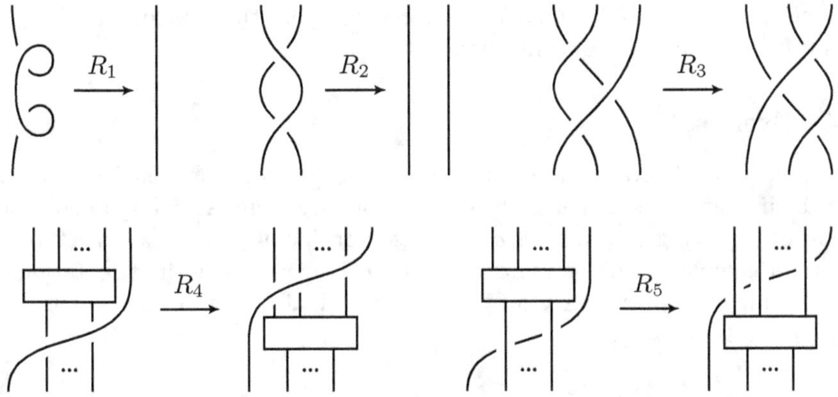

Figure 3.2: Moves on Reidemeister diagrams

Lemma 3.1. *If $\mathcal{C}$ is a braided pivotal category and two $\mathcal{C}$-colored Reidemeister diagrams $\mathcal{R}, \mathcal{R}'$ are related by a move R_i with $i = 2, 3, 4, 5$, then $\langle\mathcal{R}\rangle = \langle\mathcal{R}'\rangle$. If $\mathcal{C}$ is spherical and $\mathcal{R}, \mathcal{R}'$ have no free ends, then the same is true for $i = 1$.*

Proof. When all strands in the picture of R_i with $i = 2, 3, 4, 5$ are oriented downward, the equality $\langle\mathcal{R}\rangle = \langle\mathcal{R}'\rangle$ follows directly from the invertibility, naturality, and $\otimes$-multiplicativity of the braiding. The general case is deduced using that

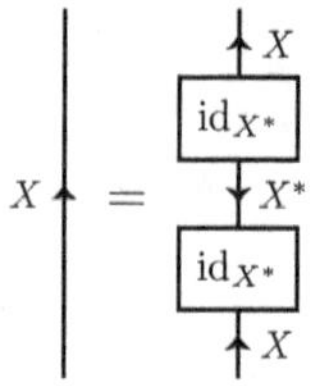

for any $X \in \mathrm{Ob}(\mathcal{C})$ together with the naturality of the braiding. Assume now that $\mathcal{C}$ is spherical and that $\mathcal{R}$, $\mathcal{R}'$ have no free ends and are related by a move R_1. Then $\mathcal{R}$ can be transformed into $\mathcal{R}'$ by a (finite) sequence of moves $R_2^{\pm 1}, R_3^{\pm 1}, R_4^{\pm 1}$ and isotopy in S^2. Indeed, pushing the over-passing branch in the picture of $\mathcal{R}$ to the left over the rest of $\mathcal{R}$ and across the point $\infty \in S^2$, we obtain $\mathcal{R}'$. We conclude using Lemma 2.9 and the first claim of the lemma. □

3.2.3 Remark

If the braiding τ is symmetric, then the boxes in Figure 3.1 carry the same morphisms. The morphisms associated with colored Reidemeister diagrams do not depend in this case on the choice of over/under-passes at the double points of the diagrams. Thus there is no need to involve over/under-passes; one can directly apply the graphical calculus to planar diagrams with double transversal intersections. The resulting graphical calculus is uninteresting from the topological viewpoint but is useful in the study of symmetric categories.

3.2.4 Exercise*

Let $\mathcal{C}$ be a braided pivotal category and $\mathcal{R}$ be a $\mathcal{C}$-colored Reidemeister diagram having a circular strand c. Let $\mathcal{R}'$ be the $\mathcal{C}$-colored Reidemeister diagram obtained from $\mathcal{R}$ by replacing the color of c by an isomorphic object of $\mathcal{C}$. Let $\mathcal{R}''$ be the $\mathcal{C}$-colored Reidemeister diagram obtained from $\mathcal{R}$ by reversing the orientation of c and replacing the color $X \in \mathrm{Ob}(\mathcal{C})$ of c by the dual object $X^* \in \mathrm{Ob}(\mathcal{C})$. Prove that $\langle \mathcal{R} \rangle = \langle \mathcal{R}' \rangle = \langle \mathcal{R}'' \rangle$.

3.2.5 Exercise*

Expand the move on Reidemeister diagrams

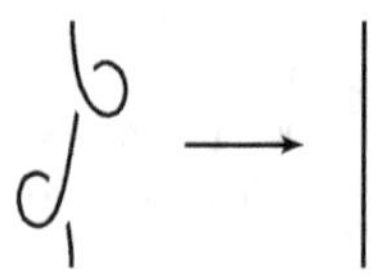

as a composition of several $R_2^{\pm 1}$ and a single R_3.

3.2.6 Exercise*

Expand the move on Reidemeister diagrams

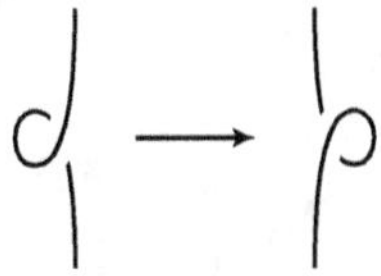

as a composition of a single R_1^{-1}, several $R_2^{\pm 1}$, and a single R_3.

3.2.7 Exercise*

Expand the move on Reidemeister diagrams

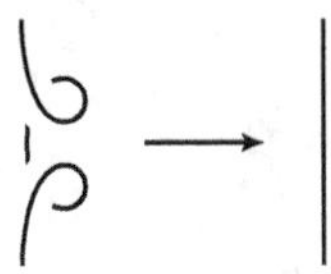

as a composition of a single R_1^{-1}, several $R_2^{\pm 1}$, and two R_3.

3.3 Twists and ribbon categories

We introduce twists in braided pivotal categories and define ribbon categories.

3.3.1 Twists

Let $\mathcal{C}$ be a braided pivotal category. The *left twist* of $\mathcal{C}$ is the family $\theta^l = \{\theta^l_X\}_{X \in \mathrm{Ob}(\mathcal{C})}$ of morphisms in $\mathcal{C}$ defined by

$\theta^l_X =$  $= (\mathrm{ev}_X \otimes \mathrm{id}_X)(\mathrm{id}_{X^*} \otimes \tau_{X,X})(\widetilde{\mathrm{coev}}_X \otimes \mathrm{id}_X) \colon X \to X.$

The *right twist* of $\mathcal{C}$ is the family $\theta^r = \{\theta^r_X\}_{X \in \mathrm{Ob}(\mathcal{C})}$ of morphisms in $\mathcal{C}$ defined by

$\theta^r_X =$ 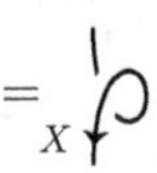 $= (\mathrm{id}_X \otimes \widetilde{\mathrm{ev}}_X)(\tau_{X,X} \otimes \mathrm{id}_{X^*})(\mathrm{id}_X \otimes \mathrm{coev}_X) \colon X \to X.$

Lemma 3.2. *The twists θ^l and θ^r are natural isomorphisms (from the identity functor $1_{\mathcal{C}}$ to itself). Their inverses are computed as follows: for any $X \in \mathrm{Ob}(\mathcal{C})$,*

$(\theta^l)^{-1}_X = (\theta^l_X)^{-1} =$ 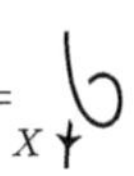 $= (\mathrm{id}_X \otimes \widetilde{\mathrm{ev}}_X)(\tau^{-1}_{X,X} \otimes \mathrm{id}_{X^*})(\mathrm{id}_X \otimes \mathrm{coev}_X)$

and

$$(\theta^r)^{-1}_X = (\theta^r_X)^{-1} = $$ 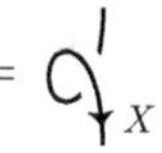 $$= (\mathrm{ev}_X \otimes \mathrm{id}_X)(\mathrm{id}_{X^*} \otimes \tau^{-1}_{X,X})(\widetilde{\mathrm{coev}}_X \otimes \mathrm{id}_X).$$

Moreover, $(\theta^r_X)^* = \theta^l_{X^*}$, $(\theta^l_X)^* = \theta^r_{X^*}$, *and*

 $= \theta^l_{X^*}$,  $= \theta^r_{X^*}$,  $= (\theta^l_{X^*})^{-1}$,  $= (\theta^r_{X^*})^{-1}$.

Proof. The naturality of the twists is a direct consequence of the naturality of τ and the dual morphism identities in Figure 2.4. For any $X \in \mathrm{Ob}(\mathcal{C})$, set

$\nu^l_X =$ 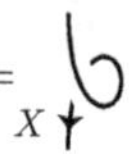.

Theorem 2.6 and Lemma 3.1 imply that

$\theta^l_X \nu^l_X =$ 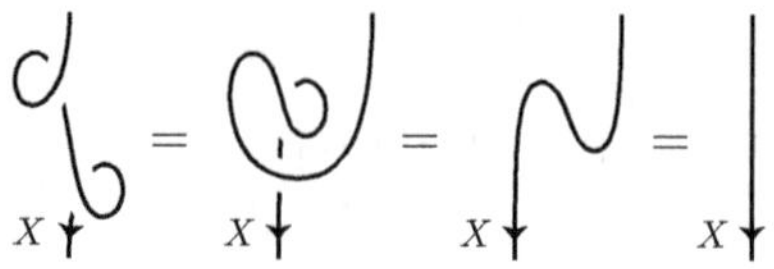 $= \mathrm{id}_X$.

The naturality of θ^l implies that $\nu_X \theta^l_X = \theta^l_X \nu_X = \mathrm{id}_X$. Thus the morphism θ^l_X is invertible and $(\theta^l_X)^{-1} = \nu_X$. Hence, the natural transformation θ^l is invertible with inverse $\{(\theta^l_X)^{-1}\}_{X \in \mathrm{Ob}(\mathcal{C})}$. The invertibility of θ^r is proved similarly. Now, denoting by $\psi = \{\psi_X\}_{X \in \mathrm{Ob}(\mathcal{C})}$ the pivotal structure of $\mathcal{C}$ and using Lemma 2.3, we obtain that for any $X \in \mathrm{Ob}(\mathcal{C})$,

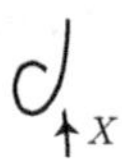
$$\begin{aligned} &= (\widetilde{\mathrm{ev}}_X \otimes \mathrm{id}_{X^*})(\mathrm{id}_X \otimes \tau_{X^*,X^*})(\mathrm{coev}_X \otimes \mathrm{id}_{X^*}) \\ &= (\mathrm{ev}_{X^*} \otimes \mathrm{id}_{X^*})(\psi_X \psi_X^{-1} \otimes \tau_{X^*,X^*})(\widetilde{\mathrm{coev}}_{X^*} \otimes \mathrm{id}_{X^*}) \\ &= (\mathrm{ev}_{X^*} \otimes \mathrm{id}_{X^*})(\mathrm{id}_X \otimes \tau_{X^*,X^*})(\widetilde{\mathrm{coev}}_{X^*} \otimes \mathrm{id}_{X^*}) = \theta^l_{X^*}. \end{aligned}$$

Then, using Theorem 2.6 and Lemma 3.1, we deduce that

$(\theta^r_X)^* =$ 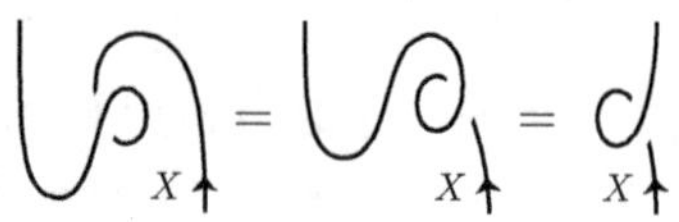 $= \theta^l_{X^*}$.

The other formulas in the statement of the lemma are proved similarly. □

It follows from the definition of the traces that

$\mathrm{tr}_r(\theta^l_X) =$  $= \mathrm{tr}_l(\theta^r_X)$. (3.4)

3.3.2 Ribbon categories

A *ribbon category* is a braided pivotal category $\mathcal{C}$ whose left twist θ^l and right twist θ^r are equal. Then $\theta = \theta^l = \theta^r$ is the *twist* of $\mathcal{C}$. By Lemma 3.2, the twist θ is self-dual in the sense that for all $X \in \mathrm{Ob}(\mathcal{C})$,

$$(\theta_X)^* = \theta_{X^*}.$$

The following result is well known (see, for example, [Sh, JS3, Tu1]).

Theorem 3.3. *If $\mathcal{C}$ is a ribbon category and two $\mathcal{C}$-colored Reidemeister diagrams $\mathcal{R}, \mathcal{R}'$ are related by isotopies and the moves $\{R_i^{\pm 1}\}_{i=1}^5$, then $\langle \mathcal{R} \rangle = \langle \mathcal{R}' \rangle$.*

Proof. Theorem 2.6 and Lemma 3.1 imply that $\langle \cdot \rangle$ is invariant under isotopies and the moves $R_2 - R_5$. Let us prove the invariance under the move R_1. Assume that the strand in the picture of R_1 is colored by $X \in \mathrm{Ob}(\mathcal{C})$. The computations in Lemma 3.2 imply that

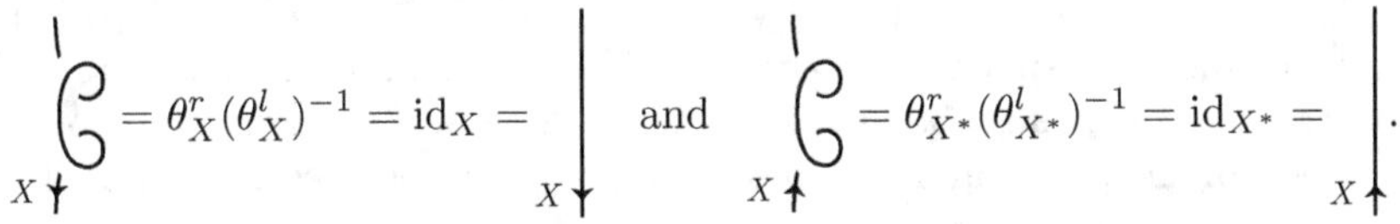

□

Corollary 3.4. *All ribbon categories are spherical.*

Proof. Let f be an endomorphism of an object X of a ribbon category $\mathcal{C}$. Theorem 3.3 and the dual morphism identities in Figure 2.4 imply that

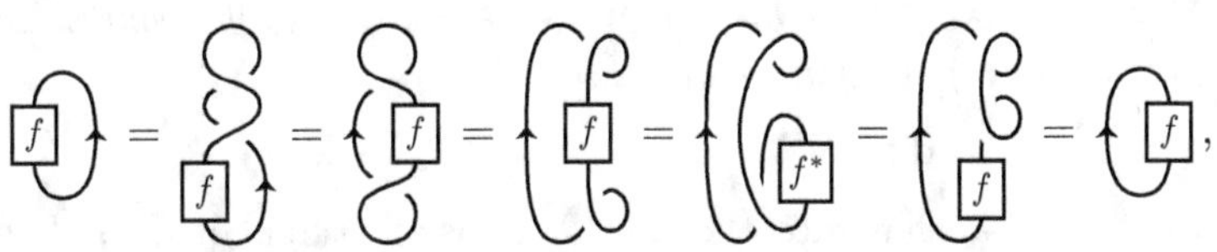

where all strands are colored by X. So, $\mathrm{tr}_r(f) = \mathrm{tr}_l(f)$. Thus $\mathcal{C}$ is spherical. □

We now use symmetries to construct a class of ribbon categories.

Lemma 3.5. *Let $\mathcal{C}$ be a monoidal category with a symmetry τ and a left duality $\{({}^\vee X, \mathrm{ev}_X)\}_{X \in \mathrm{Ob}(\mathcal{C})}$. The family*

$$\{(X^* = {}^\vee X,\, \mathrm{ev}_X \colon X^* \otimes X \to \mathbb{1},\, \widetilde{\mathrm{ev}}_X = \mathrm{ev}_X \tau_{X,X^*} \colon X \otimes X^* \to \mathbb{1})\}_{X \in \mathrm{Ob}(\mathcal{C})}$$

is a pivotal duality in $\mathcal{C}$ which turns $\mathcal{C}$ into a pivotal category. This pivotal category endowed with the symmetry τ is a ribbon category with twist $\{\mathrm{id}_X\}_{X \in \mathrm{Ob}(\mathcal{C})}$. The traces of endomorphisms in $\mathcal{C}$ and the dimensions of objects of $\mathcal{C}$ do not depend on the choice of left duality in $\mathcal{C}$.

Proof. The axioms of a pivotal category are verified directly from the definitions. For all $X \in \mathrm{Ob}(\mathcal{C})$,

$$\begin{aligned}
\theta^r_X &= (\mathrm{id}_X \otimes \widetilde{\mathrm{ev}}_X)(\tau_{X,X} \otimes \mathrm{id}_{X^*})(\mathrm{id}_X \otimes \mathrm{coev}_X)\\
&= (\mathrm{id}_X \otimes \mathrm{ev}_X)(\mathrm{id}_X \otimes \tau_{X,X^*})(\tau_{X,X} \otimes \mathrm{id}_{X^*})(\mathrm{id}_X \otimes \mathrm{coev}_X)\\
&= (\mathrm{id}_X \otimes \mathrm{ev}_X)\,\tau_{X,X\otimes X^*}(\mathrm{id}_X \otimes \mathrm{coev}_X)\\
&= (\mathrm{id}_X \otimes \mathrm{ev}_X)(\mathrm{coev}_X \otimes \mathrm{id}_X)\,\tau_{X,\mathbb{1}} = \mathrm{id}_X.
\end{aligned}$$

Since τ is a symmetry, Lemma 3.2 implies that $(\theta^l_X)^{-1} = \theta^r_X$. So, $\theta^l_X = \mathrm{id}_X$.

If $(Y, \omega\colon Y \otimes X \to \mathbb{1})$ is another left dual of $X \in \mathrm{Ob}(\mathcal{C})$, then the uniqueness of left duals implies that there is a (unique) isomorphism $\phi\colon Y \to {}^\vee X$ such that $\omega = \mathrm{ev}_X(\phi \otimes \mathrm{id}_Y)$. Then $\Omega = (\mathrm{id}_X \otimes \phi^{-1})\,\mathrm{coev}_X$ is the inverse of ω. Consequently, the trace of $f\colon X \to X$ defined using the left dual (Y,ω) of X is

$$\begin{aligned}
&\omega(\mathrm{id}_Y \otimes f)\tau_{X,Y}\Omega = \mathrm{ev}_X(\phi \otimes f)\tau_{X,Y}(\mathrm{id}_X \otimes \phi^{-1})\mathrm{coev}_X\\
&= \mathrm{ev}_X(\phi\phi^{-1} \otimes f)\tau_{X,{}^\vee X}\mathrm{coev}_X = \mathrm{ev}_X(\mathrm{id}_{{}^\vee X} \otimes f)\widetilde{\mathrm{coev}}_X = \mathrm{tr}(f).
\end{aligned}$$

Hence, the trace of f does not depend on the choice of a left dual of X. □

For instance, the spherical category $\mathrm{proj}_{\Bbbk}$ from Example 2.7.3 is obtained by the construction of Lemma 3.5 and therefore is a ribbon category. The following computation of the traces and dimensions in $\mathrm{proj}_{\Bbbk}$ will be used in Section 12.3.

Lemma 3.6. *Let $\omega\colon X \otimes_{\Bbbk} Y \to \Bbbk$ be a non-degenerate pairing between $\Bbbk$-modules X and Y (which are then projective of finite type by Lemma 1.6). For any $\Bbbk$-linear homomorphisms $f\colon X \to X$ and $g\colon Y \to Y$, we have*

$$\mathrm{tr}(f) = \omega(f \otimes_{\Bbbk} \mathrm{id}_Y)P(*_\omega) \in \Bbbk \quad \textit{and} \quad \mathrm{tr}(g) = \omega(\mathrm{id}_X \otimes_{\Bbbk} g)P(*_\omega) \in \Bbbk,$$

where $P\colon Y \otimes_{\Bbbk} X \to X \otimes_{\Bbbk} Y$ is the flip and $*_\omega \in Y \otimes_{\Bbbk} X$ is the contraction vector of ω (see Section 1.5.2). In particular,

$$\dim(X) = \dim(Y) = \omega P(*_\omega) \in \Bbbk.$$

Proof. Since ω is non-degenerate, the pair (X,ω) is a left dual of Y. By Lemma 3.5, this left dual may be used to compute $\mathrm{tr}(f)$. So,

$$\mathrm{tr}(f) = \omega(f \otimes_{\Bbbk} \mathrm{id}_Y)P\Omega,$$

where $\Omega\colon \Bbbk \to Y \otimes_{\Bbbk} X$ is the inverse of ω. We deduce the formula for $\mathrm{tr}(f)$ using the identification $\mathrm{End}_{\mathrm{proj}_{\Bbbk}}(\Bbbk) = \Bbbk$ and the equality $*_\omega = \Omega(1_{\Bbbk})$ (see Section 1.5.2). The formula for $\mathrm{tr}(g)$ is deduced from that for $\mathrm{tr}(f)$ by replacing ω with ωP. Finally, the dimensions of X and Y are computed by taking $f = \mathrm{id}_X$ and $g = \mathrm{id}_Y$. □

3.3.3 Exercise

Prove that the left twist θ^l and the right twist θ^r of a braided pivotal category $\mathcal{C}$ satisfy the following identities: for all $X, Y \in \mathrm{Ob}(\mathcal{C})$,

$$\theta^l_{X\otimes Y} = (\theta^l_X \otimes \theta^l_Y)\,\tau_{Y,X}\,\tau_{X,Y}, \quad \theta^r_{X\otimes Y} = (\theta^r_X \otimes \theta^r_Y)\,\tau_{Y,X}\,\tau_{X,Y}, \quad \theta^l_{\mathbb{1}} = \theta^r_{\mathbb{1}} = \mathrm{id}_{\mathbb{1}}.$$

3.3.4 Exercise

Prove that any braided pivotal functor $F\colon \mathcal{C}_1 \to \mathcal{C}_2$ between braided pivotal categories carries the left twist θ_1^l of $\mathcal{C}_1$ into the left twist θ_2^l of $\mathcal{C}_2$, that is, $F((\theta_1^l)_X) = (\theta_2^l)_{F(X)}$ for all $X \in \mathrm{Ob}(\mathcal{C}_1)$. Prove a similar claim for the right twist.

3.3.5 Exercise

Prove that any symmetric strong monoidal functor between pivotal categories derived as in Lemma 3.5 from symmetric categories with distinguished left duality is pivotal.

Chapter 4

Fusion categories

We introduce the classes of linear, pre-fusion, fusion, and modular categories. Recall that the symbol $\Bbbk$ denotes a commutative ring.

4.1 Linear categories

4.1.1 Linear categories and functors

A category $\mathcal{C}$ is $\Bbbk$-*linear* if for all objects X, Y of $\mathcal{C}$, the set $\mathrm{Hom}_{\mathcal{C}}(X,Y)$ is endowed with a structure of a left $\Bbbk$-module so that the composition of morphisms in $\mathcal{C}$ is $\Bbbk$-bilinear. For shortness, $\Bbbk$-linear categories are called $\Bbbk$-*categories*. Any subcategory of a $\Bbbk$-category is a $\Bbbk$-category in the obvious way.

A functor $F\colon \mathcal{C} \to \mathcal{D}$ between $\Bbbk$-categories is $\Bbbk$-*linear* if its action on the Hom-sets is $\Bbbk$-linear, that is, if for all $X, Y \in \mathrm{Ob}(\mathcal{C})$, the map

$$\mathrm{Hom}_{\mathcal{C}}(X,Y) \to \mathrm{Hom}_{\mathcal{D}}(F(X),F(Y)), \quad f \mapsto F(f)$$

is $\Bbbk$-linear. For example, the identity functor of a $\Bbbk$-category is $\Bbbk$-linear. Clearly, the composition of $\Bbbk$-linear functors is a $\Bbbk$-linear functor.

Two $\Bbbk$-categories are *isomorphic* (respectively, *equivalent*) if there is a $\Bbbk$-linear isomorphism (respectively, $\Bbbk$-linear equivalence) between them.

4.1.2 Direct sums

Let $(X_\alpha)_{\alpha\in\Lambda}$ be a finite family of objects of a $\Bbbk$-category $\mathcal{C}$. An object D of $\mathcal{C}$ is a *direct sum* of the family $(X_\alpha)_{\alpha\in\Lambda}$ if there is a family of morphisms in $\mathcal{C}$

$$(p_\alpha\colon D \to X_\alpha,\ q_\alpha\colon X_\alpha \to D)_{\alpha\in\Lambda}$$

such that

$$\mathrm{id}_D = \sum_{\alpha\in\Lambda} q_\alpha p_\alpha \quad \text{and} \quad p_\alpha q_\beta = \delta_{\alpha,\beta}\,\mathrm{id}_{X_\alpha} \quad \text{for all } \alpha, \beta \in \Lambda,$$

where $\delta_{\alpha,\beta}$ is the Kronecker symbol defined by $\delta_{\alpha,\beta} = 1$ if $\alpha = \beta$ and $\delta_{\alpha,\beta} = 0$ otherwise. The morphisms p_α and q_α are called respectively *projections* and *inclusions*. Such an object D, if it exists, is unique (up to a unique isomorphism commuting with the projections and the inclusions) and is denoted $\bigoplus_{\alpha\in\Lambda} X_\alpha$. It is easy to see that for any $Y \in \mathrm{Ob}(\mathcal{C})$, there are $\Bbbk$-linear isomorphisms

$$\mathrm{Hom}_{\mathcal{C}}(Y, \bigoplus_{\alpha\in\Lambda} X_\alpha) \simeq \bigoplus_{\alpha\in\Lambda} \mathrm{Hom}_{\mathcal{C}}(Y, X_\alpha)$$

and

$$\mathrm{Hom}_{\mathcal{C}}(\bigoplus_{\alpha\in\Lambda} X_\alpha, Y) \simeq \bigoplus_{\alpha\in\Lambda} \mathrm{Hom}_{\mathcal{C}}(X_\alpha, Y).$$

By definition, a direct sum of an empty family of objects of a $\Bbbk$-category $\mathcal{C}$ is a *zero object* of $\mathcal{C}$, that is, an object $\mathbf{0}$ of $\mathcal{C}$ such that $\mathrm{End}_{\mathcal{C}}(\mathbf{0}) = 0$. (For properties of zero objects, see Appendix C.)

A $\Bbbk$-category $\mathcal{C}$ is *additive* if any finite (possibly empty) family of objects of $\mathcal{C}$ has a direct sum in $\mathcal{C}$. We explain in Section 16.3.1 that every $\Bbbk$-category embeds in an additive $\Bbbk$-category in a canonical way.

4.1.3 Simple objects in $\Bbbk$-categories

For any object X of a $\Bbbk$-category $\mathcal{C}$, the monoid $\mathrm{End}_{\mathcal{C}}(X)$ is a $\Bbbk$-algebra with unit element id_X. It is easy to see that the following conditions on X are equivalent:

(i) the map $\Bbbk \to \mathrm{End}_{\mathcal{C}}(X)$, $k \mapsto k\,\mathrm{id}_X$ is an isomorphism of $\Bbbk$-modules;

(ii) the map $\Bbbk \to \mathrm{End}_{\mathcal{C}}(X)$, $k \mapsto k\,\mathrm{id}_X$ is an isomorphism of $\Bbbk$-algebras;

(iii) the $\Bbbk$-algebra $\mathrm{End}_{\mathcal{C}}(X)$ is isomorphic to $\Bbbk$;

(iv) the $\Bbbk$-module $\mathrm{End}_{\mathcal{C}}(X)$ is free of rank 1.

Objects of $\mathcal{C}$ satisfying these conditions are said to be *simple* (some authors use the terms "scalar" or "absolutely simple" objects). The $\Bbbk$-bilinearity of the composition of morphisms in $\mathcal{C}$ implies that all objects of $\mathcal{C}$ isomorphic to a simple object are simple.

4.1.4 Example

The category $\mathrm{Mod}_\Bbbk$ of $\Bbbk$-modules and $\Bbbk$-linear homomorphisms has a canonical structure of an additive $\Bbbk$-category: for all $\Bbbk$-modules X, Y, the set $\mathrm{Hom}_{\mathrm{Mod}_\Bbbk}(X,Y)$ of $\Bbbk$-linear homomorphisms from X to Y carries the usual left $\Bbbk$-module structure defined by

$$\Bbbk \times \mathrm{Hom}_{\mathrm{Mod}_\Bbbk}(X,Y) \to \mathrm{Hom}_{\mathrm{Mod}_\Bbbk}(X,Y), \quad (k,f) \mapsto (x \mapsto f(kx)).$$

4.1.5 Exercise

Prove that a $\Bbbk$-linear functor between $\Bbbk$-categories preserves direct sums and carries zero objects to zero objects.

4.1.6 Exercise

Prove that any quasi-inverse of a $\Bbbk$-linear equivalence between $\Bbbk$-categories is $\Bbbk$-linear.

4.1.7 Exercise

Prove that a $\Bbbk$-linear equivalence between $\Bbbk$-categories induces a bijective correspondence between isomorphism classes of simple objects in these categories.

4.2 Monoidal linear categories

4.2.1 Monoidal $\Bbbk$-categories

The definitions of Chapter 1 can be easily adapted to the setting of $\Bbbk$-categories. By a monoidal (respectively, left/right rigid, rigid, pivotal, spherical) $\Bbbk$-category, we mean a $\Bbbk$-category which is monoidal (respectively, left/right rigid, rigid, pivotal, spherical) and such that monoidal product of morphisms is $\Bbbk$-bilinear. Clearly, any monoidal subcategory of a monoidal $\Bbbk$-category is a monoidal $\Bbbk$-category. If $\mathcal{C}$ is a monoidal $\Bbbk$-category, then so are $\mathcal{C}^{\mathrm{op}}$, $\mathcal{C}^{\otimes\mathrm{op}}$, and $\mathcal{C}^{\mathrm{rev}}$ (see Section 1.2.2).

Isomorphisms and equivalences of monoidal/braided/pivotal $\Bbbk$-categories are always required to be $\Bbbk$-linear. In particular, two pivotal $\Bbbk$-categories are *equivalent* if there is a $\Bbbk$-linear pivotal functor between them which is an equivalence of the underlying categories.

It follows from the definitions that all left/right dual functors of a left/right rigid $\Bbbk$-category are $\Bbbk$-linear. In particular, given a pivotal $\Bbbk$-category $\mathcal{C}$, the dual functor $?^*\colon \mathcal{C}^{\mathrm{rev}} \to \mathcal{C}$ is $\Bbbk$-linear, and so $\mathcal{C}^{\mathrm{rev}}$ and $\mathcal{C}$ are equivalent pivotal $\Bbbk$-categories (cf. Section 1.7.5). Consequently, $\mathcal{C}^{\otimes\mathrm{op}} = (\mathcal{C}^{\mathrm{op}})^{\mathrm{rev}}$ and $\mathcal{C}^{\mathrm{op}}$ are equivalent pivotal $\Bbbk$-categories.

4.2.2 Remarks

1. Any monoidal $\Bbbk$-category $\mathcal{C}$ with simple unit object is pure in the sense of Section 1.3.3. This follows from the $\Bbbk$-linearity of the monoidal product and the equalities $\mathrm{End}_{\mathcal{C}}(\mathbb{1}) = \Bbbk\, \mathrm{id}_{\mathbb{1}}$ and $\mathrm{id}_{\mathbb{1}} \otimes \mathrm{id}_X = \mathrm{id}_X = \mathrm{id}_X \otimes \mathrm{id}_{\mathbb{1}}$ for all $X \in \mathrm{Ob}(\mathcal{C})$.

2. The left and right traces of endomorphisms in a pivotal $\Bbbk$-category are $\Bbbk$-linear. This follows from the $\Bbbk$-linearity of the monoidal product and composition of morphisms. As a consequence, the left and right dimensions of any zero object of a pivotal $\Bbbk$-category are equal to zero.

3. If a finite family $(X_\alpha)_{\alpha\in\Lambda}$ of objects of a pivotal $\Bbbk$-category has a direct sum, then

$$\dim_l\Big(\bigoplus_{\alpha\in\Lambda} X_\alpha\Big) = \sum_{\alpha\in\Lambda} \dim_l(X_\alpha) \quad \text{and} \quad \dim_r\Big(\bigoplus_{\alpha\in\Lambda} X_\alpha\Big) = \sum_{\alpha\in\Lambda} \dim_r(X_\alpha).$$

This follows from the fact that the left/right dimension of an object of a pivotal $\Bbbk$-category is the left/right trace of the identity morphism of that object, and from the $\Bbbk$-linearity and symmetry of the left/right traces of endomorphisms.

4.2.3 Convention

In the sequel, given a monoidal $\Bbbk$-category $\mathcal{C}$ with simple unit object, we identify $\mathrm{End}_\mathcal{C}(\mathbb{1})$ with $\Bbbk$ via the $\Bbbk$-algebra isomorphism $\Bbbk \to \mathrm{End}_\mathcal{C}(\mathbb{1})$, $k \mapsto k\,\mathrm{id}_\mathbb{1}$.

4.2.4 Example

The category $\mathrm{Mod}_\Bbbk$ with its canonical structure of a $\Bbbk$-category (see Example 4.1.4) and the usual monoidal product $\otimes_\Bbbk$ is a monoidal $\Bbbk$-category. The full subcategory $\mathrm{proj}_\Bbbk$ of $\mathrm{Mod}_\Bbbk$ consisting of projective $\Bbbk$-modules of finite type (see Section 2.7.3) is a spherical $\Bbbk$-category.

4.2.5 Example

For a group G, the category $G_\Bbbk$ from Example 1.1.4 has a structure of a $\Bbbk$-category defined by providing each Hom-set (which is either $\Bbbk$ or $\{0\}$) with the left $\Bbbk$-module structure given by multiplication. Given a 3-cocycle $\alpha\colon G^3 \to \Bbbk^*$ and a group homomorphism $d\colon G \to \Bbbk^*$, the pivotal category $G_\Bbbk^{\alpha,d}$ from Example 1.7.3 is a pivotal $\Bbbk$-category. This $\Bbbk$-category is not additive but has an additive version whose objects are G-graded $\Bbbk$-modules, see Appendix A.3. Exercise 4.2.7 below implies that the equivalence classes of the pivotal $\Bbbk$-categories $\{G_\Bbbk^{\alpha,d}\}_{\alpha,d}$ bijectively correspond to the orbits of the obvious diagonal action of the group of automorphisms of G on $H^3(G;\Bbbk^*) \times H^1(G;\Bbbk^*)$.

4.2.6 Exercise

Prove that all left duals and right duals of a simple object of a monoidal $\Bbbk$-category are simple.

4.2.7 Exercise*

For $i \in \{1,2\}$, let G_i be a group, let α_i be a $\Bbbk^*$-valued 3-cocycle on G_i, and let $d_i\colon G_i \to \Bbbk^*$ be a group homomorphism. Prove that the pivotal $\Bbbk$-categories $\mathcal{C}_1 = (G_1)_\Bbbk^{\alpha_1,d_1}$ and $\mathcal{C}_2 = (G_2)_\Bbbk^{\alpha_2,d_2}$ from Example 4.2.5 are equivalent if and only if there is a group isomorphism $f\colon G_1 \to G_2$ such that $d_1 = d_2 f$ and $[\alpha_1] = [f^*(\alpha_2)]$, where the square brackets stand for the cohomology class in $H^3(G,\Bbbk^*)$.

4.3 Non-degenerate categories

We introduce the class of non-degenerate monoidal $\Bbbk$-categories.

4.3.1 Non-degenerate $\Bbbk$-categories

Any pairing $e\colon X \otimes Y \to \mathbb{1}$ between objects X and Y of a monoidal $\Bbbk$-category $\mathcal{C}$ induces a $\Bbbk$-linear homomorphism

$$\mathrm{Hom}_{\mathcal{C}}(\mathbb{1}, X) \otimes_{\Bbbk} \mathrm{Hom}_{\mathcal{C}}(\mathbb{1}, Y) \to \mathrm{End}_{\mathcal{C}}(\mathbb{1}), \ \ \alpha \otimes_{\Bbbk} \beta \mapsto e(\alpha \otimes \beta)$$

where $\alpha \in \mathrm{Hom}_{\mathcal{C}}(\mathbb{1}, X)$ and $\beta \in \mathrm{Hom}_{\mathcal{C}}(\mathbb{1}, Y)$. When the unit object $\mathbb{1}$ of $\mathcal{C}$ is simple, we identify $\mathrm{End}_{\mathcal{C}}(\mathbb{1}) = \Bbbk$ (as in Section 4.1.3) and obtain a pairing in $\mathrm{Mod}_{\Bbbk}$

$$\mathrm{Hom}_{\mathcal{C}}(\mathbb{1}, X) \otimes_{\Bbbk} \mathrm{Hom}_{\mathcal{C}}(\mathbb{1}, Y) \to \Bbbk, \quad \alpha \otimes_{\Bbbk} \beta \mapsto e(\alpha \otimes \beta). \tag{4.1}$$

A monoidal $\Bbbk$-category $\mathcal{C}$ is *non-degenerate* if its unit object $\mathbb{1}$ is simple and for each non-degenerate pairing $X \otimes Y \to \mathbb{1}$ in $\mathcal{C}$, the induced pairing (4.1) is non-degenerate in the monoidal category $\mathrm{Mod}_{\Bbbk}$ (see Section 1.5.2).

Lemma 4.1. *A left rigid $\Bbbk$-category with distinguished left duality is non-degenerate if and only if its unit object is simple and the pairings* (4.1) *induced by the left evaluations are non-degenerate. The same holds with left replaced by right.*

Proof. Let $\mathcal{C}$ be a left rigid $\Bbbk$-category with left duality $\{({}^{\vee}X, \mathrm{ev}_X)\}_{X \in \mathrm{Ob}(\mathcal{C})}$. Since each pairing ev_X is non-degenerate, the 'only if' part follows from the definitions. We prove the 'if' part. Let $e\colon X \otimes Y \to \mathbb{1}$ be a non-degenerate pairing in $\mathcal{C}$. Denote by

$$\omega_e\colon \mathrm{Hom}_{\mathcal{C}}(\mathbb{1}, X) \otimes_{\Bbbk} \mathrm{Hom}_{\mathcal{C}}(\mathbb{1}, Y) \to \Bbbk, \quad \omega_Y\colon \mathrm{Hom}_{\mathcal{C}}(\mathbb{1}, {}^{\vee}Y) \otimes_{\Bbbk} \mathrm{Hom}_{\mathcal{C}}(\mathbb{1}, Y) \to \Bbbk$$

the pairings (4.1) induced by e and ev_Y, respectively. Since e is non-degenerate, the pair (X, e) is a left dual of Y. The uniqueness of duals (Section 1.6.1) implies that there is an isomorphism $\phi\colon X \to {}^{\vee}Y$ in $\mathcal{C}$ such that $e = \mathrm{ev}_Y(\phi \otimes \mathrm{id}_Y)$. Clearly, the map

$$\Phi\colon \mathrm{Hom}_{\mathcal{C}}(\mathbb{1}, X) \to \mathrm{Hom}_{\mathcal{C}}(\mathbb{1}, {}^{\vee}Y), \quad f \mapsto \phi f$$

is a $\Bbbk$-linear isomorphism. It follows from the definitions that

$$\omega_e = \omega_Y(\Phi \otimes_{\Bbbk} \mathrm{id}_{\mathrm{Hom}_{\mathcal{C}}(\mathbb{1}, Y)}).$$

Since ω_Y is non-degenerate in $\mathrm{Mod}_{\Bbbk}$ (by assumption) and Φ is a $\Bbbk$-linear isomorphism, the pairing ω_e is non-degenerate in $\mathrm{Mod}_{\Bbbk}$. Hence, the category $\mathcal{C}$ is non-degenerate. Right rigid categories are treated similarly. □

4.3.2 The case of pivotal $\Bbbk$-categories

Let $\mathcal{C}$ be a pivotal $\Bbbk$-category. Consider a tuple S of signed objects of $\mathcal{C}$ and the evaluation $\mathrm{ev}_S\colon X_{S^*} \otimes X_S \to \mathbb{1}$ from Section 2.3.2. By Section 4.3.1, ev_S induces a pairing in $\mathrm{Mod}_\Bbbk$

$$\omega_S\colon \mathrm{Hom}_\mathcal{C}(\mathbb{1}, X_{S^*}) \otimes_\Bbbk \mathrm{Hom}_\mathcal{C}(\mathbb{1}, X_S) \to \mathrm{End}_\mathcal{C}(\mathbb{1}).$$

By definition, $\omega_S(\alpha \otimes_\Bbbk \beta) = \mathrm{ev}_S(\alpha \otimes \beta)$ for $\alpha \in \mathrm{Hom}_\mathcal{C}(\mathbb{1}, X_{S^*})$, $\beta \in \mathrm{Hom}_\mathcal{C}(\mathbb{1}, X_S)$. By the isotopy invariance of the graphical calculus, $\omega_S(\alpha \otimes_\Bbbk \beta) = \omega_{S^*}(\beta \otimes_\Bbbk \alpha)$ for all S, α, β. For example, for any $\alpha \in \mathrm{Hom}_\mathcal{C}(\mathbb{1}, Y^* \otimes X)$ and $\beta \in \mathrm{Hom}_\mathcal{C}(\mathbb{1}, X^* \otimes Y)$,

$\omega_{((X,-),(Y,+))}(\alpha \otimes_\Bbbk \beta) =$ 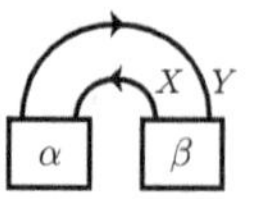

$=$ 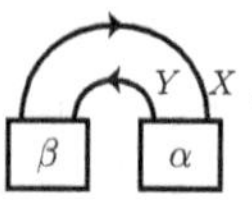

$= \omega_{((Y,-),(X,+))}(\beta \otimes_\Bbbk \alpha).$

If $\mathcal{C}$ is non-degenerate, then the pairing ω_S is non-degenerate in $\mathrm{Mod}_\Bbbk$ for any tuple S of signed objects of $\mathcal{C}$. In particular, for any $X \in \mathrm{Ob}(\mathcal{C})$, the pairing

$$\omega_{(X,+)}\colon \mathrm{Hom}_\mathcal{C}(\mathbb{1}, X^*) \otimes_\Bbbk \mathrm{Hom}_\mathcal{C}(\mathbb{1}, X) \to \mathrm{End}_\mathcal{C}(\mathbb{1}) = \Bbbk$$

is non-degenerate in $\mathrm{Mod}_\Bbbk$. By Lemma 1.6, if $\mathcal{C}$ is non-degenerate, then for any $X \in \mathrm{Ob}(\mathcal{C})$, the $\Bbbk$-module $\mathrm{Hom}_\mathcal{C}(\mathbb{1}, X)$ is projective of finite type.

It follows from Lemma 4.1 that the category $\mathcal{C}$ is non-degenerate if and only if the unit object $\mathbb{1}$ is simple and the pairing $\omega_{(X,+)}$ is non-degenerate in $\mathrm{Mod}_\Bbbk$ for all $X \in \mathrm{Ob}(\mathcal{C})$.

Lemma 4.2. *Let $\mathcal{C}$ be a non-degenerate pivotal $\Bbbk$-category. Then:*

(a) *The left and right dimensions of any simple object of $\mathcal{C}$ are invertible in $\Bbbk$;*

(b) *Any endomorphism f of a simple object i of $\mathcal{C}$ expands as*

$$f = (\dim_l(i))^{-1}\,\mathrm{tr}_l(f)\,\mathrm{id}_i = (\dim_r(i))^{-1}\,\mathrm{tr}_r(f)\,\mathrm{id}_i.$$

Proof. Let i be a simple object of $\mathcal{C}$. Consider the pair $S = ((i,+),(i,-))$. Clearly, $S^* = S$ and $X_S = i \otimes i^*$. Section 1.5.1 implies that $\mathrm{Hom}_\mathcal{C}(\mathbb{1}, i \otimes i^*) \simeq \mathrm{Hom}_\mathcal{C}(i,i)$ is a free $\Bbbk$-module of rank 1 with basis vector $\mathrm{coev}_i\colon \mathbb{1} \to i \otimes i^*$. The matrix of the pairing ω_S in this basis is the (1×1)-matrix $[\dim_r(i)]$ because

$\omega_S(\mathrm{coev}_i \otimes_\Bbbk \mathrm{coev}_i) =$ 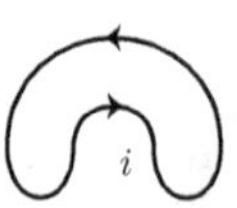

$=$ 

$= \dim_r(i) \in \mathrm{End}_\mathcal{C}(\mathbb{1}) = \Bbbk.$

The non-degeneracy of ω_S and Lemma 1.7 imply that $\dim_r(i)$ is invertible. A similar argument with $S = ((i,-),(i,+))$ shows that $\dim_l(i)$ is invertible. Since i is simple, any endomorphism f of i expands as $f = k\,\mathrm{id}_i$ with $k \in \Bbbk$. The $\Bbbk$-linearity of the left trace implies that $\mathrm{tr}_l(f) = k\,\mathrm{tr}_l(\mathrm{id}_i) = k\,\dim_l(i)$. Since $\dim_l(i)$ is invertible, we obtain $k = (\dim_l(i))^{-1}\,\mathrm{tr}_l(f)$. The second formula for f is obtained similarly by using the right trace. □

Lemma 4.2(b) implies that the left and right traces of an endomorphism f of a simple object i of a non-degenerate pivotal $\Bbbk$-category are related by

$$\dim_l(i)\operatorname{tr}_r(f) = \dim_r(i)\operatorname{tr}_l(f). \tag{4.2}$$

4.3.3 Example

The monoidal $\Bbbk$-category $\mathrm{Mod}_\Bbbk$ from Example 4.2.4 is non-degenerate because $\mathrm{Hom}_{\mathrm{Mod}_\Bbbk}(\Bbbk, X) \simeq X$ for any $\Bbbk$-module X. The category $\mathrm{proj}_\Bbbk$ is non-degenerate by the same argument.

4.3.4 Example

The pivotal $\Bbbk$-category $G_\Bbbk^{\alpha,d}$ from Example 4.2.5 is non-degenerate.

4.4 Pre-fusion categories

We introduce pre-fusion $\Bbbk$-categories.

4.4.1 Pre-fusion $\Bbbk$-categories

A *pre-fusion* $\Bbbk$*-category* is a monoidal $\Bbbk$-category $\mathcal{C}$ such that there is a set I of simple objects of $\mathcal{C}$ satisfying the following conditions:

(a) the unit object $\mathbb{1}$ of $\mathcal{C}$ belongs to I;

(b) $\mathrm{Hom}_\mathcal{C}(i,j) = 0$ for any distinct $i, j \in I$;

(c) every object of $\mathcal{C}$ is a direct sum of a finite family of elements of I.

Such a set I is called a *representative set of simple objects* of $\mathcal{C}$. This terminology is justified by the fact that every simple object of $\mathcal{C}$ is isomorphic to precisely one simple object belonging to the set I.

Let $\mathcal{C}$ be a pre-fusion $\Bbbk$-category and let I be a representative set of simple objects of $\mathcal{C}$. Condition (a) implies that $\mathcal{C}$ is pure and $\mathrm{End}_\mathcal{C}(\mathbb{1}) = \Bbbk$ (see Sections 4.2.2 and 4.2.3). Conditions (b) and (c) imply that the Hom-sets in $\mathcal{C}$ are free of finite rank and that for all $X, Y \in \mathrm{Ob}(\mathcal{C})$,

$$\mathrm{Hom}_\mathcal{C}(X,Y) \simeq \bigoplus_{i\in I} \mathrm{Hom}_\mathcal{C}(X,i) \otimes_\Bbbk \mathrm{Hom}_\mathcal{C}(i,Y). \tag{4.3}$$

By Condition (c), for each object X of $\mathcal{C}$, there is a finite family of morphisms

$$(p_\alpha\colon X \to i_\alpha,\ q_\alpha\colon i_\alpha \to X)_{\alpha\in\Lambda}$$

such that

$$i_\alpha \in I, \quad \mathrm{id}_X = \sum_{\alpha\in\Lambda} q_\alpha p_\alpha, \quad \text{and} \quad p_\alpha q_\beta = \delta_{\alpha,\beta}\,\mathrm{id}_{i_\alpha} \ \text{ for all } \alpha,\beta \in \Lambda.$$

We call such a family an *I-partition* of X. For a simple object i of $\mathcal{C}$, denote by $N_X^i \geq 0$ the number of simple objects isomorphic to i in an I-partition of X. This number is called the *multiplicity index of* i *in* X and is equal to the ranks of the free $\Bbbk$-modules $\mathrm{Hom}_{\mathcal{C}}(X,i)$ and $\mathrm{Hom}_{\mathcal{C}}(i,X)$. Clearly, the family $\{N_X^i\}_{i\in I}$ has only a finite number of non-zero terms. It follows from (4.3) that for all $X, Y \in \mathrm{Ob}(\mathcal{C})$,

$$\mathrm{rank}_{\Bbbk}(\mathrm{Hom}_{\mathcal{C}}(X,Y)) = \sum_{i\in I} N_X^i \, N_Y^i.$$

Given a simple object i of $\mathcal{C}$, an *i-partition* of $X \in \mathrm{Ob}(\mathcal{C})$ is a family of morphisms $(p_\alpha \colon X \to i,\, q_\alpha \colon i \to X)_{\alpha\in A}$ such that $(p_\alpha)_{\alpha\in A}$ is a basis of $\mathrm{Hom}_{\mathcal{C}}(X,i)$, $(q_\alpha)_{\alpha\in A}$ is a basis of $\mathrm{Hom}_{\mathcal{C}}(i,X)$, and $p_\alpha q_\beta = \delta_{\alpha,\beta}\,\mathrm{id}_i$ for all $\alpha,\beta \in A$. Note that the cardinality of the set A is equal to the multiplicity index N_X^i. For any I-partition $(p_\alpha \colon X \to i_\alpha, q_\alpha \colon i_\alpha \to X)_{\alpha\in\Lambda}$ of X and any $i \in I$, the family $(p_\alpha, q_\alpha)_{\alpha\in\Lambda,\, i_\alpha = i}$ is an i-partition of X. Conversely, a union of i-partitions of X over all $i \in I$ is an I-partition of X.

4.4.2 Properties of pre-fusion categories

We establish several useful properties of pre-fusion $\Bbbk$-categories.

Lemma 4.3. *All pre-fusion* $\Bbbk$*-categories are non-degenerate in the sense of Section* 4.3.1.

Proof. Consider a non-degenerate pairing $e \colon X \otimes Y \to \mathbb{1}$ in a pre-fusion $\Bbbk$-category $\mathcal{C}$ with inverse $E \colon \mathbb{1} \to Y \otimes X$. Pick a $\mathbb{1}$-partition $(p_\alpha \colon X \to \mathbb{1}, q_\alpha \colon \mathbb{1} \to X)_{\alpha\in A}$ of X. By definition, the families $(p_\alpha)_{\alpha\in A}$ and $(q_\alpha)_{\alpha\in A}$ are bases of the (free) $\Bbbk$-modules $\mathrm{Hom}_{\mathcal{C}}(X,\mathbb{1})$ and $\mathrm{Hom}_{\mathcal{C}}(\mathbb{1},X)$, respectively. Since the bijection (1.10) of Section 1.5.1 for $T = Z = \mathbb{1}$ is a $\Bbbk$-linear isomorphism $\mathrm{Hom}_{\mathcal{C}}(\mathbb{1},Y) \to \mathrm{Hom}_{\mathcal{C}}(X,\mathbb{1})$, the family $(p'_\alpha = (\mathrm{id}_Y \otimes p_\alpha)E \colon \mathbb{1} \to Y)_{\alpha\in A}$ is a basis of the (free) $\Bbbk$-module $\mathrm{Hom}_{\mathcal{C}}(\mathbb{1},Y)$. Now, for all $\alpha,\beta \in A$,

$$e(q_\alpha \otimes p'_\beta) = p_\beta(e \otimes \mathrm{id}_X)(\mathrm{id}_X \otimes E)q_\alpha = p_\beta q_\alpha = \delta_{\alpha,\beta}.$$

Thus the pairing

$$\mathrm{Hom}_{\mathcal{C}}(\mathbb{1},X) \otimes_{\Bbbk} \mathrm{Hom}_{\mathcal{C}}(\mathbb{1},Y) \to \Bbbk, \quad f \otimes_{\Bbbk} g \mapsto e(f \otimes g)$$

induced by e is non-degenerate (see Lemma 1.7). □

In a pivotal pre-fusion $\Bbbk$-category $\mathcal{C}$, the traces of endomorphisms and the dimensions of objects are $\otimes$-multiplicative. This follows from results of Section 2.6.1 and the fact that $\mathcal{C}$ is pure. Also, Lemmas 4.2 and 4.3 imply that the left and right dimensions of every simple object of $\mathcal{C}$ are invertible in $\mathrm{End}_{\mathcal{C}}(\mathbb{1}) = \Bbbk$.

Lemma 4.4. *A pre-fusion pivotal* $\Bbbk$*-category* $\mathcal{C}$ *is spherical if and only if* $\dim_l(i) = \dim_r(i)$ *for every simple object* i *of* $\mathcal{C}$.

Proof. Let f be an endomorphism of an object X of $\mathcal{C}$. Pick a decomposition $(p_\alpha \colon X \to i_\alpha, q_\alpha \colon i_\alpha \to X)_{\alpha\in\Lambda}$ of X as a direct sum of simple objects i_α. For $\alpha \in \Lambda$, the simplicity of i_α implies that $p_\alpha f q_\alpha = \lambda_\alpha \mathrm{id}_{i_\alpha}$ for some $\lambda_\alpha \in \Bbbk$. Then

$$\mathrm{tr}_l(f) = \mathrm{tr}_l(f \circ \mathrm{id}_X) = \mathrm{tr}_l(f \sum_{\alpha\in\Lambda} q_\alpha p_\alpha) = \mathrm{tr}_l(\sum_{\alpha\in\Lambda} p_\alpha f q_\alpha) = \sum_{\alpha\in\Lambda} \lambda_\alpha \dim_l(i_\alpha).$$

Similarly, $\mathrm{tr}_r(f) = \sum_{\alpha\in\Lambda} \lambda_\alpha \dim_r(i_\alpha)$. These formulas imply the claim of the lemma. □

Lemma 4.5. *A braided pre-fusion pivotal $\Bbbk$-category is ribbon if and only if it is spherical.*

Proof. By Lemma 3.4, all ribbon categories are spherical. Conversely, let $\mathcal{C}$ be a braided spherical pre-fusion $\Bbbk$-category. We need to prove that the left twist θ^l and the right twist θ^r of $\mathcal{C}$ are equal. Notice first that for any simple object i of $\mathcal{C}$, Lemma 4.2(b) and Formula (3.4) yield

$$\theta^l_i = (\dim(i))^{-1}\,\mathrm{tr}(\theta^l_i)\,\mathrm{id}_i = (\dim(i))^{-1}\,\mathrm{tr}(\theta^r_i)\,\mathrm{id}_i = \theta^r_i.$$

Now, let $X \in \mathrm{Ob}(\mathcal{C})$. Pick a representative set I of simple objects of $\mathcal{C}$ and consider an I-partition $(p_\alpha \colon X \to i_\alpha, q_\alpha \colon i_\alpha \to X)_{\alpha\in\Lambda}$ of X. The naturality of the twists and the formula $\mathrm{id}_X = \sum_{\alpha\in\Lambda} q_\alpha p_\alpha$ imply that

$$\theta^l_X = \sum_{\alpha\in\Lambda} q_\alpha p_\alpha \theta^l_X = \sum_{\alpha\in\Lambda} q_\alpha \theta^l_{i_\alpha} p_\alpha = \sum_{\alpha\in\Lambda} q_\alpha \theta^r_{i_\alpha} p_\alpha = \sum_{\alpha\in\Lambda} q_\alpha p_\alpha \theta^r_X = \theta^r_X.$$

Hence, $\mathcal{C}$ is ribbon. □

In Appendix C (Theorem C.6) we show that if $\Bbbk$ is a field, then any additive pre-fusion $\Bbbk$-category is abelian.

4.4.3 Example

Let $\mathrm{vect}_\Bbbk$ be the full subcategory of $\mathrm{proj}_\Bbbk$ whose objects are free $\Bbbk$-modules of finite rank. The structure of a spherical $\Bbbk$-category in $\mathrm{proj}_\Bbbk$ (see Example 4.2.4) restricts to $\mathrm{vect}_\Bbbk$ and turns it into an additive spherical $\Bbbk$-category. The category $\mathrm{vect}_\Bbbk$ is a pre-fusion $\Bbbk$-category with the singleton $\{\Bbbk\}$ as a representative set of simple objects. The category $\mathrm{proj}_\Bbbk$ itself may be pre-fusion or not depending on $\Bbbk$. For instance, if $\Bbbk$ is a field, then $\mathrm{proj}_\Bbbk = \mathrm{vect}_\Bbbk$ is a pre-fusion $\Bbbk$-category. If $\Bbbk = \mathbb{Z}/6\mathbb{Z}$, then the $\Bbbk$-modules $\Bbbk/2\Bbbk \simeq \mathbb{Z}/2\mathbb{Z}$ and $\Bbbk/3\Bbbk \simeq \mathbb{Z}/3\mathbb{Z}$ are projective of finite type because their direct sum is isomorphic to $\Bbbk$. Neither of these two modules is simple or splits as a direct sum of simple objects of $\mathrm{proj}_\Bbbk$. So, in this case, $\mathrm{proj}_\Bbbk$ is not pre-fusion.

4.4.4 Example

Given a group G, a 3-cocycle $\alpha\colon G^3 \to \Bbbk^*$, and a group homomorphism $d\colon G \to \Bbbk^*$, the pivotal $\Bbbk$-category $G_{\Bbbk}^{\alpha,d}$ from Example 4.2.5 is pre-fusion because all its objects are simple.

4.4.5 Exercise*

Let F, G be $\Bbbk$-linear functors from a pre-fusion $\Bbbk$-category $\mathcal{C}$ to a $\Bbbk$-category $\mathcal{D}$. Let I be a representative set of simple objects of $\mathcal{C}$. Prove that natural transformations $F \to G$ are uniquely determined by their restrictions to I. Moreover, any family $\{\alpha_i\colon F(i) \to G(i)\}_{i\in I}$ of morphisms in $\mathcal{D}$ extends to a natural transformation $\alpha\colon F \to G$.

4.4.6 Exercise*

Let $\mathcal{C}$ be a pre-fusion $\Bbbk$-category and I be a representative set of simple objects of $\mathcal{C}$. The *graduator* $A_{\mathcal{C}}$ of $\mathcal{C}$ is the group with generators $\{[i]\}_{i\in I}$ and defining relations $[i_1]\,[i_2]\cdots[i_m] = 1$ whenever $i_1, \ldots, i_m \in I$ with $m \geq 1$ satisfy $\mathrm{Hom}_{\mathcal{C}}(\mathbb{1}, i_1 \otimes i_2 \otimes \cdots \otimes i_m) \neq 0$. Prove that the isomorphism class of $A_{\mathcal{C}}$ does not depend on the choice of I. Prove that there is a bijective correspondence between group homomorphisms $A_{\mathcal{C}} \to \Bbbk^*$ and strong monoidal natural automorphisms of the identity endofunctor $1_{\mathcal{C}}\colon \mathcal{C} \to \mathcal{C}$. Prove that the graduator of the pre-fusion $\Bbbk$-category $G_{\Bbbk}^{\alpha,d}$ from Example 4.4.4 is isomorphic to G.

4.5 Fusion and modular categories

We introduce and study fusion and modular $\Bbbk$-categories.

4.5.1 Fusion categories

A *fusion $\Bbbk$-category* is a rigid pre-fusion $\Bbbk$-category which has only a finite number of isomorphism classes of simple objects. Fusion categories arise in various areas of mathematics including the representation theory of quantum groups, the theory of integrable modules over affine algebras, and the theory of subfactors. For a detailed study of fusion categories, see [ENO, EGNO].

We focus now on pivotal fusion $\Bbbk$-categories. The *dimension* of a pivotal fusion $\Bbbk$-category $\mathcal{C}$ is defined by

$$\dim(\mathcal{C}) = \sum_{i\in I} \dim_l(i)\dim_r(i) \in \Bbbk$$

where I is a (finite) representative set of simple objects of $\mathcal{C}$. The dimension of $\mathcal{C}$ does not depend on the choice of I because isomorphic objects of $\mathcal{C}$ have the same left/right dimensions. Note that if $\mathcal{C}$ is spherical, then

$$\dim(\mathcal{C}) = \sum_{i\in I} (\dim(i))^2.$$

Two pivotal fusion $\Bbbk$-categories are *isomorphic* (respectively, *equivalent*) if there is a $\Bbbk$-linear pivotal isomorphism (respectively, equivalence) between them. Lemma 2.8 implies that equivalent pivotal fusion $\Bbbk$-categories have the same dimension. If $\mathcal{C}$ is a pivotal fusion $\Bbbk$-category, then so are its opposites

$$\mathcal{C}^{\mathrm{op}} = (\mathcal{C}^{\mathrm{op}}, \otimes, \mathbb{1}), \quad \mathcal{C}^{\otimes\mathrm{op}} = (\mathcal{C}, \otimes^{\mathrm{op}}, \mathbb{1}), \quad \mathcal{C}^{\mathrm{rev}} = (\mathcal{C}^{\mathrm{op}}, \otimes^{\mathrm{op}}, \mathbb{1}).$$

By Section 4.2.1, $\mathcal{C}^{\mathrm{rev}}$ is equivalent to $\mathcal{C}$, and $\mathcal{C}^{\mathrm{op}}$ is equivalent to $\mathcal{C}^{\otimes\mathrm{op}}$. The dimensions of $\mathcal{C}^{\mathrm{op}}$, $\mathcal{C}^{\otimes\mathrm{op}}$, and $\mathcal{C}^{\mathrm{rev}}$ are all equal to $\dim(\mathcal{C})$.

4.5.2 Anomaly free categories and modularity

Let $\mathcal{C}$ be a ribbon fusion $\Bbbk$-category and let θ be the twist of $\mathcal{C}$. For any simple object i of $\mathcal{C}$, there is a unique invertible element v_i of $\Bbbk$, called the *twist scalar* of i, such that $\theta_i = v_i \,\mathrm{id}_i$. Pick a representative set I of simple objects of $\mathcal{C}$ and set

$$\Delta_\pm = \sum_{i\in I} \dim(i)\,\mathrm{tr}(\theta_i^{\pm 1}) = \sum_{i\in I} v_i^{\pm 1}(\dim(i))^2 \in \Bbbk\,.$$

The naturality of θ and the symmetry of tr imply that the scalars Δ_+, Δ_- do not depend on the choice of I. The category $\mathcal{C}$ is said to be *anomaly free* if $\Delta_+ = \Delta_-$.

Next, for any $i,j\in I$, set

$$S_{i,j} = \mathrm{tr}(\tau_{j,i}\tau_{i,j}) \in \mathrm{End}_{\mathcal{C}}(\mathbb{1}) = \Bbbk,$$

where τ is the braiding of $\mathcal{C}$. The matrix $[S_{i,j}]_{i,j\in I}$ is called the *S-matrix* of $\mathcal{C}$. Since the trace is symmetric, $S_{i,j} = S_{j,i}$ for all $i,j\in I$. Also, for all $i\in I$,

$$S_{\mathbb{1},i} = S_{i,\mathbb{1}} = \mathrm{tr}(\tau_{\mathbb{1},i}\tau_{i,\mathbb{1}}) = \mathrm{tr}(\mathrm{id}_i) = \dim(i).$$

Up to a bijection between the sets of indices, the S-matrix of $\mathcal{C}$ does not depend on the choice of I.

The category $\mathcal{C}$ is said to be *modular* if its S-matrix is invertible over $\Bbbk$. The modularity of $\mathcal{C}$ implies that Δ_+, Δ_-, and $\dim(\mathcal{C})$ are invertible in $\Bbbk$ and $\dim(\mathcal{C}) = \Delta_+\Delta_-$, see [Tu1]. Two modular categories are *isomorphic* (respectively, *equivalent*) if there is a $\Bbbk$-linear braided pivotal isomorphism (respectively, equivalence) between them. Modular categories play a fundamental role in topological field theory, see [Tu1] and Section 17.1.1.

In the next lemma, we state two useful properties of the S-matrix. Note first that any representative set I of simple objects of a pivotal pre-fusion $\Bbbk$-category carries a canonical involution $I \to I, i \mapsto i^\circ$ where i° is the unique element of I isomorphic to the (simple) object i^* dual to $i\in I$. The involutivity of the map $i \mapsto i^\circ$ follows from the fact that $i^{**} \simeq i$ by Lemma 2.2.

Lemma 4.6. *Let $\mathcal{C}$ be a modular $\Bbbk$-category, let I be a representative set of simple objects of $\mathcal{C}$, and let $[S_{i,j}]_{i,j\in I}$ be the S-matrix of $\mathcal{C}$. Then:*

(a) *For all $i, r \in I$,*

$$\sum_{j\in I} S_{i,j}S_{j,r} = \dim(\mathcal{C})\,\delta_{i^\circ,r} = \dim(\mathcal{C})\,\delta_{i,r^\circ}.$$

(b) *For all $i \in I$,*

$$\sum_{j\in I} v_i v_j \dim(j) S_{i,j} = \dim(\mathcal{C})\,\dim(i).$$

Proof. Both formulas are well known, see, for example, Formulas (3.8.a) and (3.8.d) in [Tu1, Chapter II]. □

4.5.3 Example

The spherical pre-fusion $\Bbbk$-category $\mathrm{vect}_\Bbbk$ from Example 4.4.3 is fusion and its dimension is $\dim(\mathrm{vect}_\Bbbk) = 1_\Bbbk \in \Bbbk$. The symmetry in $\mathrm{vect}_\Bbbk$ defined by the flips (see Example 3.1.4) turns $\mathrm{vect}_\Bbbk$ into an anomaly free modular $\Bbbk$-category with S-matrix $[1_\Bbbk]$ of size 1×1 and with $\Delta_+ = \Delta_- = 1_\Bbbk$.

4.5.4 Example

The pivotal $\Bbbk$-category $G_\Bbbk^{\alpha,d}$ from Example 4.4.4 is fusion if and only if G is finite, and then $\dim(G_\Bbbk^{\alpha,d}) = \mathrm{card}(G) \cdot 1_\Bbbk$. If G is non-abelian, then the category $G_\Bbbk^{\alpha,d}$ has no braidings and can not be modular.

4.5.5 Example

Further examples of modular categories arise in the theory of representations of quantum groups at roots of unity.

4.5.6 Exercise

Prove that equivalent pivotal fusion $\Bbbk$-categories have equal dimension.

4.6 Enriched graphical calculus

For pre-fusion pivotal $\Bbbk$-categories, we enrich the graphical calculus of Chapter 2.

4.6.1 Curvilinear boxes

Let $\mathcal{C}$ be a pivotal pre-fusion $\Bbbk$-category. Consider a simple object i of $\mathcal{C}$ and an i-partition $(p_\alpha : X \to i, q_\alpha : i \to X)_{\alpha \in A}$ of an object X of $\mathcal{C}$. Consider a (finite) formal sum of $\mathcal{C}$-colored Penrose diagrams

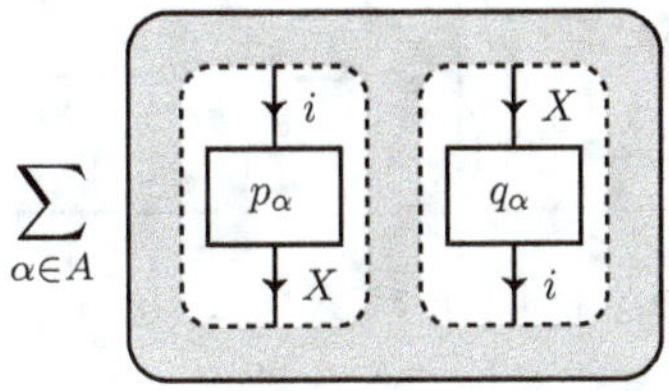

(4.4)

where the gray area represents a part of these diagrams independent of $\alpha \in A$ (and, in particular, not involving p_α, q_α). By the Penrose graphical calculus and the $\Bbbk$-linearity of $\mathcal{C}$, the sum (4.4) represents a morphism in $\mathcal{C}$. Since the tensor

$$\sum_{\alpha \in A} p_\alpha \otimes q_\alpha \in \mathrm{Hom}_{\mathcal{C}}(X, i) \otimes_{\Bbbk} \mathrm{Hom}_{\mathcal{C}}(i, X) \tag{4.5}$$

does not depend on the choice of the i-partition of X, the morphism (4.4) in $\mathcal{C}$ also does not depend on this choice. We therefore can eliminate the $\mathcal{C}$-colors p_α, q_α of the two boxes, keeping in mind only the order of the boxes and the fact that they jointly stand for the tensor (4.5). We will graphically represent this pair of boxes by two curvilinear boxes (a semi-disk and a compressed rectangle) standing respectively for p_α and q_α where α runs over A:

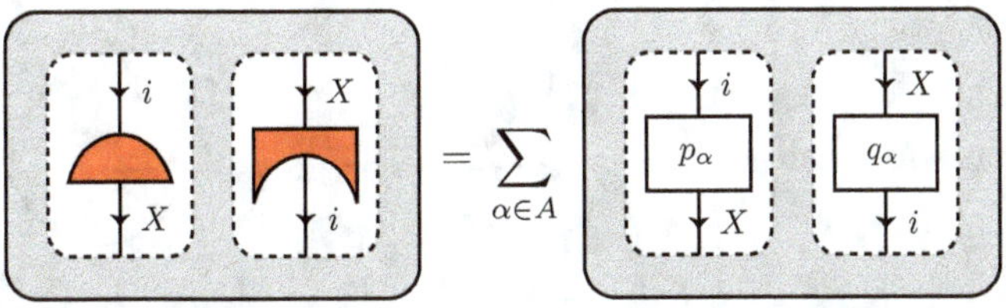

(4.6)

The gray areas in the picture are the same as above. We will also use similar notation obtained from (4.6) by reorienting the X-labeled arcs upward and replacing $(p_\alpha, q_\alpha)_{\alpha \in A}$ with an i-partition of X^*, or by reorienting the i-labeled arcs upward and replacing $(p_\alpha, q_\alpha)_{\alpha \in A}$ with an i^*-partition of X. We will allow several arcs to be attached to the bottom of the semi-disk and to the top of the compressed rectangle in (4.6). The number of these arcs, their directions (up/down), and their $\mathcal{C}$-labels should be the same. For example,

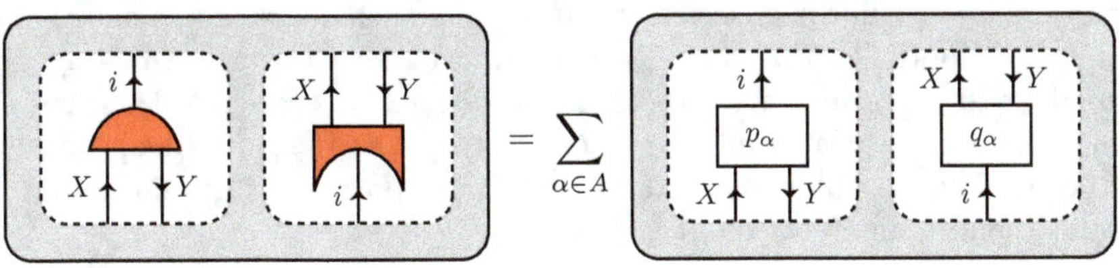

where $(p_\alpha, q_\alpha)_{\alpha\in A}$ is any i^*-partition of $X^* \otimes Y$. We will allow to erase i-labeled arcs for $i = \mathbb{1}$. In particular,

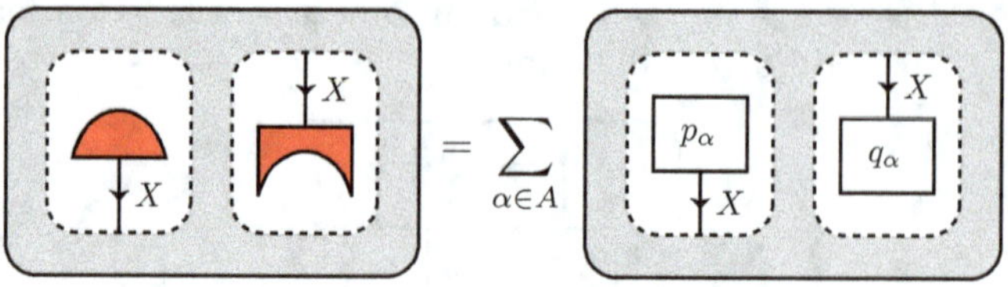

where $(p_\alpha, q_\alpha)_{\alpha\in A}$ is any $\mathbb{1}$-partition of X. Furthermore, in accordance with the isotopy invariance,

will stand for

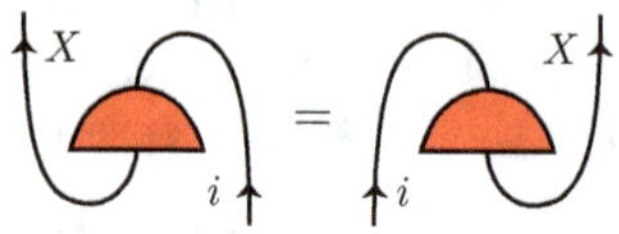

and

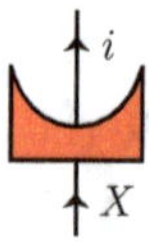

will stand for

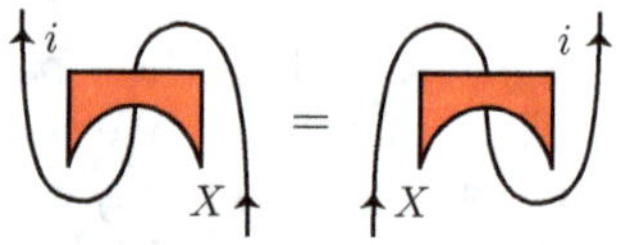

Similar notation will be applied when the X-labeled or i-labeled arcs are oriented downward and when there are several arcs attached to the horizontal sides of the curvilinear boxes. For example,

will stand for

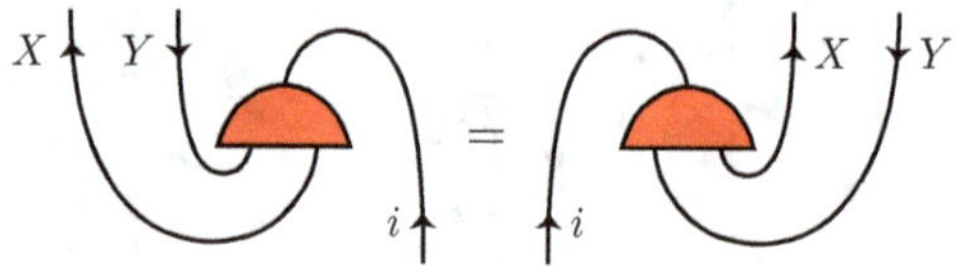

and

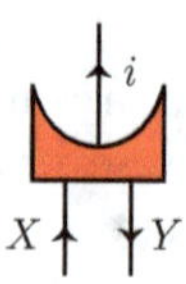

will stand for

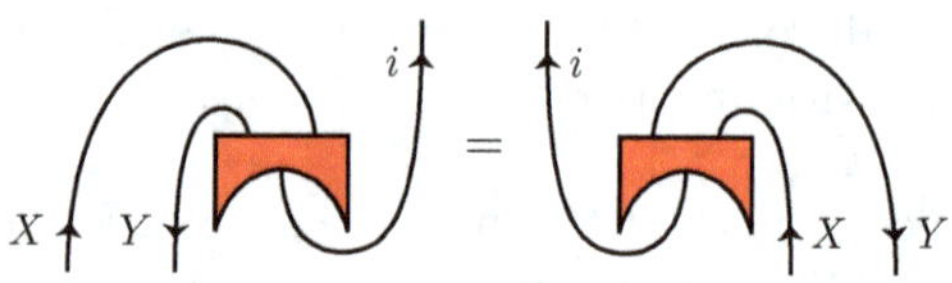

More generally, we will consider $\mathcal{C}$-colored Penrose diagrams incorporating several pairs of curvilinear boxes as above. We call such diagrams *$\mathcal{C}$-colored quasi-diagrams* and apply the procedure described above to each pair. In this way, every $\mathcal{C}$-colored quasi-diagram represents a morphism in $\mathcal{C}$. Theorem 2.6 implies that this morphism is invariant under ambient isotopies of the quasi-diagram in the plane. To distinguish different pairs of curvilinear boxes in a quasi-diagram, we will paint them in different colors.

4.6.2 Example

Let $\mathcal{C}$ be a pivotal pre-fusion $\Bbbk$-category. Consider the $\mathcal{C}$-colored quasi-diagram

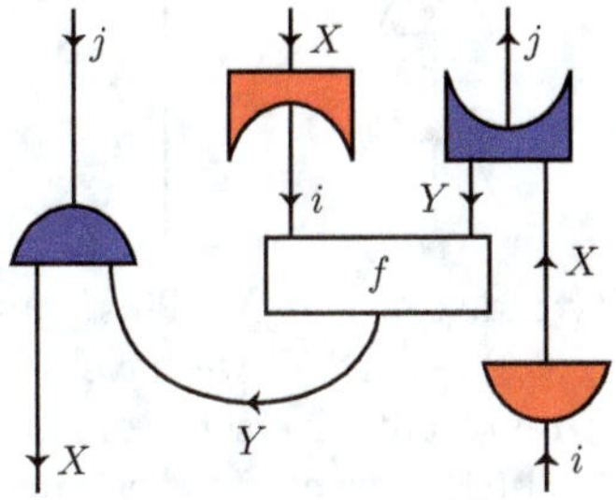

where X, Y are objects of $\mathcal{C}$, i, j are simple objects of $\mathcal{C}$, and $f\colon Y \to i \otimes Y$ is a morphism in $\mathcal{C}$. By the conventions above, this quasi-diagram represents the same morphism in $\mathcal{C}$ as the quasi-diagram

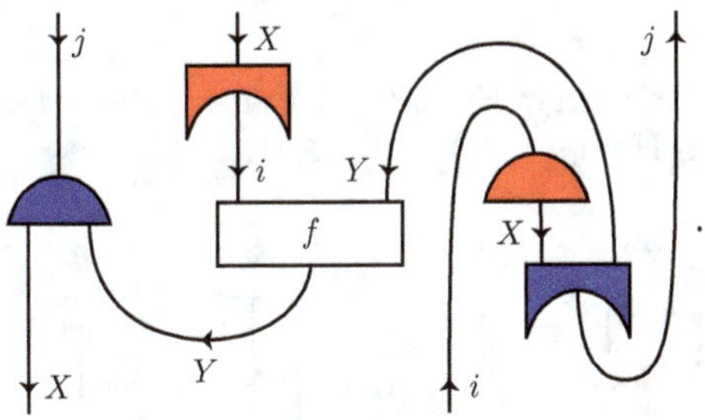

By definition, this morphism is equal to

$$\sum_{\substack{\alpha\in A\\ \beta\in B}}$$ 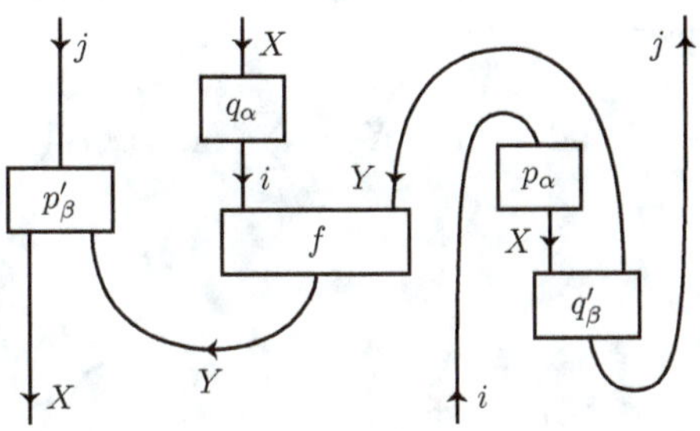 $$: X \otimes i^* \to j \otimes X \otimes j^*,$$

where $(p_\alpha, q_\alpha)_{\alpha\in A}$ is an i-partition of X and $(p'_\beta, q'_\beta)_{\beta\in B}$ is a j-partition of $X \otimes Y^*$.

4.6.3 Properties

For any object X of a pivotal pre-fusion $\Bbbk$-category $\mathcal{C}$ and any simple object i of $\mathcal{C}$, we have

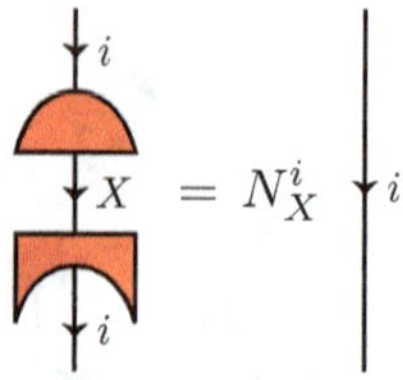

(4.7)

where N_X^i is the multiplicity index of i in X. This equality follows from the fact that given an i-partition $(p_\alpha, q_\alpha)_{\alpha\in A}$ of X, we have $p_\alpha q_\alpha = \mathrm{id}_i$ for all $\alpha \in A$ and $\mathrm{card}(A) = N_X^i$. Next, pick a representative set I of simple objects of $\mathcal{C}$. Since the union of i-partitions of $X \in \mathrm{Ob}(\mathcal{C})$ over all $i \in I$ is an I-partition of X, we have

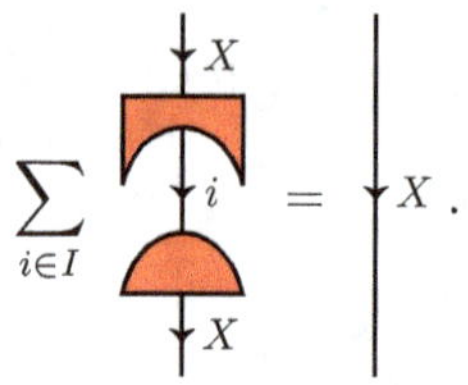

(4.8)

This formula and the fact that $\mathrm{Hom}_{\mathcal{C}}(\mathbb{1}, i) = 0 = \mathrm{Hom}_{\mathcal{C}}(i, \mathbb{1})$ for all $i \in I \setminus \{\mathbb{1}\}$ imply that for any $f \in \mathrm{Hom}_{\mathcal{C}}(\mathbb{1}, X)$ and $g \in \mathrm{Hom}_{\mathcal{C}}(X, \mathbb{1})$,

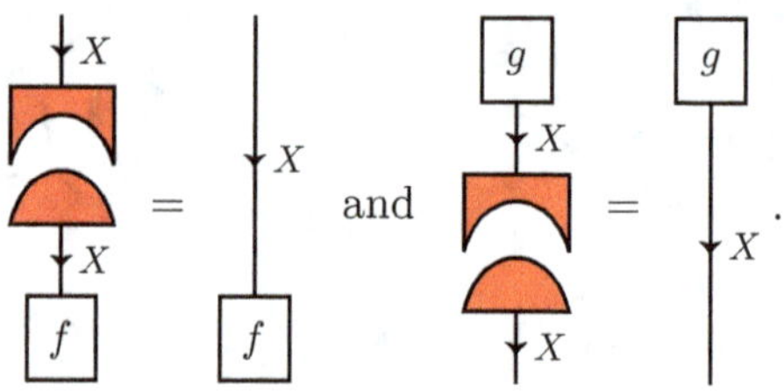

(4.9)

Also, Formulas (4.7) and (4.8) together with the isotopy invariance of the enriched graphical calculus imply the identities

$$\dim_l(X) = \sum_{i\in I} \dim_l(i) N_X^i \quad \text{and} \quad \dim_r(X) = \sum_{i\in I} \dim_r(i) N_X^i. \tag{4.10}$$

Indeed,

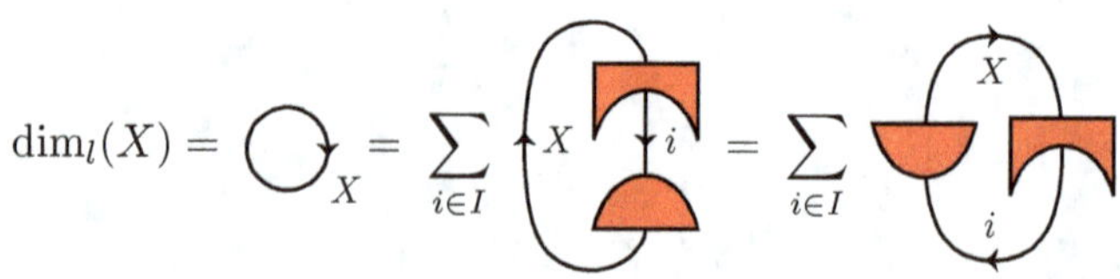

$= \sum_{i\in I}$ 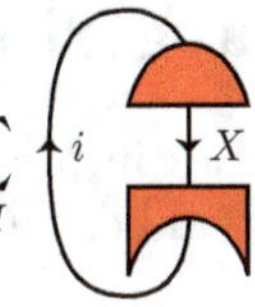 $= \sum_{i\in I} N_X^i$  $= \sum_{i\in I} \dim_l(i) N_X^i$

and similarly for the right trace. As an application, we prove the following claim.

Lemma 4.7. *Let I be a representative set of simple objects of a pivotal fusion $\Bbbk$-category $\mathcal{C}$. Then for any $X, Y \in \mathrm{Ob}(\mathcal{C})$,*

$$\sum_{m,n\in I} \dim_l(m) \dim_l(n) N^{\mathbb{1}}_{X\otimes m\otimes Y\otimes n} = \dim_r(X) \dim_r(Y) \dim(\mathcal{C})$$

and similarly with the subscripts l, r exchanged.

Proof. Using the identity $\dim_l(n) = \dim_r(n^*)$, the $\otimes$-multiplicativity of the dimensions, and (4.10) we obtain that

$$\begin{aligned}
&\sum_{m,n\in I} \dim_l(m) \dim_l(n) N^{\mathbb{1}}_{X\otimes m\otimes Y\otimes n} = \sum_{m\in I} \dim_l(m) \sum_{n\in I} \dim_r(n^*) N^{n^*}_{X\otimes m\otimes Y} \\
&= \sum_{m\in I} \dim_l(m) \sum_{k\in I} \dim_r(k) N^{k}_{X\otimes m\otimes Y} = \sum_{m\in I} \dim_l(m) \dim_r(X\otimes m\otimes Y) \\
&= \dim_r(X) \dim_r(Y) \sum_{m\in I} \dim_l(m) \dim_r(m) = \dim_r(X) \dim_r(Y) \dim(\mathcal{C}).
\end{aligned}$$

A similar proof works when the subscripts l, r are exchanged. □

4.6.4 The case of braided categories

As in Section 3.2.1, the enriched graphical calculus in a pivotal pre-fusion $\Bbbk$-category can be extended to incorporate a braiding in this category. The naturality of the braiding implies the invariance under the following moves preserving orientations and colors of the strands:

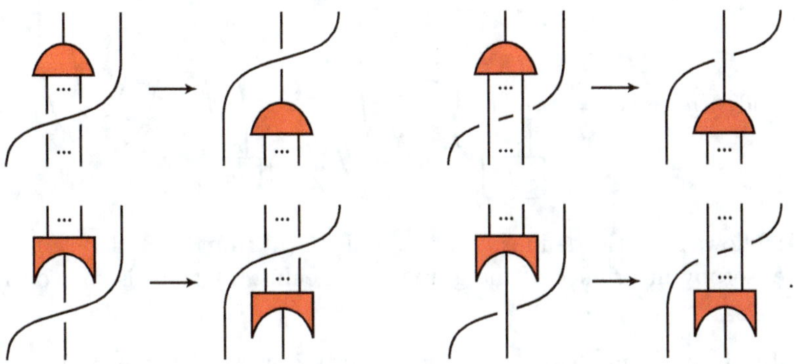

Lemma 3.1 and Theorem 3.3 directly extend to this setting.

4.7 Computations in pre-fusion categories

We establish three lemmas providing useful identities in pre-fusion $\Bbbk$-categories. The reader may skip this section at the first reading and come back to it whenever necessary.

Lemma 4.8. *Let S be a tuple of signed object of a pivotal pre-fusion $\Bbbk$-category $\mathcal{C}$ and let*

$$\omega_S \colon \mathrm{Hom}_{\mathcal{C}}(\mathbb{1}, X_{S^*}) \otimes_{\Bbbk} \mathrm{Hom}_{\mathcal{C}}(\mathbb{1}, X_S) \to \mathrm{End}_{\mathcal{C}}(\mathbb{1}) = \Bbbk$$

be the pairing determined by S, see Section 4.3.2. Then the contraction vector $*_{\omega_S} \in \mathrm{Hom}_{\mathcal{C}}(\mathbb{1}, X_S) \otimes_{\Bbbk} \mathrm{Hom}_{\mathcal{C}}(\mathbb{1}, X_{S^*})$ of ω_S is computed by

$*_{\omega_S} =$ 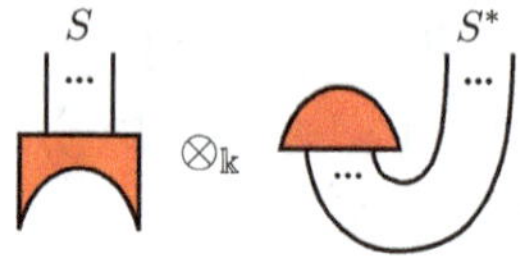 , (4.11)

where the arcs are colored and oriented so that S is the tuple of signed objects determined by the horizontal side of the curvilinear boxes (as in Section 2.2.2).

Proof. Set $E = \mathrm{Hom}_{\mathcal{C}}(\mathbb{1}, X_S)$, $F = \mathrm{Hom}_{\mathcal{C}}(\mathbb{1}, X_{S^*})$ and let $*_S \in E \otimes_{\Bbbk} F$ be the vector defined by the right-hand side of (4.11). For any $f \in E$,

$(\mathrm{id}_E \otimes_{\Bbbk} \omega_S)(*_S \otimes_{\Bbbk} f) \overset{\text{(i)}}{=}$ 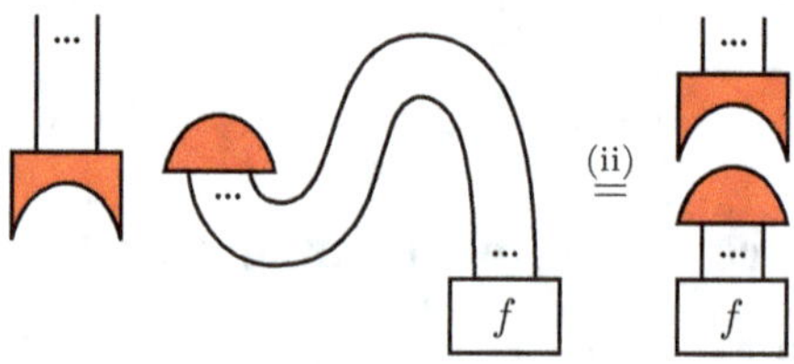 $\overset{\text{(iii)}}{=} f.$

Here, (i) follows from the definitions of ω_S and $*_S$, (ii) from the isotopy invariance of the graphical calculus, and (iii) from Formula (4.9). For any $g \in F$,

$(\omega_S \otimes_{\Bbbk} \mathrm{id}_F)(g \otimes_{\Bbbk} *_S) \overset{\text{(i)}}{=}$ 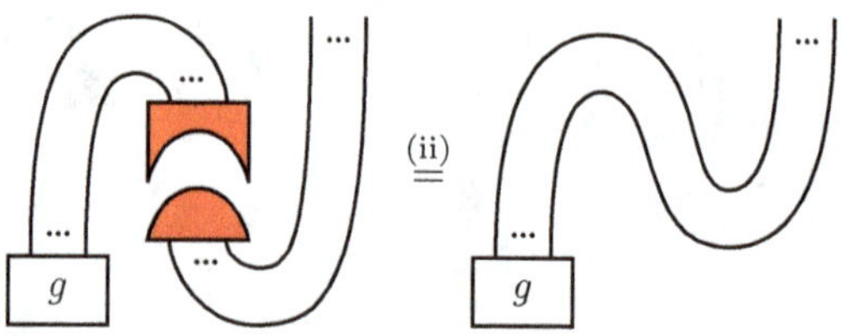 $\overset{\text{(iii)}}{=} g.$

Here, (i) follows from the definitions of ω_S and $*_S$, (ii) from Formula (4.9), and (iii) from the isotopy invariance of the graphical calculus. Thus $*_S$ is the contraction vector of ω_S. □

In the next lemma, each equality has gray areas which are supposed to be the same on both sides.

Lemma 4.9. *Let $\mathcal{C}$ be a pivotal pre-fusion $\Bbbk$-category and let I be a representative set of simple objects of $\mathcal{C}$. Then:*

(a) *For any $i, j \in I$,*

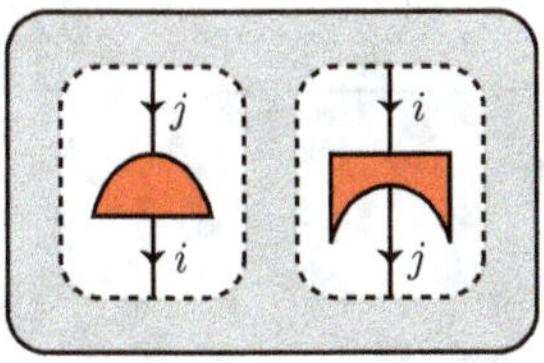

$= \delta_{i,j}$

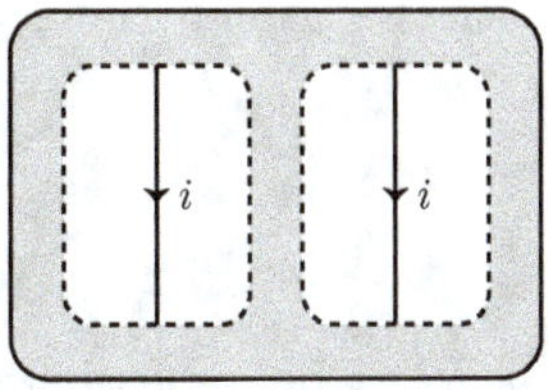

,

where $\delta_{i,j} = 1$ if $i = j$ and $\delta_{i,j} = 0$ otherwise. Also, for any $i, j \in I$,

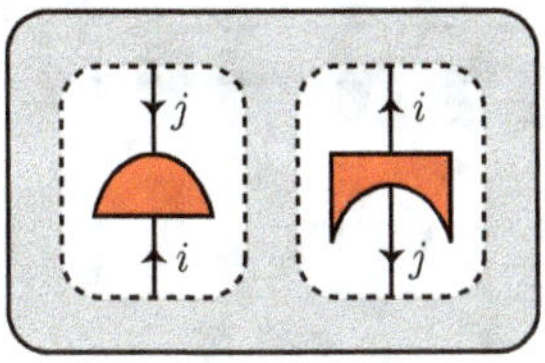

$= \delta_{i^*,j}$

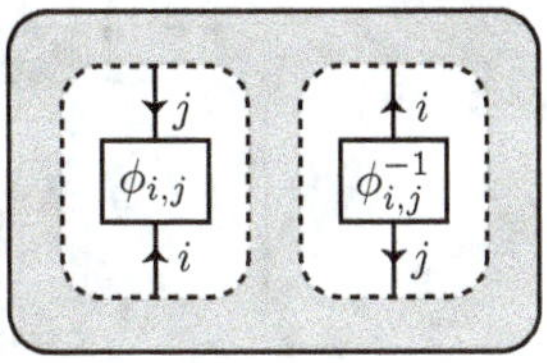

,

where $\delta_{i^*,j} = 1$ and $\phi_{i,j} \colon i^* \to j$ is any isomorphism if $i^* \simeq j$, and $\delta_{i^*,j} = 0$ otherwise.

(b) *For any $i, j \in I$,*

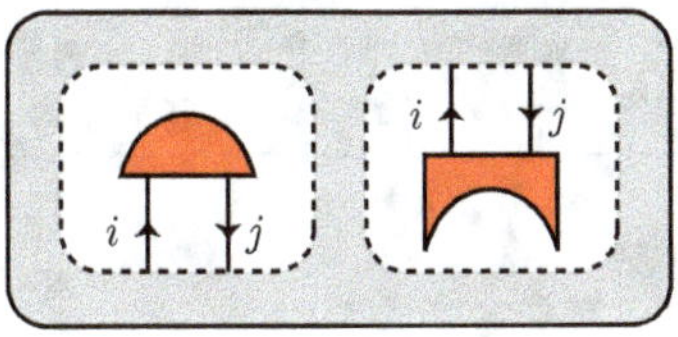

$$= \frac{\delta_{i,j}}{\dim_l(i)}$$

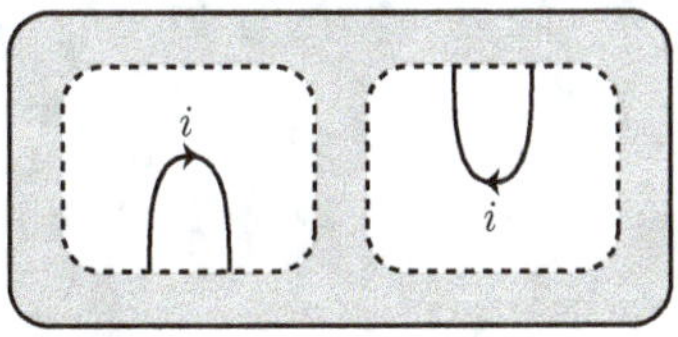

.

(c) *For any $i, j \in I$,*

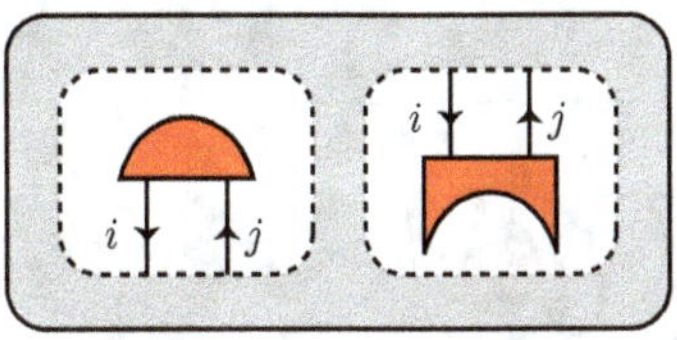

$$= \frac{\delta_{i,j}}{\dim_r(i)}$$

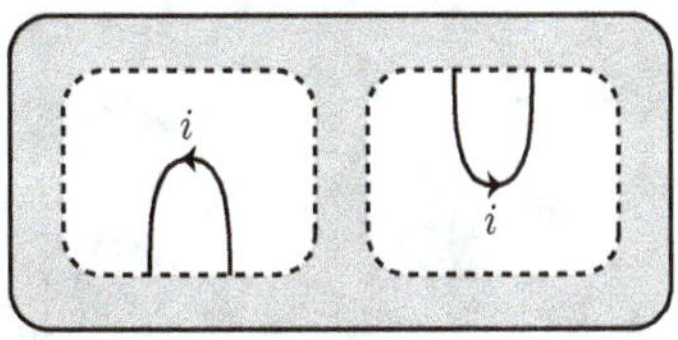

.

(d) *For any $X \in \mathrm{Ob}(\mathcal{C})$ and any simple object i of $\mathcal{C}$,*

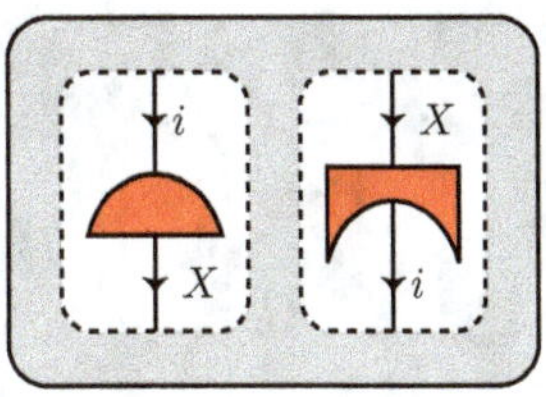

$= \dim_l(i)$

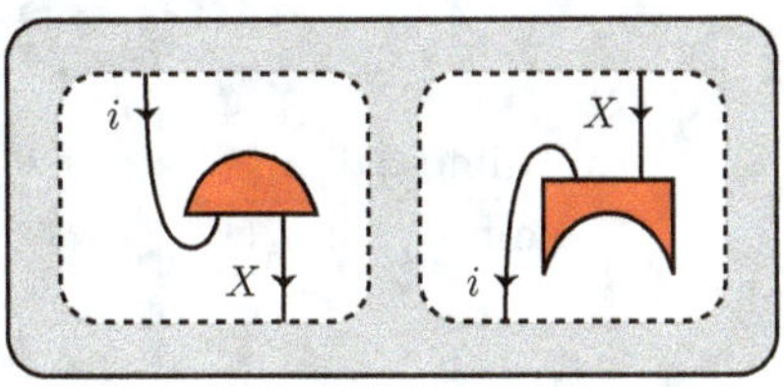

$= \dim_r(i)$ 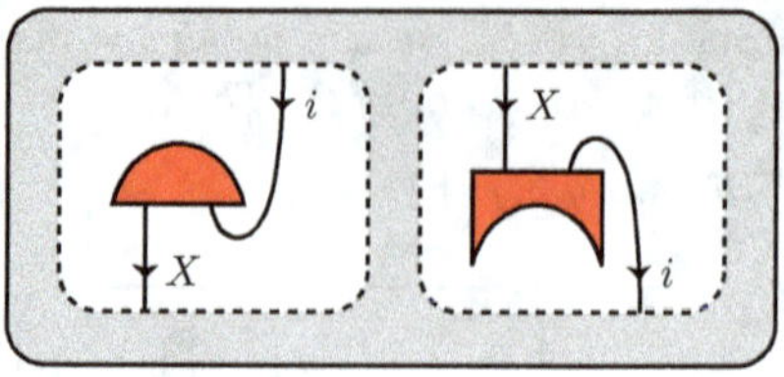

.

(e) *For any* $X \in \mathrm{Ob}(\mathcal{C})$ *and any simple objects* i, j, k *of* $\mathcal{C}$,

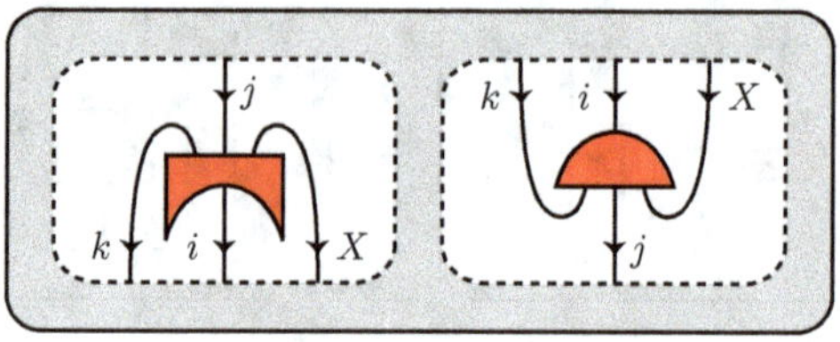

$= \dfrac{\dim_r(i)\dim_r(k)}{\dim_r(j)\dim_l(k)}$ 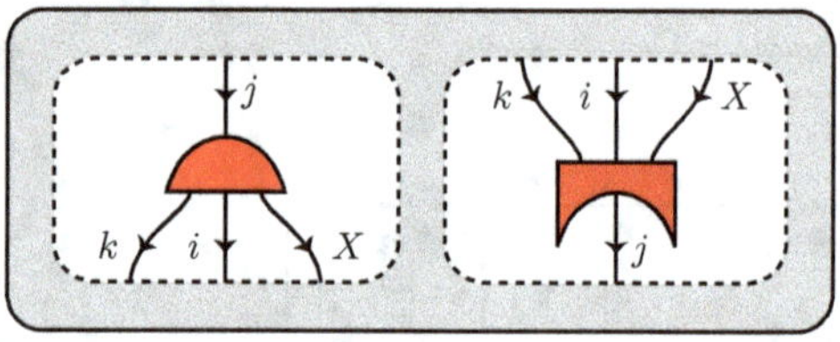

.

(f) *For any* $X \in \mathrm{Ob}(\mathcal{C})$ *and any simple objects* i, j, k *of* $\mathcal{C}$,

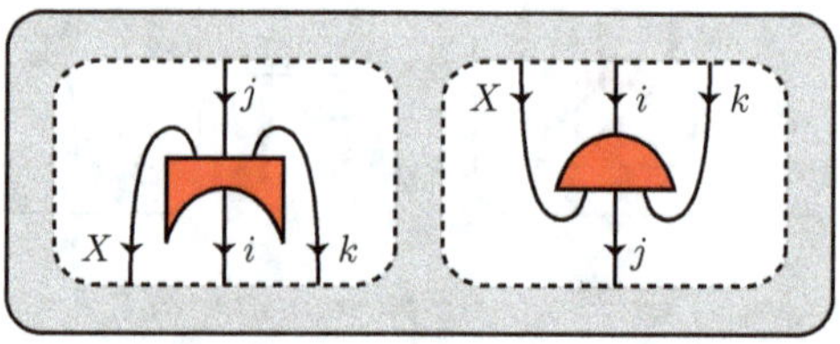

$= \dfrac{\dim_l(i)\dim_l(k)}{\dim_l(j)\dim_r(k)}$ 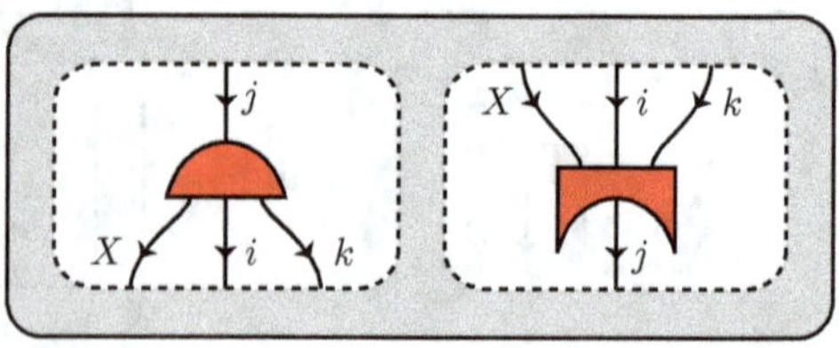

.

(g) *For any $X, Y \in \mathrm{Ob}(\mathcal{C})$ and any simple object i of $\mathcal{C}$,*

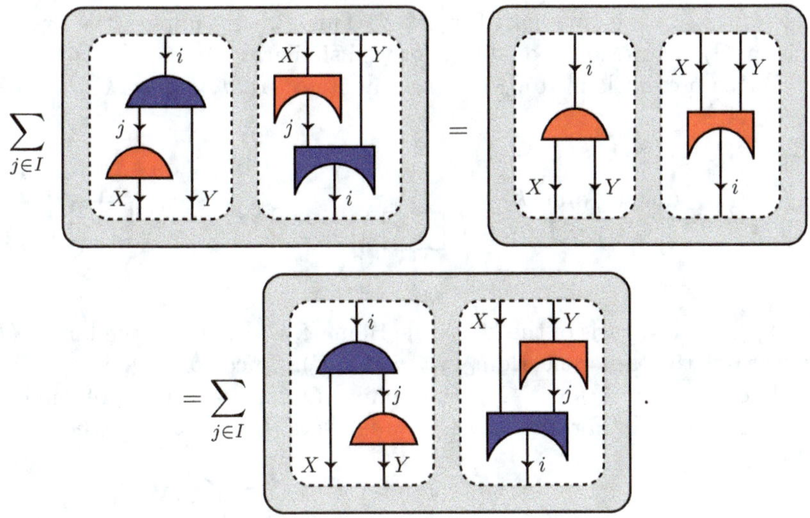

Proof. The first assertion of Claim (a) holds because $\mathrm{Hom}_{\mathcal{C}}(i,j) = 0$ if $i \neq j$ and $(\mathrm{id}_i, \mathrm{id}_i)$ is an i-partition of i. The second assertion of Claim (a) holds because $\mathrm{Hom}_{\mathcal{C}}(i^*, j) = 0$ if $i^* \not\simeq j$ and $(\phi_{i,j}, \phi_{i,j}^{-1})$ is a j-partition of i^* if $i^* \simeq j$. Claim (b) holds because

$$\mathrm{Hom}_{\mathcal{C}}(i^* \otimes j, \mathbb{1}) \simeq \mathrm{Hom}_{\mathcal{C}}(j, i) = 0$$

if $i \neq j$ and $(\mathrm{ev}_i, \dim_l(i)^{-1}\widetilde{\mathrm{coev}}_i)$ is a $\mathbb{1}$-partition of $i^* \otimes i$. Claim (c) is proved similarly.

Let us prove the first equality of Claim (d). Pick a $\mathbb{1}$-partition $(p_\alpha, q_\alpha)_{\alpha \in A}$ of $i^* \otimes X$. For each $\alpha \in A$, set

$P_\alpha = \dim_l(i)$ 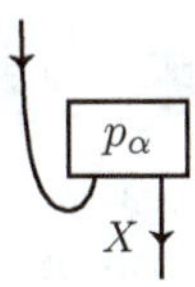 and $Q_\alpha =$ 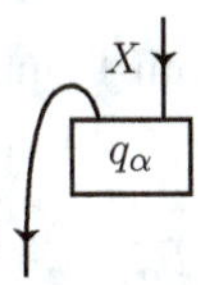.

Since $(p_\alpha)_{\alpha \in A}$ is a basis of the $\Bbbk$-module $\mathrm{Hom}_{\mathcal{C}}(i^* \otimes X, \mathbb{1})$ and $\dim_l(i)$ is invertible in $\Bbbk$ by Lemma 4.2(a), the family $(P_\alpha)_{\alpha \in A}$ is a basis of the $\Bbbk$-module $\mathrm{Hom}_{\mathcal{C}}(X, i)$. Since $(q_\alpha)_{\alpha \in A}$ is a basis of the $\Bbbk$-module $\mathrm{Hom}_{\mathcal{C}}(\mathbb{1}, i^* \otimes X)$, the family $(Q_\alpha)_{\alpha \in A}$ is a basis of the $\Bbbk$-module $\mathrm{Hom}_{\mathcal{C}}(i, X)$. Since the object i of $\mathcal{C}$ is simple, for any $\alpha, \beta \in A$, we have $P_\alpha Q_\beta = \lambda_{\alpha,\beta}\, \mathrm{id}_i$ where

$\lambda_{\alpha,\beta} \overset{\text{(i)}}{=} \dfrac{\mathrm{tr}_l(P_\alpha Q_\beta)}{\dim_l(i)} \overset{\text{(ii)}}{=}$ 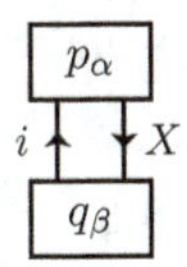 $\overset{\text{(iii)}}{=} \delta_{\alpha,\beta}.$

Here (i) follows from Lemma 4.2(b), (ii) from the definitions of P_α, Q_β and the isotopy invariance of the graphical calculus, and (iii) from the equality $p_\alpha q_\beta = \delta_{\alpha,\beta}$. Thus $(P_\alpha, Q_\alpha)_{\alpha\in A}$ is an i-partition of X. This directly implies the first equality of Claim (d). The second equality is proved similarly.

Let us prove Claim (e). Pick an i-partition $(p_\alpha, q_\alpha)_{\alpha\in A}$ of $k^* \otimes j \otimes X^*$. For any $\alpha \in A$, set

$$P_\alpha = \frac{\dim_r(j)\dim_l(k)}{\dim_r(i)\dim_r(k)}$$ 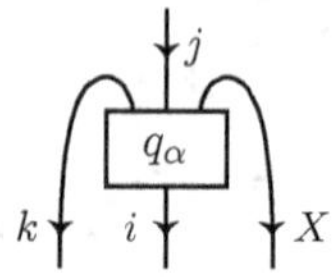 and $Q_\alpha =$ 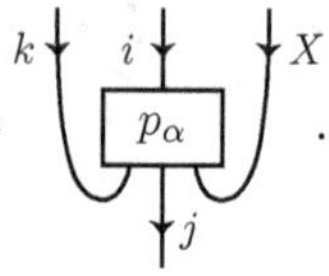.

Since $(q_\alpha)_{\alpha\in A}$ is a basis of the $\Bbbk$-module $\mathrm{Hom}_\mathcal{C}(i, k^* \otimes j \otimes X^*)$, the family $(P_\alpha)_{\alpha\in A}$ is a basis of the $\Bbbk$-module $\mathrm{Hom}_\mathcal{C}(k \otimes i \otimes X, j)$. Since $(p_\alpha)_{\alpha\in A}$ is a basis of the $\Bbbk$-module $\mathrm{Hom}_\mathcal{C}(k^* \otimes j \otimes X^*, i)$, the family $(Q_\alpha)_{\alpha\in A}$ is a basis of the $\Bbbk$-module $\mathrm{Hom}_\mathcal{C}(j, k \otimes i \otimes X)$. For any $\alpha, \beta \in A$, we have $P_\alpha Q_\beta = \lambda_{\alpha,\beta}\,\mathrm{id}_j$ where

$$\lambda_{\alpha,\beta} \overset{\text{(i)}}{=} \frac{\mathrm{tr}_r(P_\alpha Q_\beta)}{\dim_r(j)} \overset{\text{(ii)}}{=} \frac{\dim_l(k)}{\dim_r(i)\dim_r(k)}$$ 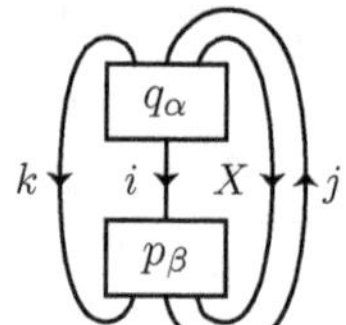

$$\overset{\text{(iii)}}{=} \frac{1}{\dim_r(i)}$$ 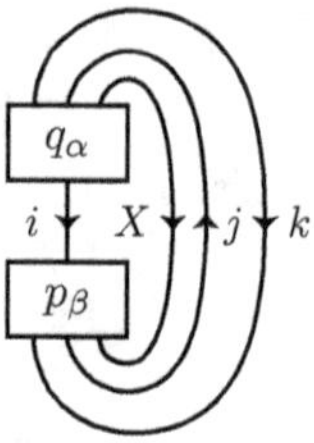 $$\overset{\text{(iv)}}{=} \frac{1}{\dim_r(i)}$$ 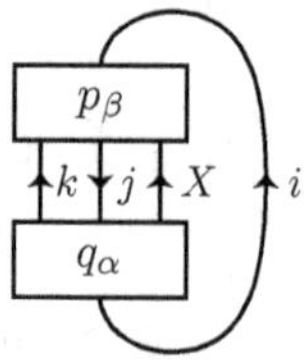 $$\overset{\text{(v)}}{=} \delta_{\alpha,\beta}.$$

Here (i) follows from Lemma 4.2(b), (ii) from the definitions of P_α and Q_β, (iii) from (4.2) applied to an endomorphism of k^*, (iv) from the isotopy invariance of the graphical calculus, (v) from the equality $p_\beta q_\alpha = \delta_{\alpha,\beta}\,\mathrm{id}_i$ and the definition of $\dim_r(i)$. Thus $(P_\alpha, Q_\alpha)_{\alpha\in A}$ is a j-partition of $k \otimes i \otimes X$. This directly implies Claim (e).

Claim (f) is obtained by applying Claim (e) to the category $\mathcal{C}^{\mathrm{op}}$. Claim (g) follows from the existence of $\Bbbk$-linear isomorphisms

$$\bigoplus_{j\in I} \mathrm{Hom}_\mathcal{C}(X, j) \otimes_\Bbbk \mathrm{Hom}_\mathcal{C}(j \otimes Y, i) \simeq \mathrm{Hom}_\mathcal{C}(X \otimes Y, i)$$
$$\simeq \bigoplus_{j\in I} \mathrm{Hom}_\mathcal{C}(X \otimes j, i) \otimes_\Bbbk \mathrm{Hom}_\mathcal{C}(Y, j)$$

induced by the isomorphisms (4.3). □

Lemma 4.10. *Let I be a representative set of simple object of a pivotal fusion $\Bbbk$-category and let $(a_i)_{i\in I}$ be a family of elements of $\Bbbk$. For any $j,k \in I$, define morphisms $f_{j,k}\colon j^* \otimes j \to k^* \otimes k$ and $g_{j,k}\colon j^* \otimes j \to k^* \otimes k$ by*

$f_{j,k} = a_k$ 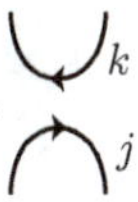 *and* $g_{j,k} = \sum_{i\in I} a_i$ 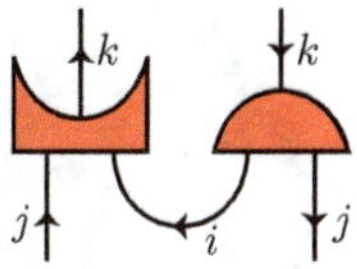 .

The equality $f_{j,k} = g_{j,k}$ holds for all $j,k \in I$ if and only if $a_i = a_{\mathbb{1}} \dim_r(i)$ for all $i \in I$.

Proof. Pick any $j,k \in I$. By duality, $f_{j,k} = g_{j,k}$ if and only if

a_k 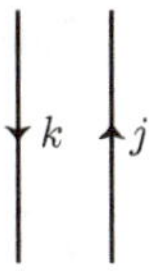 $= \sum_{i\in I} a_i$ 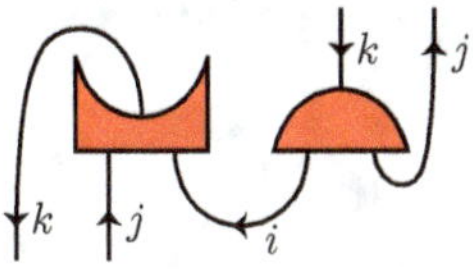 .

Lemma 4.9(e) implies that

$\sum_{i\in I} a_i$ 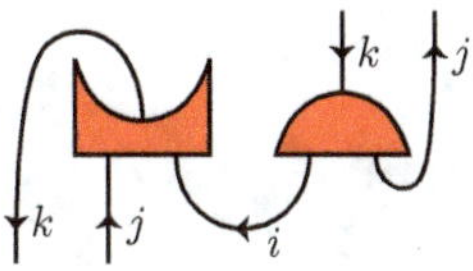 $= \sum_{i\in I} a_i \dfrac{\dim_r(k)}{\dim_r(i)}$ 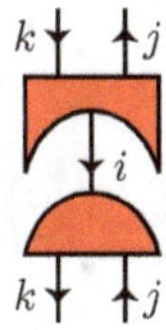 .

Therefore $f_{j,k} = g_{j,k}$ if and only if

$a_k \, \mathrm{id}_{k\otimes j^*} = \sum_{i\in I} a_i \dfrac{\dim_r(k)}{\dim_r(i)}$ 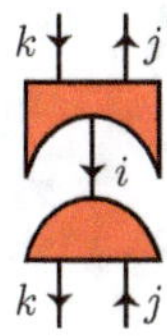 . (4.12)

If $a_i = a_{\mathbb{1}} \dim_r(i)$ for all $i \in I$, then Formula (4.8) implies (4.12) for all $j,k \in I$. Conversely, assume that (4.12) holds for all $j,k \in I$. For any $i \in I$, apply (4.12) to $k = \mathbb{1}$ and $j = i^*$. The resulting equality and Lemma 4.9(a) imply that $a_i = a_{\mathbb{1}} \dim_r(i)$. □

Chapter 5

The center of a monoidal category

We review the Drinfeld–Joyal–Street center of a monoidal category.

5.1 Half-braidings and the center

Categorifying the notion of the center of a monoid one obtains the notion of the center of a monoidal category. It is formulated in terms of half-braidings which we define first.

5.1.1 Half-braidings

A *half-braiding* of a monoidal category $\mathcal{C}$ is a pair (A, σ), where $A \in \mathrm{Ob}(\mathcal{C})$ and

$$\sigma = \{\sigma_X \colon A \otimes X \to X \otimes A\}_{X \in \mathrm{Ob}(\mathcal{C})}$$

is a natural isomorphism between the functors $A\otimes ? = \otimes(A, ?)$ and $?\otimes A = \otimes(?, A)$ which is $\otimes$-*multiplicative* in the sense that for all $X, Y \in \mathrm{Ob}(\mathcal{C})$, we have

$$\sigma_{X \otimes Y} = (\mathrm{id}_X \otimes \sigma_Y)(\sigma_X \otimes \mathrm{id}_Y). \tag{5.1}$$

The naturality of σ means that $\sigma_Y(\mathrm{id}_A \otimes f) = (f \otimes \mathrm{id}_A)\sigma_X$ for all morphisms $f \colon X \to Y$ in $\mathcal{C}$. The definition of a half-braiding σ implies that $\sigma_{\mathbb{1}} = \mathrm{id}_A$. We call A the *underlying object* of the half-braiding (A, σ).

5.1.2 The center

The *center* $\mathcal{Z}(\mathcal{C})$ of a monoidal category $\mathcal{C}$ is a braided category defined as follows. The objects of $\mathcal{Z}(\mathcal{C})$ are half-braidings of $\mathcal{C}$. A morphism $(A, \sigma) \to (B, \rho)$ in $\mathcal{Z}(\mathcal{C})$

is a morphism $f\colon A \to B$ in $\mathcal{C}$ such that for all $X \in \mathrm{Ob}(\mathcal{C})$,

$$\rho_X(f \otimes \mathrm{id}_X) = (\mathrm{id}_X \otimes f)\sigma_X \colon A \otimes X \to X \otimes B.$$

The unit object of $\mathcal{Z}(\mathcal{C})$ is $\mathbb{1}_{\mathcal{Z}(\mathcal{C})} = (\mathbb{1}, \{\mathrm{id}_X\}_{X \in \mathrm{Ob}(\mathcal{C})})$, and the monoidal product is

$$(A, \sigma) \otimes (B, \rho) = \big(A \otimes B, (\sigma \otimes \mathrm{id}_B)(\mathrm{id}_A \otimes \rho)\big).$$

The braiding τ in $\mathcal{Z}(\mathcal{C})$ is defined by

$$\tau_{(A,\sigma),(B,\rho)} = \sigma_B \colon (A, \sigma) \otimes (B, \rho) \to (B, \rho) \otimes (A, \sigma).$$

The category $\mathcal{Z}(\mathcal{C})$ is pure (since it is braided) and

$$\mathrm{End}_{\mathcal{Z}(\mathcal{C})}(\mathbb{1}_{\mathcal{Z}(\mathcal{C})}) = \{\alpha \in \mathrm{End}_{\mathcal{C}}(\mathbb{1}) \,|\, \alpha \cdot \mathrm{id}_X = \mathrm{id}_X \cdot \alpha \text{ for all } X \in \mathrm{Ob}(\mathcal{C})\},$$

where the left and right actions of $\mathrm{End}_{\mathcal{C}}(\mathbb{1})$ on the morphisms of $\mathcal{C}$ are defined in Section 1.3.2. In particular, if $\mathcal{C}$ is pure, then $\mathrm{End}_{\mathcal{Z}(\mathcal{C})}(\mathbb{1}_{\mathcal{Z}(\mathcal{C})}) = \mathrm{End}_{\mathcal{C}}(\mathbb{1})$.

The *forgetful functor* $\mathcal{Z}(\mathcal{C}) \to \mathcal{C}$ carries every half-braiding to its underlying object and acts as the identity on the morphisms. The forgetful functor is strict monoidal and reflects isomorphisms: a morphism in $\mathcal{Z}(\mathcal{C})$ carried to an isomorphism in $\mathcal{C}$ is itself an isomorphism.

5.1.3 Remark

The above-defined half-braidings of a monoidal category $\mathcal{C}$ are sometimes called *left half-braidings* and $\mathcal{Z}(\mathcal{C})$ is called the *left center* of $\mathcal{C}$. Likewise, one defines a *right half-braiding* of $\mathcal{C}$ to be a pair (A, σ), where $A \in \mathrm{Ob}(\mathcal{C})$ and

$$\sigma = \{\sigma_X \colon X \otimes A \to A \otimes X\}_{X \in \mathrm{Ob}(\mathcal{C})}$$

is a natural isomorphism satisfying the identity $\sigma_{X \otimes Y} = (\sigma_X \otimes \mathrm{id}_Y)(\mathrm{id}_X \otimes \sigma_Y)$. Right half-braidings form a monoidal category $\mathcal{Z}'(\mathcal{C})$, the *right center* of $\mathcal{C}$, with braiding $\tau'_{(A,\sigma),(B,\rho)} = \rho_A$. The left and right centers determine each other: $\mathcal{Z}'(\mathcal{C})$ is isomorphic as a braided category to the mirror of $\mathcal{Z}(\mathcal{C})$, see Example 3.1.7. The isomorphism acts on objects by $(A, \sigma) \mapsto (A, \sigma^{-1})$ and acts on morphisms as the identity. In the sequel, we consider only left half-braidings and the left center.

5.1.4 Example

Let G be a monoid and $\alpha\colon G^3 \to \Bbbk^*$ be a 3-cocycle. Half-braidings in the monoidal category $G_{\Bbbk}^{\alpha}$ of Example 1.2.6 bijectively correspond to pairs (a, σ), where a is an element of the center of G (i.e., an element of G commuting with all elements of G) and $\sigma\colon G \to \Bbbk^*$ is a map such that for all $g, h \in G$,

$$\sigma(gh) = \sigma(g)\,\sigma(h)\,\alpha(a, g, h)^{-1}\,\alpha(g, a, h)\,\alpha(g, h, a)^{-1}.$$

When $\alpha = 1$, such a map $\sigma\colon G \to \Bbbk^*$ is just a monoid homomorphism.

5.1.5 Exercise*

Let $\mathcal{C}$ be a right rigid category, $A \in \mathrm{Ob}(\mathcal{C})$, and $\sigma = \{\sigma_X : A \otimes X \to X \otimes A\}_{X \in \mathrm{Ob}(\mathcal{C})}$ be a natural transformation satisfying $\sigma_{\mathbb{1}} = \mathrm{id}_A$ and (5.1) for all $X, Y \in \mathrm{Ob}(\mathcal{C})$. Prove that the pair (A, σ) is a half-braiding of $\mathcal{C}$ and that for any $X \in \mathrm{Ob}(\mathcal{C})$, the inverse of σ_X is computed from any right dual $(X^\vee, \widetilde{\mathrm{ev}}_X)$ of X by

$\sigma_X^{-1} =$ 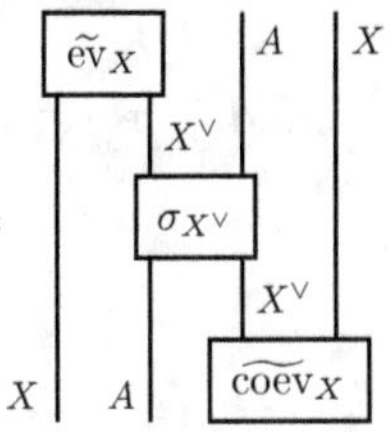 $: X \otimes A \to A \otimes X,$

where $\widetilde{\mathrm{coev}}_X$ is the inverse of the pairing $\widetilde{\mathrm{ev}}_X$.

5.2 Centers of rigid and pivotal categories

5.2.1 The center of a rigid category

Let $\mathcal{C}$ be a rigid category. We derive from a left duality $\{({}^\vee X, \mathrm{ev}_X)\}_{X \in \mathrm{Ob}(\mathcal{C})}$ and a right duality $\{(X^\vee, \widetilde{\mathrm{ev}}_X)\}_{X \in \mathrm{Ob}(\mathcal{C})}$ in $\mathcal{C}$ a left duality and a right duality in the center $\mathcal{Z}(\mathcal{C})$ of $\mathcal{C}$. For any $X \in \mathrm{Ob}(\mathcal{C})$, denote by coev_X and $\widetilde{\mathrm{coev}}_X$ the inverses of the pairings ev_X and $\widetilde{\mathrm{ev}}_X$, respectively. Consider a half-braiding (A, σ) of $\mathcal{C}$. For any $X \in \mathrm{Ob}(\mathcal{C})$, set

$({}^\dagger\sigma)_X =$ 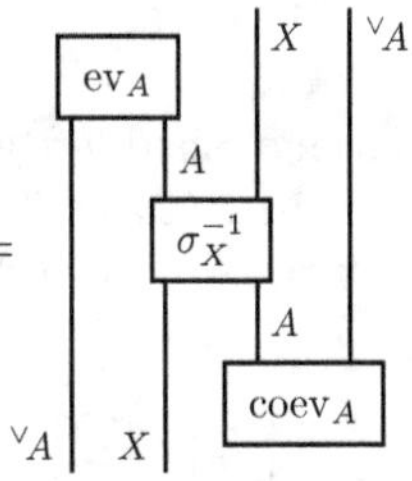 and $(\sigma^\dagger)_X =$ 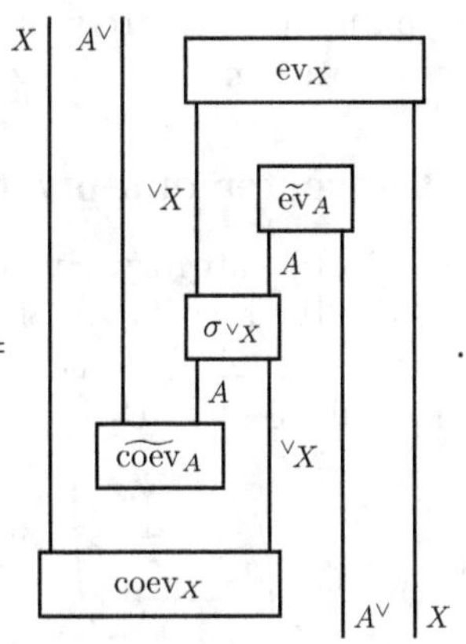 .

The duality axioms, the naturality of σ, and the formula $\sigma_{\mathbb{1}} = \mathrm{id}_A$ imply that the families

$$
\begin{aligned}
{}^\dagger\sigma &= \{({}^\dagger\sigma)_X : {}^\vee A \otimes X \to X \otimes {}^\vee A\}_{X \in \mathrm{Ob}(\mathcal{C})},\\
\sigma^\dagger &= \{(\sigma^\dagger)_X : A^\vee \otimes X \to X \otimes A^\vee\}_{X \in \mathrm{Ob}(\mathcal{C})},
\end{aligned}
$$

are natural isomorphisms with inverses

$(^{\dagger}\sigma)_X^{-1} =$ 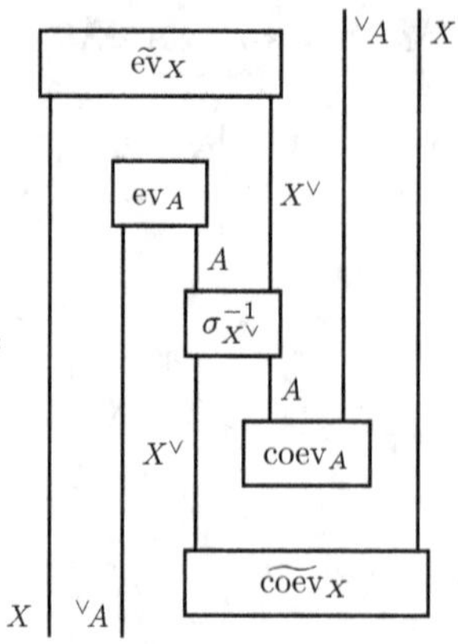

and $(\sigma^{\dagger})_X^{-1} =$ 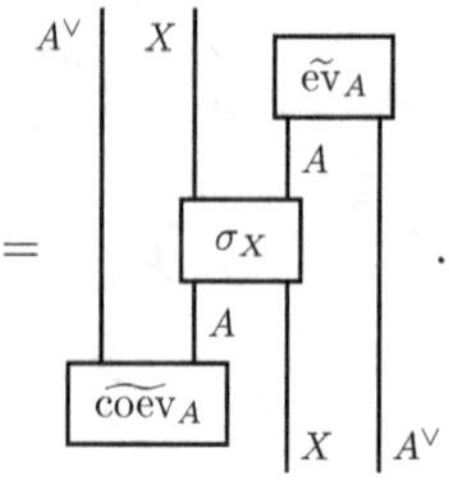

.

It follows form (5.1) that the pairs $({}^{\vee}A, {}^{\dagger}\sigma)$ and $(A^{\vee}, \sigma^{\dagger})$ are half-braidings of $\mathcal{C}$. Furthermore, ev_A is a morphism in $\mathcal{Z}(\mathcal{C})$ from $({}^{\vee}A, {}^{\dagger}\sigma) \otimes (A, \sigma)$ to $\mathbb{1}_{\mathcal{Z}(\mathcal{C})}$ and $\widetilde{\mathrm{ev}}_A$ is a morphism in $\mathcal{Z}(\mathcal{C})$ from $(A, \sigma) \otimes (A^{\vee}, \sigma^{\dagger})$ to $\mathbb{1}_{\mathcal{Z}(\mathcal{C})}$. These morphisms are non-degenerate pairings in $\mathcal{Z}(\mathcal{C})$ with inverses coev_A and $\widetilde{\mathrm{coev}}_A$, respectively. Thus the family

$$\big\{\big({}^{\vee}(A,\sigma) = ({}^{\vee}A, {}^{\dagger}\sigma),\ \mathrm{ev}_{(A,\sigma)} = \mathrm{ev}_A\big)\big\}_{(A,\sigma)\in \mathrm{Ob}(\mathcal{Z}(\mathcal{C}))} \tag{5.2}$$

is a left duality in $\mathcal{Z}(\mathcal{C})$, and the family

$$\big\{\big((A,\sigma)^{\vee} = (A^{\vee}, \sigma^{\dagger}),\ \widetilde{\mathrm{ev}}_{(A,\sigma)} = \widetilde{\mathrm{ev}}_A\big)\big\}_{(A,\sigma)\in \mathrm{Ob}(\mathcal{Z}(\mathcal{C}))} \tag{5.3}$$

is a right duality in $\mathcal{Z}(\mathcal{C})$. In particular, the category $\mathcal{Z}(\mathcal{C})$ is rigid. Note that the dualities (5.2) and (5.3) are lifts along the forgetful functor $\mathcal{Z}(\mathcal{C}) \to \mathcal{C}$ of the left and right dualities in $\mathcal{C}$ respectively (see Section 1.6.4).

5.2.2 The center of a pivotal category

Let $\mathcal{C}$ be a pivotal category. By definition, $\mathcal{C}$ has distinguished left and right dualities. Consider the associated left and right dualities in $\mathcal{Z}(\mathcal{C})$ as in Section 5.2.1. The half-braidings $(A^*, {}^{\dagger}\sigma)$ and $(A^*, \sigma^{\dagger})$ associated with a half-braiding (A, σ) of $\mathcal{C}$ are given for all $X \in \mathrm{Ob}(\mathcal{C})$ by

$({}^{\dagger}\sigma)_X =$ 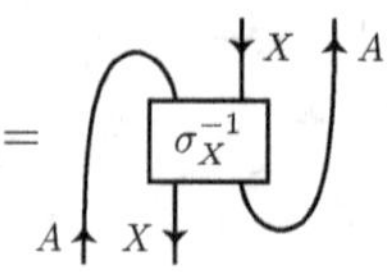

and $(\sigma^{\dagger})_X =$ 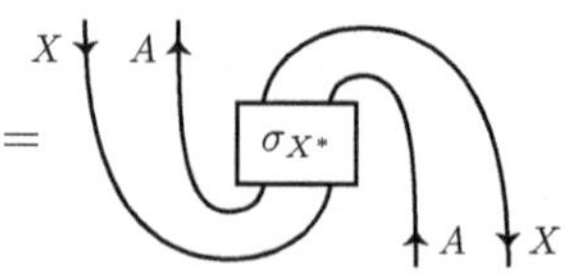

,

with inverses

$({}^{\dagger}\sigma)_X^{-1} =$ 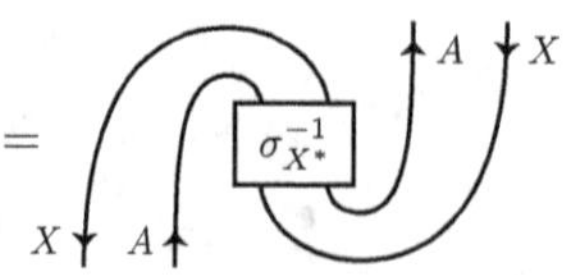

and $(\sigma^{\dagger})_X^{-1} =$ 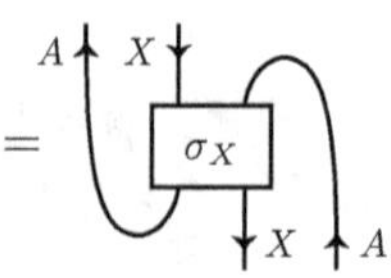

.

Lemma 5.1. $^\dagger\sigma = \sigma^\dagger$.

Proof. The isotopy invariance of the graphical calculus (Theorem 2.6) and the invertibility of σ imply that for any $X \in \mathrm{Ob}(\mathcal{C})$,

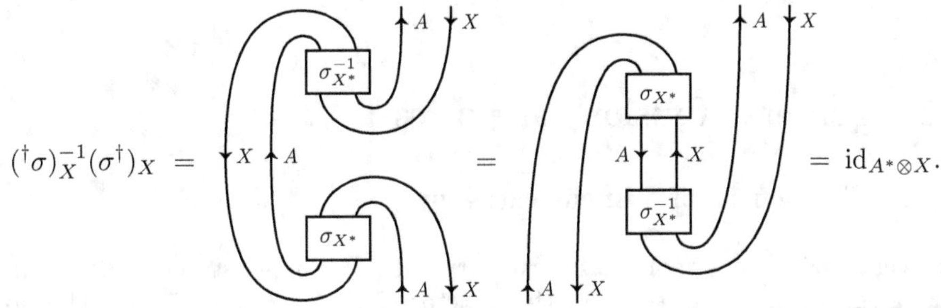

Therefore $(^\dagger\sigma)_X = (\sigma^\dagger)_X$. □

Lemma 5.1 implies that the family

$$\left\{\left((A,\sigma)^* = (A^*,\sigma^\dagger),\ \mathrm{ev}_{(A,\sigma)} = \mathrm{ev}_A,\ \widetilde{\mathrm{ev}}_{(A,\sigma)} = \widetilde{\mathrm{ev}}_A\right)\right\}_{(A,\sigma)\in \mathrm{Ob}(\mathcal{Z}(\mathcal{C}))} \tag{5.4}$$

is a pivotal duality in $\mathcal{Z}(\mathcal{C})$. It turns $\mathcal{Z}(\mathcal{C})$ into a pivotal category. The left and right coevaluations in $\mathcal{Z}(\mathcal{C})$ are computed by

$$\mathrm{coev}_{(A,\sigma)} = \mathrm{coev}_A \quad \text{and} \quad \widetilde{\mathrm{coev}}_{(A,\sigma)} = \widetilde{\mathrm{coev}}_A.$$

This implies that the (strict monoidal) forgetful functor $\mathcal{U}\colon \mathcal{Z}(\mathcal{C}) \to \mathcal{C}$ is strictly pivotal (see Section 1.7.5). By Lemma 2.8, the left/right traces of endomorphisms in $\mathcal{Z}(\mathcal{C})$ and the left/right dimensions of objects of $\mathcal{Z}(\mathcal{C})$ are the same as in $\mathcal{C}$. Consequently, if $\mathcal{C}$ is spherical, then so is $\mathcal{Z}(\mathcal{C})$.

The definition of the braiding in $\mathcal{Z}(\mathcal{C})$ (see Section 5.1.2) implies that the left twist θ^l and the right twist θ^r of $\mathcal{Z}(\mathcal{C})$ (defined in Section 3.3.1) are computed by

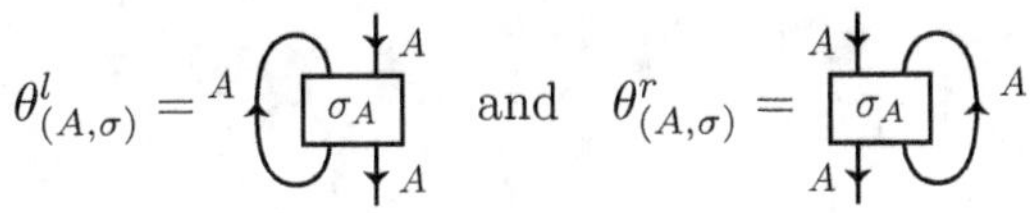

for any half-braiding (A,σ) of $\mathcal{C}$. Consequently, $\mathcal{Z}(\mathcal{C})$ is ribbon if and only if

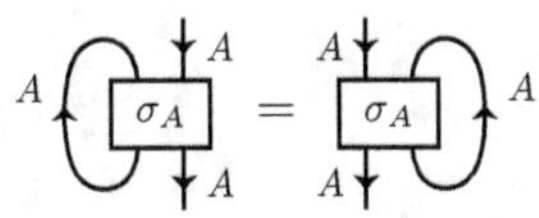

for all half braidings (A,σ) of $\mathcal{C}$.

5.2.3 Exercise*

Prove that the centers of equivalent monoidal categories are equivalent braided categories.

5.2.4 Exercise

Prove that the centers of equivalent pivotal categories are equivalent braided pivotal categories.

5.3 Centers of fusion categories

5.3.1 The center of a linear category

The center $\mathcal{Z}(\mathcal{C})$ of a monoidal $\Bbbk$-category $\mathcal{C}$ has a unique structure of a $\Bbbk$-linear category such that the forgetful functor $\mathcal{Z}(\mathcal{C}) \to \mathcal{C}$ is $\Bbbk$-linear. This follows from the fact that for all half-braidings (A, σ) and (B, ρ) of $\mathcal{C}$, the set $\mathrm{Hom}_{\mathcal{Z}(\mathcal{C})}((A, \sigma), (B, \rho))$ is a submodule of the $\Bbbk$-module $\mathrm{Hom}_{\mathcal{C}}(A, B)$. We always endow $\mathcal{Z}(\mathcal{C})$ with this $\Bbbk$-linear structure. This turns $\mathcal{Z}(\mathcal{C})$ into a braided monoidal $\Bbbk$-category. If $\mathcal{C}$ is additive, then so is $\mathcal{Z}(\mathcal{C})$.

5.3.2 The center of a pre-fusion category

By Sections 5.2.2 and 5.3.1, the center of a pivotal $\Bbbk$-category $\mathcal{C}$ is a braided pivotal $\Bbbk$-category. We now show that if $\mathcal{C}$ is spherical and pre-fusion, then $\mathcal{Z}(\mathcal{C})$ is ribbon.

Lemma 5.2. *The center of a spherical pre-fusion $\Bbbk$-category is ribbon.*

Proof. Let (A, σ) be a half-braiding of $\mathcal{C}$. For any simple object i of $\mathcal{C}$ and any morphisms $p\colon A \to i$ and $q\colon i \to A$ in $\mathcal{C}$,

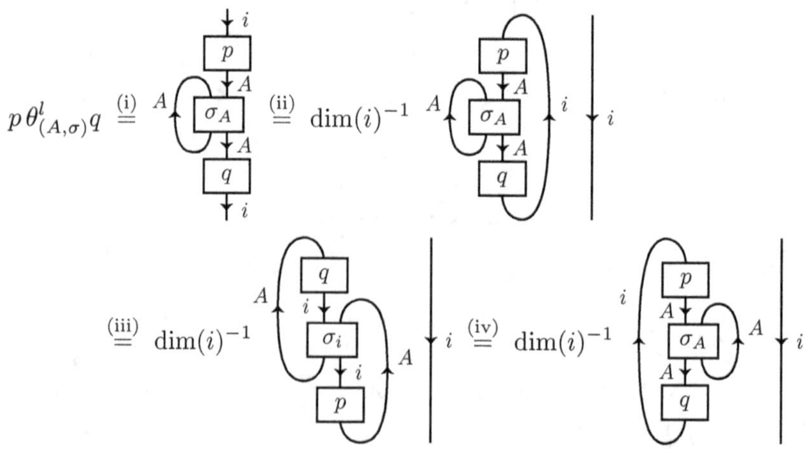

$\overset{(v)}{=}$ 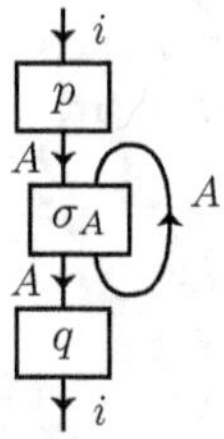

$\overset{(vi)}{=} \; p\,\theta^r_{(A,\sigma)}q\,.$

Here (i) and (vi) follow from the computation of the twists of $\mathcal{Z}(\mathcal{C})$ given at the end of Section 5.2.2, (ii) and (v) from Lemma 4.2(b), and (iii) and (iv) from the naturality of σ. Now consider an I-partition $(p_\alpha\colon A \to i_\alpha, q_\alpha\colon i_\alpha \to A)_{\alpha\in\Lambda}$ of A, where I is a representative set of simple objects of $\mathcal{C}$. Since $\mathrm{id}_A = \sum_{\alpha\in\Lambda} q_\alpha p_\alpha$ and $\mathrm{Hom}_{\mathcal{C}}(i_\alpha, i_\beta) = 0$ for $i_\alpha \neq i_\beta$, any morphism $f \in \mathrm{End}_{\mathcal{C}}(A)$ expands as

$$f = \sum_{\alpha,\beta\in\Lambda,\, i_\alpha = i_\beta} q_\alpha(p_\alpha f q_\beta)p_\beta.$$

Thus

$$\theta^l_{(A,\sigma)} = \sum_{\alpha,\beta\in\Lambda,\, i_\alpha = i_\beta} q_\alpha(p_\alpha \theta^l_{(A,\sigma)} q_\beta)p_\beta = \sum_{\alpha,\beta\in\Lambda,\, i_\alpha = i_\beta} q_\alpha(p_\alpha \theta^r_{(A,\sigma)} q_\beta)p_\beta = \theta^r_{(A,\sigma)}.$$

Hence, $\mathcal{Z}(\mathcal{C})$ is ribbon. □

5.3.3 The center of a pivotal fusion category

We state two fundamental theorems of Michael Müger on the centers of additive pivotal fusion categories, see [Mü, Theorems 1.2 and 3.16].

Theorem 5.3. *Let $\Bbbk$ be an algebraically closed field and let $\mathcal{C}$ be an additive pivotal fusion $\Bbbk$-category with* $\dim(\mathcal{C}) \neq 0$. *Then $\mathcal{Z}(\mathcal{C})$ is an additive pivotal braided fusion $\Bbbk$-category and* $\dim(\mathcal{Z}(\mathcal{C})) = (\dim(\mathcal{C}))^2$.

If the category $\mathcal{C}$ in Theorem 5.3 is spherical, then its center $\mathcal{Z}(\mathcal{C})$ is ribbon by Lemma 5.2. The next theorem yields a stronger result.

Theorem 5.4. *If, under the assumptions of Theorem* 5.3, *the category $\mathcal{C}$ is spherical, then $\mathcal{Z}(\mathcal{C})$ is an additive anomaly free modular $\Bbbk$-category with* $\Delta_+ = \Delta_- = \dim(\mathcal{C})$.

If the field $\Bbbk$ in Theorem 5.3 has characteristic zero, then the condition $\dim(\mathcal{C}) \neq 0$ follows from the other conditions on $\mathcal{C}$ and can be dropped, see [ENO]. If $\Bbbk$ is a field of positive characteristic, then it may happen that $\dim(\mathcal{C}) = 0$, see examples in Appendix A.3. The following theorem from [BV3] ensures that in this case the center of $\mathcal{C}$ is not a pre-fusion category.

Theorem 5.5. *If $\Bbbk$ is a field and $\mathcal{C}$ is an additive pivotal fusion $\Bbbk$-category with* $\dim(\mathcal{C}) = 0$, *then the $\Bbbk$-category $\mathcal{Z}(\mathcal{C})$ is not pre-fusion.*

We prove Theorems 5.3–5.5 in Section 9.5 using the theory of Hopf monads. These theorems may be generalized, in a weaker form, to arbitrary commutative ground rings, see [BV3].

Part II

Hopf Algebras and Monads

Chapter 6

Hopf algebras in braided categories

We define and study Hopf algebras in braided categories. This leads us to a useful re-formulation of the notion of a modular category and paves the way for the study of Hopf monads in the next chapters.

6.1 Algebras, coalgebras, and bialgebras

We review categorical versions of the notions of an algebra, a module over an algebra, a coalgebra, and a bialgebra.

6.1.1 Algebras and modules

Let $\mathcal{C}$ be a monoidal category. An *algebra in* $\mathcal{C}$ is an object A of $\mathcal{C}$ endowed with morphisms $m\colon A \otimes A \to A$ (the product) and $u\colon \mathbb{1} \to A$ (the unit) satisfying the following associativity and unitality axioms:

$$m(m \otimes \mathrm{id}_A) = m(\mathrm{id}_A \otimes m) \quad \text{and} \quad m(\mathrm{id}_A \otimes u) = \mathrm{id}_A = m(u \otimes \mathrm{id}_A).$$

We depict the product and the unit as

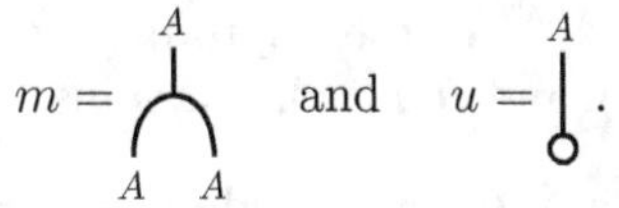

The axioms above have the following graphical interpretation:

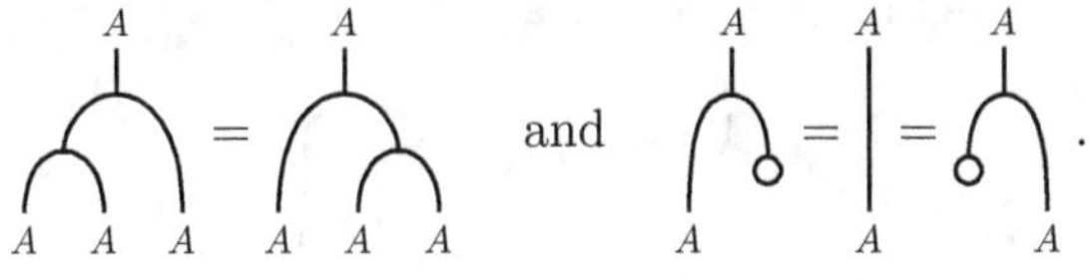

For an algebra $A = (A, m, u)$ in $\mathcal{C}$, a (left) *A-module* is a pair (M, r), where $M \in \mathrm{Ob}(\mathcal{C})$ and $r\colon A \otimes M \to M$ is a morphism in $\mathcal{C}$ called the *action*, such that

$$r(m \otimes \mathrm{id}_M) = r(\mathrm{id}_A \otimes r) \quad \text{and} \quad r(u \otimes \mathrm{id}_M) = \mathrm{id}_M.$$

Graphically, these conditions are depicted as

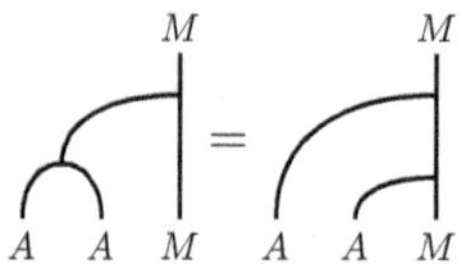

and
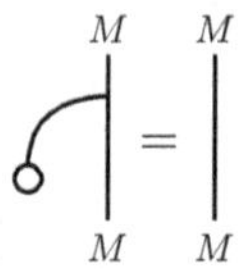

, where $r =$

.

For example, the pair (A, m) is a (left) A-module. One can similarly introduce right A-modules, but we will not use them. From now on, by an A-module, we mean a left A-module.

An *A-linear morphism* between two A-modules (M, r) and (N, s) is a morphism $f\colon M \to N$ in $\mathcal{C}$ such that $fr = s(\mathrm{id}_A \otimes f)$, that is, pictorially,

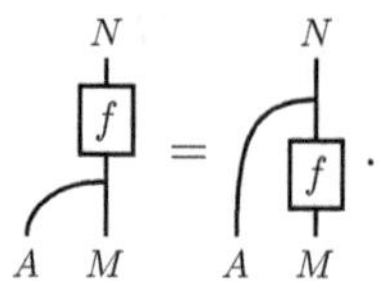

.

We let $\mathrm{mod}_A = \mathrm{mod}_A^{\mathcal{C}}$ be the category of A-modules and A-linear morphisms with composition inherited from $\mathcal{C}$. The *forgetful functor* $\mathcal{U}_A\colon \mathrm{mod}_A \to \mathcal{C}$ carries any A-module (M, r) to M and any A-linear morphism to itself.

An *algebra morphism* between algebras (A, m, u) and (A', m', u') in $\mathcal{C}$ is a morphism $\phi\colon A \to A'$ in $\mathcal{C}$ such that $\phi m = m'(\phi \otimes \phi)$ and $\phi u = u'$. Such a ϕ induces a functor $\phi^*\colon \mathrm{mod}_{A'} \to \mathrm{mod}_A$ by

$$\phi^*(M, r) = (M, r(\phi \otimes \mathrm{id}_M)) \quad \text{and} \quad \phi^*(f) = f$$

for any A'-module (M, r) and any A'-linear morphism f. Clearly, $\mathcal{U}_A\, \phi^* = \mathcal{U}_{A'}$.

6.1.2 Coalgebras

A *coalgebra* in a monoidal category $\mathcal{C}$ is an algebra in the monoidal category $\mathcal{C}^{\mathrm{op}} = (\mathcal{C}^{\mathrm{op}}, \otimes, \mathbb{1})$. In other words, a coalgebra in $\mathcal{C}$ is an object A of $\mathcal{C}$ endowed with morphisms $\Delta\colon A \to A \otimes A$ (the coproduct) and $\varepsilon\colon A \to \mathbb{1}$ (the counit) satisfying the coassociativity and counitality axioms:

$$(\Delta \otimes \mathrm{id}_A)\Delta = (\mathrm{id}_A \otimes \Delta)\Delta \quad \text{and} \quad (\mathrm{id}_A \otimes \varepsilon)\Delta = \mathrm{id}_A = (\varepsilon \otimes \mathrm{id}_A)\Delta.$$

We depict the coproduct and the counit as

$\Delta =$
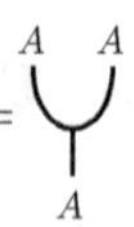

and $\varepsilon =$

.

The axioms above are depicted as

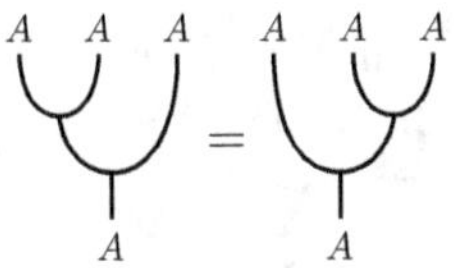

and

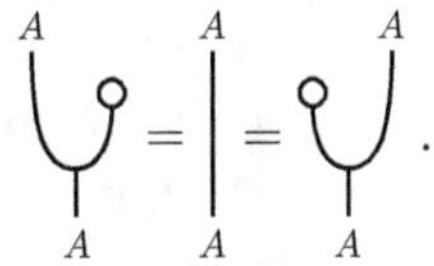

A *coalgebra morphism* between coalgebras (A, Δ, ε) and $(A', \Delta', \varepsilon')$ in $\mathcal{C}$ is a morphism $\psi\colon A \to A'$ in $\mathcal{C}$ such that $\Delta'\psi = (\psi \otimes \psi)\Delta$ and $\varepsilon'\psi = \varepsilon$.

One can define comodules over coalgebras in a way opposite to modules over algebras but we will not study comodules.

6.1.3 Bialgebras

To define bialgebras in a monoidal category, we need compatibility conditions between multiplication and comultiplication, and the formulation of one of the conditions requires a braiding. A *bialgebra* in a braided category $\mathcal{C}$ with braiding τ is a tuple $(A, m, u, \Delta, \varepsilon)$, where $A \in \mathrm{Ob}(\mathcal{C})$, (m, u) is an algebra structure in A, and (Δ, ε) is a coalgebra structure in A such that

$$\Delta m = (m \otimes m)(\mathrm{id}_A \otimes \tau_{A,A} \otimes \mathrm{id}_A)(\Delta \otimes \Delta), \quad \Delta u = u \otimes u, \tag{6.1}$$

$$\varepsilon m = \varepsilon \otimes \varepsilon, \quad \varepsilon u = \mathrm{id}_{\mathbb{1}}. \tag{6.2}$$

Pictorially,

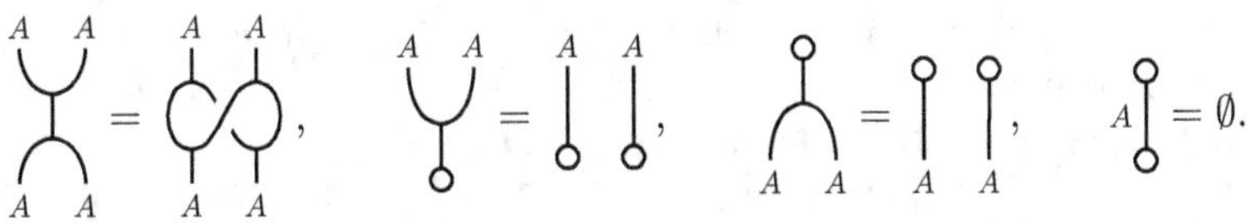

For a bialgebra $A = (A, m, u, \Delta, \varepsilon)$ in $\mathcal{C}$, the category mod_A of A-modules has a canonical structure of a monoidal category. Its unit object is the pair $(\mathbb{1}, \varepsilon)$. Its monoidal product is given on the objects by

$$(M, r) \otimes (N, s) = (M \otimes N, t)$$

where

$$t = (r \otimes s)(\mathrm{id}_A \otimes \tau_{A,M} \otimes \mathrm{id}_N)(\Delta \otimes \mathrm{id}_{M \otimes N}) =$$

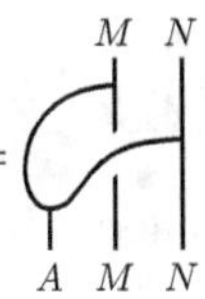

and on the morphisms by the monoidal product in $\mathcal{C}$. It is clear that the forgetful functor $\mathrm{mod}_A \to \mathcal{C}$ is strict monoidal.

A *bialgebra morphism* $\phi\colon A \to A'$ between two bialgebras in $\mathcal{C}$ is a morphism in $\mathcal{C}$ which is both an algebra morphism and a coalgebra morphism. Then the induced functor $\phi^*\colon \mathrm{mod}_{A'} \to \mathrm{mod}_A$ is strict monoidal.

6.1.4 Example

Algebras, modules over algebras, coalgebras, and bialgebras in the symmetric monoidal category $\mathrm{Mod}_{\Bbbk}$ of $\Bbbk$-modules are the usual $\Bbbk$-algebras, modules over $\Bbbk$-algebras, $\Bbbk$-coalgebras, and $\Bbbk$-bialgebras, respectively.

6.1.5 Example

Algebras, modules over algebras, coalgebras, and bialgebras in the symmetric monoidal category $\mathrm{SMod}_{\Bbbk}$ of super $\Bbbk$-modules (see Example 3.1.5) are the usual super $\Bbbk$-algebras, super modules over super $\Bbbk$-algebras, super $\Bbbk$-coalgebras, and super $\Bbbk$-bialgebras, respectively.

6.1.6 Exercise

Let $\mathcal{C}$ be a monoidal category, (A, m, u) be an algebra in $\mathcal{C}$, and (C, Δ, ε) be a coalgebra in $\mathcal{C}$. Prove that the set $\mathrm{Hom}_{\mathcal{C}}(C, A)$ is a monoid with (associative) product $*$ defined by $f * g = m(f \otimes g)\Delta$ for all $f, g \in \mathrm{Hom}_{\mathcal{C}}(C, A)$ and with unit $u\varepsilon$. This monoid is called the *convolution monoid* of C and A.

6.1.7 Exercise

Show that in the definition of a bialgebra $(A, m, u, \Delta, \varepsilon)$, the conditions (6.1) and (6.2) can be reformulated by saying that $\Delta\colon A \to A \otimes A$ and $\varepsilon\colon A \to \mathbb{1}$ are algebra morphisms or, equivalently, that $m\colon A \otimes A \to A$ and $u\colon \mathbb{1} \to A$ are coalgebra morphisms, where $\mathbb{1}$ has the trivial product, coproduct, unit, and counit all equal to $\mathrm{id}_{\mathbb{1}}$ and $A \otimes A$ has the algebra structure $(\widetilde{m}, \widetilde{u})$ and the coalgebra structure $(\widetilde{\Delta}, \widetilde{\varepsilon})$ defined by

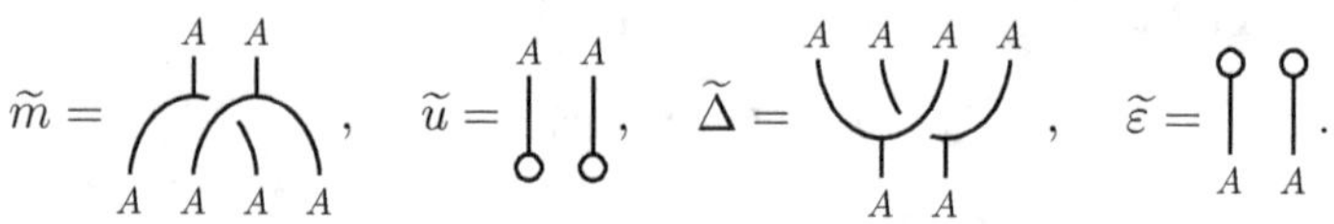

6.1.8 Exercise

Let A be an algebra in a monoidal $\Bbbk$-category $\mathcal{C}$. Prove that:

(a) The category mod_A has a unique structure of a $\Bbbk$-category such that the forgetful functor $\mathrm{mod}_A \to \mathcal{C}$ is $\Bbbk$-linear;

(b) If $\mathcal{C}$ is additive, then so is mod_A;

(c) If $\mathcal{C}$ is braided and A is a bialgebra in $\mathcal{C}$, then mod_A is a monoidal $\Bbbk$-category.

6.2 Hopf algebras

We review a categorical version of antipodes and Hopf algebras.

6.2.1 Antipodes

Let $A = (A, m, u, \Delta, \varepsilon)$ be a bialgebra in a braided category $\mathcal{C}$. An *antipode* of A is a morphism $S\colon A \to A$ in $\mathcal{C}$ such that

$$m(S \otimes \mathrm{id}_A)\Delta = u\varepsilon = m(\mathrm{id}_A \otimes S)\Delta.$$

Pictorially,

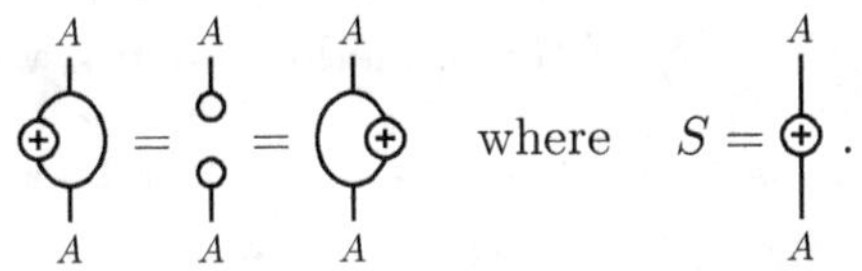

These conditions may be rephrased by saying that the antipode is a (two-sided) inverse of id_A in the convolution monoid $\mathrm{Hom}_{\mathcal{C}}(A, A)$, see Exercise 6.1.6. As a consequence, if an antipode exists, then it is unique.

If A has an antipode, then it is anti-multiplicative in the sense that

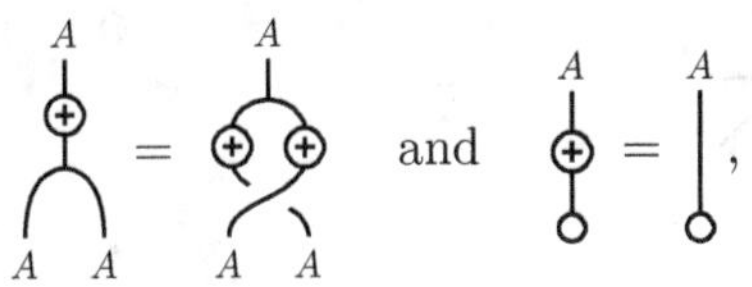

and anti-comultiplicative in the sense that

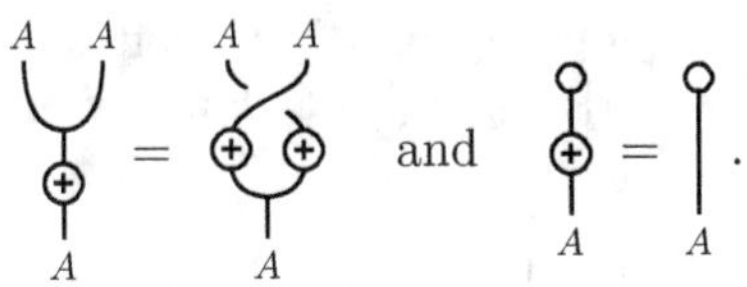

These equalities may be proved using the convolution monoids (see Exercise 6.1.6) and the uniqueness of an inverse in a monoid. In particular, the anti-multiplicativity of S follows from the fact that both Sm and $m(S \otimes S)\tau_{A,A}$ are inverse to m in the convolution monoid $\mathrm{Hom}_{\mathcal{C}}(A \otimes A, A)$. Here τ is the braiding of $\mathcal{C}$ and $A \otimes A$ carries the coalgebra structure from Exercise 6.1.7.

When the antipode S is invertible, we depict its inverse $S^{-1}\colon A \to A$ as

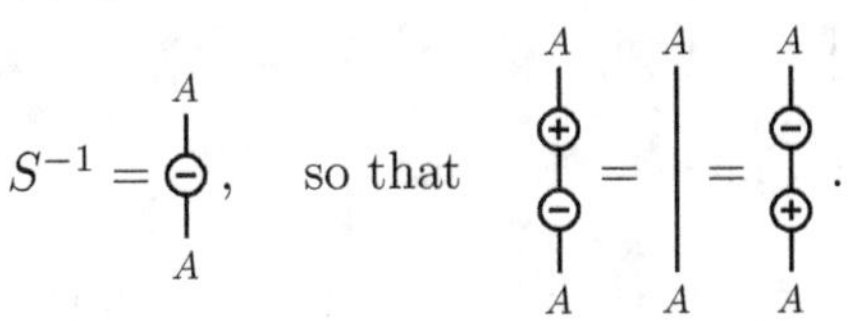

6.2.2 Hopf algebras

A *Hopf algebra* in a braided category $\mathcal{C}$ is a bialgebra in $\mathcal{C}$ which has an invertible antipode. A *Hopf algebra morphism* between two Hopf algebras in $\mathcal{C}$ is a bialgebra morphism between them. Such a morphism necessarily commutes with the antipodes. Hopf algebras naturally come up in the study of rigid categories as is clear from the following lemma.

Lemma 6.1. *Let A be a bialgebra in a braided rigid category $\mathcal{C}$. The monoidal category* mod_A *is rigid if and only if A is a Hopf algebra. Moreover, if A is a Hopf algebra, then any left/right duality in $\mathcal{C}$ has a unique lift in* mod_A *along the forgetful functor* $\mathrm{mod}_A \to \mathcal{C}$ *(in the sense of Section* 1.6.4*).*

Proof. Suppose first that mod_A is rigid. Pick a left dual $({}^\vee(A,m), \mathrm{ev}_{(A,m)})$ and a right dual $((A,m)^\vee, \widetilde{\mathrm{ev}}_{(A,m)})$ of the A-module (A,m), where $m\colon A\otimes A \to A$ is the product of A. Let $\mathrm{coev}_{(A,m)}$ and $\widetilde{\mathrm{coev}}_{(A,m)}$ be the inverses of the pairings $\mathrm{ev}_{(A,m)}$ and $\widetilde{\mathrm{ev}}_{(A,m)}$, respectively. The A-linearity of these four morphisms implies that the following formulas, where β is the action of ${}^\vee(A,m)$ and γ is the action of $(A,m)^\vee$, define an antipode S of A and its inverse:

$S =$ 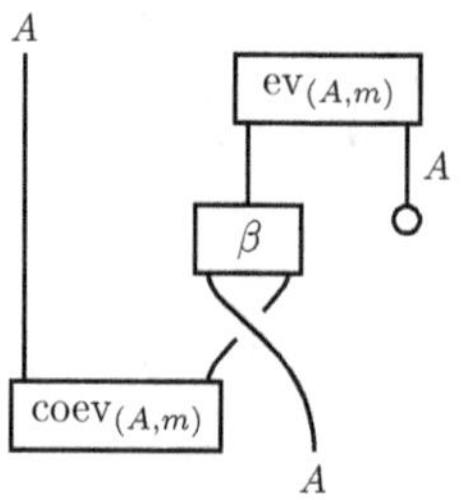 and $S^{-1} =$ 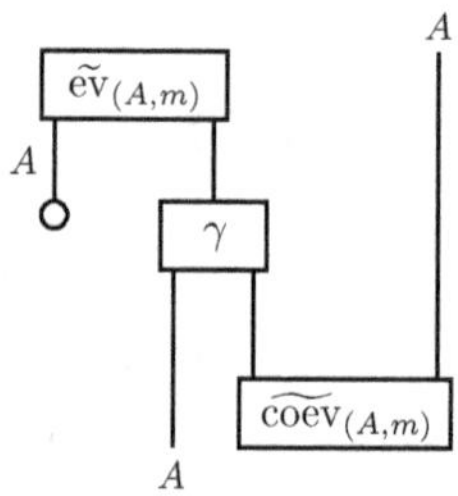 .

Thus, A is a Hopf algebra.

Conversely, let A be a Hopf algebra. Fix a left duality $\{({}^\vee X, \mathrm{ev}_X)\}_{X\in \mathrm{Ob}(\mathcal{C})}$ and a right duality $\{(X^\vee, \widetilde{\mathrm{ev}}_X)\}_{X\in\mathrm{Ob}(\mathcal{C})}$ in $\mathcal{C}$. For any A-module (M,r), set

${}^\circ r =$ 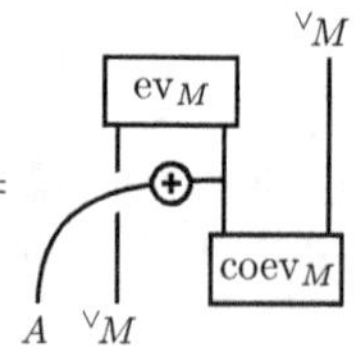 and $r^\circ =$ 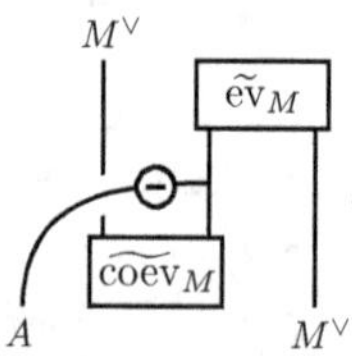 .

Then it follows from the axioms of an antipode that $(({}^\vee M, {}^\circ r), \mathrm{ev}_M)$ is a left dual of (M,r) and $((M^\vee, r^\circ), \widetilde{\mathrm{ev}}_M)$ is a right dual of (M,r). Thus, the family

$$\{({}^\vee(M,r) = ({}^\vee M, {}^\circ r),\ \mathrm{ev}_{(M,r)} = \mathrm{ev}_M)\}_{(M,r)\in\mathrm{Ob}(\mathrm{mod}_A)} \tag{6.3}$$

is a left duality in mod_A, and the family

$$\{((M,r)^\vee = (M^\vee, r^\circ),\ \widetilde{\mathrm{ev}}_{(M,r)} = \widetilde{\mathrm{ev}}_M)\}_{(M,r)\in\mathrm{Ob}(\mathrm{mod}_A)} \tag{6.4}$$

is a right duality in mod_A. In particular, the monoidal category mod_A is rigid. Consider now the forgetful functor $\mathcal{U}_A \colon \operatorname{mod}_A \to \mathcal{C}$. Clearly, the left duality (6.3) in mod_A is a lift along $\mathcal{U}_A$ of the chosen left duality in $\mathcal{C}$. Let us prove that such a lift is unique. To this end, consider another such lift

$$\{((M,r)^\dagger, e_{(M,r)})\}_{(M,r)\in\operatorname{Ob}(\operatorname{mod}_A)}.$$

By the definition of a lift (see Section 1.6.4), for any A-module (M,r),

$$\mathcal{U}_A((M,r)^\dagger) = {}^\vee M \quad \text{and} \quad (e_{(M,r)})^{\mathcal{U}_A} = \operatorname{ev}_M.$$

Then there is an action $a_{(M,r)}$ of A on ${}^\vee M$ such that $(M,r)^\dagger = ({}^\vee M, a_{(M,r)})$. Denoting by $\operatorname{coev}_{(M,r)}$ the inverse of the pairing $\operatorname{ev}_{(M,r)}$, the A-linearity of the morphism

$$(e_{(M,r)} \otimes \operatorname{id}_{{}^\vee(M,r)})(\operatorname{id}_{(M,r)^\dagger} \otimes \operatorname{coev}_{(M,r)}) = \operatorname{id}_{{}^\vee M} \colon ({}^\vee M, a_{(M,r)}) \to ({}^\vee M, {}^\circ r)$$

gives that $a_{(M,r)} = {}^\circ r$, and so $(M,r)^\dagger = {}^\vee(M,r)$. Also, the strict monoidality of $\mathcal{U}_A$ and the fact that $\mathcal{U}_A$ acts as the identity on morphisms imply that $(e_{(M,r)})^{\mathcal{U}_A} = e_{(M,r)}$, and so $e_{(M,r)} = \operatorname{ev}_M$. Consequently, the left duality (6.3) in mod_A is the unique lift along $\mathcal{U}_A$ of the chosen left duality in $\mathcal{C}$. Similarly, the right duality (6.4) in mod_A is the unique lift along $\mathcal{U}_A$ of the chosen right duality in $\mathcal{C}$. □

6.2.3 Involutory Hopf algebras

Let $\mathcal{C}$ be a braided pivotal category. A Hopf algebra A in $\mathcal{C}$ is *involutory* if its antipode S satisfies $S^2 = \theta_A^r$, where θ^r is the right twist of $\mathcal{C}$ (see Section 3.3.1). Graphically,

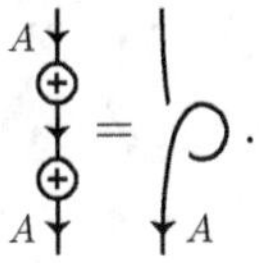

Given an involutory Hopf algebra A in $\mathcal{C}$, the category mod_A carries a unique structure of a pivotal category so that the forgetful functor $\operatorname{mod}_A \to \mathcal{C}$ is strictly pivotal. The corresponding pivotal duality is computed by

$$\{((M,r)^* = (M^*, r^\dagger),\, \operatorname{ev}_{(M,r)} = \operatorname{ev}_M,\, \widetilde{\operatorname{ev}}_{(M,r)} = \widetilde{\operatorname{ev}}_M)\}_{(M,r)\in\operatorname{Ob}(\operatorname{mod}_A)}$$

where

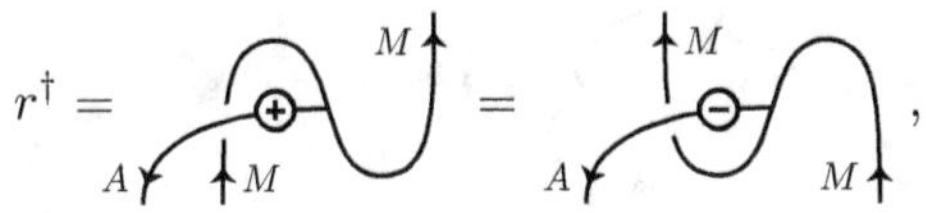

The latter equality holds because the condition $S^2 = \theta_A^r$ guarantees that the morphisms ${}^\circ r$ and r° from the proof of Lemma 6.1 are equal. (This may also be deduced from Theorem 8.10, see Example 8.4.3). Since the forgetful functor $\mathrm{mod}_A \to \mathcal{C}$ is pivotal, strict monoidal, and acts as the identity on morphisms, Lemma 2.8 implies that the left/right trace of an endomorphism f in mod_A coincides with the left/right trace of f viewed as an endomorphism in $\mathcal{C}$. Consequently, if $\mathcal{C}$ is spherical, then so is mod_A.

6.2.4 Example

Hopf algebras in the symmetric monoidal category $\mathrm{Mod}_\Bbbk$ of $\Bbbk$-modules are the usual Hopf $\Bbbk$-algebras.

6.2.5 Example

Hopf algebras in the symmetric monoidal category $\mathrm{SMod}_\Bbbk$ of super $\Bbbk$-modules (see Example 3.1.5) are the usual super Hopf $\Bbbk$-algebras.

6.2.6 Example

Recall from Example 1.7.2 the pivotal category $\mathrm{proj}_\Bbbk$ of projective $\Bbbk$-modules of finite type. Restricting the symmetry in $\mathrm{Mod}_\Bbbk$ to $\mathrm{proj}_\Bbbk$ we turn the latter into a symmetric category. A Hopf algebra in $\mathrm{proj}_\Bbbk$ is involutory in the sense of Section 6.2.3 if and only if its antipode is an involution.

6.2.7 Exercise*

Prove that a monoid G is a group if and only if the map

$$G \times G \to G \times G, \quad (g, h) \mapsto (g, gh)$$

is bijective. Similarly, prove that a bialgebra A in a braided category is a Hopf algebra if and only if both morphisms

$H_l =$  $: A \otimes A \to A \otimes A$

and

$H_r =$  $: A \otimes A \to A \otimes A$

are invertible. These morphisms are called respectively the *left and right fusion operators* of A.

6.3 Integrals, cointegrals, and bialgebra pairings

We discuss integrals, cointegrals, and bialgebra pairings for bialgebras in braided categories.

6.3.1 Integrals and cointegrals

Consider a bialgebra $A = (A, m, u, \Delta, \varepsilon)$ in a braided category $\mathcal{C}$. A *left integral* of A is a morphism $\Lambda\colon \mathbb{1} \to A$ such that

$$m(\mathrm{id}_A \otimes \Lambda) = \Lambda\,\varepsilon, \quad \text{i.e.,}$$

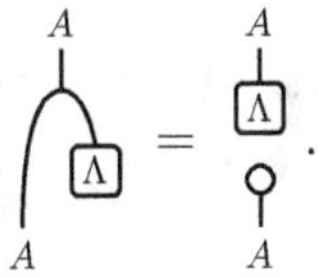

Similarly, a *right integral* of A is a morphism $\Lambda\colon \mathbb{1} \to A$ such that

$$m(\Lambda \otimes \mathrm{id}_A) = \Lambda\,\varepsilon, \quad \text{i.e.,}$$

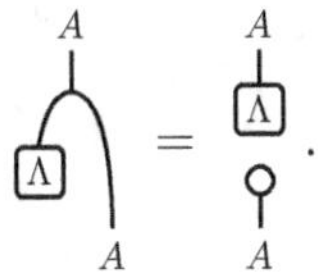

A *left cointegral* of A is a morphism $\lambda\colon A \to \mathbb{1}$ such that

$$(\mathrm{id}_A \otimes \lambda)\Delta = u\,\lambda, \quad \text{i.e.,}$$

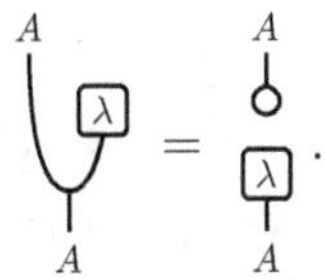

Similarly, a *right cointegral* of A is a morphism $\lambda\colon A \to \mathbb{1}$ such that

$$(\lambda \otimes \mathrm{id}_A)\Delta = u\,\lambda, \quad \text{i.e.,}$$

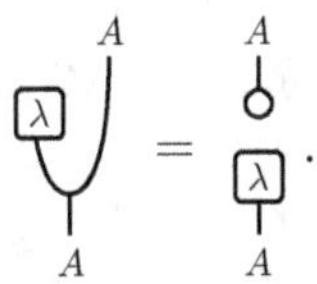

We note two properties of (co)integrals of the bialgebra A. First, if $\mathcal{C}$ is a $\Bbbk$-category, then all scalar multiples of a left/right integral of A are left/right integrals of A and similarly for cointegrals. This follows from the $\Bbbk$-bilinearity of the composition and of the monoidal product in $\mathcal{C}$. Second, suppose that A is a Hopf algebra with antipode S. In this case, a morphism $\Lambda\colon \mathbb{1} \to A$ is a left/right integral of A if and only if $S\Lambda$ is a right/left integral of A. Similarly, a morphism $\lambda\colon A \to \mathbb{1}$ is a left/right cointegral of A if and only if λS is a right/left cointegral of A. These properties follow from the anti-multiplicativity and anti-comultiplicativity of S.

6.3.2 Bialgebra pairings

If A is a $\Bbbk$-bialgebra (i.e., a bialgebra in the category of $\Bbbk$-modules), then $A^\star = \mathrm{Hom}_\Bbbk(A, \Bbbk)$ is a $\Bbbk$-algebra with unit the counit of A and product $fg = (g \otimes f)\Delta$ for $f, g \in A^\star$, where Δ is the coproduct of A. To study a $\Bbbk$-bialgebra A, one often considers bilinear pairings $A \otimes_\Bbbk A \to \Bbbk$ whose left and right adjoint maps $A \to A^\star$ are algebra homomorphisms. Such pairings are generalized to bialgebras in braided categories as follows.

Let $A = (A, m, u, \Delta, \varepsilon)$ be a bialgebra in a braided category $\mathcal{C}$. A *bialgebra pairing* for A is a morphism $\omega\colon A \otimes A \to \mathbb{1}$ in $\mathcal{C}$ such that

$$\omega(m \otimes \mathrm{id}_A) = \omega(\mathrm{id}_A \otimes \omega \otimes \mathrm{id}_A)(\mathrm{id}_{A \otimes A} \otimes \Delta), \qquad \omega(u \otimes \mathrm{id}_A) = \varepsilon,$$
$$\omega(\mathrm{id}_A \otimes m) = \omega(\mathrm{id}_A \otimes \omega \otimes \mathrm{id}_A)(\Delta \otimes \mathrm{id}_{A \otimes A}), \qquad \omega(\mathrm{id}_A \otimes u) = \varepsilon.$$

Graphically:

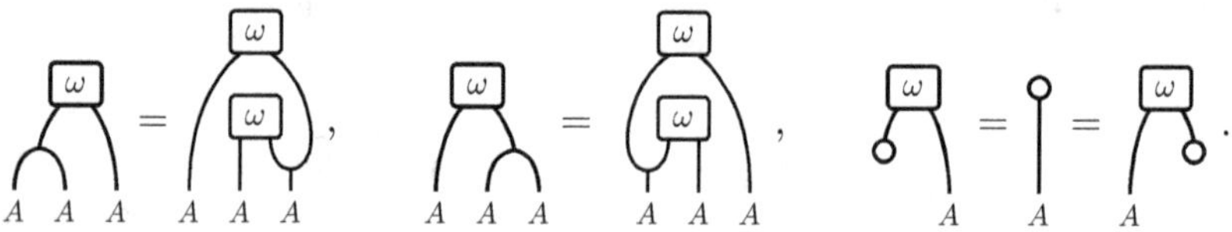

Bialgebra pairings for Hopf algebras are called *Hopf pairings*. The following lemma provides useful sufficient conditions for a Hopf pairing to be non-degenerate (in the sense of Section 1.5.1).

Lemma 6.2. *Let A be a Hopf algebra in a braided category $\mathcal{C}$. Let $\omega\colon A \otimes A \to \mathbb{1}$ be a Hopf pairing for A and let $\Lambda, \Lambda'\colon \mathbb{1} \to A$ be morphisms in $\mathcal{C}$ such that:*

(a) *$\lambda = \omega(\Lambda \otimes \mathrm{id}_A)$ and $\lambda' = \omega(\mathrm{id}_A \otimes \Lambda')$ are left cointegrals of A;*

(b) *$\omega(\Lambda \otimes \Lambda') = \mathrm{id}_{\mathbb{1}}$.*

Then the pairing ω is non-degenerate and Λ, Λ' are right integrals of A.

Proof. Let us depict ω, Λ, λ, Λ', λ' as follows:

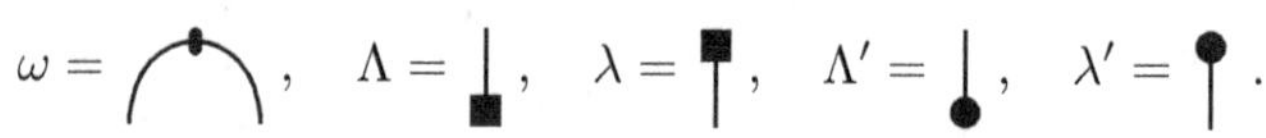

By definition,

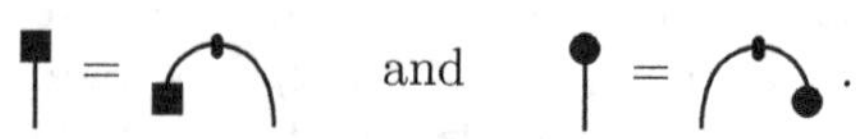

Let $S\colon A \to A$ be the antipode of A. Set

$$\Omega = (S \otimes \mathrm{id}_A \otimes \omega)(\mathrm{id}_A \otimes \Delta\Lambda \otimes \mathrm{id}_A)\Delta\Lambda' =$$ 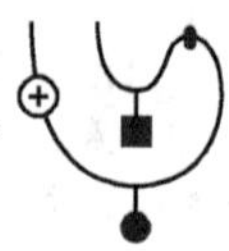 $$: \mathbb{1} \to A \otimes A\,.$$

We claim that Ω is an inverse of the pairing ω. Indeed,

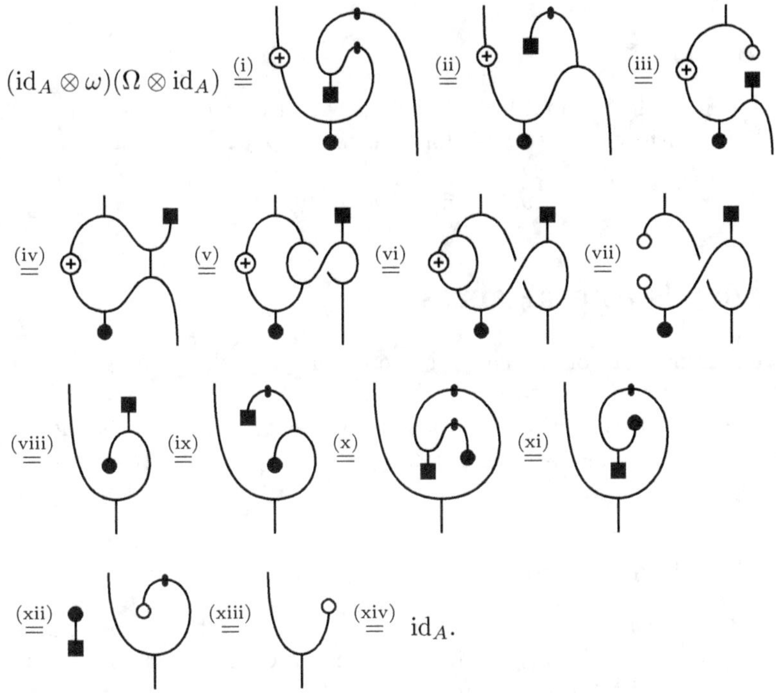

Here (i) follows from the definition of Ω, (ii) from the fact that ω is a Hopf pairing, (iii) from the definition of λ and the unitality of the product of A, (iv) from the fact that λ is a left cointegral, (v) from the multiplicativity of the coproduct of A, (vi) from the (co)associativity of the (co)product of A, (vii) from the definition of an antipode, (viii) from the (co)unitality of the (co)product of A and the naturality of the braiding, (ix) from the definition of λ, (x) from the fact that ω is a Hopf pairing, (xi) from the definition of λ', (xii) from the fact that λ' is a left cointegral, (xiii) from the facts that $\lambda'\Lambda = \omega(\Lambda \otimes \Lambda') = \mathrm{id}_{\mathbb{1}}$ and that ω is a Hopf pairing, and (xiv) from the counitality of the coproduct of A. One similarly shows that $(\omega \otimes \mathrm{id}_A)(\mathrm{id}_A \otimes \Omega) = \mathrm{id}_A$. Thus ω is non-degenerate with inverse Ω. Next, we have:

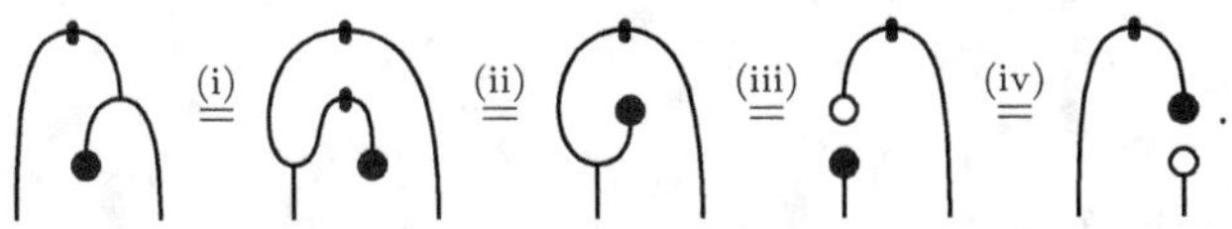

Here (i) follows from the fact that ω is a Hopf pairing, (ii) from the definition of λ', (iii) from the fact that λ' is a left cointegral, and (iv) from the definition of λ' and the fact that ω is a Hopf pairing. Since Ω is the inverse of ω, we obtain

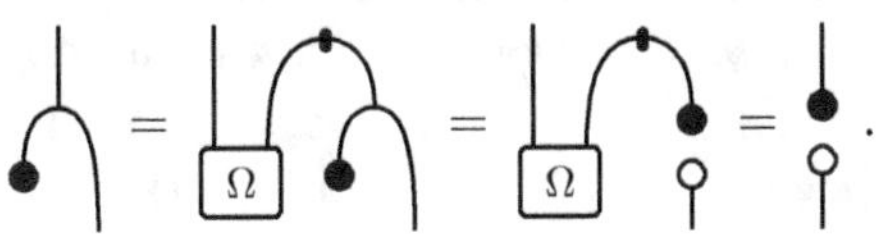

Thus Λ' is a right integral of A. That Λ is a right integral is proved similarly. □

6.3.3 Exercise*

Let A be a Hopf algebra in a braided category $\mathcal{C}$. Prove that the antipode S of A is self-adjoint with respect to any Hopf pairing ω for A in the sense that

$$\omega(S \otimes \mathrm{id}_A) = \omega(\mathrm{id}_A \otimes S).$$

6.4 Coends as coalgebras

In the remaining part of the chapter, we discuss coends of categories and their Hopf algebra structure due to Majid [Maj].

Throughout this section, $\mathcal{C}$ is a left rigid category with distinguished left duality $\{({}^\vee X, \mathrm{ev}_X)\}_{X \in \mathrm{Ob}(\mathcal{C})}$. As usual, we denote by coev_X the inverse of the pairing ev_X.

6.4.1 Coends of categories

By a *dinatural transformation in* $\mathcal{C}$ *to an object* $D \in \mathrm{Ob}(\mathcal{C})$, we mean a family $d = \{d_X : {}^\vee X \otimes X \to D\}_{X \in \mathrm{Ob}(\mathcal{C})}$ of morphisms in $\mathcal{C}$ such that for any morphism $f : X \to Y$ in $\mathcal{C}$, the following diagram commutes:

$$\begin{array}{ccc} {}^\vee Y \otimes X & \xrightarrow{\mathrm{id}_{{}^\vee Y} \otimes f} & {}^\vee Y \otimes Y \\ {\scriptstyle {}^\vee f \otimes \mathrm{id}_X} \downarrow & & \downarrow {\scriptstyle d_Y} \\ {}^\vee X \otimes X & \xrightarrow{d_X} & D\,. \end{array}$$

A *coend* of $\mathcal{C}$ is a pair (C, ρ) where $C \in \mathrm{Ob}(\mathcal{C})$ and $\rho = \{\rho_X : {}^\vee X \otimes X \to C\}_{X \in \mathrm{Ob}(\mathcal{C})}$ is a dinatural transformation which is *universal* in the sense that for any dinatural transformation d in $\mathcal{C}$ to any $D \in \mathrm{Ob}(\mathcal{C})$, there is a unique morphism $\varphi : C \to D$ in $\mathcal{C}$ such that for all $X \in \mathrm{Ob}(\mathcal{C})$,

$$d_X = \varphi \rho_X. \tag{6.5}$$

We will often write C for (C, ρ), specifying ρ when necessary.

The category $\mathcal{C}$ may have no coends. If $\mathcal{C}$ has a coend, then it is unique up to unique isomorphism in $\mathcal{C}$ commuting with the universal dinatural transformations. This follows directly from the universality condition in the definition of a coend. Note also the following property of coends. Consider an object $D \in \mathrm{Ob}(\mathcal{C})$ and a family

$$d = \{d_{X,Y} : {}^\vee X \otimes X \otimes {}^\vee Y \otimes Y \to D\}_{X,Y \in \mathrm{Ob}(\mathcal{C})}$$

of morphisms in $\mathcal{C}$ which is dinatural in X and Y in the sense that for any morphisms $f\colon X \to A$ and $g\colon Y \to B$ in $\mathcal{C}$, the following diagram commutes:

$$\begin{array}{ccc} {}^\vee\!A \otimes X \otimes {}^\vee\!B \otimes Y & \xrightarrow{\mathrm{id}_{{}^\vee\!A} \otimes f \otimes \mathrm{id}_{{}^\vee\!B} \otimes g} & {}^\vee\!A \otimes A \otimes {}^\vee\!B \otimes B \\ {\scriptstyle {}^\vee\! f \otimes \mathrm{id}_X \otimes {}^\vee\! g \otimes \mathrm{id}_Y} \big\downarrow & & \big\downarrow {\scriptstyle d_{A,B}} \\ {}^\vee\!X \otimes X \otimes {}^\vee\!Y \otimes Y & \xrightarrow{\quad d_{X,Y} \quad} & D\,. \end{array}$$

The Fubini theorem for coends (Theorem B.3) implies that for any coend (C, ρ) of $\mathcal{C}$, there is a unique morphism $\varphi\colon C \otimes C \to D$ in $\mathcal{C}$ such that for all $X, Y \in \mathrm{Ob}(\mathcal{C})$,

$$d_{X,Y} = \varphi(\rho_X \otimes \rho_Y). \tag{6.6}$$

For more on coends, see Appendix B.

6.4.2 The coalgebra structure

Suppose that $\mathcal{C}$ has a coend (C, ρ). We represent ρ pictorially as follows:

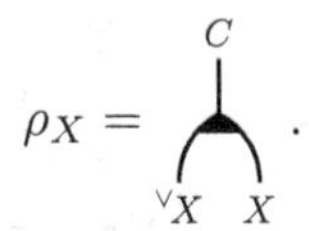

The factorization property (6.5) implies that there are unique morphisms $\Delta\colon C \to C \otimes C$ and $\varepsilon\colon C \to \mathbb{1}$ in $\mathcal{C}$ such that for any $X \in \mathrm{Ob}(\mathcal{C})$,

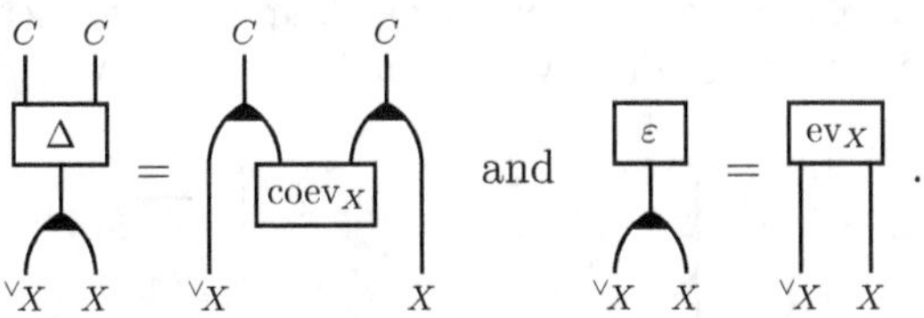

The morphisms Δ and ε are well defined because the right-hand sides of the equalities above are dinatural in X (by the identities (2.2) and the dinaturality of ρ) and so factorize through ρ.

It turns out that C is a coalgebra in $\mathcal{C}$ with coproduct Δ and counit ε. The axioms of a coalgebra directly follow from the uniqueness of the factorization in the definition of a coend. For example, the counitality condition $(\mathrm{id}_C \otimes \varepsilon)\Delta = \mathrm{id}_C$ holds because for any $X \in \mathrm{Ob}(\mathcal{C})$,

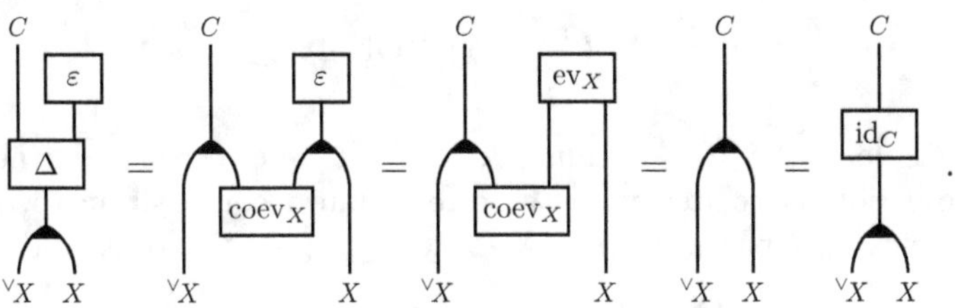

The other axioms of a coalgebra are verified similarly.

6.4.3 The universal coaction

Let (C,ρ) be a coend of $\mathcal{C}$. The *universal coaction* of C is the natural transformation

$$\delta = \{\delta_X \colon X \to X \otimes C\}_{X \in \mathrm{Ob}(\mathcal{C})}$$

defined by

$$\delta_X = (\mathrm{id}_X \otimes \rho_X)(\mathrm{coev}_X \otimes \mathrm{id}_X) =$$

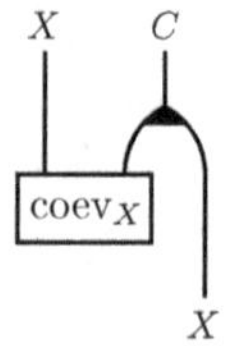

and depicted as $\delta_X =$

.

The universal coaction inherits from the universal dinatural transformation ρ the following factorization property: for any $D \in \mathrm{Ob}(\mathcal{C})$ and any natural transformation $\alpha = \{\alpha_X \colon X \to X \otimes D\}_{X \in \mathrm{Ob}(\mathcal{C})}$, there is a unique morphism $\varphi \colon C \to D$ such that $(\mathrm{id}_X \otimes \varphi)\delta_X = \alpha_X$ for all $X \in \mathrm{Ob}(\mathcal{C})$, i.e.,

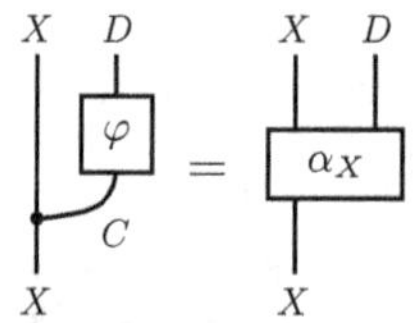

.

This property of δ implies that the morphisms Δ and ε introduced in Section 6.4.2 are uniquely determined by the following equalities: for any $X \in \mathrm{Ob}(\mathcal{C})$,

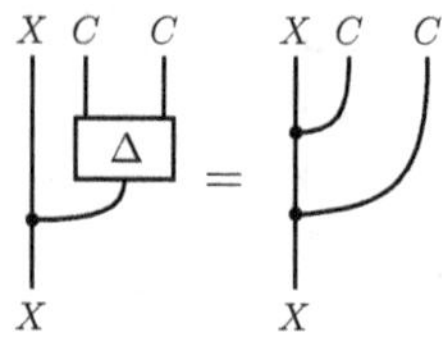

and

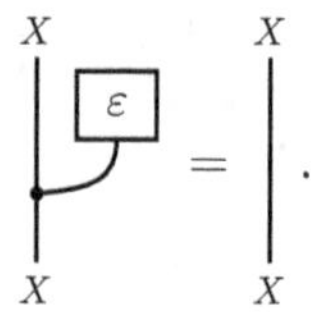

.

6.4.4 A special case

Suppose that $\mathcal{C}$ is an additive fusion $\Bbbk$-category (with distinguished left duality). Then $\mathcal{C}$ has a coend (C,ρ) constructed as follows. Pick a (finite) representative set I of simple objects of $\mathcal{C}$ and set

$$C = \bigoplus_{i \in I} {}^\vee i \otimes i \in \mathrm{Ob}(\mathcal{C}). \tag{6.7}$$

For each $i \in I$, let $\rho_i \colon {}^\vee i \otimes i \to C$ and $\nu_i \colon C \to {}^\vee i \otimes i$ be respectively the inclusion and the projection associated with the direct sum decomposition (6.7). For any $X \in \mathrm{Ob}(\mathcal{C})$, pick an I-partition $(p_\alpha \colon X \to i_\alpha,\ q_\alpha \colon i_\alpha \to X)_\alpha$ of X and set

$$\rho_X = \sum_\alpha \rho_{i_\alpha}\big({}^\vee(q_\alpha) \otimes p_\alpha\big) \colon {}^\vee X \otimes X \to C.$$

Note that for $X = i \in I$, we have $\rho_X = \rho_i$ because the pair $(\mathrm{id}_i, \mathrm{id}_i)$ is an I-partition of i. It is easy to check that the family $\rho = \{\rho_X : {}^{\vee}X \otimes X \to C\}_{X \in \mathrm{Ob}(\mathcal{C})}$ is a well-defined universal dinatural transformation.

For any dinatural transformation $d = \{d_X : {}^{\vee}X \otimes X \to D\}_{X \in \mathrm{Ob}(\mathcal{C})}$ with $D \in \mathrm{Ob}(\mathcal{C})$, the unique morphism $\varphi : C \to D$ satisfying $d = \varphi\rho$ is computed by

$$\varphi = \sum_{i \in I} d_i \nu_i : C \to D. \tag{6.8}$$

Similarly, given a family of morphisms in $\mathcal{C}$

$$d = \{d_{X,Y} : {}^{\vee}X \otimes X \otimes {}^{\vee}Y \otimes Y \to D\}_{X,Y \in \mathrm{Ob}(\mathcal{C})}$$

which is dinatural in both X and Y (see Section 6.4.1), the unique morphism $\varphi : C \otimes C \to D$ such that $d = \varphi(\rho \otimes \rho)$ is computed by

$$\varphi = \sum_{i,j \in I} d_{i,j}(\nu_i \otimes \nu_j) : C \otimes C \to D. \tag{6.9}$$

Formulas (6.8) and (6.9) imply that the coproduct $\Delta : C \to C \otimes C$ and counit $\varepsilon : C \to \mathbb{1}$ of the coend C of $\mathcal{C}$ are computed by

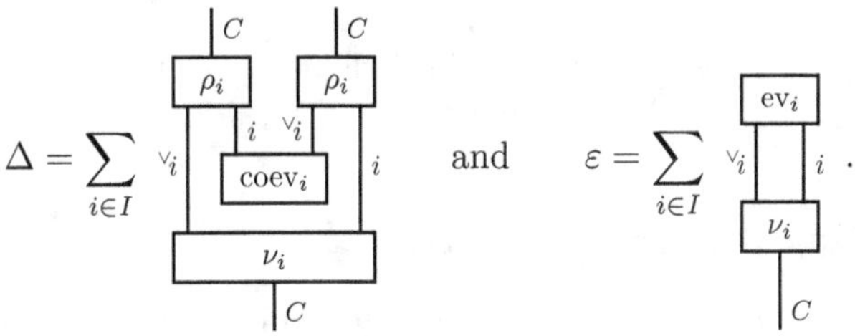

6.4.5 Exercise

Let (C, ρ) and (C', ρ') be two coends of $\mathcal{C}$. Prove that the (unique) isomorphism $C \to C'$ in $\mathcal{C}$ commuting with ρ and ρ' is a coalgebra isomorphism.

6.4.6 Exercise*

Prove that if $\mathcal{C}$ has a coend for one choice of a left duality in $\mathcal{C}$, then $\mathcal{C}$ has a coend for all choices of a left duality.

6.4.7 Exercise*

Let A be a projective Hopf $\Bbbk$-algebra of finite type, that is, a Hopf algebra in the symmetric pivotal category $\mathrm{proj}_{\Bbbk}$ of projective $\Bbbk$-modules of finite type (see Example 1.7.2). Consider the category mod_A of A-modules whose underlying $\Bbbk$-modules are projective of finite type. By Lemma 6.1, mod_A is a rigid category with distinguished duality. Let C be the A-module whose underlying $\Bbbk$-module is $A^* = \mathrm{Hom}_\Bbbk(A, \Bbbk)$ and whose action is the coadjoint action $\cdot$ of A defined by

$$(h \cdot f)(a) = f(S(h_{(1)})ah_{(2)})$$

for all $f \in A^*$ and $a, h \in A$. Here, S is the antipode of A and we write the coproduct of A as $h \mapsto h_{(1)} \otimes h_{(2)}$, leaving the summation implicit. Consider the family $\rho = \{\rho_X : {}^\vee X \otimes_\Bbbk X \to C\}_{X \in \mathrm{Ob}(\mathrm{mod}_A)}$ of homomorphisms defined by

$$(\rho_X(g \otimes_\Bbbk x))(a) = g(ax)$$

for all $g \in {}^\vee X = \mathrm{Hom}_\Bbbk(X, \Bbbk)$, $x \in X$, and $a \in A$. Prove that (C, ρ) is a coend of mod_A and that its coproduct Δ and counit ε are computed by

$$\Delta(f)(a \otimes_\Bbbk b) = f(ab) \quad \text{and} \quad \varepsilon(f) = f(1_A)$$

for all $f \in C$ and $a, b \in A$.

6.5 Coends as Hopf algebras

In this section, the category $\mathcal{C}$ is a braided rigid category with distinguished left duality $\{({}^\vee X, \mathrm{ev}_X)\}_{X \in \mathrm{Ob}(\mathcal{C})}$. We suppose that $\mathcal{C}$ has a coend (C, ρ).

6.5.1 The Hopf algebra structure

We extend the coalgebra structure in C defined in Section 6.4.2 to a Hopf algebra structure. The factorization property (6.6) implies the existence of a unique morphism $m : C \otimes C \to C$ such that for all $X, Y \in \mathrm{Ob}(\mathcal{C})$,

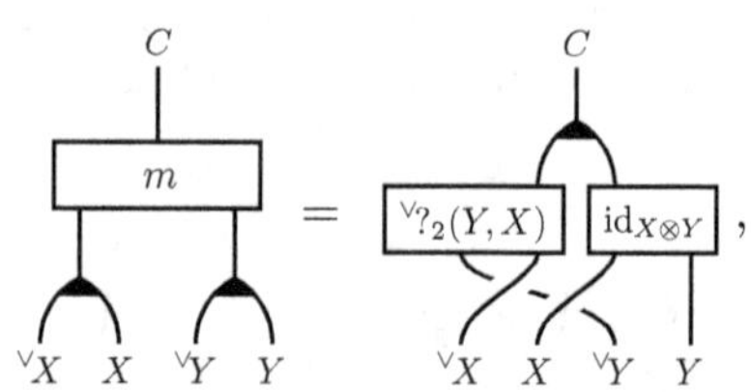

where ${}^\vee ?_2$ is the monoidal constraint of the left dual functor associated with the distinguished left duality in $\mathcal{C}$ (see Section 1.6.3). Indeed, the naturality of ${}^\vee ?_2$, the naturality of the braiding, and the dinaturality of ρ imply that the right-hand side is dinatural in X and Y. Next, set

$$u = \rho_\mathbb{1} \, \mathrm{coev}_\mathbb{1} : \mathbb{1} \to C,$$

where, as usual, coev_X is the inverse of the pairing ev_X. Finally, there is a unique morphism $S\colon C \to C$ in $\mathcal{C}$ such that for any $X \in \mathrm{Ob}(\mathcal{C})$,

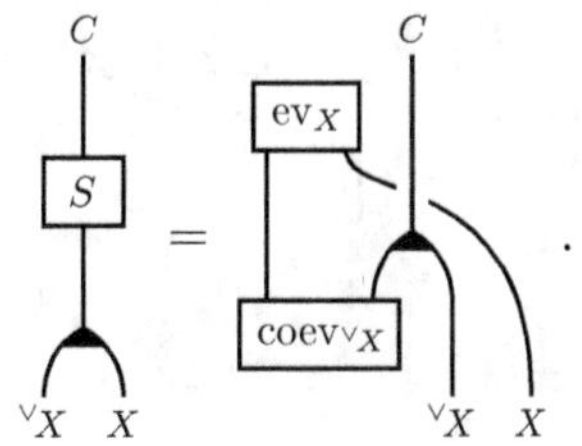

The morphism S is well defined because the right-hand side of the equality above is dinatural in X (by the identities (2.2) and the naturality of the braiding) and so factorizes through ρ.

The coalgebra C endowed with product m, unit u, and antipode S is a Hopf algebra in $\mathcal{C}$. The axioms of a Hopf algebra are deduced from the uniqueness of the factorization in the definition of a coend. The inverse of S is computed from an arbitrary right duality $\{(X^\vee, \widetilde{\mathrm{ev}}_X)\}_{X\in\mathrm{Ob}(\mathcal{C})}$ in $\mathcal{C}$ via the identity

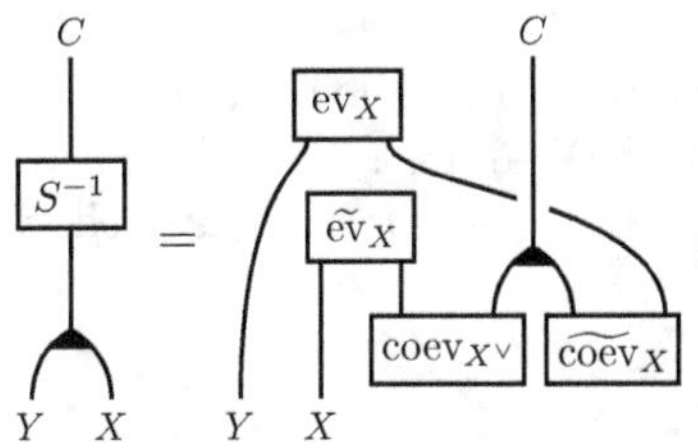

where X, Y run over $\mathrm{Ob}(\mathcal{C})$ and $\widetilde{\mathrm{coev}}_X$ is the inverse of the pairing $\widetilde{\mathrm{ev}}_X$.

Note that $u = \delta_{\mathbb{1}}$, where δ is the universal coaction of C defined in Section 6.4.3. Besides the factorization property stated there, the coaction δ has the following factorization property which is a direct consequence of (6.6): for any $D \in \mathrm{Ob}(\mathcal{C})$ and any natural transformation $\alpha = \{\alpha_{X,Y}\colon X\otimes Y \to X\otimes Y\otimes D\}_{X,Y\in\mathrm{Ob}(\mathcal{C})}$, there is a unique morphism $\varphi\colon C\otimes C \to D$ such that for all $X, Y \in \mathrm{Ob}(\mathcal{C})$,

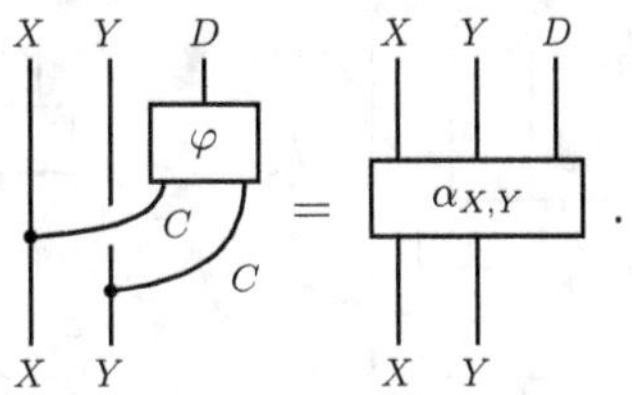

These factorization properties of δ imply that the morphisms m and $S^{\pm 1}$ above

are uniquely determined by the following equalities: for any $X, Y \in \mathrm{Ob}(\mathcal{C})$,

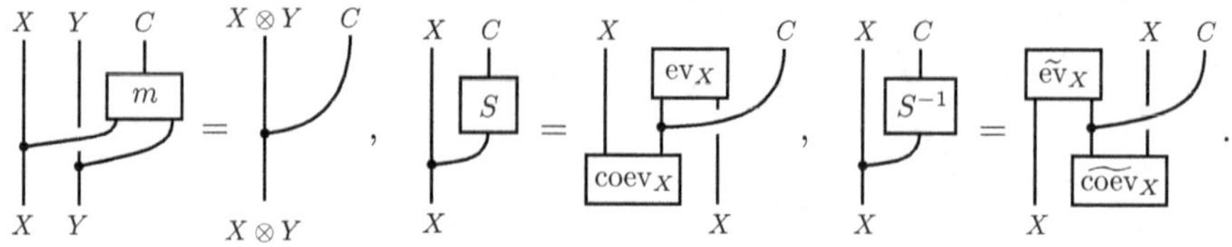

6.5.2 The canonical pairing

The factorization property (6.6) implies that there is a unique morphism $\omega\colon C \otimes C \to \mathbb{1}$ such that for all $X, Y \in \mathrm{Ob}(\mathcal{C})$,

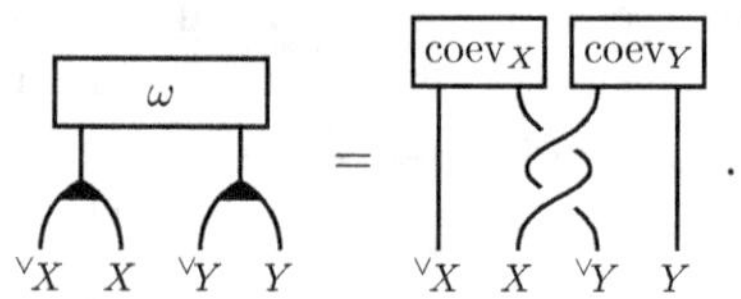

Indeed, the naturality of the braiding and the identities (2.2) imply that the right-hand side is dinatural in both X and Y. The morphism ω is called the *canonical pairing* of C. In terms of the universal coaction, ω is characterized by the condition that for all $X, Y \in \mathrm{Ob}(\mathcal{C})$,

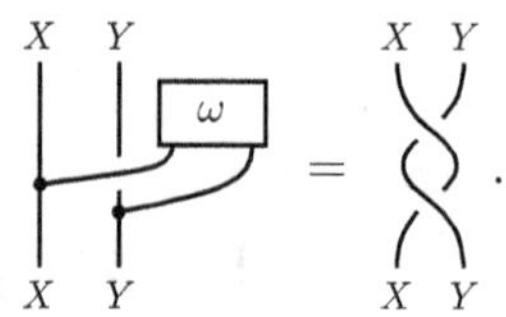

Note that ω is a Hopf pairing for the Hopf algebra C. Indeed, it follows from the definitions of the product m and the coproduct Δ of C that

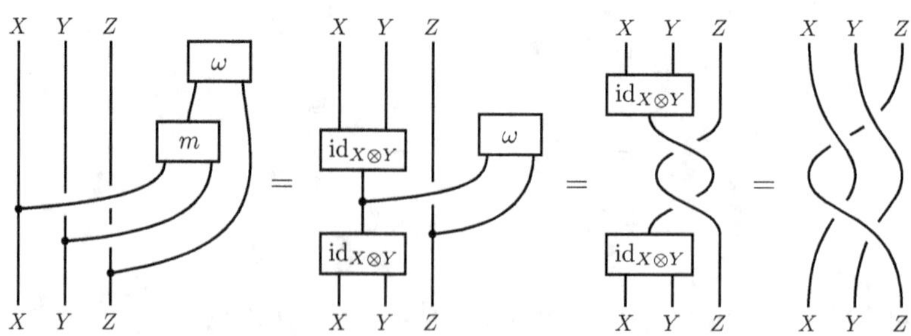

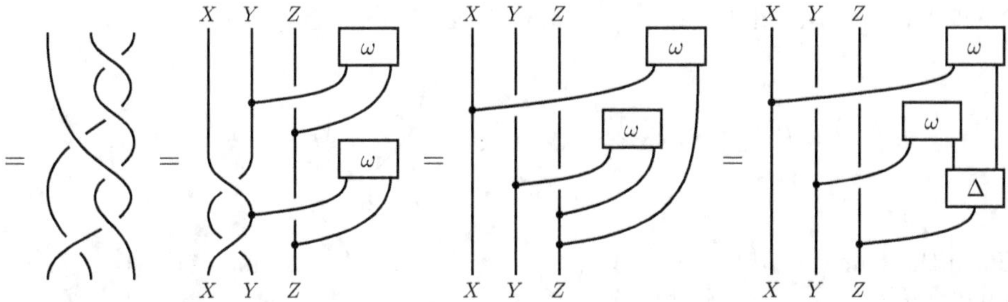

for all $X, Y, Z \in \mathrm{Ob}(\mathcal{C})$. The uniqueness of factorization through the universal coaction implies that

$$\omega(m \otimes \mathrm{id}_C) = \omega(\mathrm{id}_C \otimes \omega \otimes \mathrm{id}_C)(\mathrm{id}_{C \otimes C} \otimes \Delta).$$

All the other axioms of a Hopf pairing are verified similarly.

6.5.3 Computation of the center

We outline a computation of the center $\mathcal{Z}(\mathcal{C})$ of $\mathcal{C}$ (see Section 5.1.2) in terms of the Hopf algebra C. To this end we provide the monoidal category mod_C of C-modules with a braiding and construct a braided strict monoidal isomorphism $\mathrm{mod}_C \simeq \mathcal{Z}(\mathcal{C})$. This construction will not be used in the sequel, and consequently we skip all the proofs. However, this construction motivates the content of the next three chapters which culminate in a more general computation of the center of (possibly non-braided) monoidal categories.

For any left C-modules $(M, r\colon C \otimes M \to M)$ and $(N, s\colon C \otimes N \to N)$, set

$$c_{(M,r),(N,s)} = (\mathrm{id}_N \otimes r)(\delta_N \otimes \mathrm{id}_M)\tau_{M,N} =$$  ,

where τ is the braiding of $\mathcal{C}$ and δ is the universal coaction of C. Then the family

$$\{c_{(M,r),(N,s)}\colon (M,r) \otimes (N,s) \to (N,s) \otimes (M,r)\}_{(M,r),(N,s) \in \mathrm{Ob}(\mathrm{mod}_C)}$$

is a braiding of mod_C with inverse computed by

$$(c_{(M,r),(N,s)})^{-1} =$$ 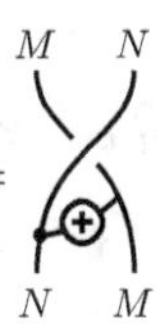 .

Next, for any C-module (M,r) and any object X of $\mathcal{C}$, set

$$\sigma_X^{(M,r)} = (\mathrm{id}_X \otimes r)(\delta_X \otimes \mathrm{id}_M)\tau_{M,X} =$$

$$: M \otimes X \to X \otimes M.$$

Then the pair $\big(M, \sigma^{(M,r)} = \{\sigma_X^{(M,r)}\}_{X \in \mathrm{Ob}(\mathcal{C})}\big)$ is a half-braiding of $\mathcal{C}$. We define a functor $K\colon \mathrm{mod}_C \to \mathcal{Z}(\mathcal{C})$ by

$$K(M,r) = (M, \sigma^{(M,r)}) \quad \text{and} \quad K(f) = f$$

for any C-module (M,r) and any C-linear morphism f between C-modules. It turns out that K is a braided strict monoidal isomorphism. Its inverse is given by

$$K^{-1}(M,\sigma) = (M, r^{(M,\sigma)}) \quad \text{and} \quad K^{-1}(f) = f$$

for any half-braiding (M,σ) and any morphism f in $\mathcal{Z}(\mathcal{C})$. Here

$$r^{(M,\sigma)} = (\mathrm{id}_M \otimes \mathrm{ev}_M)(\varphi^{(M,\sigma)} \otimes \mathrm{id}_M) =$$

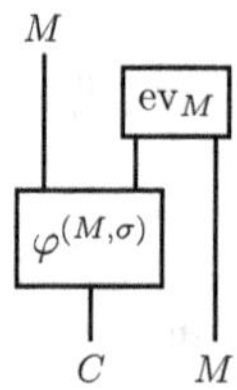

$$: C \otimes M \to M,$$

where $\varphi^{(M,\sigma)}\colon C \to M \otimes {}^\vee M$ is the unique morphism such that, for all $X \in \mathrm{Ob}(\mathcal{C})$,

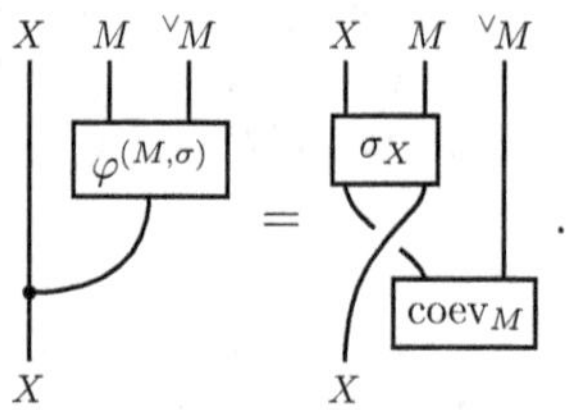

6.5.4 Exercise

Prove the following assertions.

(a) The universal coaction $\delta_C\colon C \to C \otimes C$ of C on itself is computed by

$$\delta_C = (\mathrm{id}_C \otimes m)(\tau_{C,C} \otimes \mathrm{id}_C)(S \otimes \Delta)\Delta, \quad \text{that is,}$$

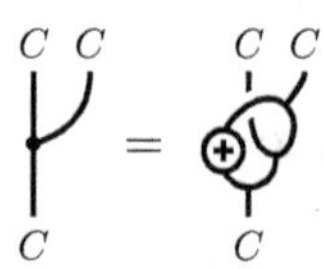

(b) The canonical pairing $\omega\colon C \otimes C \to \mathbb{1}$ satisfies $\omega\tau_{C,C}(S \otimes S) = \omega$.

(c) The braiding in $\mathcal{C}$ is symmetric if and only if $\omega = \varepsilon \otimes \varepsilon$, where $\varepsilon\colon C \to \mathbb{1}$ is the counit of C.

6.5.5 Exercise

Let (C, ρ) and (C', ρ') be two coends of $\mathcal{C}$. Prove that the (unique) isomorphism $C \to C'$ in $\mathcal{C}$ commuting with ρ and ρ' is a Hopf algebra isomorphism preserving the canonical pairing.

6.6 Special cases

We discuss several important special cases of the previous constructions.

6.6.1 The case of pivotal fusion categories

Let $\mathcal{C}$ be an additive pivotal fusion $\Bbbk$-category and let I be a representative set of simple objects of $\mathcal{C}$. By Section 6.4.4, the category $\mathcal{C}$ has a coend (C, ρ) where $C = \bigoplus_{i\in I} i^* \otimes i$. For any $X \in \mathrm{Ob}(\mathcal{C})$, we can represent the morphism ρ_X using the enriched graphical calculus of Section 4.6:

$$\rho_X = \sum_{i\in I}$$
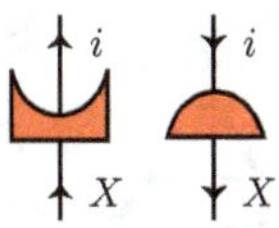

$$: X^* \otimes X \to C. \qquad (6.10)$$

In this figure and in the sequel, we drop the inclusions $i^* \otimes i \to C$ associated with the direct sum decomposition $C = \bigoplus_{i\in I} i^* \otimes i$. We will similarly drop the projections $C \to i^* \otimes i$ associated with this decomposition.

Lemma 6.3. $\dim_l(C) = \dim_r(C) = \dim(\mathcal{C})$.

Proof. Using the properties of the left and right dimensions (see Sections 2.6.2 and 4.2.2), we obtain:

$$\begin{aligned} \dim_l(C) &= \dim_l\Big(\bigoplus_{i\in I} i^* \otimes i\Big) = \sum_{i\in I} \dim_l(i^* \otimes i) \\ &= \sum_{i\in I} \dim_l(i^*) \dim_l(i) = \sum_{i\in I} \dim_r(i) \dim_l(i) = \dim(\mathcal{C}). \end{aligned}$$

The equality $\dim_r(C) = \dim(\mathcal{C})$ is proved similarly. $\square$

6.6.2 The case of braided categories

Let $\mathcal{C}$ be an additive braided pivotal fusion $\Bbbk$-category and let I be a representative set of simple objects of $\mathcal{C}$. Consider the coend (C, ρ) of $\mathcal{C}$, where $C = \bigoplus_{i\in I} i^* \otimes i$ and ρ is given by (6.10), see Section 6.6.1. By Section 6.5.1, C is a Hopf algebra in $\mathcal{C}$. Using the definitions of Section 6.5 and Formulas (6.8), (6.9), (6.10), we can graphically compute the structural morphisms and canonical pairing of C as follows:

$$\Delta = \sum_{i \in I}$$ 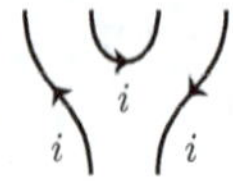 $$: C \to C \otimes C,$$

$$\varepsilon = \sum_{i \in I}$$  $$: C \to \mathbb{1},$$

$$m = \sum_{i,j,k \in I}$$ 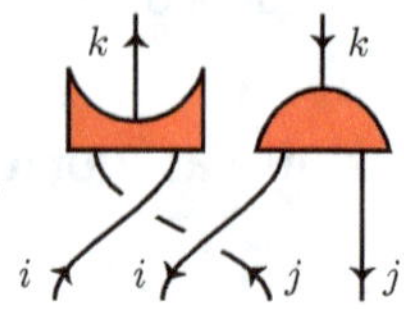 $$: C \otimes C \to C,$$

$$u =$$  $$= \widetilde{\mathrm{coev}}_{\mathbb{1}} \colon \mathbb{1} \to \mathbb{1}^* \otimes \mathbb{1} \hookrightarrow C,$$

$$S = \sum_{i \in I}$$ 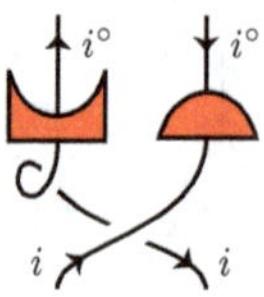 $$: C \to C,$$

$$S^{-1} = \sum_{i \in I}$$ 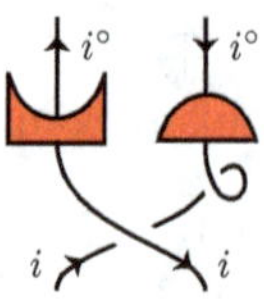 $$: C \to C,$$

$$\omega = \sum_{i,j \in I}$$  $$: C \otimes C \to \mathbb{1}.$$

As above, we drop in our pictures the inclusions and projections associated with the direct sum decomposition $C = \bigoplus_{i \in I} i^* \otimes i$. In the expressions for $S^{\pm 1}$, the symbol i° with $i \in I$ stands for the unique element of I isomorphic to i^*.

Lemma 6.4. *The morphism*

$$\Lambda_r = \sum_{i \in I} \dim_r(i)\, \rho_i\, \widetilde{\mathrm{coev}}_i = \sum_{i \in I} \dim_r(i)$$  $$: \mathbb{1} \to C$$

is a right integral of C, and all right integrals of C are scalar multiples of Λ_r.

Proof. For $i \in I$, the $\Bbbk$-module $\mathrm{Hom}_{\mathcal{C}}(\mathbb{1}, i^* \otimes i) \simeq \mathrm{Hom}_{\mathcal{C}}(i, i)$ is free of rank 1 with basis $(\widetilde{\mathrm{coev}}_i \colon \mathbb{1} \to i^* \otimes i)$. By Section 6.4.4, the morphism $\rho_i \colon i^* \otimes i \to C$ is the inclusion associated with the direct sum decomposition $C = \bigoplus_{i\in I} i^* \otimes i$. Therefore the $\Bbbk$-module

$$\mathrm{Hom}_{\mathcal{C}}(\mathbb{1}, C) = \mathrm{Hom}_{\mathcal{C}}\big(\mathbb{1}, \bigoplus_{i\in I} i^* \otimes i\big) \simeq \bigoplus_{i\in I} \mathrm{Hom}_{\mathcal{C}}(\mathbb{1}, i^* \otimes i)$$

is free with basis $(e_i = \rho_i\, \widetilde{\mathrm{coev}}_i \colon \mathbb{1} \to C)_{i\in I}$. Thus, each morphism $f\colon \mathbb{1} \to C$ in $\mathcal{C}$ expands uniquely as $f = \sum_{i\in I} a_i e_i$ with $a_i \in \Bbbk$. Using the description above of the counit ε and product m of C, we obtain

$$f\varepsilon = \sum_{j,k\in I} a_k$$ 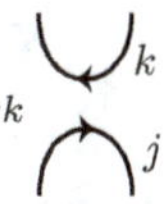

and

$$m(f \otimes \mathrm{id}_C) = \sum_{i,j,k\in I} a_i$$ 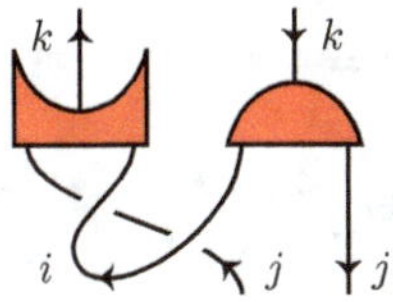

$$= \sum_{i,j,k\in I} a_i$$ 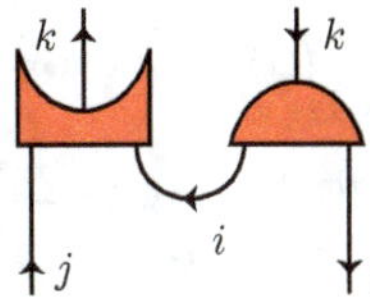

.

Lemma 4.10 implies that f is a right integral for C if and only if $a_i = a_{\mathbb{1}} \dim_r(i)$ for all $i \in I$, that is, if and only if $f = a_{\mathbb{1}} \Lambda_r$. This implies the lemma. $\square$

6.6.3 The case of ribbon categories

Let $\mathcal{C}$ be a ribbon category having a coend (C, ρ). Let $\theta = \{\theta_X \colon X \to X\}_{X\in \mathrm{Ob}(\mathcal{C})}$ be the twist of $\mathcal{C}$ (see Section 3.3.2). The factorization property (6.5) implies the existence of unique morphisms $\theta_+\colon C \to \mathbb{1}$ and $\theta_-\colon C \to \mathbb{1}$ in $\mathcal{C}$ such that for any $X \in \mathrm{Ob}(\mathcal{C})$,

$$\theta_+\rho_X = \mathrm{ev}_X(\mathrm{id}_{X^*} \otimes \theta_X) \quad \text{and} \quad \theta_-\rho_X = \mathrm{ev}_X(\mathrm{id}_{X^*} \otimes \theta_X^{-1}).$$

Indeed, the right-hand sides of these equalities are dinatural in X (by (2.2) and the naturality of θ) and so factorize through ρ. Graphically,

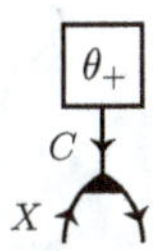

$=$ 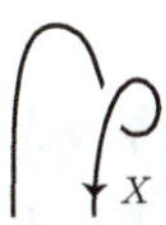

and 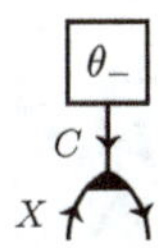

$=$ 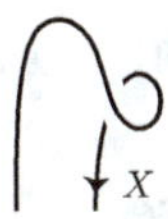

.

The morphisms $\theta_\pm$ are called the *twist forms* of C.

Lemma 6.5. *Let $\mathcal{C}$ be an additive ribbon fusion $\Bbbk$-category such that $\dim(\mathcal{C})$ is not a zero divisor of $\Bbbk$. Let $C \in \mathrm{Ob}(\mathcal{C})$ be a coend of $\mathcal{C}$ with counit $\varepsilon\colon C \to \mathbb{1}$ and twist forms $\theta_\pm\colon C \to \mathbb{1}$. Then the scalars $\Delta_\pm \in \Bbbk$ introduced in Section 4.5.2 are computed by*

$$\Delta_\pm = \theta_\pm \Lambda,$$

where $\Lambda\colon \mathbb{1} \to C$ is any right integral of C such that $\varepsilon\Lambda = \dim(\mathcal{C})$.

Proof. Since all coends of $\mathcal{C}$ are isomorphic to each other, we can assume that $C = \bigoplus_{i\in I} i^* \otimes i$, where I is a representative set of simple objects of $\mathcal{C}$, and that ρ is given by (6.10). By Lemma 6.4, all right integrals of C are scalar multiples of

$$\Lambda_r = \sum_{i\in I} \dim(i)\, \rho_i \widetilde{\mathrm{coev}}_i \colon \mathbb{1} \to C.$$

In particular, $\Lambda = k\,\Lambda_r$ for some $k \in \Bbbk$. Then

$$\begin{aligned}\dim(\mathcal{C}) = \varepsilon\Lambda = k\,\varepsilon\Lambda_r &= k \sum_{i\in I} \dim(i)\, \varepsilon\rho_i \widetilde{\mathrm{coev}}_i \\ &= k \sum_{i\in I} \dim(i)\, \mathrm{ev}_i \widetilde{\mathrm{coev}}_i = k \sum_{i\in I} (\dim(i))^2 = k \dim(\mathcal{C}).\end{aligned}$$

Since $\dim(\mathcal{C})$ is not a zero divisor, $k = 1$ and $\Lambda = \Lambda_r$. Therefore,

$$\begin{aligned}\theta_\pm\Lambda = \theta_\pm\Lambda_r &= \sum_{i\in I} \dim(i)\, \theta_\pm \rho_i \widetilde{\mathrm{coev}}_i \\ &= \sum_{i\in I} \dim(i)\, \mathrm{ev}_i(\mathrm{id}_{i^*} \otimes \theta_i^{\pm 1}) \widetilde{\mathrm{coev}}_i = \sum_{i\in I} \dim(i)\, \mathrm{tr}(\theta_i^{\pm 1}) = \Delta_\pm.\end{aligned}$$

□

6.6.4 Exercise*

Let $\mathcal{C}$ be a braided pivotal category having a coend (C, ρ). Prove that $\dim_l(C) = \dim_r(C)$ and $\theta^l_C = \theta^r_C$, where θ^l and θ^r are the left and right twists of $\mathcal{C}$ (see Section 3.3.1).

6.6.5 Exercise

Let $\mathcal{C}$ be a braided pivotal category having a coend C. Prove that the Hopf algebra C is involutory. Consequently, by Sections 5.2.2 and 6.2.3, the categories $\mathcal{Z}(\mathcal{C})$ and mod_C are pivotal. Prove that the isomorphism $\mathrm{mod}_C \simeq \mathcal{Z}(\mathcal{C})$ from Section 6.5.3 is strictly pivotal.

6.6.6 Exercise

Let $\mathcal{C}$ be a ribbon category having a coend C. Prove that the twist forms $\theta_\pm \colon C \to \mathbb{1}$ from Section 6.6.3 are characterized in terms of the universal coaction of C (see Section 6.4.3) as follows: for any $X \in \mathrm{Ob}(\mathcal{C})$,

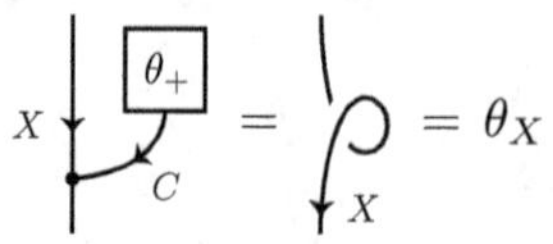 and 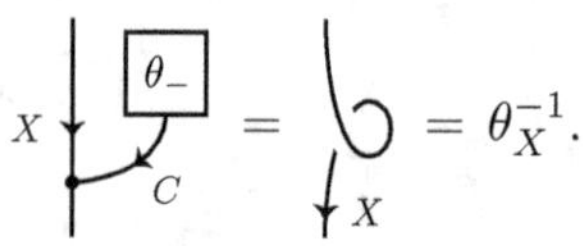

6.6.7 Exercise

Let $\mathcal{C}$ be a ribbon category having a coend C. Prove that

$$(\theta_+ \otimes \theta_-)\Delta = \varepsilon = (\theta_- \otimes \theta_+)\Delta, \quad (\theta_\pm \otimes \mathrm{id}_C)\Delta = \varepsilon = (\mathrm{id}_C \otimes \theta_\pm)\Delta,$$
$$\theta_\pm u = \mathrm{id}_{\mathbb{1}}, \quad \theta_\pm S = \theta_\pm, \quad m\theta_- = (\theta_- \otimes \omega \otimes \theta_-)(\Delta \otimes \Delta)$$

where Δ, ε, m, u, S, $\theta_\pm$, and ω are the coproduct, counit, product, unit, antipode, twist forms, and canonical pairing of C. The first two equalities mean that θ_+ and θ_- are inverse to each other in the convolution monoid $\mathrm{Hom}_{\mathcal{C}}(C, \mathbb{1})$, see Exercise 6.1.6.

6.7 Modularity re-examined

Following Volodymyr Lyubashenko [Lyu, KL], we call a braided rigid category *modularmodular* if it has a coend and the canonical pairing of the coend is non-degenerate. This condition does not depend on the choice of a coend, see Exercise 6.5.5. The next theorem shows that for additive ribbon fusion $\Bbbk$-categories, this definition of modularity is equivalent to that of Section 4.5.2.

Theorem 6.6. *An additive ribbon fusion $\Bbbk$-category $\mathcal{C}$ is modular in the sense of Lyubashenko if and only if the S-matrix of $\mathcal{C}$ is invertible.*

Proof. Let I be a representative set of simple objects of $\mathcal{C}$. Consider the coend (C, ρ) of $\mathcal{C}$, where $C = \bigoplus_{i\in I} i^* \otimes i$ and ρ is given by (6.10). Recall from Section 6.4.4 that for each $i \in I$, the morphism $\rho_i \colon i^* \otimes i \to C$ is the inclusion associated with the direct sum decomposition $C = \bigoplus_{i\in I} i^* \otimes i$. Let $\nu_i \colon C \to i^* \otimes i$ be the projection associated with this direct sum decomposition. Consider the bases

$$(e_i = \rho_i \widetilde{\mathrm{coev}}_i \colon \mathbb{1} \to C)_{i\in I} \quad \text{and} \quad (f_i = \dim(i)^{-1} \mathrm{ev}_i \nu_i \colon C \to \mathbb{1})_{i \in I}$$

of the $\Bbbk$-modules $\mathrm{Hom}_{\mathcal{C}}(\mathbb{1}, C)$ and $\mathrm{Hom}_{\mathcal{C}}(C, \mathbb{1})$, respectively (cf. the proof of Lemma 6.4). It follows from the identities $\nu_j \rho_i = \delta_{i,j}\, \mathrm{id}_{i^*\otimes i}$ and $\mathrm{ev}_i \widetilde{\mathrm{coev}}_i = \dim(i)$ that for any $i, j \in I$,

$$f_j e_i = \delta_{i,j}. \tag{6.11}$$

As a consequence, for any $\alpha \in \mathrm{Hom}_{\mathcal{C}}(\mathbb{1}, C)$ and $\beta \in \mathrm{Hom}_{\mathcal{C}}(C, \mathbb{1})$, we have

$$\alpha = \sum_{k\in I} (f_k \alpha)\, e_k, \quad \beta = \sum_{k \in I} (\beta e_k)\, f_k, \quad \text{and} \quad \beta\alpha = \sum_{k\in I} (\beta e_k)(f_k \alpha). \tag{6.12}$$

We provide C with the structure of a Hopf algebra in $\mathcal{C}$ as in Section 6.5.1. Let $S \colon C \to C$ be the antipode of C. We claim that for all $i \in I$,

$$S e_i = e_{i^\circ} \quad \text{and} \quad f_i S = f_{i^\circ}, \tag{6.13}$$

where $I \to I$, $i \mapsto i^\circ$ is the involution from Section 4.5.2. Indeed, dropping as usual the inclusions and projections associated with the direct sum decomposition $C = \bigoplus_{i\in I} i^* \otimes i$ and picking an isomorphism $\alpha\colon i^* \to i^\circ$, we obtain

$Se_i \overset{\text{(i)}}{=}$ 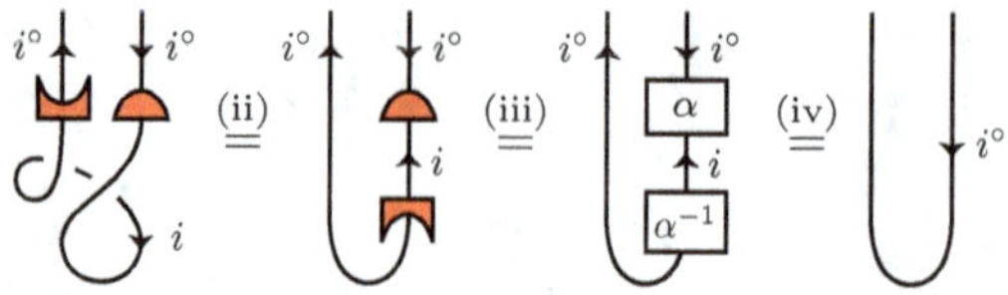 $\overset{\text{(v)}}{=} e_{i^\circ}.$

Here (i) follows from the computation of S in Section 6.6.2, (ii) from Exercise 3.2.7 and Theorem 3.3, (iii) from the fact that (α, α^{-1}) is an i°-partition of i^*, (iv) from the identity $\alpha\alpha^{-1} = \mathrm{id}_{i^\circ}$, and (v) from the definition of e_{i°. Applying the second formula in (6.12) to $\beta = f_i S$, we obtain

$$f_i S = \sum_{k\in I} (f_i S e_k)\, f_k = \sum_{k\in I} (f_i e_{k^\circ})\, f_k = \sum_{k\in I} \delta_{i,k^\circ}\, f_k = f_{i^\circ}.$$

Recall from Section 4.5.2 that the S-matrix $[S_{i,j}]_{i,j\in I}$ of $\mathcal{C}$ is defined by

$$S_{i,j} = \mathrm{tr}(\tau_{j,i}\tau_{i,j}) \in \mathrm{End}_{\mathcal{C}}(\mathbb{1}) = \Bbbk,$$

where τ is the braiding of $\mathcal{C}$. This matrix is symmetric and is related to the canonical pairing $\omega\colon C \otimes C \to \mathbb{1}$ as follows: for all $i, j \in I$,

$$\omega(e_i \otimes e_{j^\circ}) = S_{i,j}. \tag{6.14}$$

Indeed,

$\omega(e_i \otimes e_{j^\circ}) \overset{\text{(i)}}{=}$ 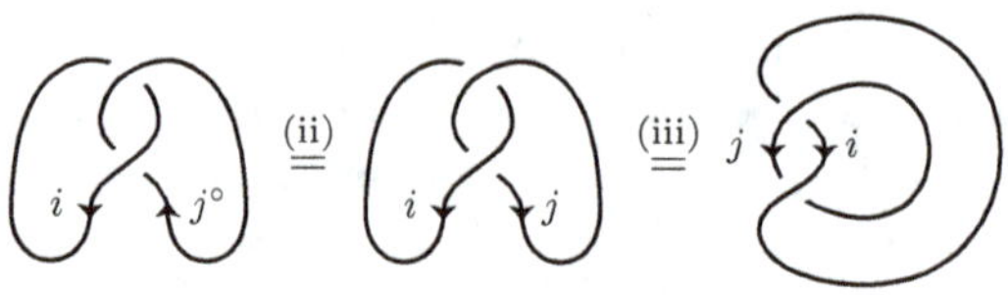

$\overset{\text{(iv)}}{=} \mathrm{tr}(\tau_{j,i}\tau_{i,j}) = S_{i,j}.$

Here (i) follows from the computation of ω in Section 6.6.2, (ii) from the fact that $j^\circ \simeq j^*$ (see Exercise 3.2.4), (iii) from the fact that $\mathcal{C}$ is spherical (as are all ribbon categories), and (iv) from the definition of tr.

We now prove the claim of the theorem. Assume first that the pairing ω is non degenerate and let $\Omega\colon \mathbb{1} \to C \otimes C$ be its inverse. For any $i, j \in I$, set

$$T_{i,j} = (f_{i^\circ} \otimes f_j)\Omega \in \mathrm{End}_{\mathcal{C}}(\mathbb{1}) = \Bbbk.$$

We have:

$$\begin{aligned}\delta_{i,j} &\overset{\text{(i)}}{=} f_j e_i \overset{\text{(ii)}}{=} f_j(\omega \otimes \mathrm{id}_C)(\mathrm{id}_C \otimes \Omega)e_i \overset{\text{(iii)}}{=} \omega(e_i \otimes S)(S^{-1} \otimes f_j)\,\Omega \\ &\overset{\text{(iv)}}{=} \sum_{k\in I} \omega(e_i \otimes S e_k)(f_k S^{-1} \otimes f_j)\,\Omega \overset{\text{(v)}}{=} \sum_{k\in I} \omega(e_i \otimes e_{k^\circ})(f_{k^\circ} \otimes f_j)\,\Omega \overset{\text{(vi)}}{=} \sum_{k\in I} S_{i,k} T_{k,j}.\end{aligned}$$

Here (i) follows from (6.11), (ii) from the assumption that Ω is the inverse of the pairing ω, (iii) from the equality $SS^{-1} = \mathrm{id}_C$, (iv) from the third formula in (6.12) applied to $\beta = \omega(e_i \otimes S)$ and $\alpha = (S^{-1} \otimes f_j)\Omega$, (v) from (6.13), and (vi) from (6.14) and the definition of $T_{k,j}$. Therefore the matrices $[S_{i,j}]_{i,j\in I}$ and $[T_{i,j}]_{i,j\in I}$ are inverse to each other, and so the matrix $[S_{i,j}]_{i,j\in I}$ is invertible.

Conversely, suppose that the matrix $[S_{i,j}]_{i,j\in I}$ is invertible, and let $[T_{i,j}]_{i,j\in I}$ be its inverse (over $\Bbbk$). We verify below that the morphisms

$$\Lambda = \sum_{i\in I} T_{\mathbb{1},i^\circ} e_i \colon \mathbb{1} \to C \quad \text{and} \quad \Lambda' = \dim(\mathcal{C})\,\Lambda \colon \mathbb{1} \to C$$

satisfy the conditions of Lemma 6.2. By this lemma, ω is non-degenerate.

To verify the conditions of Lemma 6.2, set

$$\lambda = \omega(\Lambda \otimes \mathrm{id}_C) \colon C \to \mathbb{1} \quad \text{and} \quad \lambda' = \omega(\mathrm{id}_C \otimes \Lambda') \colon C \to \mathbb{1}.$$

We have:

$$\begin{aligned}\lambda &\overset{\text{(i)}}{=} \sum_{k\in I} \omega(\Lambda \otimes e_k)\, f_k \overset{\text{(ii)}}{=} \sum_{i,k\in I} T_{\mathbb{1},i^\circ}\, \omega(e_i \otimes e_k)\, f_k \\ &\overset{\text{(iii)}}{=} \sum_{k\in I} \Big(\sum_{i\in I} T_{\mathbb{1},i}\, S_{i,k^\circ} \Big) f_k \overset{\text{(iv)}}{=} \sum_{k\in I} \delta_{\mathbb{1},k^\circ}\, f_k = f_{\mathbb{1}}.\end{aligned}$$

Here (i) follows from the first formula in (6.12), (ii) from the definition of Λ, (iii) from (6.14), and (iv) from the assumption that $[T_{i,j}]_{i,j\in I}$ is the inverse of $[S_{i,j}]_{i,j\in I}$. A similar computation yields that $\omega(\mathrm{id}_C \otimes \Lambda) = f_{\mathbb{1}}$, and so $\lambda' = \dim(\mathcal{C})\, f_{\mathbb{1}}$. Observe next that $f_{\mathbb{1}}$ is a left cointegral of C: the definition of $f_{\mathbb{1}}$ and the computation of the coproduct Δ and unit u of C in Section 6.6.2, imply that

$(\mathrm{id}_C \otimes f_{\mathbb{1}})\Delta \;=\; \dim(\mathbb{1})$ 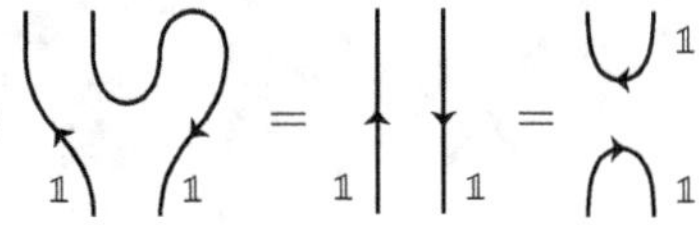 $= u f_{\mathbb{1}}$.

Hence λ and λ' are left cointegrals of C. Furthermore,

$$\lambda\Lambda = \sum_{i\in I} T_{\mathbb{1},i^\circ}\, f_{\mathbb{1}} e_i = \sum_{i\in I} T_{\mathbb{1},i^\circ}\, \delta_{\mathbb{1},i} = T_{\mathbb{1},\mathbb{1}}$$

and so

$$\omega(\Lambda \otimes \Lambda') = \omega(\mathrm{id}_C \otimes \Lambda')\Lambda = \lambda'\Lambda = \dim(\mathcal{C})\, \lambda\Lambda = \dim(\mathcal{C})\, T_{\mathbb{1},\mathbb{1}}.$$

Now, Lemma 4.6(a) implies that $T_{i,j} = (\dim(\mathcal{C}))^{-1} S_{i,j^\circ}$ for all $i,j \in I$. Therefore

$$\omega(\Lambda \otimes \Lambda') = \dim(\mathcal{C})\, T_{\mathbb{1},\mathbb{1}} = S_{\mathbb{1},\mathbb{1}^\circ} = S_{\mathbb{1},\mathbb{1}} = \dim(\mathbb{1}) = 1_{\Bbbk}. \qquad \square$$

Chapter 7

Monads and bimonads

The notion of a monad (or a triple) on a category goes back to Roger Godement [Go] and was later developed by Samuel Eilenberg, John Moore, Jonathan Beck, and others. Combining the theory of monads with that of monoidal categories, Ieke Moerdijk [Mo] formulated the notion of a bimonad. In this chapter, we outline the basics of the theory of monads and bimonads.

7.1 Monads

We define monads and show how to construct them from algebras.

7.1.1 Endofunctors

An *endofunctor* of a category $\mathcal{C}$ is a functor from $\mathcal{C}$ to itself. For any endofunctor F of $\mathcal{C}$, we set $F^0 = 1_{\mathcal{C}} \colon \mathcal{C} \to \mathcal{C}$ and, for each integer $n \geq 1$, we set

$$F^n = \underbrace{F \circ \cdots \circ F}_{n \text{ times}} \colon \mathcal{C} \to \mathcal{C}.$$

7.1.2 Monads

A *monad* on a category $\mathcal{C}$ is a triple (T, μ, η) consisting of an endofunctor T of $\mathcal{C}$ and two natural transformations

$$\mu = \{\mu_X \colon T^2(X) \to T(X)\}_{X \in \mathrm{Ob}(\mathcal{C})} \quad \text{and} \quad \eta = \{\eta_X \colon X \to T(X)\}_{X \in \mathrm{Ob}(\mathcal{C})},$$

called respectively the *product* and the *unit* of T, such that for all $X \in \mathrm{Ob}(\mathcal{C})$, the diagrams

$$
\begin{array}{ccc}
T^3(X) & \xrightarrow{T(\mu_X)} & T^2(X) \\
{\scriptstyle \mu_{T(X)}}\downarrow & & \downarrow{\scriptstyle \mu_X} \\
T^2(X) & \xrightarrow[\mu_X]{} & T(X)
\end{array}
\quad \text{and} \quad
\begin{array}{ccccc}
 & & T^2(X) & & \\
 & {\scriptstyle \eta_{T(X)}}\nearrow & & \searrow{\scriptstyle \mu_X} & \\
T(X) & & \xrightarrow{\mathrm{id}_{T(X)}} & & T(X) \\
 & {\scriptstyle T(\eta_X)}\searrow & & \nearrow{\scriptstyle \mu_X} & \\
 & & T^2(X) & &
\end{array}
$$

commute, that is,

$$\mu_X T(\mu_X) = \mu_X \mu_{T(X)} \quad \text{and} \quad \mu_X \eta_{T(X)} = \mathrm{id}_{T(X)} = \mu_X T(\eta_X).$$

This commutativity condition is usually expressed by saying that the product μ of T is *associative* and *unital.*

For example, the identity functor $1_{\mathcal{C}} \colon \mathcal{C} \to \mathcal{C}$ with the identity natural transformation as the product and the unit is a monad on $\mathcal{C}$. It is called the *trivial monad.*

7.1.3 Modules

Let $T = (T, \mu, \eta)$ be a monad on a category $\mathcal{C}$. An *action* of T on an object M of $\mathcal{C}$ is a morphism $r \colon T(M) \to M$ in $\mathcal{C}$ such that the diagrams

$$
\begin{array}{ccc}
T^2(M) & \xrightarrow{T(r)} & T(M) \\
{\scriptstyle \mu_M}\downarrow & & \downarrow{\scriptstyle r} \\
T(M) & \xrightarrow[r]{} & M
\end{array}
\quad \text{and} \quad
\begin{array}{ccc}
 & T(M) & \\
{\scriptstyle \eta_M}\nearrow & & \searrow{\scriptstyle r} \\
M & \xrightarrow[\mathrm{id}_M]{} & M
\end{array}
$$

commute, that is,

$$rT(r) = r\mu_M \quad \text{and} \quad r\eta_M = \mathrm{id}_M.$$

Such a pair (M, r) is called a *T-module.* (In the literature, T-modules are also called T-algebras.) A *morphism of T-modules* $(M, r) \to (N, s)$ is a morphism $f \colon M \to N$ in $\mathcal{C}$ which is *T-linear* in the sense that

$$fr = sT(f) \colon T(M) \to N.$$

We denote by $\mathcal{C}^T$ the *category of T-modules.* Its objects are T-modules in $\mathcal{C}$ and its morphisms are morphisms of T-modules with composition induced by that in $\mathcal{C}$. The *forgetful functor* $U_T \colon \mathcal{C}^T \to \mathcal{C}$ is defined by

$$U_T(M, r) = M \quad \text{and} \quad U_T(f) = f$$

for any T-module (M, r) and any T-linear morphism f.

The functor U_T is faithful, meaning that for all T-modules $(M,r),(N,s)$, the map $\mathrm{Hom}_{\mathcal{C}^T}((M,r),(N,s)) \to \mathrm{Hom}_{\mathcal{C}}(M,N)$ induced by U_T is injective. Also, the functor U_T reflects isomorphisms, meaning that a morphism in $\mathcal{C}^T$ is an isomorphism if and only if its image under U_T is an isomorphism in $\mathcal{C}$.

The *free module functor* $F_T\colon \mathcal{C} \to \mathcal{C}^T$ is defined by

$$F_T(X) = (T(X), \mu_X) \quad \text{and} \quad F_T(f) = T(f)$$

for any object X of $\mathcal{C}$ and any morphism f in $\mathcal{C}$. Note that the action $r\colon T(M) \to M$ of any T-module (M,r) induces a T-linear morphism $r\colon F_T(M) \to (M,r)$.

7.1.4 Example

Let (A,m,u) be an algebra in a monoidal category $\mathcal{C}$. Define an endofunctor $T = A\otimes ?$ of $\mathcal{C}$ by

$$T(X) = A \otimes X \quad \text{and} \quad T(f) = \mathrm{id}_A \otimes f$$

for any $X \in \mathrm{Ob}(\mathcal{C})$ and any morphism f in $\mathcal{C}$. Then T is a monad on $\mathcal{C}$ with product μ and unit η given for any $X \in \mathrm{Ob}(\mathcal{C})$ by

$\mu_X = m \otimes \mathrm{id}_X =$ 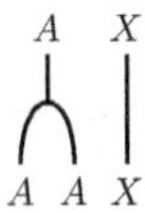 and $\eta_X = u \otimes \mathrm{id}_X =$  .

Clearly, $\mathcal{C}^T = \mathrm{mod}_A^{\mathcal{C}}$ is the category of (left) A-modules (see Section 6.1.1). This construction exhibits monads as generalizations of algebras.

7.1.5 Remark

A category is said to be *small* if its class of objects is a set. For a small category $\mathcal{C}$, the natural transformations between any two endofunctors of $\mathcal{C}$ form a set. Thus, we can consider the category $\mathrm{End}(\mathcal{C})$ whose objects are endofunctors of $\mathcal{C}$ and whose morphisms are natural transformations between endofunctors (with the usual composition). We provide $\mathrm{End}(\mathcal{C})$ with a monoidal product using composition of endofunctors as follows. For endofunctors F,G of $\mathcal{C}$, set $F \otimes G = FG\colon \mathcal{C} \to \mathcal{C}$. For natural transformations $\alpha\colon F \to F'$ and $\beta\colon G \to G'$ of endofunctors of $\mathcal{C}$ and for any $X \in \mathrm{Ob}(\mathcal{C})$, the diagram

$$\begin{array}{ccc} FG(X) & \xrightarrow{\ \alpha_{G(X)}\ } & F'G(X) \\ {\scriptstyle F(\beta_X)}\downarrow & & \downarrow{\scriptstyle F'(\beta_X)} \\ FG'(X) & \xrightarrow{\ \alpha_{G'(X)}\ } & F'G'(X) \end{array}$$

commutes, as follows from the naturality of α. Set

$$(\alpha \otimes \beta)_X = \alpha_{G'(X)} F(\beta_X) = F'(\beta_X)\, \alpha_{G(X)}\colon FG(X) \to F'G'(X).$$

This defines a natural transformation $\alpha \otimes \beta \colon F \otimes G \to F' \otimes G'$. The resulting monoidal product $\otimes$ turns $\mathrm{End}(\mathcal{C})$ into a strict monoidal category whose unit object is the identity functor $1_{\mathcal{C}} \colon \mathcal{C} \to \mathcal{C}$. It is easy to check that a monad on a small category $\mathcal{C}$ is nothing but an algebra in the monoidal category $\mathrm{End}(\mathcal{C})$.

7.1.6 Exercise*

Let $T = (T, \mu, \eta)$ and $T' = (T', \mu', \eta')$ be monads on a category $\mathcal{C}$. A *monad morphism* from T to T' is a natural transformation $f = \{f_X \colon T(X) \to T'(X)\}_{X \in \mathrm{Ob}(\mathcal{C})}$ such that, for all $X \in \mathrm{Ob}(\mathcal{C})$,

$$f_X \mu_X = \mu'_X f_{T'(X)} T(f_X) \quad \text{and} \quad f_X \eta_X = \eta'_X.$$

Prove that there is a bijective correspondence between

(i) monad morphisms from T to T';

(ii) functors $F \colon \mathcal{C}^{T'} \to \mathcal{C}^T$ such that $U_T F = U_{T'}$.

7.2 Monads versus adjunctions

We discuss the relationships between adjunctions and monads. We begin by defining adjunctions between categories.

7.2.1 Adjunctions

Observe first that any pair of functors $F \colon \mathcal{A} \to \mathcal{C}$ and $G \colon \mathcal{B} \to \mathcal{C}$ between categories $\mathcal{A}$, $\mathcal{B}$, $\mathcal{C}$ induces a functor

$$H = \mathrm{Hom}_{\mathcal{C}}(F, G) \colon \mathcal{A}^{\mathrm{op}} \times \mathcal{B} \to \mathrm{Set},$$

where Set is the category of sets and maps between sets. This functor is defined as follows. For an object $(A, B) \in \mathrm{Ob}(\mathcal{A}^{\mathrm{op}} \times \mathcal{B}) = \mathrm{Ob}(\mathcal{A}) \times \mathrm{Ob}(\mathcal{B})$, set

$$H(A, B) = \mathrm{Hom}_{\mathcal{C}}(F(A), G(B)).$$

For a morphism

$$(f, g) \in \mathrm{Hom}_{\mathcal{A}^{\mathrm{op}} \times \mathcal{B}}((A, B), (A', B')) = \mathrm{Hom}_{\mathcal{A}}(A', A) \times \mathrm{Hom}_{\mathcal{B}}(B, B'),$$

the map

$$H(f, g) \colon \mathrm{Hom}_{\mathcal{C}}(F(A), G(B)) \to \mathrm{Hom}_{\mathcal{C}}(F(A'), G(B'))$$

carries any $\alpha \in \mathrm{Hom}_{\mathcal{C}}(F(A), G(B))$ to $G(g)\alpha F(f)$.

Lemma 7.1. *Let $F\colon \mathcal{C} \to \mathcal{D}$ and $U\colon \mathcal{D} \to \mathcal{C}$ be functors between categories $\mathcal{C}$ and $\mathcal{D}$. There is a bijective correspondence between natural isomorphisms*

$$\Psi = \{\Psi_{X,Y}\colon \mathrm{Hom}_{\mathcal{D}}(F(X), Y) \to \mathrm{Hom}_{\mathcal{C}}(X, U(Y))\}_{X\in \mathrm{Ob}(\mathcal{C}), Y\in \mathrm{Ob}(\mathcal{D})}$$

from $\mathrm{Hom}_{\mathcal{D}}(F, 1_{\mathcal{D}})$ to $\mathrm{Hom}_{\mathcal{C}}(1_{\mathcal{C}}, U)$ and pairs (η, ε) of natural transformations

$$\eta = \{\eta_X\colon X \to UF(X)\}_{X\in \mathrm{Ob}(\mathcal{C})}\colon 1_{\mathcal{C}} \to UF,$$
$$\varepsilon = \{\varepsilon_Y\colon FU(Y) \to Y\}_{Y\in \mathrm{Ob}(\mathcal{D})}\colon FU \to 1_{\mathcal{D}},$$

such that, for all $X \in \mathrm{Ob}(\mathcal{C})$ and $Y \in \mathrm{Ob}(\mathcal{D})$,

$$\varepsilon_{F(X)}F(\eta_X) = \mathrm{id}_{F(X)} \quad \text{and} \quad U(\varepsilon_Y)\,\eta_{U(Y)} = \mathrm{id}_{U(Y)}. \tag{7.1}$$

Proof. Consider a natural isomorphism $\Psi = \{\Psi_{X,Y}\}_{X\in \mathrm{Ob}(\mathcal{C}), Y\in \mathrm{Ob}(\mathcal{D})}$ as in the lemma. By the definition of a natural isomorphism, all the maps $\Psi_{X,Y}$ are bijections such that for any objects $X, X' \in \mathrm{Ob}(\mathcal{C})$, $Y, Y' \in \mathrm{Ob}(\mathcal{D})$ and any morphisms $f \in \mathrm{Hom}_{\mathcal{C}}(X', X)$, $g \in \mathrm{Hom}_{\mathcal{D}}(Y, Y')$, and $\alpha \in \mathrm{Hom}_{\mathcal{D}}(F(X), Y)$, we have

$$\Psi_{X',Y'}(g \circ \alpha \circ F(f)) = U(g) \circ \Psi_{X,Y}(\alpha) \circ f.$$

For $X \in \mathrm{Ob}(\mathcal{C})$ and $Y \in \mathrm{Ob}(\mathcal{D})$, set

$$\eta_X = \Psi_{X,F(X)}(\mathrm{id}_{F(X)}) \quad \text{and} \quad \varepsilon_Y = (\Psi_{U(Y),Y})^{-1}(\mathrm{id}_{U(Y)}).$$

The naturality of Ψ implies that the families

$$\eta = \{\eta_X\colon X \to UF(X)\}_{X\in \mathrm{Ob}(\mathcal{C})} \quad \text{and} \quad \varepsilon = \{\varepsilon_Y\colon FU(Y) \to Y\}_{Y\in \mathrm{Ob}(\mathcal{D})}$$

are natural transformations. Furthermore, for any $X \in \mathrm{Ob}(\mathcal{C})$, the naturality of Ψ and the definitions of η and ε imply that

$$\Psi_{X,F(X)}(\varepsilon_{F(X)}F(\eta_X)) = \Psi_{UF(X),F(X)}(\varepsilon_{F(X)})\eta_X = \Psi_{X,F(X)}(\mathrm{id}_{F(X)}).$$

Since $\Psi_{X,F(X)}$ is injective, we deduce that $\varepsilon_{F(X)}F(\eta_X) = \mathrm{id}_{F(X)}$. Also, for any $Y \in \mathrm{Ob}(\mathcal{D})$,

$$U(\varepsilon_Y)\eta_{U(Y)} = U(\varepsilon_Y)\Psi_{U(Y),FU(Y)}(\mathrm{id}_{FU(Y)}) = \Psi_{U(Y)Y}(\varepsilon_Y) = \mathrm{id}_{U(Y)}.$$

Hence, η and ε satisfy (7.1).

Conversely, let $\eta\colon 1_{\mathcal{C}} \to UF$ and $\varepsilon\colon FU \to 1_{\mathcal{D}}$ be natural transformations satisfying (7.1). For any $X \in \mathrm{Ob}(\mathcal{C})$, $Y \in \mathrm{Ob}(\mathcal{D})$, $f \in \mathrm{Hom}_{\mathcal{D}}(F(X), Y)$, and $g \in \mathrm{Hom}_{\mathcal{C}}(X, U(Y))$, set

$$\Psi_{X,Y}(f) = U(f)\eta_X \quad \text{and} \quad \Phi_{X,Y}(g) = \varepsilon_Y F(g).$$

It follows from (7.1) and the naturality of η and ε that

$$\Psi_{X,Y}(\Phi_{X,Y}(g)) = U(\varepsilon_Y F(g))\eta_X = U(\varepsilon_Y)\eta_{U(Y)}g = g,$$
$$\Phi_{X,Y}(\Psi_{X,Y}(f)) = \varepsilon_Y F(U(f)\eta_X) = f\varepsilon_{F(X)}F(\eta_X) = f.$$

Thus, the map

$$\Psi_{X,Y} \colon \mathrm{Hom}_{\mathcal{D}}(F(X),Y) \to \mathrm{Hom}_{\mathcal{C}}(X,U(Y))$$

is bijective. The family $\Psi = \{\Psi_{X,Y}\}_{X\in \mathrm{Ob}(\mathcal{C}),Y\in\mathrm{Ob}(\mathcal{D})}$ is natural since, for any $f \in \mathrm{Hom}_{\mathcal{C}}(X',X)$, $g \in \mathrm{Hom}_{\mathcal{D}}(Y,Y')$, and $\alpha \in \mathrm{Hom}_{\mathcal{D}}(F(X),Y)$, the naturality of η implies that

$$\Psi_{X',Y'}(g\alpha F(f)) = U(g\alpha F(f))\eta_{X'} = U(g)U(\alpha)\eta_X f = U(g)\Psi_{X,Y}(\alpha)f.$$

Finally, the above constructions yield mutually inverse bijections since

$$U(f)\Psi_{X,F(X)}(\mathrm{id}_{F(X)}) = \Psi_{X,Y}(f), \quad U(\mathrm{id}_{F(X)})\eta_X = \eta_X, \quad \varepsilon_Y F(\mathrm{id}_{U(Y)}) = \varepsilon_Y$$

for all $X \in \mathrm{Ob}(\mathcal{C})$, $Y \in \mathrm{Ob}(\mathcal{D})$, and $f \in \mathrm{Hom}_{\mathcal{D}}(F(X),Y)$. □

An *adjunction* between categories $\mathcal{C}$ and $\mathcal{D}$ is a pair of functors

$$(F\colon \mathcal{C} \to \mathcal{D},\ U\colon \mathcal{D} \to \mathcal{C}) \tag{7.2}$$

together with a natural isomorphism $\mathrm{Hom}_{\mathcal{D}}(F,1_{\mathcal{D}}) \to \mathrm{Hom}_{\mathcal{C}}(1_{\mathcal{C}},U)$. We denote such an adjunction by $F: \mathcal{C} \rightleftarrows \mathcal{D} : U$ or, shorter, by $F \dashv U$. By Lemma 7.1, an adjunction between $\mathcal{C}$ and $\mathcal{D}$ may be equivalently defined as a pair of functors (7.2) together with natural transformations $\eta\colon 1_{\mathcal{C}} \to UF$ and $\varepsilon\colon FU \to 1_{\mathcal{D}}$ satisfying (7.1). One calls η the *unit* of the adjunction and ε the *counit* of the adjunction.

Adjunctions can be composed as follows: the *composition* of an adjunction $F: \mathcal{C} \rightleftarrows \mathcal{D} : U$ with unit η and counit ε and an adjunction $F' : \mathcal{D} \rightleftarrows \mathcal{E} : U'$ with unit η' and counit ε', is the adjunction $F'F : \mathcal{C} \rightleftarrows \mathcal{E} : UU'$ with unit and counit defined respectively by

$$\eta'' = \{U(\eta'_{F(X)})\eta_X\}_{X\in\mathrm{Ob}(\mathcal{C})} \quad \text{and} \quad \varepsilon'' = \{\varepsilon'_Z F'(\varepsilon_{U'(Z)})\}_{Z\in\mathrm{Ob}(\mathcal{E})}.$$

Note that if $F : \mathcal{C} \rightleftarrows \mathcal{D} : U$ is an adjunction with unit η and counit ε, then $U\colon \mathcal{D}^{\mathrm{op}} \rightleftarrows \mathcal{C}^{\mathrm{op}} : F$ is an adjunction with unit ε and counit η.

A *left adjoint* of a functor $U\colon \mathcal{D} \to \mathcal{C}$ is a functor $F\colon \mathcal{C} \to \mathcal{D}$ such that there is a natural isomorphism $\mathrm{Hom}_{\mathcal{D}}(F,1_{\mathcal{D}}) \to \mathrm{Hom}_{\mathcal{C}}(1_{\mathcal{C}},U)$ making the pair (F,U) an adjunction. A *right adjoint* of a functor $F\colon \mathcal{C} \to \mathcal{D}$ is a functor $U\colon \mathcal{D} \to \mathcal{C}$ such that there is a natural isomorphism $\mathrm{Hom}_{\mathcal{D}}(F,1_{\mathcal{D}}) \to \mathrm{Hom}_{\mathcal{C}}(1_{\mathcal{C}},U)$ making the pair (F,U) an adjunction.

Lemma 7.2. *A left (respectively, right) adjoint of a given functor, if it exists, is unique up to natural isomorphism.*

Proof. We prove the lemma for the left adjoints, the right adjoints are treated similarly. Assume that $F,F'\colon \mathcal{C} \to \mathcal{D}$ are two left adjoints of a functor $U\colon \mathcal{D} \to \mathcal{C}$. Let η, ε be the unit and counit of the adjunction $F \dashv U$ and let η', ε' be the unit and counit of the adjunction $F' \dashv U$. For any $X \in \mathrm{Ob}(\mathcal{C})$, set

$$\alpha_X = \varepsilon_{F'(X)}F(\eta'_X)\colon F(X) \to F'(X)$$

and

$$\beta_X = \varepsilon'_{F(X)} F'(\eta_X) \colon F'(X) \to F(X).$$

We have:

$$\begin{aligned}
\alpha_X \beta_X &= \varepsilon_{F'(X)} F(\eta'_X) \varepsilon'_{F(X)} F'(\eta_X) \\
&\overset{\text{(i)}}{=} \varepsilon'_{F'(X)} F'U(\varepsilon_{F'(X)}) F'UF(\eta'_X) F'(\eta_X) \\
&\overset{\text{(ii)}}{=} \varepsilon'_{F'(X)} F'(U(\varepsilon_{F'(X)}) \eta_{UF'(X)}) F'(\eta'_X) \\
&\overset{\text{(iii)}}{=} \varepsilon'_{F'(X)} F'(\eta'_X) \overset{\text{(iv)}}{=} \mathrm{id}_{F'(X)}.
\end{aligned}$$

Here, (i) follows from the naturality of ε', (ii) from the naturality of η, and (iii) and (iv) from (7.1). A similar argument proves that $\beta_X \alpha_X = \mathrm{id}_{F(X)}$. Thus α_X is an isomorphism. The naturality of η' and ε implies that the family $\alpha = \{\alpha_X\}_{X \in \mathrm{Ob}(\mathcal{C})}$ is a natural isomorphism from F to F'. □

Note that the composition of two functors having a left (respectively, right) adjoint has a left (respectively, right) adjoint. This follows from the fact that the composition of two adjunctions is an adjunction.

7.2.2 From monads to adjunctions and back

If T is a monad on a category $\mathcal{C}$, then the forgetful functor $U_T \colon \mathcal{C}^T \to \mathcal{C}$ and the free module functor $F_T \colon \mathcal{C} \to \mathcal{C}^T$ (see Section 7.1.3) give rise to an adjunction $F_T \dashv U_T$ whose unit is the unit $\eta \colon 1_\mathcal{C} \to T$ of T and whose counit $\varepsilon \colon F_T U_T \to 1_{\mathcal{C}^T}$ is given by $\varepsilon_{(M,r)} = r$ for any T-module (M,r). Condition (7.1) in this case is just a reformulation of the identities $\mu_X T(\eta_X) = \mathrm{id}_{T(X)}$ and $r\eta_M = \mathrm{id}_M$ for all $X \in \mathrm{Ob}(\mathcal{C})$ and all T-modules (M,r). The adjunction $F_T \dashv U_T$ is called *the adjunction of* T.

Conversely, let $\mathcal{C}, \mathcal{D}$ be categories and let $F : \mathcal{C} \rightleftarrows \mathcal{D} : U$ be an adjunction with unit $\eta \colon 1_\mathcal{C} \to UF$ and counit $\varepsilon \colon FU \to 1_\mathcal{D}$. It follows from (7.1) that the functor $T = UF \colon \mathcal{C} \to \mathcal{C}$ is a monad with unit η and product

$$\mu = \{\mu_X = U(\varepsilon_{F(X)})\}_{X \in \mathrm{Ob}(\mathcal{C})}.$$

Moreover, there is a unique functor $K \colon \mathcal{D} \to \mathcal{C}^T$ such that $U_T K = U$ and $KF = F_T$. This functor is computed by $K(Y) = (U(Y), U(\varepsilon_Y))$ for any object Y of $\mathcal{D}$ and $K(g) = U(g)$ for any morphism g in $\mathcal{D}$. The monad T and the functor K are called respectively the *monad* and the *comparison functor* of the adjunction $F \dashv U$. An adjunction is *monadic* if its comparison functor is an equivalence of categories.

7.2.3 Example

For any monad T on a category $\mathcal{C}$, the monad of the adjunction $F_T\colon \mathcal{C} \rightleftarrows \mathcal{C}^T : U_T$ is the monad T itself. The comparison functor in this case is the identity functor $\mathcal{C}^T \to \mathcal{C}^T$ and so the adjunction in question is monadic.

7.2.4 Example

Let M be an object of a monoidal category $\mathcal{C}$ having a left dual $({}^\vee\! M, \mathrm{ev}_M)$. Let $\mathrm{coev}_M\colon \mathbb{1} \to M \otimes {}^\vee\! M$ be the inverse of $\mathrm{ev}_M\colon {}^\vee\! M \otimes M \to \mathbb{1}$. Then the functor $? \otimes M\colon \mathcal{C} \to \mathcal{C}$ is a left adjoint of the functor $? \otimes {}^\vee\! M\colon \mathcal{C} \to \mathcal{C}$. The unit η and counit ε of the adjunction $(? \otimes M) \dashv (? \otimes {}^\vee\! M)$ are given for all $X \in \mathrm{Ob}(\mathcal{C})$ by

$$\eta_X = \mathrm{id}_X \otimes \mathrm{coev}_M \quad \text{and} \quad \varepsilon_X = \mathrm{id}_X \otimes \mathrm{ev}_M.$$

The monad of this adjunction is the monad $? \otimes A$, where $A = M \otimes {}^\vee\! M$ is the algebra in $\mathcal{C}$ with product $\mathrm{id}_M \otimes \mathrm{ev}_M \otimes \mathrm{id}_{{}^\vee\! M}$ and unit coev_M (see Example 7.1.4). Note that for any $X, Y \in \mathrm{Ob}(\mathcal{C})$, the bijection

$$\mathrm{Hom}_{\mathcal{C}}(X \otimes M, Y) \to \mathrm{Hom}_{\mathcal{C}}(X, Y \otimes {}^\vee\! M), \quad f \mapsto (f \otimes \mathrm{id}_{{}^\vee\! M})(\mathrm{id}_X \otimes \mathrm{coev}_M)$$

associated with the adjunction $(? \otimes M) \dashv (? \otimes {}^\vee\! M)$ is nothing but the inverse of the bijection (1.11) induced by the non-degenerate pairing ev_M.

7.2.5 Exercise

Prove that the counit of an adjunction $F \dashv U$ is an isomorphism if and only if the functor U is fully faithful.

7.2.6 Exercise

Consider arbitrary functors $A\colon \mathcal{C} \to \mathcal{D}$ and $B\colon \mathcal{D} \to \mathcal{C}$ between categories $\mathcal{C}$ and $\mathcal{D}$. Prove that the following conditions are equivalent:

(i) A is an equivalence of categories and B is a quasi-inverse of A;

(ii) A is a left adjoint of B and both A and B are fully faithful;

(iii) A is a left adjoint of B and the unit and counit of the adjunction $A \dashv B$ are isomorphisms;

(iv) A is a right adjoint of B and both A and B are fully faithful;

(v) A is a right adjoint of B and the unit and counit of the adjunction $B \dashv A$ are isomorphisms.

7.3 Semisimple and separable monads

We recall standard terminology concerning morphisms in categories. Then we introduce free modules and projective modules over monads and define semisimple and separable monads.

7.3.1 Retractions and sections

An object Y of a category $\mathcal{C}$ is a *retract* of an object X of $\mathcal{C}$ if there are morphisms $r\colon X \to Y$ and $s\colon Y \to X$ in $\mathcal{C}$ such that $rs = \mathrm{id}_Y$. The morphism r is then called a *left inverse* or a *retraction* of s while s is called a *right inverse* or a *section* of r. For example, any isomorphism in $\mathcal{C}$ is both a retraction and a section of the inverse isomorphism.

A morphism $f\colon X \to Y$ in a category $\mathcal{C}$ is an *epimorphism* if for any distinct morphisms g and h from Y to the same object of $\mathcal{C}$, the morphisms gf and hf are distinct. A morphism $f\colon X \to Y$ in $\mathcal{C}$ is a *monomorphism* if for any distinct morphisms g and h from the same object of $\mathcal{C}$ to X, the morphisms fg and fh are distinct. It is easy to see that all retractions in $\mathcal{C}$ are epimorphisms and all sections in $\mathcal{C}$ are monomorphisms.

7.3.2 Free and projective modules

Let $T = (T, \mu, \eta)$ be a monad on a category $\mathcal{C}$. A T-module is *free* if it is isomorphic in the category $\mathcal{C}^T$ of T-modules to the T-module $F_T(X) = (T(X), \mu_X)$ for some $X \in \mathrm{Ob}(\mathcal{C})$. A T-module is *projective* if it is a retract of a free T-module in $\mathcal{C}^T$. The following lemma gives a useful characterization of projective T-modules.

Lemma 7.3. *A T-module (M, r) is projective if and only if the T-linear morphism $r\colon F_T(M) \to (M, r)$ is a retraction in $\mathcal{C}^T$.*

Proof. If r is a retraction in $\mathcal{C}^T$, then (M, r) is a retract of the free module $F_T(M)$, and so is projective. Conversely, assume that (M, r) is projective. Then (M, r) is a retract of a T-module isomorphic to $F_T(X)$ for some $X \in \mathrm{Ob}(\mathcal{C})$. This implies that (M, r) is a retract of $F_T(X)$. Let $p\colon T(X) \to M$ and $q\colon M \to T(X)$ be T-linear morphisms such that $pq = \mathrm{id}_M$. Set $s = T(p\eta_X)q\colon M \to T(M)$. Then

$$\mu_M T(s) \overset{\text{(i)}}{=} \mu_M T^2(p\eta_X)T(q) \overset{\text{(ii)}}{=} T(p\eta_X)\mu_X T(q) \overset{\text{(iii)}}{=} T(p\eta_X)qr \overset{\text{(iv)}}{=} sr.$$

Here, (i) and (iv) follow from the definition of s, (ii) from the naturality of μ, and (iii) from the T-linearity of q. Thus, s is T-linear. Also,

$$rs \overset{\text{(i)}}{=} rT(p)T(\eta_X)q \overset{\text{(ii)}}{=} p\mu_X T(\eta_X)q \overset{\text{(iii)}}{=} pq \overset{\text{(iv)}}{=} \mathrm{id}_M.$$

Here, (i) follows from the definition of s, (ii) from the T-linearity of p, (iii) from the unitality of μ, and (iv) from the assumptions. Thus, s is a section of r. Consequently, r is a retraction in $\mathcal{C}^T$. □

7.3.3 Semisimplicity and separability

A monad T is *semisimple* if all T-modules are projective. A monad T is *separable* if it is semisimple and T-linear sections of T-actions (see Lemma 7.3) can be chosen in a natural way. More precisely, a monad T on a category $\mathcal{C}$ is separable if there is a natural transformation

$$\nu = \{\nu_{(M,r)} \colon (M,r) \to (T(M), \mu_M)\}_{(M,r)\in \mathrm{Ob}(\mathcal{C}^T)}$$

from the identity functor $1_{\mathcal{C}^T}$ to the functor $F_T U_T$ such that $r\nu_{(M,r)} = \mathrm{id}_M$ for every T-module (M,r).

Lemma 7.4. *A monad $T = (T, \mu, \eta)$ on a category $\mathcal{C}$ is separable if and only if there is a natural transformation $\gamma = \{\gamma_X \colon X \to T^2(X)\}_{X\in \mathrm{Ob}(\mathcal{C})}$ such that*

$$T(\mu_X)\gamma_{T(X)} = \mu_{T(X)} T(\gamma_X) \quad \textit{and} \quad \mu_X \gamma_X = \eta_X$$

for all $X \in \mathrm{Ob}(\mathcal{C})$.

Proof. Assume that T is separable, and let ν be a natural transformation as in the definition of a separable monad. Consider the natural transformation

$$\gamma = \{\gamma_X = \nu_{F_T(X)}\eta_X \colon X \to T^2(X)\}_{X\in \mathrm{Ob}(\mathcal{C})} \colon 1_{\mathcal{C}} \to T^2.$$

Then for any $X \in \mathrm{Ob}(\mathcal{C})$, we have

$$\begin{aligned} T(\mu_X)\gamma_{T(X)} &\overset{\text{(i)}}{=} T(\mu_X)\nu_{F_T(T(X))}\eta_{T(X)} \overset{\text{(ii)}}{=} \nu_{F_T(X)}\mu_X \eta_{T(X)} \overset{\text{(iii)}}{=} \nu_{F_T(X)} \\ &\overset{\text{(iv)}}{=} \nu_{F_T(X)}\mu_X T(\eta_X) \overset{\text{(v)}}{=} \mu_{T(X)} T(\nu_{F_T(X)}) T(\eta_X) \overset{\text{(vi)}}{=} \mu_{T(X)} T(\gamma_X). \end{aligned}$$

Here, (i) and (vi) follow from the definition of γ, (ii) from the naturality of ν, (iii) and (iv) from the unitality of μ, and (v) from the T-linearity of ν. Also,

$$\mu_X \gamma_X \overset{\text{(i)}}{=} \mu_X \nu_{F_T(X)}\eta_X \overset{\text{(ii)}}{=} \eta_X.$$

Here, (i) follows from the definition of γ and (ii) from the fact $\nu_{F_T(X)}$ is a section of the action μ_X of $F_T(X)$.

Conversely, assume the existence of a natural transformation γ satisfying the conditions of the lemma. For a T-module (M,r), set

$$\nu_{(M,r)} = T(r)\gamma_M \colon M \to T(M).$$

Then

$$r\nu_{(M,r)} \overset{\text{(i)}}{=} rT(r)\gamma_M \overset{\text{(ii)}}{=} r\mu_M \gamma_M \overset{\text{(iii)}}{=} r\eta_M \overset{\text{(iv)}}{=} \mathrm{id}_M.$$

Here, (i) follows from the definition of $\nu_{(M,r)}$, (ii) and (iv) from the definition of a T-module, and (iii) from the assumptions on γ. Also,

$$\begin{aligned} \nu_{(M,r)} r &\overset{\text{(i)}}{=} T(r)\gamma_M r \overset{\text{(ii)}}{=} T(rT(r))\gamma_{T(M)} \overset{\text{(iii)}}{=} T(r)T(\mu_M)\gamma_{T(M)} \\ &\overset{\text{(iv)}}{=} T(r)\mu_{T(M)} T(\gamma_M) \overset{\text{(v)}}{=} \mu_M T(T(r)\gamma_M)) \overset{\text{(vi)}}{=} \mu_M T(\nu_{(M,r)}). \end{aligned}$$

Here, (i) and (vi) follow from the definition of $\nu_{(M,r)}$, (ii) from the naturality of γ, (iii) from the definition of a T-module, (iv) from the assumptions on γ, and (v) from the naturality of μ. Thus $\nu_{(M,r)}$ is a T-linear section of $r\colon F_T(M) \to (M,r)$. Finally, for any morphism $f\colon (M,r) \to (N,s)$ in $\mathcal{C}^T$, we have

$$\nu_{(N,s)}f \overset{\text{(i)}}{=} T(s)\gamma_N f \overset{\text{(ii)}}{=} T(sT(f))\gamma_M \overset{\text{(iii)}}{=} T(fr)\gamma_M \overset{\text{(iv)}}{=} T(f)\nu_{(M,r)}.$$

Here, (i) and (iv) follow from the definition of ν, (ii) from the naturality of γ, and (iii) from the T-linearity of f. Thus, $\nu = \{\nu_{(M,r)}\}_{(M,r)\in \mathrm{Ob}(\mathcal{C}^T)}$ is a natural transformation. □

For a separable monad T on $\mathcal{C}$, we can describe the morphisms of $\mathcal{C}^T$ as follows.

Lemma 7.5. *Let $T = (T,\mu,\eta)$ be a separable monad on a category $\mathcal{C}$, and let γ be as in Lemma 7.4. For any T-modules (M,r) and (N,s), the map*

$$\pi\colon \mathrm{Hom}_{\mathcal{C}}(M,N) \to \mathrm{Hom}_{\mathcal{C}}(M,N), \quad f \mapsto \pi(f) = sT(fr)\gamma_M$$

is a projector with image $\mathrm{Hom}_{\mathcal{C}^T}((M,r),(N,s))$.

Proof. For any $f \in \mathrm{Hom}_{\mathcal{C}}(M,N)$, we have

$$\begin{aligned}\pi(f)r &\overset{\text{(i)}}{=} sT(fr)\gamma_M r \overset{\text{(ii)}}{=} sT(frT(r))\gamma_{T(M)}\\ &\overset{\text{(iii)}}{=} sT(fr)T(\mu_M)\gamma_{T(M)} \overset{\text{(iv)}}{=} sT(fr)\mu_{T(M)}T(\gamma_M)\\ &\overset{\text{(v)}}{=} s\mu_N T(T(fr)\gamma_M) \overset{\text{(vi)}}{=} sT(s)T(T(fr)\gamma_M) \overset{\text{(vii)}}{=} sT(\pi(f)).\end{aligned}$$

Here, (i) and (vii) follow from the definition of π, (ii) from the naturality of γ, (iii) and (vi) from the definition of a T-module, (iv) from the properties of γ stated in Lemma 7.4, and (v) from the naturality of μ. Thus, $\pi(f) \in \mathrm{Hom}_{\mathcal{C}^T}((M,r),(N,s))$.

For any $f \in \mathrm{Hom}_{\mathcal{C}^T}((M,r),(N,s))$, we have

$$\begin{aligned}\pi(f) &\overset{\text{(i)}}{=} sT(fr)\gamma_M \overset{\text{(ii)}}{=} sT(s)T^2(f)\gamma_M\\ &\overset{\text{(iii)}}{=} sT(s)\gamma_N f \overset{\text{(iv)}}{=} s\mu_N\gamma_N f \overset{\text{(v)}}{=} s\eta_N f \overset{\text{(vi)}}{=} f.\end{aligned}$$

Here, (i) follows from the definition of π, (ii) from the T-linearity of f, (iii) from the naturality of γ, (iv) and (vi) from the definition of a T-module, and (v) from the properties of γ stated in Lemma 7.4. Hence, π is a projector onto $\mathrm{Hom}_{\mathcal{C}^T}((M,r),(N,s))$. □

7.3.4 Exercise*

Let $T = (T,\mu,\eta)$ be a monad on a category $\mathcal{C}$. Prove that the full subcategory $\mathcal{F}$ of $\mathcal{C}^T$ whose objects are free T-modules is equivalent to the *Kleisli category* $\mathcal{K}$ of T

defined as follows. The objects of $\mathcal{K}$ are the objects of $\mathcal{C}$. The set of morphisms in $\mathcal{K}$ from an object X to an object Y is $\mathrm{Hom}_{\mathcal{C}}(X, T(Y))$. The identity morphism of $X \in \mathrm{Ob}(\mathcal{K})$ is η_X. The composition of $f \in \mathrm{Hom}_{\mathcal{K}}(X, Y)$ and $g \in \mathrm{Hom}_{\mathcal{K}}(Y, Z)$ is the morphism

$$\mu_Z T(g) f \in \mathrm{Hom}_{\mathcal{C}}(X, T(Z)) = \mathrm{Hom}_{\mathcal{K}}(X, Z).$$

7.3.5 Exercise*

Show that separable monads generalize separable algebras. More precisely, consider an algebra A in a monoidal category $\mathcal{C}$ (see Section 6.1.1). Then A carries left and right actions of itself given by the product $m\colon A \otimes A \to A$ of A. The object $A \otimes A$ of $\mathcal{C}$ also carries a left action of A given by

$$m \otimes \mathrm{id}_A \colon A \otimes A \otimes A \to A \otimes A$$

and a right action of A given by

$$\mathrm{id}_A \otimes m \colon A \otimes A \otimes A \to A \otimes A.$$

Clearly, the morphism m is both left and right A-linear. The algebra A is said to be *separable* if m has a section which is both left and right A-linear. Prove that the algebra A is separable if and only if the monad $A\otimes ?$ from Example 7.1.4 is separable.

7.3.6 Exercise*

Prove that a right adjoint functor preserves monomorphisms and a left adjoint functor preserves epimorphisms.

7.4 Linear monads

We introduce and study linear monads on a $\Bbbk$-category.

7.4.1 Linear monads and adjunctions

A monad on a $\Bbbk$-category $\mathcal{C}$ is $\Bbbk$-*linear* if it is $\Bbbk$-linear as an endofunctor of $\mathcal{C}$ (see Section 4.1.1).

Lemma 7.6. *Let T be a $\Bbbk$-linear monad on a $\Bbbk$-category $\mathcal{C}$. Then $\mathcal{C}^T$ has a unique structure of a $\Bbbk$-category making the forgetful functor $U_T\colon \mathcal{C}^T \to \mathcal{C}$ $\Bbbk$-linear. Furthermore, the free module functor $F_T\colon \mathcal{C} \to \mathcal{C}^T$ is $\Bbbk$-linear.*

Proof. Let (M, r) and (N, s) be two T-modules. Since T is $\Bbbk$-linear and the composition in $\mathcal{C}$ is $\Bbbk$-bilinear, the set

$$\mathrm{Hom}_{\mathcal{C}^T}((M, r), (N, s)) = \{f \in \mathrm{Hom}_{\mathcal{C}}(M, N) \mid fr = sT(f)\}$$

is a $\Bbbk$-submodule of the $\Bbbk$-module $\mathrm{Hom}_{\mathcal{C}}(M, N)$. This turns $\mathrm{Hom}_{\mathcal{C}^T}((M,r),(N,s))$ into a $\Bbbk$-module. Then the composition in $\mathcal{C}^T$ is $\Bbbk$-bilinear (since it is the restriction of the composition in $\mathcal{C}$). Thus, $\mathcal{C}^T$ is a $\Bbbk$-category. The $\Bbbk$-linearity of the functors U_T and F_T is obvious. The uniqueness follows from the fact that if U_T is $\Bbbk$-linear, then the inclusion $\mathrm{Hom}_{\mathcal{C}^T}((M,r),(N,s)) \hookrightarrow \mathrm{Hom}_{\mathcal{C}}(M,N)$ is a $\Bbbk$-linear homomorphism. □

An adjunction between $\Bbbk$-categories is $\Bbbk$-*linear* if its underlying functors are $\Bbbk$-linear. For example, Lemma 7.6 implies that the adjunction of a $\Bbbk$-linear monad (see Section 7.2.2) is $\Bbbk$-linear. Conversely, the monad of a $\Bbbk$-linear adjunction is $\Bbbk$-linear (since the composition of $\Bbbk$-linear functors is $\Bbbk$-linear).

Lemma 7.7. *A left (respectively right) adjoint of a $\Bbbk$-linear functor is $\Bbbk$-linear.*

Proof. Let $\mathcal{C}$ and $\mathcal{D}$ be $\Bbbk$-categories and let $F\colon \mathcal{C} \rightleftarrows \mathcal{D} : U$ be an adjunction with unit η and counit ε. Assume that U is $\Bbbk$-linear. For any $\alpha \in \Bbbk$ and any morphisms $f, g\colon X \to Y$ in $\mathcal{C}$, we have:

$$
\begin{aligned}
&F(f+\alpha g) - F(f) - \alpha F(g)\\
&\overset{\text{(i)}}{=} \big(F(f+\alpha g) - F(f) - \alpha F(g)\big)\varepsilon_{F(X)}F(\eta_X)\\
&\overset{\text{(ii)}}{=} \varepsilon_{F(Y)}FU\big(F(f+\alpha g) - F(f) - \alpha F(g)\big)F(\eta_X)\\
&\overset{\text{(iii)}}{=} \varepsilon_{F(Y)}F\big(UF(f+\alpha g)\eta_X - UF(f)\eta_X - \alpha UF(g)\eta_X\big)\\
&\overset{\text{(iv)}}{=} \varepsilon_{F(Y)}F(\eta_Y)F\big((f+\alpha g) - f - \alpha g\big) \overset{\text{(v)}}{=} F(0).
\end{aligned}
$$

Here, (i) follows from (7.1), (ii) from the naturality of ε, (iii) from the $\Bbbk$-linearity of U and the $\Bbbk$-bilinearity of the composition in $\mathcal{C}$, (iv) from the naturality of η, and (v) from (7.1) and the $\Bbbk$-linearity of $\mathcal{C}$. For $\alpha = 0$ and $f = g = 0$, this computation gives $F(0) = 0$. Thus, for any α, f, g as above, $F(f + \alpha g) = F(f) + \alpha F(g)$. Hence, the functor F is $\Bbbk$-linear. A similar argument shows that if F is $\Bbbk$-linear, then so is U. □

Lemma 7.7 implies that an adjunction between $\Bbbk$-categories is $\Bbbk$-linear whenever one of its underlying functors is $\Bbbk$-linear.

7.4.2 Properties

We state several properties of linear monads. We use the standard notions of a (co)kernel of a morphism, of a functor preserving/creating (co)kernels, and of an abelian category. These notions are recalled in Appendix C.

Lemma 7.8. *Let T be a $\Bbbk$-linear monad on a $\Bbbk$-category $\mathcal{C}$ having a zero object. Then:*

(a) *The $\Bbbk$-category $\mathcal{C}^T$ has a zero object;*

(b) *The forgetful functor* $U_T \colon \mathcal{C}^T \to \mathcal{C}$ *creates kernels;*

(c) *If* T *preserves cokernels, then* U_T *creates cokernels;*

(d) *If* $\mathcal{C}$ *is additive, then* $\mathcal{C}^T$ *is additive;*

(e) *If* $\mathcal{C}$ *is abelian and* T *preserves cokernels, then* $\mathcal{C}^T$ *is abelian.*

Proof. Let μ and η be the product and unit of T, respectively. To prove (a), pick a zero object $\mathbf{0} \in \mathrm{Ob}(\mathcal{C})$. The zero morphism $0 \colon T(\mathbf{0}) \to \mathbf{0}$ is a T-action since

$$0T(0) = 0 = 0\mu_{T(0)} \quad \text{and} \quad 0\eta_0 = 0 = \mathrm{id}_{\mathbf{0}}.$$

Furthermore, $\mathrm{End}_{\mathcal{C}^T}((\mathbf{0},0)) \subset \mathrm{End}_{\mathcal{C}}(\mathbf{0}) = \{0\}$, and so $\mathrm{End}_{\mathcal{C}^T}((\mathbf{0},0)) = \{0\}$. Hence, $(\mathbf{0},0)$ is a zero object of $\mathcal{C}^T$.

Let us prove (b). Let $f \colon (M,r) \to (N,s)$ be a morphism in $\mathcal{C}^T$ such that the morphism $U_T(f) = f \colon M \to N$ has a kernel $q \colon K \to M$ in $\mathcal{C}$. We need to prove that there is a unique T-action $k \colon T(K) \to K$ such that q is a T-linear morphism from (K,k) to (M,r), and that this morphism is a kernel of f. The T-linearity of f and $\Bbbk$-linearity of T imply that

$$frT(q) = sT(f)T(q) = sT(fq) = sT(0) = s0 = 0.$$

Thus the morphism $rT(q)$ factors uniquely through the kernel q of f, i.e., there is a unique morphism $k \colon T(K) \to K$ in $\mathcal{C}$ such that $rT(q) = qk$. The fact that r is a T-action and the naturality of μ and η imply that

$$\begin{aligned} qkT(k) &= rT(q)T(k) = rT(qk) = rT(rT(q)) \\ &= rT(r)T^2(q) = r\mu_M T^2(q) = rT(q)\mu_K = qk\mu_K \end{aligned}$$

and

$$qk\eta_K = rT(q)\eta_K = r\eta_M q = \mathrm{id}_M q = q = q\,\mathrm{id}_K.$$

Since q, being a kernel, is a monomorphism, we deduce that $kT(k) = k\mu_K$ and $k\eta_K = \mathrm{id}_K$. Thus $k \colon T(K) \to K$ is a T-action, and the equality $rT(q) = qk$ means that q is a T-linear morphism from (K,k) to (M,r). If $k' \colon T(K) \to K$ is another T-action such that q is a T-linear morphism from (K,k') to (M,r), then $qk' = rT(q) = qk$, and so $k = k'$. It remains to show that $q \colon (K,k) \to (M,r)$ is a kernel of f. Clearly, $fq = 0$. Let $g \colon (Z,z) \to (M,r)$ be a morphism in $\mathcal{C}^T$ such that $fg = 0$ in $\mathcal{C}^T$. Then $fg = 0$ in $\mathcal{C}$ and so there is a unique morphism $h \colon Z \to K$ in $\mathcal{C}$ such that $g = qh$. The T-linearity of g and q implies that

$$qhz = gz = rT(g) = rT(qh) = rT(q)T(h) = qkT(h).$$

Since q is a monomorphism, we deduce that $hz = kT(h)$. Therefore, h is a T-linear morphism from (Z,z) to (K,k) and the equality $g = qh$ in $\mathcal{C}$ implies the same equality in $\mathcal{C}^T$. If $h' \colon (Z,z) \to (K,k)$ is a T-linear morphism satisfying $g = qh'$

in $\mathcal{C}^T$, then $g = qh'$ in $\mathcal{C}$, and so $h = h'$ by the uniqueness of h. We conclude that $q\colon (K,k) \to (M,r)$ is a kernel of f.

Let us prove (c). Let $f\colon (M,r) \to (N,s)$ be a morphism in $\mathcal{C}^T$ such that $U_T(f) = f\colon M \to N$ has a cokernel $p\colon N \to C$ in $\mathcal{C}$. We need to prove that there is a unique T-action $c\colon T(C) \to C$ such that p is a T-linear morphism from (N,s) to (C,c), and that this morphism is a cokernel of f. Since the functor T preserves cokernels, $T(p)$ is a cokernel of $T(f)$ and $T^2(p)$ is a cokernel of $T^2(f)$. In particular, the morphisms p, $T(p)$, $T^2(p)$ are epimorphisms. The T-linearity of f implies that

$$psT(f) = pfr = 0r = 0.$$

Thus, since $T(p)$ is a cokernel of $T(f)$, there is a unique morphism $c\colon T(C) \to C$ such that $ps = cT(p)$. The fact that s is a T-action and the naturality of μ and η imply that

$$\begin{aligned} cT(c)T^2(p) &= cT(cT(p)) = cT(ps) = cT(p)T(s) \\ &= psT(s) = ps\mu_N = cT(p)\mu_N = c\mu_C T^2(p) \end{aligned}$$

and

$$c\eta_C p = cT(p)\eta_N = ps\eta_N = p\,\mathrm{id}_N = p = \mathrm{id}_C\, p.$$

Since $T^2(p)$ and p are epimorphisms, we deduce that $cT(c) = c\mu_C$ and $c\eta_C = \mathrm{id}_C$. Thus $c\colon T(C) \to C$ is a T-action, and the equality $ps = cT(p)$ means that p is a T-linear morphism from (N,s) to (C,c). If $c'\colon T(C) \to C$ is another T-action such that p is a T-linear morphism from (N,s) to (C,c'), then $c' = c$ because

$$c'T(p) = ps = cT(p)$$

and $T(p)$ is an epimorphism. It remains to show that $p\colon (N,s) \to (C,c)$ is a cokernel of f. Clearly, $pf = 0$. Let $g\colon (N,s) \to (Z,z)$ be a morphism in $\mathcal{C}^T$ such that $gf = 0$ in $\mathcal{C}^T$. Then $gf = 0$ in $\mathcal{C}$ and so there exists a unique morphism $h\colon C \to Z$ in $\mathcal{C}$ such that $g = hp$. The T-linearity of p and g implies that

$$hcT(p) = hps = gs = zT(g) = zT(hp) = zT(h)T(p).$$

Since $T(p)$ is an epimorphism, we deduce that $hc = zT(h)$. Therefore, h is a T-linear morphism from (C,c) to (Z,z) and the equality $g = hp$ in $\mathcal{C}$ implies the same equality in $\mathcal{C}^T$. If $h'\colon (C,c) \to (Z,z)$ is a T-linear morphism satisfying $g = h'p$ in $\mathcal{C}^T$, then $g = h'p$ in $\mathcal{C}$, and so $h = h'$ by the uniqueness of h. We conclude that $p\colon (N,s) \to (C,c)$ is a cokernel of f.

Let us prove (d). Let $(M_\alpha, r_\alpha)_{\alpha\in\Lambda}$ be a finite family of T-modules. Set $M = \bigoplus_{\alpha\in\Lambda} M_\alpha \in \mathrm{Ob}(\mathcal{C})$ and consider the associated projections and inclusions $(p_\alpha\colon M \to M_\alpha, q_\alpha\colon M_\alpha \to M)_{\alpha\in\Lambda}$ such that

$$\mathrm{id}_M = \sum_{\alpha\in\Lambda} q_\alpha p_\alpha \quad \text{and} \quad p_\alpha q_\beta = \delta_{\alpha,\beta}\,\mathrm{id}_{M_\alpha}$$

for all $\alpha, \beta \in \Lambda$. Set

$$r = \sum_{\alpha \in \Lambda} q_\alpha r_\alpha T(p_\alpha)\colon T(M) \to M.$$

The fact that each r_α is a T-action and the naturality of μ and η imply that

$$\begin{aligned} rT(r) &= \sum_{\alpha,\beta \in \Lambda} q_\alpha r_\alpha T(p_\alpha) T(q_\beta) T(r_\beta) T^2(p_\beta) = \sum_{\alpha \in \Lambda} q_\alpha r_\alpha T(r_\alpha) T^2(p_\alpha) \\ &= \sum_{\alpha \in \Lambda} q_\alpha r_\alpha \mu_{M_\alpha} T^2(p_\alpha) = \sum_{\alpha \in \Lambda} q_\alpha r_\alpha T(p_\alpha) \mu_M = r\mu_M \end{aligned}$$

and

$$r\eta_M = \sum_{\alpha \in \Lambda} q_\alpha r_\alpha T(p_\alpha) \eta_M = \sum_{\alpha \in \Lambda} q_\alpha r_\alpha \eta_{M_\alpha} p_\alpha = \sum_{\alpha \in \Lambda} q_\alpha p_\alpha = \mathrm{id}_M.$$

Thus (M, r) is a T-module. Furthermore, for any $\alpha \in \Lambda$,

$$p_\alpha r = \sum_{\beta \in \Lambda} p_\alpha q_\beta r_\beta T(p_\beta) = r_\alpha T(p_\alpha)$$

and

$$rT(q_\alpha) = \sum_{\beta \in \Lambda} q_\beta r_\beta T(p_\beta q_\alpha) = q_\alpha r_\alpha.$$

Thus p_α and q_α are T-linear. It follows that $(M, r) = \bigoplus_{\alpha \in \Lambda} (M_\alpha, r_\alpha)$ in $\mathcal{C}^T$. Hence the category $\mathcal{C}^T$ is additive.

We now prove (e). By Lemma 7.6 and (d), $\mathcal{C}^T$ is an additive $\Bbbk$-category and U_T is $\Bbbk$-linear. Since $\mathcal{C}$ is abelian, kernels and cokernels exist in $\mathcal{C}$. By (b) and (c), U_T creates kernels and cokernels. Lemma C.2(e)–(f) implies that kernels and cokernels exist in $\mathcal{C}^T$ and that U_T preserves them. Then, by Lemma C.2(c)–(d), the functor U_T preserves monomorphisms and epimorphisms.

Let us prove that in $\mathcal{C}^T$, any monomorphism $f\colon (M, r) \to (N, s)$ is a kernel of a morphism. Since U_T preserves monomorphisms, $f = U_T(f)\colon M \to N$ is a monomorphism in $\mathcal{C}$. Let $p\colon N \to C$ be a cokernel of $f\colon M \to N$ in $\mathcal{C}$. Since U_T creates cokernels, there is a (unique) T-action $c\colon T(C) \to C$ such that p is a T-linear morphism from (N, s) to (C, c) and is a cokernel of f in $\mathcal{C}^T$. Since $f\colon M \to N$ is a monomorphism and $\mathcal{C}$ is abelian, Lemma C.3(a) implies that $f\colon M \to N$ is a kernel of its cokernel $p\colon N \to C$. Since U_T creates kernels, we obtain that there is a (unique) T-action $r'\colon T(M) \to M$ such that f is a T-linear morphism from (M, r') to (N, s) and is a kernel of $p\colon (N, s) \to (C, c)$ in $\mathcal{C}^T$. The T-linearity of f implies that

$$fr' = sT(f) = fr.$$

Thus $r = r'$ since f is a monomorphism in $\mathcal{C}$. We conclude that $f\colon (M, r) \to (N, s)$ is a kernel of $p\colon (N, s) \to (C, c)$ in $\mathcal{C}^T$.

It remains to prove that in $\mathcal{C}^T$, any epimorphism $f\colon (M,r) \to (N,s)$ is a cokernel of a morphism. Since U_T preserves epimorphisms, $f = U_T(f)\colon M \to N$ is an epimorphism in $\mathcal{C}$. By assumption, T preserves cokernels and so Lemma C.2(b) implies that $T(f)\colon T(M) \to T(N)$ is an epimorphism in $\mathcal{C}$. Next, let $q\colon K \to M$ be a kernel of $f\colon M \to N$ in $\mathcal{C}$. Since U_T creates kernels, there is a (unique) T-action $k\colon T(K) \to K$ such that q is a T-linear morphism from (K,k) to (M,r) and is a kernel of f in $\mathcal{C}^T$. Since $f\colon M \to N$ is an epimorphism in $\mathcal{C}$ and $\mathcal{C}$ is abelian, Lemma C.3(b) implies that $f\colon M \to N$ is a cokernel in $\mathcal{C}$ of its kernel $q\colon K \to M$. Since U_T creates cokernels, there is a (unique) T-action $s'\colon T(N) \to N$ such that f is a T-linear morphism from (M,r) to (N,s') and is a cokernel of $q\colon (K,k) \to (M,r)$ in $\mathcal{C}^T$. The T-linearity of f implies that

$$sT(f) = fr = s'T(f).$$

Since $T(f)$ is an epimorphism in $\mathcal{C}$, we deduce that $s = s'$. We conclude that $f\colon (M,r) \to (N,s)$ is a cokernel of $q\colon (K,k) \to (M,r)$ in $\mathcal{C}^T$. □

7.4.3 Exercise*

Let (M,r) be a module over a $\Bbbk$-linear monad T on a $\Bbbk$-category $\mathcal{C}$. Prove that if M is simple, then so is (M,r).

7.5 Comonoidal functors and bimonads

We introduce comonoidal functors and use them to define bimonads.

7.5.1 Comonoidal functors and natural transformations

A comonoidal functor between monoidal categories $\mathcal{C} = (\mathcal{C}, \otimes, \mathbb{1})$ and $\mathcal{D} = (\mathcal{D}, \otimes, \mathbb{1})$ is a monoidal functor from $\mathcal{C}^{\mathrm{op}} = (\mathcal{C}^{\mathrm{op}}, \otimes, \mathbb{1})$ to $\mathcal{D}^{\mathrm{op}} = (\mathcal{D}^{\mathrm{op}}, \otimes, \mathbb{1})$. For convenience of the reader, we give here a more detailed definition. A *comonoidal functor* from $\mathcal{C}$ to $\mathcal{D}$ is a functor $F\colon \mathcal{C} \to \mathcal{D}$ endowed with a morphism $F_0\colon F(\mathbb{1}) \to \mathbb{1}$ in $\mathcal{D}$ and a natural transformation

$$F_2 = \{F_2(X,Y)\colon F(X \otimes Y) \to F(X) \otimes F(Y)\}_{X,Y \in \mathrm{Ob}(\mathcal{C})}$$

between the functors $F\otimes\colon \mathcal{C} \times \mathcal{C} \to \mathcal{D}$ and $F \otimes F = \otimes(F \times F)\colon \mathcal{C} \times \mathcal{C} \to \mathcal{D}$ such that for all $X, Y, Z \in \mathrm{Ob}(\mathcal{C})$, the following two diagrams commute:

$$\begin{array}{ccc} F(X \otimes Y \otimes Z) & \xrightarrow{F_2(X,Y\otimes Z)} & F(X) \otimes F(Y \otimes Z) \\ {\scriptstyle F_2(X\otimes Y,Z)}\downarrow & & \downarrow{\scriptstyle \mathrm{id}_{F(X)}\otimes F_2(Y,Z)} \\ F(X \otimes Y) \otimes F(Z) & \xrightarrow[F_2(X,Y)\otimes \mathrm{id}_{F(Z)}]{} & F(X) \otimes F(Y) \otimes F(Z), \end{array} \tag{7.3}$$

$$
\begin{array}{ccc}
F(X) & \xrightarrow{F_2(X,\mathbb{1})} & F(X)\otimes F(\mathbb{1}) \\
{\scriptstyle F_2(\mathbb{1},X)}\downarrow & \searrow^{\mathrm{id}_{F(X)}} & \downarrow{\scriptstyle \mathrm{id}_{F(X)}\otimes F_0} \\
F(\mathbb{1})\otimes F(X) & \xrightarrow[F_0\otimes \mathrm{id}_{F(X)}]{} & F(X).
\end{array}
\tag{7.4}
$$

Recall that the naturality of F_2 means that for arbitrary morphisms $f\colon X \to X'$ and $g\colon Y \to Y'$ in $\mathcal{C}$, the following diagram commutes:

$$
\begin{array}{ccc}
F(X\otimes Y) & \xrightarrow{F_2(X,Y)} & F(X)\otimes F(Y) \\
{\scriptstyle F(f\otimes g)}\downarrow & & \downarrow{\scriptstyle F(f)\otimes F(g)} \\
F(X'\otimes Y') & \xrightarrow[F_2(X',Y')]{} & F(X')\otimes F(Y').
\end{array}
$$

The morphism F_0 and the natural transformation F_2 are the *comonoidal constraints* of F. A comonoidal functor (F, F_2, F_0) is *strong* (respectively, *strict*) if F_2 and F_0 are isomorphisms (respectively, identities). Then (F, F_2^{-1}, F_0^{-1}) is a strong (respectively, strict) monoidal functor in the sense of Section 1.4.1.

The composition $GF\colon \mathcal{C} \to \mathcal{E}$ of two comonoidal functors $F\colon \mathcal{C} \to \mathcal{D}$ and $G\colon \mathcal{D} \to \mathcal{E}$ is a comonoidal functor with

$$(GF)_0 = G_0\, G(F_0) \quad \text{and} \quad (GF)_2(X,Y) = G_2(F(X),F(Y))\, G(F_2(X,Y))$$

for all $X, Y \in \mathrm{Ob}(\mathcal{C})$.

A natural transformation φ from a comonoidal functor $F\colon \mathcal{C} \to \mathcal{D}$ to a comonoidal functor $G\colon \mathcal{C} \to \mathcal{D}$ is *comonoidal* if

$$G_0\varphi_{\mathbb{1}} = F_0 \quad \text{and} \quad G_2(X,Y)\varphi_{X\otimes Y} = (\varphi_X \otimes \varphi_Y)F_2(X,Y)$$

for all $X, Y \in \mathrm{Ob}(\mathcal{C})$. Note that the composition of comonoidal natural transformations is a comonoidal natural transformation.

7.5.2 Bimonads

A *bimonad* on a monoidal category $\mathcal{C} = (\mathcal{C}, \otimes, \mathbb{1})$ is a monad (T, μ, η) on $\mathcal{C}$ such that the functor $T\colon \mathcal{C} \to \mathcal{C}$ is comonoidal and the natural transformations $\mu\colon T^2 \to T$ and $\eta\colon 1_{\mathcal{C}} \to T$ are comonoidal. The comonoidality of the functor T means that it is endowed with a morphism $T_0\colon T(\mathbb{1}) \to \mathbb{1}$ in $\mathcal{C}$ and a natural transformation

$$T_2 = \{T_2(X,Y)\colon T(X\otimes Y) \to T(X)\otimes T(Y)\}_{X,Y\in \mathrm{Ob}(\mathcal{C})}$$

such that the diagrams obtained by substituting $F = T$ in (7.3) and (7.4) commute for all $X, Y, Z \in \mathrm{Ob}(\mathcal{C})$. Using the definition of the comonoidal constraints of T^2

(see Section 7.5.1), the comonoidality of μ means that the diagram

$$\begin{array}{ccccc} T^2(X\otimes Y) & \xrightarrow{\mu_{X\otimes Y}} & T(X\otimes Y) & \xrightarrow{T_2(X,Y)} & T(X)\otimes T(Y) \\ {\scriptstyle T(T_2(X,Y))}\downarrow & & & & \uparrow{\scriptstyle \mu_X\otimes\mu_Y} \\ T(T(X)\otimes T(Y)) & & \xrightarrow[T_2(T(X),T(Y))]{} & & T^2(X)\otimes T^2(Y) \end{array}$$

commutes for all $X, Y \in \mathrm{Ob}(\mathcal{C})$ and that the diagram

$$\begin{array}{ccc} T^2(\mathbb{1}) & \xrightarrow{\mu_{\mathbb{1}}} & T(\mathbb{1}) \\ {\scriptstyle T(T_0)}\downarrow & & \downarrow{\scriptstyle T_0} \\ T(\mathbb{1}) & \xrightarrow[T_0]{} & \mathbb{1} \end{array}$$

commutes. The comonoidality of η means that the diagram

$$\begin{array}{ccc} & T(X\otimes Y) & \\ {\scriptstyle \eta_{X\otimes Y}}\nearrow & & \searrow{\scriptstyle T_2(X,Y)} \\ X\otimes Y & \xrightarrow[\eta_X\otimes\eta_Y]{} & T(X)\otimes T(Y) \end{array}$$

commutes for all $X, Y \in \mathrm{Ob}(\mathcal{C})$ and that the diagram

$$\begin{array}{ccc} & T(\mathbb{1}) & \\ {\scriptstyle \eta_{\mathbb{1}}}\nearrow & & \searrow{\scriptstyle T_0} \\ \mathbb{1} & \xrightarrow[\mathrm{id}_{\mathbb{1}}]{} & \mathbb{1} \end{array}$$

commutes.

The notion of a bimonad is motivated by the following fact due to Ieke Moerdijk (see [Mo, Theorem 7.1]): for any monad $T = (T, \mu, \eta)$ on a monoidal category $\mathcal{C}$, there is a bijective correspondence between:

(i) bimonad structures on T;

(ii) monoidal structures on the category $\mathcal{C}^T$ of T-modules such that the forgetful functor $U_T : \mathcal{C}^T \to \mathcal{C}$ is strict monoidal.

Indeed, a bimonad structure on T determines a monoidal product on $\mathcal{C}^T$:

$$(M, r) \otimes (N, s) = \big(M \otimes N, (r \otimes s)T_2(M, N)\big)$$

for any T-modules $(M, r), (N, s)$, and the monoidal product of two morphisms in $\mathcal{C}^T$ is their monoidal product in $\mathcal{C}$. This turns $\mathcal{C}^T$ into a monoidal category with monoidal unit $\mathbb{1}_{\mathcal{C}^T} = (\mathbb{1}, T_0)$ and with monoidal constraints induced by that of $\mathcal{C}$ in the obvious way. Clearly, the functor $U_T \colon \mathcal{C}^T \to \mathcal{C}$ is strict monoidal.

Conversely, a monoidal structure on $\mathcal{C}^T$ such that U_T is strict monoidal determines a bimonad structure on T as follows. The strict monoidality of U_T implies that $\mathbb{1}_{\mathcal{C}^T} = (\mathbb{1}, r_0)$ and

$$(T(X), \mu_X) \otimes (T(Y), \mu_Y) = (T(X) \otimes T(Y), r_{X,Y})$$

for all $X, Y \in \mathrm{Ob}(\mathcal{C})$, where r_0 and $r_{X,Y}$ are actions of T. Then the comonoidal constraints of T are given by $T_0 = r_0$ and

$$T_2 = \{T_2(X,Y) = r_{X,Y} T(\eta_X \otimes \eta_Y) \colon T(X \otimes Y) \to T(X) \otimes T(Y)\}_{X,Y \in \mathrm{Ob}(\mathcal{C})}.$$

A bimonad is $\Bbbk$-*linear* (respectively, *semisimple*, *separable*) if it is $\Bbbk$-linear (respectively, semisimple, separable) as a monad.

Lemma 7.9. *Let T be a monad on a monoidal $\Bbbk$-category $\mathcal{C}$. The Moerdijk bijective correspondence above induces a bijective correspondence between:*

(i) *$\Bbbk$-linear bimonad structures on T;*

(ii) *$\Bbbk$-linear monoidal structures on $\mathcal{C}^T$ such that U_T is $\Bbbk$-linear strict monoidal.*

Proof. Assume that $\mathcal{C}^T$ has a structure of a monoidal $\Bbbk$-category such that U_T is $\Bbbk$-linear strict monoidal. By the Moerdijk correspondence, the monad T is a bimonad. By Lemma 7.7, the free module functor $F_T \colon \mathcal{C} \to \mathcal{C}^T$, being a left adjoint of U_T, is $\Bbbk$-linear. Consequently, $T = U_T F_T$ is $\Bbbk$-linear. Conversely, consider a $\Bbbk$-linear bimonad structure on T. We endow $\mathcal{C}^T$ with its unique structure of a monoidal category making U_T strict monoidal (provided by the above correspondence) and with its unique structure of a $\Bbbk$-category making U_T $\Bbbk$-linear (provided by Lemma 7.6). The $\Bbbk$-bilinearity of the monoidal product of morphisms in $\mathcal{C}$ and the fact that U_T is $\Bbbk$-linear and acts as the identity on morphisms imply that the monoidal product of morphisms in $\mathcal{C}^T$ is $\Bbbk$-bilinear. Hence, $\mathcal{C}^T$ is a monoidal $\Bbbk$-category and U_T is $\Bbbk$-linear strict monoidal. Clearly, these constructions yield mutually inverse bijections. □

7.5.3 Example

For a monoidal category $\mathcal{C}$, the trivial monad $1_{\mathcal{C}} \colon \mathcal{C} \to \mathcal{C}$ (see Section 7.1.2) endowed with the identity comonoidal constraints is a bimonad on $\mathcal{C}$. It is called the *trivial bimonad*.

7.5.4 Example

Let A be a bialgebra in a braided category $\mathcal{C}$. The monad $T = A\otimes ?$ from Example 7.1.4 is a bimonad on $\mathcal{C}$ with comonoidal constraints

$T_0 =$  and $T_2(X,Y) =$ 

for any $X, Y \in \mathrm{Ob}(\mathcal{C})$. Furthermore, $\mathcal{C}^T = \mathrm{mod}_A^{\mathcal{C}}$ as monoidal categories and $U_T = \mathcal{U}_A : \mathrm{mod}_A^{\mathcal{C}} \to \mathcal{C}$ (see Section 6.1.3).

7.5.5 Example

Let $((A,\sigma), m, u, \Delta, \varepsilon)$ be a bialgebra in the center $\mathcal{Z}(\mathcal{C})$ of a monoidal category $\mathcal{C}$. Here (A,σ) is a half-braiding of $\mathcal{C}$ and $m, u, \Delta, \varepsilon$ are respectively the product, the unit, the coproduct, and the counit of the bialgebra. By Example 7.5.4, this bialgebra defines a bimonad on $\mathcal{Z}(\mathcal{C})$. It also defines a bimonad $T = A\otimes_\sigma ?$ on $\mathcal{C}$ as follows. As a functor from $\mathcal{C}$ to itself, T is given by

$$T(X) = A \otimes X \quad \text{and} \quad T(f) = \mathrm{id}_A \otimes f$$

for any $X \in \mathrm{Ob}(\mathcal{C})$ and any morphism f in $\mathcal{C}$. The comonoidal constraints of T are given by

$T_0 = \varepsilon$ and $T_2(X,Y) = (\mathrm{id}_A \otimes \sigma_X)(\Delta \otimes \mathrm{id}_X) \otimes \mathrm{id}_Y =$ 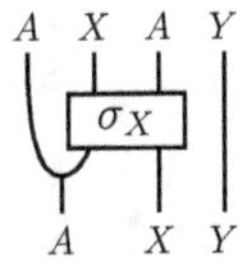

for any $X, Y \in \mathrm{Ob}(\mathcal{C})$. The product μ and the unit η of T are given by

$\mu_X = m \otimes \mathrm{id}_X =$ 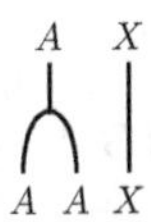

and $\eta_X = u \otimes \mathrm{id}_X =$ 

for any $X \in \mathrm{Ob}(\mathcal{C})$. This construction generalizes Example 7.5.4: if $\mathcal{C}$ is braided with braiding τ and if $(A, m, u, \Delta, \varepsilon)$ is a bialgebra in $\mathcal{C}$, then $((A, \tau_{A,-}), m, u, \Delta, \varepsilon)$ is a bialgebra in $\mathcal{Z}(\mathcal{C})$ and $A\otimes_{\tau_{A,-}} ? = A\otimes ?$ as bimonads on $\mathcal{C}$.

7.5.6 Remark

If a monoidal category $\mathcal{C}$ is small, then we can consider the strict monoidal category $\mathrm{End}_c(\mathcal{C})$ of comonoidal endofunctors of $\mathcal{C}$. The objects of $\mathrm{End}_c(\mathcal{C})$ are the comonoidal endofunctors of $\mathcal{C}$. The morphisms of $\mathrm{End}_c(\mathcal{C})$ are the comonoidal natural transformations (with the obvious composition). The monoidal product of the objects is given by the composition of comonoidal endofunctors. The monoidal product of the morphisms is defined in the same way as in $\mathrm{End}(\mathcal{C})$, see Section 7.1.5. It is easy to check that a bimonad on $\mathcal{C}$ is nothing but an algebra in the monoidal category $\mathrm{End}_c(\mathcal{C})$.

7.5.7 Exercise

Let $T = (T, \mu, \eta)$ be a bimonad on a monoidal category $\mathcal{C} = (\mathcal{C}, \otimes, \mathbb{1})$. Consider the opposite monoidal category $\mathcal{C}^{\otimes\mathrm{op}} = (\mathcal{C}, \otimes^{\mathrm{op}}, \mathbb{1})$ and the opposite comonoidal functor $T^{\otimes\mathrm{op}} \colon \mathcal{C}^{\otimes\mathrm{op}} \to \mathcal{C}^{\otimes\mathrm{op}}$ which is the same functor T with comonoidal constraints defined for all $X, Y \in \mathrm{Ob}(\mathcal{C})$ by

$$(T^{\otimes\mathrm{op}})_2(X,Y) = T_2(Y,X) \quad \text{and} \quad (T^{\otimes\mathrm{op}})_0 = T_0.$$

Prove that $T^{\otimes\mathrm{op}} = (T^{\otimes\mathrm{op}}, \mu, \eta)$ is a bimonad on $\mathcal{C}^{\otimes\mathrm{op}}$.

7.5.8 Exercise

Let T and T' be bimonads on a category $\mathcal{C}$. A *bimonad morphism* $T \to T'$ is a monad morphism from T to T' (see Exercise 7.1.6) which is comonoidal. Prove that there is a bijective correspondence between bimonad morphisms $T \to T'$ and strict monoidal functors $F \colon \mathcal{C}^{T'} \to \mathcal{C}^T$ such that $U_T F = U_{T'}$ as monoidal functors.

7.5.9 Exercise

Consider monoidal categories $\mathcal{B}, \mathcal{C}, \mathcal{D}, \mathcal{E}$ and comonoidal functors

$$\mathcal{B} \xrightarrow{E} \mathcal{C} \underset{G}{\overset{F}{\rightrightarrows}} \mathcal{D} \xrightarrow{H} \mathcal{E}.$$

Let $\varphi = \{\varphi_X \colon F(X) \to G(X)\}_{X \in \mathrm{Ob}(\mathcal{C})}$ be a comonoidal transformation from F to G. Prove that the natural transformations

$$\varphi_E = \{\varphi_{E(B)} \colon FE(B) \to GE(B)\}_{B \in \mathrm{Ob}(\mathcal{B})} \colon FE \to GE$$

and

$$H(\varphi) = \{H(\varphi_X) \colon HF(X) \to HG(X)\}_{X \in \mathrm{Ob}(\mathcal{C})} \colon HF \to HG$$

are comonoidal, where the functors FE, GE, HF, HG are endowed with the composition comonoidal structure (see Section 7.5.1).

7.6 Bimonads versus adjunctions

By Section 7.2.2, monads are closely related to adjunctions. In this section, we introduce comonoidal adjunctions and discuss their relations with bimonads.

7.6.1 Comonoidal adjunctions

A *comonoidal adjunction* is an adjunction $F : \mathcal{C} \rightleftarrows \mathcal{D} : U$ between monoidal categories $\mathcal{C}, \mathcal{D}$ such that the functors F and U are comonoidal and both the unit $\eta\colon 1_{\mathcal{C}} \to UF$ and the counit $\varepsilon\colon FU \to 1_{\mathcal{D}}$ are comonoidal natural transformations. The following lemma describes the lifts of adjunctions to comonoidal adjunctions.

Lemma 7.10. *Let $F : \mathcal{C} \rightleftarrows \mathcal{D} : U$ be an adjunction between monoidal categories $\mathcal{C}$ and $\mathcal{D}$. Lifts of this adjunction to a comonoidal adjunction bijectively correspond to lifts of U to a strong monoidal functor $\mathcal{D} \to \mathcal{C}$.*

Proof. Let $\eta\colon 1_{\mathcal{C}} \to UF$ and $\varepsilon\colon FU \to 1_{\mathcal{D}}$ be the unit and counit of our adjunction. Assume that it is lifted to a comonoidal adjunction. It follows then from the definition of the comonoidal constraints of the composition of comonoidal functors (see Section 7.5.1), the definition of a comonoidal transformation, and (7.1) that the comonoidal constraints U_0 and U_2 are invertible with inverses

$$U_0^{-1} = U(F_0)\eta_{\mathbb{1}}$$

and, for all $X, Y \in \mathrm{Ob}(\mathcal{D})$,

$$U_2(X,Y)^{-1} = U(\varepsilon_X \otimes \varepsilon_Y)\, U\big(F_2(U(X), U(Y))\big)\, \eta_{U(X)\otimes U(Y)}.$$

Thus, (U, U_2, U_0) is a strong comonoidal functor $\mathcal{D} \to \mathcal{C}$ and (U, U_2^{-1}, U_0^{-1}) is a strong monoidal functor $\mathcal{D} \to \mathcal{C}$.

Conversely, assume that U is lifted to a strong monoidal functor (U, U_2, U_0). Then (U, U_2^{-1}, U_0^{-1}) is a strong comonoidal functor. Set

$$F_0 = \varepsilon_{\mathbb{1}} F(U_0) \quad \text{and} \quad F_2(X,Y) = \varepsilon_{FX\otimes FY}\, F\big(U_2(F(X), F(Y))\big)\, F(\eta_X \otimes \eta_Y)$$

for all $X, Y \in \mathrm{Ob}(\mathcal{C})$. It follows from the monoidality of U and (7.1) that (F, F_2, F_0) is comonoidal and that the natural transformations η and ε are comonoidal. This gives a lift of the adjunction $F : \mathcal{C} \rightleftarrows \mathcal{D} : U$ to a comonoidal adjunction.

One directly checks that the above two lifting procedures are inverse to each other. This concludes the proof of the lemma. □

7.6.2 From bimonads to comonoidal adjunctions and back

The adjunction $F_T \dashv U_T$ of any bimonad T (see Section 7.2.2) is a comonoidal adjunction because the functor U_T is strict monoidal and Lemma 7.10 applies. Since the strict monoidal functor U_T acts as the identity on morphisms, the formulas in the proof of Lemma 7.10 give that for all $X, Y \in \mathrm{Ob}(\mathcal{C})$,

$$(F_T)_0 = T_0 \quad \text{and} \quad (F_T)_2(X,Y) = T_2(X,Y).$$

Conversely, consider an arbitrary comonoidal adjunction $F : \mathcal{C} \rightleftarrows \mathcal{D} : U$. We turn the monad $T = UF$ into a comonoidal functor by providing it with the composition of the comonoidal structures of U and F, that is, by setting

$$T_0 = U_0 U(F_0) \quad \text{and} \quad T_2(X,Y) = U_2(F(X), F(Y))\, U(F_2(X,Y))$$

for all $X, Y \in \mathrm{Ob}(\mathcal{C})$. The product μ and counit η of T (defined in Section 7.2.2) are then comonoidal natural transformations (see Exercise 7.5.9). Thus, T is a bimonad on $\mathcal{C}$, called the *bimonad of the comonoidal adjunction* $F \dashv U$. Note that by Lemma 7.10, the comonoidal constraints U_2 and U_0 of U are invertible. Consequently, the comparison functor $K \colon \mathcal{D} \to \mathcal{C}^T$ is strong monoidal with monoidal constraints

$$K_0 = U_0^{-1} \quad \text{and} \quad K_2(X,Y) = U_2(X,Y)^{-1}$$

for all $X, Y \in \mathrm{Ob}(\mathcal{D})$. Clearly, $U_T K = U$ as monoidal functors and $KF = F_T$ as comonoidal functors.

7.7 Quasitriangular bimonads

We describe braidings on the categories of modules over bimonads in terms of R-matrices introduced in [BV1].

7.7.1 R-matrices

Let $T = (T, \mu, \eta)$ be a bimonad on a monoidal category $\mathcal{C}$. An R-*matrix* for T is a natural transformation

$$R = \{R_{X,Y} \colon X \otimes Y \to T(Y) \otimes T(X)\}_{X,Y \in \mathrm{Ob}(\mathcal{C})} \tag{7.5}$$

such that for all $X, Y, Z \in \mathrm{Ob}(\mathcal{C})$, the following four diagrams commute:

$$\begin{array}{ccc} T(X \otimes Y) & \xrightarrow{T(R_{X,Y})} & T(T(Y) \otimes T(X)) \\ {\scriptstyle T_2(X,Y)} \downarrow & & \downarrow {\scriptstyle T_2(T(Y),T(X))} \\ T(X) \otimes T(Y) & & T^2(Y) \otimes T^2(X) \\ {\scriptstyle R_{T(X),T(Y)}} \downarrow & & \downarrow {\scriptstyle \mu_Y \otimes \mu_X} \\ T^2(Y) \otimes T^2(X) & \xrightarrow[\mu_Y \otimes \mu_X]{} & T(Y) \otimes T(X), \end{array} \tag{7.6}$$

$$\begin{array}{ccc} X \otimes Y \otimes Z & \xrightarrow{R_{X,Y \otimes Z}} & T(Y \otimes Z) \otimes T(X) \\ & & \downarrow {\scriptstyle T_2(Y,Z) \otimes \mathrm{id}_{T(X)}} \\ {\scriptstyle R_{X,Y} \otimes \mathrm{id}_Z} \downarrow & & T(Y) \otimes T(Z) \otimes T(X) \\ & & \uparrow {\scriptstyle \mathrm{id}_{T(Y) \otimes T(Z)} \otimes \mu_X} \\ T(Y) \otimes T(X) \otimes Z & \xrightarrow[\mathrm{id}_{T(Y)} \otimes R_{T(X),Z}]{} & T(Y) \otimes T(Z) \otimes T^2(X), \end{array} \tag{7.7}$$

$$\begin{CD}
X \otimes Y \otimes Z @>{R_{X\otimes Y,Z}}>> T(Z) \otimes T(X \otimes Y) \\
@| @VV{\mathrm{id}_{T(Z)}\otimes T_2(X,Y)}V \\
@. T(Z) \otimes T(X) \otimes T(Y) \\
@V{\mathrm{id}_X \otimes R_{Y,Z}}VV @AA{\mu_Z \otimes \mathrm{id}_{T(X)\otimes T(Y)}}A \\
X \otimes T(Z) \otimes T(Y) @>>{R_{X,T(Z)}\otimes \mathrm{id}_{T(Y)}}> T^2(Z) \otimes T(X) \otimes T(Y),
\end{CD} \tag{7.8}$$

$$\begin{array}{ccc}
X & \xrightarrow{R_{\mathbb{1},X}} & T(X) \otimes T(\mathbb{1}) \\
{\scriptstyle R_{X,\mathbb{1}}}\downarrow & \searrow^{\eta_X} & \downarrow{\scriptstyle \mathrm{id}_{T(X)}\otimes T_0} \\
T(\mathbb{1}) \otimes T(X) & \xrightarrow[T_0 \otimes \mathrm{id}_{T(X)}]{} & T(X).
\end{array} \tag{7.9}$$

Here, the naturality of R means that for arbitrary morphisms $f\colon X \to X'$ and $g\colon Y \to Y'$ in $\mathcal{C}$, the following diagram commutes:

$$\begin{CD}
X \otimes Y @>{f\otimes g}>> X' \otimes Y' \\
@V{R_{X,Y}}VV @VV{R_{X',Y'}}V \\
T(Y) \otimes T(X) @>>{T(g)\otimes T(f)}> T(Y') \otimes T(X').
\end{CD}$$

A natural transformation (7.5) is said to be $*$-*invertible* if there exists a natural transformation

$$L = \{L_{X,Y}\colon X \otimes Y \to T(Y) \otimes T(X)\}_{X,Y \in \mathrm{Ob}(\mathcal{C})}$$

such that for all $X, Y \in \mathrm{Ob}(\mathcal{C})$, the following diagram commutes:

$$\begin{CD}
T(Y) \otimes T(X) @<{L_{X,Y}}<< X \otimes Y @>{R_{X,Y}}>> T(Y) \otimes T(X) \\
@V{R_{T(Y),T(X)}}VV @VV{\eta_X \otimes \eta_Y}V @VV{L_{T(Y),T(X)}}V \\
T^2(X) \otimes T^2(Y) @>>{\mu_X \otimes \mu_Y}> T(X) \otimes T(Y) @<<{\mu_X \otimes \mu_Y}< T^2(X) \otimes T^2(Y).
\end{CD}$$

Such an L is then unique (this is easily shown using the same arguments as in the proof of the uniqueness of the inverse of a non-degenerate pairing). It is called the $*$-*inverse* of R and is denoted R^{*-1}.

A bimonad endowed with a $*$-invertible R-matrix is said to be *quasitriangular*. The notions of an R-matrix and a quasitriangular bimonad are motivated by the following theorem.

Theorem 7.11. *Any $*$-invertible* R*-matrix R for a bimonad T on a monoidal category $\mathcal{C}$ determines a braiding τ in the monoidal category $\mathcal{C}^T$ of T-modules by*

$$\tau_{(M,r),(N,s)} = (s \otimes r)R_{M,N} \colon (M,r) \otimes (N,s) \to (N,s) \otimes (M,r)$$

for any T-modules (M,r) and (N,s). This induces a bijective correspondence between $*$-invertible R*-matrices for T and braidings in $\mathcal{C}^T$.*

Proof. Let μ and η be respectively the product and unit of T. We first prove that τ is a braiding in $\mathcal{C}^T$. Pick T-modules $(M,r),(N,s)$. We have:

$$\begin{aligned}
&\tau_{(M,r),(N,s)}(r \otimes s)T_2(M,N)\\
&\overset{\text{(i)}}{=} (s \otimes r)R_{M,N}(r \otimes s)T_2(M,N)\\
&\overset{\text{(ii)}}{=} (sT(s) \otimes rT(r))R_{T(M),T(N)}T_2(M,N)\\
&\overset{\text{(iii)}}{=} (s\mu_N \otimes r\mu_M)R_{T(M),T(N)}T_2(M,N)\\
&\overset{\text{(iv)}}{=} (s\mu_N \otimes r\mu_M)\, T_2(T(N),T(M))\, T(R_{M,N})\\
&\overset{\text{(v)}}{=} (sT(s) \otimes rT(r))\, T_2(T(N),T(M))\, T(R_{M,N})\\
&\overset{\text{(vi)}}{=} (s \otimes r)T_2(N,M)T((s \otimes r)R_{M,N})\\
&\overset{\text{(vii)}}{=} (s \otimes r)T_2(N,M)T(\tau_{(M,r),(N,s)}).
\end{aligned}$$

Here, (i) and (vii) follow from the definition of $\tau_{(M,r),(N,s)}$, (ii) from the naturality of R, (iii) and (v) from the fact that r and s are T-actions, (iv) from the commutativity of (7.6), and (vi) from the naturality of T_2. This proves that the morphism $\tau_{(M,r),(N,s)}$ is T-linear. To prove that it is invertible in $\mathcal{C}^T$, it suffices to prove that it is invertible in $\mathcal{C}$ (because the forgetful functor $\mathcal{C}^T \to \mathcal{C}$ reflects isomorphisms, see Section 7.1.3). To this end, set

$$\xi = (r \otimes s)R^{*-1}_{N,M} \colon N \otimes M \to M \otimes N.$$

Then

$$\begin{aligned}
\xi\,\tau_{(M,r),(N,s)} &\overset{\text{(i)}}{=} (r \otimes s)R^{*-1}_{N,M}(s \otimes r)R_{M,N}\\
&\overset{\text{(ii)}}{=} (rT(r) \otimes sT(s))R^{*-1}_{T(N),T(M)}R_{M,N}\\
&\overset{\text{(iii)}}{=} (r\mu_M \otimes s\mu_N)R^{*-1}_{T(N),T(M)}R_{M,N}\\
&\overset{\text{(iv)}}{=} r\eta_M \otimes s\eta_N \overset{\text{(v)}}{=} \mathrm{id}_M \otimes \mathrm{id}_N \overset{\text{(vi)}}{=} \mathrm{id}_{M\otimes N}.
\end{aligned}$$

Here, (i) follows from the definitions of $\tau_{(M,r),(N,s)}$ and ξ, (ii) from the naturality of R^{*-1}, (iii) and (v) from the fact that r and s are T-actions, (iv) from the definition of R^{*-1}, and (vi) from the functoriality of $\otimes$. One proves similarly that

$$\tau_{(M,r),(N,s)}\,\xi = \mathrm{id}_{N\otimes M}.$$

Thus, $\tau_{(M,r),(N,s)}$ is invertible with inverse ξ. Consequently,

$$\tau = \{\tau_{(M,r),(N,s)}\colon (M,r)\otimes(N,s)\to(N,s)\otimes(M,r)\}_{(M,r),(N,s)\in \mathrm{Ob}(\mathcal{C}^T)}$$

is a family of isomorphisms in $\mathcal{C}^T$. Moreover, for any morphisms $f\colon (M,r)\to(A,a)$ and $g\colon (N,s)\to(B,b)$ in $\mathcal{C}^T$, we have

$$\begin{aligned}(g\otimes f)\tau_{(M,r),(N,s)} &\overset{\text{(i)}}{=} (gs\otimes fr)R_{M,N} \overset{\text{(ii)}}{=} (bT(g)\otimes aT(f))R_{M,N}\\ &\overset{\text{(iii)}}{=} (b\otimes a)R_{A,B}(f\otimes g) \overset{\text{(iv)}}{=} \tau_{(A,a),(B,b)}(f\otimes g).\end{aligned}$$

Here, (i) and (iv) follow from the definition of τ, (ii) from the T-linearity of f and g, and (iii) from the naturality of R. Thus the family τ is natural. Now, for any T-modules $(M,r),(N,s),(P,t)$, we have:

$$\begin{aligned}\tau_{(M,r),(N,s)\otimes(P,t)} &\overset{\text{(i)}}{=} ((s\otimes t)T_2(N,P)\otimes r)R_{M,N\otimes P}\\ &\overset{\text{(ii)}}{=} (s\otimes t\otimes r\mu_M)(\mathrm{id}_{T(N)}\otimes R_{T(M),P})(R_{M,N}\otimes \mathrm{id}_P)\\ &\overset{\text{(iii)}}{=} (s\otimes t\otimes rT(r))(\mathrm{id}_{T(N)}\otimes R_{T(M),P})(R_{M,N}\otimes \mathrm{id}_P)\\ &\overset{\text{(iv)}}{=} (\mathrm{id}_N\otimes (t\otimes r)R_{M,P})((s\otimes r)R_{M,N}\otimes \mathrm{id}_P)\\ &\overset{\text{(v)}}{=} (\mathrm{id}_{(N,s)}\otimes \tau_{(M,r),(P,t)})(\tau_{(M,r),(N,s)}\otimes \mathrm{id}_{(P,t)}).\end{aligned}$$

Here, (i) and (v) follow from the definitions of τ and the monoidal product in $\mathcal{C}^T$, (ii) from the commutativity of (7.7), (iii) from the fact that r is a T-action, and (iv) from the naturality of R. Thus τ satisfies (3.1). Similarly, using (7.7), one proves that τ satisfies (3.2). Hence τ is a braiding in $\mathcal{C}^T$.

Conversely, let τ be an arbitrary braiding in $\mathcal{C}^T$. For any $X,Y\in \mathrm{Ob}(\mathcal{C})$, set

$$R_{X,Y} = \tau_{F_T(X),F_T(Y)}(\eta_X\otimes\eta_Y)\colon X\otimes Y\to T(Y)\otimes T(X).$$

The T-linearity of $\tau_{F_T(X),F_T(Y)}$ implies that $R=\{R_{X,Y}\}_{X,Y\in\mathrm{Ob}(\mathcal{C})}$ satisfies (7.6). Similarly, (3.1), (3.2), and (3.3) for τ applied to free T-modules imply that R satisfies (7.7), (7.8), and (7.9). Thus R is an R-matrix for T. The invertibility of τ implies that R is $*$-invertible with $*$-inverse

$$R^{*-1}_{X,Y} = \tau^{-1}_{F_T(Y),F_T(X)}(\eta_X\otimes\eta_Y)\colon X\otimes Y\to T(Y)\otimes T(X).$$

Finally, if τ is the braiding in $\mathcal{C}^T$ associated to a $*$-invertible R-matrix R for T, then the R-matrix associated with τ is computed for all $X,Y\in\mathrm{Ob}(\mathcal{C})$ by

$$\begin{aligned}\tau_{F_T(X),F_T(Y)}(\eta_X\otimes\eta_Y) &\overset{\text{(i)}}{=} (\mu_Y\otimes\mu_X)R_{T(X),T(Y)}(\eta_X\otimes\eta_Y)\\ &\overset{\text{(ii)}}{=} (\mu_Y T(\eta_Y)\otimes\mu_X T(\eta_X))R_{X,Y} \overset{\text{(iii)}}{=} R_{X,Y}.\end{aligned}$$

Here, (i) follows from the definition of τ, (ii) from the naturality of R, and (iii) from the unitality of μ. Conversely, if R is the R-matrix for T associated with a braiding τ in $\mathcal{C}^T$, then the braiding associated with R is given for all T-modules (M,r) and (N,s) by

$$\begin{aligned}(s\otimes r)R_{M,N} &\overset{\text{(i)}}{=} (s\otimes r)\tau_{F_T(M),F_T(N)}(\eta_M\otimes\eta_N)\\ &\overset{\text{(ii)}}{=} \tau_{(M,r),(N,s)}(r\eta_M\otimes s\eta_N) \overset{\text{(iii)}}{=} \tau_{(M,r),(N,s)}.\end{aligned}$$

Here, (i) follows from the definition of R, (ii) from the naturality of τ since $r\colon F_T(M)\to(M,r)$ and $s\colon F_T(N)\to(N,s)$ are T-linear morphisms, and (iii) from the fact that r and s are T-actions. Consequently, the above constructions yield mutually inverse bijections. □

7.7.2 Exercise

Prove that if R is a $*$-invertible R-matrix for a bimonad on a monoidal category, then R^{*-1} also is a $*$-invertible R-matrix for this bimonad. Prove that the braiding associated with R^{*-1} is the mirror of the braiding associated with R (see Example 3.1.7).

7.7.3 Exercise

Let T be a bimonad on a monoidal category $\mathcal{C}$ and let

$$R=\{R_{X,Y}\colon X\otimes Y\to T(Y)\otimes T(X)\}_{X,Y\in\mathrm{Ob}(\mathcal{C})}$$

be a $*$-invertible natural transformation satisfying (7.6), (7.7), and (7.8). Prove that R is an R-matrix for T, i.e., that the axiom (7.9) is also met.

7.7.4 Exercise

Let $T=(T,\mu,\eta)$ be a bimonad on a monoidal category $\mathcal{C}$ and let R be an R-matrix for T (not necessarily $*$-invertible). Prove that the family

$$\{\tau_{(M,r),(N,s)}=(s\otimes r)R_{M,N}\}_{(M,r),(N,s)\in\mathrm{Ob}(\mathcal{C}^T)}$$

is a weak braiding in $\mathcal{C}^T$ in the sense of Exercise 3.1.12. Deduce that R satisfies the following Yang–Baxter equation:

$$\begin{aligned}&(\mu_Z\otimes\mu_Y\otimes\mu_X)(R_{T(Y),T(Z)}\otimes\mathrm{id}_{T^2(X)})(\mathrm{id}_{T(Y)}\otimes R_{T(X),Z})(R_{X,Y}\otimes\mathrm{id}_Z)\\ &\quad=(\mu_Z\otimes\mu_Y\otimes\mu_X)(\mathrm{id}_{T^2(Z)}\otimes R_{T(X),T(Y)})(R_{X,T(Z)}\otimes\mathrm{id}_{T(Y)})(\mathrm{id}_X\otimes R_{Y,Z})\end{aligned}$$

for all $X, Y, Z \in \mathrm{Ob}(\mathcal{C})$. In the graphical form,

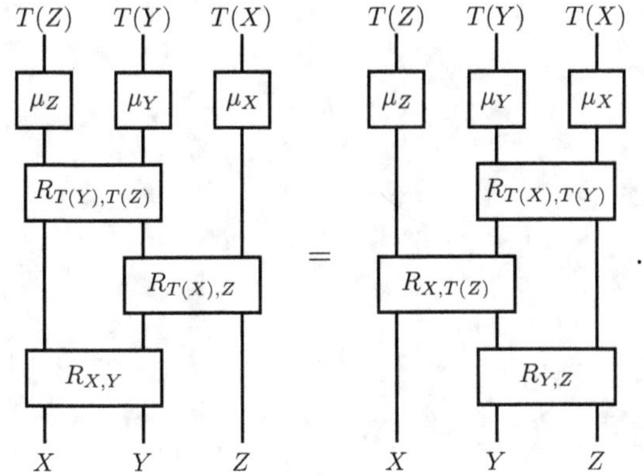

Chapter 8

Hopf monads

We present in this chapter the theory of Hopf monads introduced by the second named author jointly with Alain Bruguières and Steve Lack, see [BV1, BLV].

8.1 Fusion operators and Hopf monads

We define fusion operators of bimonads and then use these operators to define Hopf monads.

8.1.1 Fusion operators of a bimonad

Consider a bimonad $T = (T, \mu, \eta)$ on a monoidal category $\mathcal{C}$. For any $X, Y \in \mathrm{Ob}(\mathcal{C})$, set

$$H^l_{X,Y} = (\mathrm{id}_{T(X)} \otimes \mu_Y)\, T_2(X, T(Y)) =$$

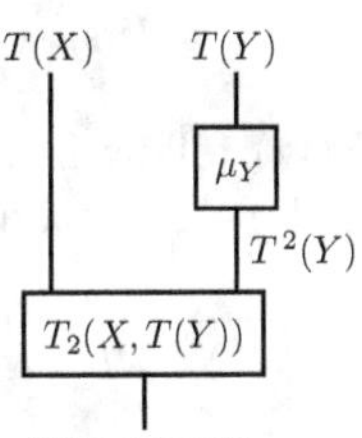

and

$$H^r_{X,Y} = (\mu_X \otimes \mathrm{id}_{T(Y)})\, T_2(T(X), Y) =$$

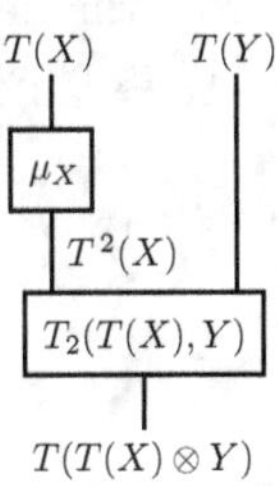

.

The naturality of T_2 and μ implies that the families

$$H^l = \{H^l_{X,Y} \colon T(X \otimes T(Y)) \to T(X) \otimes T(Y)\}_{X,Y \in \mathrm{Ob}(\mathcal{C})}$$

and

$$H^r = \{H^r_{X,Y} \colon T(T(X) \otimes Y) \to T(X) \otimes T(Y)\}_{X,Y \in \mathrm{Ob}(\mathcal{C})}$$

are natural transformations. They are called respectively the *left fusion operator* and the *right fusion operator* of T.

The next lemma yields several useful identities involving the left fusion operator H^l. Similar identities for the right fusion operator H^r may be deduced from this lemma using the opposite bimonad, see Exercise 8.1.7.

Lemma 8.1. *For any $X, Y, Z \in \mathrm{Ob}(\mathcal{C})$, we have:*

(a) $H^l_{X,Y} T(\mathrm{id}_X \otimes \mu_Y) = (\mathrm{id}_{T(X)} \otimes \mu_Y) H^l_{X,T(Y)}$;

(b) $H^l_{X,Y} T(\mathrm{id}_X \otimes \eta_Y) = T_2(X, Y)$;

(c) $(T_2(X,Y) \otimes \mathrm{id}_{T(Z)}) H^l_{X \otimes Y, Z} = (\mathrm{id}_{T(X)} \otimes H^l_{Y,Z}) T_2(X, Y \otimes T(Z))$;

(d) $(T_0 \otimes \mathrm{id}_{T(X)}) H^l_{\mathbb{1},X} = \mu_X$;

(e) $(\mathrm{id}_{T(X)} \otimes H^l_{Y,Z}) H^l_{X, Y \otimes T(Z)} = (H^l_{X,Y} \otimes \mathrm{id}_{T(Z)}) H^l_{X \otimes T(Y), Z} T(\mathrm{id}_X \otimes H^l_{Y,Z})$.

Proof. All the identities of the lemma follow from the associativity and unitality of μ (see Section 7.1.2) and from the comonoidality of T and μ (see Section 7.5.2). We first prove (a):

$$\begin{aligned} H^l_{X,Y} T(\mathrm{id}_X \otimes \mu_Y) &\overset{\text{(i)}}{=} (\mathrm{id}_{T(X)} \otimes \mu_Y) T_2(X, T(Y)) T(\mathrm{id}_X \otimes \mu_Y) \\ &\overset{\text{(ii)}}{=} (\mathrm{id}_{T(X)} \otimes \mu_Y T(\mu_Y)) T_2(X, T^2(Y)) \\ &\overset{\text{(iii)}}{=} (\mathrm{id}_{T(X)} \otimes \mu_Y \mu_{T(Y)}) T_2(X, T^2(Y)) \overset{\text{(iv)}}{=} (\mathrm{id}_{T(X)} \otimes \mu_Y) H^l_{X,T(Y)}. \end{aligned}$$

Here, (i) and (iv) follow from the definition of H^l, (ii) from the naturality of T_2, and (iii) from the associativity of μ. Next, we prove (b):

$$\begin{aligned} H^l_{X,Y} T(\mathrm{id}_X \otimes \eta_Y) &\overset{\text{(i)}}{=} (\mathrm{id}_{T(X)} \otimes \mu_Y) T_2(X, T(Y)) T(\mathrm{id}_X \otimes \eta_Y) \\ &\overset{\text{(ii)}}{=} (\mathrm{id}_{T(X)} \otimes \mu_Y T(\eta_Y)) T_2(X, Y) \overset{\text{(iii)}}{=} T_2(X, Y). \end{aligned}$$

Here, (i) follows from the definition of H^l, (ii) from the naturality of T_2, and (iii) from the unitality of μ. Now, we prove (c):

$$\begin{aligned} (T_2(X,Y) \otimes \mathrm{id}_{T(Z)}) H^l_{X \otimes Y, Z} &\overset{\text{(i)}}{=} (T_2(X,Y) \otimes \mu_Z) T_2(X \otimes Y, T(Z)) \\ &\overset{\text{(ii)}}{=} (\mathrm{id}_{T(X)} \otimes (\mathrm{id}_{T(Y)} \otimes \mu_Z) T_2(Y, Z)) T_2(X, Y \otimes T(Z)) \\ &\overset{\text{(iii)}}{=} (\mathrm{id}_{T(X)} \otimes H^l_{Y,Z}) T_2(X, Y \otimes T(Z)). \end{aligned}$$

Here, (i) and (iii) follow from the definition of H^l and (ii) from the coassociativity (7.3) of T_2. Formula (d) follows from the equalities

$$(T_0 \otimes \mathrm{id}_{T(X)})H^l_{\mathbb{1},X} \overset{\text{(i)}}{=} (T_0 \otimes \mu_X)T_2(\mathbb{1}, T(X)) \overset{\text{(ii)}}{=} \mu_X.$$

Here, (i) follows from the definition of H^l and (ii) from the counitality (7.4) of T_2. To prove (e), observe that

$$H^l_{Y,Z}\,\mu_{Y\otimes T(Z)} = (\mu_Y \otimes \mu_Z)T_2(T(Y), T(Z))T(H^l_{Y,Z}). \tag{8.1}$$

Indeed,

$$\begin{aligned}
H^l_{Y,Z}\,\mu_{Y\otimes T(Z)} &\overset{\text{(i)}}{=} (\mathrm{id}_{T(Y)} \otimes \mu_Z)T_2(Y, T(Z))\mu_{Y\otimes T(Z)}\\
&\overset{\text{(ii)}}{=} (\mu_Y \otimes \mu_Z\mu_{T(Z)})T_2(T(Y), T^2(Z))T(T_2(Y, T(Z)))\\
&\overset{\text{(iii)}}{=} (\mu_Y \otimes \mu_Z T(\mu_Z))T_2(T(Y), T^2(Z))T(T_2(Y, T(Z)))\\
&\overset{\text{(iv)}}{=} (\mu_Y \otimes \mu_Z)T_2(T(Y), T(Z))T(H^l_{Y,Z}).
\end{aligned}$$

Here, (i) follows from the definition of H^l, (ii) from the comonoidality of μ, (iii) from the associativity of μ, and (iv) from the naturality of T_2 and the definition of H^l. Then,

$$\begin{aligned}
&(\mathrm{id}_{T(X)} \otimes H^l_{Y,Z})H^l_{X,Y\otimes T(Z)}\\
&\overset{\text{(i)}}{=} (\mathrm{id}_{T(X)} \otimes H^l_{Y,Z}\,\mu_{Y\otimes T(Z)})T_2(X, T(Y \otimes T(Z)))\\
&\overset{\text{(ii)}}{=} (\mathrm{id}_{T(X)} \otimes (\mu_Y \otimes \mu_Z)T_2(T(Y), T(Z))T(H^l_{Y,Z}))T_2(X, T(Y \otimes T(Z)))\\
&\overset{\text{(iii)}}{=} (\mathrm{id}_{T(X)} \otimes \mu_Y \otimes \mu_Z)\big(\mathrm{id}_{T(X)} \otimes T_2(T(Y), T(Z))\big)\\
&\qquad \circ\, T_2(X, T(Y) \otimes T(Z)))\,T(\mathrm{id}_X \otimes H^l_{Y,Z})\\
&\overset{\text{(iv)}}{=} (\mathrm{id}_{T(X)} \otimes \mu_Y \otimes \mu_Z)\big(T_2(X, T(Y)) \otimes \mathrm{id}_{T^2(Z)}\big)\\
&\qquad \circ\, T_2(X \otimes T(Y), T(Z)))\,T(\mathrm{id}_X \otimes H^l_{Y,Z})\\
&\overset{\text{(v)}}{=} (H^l_{X,Y} \otimes \mathrm{id}_{T(Z)})H^l_{X\otimes T(Y),Z}T(\mathrm{id}_X \otimes H^l_{Y,Z}).
\end{aligned}$$

Here, (i) and (v) follow from the definition of H^l, (ii) from (8.1), (iii) from the naturality of T_2, and (iv) from the coassociativity (7.3) of T_2. This proves (e) and completes the proof of the lemma. □

8.1.2 Hopf monads

A *left* (respectively, *right*) *Hopf monad* on a monoidal category $\mathcal{C}$ is a bimonad on $\mathcal{C}$ whose left (respectively, right) fusion operator is invertible as a natural transformation. A *Hopf monad* on a monoidal category $\mathcal{C}$ is a bimonad on $\mathcal{C}$ whose left and right fusion operators are invertible.

8.1.3 Example

The trivial bimonad on a monoidal category (see Example 7.5.3) is a Hopf monad since its fusion operators are the identity natural transformations and are invertible. It is called the *trivial Hopf monad.*

8.1.4 Example

Let A be a bialgebra in a braided category $\mathcal{C}$ and let $T = A\otimes ?$ be the bimonad on $\mathcal{C}$ of Example 7.5.4. Then the left/right fusion operators of T are given for all $X, Y \in \mathrm{Ob}(\mathcal{C})$ by

$H^l_{X,Y} =$ 

and $H^r_{X,Y} =$ 

.

It follows from Exercise 6.2.7 that T is a Hopf monad if and only if A is a Hopf algebra. The inverses of the fusion operators are computed then by

$(H^l_{X,Y})^{-1} =$ 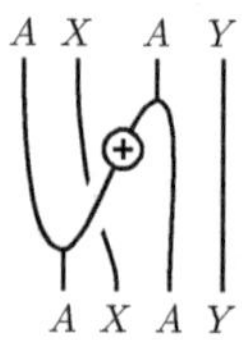

and $(H^r_{X,Y})^{-1} =$ 

for all $X, Y \in \mathrm{Ob}(\mathcal{C})$.

8.1.5 Example

The bimonad $A\otimes_\sigma ?$ on $\mathcal{C}$ in Example 7.5.5 is a Hopf monad if and only if the bialgebra $((A,\sigma), m, u, \Delta, \varepsilon)$ in $\mathcal{Z}(\mathcal{C})$ is a Hopf algebra. This generalizes Example 8.1.4. Indeed, if $\mathcal{C}$ is braided with braiding τ, then a bialgebra $(A, m, u, \Delta, \varepsilon)$ in $\mathcal{C}$ is a Hopf algebra if and only if the bialgebra $((A, \tau_{A,-}), m, u, \Delta, \varepsilon)$ in $\mathcal{Z}(\mathcal{C})$ is a Hopf algebra, and then $A\otimes_{\tau_{A,-}} ? = A\otimes ?$ as Hopf monads on $\mathcal{C}$.

8.1.6 Remark

Hopf monads arising from Hopf algebras as in Examples 8.1.4 and 8.1.5 are said to be *representable by Hopf algebras*. Not all Hopf monads arise in this way, see Example 9.2.4 below. We characterize Hopf monads representable by Hopf algebras in Appendix D.

8.1.7 Exercise

Let T be a bimonad on a monoidal category $\mathcal{C}$. Consider the opposite bimonad $T^{\otimes\mathrm{op}}$ from Exercise 7.5.7. Prove that the left (respectively, right) fusion operator of $T^{\otimes\mathrm{op}}$ is nothing but the right (respectively, left) fusion operator of T. Deduce that $T^{\otimes\mathrm{op}}$ is a left (respectively, right) Hopf monad if and only if T is a right (respectively, left) Hopf monad.

8.1.8 Exercise

Prove that a bimonad morphism between bimonads on a monoidal category (see Exercise 7.5.8) preserves both the left and the right fusion operators in the obvious sense.

8.2 Fusion operators and Hopf adjunctions

We study a class of comonoidal adjunctions (the so-called Hopf adjunctions) whose associated bimonads (see Section 7.6.2) are Hopf monads.

8.2.1 Fusion operators of a comonoidal adjunction

The *left and right fusion operators* of a comonoidal adjunction $F\colon \mathcal{C} \rightleftarrows \mathcal{D} : U$ are the natural transformations

$$\mathbb{H}^l = \{\mathbb{H}^l_{c,d} = (\mathrm{id}_{F(c)} \otimes \varepsilon_d)\, F_2(c, U(d))\colon F(c \otimes U(d)) \to F(c) \otimes d\}_{c\in\mathrm{Ob}(\mathcal{C}),d\in\mathrm{Ob}(\mathcal{D})},$$
$$\mathbb{H}^r = \{\mathbb{H}^r_{d,c} = (\varepsilon_d \otimes \mathrm{id}_{F(c)})\, F_2(U(d), c)\colon F(U(d) \otimes c) \to d \otimes F(c)\}_{c\in\mathrm{Ob}(\mathcal{C}),d\in\mathrm{Ob}(\mathcal{D})},$$

where ε is the counit of the adjunction.

8.2.2 Hopf adjunction

A *left* (respectively, *right*) *Hopf adjunction* is a comonoidal adjunction whose left (respectively, right) fusion operator is invertible. A *Hopf adjunction* is a comonoidal adjunction whose left and right fusion operators are invertible.

Lemma 8.2. *The bimonad of a left (respectively, right) Hopf adjunction is a left (respectively, right) Hopf monad. Consequently, the bimonad of a Hopf adjunction is a Hopf monad.*

Proof. We consider the left case; the right case is similar. Let $F\colon \mathcal{C} \rightleftarrows \mathcal{D} : U$ be a left Hopf adjunction and let $T = UF$ be the associated bimonad (see Section 7.6.2). The left fusion operator H^l of T is computed from the left fusion operator $\mathbb{H}^l$ of the adjunction as follows: for all $X, Y \in \mathrm{Ob}(\mathcal{C})$,

$$H^l_{X,Y} = U_2(F(X), F(Y))\, U(\mathbb{H}^l_{X,F(Y)}). \tag{8.2}$$

Indeed, denoting by μ the product of T and by ε the counit of $F \dashv U$, we obtain

$$\begin{aligned} H^l_{X,Y} &\overset{\text{(i)}}{=} (\mathrm{id}_{T(X)} \otimes \mu_Y)\, T_2(X, T(Y)) \\ &\overset{\text{(ii)}}{=} (\mathrm{id}_{UF(X)} \otimes U(\varepsilon_{F(Y)}))\, U_2(F(X), FUF(Y))\, U(F_2(X, UF(Y))) \\ &\overset{\text{(iii)}}{=} U_2(F(X), F(Y))\, U\big((\mathrm{id}_{F(X)} \otimes \varepsilon_{F(Y)}) F_2(X, UF(Y))\big) \\ &\overset{\text{(iv)}}{=} U_2(F(X), F(Y))\, U(\mathbb{H}^l_{X,F(Y)}). \end{aligned}$$

Here, (i) follows from the definition of H^l (see Section 8.1.1), (ii) the definitions of μ and of the comonoidal constraints T_2 of T (see Sections 7.2.2 and 7.6.2), (iii) from the naturality of U_2, and (iv) from the definition of $\mathbb{H}^l$ (see Section 8.2.1). Since U_2 is invertible by Lemma 7.10 and $\mathbb{H}^l$ is invertible by the assumptions, the identity (8.2) implies that H^l is invertible. Hence, T is a left Hopf monad. □

The next theorem characterizes Hopf monads in terms of Hopf adjunctions.

Theorem 8.3. *A bimonad T on a monoidal category is a left (respectively, right) Hopf monad if and only if the comonoidal adjunction $F_T \dashv U_T$ of T is a left (respectively, right) Hopf adjunction. Consequently, a bimonad T is a Hopf monad if and only if the comonoidal adjunction $F_T \dashv U_T$ is a Hopf adjunction.*

Proof. We consider the left case; the right case is similar. Since T is the bimonad of the comonoidal adjunction $F_T \dashv U_T$, the 'if' part of the theorem follows from Lemma 8.2. Conversely, let $T = (T, \mu, \eta)$ be a left Hopf monad on a monoidal category $\mathcal{C}$. Then the left fusion operator H^l of T is invertible. Consider the left fusion operator $\mathbb{H}^l$ of the comonoidal adjunction $F_T \dashv U_T$. Pick an object X of $\mathcal{C}$ and a T-module (M, r). It follows from the definitions of the counit of $F_T \dashv U_T$ (see Section 7.2.2) and the comonoidal structure of F_T (see Section 7.6.2) that

$$\mathbb{H}^l_{X,(M,r)} = (\mathrm{id}_{T(X)} \otimes r) T_2(X, M). \tag{8.3}$$

Let us prove that $\mathbb{H}^l_{X,(M,r)}$ is invertible with inverse

$$\mathbb{L}_{X,(M,r)} = T(\mathrm{id}_X \otimes r)(H^l_{X,M})^{-1}(\mathrm{id}_{T(X)} \otimes \eta_M) \colon T(X) \otimes M \to T(X \otimes M).$$

First, we have:

$$\begin{aligned} \mathbb{H}^l_{X,(M,r)} \mathbb{L}_{X,(M,r)} &\overset{\text{(i)}}{=} (\mathrm{id}_{T(X)} \otimes r) T_2(X, M) T(\mathrm{id}_X \otimes r)(H^l_{X,M})^{-1}(\mathrm{id}_{T(X)} \otimes \eta_M) \\ &\overset{\text{(ii)}}{=} (\mathrm{id}_{T(X)} \otimes rT(r)) T_2(X, T(M))(H^l_{X,M})^{-1}(\mathrm{id}_{T(X)} \otimes \eta_M) \\ &\overset{\text{(iii)}}{=} (\mathrm{id}_{T(X)} \otimes r\mu_M) T_2(X, T(M))(H^l_{X,M})^{-1}(\mathrm{id}_{T(X)} \otimes \eta_M) \\ &\overset{\text{(iv)}}{=} (\mathrm{id}_{T(X)} \otimes r) H^l_{X,M}(H^l_{X,M})^{-1}(\mathrm{id}_{T(X)} \otimes \eta_M) \\ &\overset{\text{(v)}}{=} \mathrm{id}_{T(X)} \otimes r\eta_M \overset{\text{(vi)}}{=} \mathrm{id}_{T(X) \otimes M}. \end{aligned}$$

Here, (i) follows from (8.3) and the definition of $\mathbb{L}_{X,(M,r)}$, (ii) from the naturality of T_2, (iii) and (vi) from the fact that r is a T-action, (iv) from the definition of H^l, and (v) from the invertibility of H^l. Second, we have:

$$\begin{aligned}
\mathbb{L}_{X,(M,r)}\,\mathbb{H}^l_{X,(M,r)} &\overset{\text{(i)}}{=} T(\mathrm{id}_X \otimes r)(H^l_{X,M})^{-1}(\mathrm{id}_{T(X)} \otimes \eta_M r)T_2(X,M)\\
&\overset{\text{(ii)}}{=} T(\mathrm{id}_X \otimes r)(H^l_{X,M})^{-1}(\mathrm{id}_{T(X)} \otimes T(r)\eta_{T(M)})T_2(X,M)\\
&\overset{\text{(iii)}}{=} T(\mathrm{id}_X \otimes rT(r))(H^l_{X,T(M)})^{-1}(\mathrm{id}_{T(X)} \otimes \eta_{T(M)})T_2(X,M)\\
&\overset{\text{(iv)}}{=} T(\mathrm{id}_X \otimes r\mu_M)(H^l_{X,T(M)})^{-1}(\mathrm{id}_{T(X)} \otimes \eta_{T(M)})T_2(X,M)\\
&\overset{\text{(v)}}{=} T(\mathrm{id}_X \otimes r)(H^l_{X,M})^{-1}(\mathrm{id}_{T(X)} \otimes \mu_M\eta_{T(M)})T_2(X,M)\\
&\overset{\text{(vi)}}{=} T(\mathrm{id}_X \otimes r)(H^l_{X,M})^{-1}(\mathrm{id}_{T(X)} \otimes \mu_M T(\eta_M))T_2(X,M)\\
&\overset{\text{(vii)}}{=} T(\mathrm{id}_X \otimes r)(H^l_{X,M})^{-1}H^l_{X,M}T(\mathrm{id}_X \otimes \eta_M)\\
&\overset{\text{(viii)}}{=} T(\mathrm{id}_X \otimes r\eta_M) \overset{\text{(ix)}}{=} \mathrm{id}_{T(X\otimes M)}.
\end{aligned}$$

Here, (i) follows from (8.3) and the definition of $\mathbb{L}_{X,(M,r)}$, (ii) from the naturality of η, (iii) from the naturality of H^l, (iv) and (ix) from the fact that r is a T-action, (v) from Lemma 8.1(a), (vi) from the unitality of μ, (vii) from the naturality of T_2 and the definition of H^l, and (viii) from the invertibility of H^l. Thus $\mathbb{H}^l_{X,(M,r)}$ is invertible. Hence, $F_T \dashv U_T$ is a left Hopf adjunction. □

Theorem 8.4. *A comonoidal adjunction between left (respectively, right) rigid categories is a left (respectively, right) Hopf adjunction. Consequently, a comonoidal adjunction between rigid categories is a Hopf adjunction.*

Proof. We consider the left case; the right case is similar. Let $\mathcal{C}$ and $\mathcal{D}$ be left rigid categories with left dualities $\{({}^\vee c, \mathrm{ev}_c)\}_{c\in\mathrm{Ob}(\mathcal{C})}$ and $\{({}^\vee d, \mathrm{ev}_d)\}_{d\in\mathrm{Ob}(\mathcal{D})}$. Let $F\colon \mathcal{C} \rightleftarrows \mathcal{D} : U$ be a comonoidal adjunction with unit η and counit ε. Pick $c \in \mathrm{Ob}(\mathcal{C})$ and $d \in \mathrm{Ob}(\mathcal{D})$. For any $e \in \mathrm{Ob}(\mathcal{D})$, consider the composition

$$h_e\colon \mathrm{Hom}_{\mathcal{D}}(F(c)\otimes d, e) \to \mathrm{Hom}_{\mathcal{D}}(F(c\otimes U(d)), e)$$

of the following five bijections:

$$\begin{aligned}
\mathrm{Hom}_{\mathcal{D}}(F(c)\otimes d, e) &\overset{(A)}{\simeq} \mathrm{Hom}_{\mathcal{D}}(F(c), e\otimes {}^\vee d)\\
&\overset{(B)}{\simeq} \mathrm{Hom}_{\mathcal{C}}(c, U(e\otimes {}^\vee d)) \overset{(C)}{\simeq} \mathrm{Hom}_{\mathcal{C}}(c, U(e)\otimes {}^\vee U(d))\\
&\overset{(D)}{\simeq} \mathrm{Hom}_{\mathcal{C}}(c\otimes U(d), U(e)) \overset{(E)}{\simeq} \mathrm{Hom}_{\mathcal{D}}(F(c\otimes U(d)), e).
\end{aligned}$$

Here, the bijections (A) and (D) are respectively induced by the non-degenerate pairings ev_d and $\mathrm{ev}_{U(d)}$ (see Section 1.5.1), (B) and (E) are induced by the adjunction $F \dashv U$ (see Section 7.2.1), and (C) comes from the fact that the functor U, being strong monoidal by Lemma 7.10, induces isomorphisms

$$U(e \otimes {}^\vee d) \simeq U(e) \otimes U({}^\vee d) \quad \text{and} \quad U({}^\vee d) \simeq {}^\vee U(d)$$

(see Section 1.6.4). Denoting by (U_2, U_0) the comonoidal constraints of U, the latter isomorphisms are computed by

$$U_2(e, {}^\vee d) \colon U(e \otimes {}^\vee d) \to U(e) \otimes U({}^\vee d)$$

and

$$(U_0 U(\mathrm{ev}_d) U_2({}^\vee d, d)^{-1} \otimes \mathrm{id}_{{}^\vee U(d)})(\mathrm{id}_{U({}^\vee d)} \otimes \mathrm{coev}_{U(d)}) \colon U({}^\vee d) \to {}^\vee U(d).$$

Thus, the composition of (C) and (D) carries $\beta \in \mathrm{Hom}_{\mathcal{C}}(c, U(e \otimes {}^\vee d))$ to

$$\begin{aligned}
&(\mathrm{id}_{U(e)} \otimes U_0 U(\mathrm{ev}_d) U_2({}^\vee d, d)^{-1})(U_2(e, {}^\vee d)\beta \otimes \mathrm{id}_{U(d)}) \\
&\overset{\text{(i)}}{=} (\mathrm{id}_{U(e)} \otimes U_0 U(\mathrm{ev}_d)) U_2(e, {}^\vee d \otimes d) U_2(e \otimes {}^\vee d, d)^{-1}(\beta \otimes \mathrm{id}_{U(d)}) \\
&\overset{\text{(ii)}}{=} (\mathrm{id}_{U(e)} \otimes U_0) U_2(e, \mathbb{1}) U(\mathrm{id}_e \otimes \mathrm{ev}_d) U_2(e \otimes {}^\vee d, d)^{-1}(\beta \otimes \mathrm{id}_{U(d)}) \\
&\overset{\text{(iii)}}{=} U(\mathrm{id}_e \otimes \mathrm{ev}_d) U_2(e \otimes {}^\vee d, d)^{-1}(\beta \otimes \mathrm{id}_{U(d)}) \in \mathrm{Hom}_{\mathcal{C}}(c \otimes U(d), U(e)).
\end{aligned}$$

Here, (i) follows from the invertibility and coassociativity (7.3) of U_2, (ii) from the naturality of U_2, and (iii) from the counitality (7.4) of U_2. It follows from Section 1.5.1 and the proof of Lemma 7.1 that the composition of the bijections (A) and (B) carries any $\alpha \in \mathrm{Hom}_{\mathcal{D}}(F(c) \otimes d, e)$ to

$$U((\alpha \otimes \mathrm{id}_{{}^\vee d})(\mathrm{id}_{F(c)} \otimes \mathrm{coev}_d))\eta_c \in \mathrm{Hom}_{\mathcal{C}}(c, U(e \otimes {}^\vee d)).$$

Thus, the composition of the bijections (A)-(D) carries α to

$$\begin{aligned}
&U(\mathrm{id}_e \otimes \mathrm{ev}_d) U_2(e \otimes {}^\vee d, d)^{-1}\big(U((\alpha \otimes \mathrm{id}_{{}^\vee d})(\mathrm{id}_{F(c)} \otimes \mathrm{coev}_d))\eta_c \otimes \mathrm{id}_{U(d)}\big) \\
&\overset{\text{(i)}}{=} U\big((\alpha \otimes \mathrm{ev}_d)(\mathrm{id}_{F(c)} \otimes \mathrm{coev}_d \otimes \mathrm{id}_d)\big) U_2(F(c), d)^{-1}(\eta_c \otimes \mathrm{id}_{U(d)}) \\
&\overset{\text{(ii)}}{=} U(\alpha) U_2(F(c), d)^{-1}(\eta_c \otimes \mathrm{id}_{U(d)}) \in \mathrm{Hom}_{\mathcal{C}}(c \otimes U(d), U(e)).
\end{aligned}$$

Here, (i) follows from the naturality of U_2 and (ii) from the fact that coev_d is the inverse of the pairing ev_d. Consequently,

$$\begin{aligned}
h_e(\alpha) &= \varepsilon_e F\big(U(\alpha) U_2(F(c), d)^{-1}(\eta_c \otimes \mathrm{id}_{U(d)})\big) \\
&\overset{\text{(i)}}{=} \alpha \varepsilon_{F(c) \otimes d} F(U_2(F(c), d)^{-1}) F(\eta_c \otimes \mathrm{id}_{U(d)}) \\
&\overset{\text{(ii)}}{=} \alpha(\varepsilon_{F(c)} \otimes \varepsilon_d) F_2(UF(c), U(d)) F(\eta_c \otimes \mathrm{id}_{U(d)}) \\
&\overset{\text{(iii)}}{=} \alpha(\varepsilon_{F(c)} F(\eta_c) \otimes \varepsilon_d) F_2(c, U(d)) \\
&\overset{\text{(iv)}}{=} \alpha(\mathrm{id}_{F(c)} \otimes \varepsilon_d) F_2(c, U(d)) \overset{\text{(v)}}{=} \alpha\, \mathbb{H}^l_{c,d}.
\end{aligned}$$

Here, (i) follows from the naturality of ε, (ii) from the comonoidality of ε, (iii) from the naturality of F_2, (iv) from (7.1), and (v) from the definition of the left fusion operator $\mathbb{H}^l$ of $F \dashv U$. We conclude that for any $e \in \mathrm{Ob}(\mathcal{D})$, the map

$$h_e \colon \mathrm{Hom}_{\mathcal{D}}(F(c) \otimes d, e) \to \mathrm{Hom}_{\mathcal{D}}(F(c \otimes U(d)), e), \quad \alpha \mapsto \alpha\, \mathbb{H}^l_{c,d}$$

is a bijection. Thus, $\mathbb{H}^l_{c,d}$ is invertible (see Exercise 1.1.11). Hence, the comonoidal adjunction $F \dashv U$ is a left Hopf adjunction. □

The next corollary follows directly from Lemma 8.2 and Theorem 8.4.

Corollary 8.5. *The bimonad of a comonoidal adjunction between left (respectively, right) rigid categories is a left (respectively, right) Hopf monad. Consequently, the bimonad of a comonoidal adjunction between rigid categories is a Hopf monad.*

8.3 Hopf monads on rigid categories

We study antipodes and adjoint functors for Hopf monads on rigid categories.

8.3.1 Antipodes

Let $T = (T, \mu, \eta)$ be a bimonad on a monoidal category $\mathcal{C}$. If $\mathcal{C}$ has a distinguished left duality $\{({}^\vee X, \mathrm{ev}_X)\}_{X \in \mathrm{Ob}(\mathcal{C})}$, then a *left antipode* for T is a natural transformation

$$s^l = \{s^l_X \colon T({}^\vee T(X)) \to {}^\vee X\}_{X \in \mathrm{Ob}(\mathcal{C})}$$

such that for all $X \in \mathrm{Ob}(\mathcal{C})$,

$$T_0\, T(\mathrm{ev}_X)\, T({}^\vee \eta_X \otimes \mathrm{id}_X) = \mathrm{ev}_{T(X)}\, (s^l_{T(X)} T({}^\vee \mu_X) \otimes \mathrm{id}_{T(X)})\, T_2({}^\vee T(X), X) \tag{8.4}$$

and

$$(\eta_X \otimes \mathrm{id}_{{}^\vee X})\, \mathrm{coev}_X T_0 = (\mu_X \otimes s^l_X)\, T_2(T(X), {}^\vee T(X))\, T(\mathrm{coev}_{T(X)}). \tag{8.5}$$

If $\mathcal{C}$ has a distinguished right duality $\{(X^\vee, \widetilde{\mathrm{ev}}_X)\}_{X \in \mathrm{Ob}(\mathcal{C})}$, then a *right antipode* for T is a natural transformation

$$s^r = \{s^r_X \colon T(T(X)^\vee) \to X^\vee\}_{X \in \mathrm{Ob}(\mathcal{C})}$$

such that for all $X \in \mathrm{Ob}(\mathcal{C})$,

$$T_0\, T(\widetilde{\mathrm{ev}}_X)\, T(\mathrm{id}_X \otimes \eta_X^\vee) = \widetilde{\mathrm{ev}}_{T(X)}\, (\mathrm{id}_{T(X)} \otimes s^r_{T(X)}\, T(\mu_X^\vee))\, T_2(X, T(X)^\vee) \tag{8.6}$$

and

$$(\mathrm{id}_{X^\vee} \otimes \eta_X)\, \widetilde{\mathrm{coev}}_X\, T_0 = (s^r_X \otimes \mu_X)\, T_2(T(X)^\vee, T(X))\, T(\widetilde{\mathrm{coev}}_{T(X)}). \tag{8.7}$$

The left and right antipodes play in the study of bimonads the role of an antipode of a bialgebra and its inverse (see Section 6.2.1). The next lemma shows that the antipodes for bimonads are anti-multiplicative (as are antipodes for bialgebras).

Lemma 8.6. *A left antipode s^l for T satisfies*

$$s^l_X \mu_{{}^\vee T(X)} = s^l_X T(s^l_{T(X)}) T^2({}^\vee\mu_X) \quad \textit{and} \quad s^l_X \eta_{{}^\vee T(X)} = {}^\vee\eta_X$$

for all $X \in \mathrm{Ob}(\mathcal{C})$. Similarly, a right antipode s^r for T satisfies

$$s^r_X \mu_{T(X)^\vee} = s^r_X T(s^r_{T(X)}) T^2(\mu^\vee_X) \quad \textit{and} \quad s^r_X \eta_{T(X)^\vee} = \eta^\vee_X$$

for all $X \in \mathrm{Ob}(\mathcal{C})$.

We prove Lemma 8.6 in Section 8.3.2. Recall from Lemma 6.1 that a bialgebra in a braided rigid category is a Hopf algebra if and only if its category of modules is rigid. The following theorem generalizes this fact to Hopf monads.

Theorem 8.7. *Let T be a bimonad on a left (respectively, right) rigid category $\mathcal{C}$ with distinguished left (respectively, right) duality. Then the following three conditions are equivalent:*

(i) *T is a left (respectively, right) Hopf monad;*

(ii) *T has a left (respectively, right) antipode;*

(iii) *The monoidal category $\mathcal{C}^T$ of T-modules is left (respectively, right) rigid.*

If these conditions hold, then

(a) *The left (respectively, right) antipode of T is unique;*

(b) *There is a unique left (respectively, right) duality in $\mathcal{C}^T$ which is a lift (in the sense of Section 1.6.4) of the left (respectively, right) distinguished duality in $\mathcal{C}$ along the forgetful functor $\mathcal{C}^T \to \mathcal{C}$.*

We prove Theorem 8.7 in Section 8.3.3. In this theorem the category $\mathcal{C}^T$ carries the unique structure of a monoidal category such that the forgetful functor $\mathcal{C}^T \to \mathcal{C}$ is strict monoidal, see Section 7.5.2. It follows from Section 8.3.3 that the left duality in $\mathcal{C}^T$ produced by Theorem 8.7(b) is computed from the left antipode s^l of T and the distinguished left duality $\{({}^\vee X, \mathrm{ev}_X)\}_{X \in \mathrm{Ob}(\mathcal{C})}$ in $\mathcal{C}$ as follows:

$$\big\{\big({}^\vee(M,r) = ({}^\vee M, s^l_M T({}^\vee r)),\ \mathrm{ev}_{(M,r)} = \mathrm{ev}_M\big)\big\}_{(M,r) \in \mathrm{Ob}(\mathcal{C}^T)}.$$

Likewise, the right duality in $\mathcal{C}^T$ produced by Theorem 8.7(b) is computed from the right antipode s^r of T and the distinguished right duality $\{(X^\vee, \widetilde{\mathrm{ev}}_X)\}_{X \in \mathrm{Ob}(\mathcal{C})}$ in $\mathcal{C}$ as follows:

$$\big\{\big((M,r)^\vee = (M^\vee, s^r_M T(r^\vee)),\ \widetilde{\mathrm{ev}}_{(M,r)} = \widetilde{\mathrm{ev}}_M\big)\big\}_{(M,r) \in \mathrm{Ob}(\mathcal{C}^T)}.$$

8.3.2 Proof of Lemma 8.6

We consider the case of a left antipode s^l for T; the case of a right antipode is similar. Pick $X \in \mathrm{Ob}(\mathcal{C})$. First, we have:

$$
\begin{aligned}
(\mathrm{id}_{T(X)} \otimes {}^\vee\eta_X)\mathrm{coev}_{T(X)} &\overset{\text{(i)}}{=} (\eta_X \otimes \mathrm{id}_{{}^\vee X})\mathrm{coev}_X T_0 \eta_{\mathbb{1}} \\
&\overset{\text{(ii)}}{=} (\mu_X \otimes s^l_X) T_2(T(X), {}^\vee T(X)) T(\mathrm{coev}_{T(X)})\eta_{\mathbb{1}} \\
&\overset{\text{(iii)}}{=} (\mu_X \otimes s^l_X) T_2(T(X), {}^\vee T(X))\eta_{T(X)\otimes {}^\vee T(X)}\mathrm{coev}_{T(X)} \\
&\overset{\text{(iv)}}{=} (\mu_X \eta_{T(X)} \otimes s^l_X \eta_{{}^\vee T(X)})\mathrm{coev}_{T(X)} \overset{\text{(v)}}{=} (\mathrm{id}_{T(X)} \otimes s^l_X \eta_{{}^\vee T(X)})\mathrm{coev}_{T(X)}.
\end{aligned}
$$

Here, (i) follows from (2.2) and the comonoidality of η, (ii) from (8.5), (iii) from the naturality of η, (iv) from the comonoidality of η, and (v) from the unitality of μ. Then, composing on the left with $\mathrm{ev}_{T(X)} \otimes \mathrm{id}_{{}^\vee T(X)}$ and using that $\mathrm{coev}_{T(X)}$ is the inverse of the pairing $\mathrm{ev}_{T(X)}$, we obtain that $s^l_X \eta_{{}^\vee T(X)} = {}^\vee\eta_X$.

We now prove the first equality of the lemma. To simplify the notation, set

$$A = T(X), \quad B = T^2({}^\vee A), \quad \nu = s^l_A T({}^\vee\mu_X), \quad \mathcal{L} = s^l_X \mu_{{}^\vee A}, \quad \mathcal{R} = s^l_X T(\nu).$$

We need to prove that $\mathcal{L} = \mathcal{R}$. To this end, consider the morphisms

$$e = \mathrm{ev}_A\big(\nu T(\nu) \otimes \mathrm{id}_A\big) \colon B \otimes A \to \mathbb{1}$$

and

$$c = (\mu_X T(\mu_X) \otimes \mathrm{id}_B)\,(T^2)_2(A, {}^\vee A)\,T^2(\mathrm{coev}_A) \colon T^2(\mathbb{1}) \to A \otimes B.$$

We claim that

$$(\mathrm{id}_A \otimes \mathcal{L})c = (\mathrm{id}_A \otimes \mathcal{R})c, \tag{8.8}$$

$$(e \otimes \mathrm{id}_B)(\mathrm{id}_B \otimes c)(T^2)_2({}^\vee A, \mathbb{1}) = \mathrm{id}_B. \tag{8.9}$$

These two formulas will imply the equality $\mathcal{L} = \mathcal{R}$ because

$$\mathcal{L} \overset{\text{(i)}}{=} (e \otimes \mathcal{L})(\mathrm{id}_B \otimes c)(T^2)_2({}^\vee A, \mathbb{1}) \overset{\text{(ii)}}{=} (e \otimes \mathcal{R})(\mathrm{id}_B \otimes c)(T^2)_2({}^\vee A, \mathbb{1}) \overset{\text{(iii)}}{=} \mathcal{R},$$

where (i) and (iii) follow from (8.9) and (ii) from (8.8).

We now prove (8.8):

$$
\begin{aligned}
&(\mathrm{id}_A \otimes \mathcal{L})c \\
&\overset{\text{(i)}}{=} (\mu_X T(\mu_X) \otimes s^l_X \mu_{{}^\vee A})\,(T^2)_2(A, {}^\vee A)\,T^2(\mathrm{coev}_A) \\
&\overset{\text{(ii)}}{=} (\mu_X \mu_A \otimes s^l_X \mu_{{}^\vee A})\,(T^2)_2(A, {}^\vee A)\,T^2(\mathrm{coev}_A) \\
&\overset{\text{(iii)}}{=} (\mu_X \otimes s^l_X)\,T_2(A, {}^\vee A)\,\mu_{A \otimes {}^\vee A}\,T^2(\mathrm{coev}_A)
\end{aligned}
$$

$$
\begin{aligned}
&\overset{\text{(iv)}}{=} (\mu_X \otimes s^l_X)\, T_2(A, {}^\vee\!A)\, T(\mathrm{coev}_A)\, \mu_{\mathbb{1}}\\
&\overset{\text{(v)}}{=} (\eta_X \otimes \mathrm{id}_{{}^\vee\!X})\mathrm{coev}_X T_0 \mu_{\mathbb{1}} \overset{\text{(vi)}}{=} (\eta_X \otimes \mathrm{id}_{{}^\vee\!X})\mathrm{coev}_X T_0 T(T_0)\\
&\overset{\text{(vii)}}{=} (\mu_X \otimes s^l_X)\, T_2(A, {}^\vee\!A)\, T(\mathrm{coev}_A T_0)\\
&\overset{\text{(viii)}}{=} (\mu_X \otimes s^l_X)\, T_2(A, {}^\vee\!A)\, T\big((\mu_X \eta_A \otimes \mathrm{id}_{{}^\vee\!A})\mathrm{coev}_A T_0\big)\\
&\overset{\text{(ix)}}{=} (\mu_X \otimes s^l_X)\, T_2(A, {}^\vee\!A)\, T\big((\mu_X \mu_A \otimes s^l_A)\, T_2(T(A), {}^\vee T(A))\, T(\mathrm{coev}_{T(A)})\big)\\
&\overset{\text{(x)}}{=} (\mu_X \otimes s^l_X)\, T_2(A, {}^\vee\!A)\, T\big((\mu_X T(\mu_X) \otimes s^l_A)\, T_2(T(A), {}^\vee T(A))\, T(\mathrm{coev}_{T(A)})\big)\\
&\overset{\text{(xi)}}{=} (\mu_X \otimes s^l_X)\, T_2(A, {}^\vee\!A)\, T\big((\mu_X \otimes s^l_A T({}^\vee\mu_X))\, T_2(A, {}^\vee\!A)\, T(\mathrm{coev}_A)\big)\\
&\overset{\text{(xii)}}{=} \big(\mu_X T(\mu_X) \otimes s^l_X T(s^l_A) T^2({}^\vee\mu_X)\big)\, T_2(T(A), T({}^\vee\!A)) T(T_2(A, {}^\vee\!A))\, T^2(\mathrm{coev}_A)\\
&\overset{\text{(xiii)}}{=} \big(\mu_X T(\mu_X) \otimes s^l_X T(s^l_A) T^2({}^\vee\mu_X)\big)\, (T^2)_2(A, {}^\vee\!A)\, T^2(\mathrm{coev}_A)\\
&\overset{\text{(xiv)}}{=} (\mathrm{id}_A \otimes \mathcal{R})c.
\end{aligned}
$$

Here, (i) follows from the definitions of c and $\mathcal{L}$, (ii) and (x) from the associativity of μ, (iii) and (vi) from the comonoidality of μ, (iv) from the naturality of μ, (v), (vii), and (ix) from (8.5), (viii) from the unitality of μ, (xi) from the naturality of T_2 and (2.2), (xii) from the naturality of T_2, (xiii) from the definition of $(T^2)_2$ (see Section 7.5.1), and (xiv) from the definitions of c and $\mathcal{R}$.

It remains to prove (8.9). First, we have:

$$
\mathrm{ev}_A(\nu \otimes \mu_X)\, T_2({}^\vee\!A, A) = T_0\, T(\mathrm{ev}_A). \tag{8.10}
$$

Indeed,

$$
\begin{aligned}
\mathrm{ev}_A(\nu \otimes \mu_X)\, T_2(A, {}^\vee\!A) &\overset{\text{(i)}}{=} \mathrm{ev}_{T(A)}({}^\vee\mu_X s^l_A T({}^\vee\mu_X) \otimes \mathrm{id}_{T(A)})\, T_2({}^\vee\!A, A)\\
&\overset{\text{(ii)}}{=} \mathrm{ev}_{T(A)}(s^l_{T(A)} T({}^\vee T(\mu_X) {}^\vee\mu_X) \otimes \mathrm{id}_{T(A)})\, T_2({}^\vee\!A, A)\\
&\overset{\text{(iii)}}{=} \mathrm{ev}_{T(A)}(s^l_{T(A)} T({}^\vee\mu_A {}^\vee\mu_X) \otimes \mathrm{id}_{T(A)})\, T_2({}^\vee\!A, A)\\
&\overset{\text{(iv)}}{=} \mathrm{ev}_{T(A)}(s^l_{T(A)} T({}^\vee\mu_A) \otimes \mathrm{id}_{T(A)})\, T_2({}^\vee T(A), A) T({}^\vee\mu_X \otimes \mathrm{id}_A)\\
&\overset{\text{(v)}}{=} T_0\, T(\mathrm{ev}_A)\, T({}^\vee\eta_A {}^\vee\mu_X \otimes \mathrm{id}_A) \overset{\text{(vi)}}{=} T_0\, T(\mathrm{ev}_A).
\end{aligned}
$$

Here, (i) follows from (2.2) and the definition of ν, (ii) from the naturality of s^l, (iii) from the associativity of μ, (iv) from the naturality of T_2, (v) from (8.4), and (vi) from the unitality of μ. Second, we have:

$$
\mathrm{ev}_A\big(\nu T(\nu) \otimes \mu_X T(\mu_X)\big)\, (T^2)_2({}^\vee\!A, A) = (T^2)_0 T^2(\mathrm{ev}_A). \tag{8.11}
$$

Indeed,

$$
\begin{aligned}
&\mathrm{ev}_A\big(\nu T(\nu)\otimes\mu_X T(\mu_X)\big)\,(T^2)_2({}^\vee\!A,A)\\
&\overset{\text{(i)}}{=}\mathrm{ev}_A\big(\nu T(\nu)\otimes\mu_X T(\mu_X)\big)\,T_2(T({}^\vee\!A),T(A))\,T(T_2({}^\vee\!A,A))\\
&\overset{\text{(ii)}}{=}\mathrm{ev}_A(\nu\otimes\mu_X)\,T_2({}^\vee\!A,A)\,T\big((\nu\otimes\mu_X)\,T_2({}^\vee\!A,A)\big)\\
&\overset{\text{(iii)}}{=}T_0\,T\big(\mathrm{ev}_A(\nu\otimes\mu_X)\,T_2({}^\vee\!A,A)\big)\overset{\text{(iv)}}{=}T_0\,T(T_0)\,T^2(\mathrm{ev}_A)\overset{\text{(v)}}{=}(T^2)_0\,T^2(\mathrm{ev}_A).
\end{aligned}
$$

Here, (i) follows from the definition of $(T^2)_2$ (see Section 7.5.1), (ii) from the naturality of T_2, (iii) and (iv) from (8.10), and (v) from the definition of $(T^2)_0$ (see Section 7.5.1). Finally, we have:

$$
\begin{aligned}
&(e\otimes\mathrm{id}_B)(\mathrm{id}_B\otimes c)(T^2)_2({}^\vee\!A,\mathbb{1})\\
&\overset{\text{(i)}}{=}\big(\mathrm{ev}_A(\nu T(\nu)\otimes\mu_X T(\mu_X))\otimes\mathrm{id}_B\big)\\
&\qquad\circ\big(\mathrm{id}_B\otimes(T^2)_2(A,{}^\vee\!A)\,T^2(\mathrm{coev}_A)\big)\,(T^2)_2({}^\vee\!A,\mathbb{1})\\
&\overset{\text{(ii)}}{=}\big(\mathrm{ev}_A(\nu T(\nu)\otimes\mu_X T(\mu_X))\otimes\mathrm{id}_B\big)\\
&\qquad\circ\big(\mathrm{id}_B\otimes(T^2)_2(A,{}^\vee\!A)\big)\,(T^2)_2({}^\vee\!A,A\otimes{}^\vee\!A)\,T^2(\mathrm{id}_{{}^\vee\!A}\otimes\mathrm{coev}_A)\\
&\overset{\text{(iii)}}{=}\big(\mathrm{ev}_A(\nu T(\nu)\otimes\mu_X T(\mu_X))\,(T^2)_2({}^\vee\!A,A)\otimes\mathrm{id}_B\big)\\
&\qquad\circ(T^2)_2({}^\vee\!A\otimes A,{}^\vee\!A)\,T^2(\mathrm{id}_{{}^\vee\!A}\otimes\mathrm{coev}_A)\\
&\overset{\text{(iv)}}{=}\big((T^2)_0T^2(\mathrm{ev}_A)\otimes\mathrm{id}_B\big)\,(T^2)_2({}^\vee\!A\otimes A,{}^\vee\!A)\,T^2(\mathrm{id}_{{}^\vee\!A}\otimes\mathrm{coev}_A)\\
&\overset{\text{(v)}}{=}\big((T^2)_0\otimes\mathrm{id}_B\big)\,(T^2)_2(\mathbb{1},{}^\vee\!A)\,T^2\big((\mathrm{ev}_A\otimes\mathrm{id}_{{}^\vee\!A})(\mathrm{id}_{{}^\vee\!A}\otimes\mathrm{coev}_A)\big)\overset{\text{(vi)}}{=}\mathrm{id}_B.
\end{aligned}
$$

Here, (i) follows from the definitions of e and c, (ii) and (v) from the naturality of $(T^2)_2$, (iii) from the coassociativity (7.3) of $(T^2)_2$, (iv) from (8.11), and (vi) from the counitality (7.4) of $(T^2)_2$ and the fact that coev_A is the inverse of the pairing ev_A. This proves (8.9) and the lemma.

8.3.3 Proof of Theorem 8.7

We consider only the left case. The right case may be deduced by considering the opposite bimonad $T^{\otimes\mathrm{op}}$, see Exercise 8.1.7. We assume that $T=(T,\mu,\eta)$ is a bimonad on a left rigid category $\mathcal{C}$ with distinguished left duality $\{({}^\vee X,\mathrm{ev}_X)\}_{X\in\mathrm{Ob}(\mathcal{C})}$. As usual, we denote by coev_X the inverse of the pairing ev_X.

We first prove the implication (iii) $\Rightarrow$ (i). Consider the comonoidal adjunction $F_T\dashv U_T$ of the bimonad T (see Section 7.6.2). Since T is the bimonad of this comonoidal adjunction and both $\mathcal{C}$ and $\mathcal{C}^T$ are left rigid, Corollary 8.5 implies that T is a left Hopf monad.

Let us prove the implication (i) $\Rightarrow$ (ii). For any $X\in\mathrm{Ob}(\mathcal{C})$, set

$$
s^l_X=\big(T_0T(\mathrm{ev}_{T(X)})(H^l_{{}^\vee T(X),X})^{-1}\otimes{}^\vee\eta_X\big)\big(\mathrm{id}_{T({}^\vee T(X))}\otimes\mathrm{coev}_{T(X)}\big),\qquad(8.12)
$$

where H^l is the left fusion operator of T (which is invertible by assumption). Let us prove that the family $s^l = \{s^l_X : T({}^\vee T(X)) \to {}^\vee X\}_{X\in \mathrm{Ob}(\mathcal{C})}$ is a left antipode for T. First, the naturality of H^l and η together with (2.2) imply that s^l is a natural transformation. Second, s^l satisfies (8.4) since

$$
\begin{aligned}
&\mathrm{ev}_{T(X)}\big(s^l_{T(X)}T({}^\vee\mu_X)\otimes \mathrm{id}_{T(X)}\big)T_2({}^\vee T(X),X)\\
&\overset{\text{(i)}}{=} T_0T(\mathrm{ev}_{T^2(X)})(H^l_{{}^\vee T^2(X),T(X)})^{-1}\big(T({}^\vee\mu_X)\otimes \eta_{T(X)}\big)T_2({}^\vee T(X),X)\\
&\overset{\text{(ii)}}{=} T_0T\big(\mathrm{ev}_{T(X)}(\mathrm{id}_{{}^\vee T(X)}\otimes \mu_X)\big)(H^l_{{}^\vee T(X),T(X)})^{-1}\\
&\qquad\circ (\mathrm{id}_{T({}^\vee T(X))}\otimes \eta_{T(X)})T_2({}^\vee T(X),X)\\
&\overset{\text{(iii)}}{=} T_0T(\mathrm{ev}_{T(X)})(H^l_{{}^\vee T(X),X})^{-1}(\mathrm{id}_{T({}^\vee T(X))}\otimes \mu_X\eta_{T(X)})T_2({}^\vee T(X),X)\\
&\overset{\text{(iv)}}{=} T_0T(\mathrm{ev}_{T(X)})(H^l_{{}^\vee T(X),X})^{-1}T_2({}^\vee T(X),X)\\
&\overset{\text{(v)}}{=} T_0T(\mathrm{ev}_{T(X)})T(\mathrm{id}_{{}^\vee T(X)}\otimes \eta_X) \overset{\text{(vi)}}{=} T_0T(\mathrm{ev}_X)T({}^\vee\eta_X\otimes \mathrm{id}_X).
\end{aligned}
$$

Here, (i) follows from the definition of s^l, Formula (2.2), and the fact that $\mathrm{coev}_{T(X)}$ is the inverse of the pairing $\mathrm{ev}_{T(X)}$, (ii) from the naturality of H^l and (2.2), (iii) from Lemma 8.1(a), (iv) from the unitality of μ, (v) from Lemma 8.1(b), and (vi) from (2.2). Third, s^l satisfies (8.5) since

$$
\begin{aligned}
&(\mu_X\otimes s^l_X)T_2(T(X),{}^\vee T(X))T(\mathrm{coev}_{T(X)})\\
&\overset{\text{(i)}}{=} \big(\mu_X\otimes T_0T(\mathrm{ev}_{T(X)})(H^l_{{}^\vee T(X),X})^{-1}\otimes {}^\vee\eta_X\big)\\
&\qquad\circ \big(T_2(T(X),{}^\vee T(X))T(\mathrm{coev}_{T(X)})\otimes \mathrm{coev}_{T(X)}\big)\\
&\overset{\text{(ii)}}{=} \big((\mu_X\otimes T_0T(\mathrm{ev}_{T(X)}))T_2(T(X),{}^\vee T(X)\otimes T(X))(H^l_{T(X)\otimes {}^\vee T(X),X})^{-1}\otimes \mathrm{id}_{{}^\vee X}\big)\\
&\qquad\circ \big(T(\mathrm{coev}_{T(X)})\otimes (\eta_X\otimes \mathrm{id}_{{}^\vee X})\mathrm{coev}_X\big)\\
&\overset{\text{(iii)}}{=} \big((\mu_X\otimes T_0)T_2(T(X),\mathbb{1})\otimes \mathrm{id}_{{}^\vee X}\big)\\
&\qquad\circ \big(T\big((\mathrm{id}_{T(X)}\otimes \mathrm{ev}_{T(X)})(\mathrm{coev}_{T(X)}\otimes \mathrm{id}_{{}^\vee T(X)})\big)\otimes \mathrm{id}_{{}^\vee X}\big)\\
&\qquad\circ \big((H^l_{\mathbb{1},X})^{-1}(\mathrm{id}_{T(\mathbb{1})}\otimes \eta_X)\otimes \mathrm{id}_{{}^\vee X}\big)(\mathrm{id}_{T(\mathbb{1})}\otimes \mathrm{coev}_X)\\
&\overset{\text{(iv)}}{=} \big(\mu_X(H^l_{\mathbb{1},X})^{-1}(\mathrm{id}_{T(\mathbb{1})}\otimes \eta_X)\otimes \mathrm{id}_{{}^\vee X}\big)(\mathrm{id}_{T(\mathbb{1})}\otimes \mathrm{coev}_X)\\
&\overset{\text{(v)}}{=} (\eta_X\otimes \mathrm{id}_{{}^\vee X})\mathrm{coev}_X\, T_0.
\end{aligned}
$$

Here, (i) follows from the definition of s^l, (ii) from Lemma 8.1(c) and (2.2), (iii) from the naturality of T_2 and H^l, (iv) from the counitality (7.4) of T_2 and the fact that $\mathrm{coev}_{T(X)}$ is the inverse of the pairing $\mathrm{ev}_{T(X)}$, and (v) from Lemma 8.1(d). Hence s^l is a left antipode for T.

Let us prove the implication (ii) $\Rightarrow$ (iii). Assume that T has a left antipode s^l. Pick a T-module (M,r). The morphism

$$a = s^l_M T({}^\vee r)\colon T({}^\vee M)\to {}^\vee M$$

is then a T-action. Indeed,

$$aT(a) \overset{\text{(i)}}{=} s^l_M T({}^\vee r) T(s^l_M) T^2({}^\vee r) \overset{\text{(ii)}}{=} s^l_M T(s^l_{T(M)}) T^2({}^\vee T(r) {}^\vee r)$$
$$\overset{\text{(iii)}}{=} s^l_M T(s^l_{T(M)}) T^2({}^\vee \mu_M {}^\vee r) \overset{\text{(iv)}}{=} s^l_M \mu_{{}^\vee T(M)} T^2({}^\vee r) \overset{\text{(v)}}{=} s^l_M T({}^\vee r) \mu_{{}^\vee M} \overset{\text{(vi)}}{=} a\mu_{{}^\vee M}.$$

Here, (i) and (vi) follow from the definition of a, (ii) from the naturality of s^l, (iii) from the fact that r is a T-action, (iv) from Lemma 8.6, and (v) from the naturality of μ. Also,

$$a\eta_{{}^\vee M} \overset{\text{(i)}}{=} s^l_M T({}^\vee r) \eta_{{}^\vee M} \overset{\text{(ii)}}{=} s^l_M \eta_{{}^\vee T(M)} {}^\vee r \overset{\text{(iii)}}{=} {}^\vee \eta_M {}^\vee r \overset{\text{(iv)}}{=} \mathrm{id}_{{}^\vee M}.$$

Here, (i) follows from the definition of a, (ii) from the naturality of η, (iii) from Lemma 8.6, and (iv) from the fact that r is a T-action. Next, ev_M is a T-linear morphism from $({}^\vee M, a) \otimes (M, r)$ to $(\mathbb{1}, T_0)$ since

$$\begin{aligned}
&\mathrm{ev}_M (a \otimes r) T_2({}^\vee M, M) \overset{\text{(i)}}{=} \mathrm{ev}_{T(M)}({}^\vee r s^l_M T({}^\vee r) \otimes \mathrm{id}_{T(M)}) T_2({}^\vee M, M)\\
&\overset{\text{(ii)}}{=} \mathrm{ev}_{T(M)}(s^l_{T(M)} T({}^\vee T(r) {}^\vee r) \otimes \mathrm{id}_{T(M)}) T_2({}^\vee M, M)\\
&\overset{\text{(iii)}}{=} \mathrm{ev}_{T(M)}(s^l_{T(M)} T({}^\vee \mu_M {}^\vee r) \otimes \mathrm{id}_{T(M)}) T_2({}^\vee M, M)\\
&\overset{\text{(iv)}}{=} \mathrm{ev}_{T(M)}(s^l_{T(M)} T({}^\vee \mu_M) \otimes \mathrm{id}_{T(M)}) T_2({}^\vee T(M), M) T({}^\vee r \otimes \mathrm{id}_M)\\
&\overset{\text{(v)}}{=} T_0 T(\mathrm{ev}_M) T({}^\vee \eta_M {}^\vee r \otimes \mathrm{id}_M) \overset{\text{(vi)}}{=} T_0 T(\mathrm{ev}_M).
\end{aligned}$$

Here, (i) follows from the definition of a and (2.2), (ii) from the naturality of s^l, (iii) and (vi) from the fact that r is a T-action, (iv) from the naturality of T_2, and (v) from (8.4). Finally, coev_M is a T-linear morphism from $(\mathbb{1}, T_0)$ to $(M, r) \otimes ({}^\vee M, a)$ since

$$\begin{aligned}
\mathrm{coev}_M T_0 &\overset{\text{(i)}}{=} (r\eta_M \otimes \mathrm{id}_{{}^\vee M}) \mathrm{coev}_M T_0\\
&\overset{\text{(ii)}}{=} (r\mu_M \otimes s^l_M)\, T_2(T(M), {}^\vee T(M)) T(\mathrm{coev}_{T(M)})\\
&\overset{\text{(iii)}}{=} (rT(r) \otimes s^l_M)\, T_2(T(M), {}^\vee T(M)) T(\mathrm{coev}_{T(M)})\\
&\overset{\text{(iv)}}{=} (r \otimes s^l_M T({}^\vee r)) T_2(M, {}^\vee M) T(\mathrm{coev}_M)\\
&\overset{\text{(v)}}{=} (r \otimes a) T_2(M, {}^\vee M) T(\mathrm{coev}_M).
\end{aligned}$$

Here, (i) and (iii) follow from the fact that r is a T-action, (ii) from (8.5), (iv) from the naturality of T_2 and (2.2), and (v) from the definition of a. This proves that the pairing $\mathrm{ev}_{(M,r)} = \mathrm{ev}_M \colon ({}^\vee M, a) \otimes (M, r) \to (\mathbb{1}, T_0)$ is non-degenerate with inverse $\mathrm{coev}_{(M,r)} = \mathrm{coev}_M$. In conclusion, the family

$$\big\{\big({}^\vee(M, r) = ({}^\vee M, s^l_M T({}^\vee r)),\ \mathrm{ev}_{(M,r)} = \mathrm{ev}_M\big)\big\}_{(M,r) \in \mathrm{Ob}(\mathcal{C}^T)} \tag{8.13}$$

is a left duality in $\mathcal{C}^T$. Hence, $\mathcal{C}^T$ is left rigid.

To prove the claims (a) and (b) of Theorem 8.7, consider a left antipode s^l of T and its associated left duality (8.13) in $\mathcal{C}^T$. Observe that if a T-module (M,r) has a left dual of the form

$$(({}^\vee M, a),\ \omega = \mathrm{ev}_M),$$

then the T-linearity of the morphism

$$(\omega \otimes \mathrm{id}_{{}^\vee M})(\mathrm{id}_{{}^\vee M} \otimes \mathrm{coev}_{(M,r)}) = \mathrm{id}_{{}^\vee M} \colon ({}^\vee M, a) \to ({}^\vee M, s^l_M T({}^\vee r))$$

means that

$$a = s^l_M T({}^\vee r). \tag{8.14}$$

We now prove the claim (a). Suppose that t^l is a left antipode for T. The proof of the implication (ii) $\Rightarrow$ (iii) above shows that for any $X \in \mathrm{Ob}(\mathcal{C})$, the pair

$$(({}^\vee T(X), t^l_{T(X)} T({}^\vee \mu_X)),\ \mathrm{ev}_{T(X)})$$

is a left dual of $(T(X), \mu_X)$. Thus, $t^l_{T(X)} T({}^\vee \mu_X) = s^l_{T(X)} T({}^\vee \mu_X)$ by (8.14). Composing on the left with ${}^\vee \eta_X$ and using the naturality of the antipodes, we get

$$t^l_X T({}^\vee T(\eta_X) {}^\vee \mu_X) = s^l_X T({}^\vee T(\eta_X) {}^\vee \mu_X).$$

Then, using the identity ${}^\vee T(\eta_X) {}^\vee \mu_X = \mathrm{id}_{{}^\vee T(X)}$ (which follows from the unitality of μ), we obtain that $s^l_X = t^l_X$. This proves (a).

Let us prove the claim (b). The strict monoidality of the forgetful functor $U_T \colon \mathcal{C}^T \to \mathcal{C}$ and the fact that U_T acts as the identity on morphisms imply that for any pairing ω in $\mathcal{C}^T$,

$$(\omega)^{U_T} = \omega. \tag{8.15}$$

This implies that the left duality (8.13) associated with s^l is a lift along U_T of the distinguished left duality in $\mathcal{C}$. Consider now a left duality in $\mathcal{C}^T$

$$\{((M,r)^\dagger,\ e_{(M,r)})\}_{(M,r) \in \mathrm{Ob}(\mathcal{C}^T)}$$

which is a lift along U_T of the distinguished left duality in $\mathcal{C}$. This means (see Section 1.6.4) that for any T-module (M,r),

$$U_T((M,r)^\dagger) = {}^\vee M \quad \text{and} \quad (e_{(M,r)})^{U_T} = \mathrm{ev}_M.$$

Thus, there is a T-action $a_{(M,r)}$ such that $(M,r)^\dagger = ({}^\vee M, a_{(M,r)})$ and $e_{(M,r)} = \mathrm{ev}_M$ by (8.15). Consequently, $a_{(M,r)} = s^l_M T({}^\vee r)$ by (8.14). Hence, the given left duality in $\mathcal{C}^T$ is nothing but the left duality (8.13) associated with s^l. This proves (b) and completes the proof of Theorem 8.7.

8.3.4 Example

Let A be a Hopf algebra in a braided rigid category $\mathcal{C}$. The antipodes of the Hopf monad $T = A \otimes ?$ on $\mathcal{C}$ from Example 8.1.4 are computed by

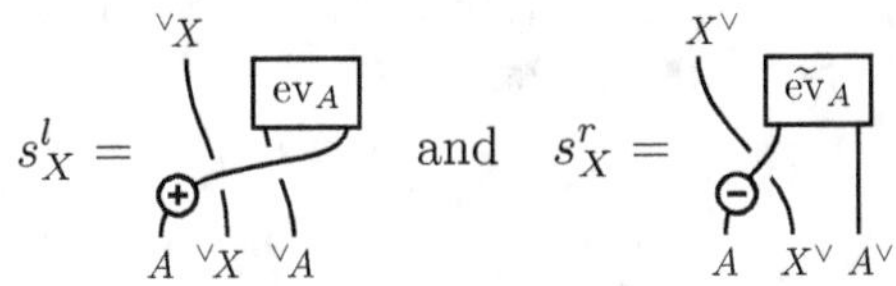

for all $X \in \mathrm{Ob}(\mathcal{C})$. This follows from the expression (8.12) of the antipodes via the inverse fusion operators (computed in Example 8.1.4). Note that the distinguished dualities in $\mathcal{C}^T = \mathrm{mod}_A$ given by Lemma 6.1 and Theorem 8.7 coincide.

8.3.5 Adjoints

We show in this section that a Hopf monad on a rigid category and its associated forgetful functor have right adjoints. We start with a lemma concerning strong monoidal functors between rigid categories.

Lemma 8.8. *If a strong monoidal functor between rigid categories has a left adjoint, then it has a right adjoint.*

Proof. Let $U: \mathcal{D} \to \mathcal{C}$ be a strong monoidal functor between rigid categories having a left adjoint $F: \mathcal{C} \to \mathcal{D}$. Fix left and right dualities in $\mathcal{C}$ and $\mathcal{D}$. Consider the functor $F^!: \mathcal{C} \to \mathcal{D}$ defined by

$$F^!(X) = F({}^\vee X)^\vee \quad \text{and} \quad F^!(f) = F({}^\vee f)^\vee$$

for all $X \in \mathrm{Ob}(\mathcal{C})$ and all morphisms f in $\mathcal{C}$. For any $X \in \mathrm{Ob}(\mathcal{C})$ and $Y \in \mathrm{Ob}(\mathcal{D})$, consider the following bijections:

$$\begin{aligned} &\mathrm{Hom}_{\mathcal{C}}(U(Y), X) \overset{(A)}{\simeq} \mathrm{Hom}_{\mathcal{C}}({}^\vee X, {}^\vee U(Y)) \overset{(B)}{\simeq} \mathrm{Hom}_{\mathcal{C}}({}^\vee X, U({}^\vee Y)) \\ &\overset{(C)}{\simeq} \mathrm{Hom}_{\mathcal{D}}(F({}^\vee X), {}^\vee Y) \overset{(D)}{\simeq} \mathrm{Hom}_{\mathcal{D}}(({}^\vee Y)^\vee, F({}^\vee X)^\vee) \overset{(E)}{\simeq} \mathrm{Hom}_{\mathcal{D}}(Y, F^!(X)). \end{aligned}$$

Here, (A) is induced by the left duality in $\mathcal{C}$ (see Section 1.5.1), (B) by the natural isomorphism $U^l(Y): U({}^\vee Y) \to {}^\vee U(Y)$ induced by U (see Section 1.6.4), (C) by the adjunction $F \dashv U$ (see Section 7.2.1), (D) by the right duality in $\mathcal{D}$ (see Section 1.5.1), and (E) by the canonical isomorphism $({}^\vee Y)^\vee \simeq Y$ (see Section 1.6.3). All these bijections are natural in X and Y. Their composition turns the functor $F^!$ into a right adjoint of U (see Section 7.2.1). □

Theorem 8.9. *Let T be a Hopf monad on a rigid category $\mathcal{C}$. Then both functors $T: \mathcal{C} \to \mathcal{C}$ and $U_T: \mathcal{C}^T \to \mathcal{C}$ have right adjoints.*

Proof. Recall the adjunction $F_T \dashv U_T$, where $F_T : \mathcal{C} \to \mathcal{C}^T$ is the free module functor, see Section 7.2.2. In particular, F_T has a right adjoint and U_T has a left adjoint. By Theorem 8.7, the category $\mathcal{C}^T$ is rigid, so that U_T is a strict monoidal functor between rigid categories. By Lemma 8.8, the functor U_T has a right adjoint. Finally, since both F_T and U_T have right adjoints, their composition $T = U_T F_T$ has a right adjoint (see Section 7.2.1). □

8.3.6 Exercise

Let T be a left Hopf monad on a left rigid category $\mathcal{C}$ with distinguished left duality. By Lemma 8.6, the left antipode s^l of T is anti-multiplicative. Prove that it is anti-comultiplicative in the following sense: for all $X, Y \in \mathrm{Ob}(\mathcal{C})$,

$$s^l_{X\otimes Y} T({}^\vee T_2(X,Y)\, {}^\vee ?_2(T(Y), T(X))) = {}^\vee ?_2(Y,X)(s^l_Y \otimes s^l_X) T_2({}^\vee T(Y), {}^\vee T(X))$$

and

$$s^l_{\mathbb{1}} T({}^\vee T_0\, {}^\vee ?_0) = T_0,$$

where ${}^\vee ?$ is the left dual functor. Establish similar facts with left replaced by right.

8.3.7 Exercise*

Let $T = (T, \mu, \eta)$ be a left Hopf monad on a left rigid category $\mathcal{C}$ with distinguished left duality. Prove that the inverse of the left fusion operator H^l of T is computed via the left antipode s^l of T as follows: for any $X, Y \in \mathrm{Ob}(\mathcal{C})$,

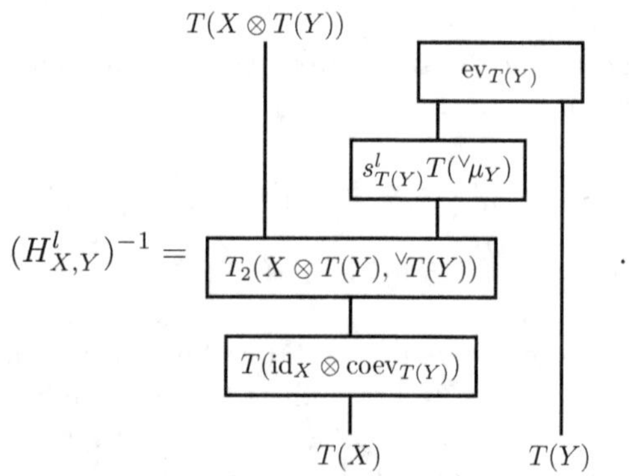

8.3.8 Exercise*

Prove that all R-matrices for a left Hopf monad on a left rigid category are $*$-invertible in the sense of Section 7.7.1. Prove the same with left replaced by right.

8.4 Involutory Hopf monads on pivotal categories

We introduce involutory Hopf monads on pivotal categories and show that their categories of modules carry a canonical pivotal structure.

8.4.1 Involutory Hopf monads

Let $\mathcal{C}$ be a pivotal category, with pivotal duality $\{(X^*, \mathrm{ev}_X, \widetilde{\mathrm{ev}}_X)\}_{X \in \mathrm{Ob}(\mathcal{C})}$. Recall that this duality induces (in the obvious way) distinguished left and right dualities in $\mathcal{C}$. A Hopf monad T on $\mathcal{C}$ is *involutory* if its left and right antipodes coincide. Then the left/right antipode is called the *antipode* of T and is denoted by

$$s = \{s_X \colon T(T(X)^*) \to X^*\}_{X \in \mathrm{Ob}(\mathcal{C})}.$$

Theorem 8.10. *Let T be a bimonad on a pivotal category $\mathcal{C}$. The following two conditions are equivalent:*

(i) *T is an involutory Hopf monad;*

(ii) *The monoidal category $\mathcal{C}^T$ has a structure of a pivotal category such that the forgetful functor $U_T \colon \mathcal{C}^T \to \mathcal{C}$ is pivotal.*

If these conditions hold, then there is a unique pivotal duality in $\mathcal{C}^T$ which makes U_T strictly pivotal or, equivalently, which is a lift along U_T of the pivotal duality in $\mathcal{C}$ (see Section 1.7.5).

The proof of Theorem 8.10 given below shows that the pivotal duality in $\mathcal{C}^T$ mentioned in the theorem is computed from the antipode s of T:

$$\big\{\big((M,r)^* = (M^*, s_M T(r^*)),\ \mathrm{ev}_{(M,r)} = \mathrm{ev}_M,\ \widetilde{\mathrm{ev}}_{(M,r)} = \widetilde{\mathrm{ev}}_M\big)\big\}_{(M,r) \in \mathrm{Ob}(\mathcal{C}^T)}.$$

Also, since U_T is pivotal, strict monoidal, and acts as the identity on morphisms, Lemma 2.8 implies that the left/right trace of any endomorphism f in $\mathcal{C}^T$ is nothing but the left/right trace of f viewed as an endomorphism in $\mathcal{C}$. Consequently, if $\mathcal{C}$ is spherical, then so is $\mathcal{C}^T$.

8.4.2 Proof of Theorem 8.10

We first prove the implication (i) $\Rightarrow$ (ii). Let s be the antipode of T. By Theorem 8.7, for any T-module (M,r), the tuples

$$\big((M^*, s_M T(r^*)),\ \mathrm{ev}_{(M,r)} = \mathrm{ev}_M\big) \quad \text{and} \quad \big((M^*, s_M T(r^*)),\ \widetilde{\mathrm{ev}}_{(M,r)} = \widetilde{\mathrm{ev}}_M\big)$$

are a left dual and a right dual of (M,r), respectively. Since the (co)evaluations in $\mathcal{C}^T$ are induced by those in $\mathcal{C}$ and the forgetful functor $U_T \colon \mathcal{C}^T \to \mathcal{C}$ acts as the identity on morphisms, the axioms (i)–(iii) of Section 1.7.1 are satisfied in $\mathcal{C}^T$. Thus the family

$$\big\{\big((M,r)^* = (M^*, s_M T(r^*)),\ \mathrm{ev}_{(M,r)} = \mathrm{ev}_M,\ \widetilde{\mathrm{ev}}_{(M,r)} = \widetilde{\mathrm{ev}}_M\big)\big\}_{(M,r)}, \tag{8.16}$$

where (M,r) runs over all T-modules, is a pivotal duality in $\mathcal{C}^T$. Also, since U_T is strict monoidal, for any pairing $\omega\colon (M,r)\otimes(N,s)\to(\mathbb{1},T_0)$ in $\mathcal{C}^T$, we have:

$$(\omega)^{U_T} = (U_T)_0^{-1} U_T(\omega)(U_T)_2((M,r),(N,s)) = \omega\colon M\otimes N \to \mathbb{1}.$$

This implies that

$$(U_T)^l(M,r) = \mathrm{id}_{M^*} = (U_T)^r(M,r)$$

for any T-module (M,r), see Section 1.6.4. Consequently, U_T is strictly pivotal.

We now prove the implication (ii) $\Rightarrow$ (i). Assume that there is a pivotal duality

$$\{((M,r)^\dagger,\, e_{(M,r)},\, \widetilde{e}_{(M,r)})\}_{(M,r)\in \mathrm{Ob}(\mathcal{C}^T)}$$

in $\mathcal{C}^T$ making U_T pivotal. In particular, $\mathcal{C}^T$ is rigid. Thus, by Theorem 8.7, the bimonad T is a Hopf monad. Let s^l and s^r be the left and right antipodes of T, respectively. We need to prove that they are equal. Pick $X\in\mathrm{Ob}(\mathcal{C})$ and set $(A,\alpha)=F_T(X)^\dagger$. Recall from Section 1.6.4 the isomorphism

$$\phi = (U_T)^l(F_T(X)) = (e_{F_T(X)}\otimes \mathrm{id}_{T(X)^*})(\mathrm{id}_A\otimes \mathrm{coev}_{T(X)})\colon A\to F_T(X)^*.$$

By Theorem 8.7, the left coevaluation $\mathrm{coev}_{T(X)}$ is a T-linear morphism from $(\mathbb{1},T_0)$ to $F_T(X)\otimes(T(X)^*, s^l_{T(X)}T(\mu_X^*))$. We deduce that ϕ is a T-linear morphism from (A,α) to $(T(X)^*, s^l_{T(X)}T(\mu_X^*))$, that is,

$$\phi\alpha = s^l_{T(X)}T(\mu_X^*)T(\phi). \tag{8.17}$$

Similarly, the isomorphism

$$\psi = (U_T)^r(F_T(X)) = (\mathrm{id}_{T(X)^*}\otimes \widetilde{e}_{F_T(X)})(\widetilde{\mathrm{coev}}_{T(X)}\otimes \mathrm{id}_A)\colon A\to F_T(X)^*$$

is a T-linear isomorphism from (A,α) to $(T(X)^*, s^r_{T(X)}T(\mu_X^*))$, that is,

$$\psi\alpha = s^r_{T(X)}T(\mu_X^*)T(\psi). \tag{8.18}$$

Now the pivotality of U_T implies that $\phi=\psi$ (see Section 1.7.5). Consequently, it follows from (8.17), (8.18), and the fact that ϕ is an isomorphism that

$$s^l_{T(X)}T(\mu_X^*) = s^r_{T(X)}T(\mu_X^*).$$

Composing on the left with η_X^*, using the naturality of the antipodes, and then the unitality of μ, we obtain that $s^l_X = s^r_X$. Hence, $s^l=s^r$ and T is involutory.

Finally, assume that the conditions (i) and (ii) hold. Consider a pivotal duality in $\mathcal{C}^T$ which makes U_T strictly pivotal or, equivalently, which is a lift along U_T of the pivotal duality in $\mathcal{C}$ (see Section 1.7.5). In particular, its associated left and right dualities in $\mathcal{C}^T$ are lifts along U_T of the distinguished left and right dualities in $\mathcal{C}$. By Theorem 8.7(b), these lifts are computed by

$$\big\{\big((M,r)^* = (M^*, s^l_M T(r^*)),\ \mathrm{ev}_{(M,r)} = \mathrm{ev}_M\big)\big\}_{(M,r)\in\mathrm{Ob}(\mathcal{C}^T)}$$

and

$$\left\{\left((M,r)^* = (M^*, s^r_M T(r^*)),\ \mathrm{ev}_{(M,r)} = \mathrm{ev}_M\right)\right\}_{(M,r)\in \mathrm{Ob}(\mathcal{C}^T)}$$

where s^l and s^r are the left and right antipodes of T. Since $s^l = s^r$ is the antipode of T, our pivotal duality in $\mathcal{C}^T$ is nothing but the pivotal duality (8.16). As shown in the proof of the implication (i) $\Rightarrow$ (ii), the pivotal duality (8.16) makes U_T strictly pivotal. Hence, there is a unique pivotal duality in $\mathcal{C}^T$ which makes U_T strictly pivotal

8.4.3 Example

Let A be a Hopf algebra in a braided pivotal category $\mathcal{C}$. Consider the Hopf monad $T = A\otimes ?$ on $\mathcal{C}$ from Example 8.1.4. By Example 8.3.4, the antipodes of T are computed by

$s^l_X =$ 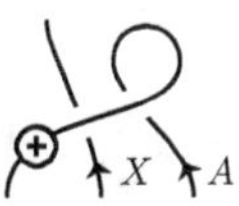

and $s^r_X =$ 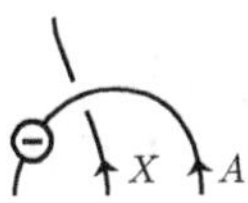

for all $X \in \mathrm{Ob}(\mathcal{C})$. Consequently, the Hopf monad T is involutory if and only if the Hopf algebra A in involutory in the sense of Section 6.2.3. Note that the pivotal dualities in $\mathcal{C}^T = \mathrm{mod}_A$ provided by Theorem 8.10 and Section 6.2.3 coincide.

8.4.4 Exercise

Let $T = (T, \mu, \eta)$ be an involutory quasitriangular Hopf monad on a pivotal category $\mathcal{C}$. By Theorems 7.11 and 8.10, the category $\mathcal{C}^T$ is braided and pivotal. Prove that it is ribbon if and only if the R-matrix R of T satisfies the following condition: for all $X \in \mathrm{Ob}(\mathcal{C})$,

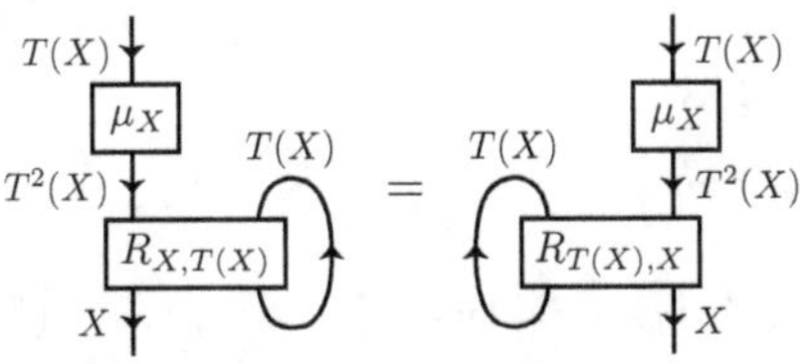

.

8.5 Semisimple Hopf monads

We study properties of semisimple Hopf monads.

8.5.1 Semisimplicity criteria

Recall the notions of semisimple and separable monads from Section 7.3.3. The following theorem reformulates these notions and establishes their equivalence in the class of Hopf monads.

Theorem 8.11. *For a Hopf monad $T = (T, \mu, \eta)$ on a monoidal category $\mathcal{C}$, the following conditions are equivalent:*

(i) *T is separable;*

(ii) *T is semisimple;*

(iii) *The monoidal unit $(\mathbb{1}, T_0)$ of $\mathcal{C}^T$ is a projective object;*

(iv) *There is a morphism $\Lambda\colon \mathbb{1} \to T(\mathbb{1})$ in $\mathcal{C}$ such that $T_0\Lambda = \mathrm{id}_\mathbb{1}$ and $\Lambda T_0 = \mu_\mathbb{1} T(\Lambda)$.*

Proof. Clearly, (i) $\Rightarrow$ (ii) $\Rightarrow$ (iii). The condition (iv) may be rephrased by saying that the T-linear morphism $T_0\colon (T(\mathbb{1}), \mu_\mathbb{1}) = F_T(\mathbb{1}) \to (\mathbb{1}, T_0)$ has a T-linear section. Thus, (iii) $\Rightarrow$ (iv) by Lemma 7.3. It remains to show that (iv) $\Rightarrow$ (i). Consider the natural transformations

$$\Omega = \{\Omega_X = (H^l_{\mathbb{1},X})^{-1}\colon T(\mathbb{1}) \otimes T(X) \to T^2(X)\}_{X \in \mathrm{Ob}(\mathcal{C})},$$

where H^l is the left fusion operator of T, and

$$\gamma = \{\gamma_X = \Omega_X(\Lambda \otimes \eta_X)\colon X \to T^2(X)\}_{X \in \mathrm{Ob}(\mathcal{C})}.$$

Claims (a) and (d) of Lemma 8.1 imply respectively that for all $X \in \mathrm{Ob}(\mathcal{C})$,

$$T(\mu_X)\Omega_{T(X)} = \Omega_X(\mathrm{id}_{T(\mathbb{1})} \otimes \mu_X), \tag{8.19}$$

$$\mu_X \Omega_X = T_0 \otimes \mathrm{id}_{T(X)}. \tag{8.20}$$

Lemma 8.1(e) and the definition of $H^l_{T(\mathbb{1}),X}$ imply that for all $X \in \mathrm{Ob}(\mathcal{C})$,

$$T(\Omega_X) = \Omega_{T(X)}(\mathrm{id}_{T(\mathbb{1})} \otimes \Omega_X)(\Omega_\mathbb{1}^{-1} \otimes \mu_X)T_2(T(\mathbb{1}), T(X)). \tag{8.21}$$

Pick $X \in \mathrm{Ob}(\mathcal{C})$. First, we have:

$$\mu_X\gamma_X \overset{\text{(i)}}{=} \mu_X\Omega_X(\Lambda \otimes \eta_X) \overset{\text{(ii)}}{=} T_0\Lambda \otimes \eta_X \overset{\text{(iii)}}{=} \eta_X.$$

Here, (i) follows from the definition of γ, (ii) from (8.20), and (iii) from the assumption on Λ. Second, we have:

$$\begin{aligned}
\mu_{T(X)}T(\gamma_X) &\overset{\text{(i)}}{=} \mu_{T(X)}T(\Omega_X)T(\Lambda \otimes \eta_X)\\
&\overset{\text{(ii)}}{=} \mu_{T(X)}\Omega_{T(X)}(\mathrm{id}_{T(\mathbb{1})} \otimes \Omega_X)(\Omega_\mathbb{1}^{-1} \otimes \mu_X)T_2(T(\mathbb{1}), T(X))T(\Lambda \otimes \eta_X)\\
&\overset{\text{(iii)}}{=} (T_0 \otimes \Omega_X)(\Omega_\mathbb{1}^{-1}T(\Lambda) \otimes \mu_X T(\eta_X))T_2(\mathbb{1}, X)\\
&\overset{\text{(iv)}}{=} \Omega_X((T_0 \otimes \mathrm{id}_{T(\mathbb{1})})\Omega_\mathbb{1}^{-1}T(\Lambda) \otimes \mu_X\eta_{T(X)})T_2(\mathbb{1}, X)\\
&\overset{\text{(v)}}{=} \Omega_X(\mu_\mathbb{1}T(\Lambda) \otimes \mu_X\eta_{T(X)})T_2(\mathbb{1}, X)\\
&\overset{\text{(vi)}}{=} \Omega_X(\Lambda T_0 \otimes \mu_X\eta_{T(X)})T_2(\mathbb{1}, X)\\
&\overset{\text{(vii)}}{=} T(\mu_X)\Omega_{T(X)}(\Lambda T_0 \otimes \eta_{T(X)})T_2(\mathbb{1}, X)\\
&\overset{\text{(viii)}}{=} T(\mu_X)\Omega_{T(X)}(\Lambda \otimes \eta_{T(X)}) \overset{\text{(ix)}}{=} T(\mu_X)\gamma_{T(X)}.
\end{aligned}$$

Here, (i) and (ix) follow from the definition of γ, (ii) from (8.21), (iii) from (8.20) and the naturality of T_2, (iv) from the unitality of μ, (v) from (8.20), (vi) from the assumptions on Λ, (vii) from (8.19), and (viii) from the counitality (7.4) of T_2. Thus, T is separable by Lemma 7.4. □

8.5.2 Remark

A Hopf algebra A in a braided category $\mathcal{C}$ gives rise to the Hopf monad $T = A\otimes ?$ on $\mathcal{C}$, see Example 8.1.4. Clearly, a morphism $\Lambda\colon \mathbb{1} \to A$ in $\mathcal{C}$ satisfies $\Lambda T_0 = \mu_{\mathbb{1}} T(\Lambda)$ (where μ is the product of T) if and only if Λ is a left integral for A (see Section 6.3.1). Theorem 8.11 and Exercise 7.3.5 imply that A is separable as an algebra in $\mathcal{C}$ if and only if A has a left integral Λ such that $\varepsilon(\Lambda) = \mathrm{id}_{\mathbb{1}}$, where ε is the counit of A. If $\Bbbk$ is a field, then Theorem 8.11 applied to the category $\mathrm{vect}_{\Bbbk}$ of finite-dimensional $\Bbbk$-vector spaces implies the well-known Maschke's semisimplicity theorem: for a finite-dimensional Hopf $\Bbbk$-algebra A, the following three conditions are equivalent:

(i) A is semisimple as a $\Bbbk$-algebra,

(ii) A is separable as a $\Bbbk$-algebra,

(iii) A has a left integral Λ such that $\varepsilon(\Lambda) = \mathrm{id}_{\mathbb{1}}$, where ε is the counit of A.

Indeed, a finite-dimensional $\Bbbk$-algebra A is semsimple if and only if the action of A on any finite-dimensional left A-module has an A-linear section, that is, if and only if the monad $A\otimes_{\Bbbk} ?$ on $\mathrm{vect}_{\Bbbk}$ is semisimple.

8.5.3 Semisimple Hopf monads on fusion categories

The following theorem exhibits a fundamental property of semisimple Hopf monads on fusion categories.

Theorem 8.12. *Let $\Bbbk$ be an algebraically closed field and let T be a semisimple $\Bbbk$-linear Hopf monad on an additive fusion $\Bbbk$-category $\mathcal{C}$. Then the category $\mathcal{C}^T$ of T-modules is an abelian fusion $\Bbbk$-category.*

Proof. We endow $\mathcal{C}^T$ with the unique structure of a monoidal $\Bbbk$-category such that the forgetful functor $U_T\colon \mathcal{C}^T \to \mathcal{C}$ is $\Bbbk$-linear and strict monoidal (see Lemma 7.9). Since $\Bbbk$ is a field, Theorem C.6 implies that $\mathcal{C}$ is abelian. By Theorem 8.9, the functor T has a right adjoint. By Lemma 7.8(a), the category $\mathcal{C}^T$ has a zero object. Thus, by Lemma C.2(b), the functor T preserves cokernels. Consequently, by Lemma 7.8(e), the $\Bbbk$-category $\mathcal{C}^T$ is abelian. It is rigid by Theorem 8.7. Next, we prove that any monomorphism $q\colon (N,s) \to (M,r)$ in $\mathcal{C}^T$ admits a retraction, i.e., a morphism $p\colon (M,r) \to (N,s)$ in $\mathcal{C}^T$ such that $pq = \mathrm{id}_N$. Being a right adjoint of the free module functor F_T, the functor U_T preserves monomorphisms (see Exercise 7.3.6). Consequently, $q = U_T(q)\colon N \to M$ is a monomorphism in $\mathcal{C}$. By Theorem C.6, it has a retraction in $\mathcal{C}$, i.e., there is a morphism $v\colon M \to N$ in $\mathcal{C}$

such that $vq = \mathrm{id}_N$. Since the Hopf monad T is semisimple, Theorem 8.11 ensures its separability. Let $\gamma = \{\gamma_X : X \to T^2(X)\}_{X \in \mathrm{Ob}(\mathcal{C})}$ be the natural transformation provided by Lemma 7.4. Set

$$p = sT(vr)\gamma_M : M \to N.$$

We claim that p is a retraction of q in $\mathcal{C}^T$. Indeed, let μ and η be the product and unit of T, respectively. We have:

$$\begin{aligned} sT(p) &\overset{\text{(i)}}{=} sT(s)T^2(vr)T(\gamma_M) \overset{\text{(ii)}}{=} s\mu_N T^2(vr)T(\gamma_M) \\ &\overset{\text{(iii)}}{=} sT(vr)\mu_{T(M)}T(\gamma_M) \overset{\text{(iv)}}{=} sT(vr\mu_M)\gamma_{T(M)} \\ &\overset{\text{(v)}}{=} sT(vrT(r))\gamma_{T(M)} \overset{\text{(vi)}}{=} sT(vr)\gamma_M r \overset{\text{(vii)}}{=} pr. \end{aligned}$$

Here, (i) and (vii) follow from the definition of p, (ii) from the fact that s is a T-action, (iii) from the naturality of μ, (iv) from the equality $\mu_{T(M)}T(\gamma_M) = T(\mu_M)\gamma_{T(M)}$ (see Lemma 7.4), (v) from the fact that r is a T-action, and (vi) from the naturality of γ. Thus, p is a morphism $(M, r) \to (N, s)$ in $\mathcal{C}^T$. Moreover, we have:

$$\begin{aligned} pq &\overset{\text{(i)}}{=} sT(vr)\gamma_M q \overset{\text{(ii)}}{=} sT(vrT(q))\gamma_N \\ &\overset{\text{(iii)}}{=} sT(vqs)\gamma_N \overset{\text{(iv)}}{=} sT(s)\gamma_N \\ &\overset{\text{(v)}}{=} s\mu_N\gamma_N \overset{\text{(vi)}}{=} s\eta_N \overset{\text{(vii)}}{=} \mathrm{id}_N. \end{aligned}$$

Here (i) follows from the definition of p, (ii) from the naturality of γ, (iii) from the T-linearity of q, (iv) from the equality $vq = \mathrm{id}_N$, (v) and (vii) from the fact that s is a T-action, and (vi) from the equality $\mu_N\gamma_N = \eta_N$ (see Lemma 7.4). Hence, p is a retraction of q in $\mathcal{C}^T$.

We now prove that $\mathcal{C}^T$ is a fusion $\Bbbk$-category. Since $\Bbbk$ is a field, all simple T-modules are indecomposable, see Section C.2.2. We prove that, conversely, all indecomposable T-modules are simple. Since $\mathcal{C}^T$ is an abelian $\Bbbk$-category in which all monomorphisms have retractions, Lemma C.4 implies that any morphism between indecomposable T-modules is either zero or an isomorphism. Consequently, the Hom-space between any two non-isomorphic indecomposable T-modules is zero, and the algebra of endomorphisms of an indecomposable T-module is a finite-dimensional division $\Bbbk$-algebra. The field $\Bbbk$ being algebraically closed, such an algebra is isomorphic to $\Bbbk$. Thus, all indecomposable T-modules are simple.

The Hom-spaces in $\mathcal{C}^T$ are vector subspaces of the Hom-spaces in $\mathcal{C}$, and the latter are finite-dimensional $\Bbbk$-vector spaces because $\mathcal{C}$ is fusion. Therefore the Hom-spaces in $\mathcal{C}^T$ are finite-dimensional $\Bbbk$-vector spaces. Lemma C.5 implies that any T-module is a direct sum of finitely many simple T-modules. Note that the monoidal unit $(\mathbb{1}, T_0)$ of $\mathcal{C}^T$ is a simple object, because $\mathbb{1}$ is a simple object of $\mathcal{C}$ and T is $\Bbbk$-linear (see Exercise 7.4.3).

To conclude that $\mathcal{C}^T$ is a fusion $\Bbbk$-category, it remains to verify that there is a finite representative set of simple objects of $\mathcal{C}^T$. Pick a (finite) representative set I of simple objects of $\mathcal{C}$. By the above, for any $i \in I$, the T-module $F_T(i)$ is a direct sum of a finite family $(S_\alpha)_{\alpha\in A_i}$ of simple T-modules. Consider the following set of simple T-modules:

$$\mathcal{J} = \{S_\alpha \,|\, \alpha \in A_i,\ i \in I\} \cup \{(\mathbb{1}, T_0)\}.$$

Since $\mathcal{J}$ is finite, it has a maximal subset J containing $(\mathbb{1}, T_0)$ and such that distinct elements of J are non-isomorphic. Clearly, J is a finite set. We claim that J is a representative set of simple objects of $\mathcal{C}^T$. Indeed, by construction, $(\mathbb{1}, T_0) \in J$. Any distinct elements A, B of J are non-isomorphic and so, by the above, $\mathrm{Hom}_{\mathcal{C}^T}(A, B) = 0$. It remains to prove that any simple T-module S is isomorphic to an element of J. We have:

$$\Bbbk \simeq \mathrm{End}_{\mathcal{C}^T}(S) \subset \mathrm{End}_{\mathcal{C}}(U_T(S)).$$

Thus, the object $U_T(S)$ is non-zero. Since $\mathcal{C}$ is fusion, $U_T(S)$ has at least one $i \in I$ as a direct summand. Using the expansion $F_T(i) \simeq \bigoplus_{\alpha\in A_i} S_\alpha$ and the adjunction bijections of $F_T \dashv U_T$, we obtain

$$\bigoplus_{\alpha\in A_i} \mathrm{Hom}_{\mathcal{C}^T}(S_\alpha, S) \simeq \mathrm{Hom}_{\mathcal{C}^T}(F_T(i), S) \simeq \mathrm{Hom}_{\mathcal{C}}(i, U_T(S)) \neq 0.$$

Thus, $\mathrm{Hom}_{\mathcal{C}^T}(S_\alpha, S) \neq 0$ for some $\alpha \in A_i$. Since S_α and S are indecomposable T-modules, we obtain (by the above) that S is isomorphic to $S_\alpha \in \mathcal{J}$. The maximality of J implies that S_α is isomorphic to an element of J, and consequently, so is S. Hence, J is a finite representative set of simple objects of $\mathcal{C}^T$. □

8.6 Coends of the categories of modules

Under certain assumptions on a Hopf monad, we compute the coend of its category of modules following [BV2]. We systematically use in this section the notion of a coend of a functor and the notions of preservation and creation of coends, see Appendix B.

8.6.1 Preservation and creation of coends

We begin with a lemma concerning monads on arbitrary categories.

Lemma 8.13. *If a monad T on a category $\mathcal{C}$ preserves coends, then the forgetful functor $U_T\colon \mathcal{C}^T \to \mathcal{C}$ creates coends.*

Proof. Let $G\colon \mathcal{E}^{\mathrm{op}} \times \mathcal{E} \to \mathcal{C}^T$ be a functor such that $U_T G\colon \mathcal{E}^{\mathrm{op}} \times \mathcal{E} \to \mathcal{C}$ has a coend (C, ρ), where $C \in \mathrm{Ob}(\mathcal{C})$ and $\rho = \{\rho_Y : U_T G(Y,Y) \to C\}_{Y\in\mathrm{Ob}(\mathcal{E})}$ is a universal dinatural transformation. We must prove that

(a) there is a unique action $\alpha\colon T(C) \to C$ of T on C such that for all $Y \in \mathrm{Ob}(\mathcal{C})$, the morphism ρ_Y is T-linear from $G(Y,Y)$ to (C,α);

(b) the pair $((C,\alpha),\rho)$ is a coend of the functor G.

Let us prove (a). For any $Y, Z \in \mathrm{Ob}(\mathcal{E})$, let

$$a_{Y,Z}\colon TU_TG(Y,Z) \to U_TG(Y,Z)$$

be the action of the T-module $G(Y,Z)$. For any $Y \in \mathrm{Ob}(\mathcal{E})$, set $a_Y = a_{Y,Y}$. Then the family $\{\rho_Y a_Y\}_{Y \in \mathrm{Ob}(\mathcal{E})}$ is a dinatural transformation from TU_TG to C. Indeed, for any morphism $f\colon Y \to Z$ in $\mathcal{E}$,

$$\begin{aligned}\rho_Z a_Z TU_TG(\mathrm{id}_Z, f) &\overset{\text{(i)}}{=} \rho_Z U_TG(\mathrm{id}_Z, f) a_{Z,Y}\\ &\overset{\text{(ii)}}{=} \rho_Y U_TG(f, \mathrm{id}_Y) a_{Z,Y} \overset{\text{(iii)}}{=} \rho_Y a_Y TU_TG(f, \mathrm{id}_Y).\end{aligned}$$

Here, the equality (i) follows from the T-linearity of $G(\mathrm{id}_Z, f)$, (ii) from the dinaturality of ρ, and (iii) from the T-linearity of $G(f,\mathrm{id}_Y)$. Now, since T preserves coends, the pair $(T(C), T(\rho))$ is a coend of the functor TU_TG. Therefore there is a unique morphism $\alpha\colon T(C) \to C$ such that for all $Y \in \mathrm{Ob}(\mathcal{E})$,

$$\alpha T(\rho_Y) = \rho_Y a_Y. \tag{8.22}$$

Let μ and η be the product and unit of T, respectively. For any $Y \in \mathrm{Ob}(\mathcal{E})$,

$$\alpha \eta_C \rho_Y \overset{\text{(i)}}{=} \alpha T(\rho_Y) \eta_{U_TG(Y,Y)} \overset{\text{(ii)}}{=} \rho_Y a_Y \eta_{U_TG(Y,Y)} \overset{\text{(iii)}}{=} \rho_Y$$

and

$$\begin{aligned}\alpha T(\alpha) T^2(\rho_Y) &\overset{\text{(iv)}}{=} \alpha T(\rho_Y) T(a_Y) \overset{\text{(v)}}{=} \rho_Y a_Y T(a_Y)\\ &\overset{\text{(vi)}}{=} \rho_Y a_Y \mu_{U_TG(Y,Y)} \overset{\text{(vii)}}{=} \alpha T(\rho_Y) \mu_{U_TG(Y,Y)} \overset{\text{(viii)}}{=} \alpha \mu_C T^2(\rho_Y).\end{aligned}$$

Here, (i) follows from the naturality of η, the equalities (ii), (iv), (v), and (vii) from (8.22), (iii) and (vi) from the fact that a_Y is a T-action, and (viii) from the naturality of μ. Since ρ and $T^2(\rho)$ are universal dinatural transformation (because T preserves coends), we conclude that $\alpha\eta_C = \mathrm{id}_C$ and $\alpha T(\alpha) = \alpha\mu_C$, that is, α is an action of T on C. Note that (8.22) means that $\rho_Y\colon G(Y,Y) \to (C,\alpha)$ is T-linear for all $Y \in \mathrm{Ob}(\mathcal{E})$. Assume that $\alpha'\colon T(C) \to C$ is another action of T on C such that $\rho_Y\colon G(Y,Y) \to (C,\alpha')$ is T-linear, that is, $\alpha' T(\rho_Y) = \rho_Y a_Y$, for all $Y \in \mathrm{Ob}(\mathcal{E})$. The uniqueness of the morphism $\alpha\colon T(C) \to C$ satisfying (8.22) for all $Y \in \mathrm{Ob}(\mathcal{E})$ implies that $\alpha' = \alpha$.

We now prove (b). By (a), the family $\{\rho_Y\colon G(Y,Y) \to (C,\alpha)\}_{Y \in \mathrm{Ob}(\mathcal{E})}$ is a dinatural transformation from G to (C,α). We need to prove that it is universal. Let d be a dinatural transformation from G to a T-module (M,r). Then $U_T(d)$ is a dinatural transformation from $U_T G$ to M and so there exists a unique morphism $\varphi\colon C \to M$ such that $\varphi\rho_Y = d_Y$ (in $\mathcal{C}$) for all $Y \in \mathrm{Ob}(\mathcal{E})$. Now, for any $Y \in \mathrm{Ob}(\mathcal{E})$,

$$\varphi\alpha T(\rho_Y) \overset{\text{(i)}}{=} \varphi\rho_Y a_Y \overset{\text{(ii)}}{=} d_Y a_Y \overset{\text{(iii)}}{=} rT(d_Y) \overset{\text{(iv)}}{=} rT(\varphi)T(\rho_Y).$$

Here, (i) follows from (8.22), (ii) and (iv) from the equality $\varphi\rho_Y = d_Y$, and (iii) from the T-linearity of d_Y. The universality of $T(\rho)$ implies that $\varphi\alpha = rT(\varphi)$. Thus φ is a morphism in $\mathcal{C}^T$ from (C,α) to (M,r) such that $\varphi\rho_Y = d_Y$ (in $\mathcal{C}^T$) for all $Y \in \mathrm{Ob}(\mathcal{E})$. If $\varphi'\colon (C,\alpha) \to (M,r)$ is another such morphism in $\mathcal{C}^T$, then $\varphi'\rho_Y = d_Y = \varphi\rho_Y$ (in $\mathcal{C}$) for all $Y \in \mathrm{Ob}(\mathcal{E})$. Since ρ is a universal dinatural tranformation from $U_T G$ to C, we have $\varphi' = \varphi$. □

Theorem 8.14. *Let T be a Hopf monad on a rigid category $\mathcal{C}$. Then the functor $T\colon \mathcal{C} \to \mathcal{C}$ preserves coends and the forgetful functor $U_T\colon \mathcal{C}^T \to \mathcal{C}$ preserves and creates coends.*

Proof. By Theorem 8.9, the functors T and U_T have right adjoints. By Lemma B.5, these functors preserve coends. Finally, by Lemma 8.13, the functor U_T creates coends. □

8.6.2 The coend of $\mathcal{C}^T$

Let $T = (T,\mu,\eta)$ be a Hopf monad on a rigid category $\mathcal{C}$ with distinguished left duality $\{({}^\vee X, \mathrm{ev}_X)\}_{X\in\mathrm{Ob}(\mathcal{C})}$. By Theorem 8.7, the category $\mathcal{C}^T$ is rigid with distinguished left duality computed from the left antipode s^l of T as follows:

$$\big\{\big({}^\vee(M,r) = ({}^\vee M, s^l_M T({}^\vee r)),\ \mathrm{ev}_{(M,r)} = \mathrm{ev}_M\big)\big\}_{(M,r)\in\mathrm{Ob}(\mathcal{C}^T)}.$$

Under an additional assumption on T we show that $\mathcal{C}^T$ has a coend and compute it. Namely, we assume the existence of the coend

$$C = \int^{X\in\mathcal{C}} {}^\vee T(X) \otimes X \in \mathrm{Ob}(\mathcal{C}) \tag{8.23}$$

with universal dinatural transformation $\varrho = \{\varrho_X\colon {}^\vee T(X)\otimes X \to C\}_{X\in\mathrm{Ob}(\mathcal{C})}$. We equip C with a T-action as follows. Since the functor $T\colon \mathcal{C}\to\mathcal{C}$ preserves coends by Theorem 8.14, the dinatural transformation

$$T(\varrho) = \{T(\varrho_X)\colon T({}^\vee T(X)\otimes X) \to T(C)\}_{X\in\mathrm{Ob}(\mathcal{C})}$$

is universal. Consider the family $d = \{d_X\colon T({}^\vee T(X)\otimes X) \to C\}_{X\in\mathrm{Ob}(\mathcal{C})}$ of morphisms in $\mathcal{C}$ defined by

$$d_X = \varrho_{T(X)}\big({}^\vee\mu_X s^l_{T(X)} T({}^\vee\mu_X) \otimes \mathrm{id}_{T(X)}\big)\, T_2\big({}^\vee T(X), X\big).$$

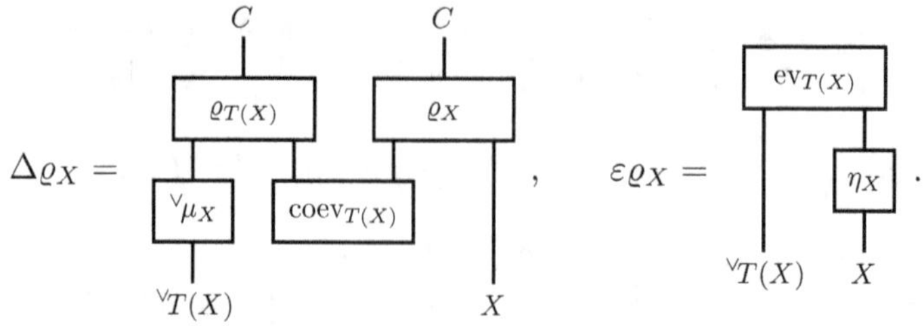

Figure 8.1: The coalgebra structure of (C, α)

Graphically:

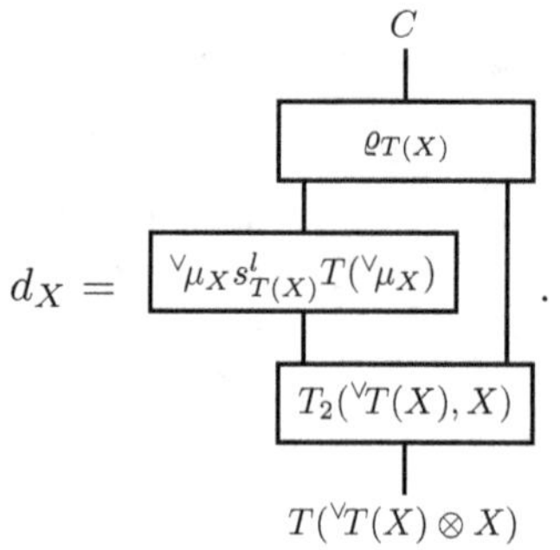

It follows from the dinaturality of ϱ and the naturality of T_2, μ, s^l that d is a dinatural transformation. Consequently, the universality of $T(\varrho)$ implies that there is a unique morphism $\alpha\colon T(C) \to C$ such that for all $X \in \mathrm{Ob}(\mathcal{C})$,

$$d_X = \alpha T(\varrho_X). \tag{8.24}$$

Theorem 8.15. *The morphism α is a T-action and the T-module (C, α) is a coend of $\mathcal{C}^T$ with universal dinatural transformation*

$$\tilde{\varrho} = \{\tilde{\varrho}_{(M,r)} = \varrho_M({}^\vee r \otimes \mathrm{id}_M)\colon {}^\vee(M,r) \otimes (M,r) \to (C,\alpha)\}_{(M,r)\in\mathrm{Ob}(\mathcal{C}^T)}.$$

The coproduct Δ and counit ε of the coend (C, α) (see Section 6.4) are uniquely determined by the equalities of Figure 8.1, where X runs over the objects of $\mathcal{C}$.

Proof. Consider the functors

$$G\colon (\mathcal{C}^T)^{\mathrm{op}} \times \mathcal{C}^T \to \mathcal{C}^T \quad \text{and} \quad H\colon (\mathcal{C}^T)^{\mathrm{op}} \times \mathcal{C} \to \mathcal{C}$$

defined by

$$G((M,r),(N,s)) = {}^\vee(M,r) \otimes (N,s), \qquad H((M,r),X) = {}^\vee M \otimes X,$$
$$G(f,g) = {}^\vee f \otimes g, \qquad H(f,h) = {}^\vee f \otimes h,$$

for all T-modules $(M,r),(N,s)$, all $X \in \mathrm{Ob}(\mathcal{C})$, all morphisms f,g in $\mathcal{C}^T$, and all morphisms h in $\mathcal{C}$. Since the forgetful functor $U_T\colon \mathcal{C}^T \to \mathcal{C}$ is strict monoidal and carries the duality in $\mathcal{C}^T$ to the duality in $\mathcal{C}$,

$$U_T G = H(1_{(\mathcal{C}^T)^{\mathrm{op}}} \times U_T)\colon (\mathcal{C}^T)^{\mathrm{op}} \times \mathcal{C}^T \to \mathcal{C}.$$

By definition, a coend of $\mathcal{C}^T$ is a coend of the functor G. By Theorem 8.14, U_T preserves and creates coends. Thus $\mathcal{C}^T$ has a coend if and only if the functor $U_T G$ has a coend. Since the free module functor $F_T\colon \mathcal{C} \to \mathcal{C}^T$ is a left adjoint of U_T, (see Section 7.2.2), Lemma B.4 implies that the functor $U_T G = H(1_{(\mathcal{C}^T)^{\mathrm{op}}} \times U_T)$ has a coend if and only if the functor $H(F_T \times 1_\mathcal{C})\colon \mathcal{C}^{\mathrm{op}} \times \mathcal{C} \to \mathcal{C}$ has a coend. Since $H(F_T \times 1_\mathcal{C}) = {}^\vee T \otimes 1_\mathcal{C}$, the coend of $H(F_T \times 1_\mathcal{C})$ exists by assumption and is (C,ϱ). Thus, the category $\mathcal{C}^T$ has a coend. Moreover, since the counit ε of the adjunction $F_T \dashv U_T$ is given by $\varepsilon_{(M,r)} = r$ for all T-modules (M,r), Lemma B.4 implies that the coend of $H(1_{(\mathcal{C}^T)^{\mathrm{op}}} \times U_T) = U_T G$ is the object $C \in \mathrm{Ob}(\mathcal{C})$ with associated universal dinatural transformation $U_T(\tilde{\varrho})$ with $\tilde{\varrho}$ as defined in the statement of the theorem. Since U_T creates coends, there is a unique T-action $\alpha\colon T(C) \to C$ such that each $U_T(\tilde{\varrho}_{(M,r)}) = \tilde{\varrho}_{(M,r)}$ becomes a T-linear morphism from

$$G((M,r),(M,r)) = {}^\vee(M,r) \otimes (M,r) = \big({}^\vee M \otimes M, (s^l_M T({}^\vee r) \otimes r) T_2({}^\vee M, M)\big)$$

to (C,α), and then the T-module (C,α) is the coend of G (i.e., of $\mathcal{C}^T$) with associated universal dinatural transformation $\tilde{\varrho}$. Let us prove that this morphism α satisfies (8.24). Pick any $X \in \mathrm{Ob}(\mathcal{C})$ and consider the T-module $F_T(X) = (T(X), \mu_X)$. The T-linearity of $\tilde{\varrho}_{F_T(X)}$ means that

$$\alpha T\big(\varrho_{T(X)}({}^\vee\mu_X \otimes \mathrm{id}_{T(X)})\big) = \varrho_{T(X)}({}^\vee\mu_X s^l_{T(X)} T({}^\vee\mu_X) \otimes \mu_X) T_2({}^\vee T(X), T(X)).$$

Composing this equality on the right with $T(\mathrm{id}_{{}^\vee T(X)} \otimes \eta_X)$ and using the naturality of T_2 and the unitality of μ, we obtain

$$\alpha T\big(\varrho_{T(X)}({}^\vee\mu_X \otimes \eta_X)\big) = \varrho_{T(X)}({}^\vee\mu_X s^l_{T(X)} T({}^\vee\mu_X) \otimes \mathrm{id}_{T(X)}) T_2({}^\vee T(X), X) = d_X.$$

This yields (8.24) since the dinaturality of ϱ and the unitality of μ imply that

$$\varrho_{T(X)}({}^\vee\mu_X \otimes \eta_X) = \varrho_X({}^\vee T(\eta_X){}^\vee\mu_X \otimes \mathrm{id}_X) = \varrho_X. \tag{8.25}$$

We now compute the coalgebra structure of the coend (C,α) of $\mathcal{C}^T$. By Section 6.4.2, the coproduct Δ is characterized as follows: for any T-module (M,r),

$$\Delta\tilde{\varrho}_{(M,r)} = (\tilde{\varrho}_{(M,r)} \otimes \tilde{\varrho}_{(M,r)})(\mathrm{id}_{{}^\vee(M,r)} \otimes \mathrm{coev}_{(M,r)} \otimes \mathrm{id}_{(M,r)}).$$

We apply this equality to the free T-module $F_T(X)$, where $X \in \mathrm{Ob}(\mathcal{C})$. Composing both sides on the right with $\mathrm{id}_{{}^\vee T(X)} \otimes \eta_X$ and using that

$$\tilde{\varrho}_{F_T(X)} = \varrho_{T(X)}({}^\vee\mu_X \otimes \mathrm{id}_{T(X)}) \quad \text{and} \quad \mathrm{coev}_{F_T(X)} = \mathrm{coev}_{T(X)},$$

we obtain the equality

$$\Delta\varrho_{T(X)}({}^{\vee}\mu_X \otimes \eta_X) = \big(\varrho_{T(X)} \otimes \varrho_{T(X)}({}^{\vee}\mu_X \otimes \eta_X)\big)({}^{\vee}\mu_X \otimes \mathrm{coev}_{T(X)} \otimes \mathrm{id}_X).$$

Therefore, using (8.25), we conclude that for any $X \in \mathrm{Ob}(\mathcal{C})$,

$$\Delta\varrho_X = (\varrho_{T(X)} \otimes \varrho_X)({}^{\vee}\mu_X \otimes \mathrm{coev}_{T(X)} \otimes \mathrm{id}_X).$$

This is the first equality in Figure 8.1. It determines Δ uniquely because the dinatural transformation ϱ is universal. Similarly, by Section 6.4.2, the counit ε is characterized by the condition that for any T-module (M, r),

$$\varepsilon\tilde{\varrho}_{(M,r)} = \mathrm{ev}_{(M,r)}.$$

We apply this equality to the free T-module $F_T(X)$, where $X \in \mathrm{Ob}(\mathcal{C})$. Composing both sides of the resulting formula on the right with $\mathrm{id}_{{}^{\vee}T(X)} \otimes \eta_X$ and using the definition of $\tilde{\varrho}_{F_T(X)}$, (8.25), and the identity $\mathrm{ev}_{F_T(X)} = \mathrm{ev}_{T(X)}$, we obtain that

$$\varepsilon\varrho_X = \mathrm{ev}_{T(X)}(\mathrm{id}_{{}^{\vee}T(X)} \otimes \eta_X).$$

This is the second equality in Figure 8.1. It determines ε uniquely because the dinatural transformation ϱ is universal. □

8.6.3 The quasitriangular case

We keep the assumptions and notation of the previous subsection and additionally assume that the Hopf monad T is quasitriangular. Then, by Theorem 7.11, the category $\mathcal{C}^T$ is braided. By Section 6.5, the coend (C, α) of $\mathcal{C}^T$ is a Hopf algebra in $\mathcal{C}^T$ carrying a canonical Hopf pairing.

Theorem 8.16. *The product m, unit u, antipode S, and canonical pairing ω of the Hopf algebra (C, α) are uniquely determined by the equalities of Figure 8.2, where μ, s^l, R are respectively the product, left antipode,* R*-matrix of T, ${}^{\vee}?_2$ is the monoidal constraint of the left dual functor in $\mathcal{C}$, and X, Y run over all objects of $\mathcal{C}$.*

Proof. Recall from Theorem 8.15 the universal dinatural transformation $\tilde{\varrho}$ associated to (C, α). By Section 6.5.1, the unit u of (C, α) is given by

$$u = \tilde{\varrho}_{(\mathbb{1},T_0)}\,\mathrm{coev}_{(\mathbb{1},T_0)} = \varrho_{\mathbb{1}}\,{}^{\vee}T_0\,\mathrm{coev}_{\mathbb{1}}.$$

This yields the expression for u in Figure 8.2. Next, pick $X, Y \in \mathrm{Ob}(\mathcal{C})$ and consider the free T-modules $F_T(X) = (T(X), \mu_X)$ and $F_T(Y) = (T(Y), \mu_Y)$. Using Formula (3.2) and Section 6.5.1, we obtain that the product m of (C, α) satisfies

$$\begin{aligned} m(\tilde{\varrho}_{F_T(X)} \otimes \tilde{\varrho}_{F_T(Y)}) &= \tilde{\varrho}_{F_T(X) \otimes F_T(Y)} \\ &\circ \big({}^{\vee}?_2(F_T(Y), F_T(X)) \otimes \mathrm{id}_{F_T(X) \otimes F_T(Y)}\big)\big(\tau_{{}^{\vee}F_T(X) \otimes F_T(X), {}^{\vee}F_T(Y)} \otimes \mathrm{id}_{F_T(Y)}\big), \end{aligned}$$

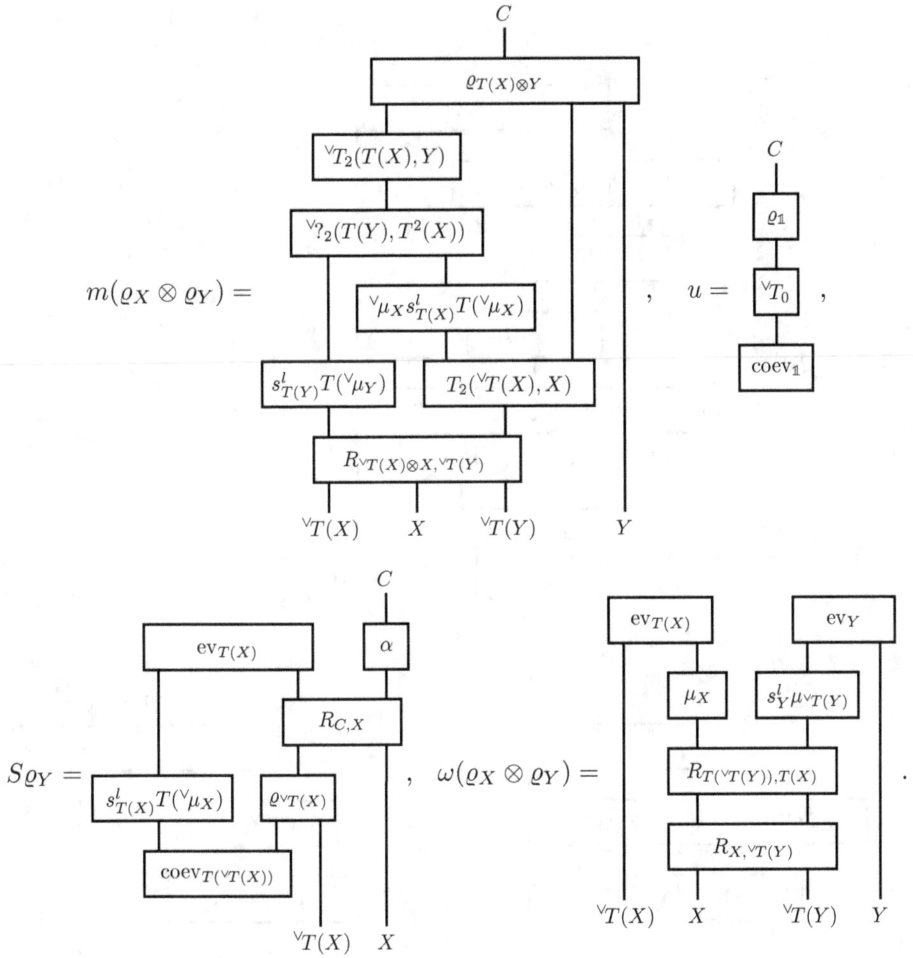

Figure 8.2: The Hopf algebra structure of (C, α)

where τ is the braiding of $\mathcal{C}^T$ determined by R and ${}^\vee ?_2$ is the monoidal constraint of the left dual functor in $\mathcal{C}^T$. Composing both sides on the right with

$$\mathrm{id}_{{}^\vee T(X)} \otimes \eta_X \otimes \mathrm{id}_{{}^\vee T(Y)} \otimes \eta_Y,$$

we obtain the equality (i) in Figure 8.3 where we use the computation of $\tilde{\varrho}$ in Theorem 8.15, the computation of the monoidal product in $\mathcal{C}^T$ via T_2 (see Section 7.5.2), the computation of the left duals in $\mathcal{C}^T$ via s^l (see Section 8.3.1), the definition of τ (see Theorem 7.11), and the equality

$${}^\vee ?_2(F_T(Y), F_T(X)) = {}^\vee ?_2(T(Y), T(X)). \tag{8.26}$$

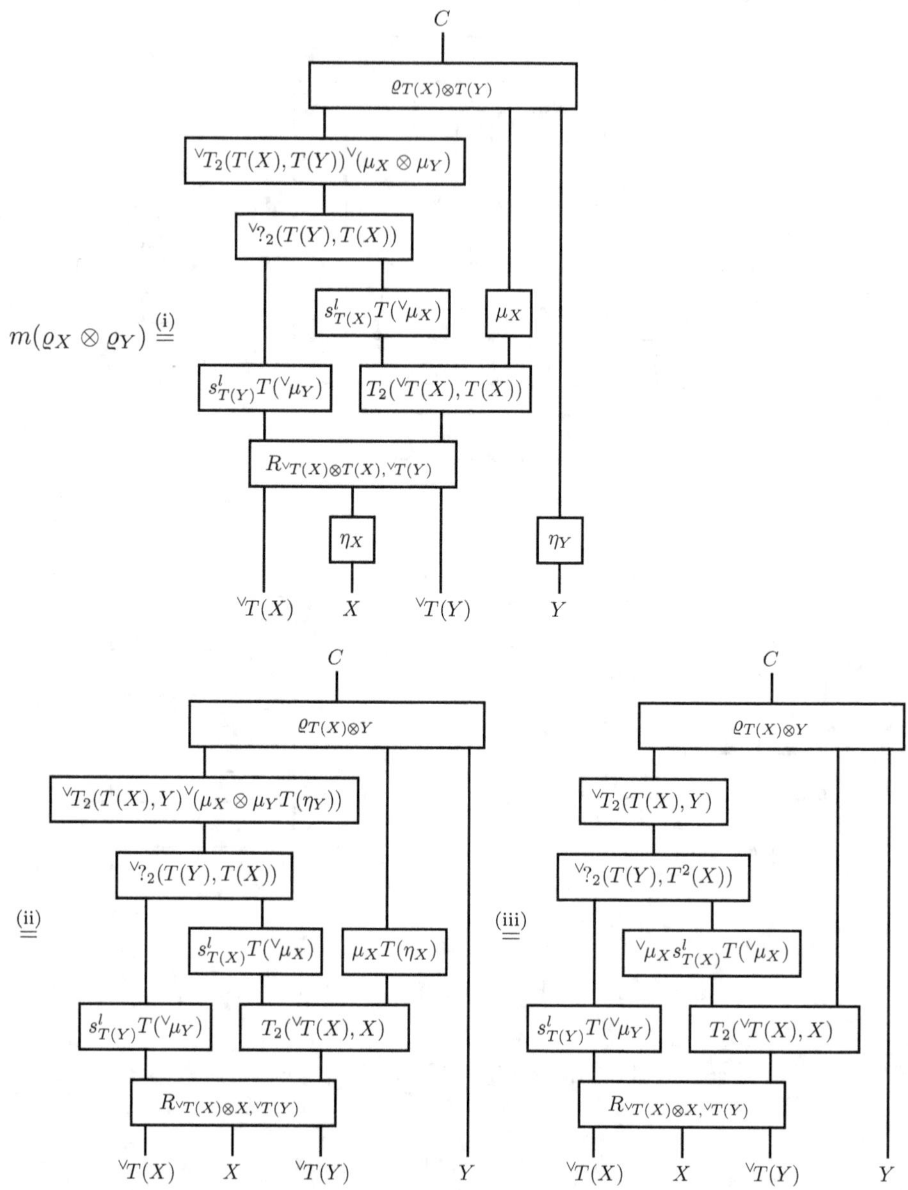

Figure 8.3: Computation of $m(\varrho_X \otimes \varrho_Y)$

The equality (8.26) relates the monoidal constraints of the left dual functors in $\mathcal{C}$

and $\mathcal{C}^T$ and is a direct consequence of the fact that the distinguished left duality in $\mathcal{C}^T$ is a lift along the forgetful functor $\mathcal{C}^T \to \mathcal{C}$ of the distinguished left duality in $\mathcal{C}$. The equality (ii) in Figure 8.3 follows from the dinaturality of ϱ and the naturality of R and T_2. The equality (iii) in Figure 8.3 follows from the unitality of μ and the naturality of ${}^{\vee}?_2$. This gives the expression for $m(\varrho_X \otimes \varrho_Y)$ in Figure 8.2. This expression determines m uniquely because the dinatural transformation ϱ is universal. The last two equalities in Figure 8.2 are obtained similarly using the definitions of S and ω given in Sections 6.5.1 and 6.5.2 as well as the anti-multiplicativity of s^l, see Lemma 8.6. □

8.6.4 The ribbon case

Let T be a quasitriangular involutory Hopf monad on a pivotal category $\mathcal{C}$ such that there is a coend

$$C = \int^{X \in \mathcal{C}} T(X)^* \otimes X \in \mathrm{Ob}(\mathcal{C})$$

with universal dinatural transformation $\varrho = \{\varrho_X : T(X)^* \otimes X \to C\}_{X \in \mathrm{Ob}(\mathcal{C})}$. By Theorems 7.11 and 8.10, the category $\mathcal{C}^T$ is braided and pivotal. Theorems 8.15 and 8.16 lift C to a Hopf algebra (C, α) in the braided pivotal category $\mathcal{C}^T$. If $\mathcal{C}^T$ is ribbon, then the twist forms θ_+ and θ_- associated with (C, α) as in Section 6.6.3 are uniquely determined by the equalities in Figure 8.4, where μ is the product of T, R is the R-matrix of T, s is the antipode of T, and X runs over all objects of $\mathcal{C}$. The proof of these equalities is similar to the proof of the equalities in Figure 8.2 in Theorem 8.16. For the second equality in Figure 8.4, additionally use that

$$\mathrm{ev}_A(\mathrm{id}_{A^*} \otimes \theta_A^{-1}) = \widetilde{\mathrm{ev}}_A \, \tau_{A^*,A}$$

for any object A of a ribbon category with braiding τ and twist θ.

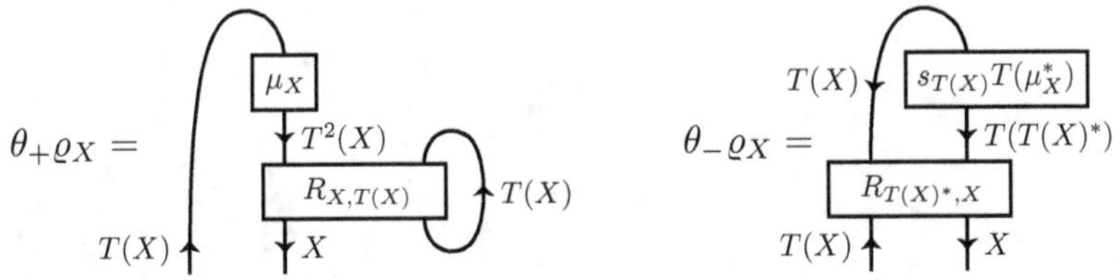

Figure 8.4: The twist forms of the Hopf algebra (C, α)

Chapter 9

Monadicity of the center

In this chapter, we show that under rather mild hypothesis, the center of a rigid category is the category of modules over a quasitriangular Hopf monad. This allows us to compute the coend of the center of a pivotal fusion category. As an application, we prove Theorems 5.3–5.5. This chapter is based on [BV2, BV3].

9.1 Centralizers of categories

We define centralizers of left rigid categories and their universal coactions.

9.1.1 Centralizers

Let $\mathcal{C}$ be a left rigid category with distinguished left duality $\{(^\vee X, \mathrm{ev}_X)\}_{X \in \mathrm{Ob}(\mathcal{C})}$. A *centralizer* of $\mathcal{C}$ is a pair (Z, ρ), where $Z\colon \mathcal{C} \to \mathcal{C}$ is an endofunctor of $\mathcal{C}$ and

$$\rho = \{\rho_{X,Y}\colon {}^\vee Y \otimes X \otimes Y \to Z(X)\}_{X,Y \in \mathrm{Ob}(\mathcal{C})}$$

is a family of morphisms in $\mathcal{C}$, such that

(i) ρ is natural in X, i.e., for all morphisms $f\colon X \to X'$ in $\mathcal{C}$ and all $Y \in \mathrm{Ob}(\mathcal{C})$, the following diagram commutes:

$$\begin{array}{ccc} {}^\vee Y \otimes X \otimes Y & \xrightarrow{\mathrm{id}_{^\vee Y} \otimes f \otimes \mathrm{id}_Y} & {}^\vee Y \otimes X' \otimes Y \\ \downarrow{\scriptstyle \rho_{X,Y}} & & \downarrow{\scriptstyle \rho_{X',Y}} \\ Z(X) & \xrightarrow{Z(f)} & Z(X'); \end{array}$$

(ii) ρ is dinatural in Y, i.e., for all morphisms $g\colon Y \to Y'$ in $\mathcal{C}$ and all $X \in \mathrm{Ob}(\mathcal{C})$, the following diagram commutes:

$$\begin{CD}
{}^{\vee}Y' \otimes X \otimes Y @>{\mathrm{id}_{{}^{\vee}Y' \otimes X} \otimes g}>> {}^{\vee}Y' \otimes X \otimes Y' \\
@V{{}^{\vee}g \otimes \mathrm{id}_{X \otimes Y}}VV @VV{\rho_{X,Y'}}V \\
{}^{\vee}Y \otimes X \otimes Y @>{\rho_{X,Y}}>> Z(X);
\end{CD}$$

(iii) for all $X \in \mathrm{Ob}(\mathcal{C})$, the dinatural transformation $\{\rho_{X,Y}\}_{Y \in \mathrm{Ob}(\mathcal{C})}$ is *universal* in the sense that for any dinatural transformation

$$d = \{d_Y : {}^{\vee}Y \otimes X \otimes Y \to D\}_{Y \in \mathrm{Ob}(\mathcal{C})}$$

with $D \in \mathrm{Ob}(\mathcal{C})$, there is a unique morphism $\varphi\colon C \to D$ in $\mathcal{C}$ such that $d_Y = \varphi\, \rho_{X,Y}$ for all $Y \in \mathrm{Ob}(\mathcal{C})$.

Sometimes, when we do not need ρ, we call the functor Z a centralizer of $\mathcal{C}$. The category $\mathcal{C}$ is *centralizable* if it has a centralizer.

In view of the definition of a coend of a functor (see Appendix B.1), the conditions (ii) and (iii) above may be rephrased by saying that for all $X \in \mathrm{Ob}(\mathcal{C})$, the pair $(Z(X), \{\rho_{X,Y}\}_{Y \in \mathrm{Ob}(\mathcal{C})})$ is a coend of the functor

$$\mathcal{C}^{\mathrm{op}} \times \mathcal{C} \to \mathcal{C}, \quad (Y, Y') \mapsto {}^{\vee}Y \otimes X \otimes Y', \quad (g, g') \mapsto {}^{\vee}g \otimes \mathrm{id}_X \otimes g' \tag{9.1}$$

where Y, Y' run over objects of $\mathcal{C}$ and g, g' run over morphisms in $\mathcal{C}$. Following [ML2], we write

$$Z(X) = \int^{Y \in \mathrm{Ob}(\mathcal{C})} {}^{\vee}Y \otimes X \otimes Y.$$

It follows from Lemma B.1 that a centralizer of $\mathcal{C}$, if it exists, is unique up to isomorphism. More precisely, if (Z, ρ) and (Z', ρ') are centralizers of $\mathcal{C}$, then there is a unique natural isomorphism $\alpha\colon Z \to Z'$ such that $\rho'_{X,Y} = \alpha_X \rho_{X,Y}$ for all $X, Y \in \mathrm{Ob}(\mathcal{C})$.

The centralizability of $\mathcal{C}$ can be verified objectwise: if for any $X \in \mathrm{Ob}(\mathcal{C})$ the functor (9.1) has a coend

$$(Z(X), \{\rho_{X,Y} : {}^{\vee}Y \otimes X \otimes Y \to Z(X)\}_{Y \in \mathrm{Ob}(\mathcal{C})}),$$

then $\mathcal{C}$ is centralizable. Indeed, by the parameters theorem for coends (Theorem B.2), the rule $X \mapsto Z(X)$ extends uniquely to a functor $Z\colon \mathcal{C} \to \mathcal{C}$ so that $\rho_{X,Y}$ is natural in X and dinatural in Y. Then the pair (Z, ρ) is a centralizer of $\mathcal{C}$.

9.1.2 The universal coaction

Let $\mathcal{C}$ be a centralizable left rigid category with distinguished left duality. The *universal coaction* associated with a centralizer (Z, ρ) of $\mathcal{C}$ is the natural transformation

$$\partial = \{\partial_{X,Y} : X \otimes Y \to Y \otimes Z(X)\}_{X,Y \in \mathrm{Ob}(\mathcal{C})}$$

defined by

$$\partial_{X,Y} = (\mathrm{id}_Y \otimes \rho_{X,Y})(\mathrm{coev}_Y \otimes \mathrm{id}_{X\otimes Y}) =$$ 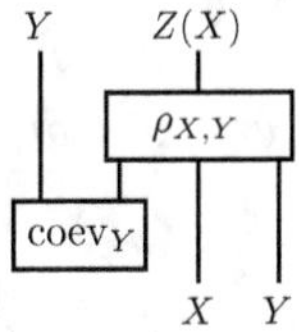 . (9.2)

We will depict the morphism $\partial_{X,Y}$ as follows:

$$\partial_{X,Y} =$$ 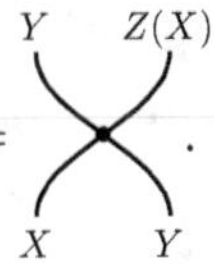 .

Note that $C = Z(\mathbb{1})$ is a coend of the category $\mathcal{C}$ and $\{\partial_{\mathbb{1},Y} \colon Y \to Y \otimes C\}_{Y\in \mathrm{Ob}(\mathcal{C})}$ is the universal coaction of C defined in Section 6.4.3. Graphically,

$$\partial_{\mathbb{1},Y} =$$ 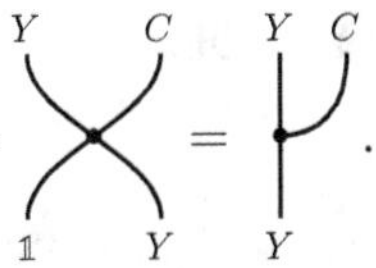 .

The universal coaction has the following useful factorization properties.

Lemma 9.1. (a) *For any $A, B \in \mathrm{Ob}(\mathcal{C})$ and any natural transformation*

$$\xi = \{\xi_Y \colon A \otimes Y \to Y \otimes B\}_{Y\in \mathrm{Ob}(\mathcal{C})},$$

there is a unique morphism $r \colon Z(A) \to B$ such that for all $Y \in \mathrm{Ob}(\mathcal{C})$,

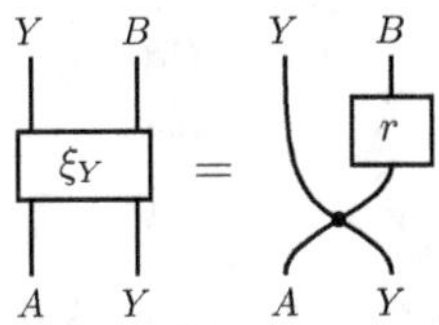 .

(b) *For any $A, B, C \in \mathrm{Ob}(\mathcal{C})$ and any natural transformation*

$$\xi = \{\xi_Y \colon A \otimes Y \otimes C \to Y \otimes B\}_{Y\in \mathrm{Ob}(\mathcal{C})},$$

there is a unique morphism $r \colon Z(A) \otimes C \to B$ such that for all $Y \in \mathrm{Ob}(\mathcal{C})$,

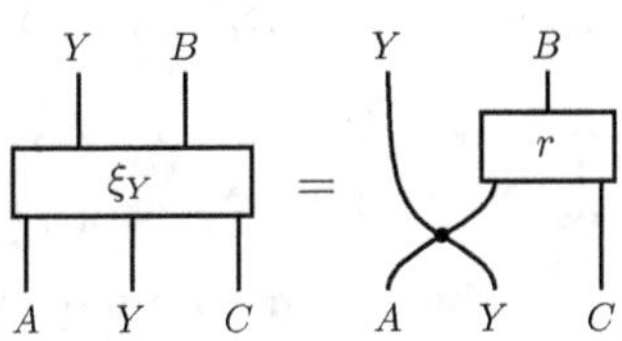 .

(c) *For any* $A, B \in \mathrm{Ob}(\mathcal{C})$ *and any natural transformation*

$$\xi = \{\xi_{Y_1,Y_2} : A \otimes Y_1 \otimes Y_2 \to Y_1 \otimes Y_2 \otimes B\}_{Y_1,Y_2 \in \mathrm{Ob}(\mathcal{C})},$$

there is a unique morphism $r\colon Z^2(A) \to B$ *such that for all* $Y_1, Y_2 \in \mathrm{Ob}(\mathcal{C})$,

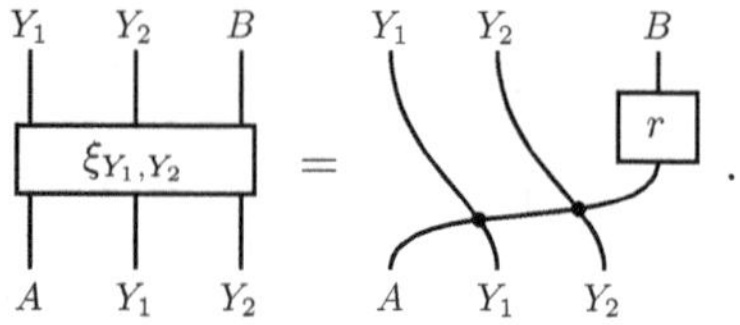

Proof. Let us prove (a). The naturality of ξ and (2.2) imply that the family

$$d = \{d_Y = (\mathrm{ev}_Y \otimes \mathrm{id}_B)(\mathrm{id}_{^\vee Y} \otimes \xi_Y)\colon {}^\vee Y \otimes A \otimes Y \to B\}_{Y \in \mathrm{Ob}(\mathcal{C})}$$

is a dinatural transformation. Since $Z(A) = \int^{Y \in \mathcal{D}} {}^\vee Y \otimes A \otimes Y$ with universal dinatural transformation $\{\rho_{A,Y}\}_{Y \in \mathrm{Ob}(\mathcal{C})}$, there exists a unique morphism $r\colon Z(A) \to B$ such that $d_Y = r\rho_{A,Y}$ for all $Y \in \mathrm{Ob}(\mathcal{C})$. The latter condition on r is equivalent to the one in (a), since for all $Y \in \mathrm{Ob}(\mathcal{C})$,

$$\xi_Y = (\mathrm{id}_Y \otimes d_Y)(\mathrm{coev}_Y \otimes \mathrm{id}_{A \otimes Y}) \quad \text{and} \quad \partial_{A,Y} = (\mathrm{id}_Y \otimes \rho_{A,Y})(\mathrm{coev}_Y \otimes \mathrm{id}_{A \otimes Y}).$$

Let us prove (b). The naturality of ξ and (2.2) imply that the family

$$\xi' = \{\xi'_Y = (\xi_Y \otimes \mathrm{id}_{^\vee C})(\mathrm{id}_{A \otimes Y} \otimes \mathrm{coev}_C)\colon A \otimes Y \to Y \otimes B \otimes {}^\vee C\}_{Y \in \mathrm{Ob}(\mathcal{C})}$$

is a natural transformation. By (a), there is a unique morphism $r'\colon Z(A) \to B \otimes {}^\vee C$ such that $\xi'_Y = (\mathrm{id}_Y \otimes r')\,\partial_{A,Y}$ for all $Y \in \mathrm{Ob}(\mathcal{C})$. Composing on the left with $\mathrm{id}_{Y \otimes B} \otimes \mathrm{ev}_C$, we obtain that

$$r = (\mathrm{id}_B \otimes \mathrm{ev}_C)(r' \otimes \mathrm{id}_C)\colon Z(A) \otimes C \to B$$

is the unique morphism such that $\xi_Y = (\mathrm{id}_Y \otimes r)\,(\partial_{A,Y} \otimes \mathrm{id}_C)$ for all $Y \in \mathrm{Ob}(\mathcal{C})$.

Let us prove (c). For each $Y_2 \in \mathrm{Ob}(\mathcal{C})$, the family

$$\{\xi_{Y_1,Y_2} : A \otimes Y_1 \otimes Y_2 \to Y_1 \otimes Y_2 \otimes B\}_{Y_1 \in \mathrm{Ob}(\mathcal{C})}$$

is a natural transformation. By (b), there is a unique morphism $r_{Y_2}\colon Z(A) \otimes Y_2 \to Y_2 \otimes B$ such that $\xi_{Y_1,Y_2} = (\mathrm{id}_{Y_1} \otimes r_{Y_2})\,(\partial_{A,Y_1} \otimes \mathrm{id}_{Y_2})$ for all $Y_1 \in \mathrm{Ob}(\mathcal{C})$. The uniqueness of the factorization in (b) implies that the family

$$\{r_{Y_2}\colon Z(A) \otimes Y_2 \to Y_2 \otimes B\}_{Y_2 \in \mathrm{Ob}(\mathcal{C})}$$

is a natural transformation. By (a), there is a morphism $r\colon Z^2(A) \to B$ such that $r_{Y_2} = (\mathrm{id}_{Y_2} \otimes r)\,\partial_{Z(A),Y_2}$ for all $Y_2 \in \mathrm{Ob}(\mathcal{C})$. Then for all $Y_1, Y_2 \in \mathrm{Ob}(\mathcal{C})$,

$$\begin{aligned}\xi_{Y_1,Y_2} &= (\mathrm{id}_{Y_1} \otimes r_{Y_2})\,(\partial_{A,Y_1} \otimes \mathrm{id}_{Y_2})\\ &= (\mathrm{id}_{Y_1 \otimes Y_2} \otimes r)\,(\mathrm{id}_{Y_1} \otimes \partial_{Z(A),Y_2})\,(\partial_{A,Y_1} \otimes \mathrm{id}_{Y_2}).\end{aligned}$$

The uniqueness of such an r follows from the uniqueness in (a) and (b). □

9.1.3 Example

Let G be a finite group. Consider the additive pivotal fusion $\Bbbk$-category $\mathcal{C} = G\text{-}\mathrm{vect}_{\Bbbk}^{1}$ of G-graded free $\Bbbk$-modules of finite rank, see Section A.2. A representative set of simple objects of $\mathcal{C}$ is $\{\Bbbk_h\}_{h\in G}$, where $\Bbbk_h$ is the G-graded $\Bbbk$-module defined by $(\Bbbk_h)_h = \Bbbk$ and $(\Bbbk_h)_g = 0$ for $g \in G\setminus\{h\}$. It is clear from Example B.1.2 that for any $X \in \mathrm{Ob}(\mathcal{C})$, the functor (9.1) has a coend

$$\bigoplus_{h\in G} \Bbbk_h^* \otimes X \otimes \Bbbk_h = \bigoplus_{g\in G}\left(\bigoplus_{\substack{h,a,b,c\in G\\ abc=g}} (\Bbbk_h^*)_a \otimes_{\Bbbk} X_b \otimes_{\Bbbk} (\Bbbk_h)_c\right) \simeq \bigoplus_{g\in G}\left(\bigoplus_{h\in G} X_{hgh^{-1}}\right).$$

Thus, $\mathcal{C}$ is centralizable with centralizer $Z\colon \mathcal{C}\to\mathcal{C}$ carrying any $X\in\mathrm{Ob}(\mathcal{C})$ to

$$Z(X) = \bigoplus_{g\in G} Z(X)_g \quad \text{where} \quad Z(X)_g = \bigoplus_{h\in G} X_{hgh^{-1}}$$

and carrying any morphism $f\colon X\to Y$ in $\mathcal{C}$ to $Z(f) = \sum_{g\in G} Z(f)_g$ where

$$Z(f)_g = \sum_{h\in G} f_{|X_{hgh^{-1}}}^{|Y_{hgh^{-1}}} \colon Z(X)_g = \bigoplus_{h\in G} X_{hgh^{-1}} \to Z(Y)_g = \bigoplus_{h\in G} Y_{hgh^{-1}}.$$

For any $X, Y \in \mathrm{Ob}(\mathcal{C})$, the associated universal coaction

$$\partial_{X,Y}\colon X\otimes Y \to Y\otimes Z(X)$$

is computed as follows: for any $g,h\in G$, $x\in X_g$, and $y\in Y_h$,

$$\partial_{X,Y}(x\otimes_{\Bbbk} y) = y\otimes_{\Bbbk} x \in Y_h\otimes_{\Bbbk} Z(X)_{h^{-1}gh} \subset Y\otimes Z(X).$$

9.2 From centralizers to the center

The aim of this section is to state two theorems relating centralizers to the center of a rigid category. The proof of these theorems is postponed to Section 9.3. As an application, we compute the coend of the center.

In this section, $\mathcal{C}$ is a centralizable rigid category with distinguished left duality $\{({}^{\vee}X, \mathrm{ev}_X)\}_{X\in\mathrm{Ob}(\mathcal{C})}$ and distinguished right duality $\{(X^{\vee}, \widetilde{\mathrm{ev}}_X)\}_{X\in\mathrm{Ob}(\mathcal{C})}$. We fix a centralizer $Z\colon\mathcal{C}\to\mathcal{C}$ of $\mathcal{C}$ with universal coaction ∂.

9.2.1 The structure of a Hopf monad

Our first aim is to turn Z into a quasitriangular Hopf monad. We define the structural morphisms of Z as follows. For any $X\in\mathrm{Ob}(\mathcal{C})$, Lemma 9.1(c) applied to $A = X$, $B = Z(X)$, and $\xi_{Y_1,Y_2} = \partial_{X,Y_1\otimes Y_2}$ yields a unique morphism

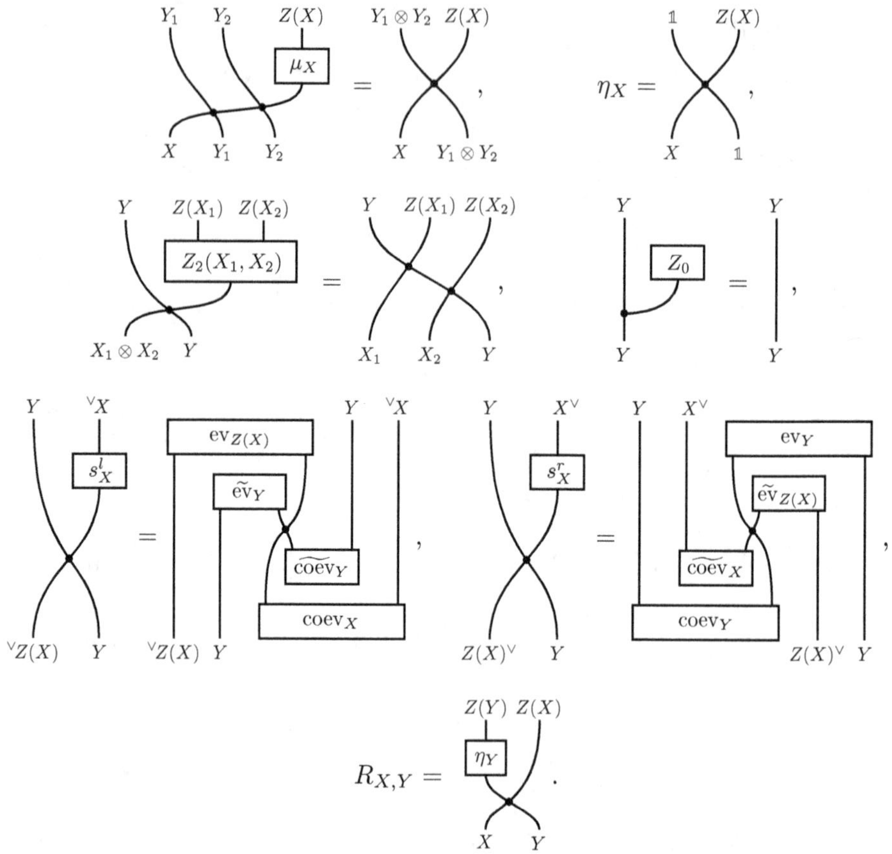

Figure 9.1: The structural morphisms of Z

$\mu_X \colon Z^2(X) \to Z(X)$ in $\mathcal{C}$ such that the first equality in Figure 9.1 holds for all $Y_1, Y_2 \in \mathrm{Ob}(\mathcal{C})$. Set

$$\mu = \{\mu_X \colon Z^2(X) \to Z(X)\}_{X \in \mathrm{Ob}(\mathcal{C})}.$$

Also, consider the family of morphisms

$$\eta = \{\eta_X = \partial_{X,\mathbb{1}} \colon X \to Z(X)\}_{X \in \mathrm{Ob}(\mathcal{C})},$$

see the second equality in Figure 9.1. For any $X_1, X_2 \in \mathrm{Ob}(\mathcal{C})$, Lemma 9.1(a) applied to $A = X_1 \otimes X_2$, $B = Z(X_1) \otimes Z(X_2)$, and

$$\xi_Y = (\partial_{X_1,Y} \otimes \mathrm{id}_{Z(X_2)})(\mathrm{id}_{X_1} \otimes \partial_{X_2,Y})$$

yields a unique morphism

$$Z_2(X_1, X_2) \colon Z(X_1 \otimes X_2) \to Z(X_1) \otimes Z(X_2)$$

such that the third equality in Figure 9.1 holds for all $Y \in \mathrm{Ob}(\mathcal{C})$. Set

$$Z_2 = \{Z_2(X_1, X_2) \colon Z(X_1 \otimes X_2) \to Z(X_1) \otimes Z(X_2)\}_{X_1, X_2 \in \mathrm{Ob}(\mathcal{C})}.$$

For $A = B = \mathbb{1}$ and $\xi_Y = \mathrm{id}_Y$, Lemma 9.1(a) gives a unique morphism

$$Z_0 \colon Z(\mathbb{1}) \to \mathbb{1}$$

such that the fourth equality in Figure 9.1 holds for all $Y \in \mathrm{Ob}(\mathcal{C})$. For any $X \in \mathrm{Ob}(\mathcal{C})$, Lemma 9.1(a) applied to $A = {}^\vee Z(X)$, $B = {}^\vee X$, and

$$\begin{aligned}\xi_Y = (\mathrm{ev}_{Z(X)} \otimes \mathrm{id}_{Y \otimes {}^\vee X})&(\mathrm{id}_{{}^\vee Z(X)} \otimes \widetilde{\mathrm{ev}}_Y \otimes \mathrm{id}_{Z(X) \otimes Y \otimes {}^\vee X}) \\ &\circ (\mathrm{id}_{{}^\vee Z(X) \otimes Y} \otimes \partial_{X, Y^\vee} \otimes \mathrm{id}_{Y \otimes {}^\vee X}) \\ &\circ (\mathrm{id}_{{}^\vee Z(X) \otimes Y \otimes X} \otimes \widetilde{\mathrm{coev}}_Y \otimes \mathrm{id}_{{}^\vee X})(\mathrm{id}_{{}^\vee Z(X) \otimes Y} \otimes \mathrm{coev}_X)\end{aligned}$$

yields a unique morphism $s^l_X \colon Z({}^\vee Z(X)) \to {}^\vee X$ such that the fifth equality in Figure 9.1 holds for all $Y \in \mathrm{Ob}(\mathcal{C})$. Set

$$s^l = \{s^l_X \colon Z({}^\vee Z(X)) \to {}^\vee X\}_{X \in \mathrm{Ob}(\mathcal{C})}.$$

Similarly, Lemma 9.1(a) applied to $A = Z(X)^\vee$, $B = X^\vee$, and

$$\begin{aligned}\xi_Y = (\mathrm{id}_{Y \otimes X^\vee} \otimes \mathrm{ev}_Y)&(\mathrm{id}_{Y \otimes X^\vee \otimes {}^\vee Y} \otimes \widetilde{\mathrm{ev}}_{Z(X)} \otimes \mathrm{id}_Y) \\ &\circ (\mathrm{id}_{Y \otimes X^\vee} \otimes \partial_{X, {}^\vee Y} \otimes \mathrm{id}_{Z(X)^\vee \otimes Y}) \\ &\circ (\mathrm{id}_Y \otimes \widetilde{\mathrm{coev}}_X \otimes \mathrm{id}_{{}^\vee Y \otimes Z(X)^\vee \otimes Y})(\mathrm{coev}_Y \otimes \mathrm{id}_{Z(X)^\vee \otimes Y})\end{aligned}$$

yields a unique morphism $s^r_X \colon Z(Z(X)^\vee) \to X^\vee$ such that the sixth equality in Figure 9.1 holds for all $Y \in \mathrm{Ob}(\mathcal{C})$. Set

$$s^r = \{s^r_X \colon Z(Z(X)^\vee) \to X^\vee\}_{X \in \mathrm{Ob}(\mathcal{C})}.$$

Finally, consider the family of morphisms

$$R = \{R_{X,Y} = (\eta_Y \otimes \mathrm{id}_{Z(X)})\partial_{X,Y} \colon X \otimes Y \to Z(Y) \otimes Z(X)\}_{X, Y \in \mathrm{Ob}(\mathcal{C})},$$

see the last equality in Figure 9.1.

Theorem 9.2. *The tuple $Z = ((Z, Z_2, Z_0), \mu, \eta)$ is a quasitriangular Hopf monad on $\mathcal{C}$ with left antipode s^l, right antipode s^r, and R-matrix R. If $\mathcal{C}$ is $\Bbbk$-linear (respectively, pivotal), then the Hopf monad Z is $\Bbbk$-linear (respectively, involutory).*

Theorem 9.2 is proved in Section 9.3.

9.2.2 The isomorphism

Theorem 9.2 together with Theorems 7.11 and 8.7 shows that the category $\mathcal{C}^Z$ of Z-modules is braided and rigid with distinguished left and right dualities. By Section 5.2.1, the center $\mathcal{Z}(\mathcal{C})$ of $\mathcal{C}$ is also a braided rigid category with distinguished left and right dualities. We now relate $\mathcal{C}^Z$ and $\mathcal{Z}(\mathcal{C})$. Any Z-module (M, r) determines a family of morphisms in $\mathcal{C}$

$$\sigma^{(M,r)} = \{\sigma_X^{(M,r)} \colon M \otimes X \to X \otimes M\}_{X \in \mathrm{Ob}(\mathcal{C})},$$

where

$$\sigma_X^{(M,r)} = (\mathrm{id}_X \otimes r)\, \partial_{M,X} = \quad (9.3)$$

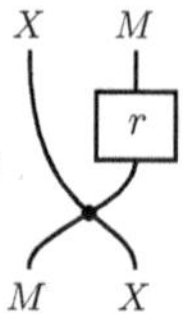

We define a functor $E \colon \mathcal{C}^Z \to \mathcal{Z}(\mathcal{C})$ by

$$E(M, r) = (M, \sigma^{(M,r)}) \quad \text{and} \quad E(f) = f$$

for all Z-modules (M, r) and all morphisms f in $\mathcal{C}^Z$. Clearly, the diagram of functors

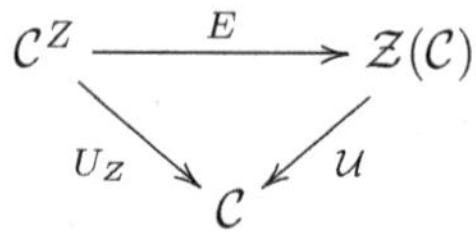

commutes, where $\mathcal{U}$ and U_Z are the forgetful functors (see Sections 5.1.2 and 7.1.3).

Theorem 9.3. (a) *The functor E is a braided strict monoidal isomorphism.*

(b) *If $\mathcal{C}$ is $\Bbbk$-linear, then both categories $\mathcal{C}^Z$ and $\mathcal{Z}(\mathcal{C})$ are $\Bbbk$-linear, and the functor E is $\Bbbk$-linear.*

(c) *If $\mathcal{C}$ is pivotal, then both categories $\mathcal{C}^Z$ and $\mathcal{Z}(\mathcal{C})$ are pivotal, and the functor E is strictly pivotal.*

Theorem 9.3 is proved in Section 9.3.

9.2.3 The coend of the center

Suppose now the existence of the coend

$$C = \int^{X \in \mathcal{C}} {}^\vee Z(X) \otimes X \in \mathrm{Ob}(\mathcal{C})$$

with universal dinatural transformation $\varrho = \{\varrho_X \colon {}^\vee Z(X) \otimes X \to C\}_{X \in \mathrm{Ob}(\mathcal{C})}$ (see Appendix B.1). We use C and ϱ to compute a coend of $\mathcal{Z}(\mathcal{C})$ as follows. Let μ and s^l be respectively the product and left antipode of Z (see Section 9.2.1). By Section 8.6.2, there is a unique morphism $\alpha \colon Z(C) \to C$ such that

$$\alpha Z(\varrho_X) = \varrho_{Z(X)} \left({}^\vee\mu_X s^l_{Z(X)} Z({}^\vee\mu_X) \otimes \mathrm{id}_{Z(X)} \right) Z_2({}^\vee Z(X), X)$$

for all $X \in \mathrm{Ob}(\mathcal{C})$. We define a family

$$\sigma = \{\sigma_X \colon C \otimes X \to X \otimes C\}_{X \in \mathrm{Ob}(\mathcal{C})}$$

of morphisms in $\mathcal{C}$ by

$$\sigma_X = (\mathrm{id}_X \otimes \alpha)\, \partial_{C,X} =$$ 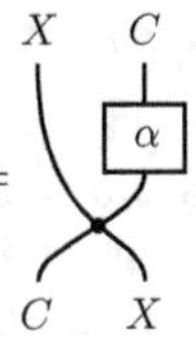 .

Theorem 9.4. *The pair (C, σ) is a coend of $\mathcal{Z}(\mathcal{C})$. Moreover, the product m, coproduct Δ, unit u, counit ε, antipode S, and canonical pairing ω of the Hopf algebra (C, σ) are uniquely determined by the equalities in Figures* 8.1 *and* 8.2, *where $T = Z$ and X, Y run over all objects of $\mathcal{C}$.*

Proof. By Theorem 8.15, α is a Z-action and the Z-module (C, α) is a coend of $\mathcal{C}^Z$ with associated universal dinatural transformation

$$\tilde{\varrho} = \{\tilde{\varrho}_{(M,r)} = \varrho_M({}^\vee r \otimes \mathrm{id}_M) \colon {}^\vee(M, r) \otimes (M, r) \to (C, \alpha)\}_{(M,r) \in \mathrm{Ob}(\mathcal{C}^Z)}.$$

Consider the strict monoidal isomorphism $E \colon \mathcal{C}^Z \to \mathcal{Z}(\mathcal{C})$ from Theorem 9.3. It follows from the proof of Theorem 9.3 that the functor E carries the distinguished left duality in $\mathcal{C}^Z$ to the distinguished left duality in $\mathcal{Z}(\mathcal{C})$. In particular,

$$E({}^\vee E^{-1}(M, \gamma) \otimes E(M, \gamma)) = {}^\vee(M, \gamma) \otimes (M, \gamma)$$

for all $(M, \gamma) \in \mathcal{Z}(\mathcal{C})$. Consequently, $E(C, \alpha) = (C, \sigma)$ is a coend of $\mathcal{Z}(\mathcal{C})$ with associated universal dinatural transformation

$$\{E(\tilde{\varrho}_{E^{-1}(M,\gamma)}) = \tilde{\varrho}_{E^{-1}(M,\gamma)} \colon {}^\vee(M, \gamma) \otimes (M, \gamma) \to (C, \sigma)\}_{(M,\gamma) \in \mathrm{Ob}(\mathcal{Z}(\mathcal{C}))}.$$

This and the fact that the functor E is braided strict monoidal, preserves the distinguished left dualities, and acts as the identity on morphisms imply that the structural morphisms of the coend (C, α) of $\mathcal{C}^Z$ and of the coend (C, σ) of $\mathcal{Z}(\mathcal{C})$ are the same (see their definitions in Sections 6.4.2, 6.5.1, and 6.5.2). By Theorems 8.15 and 8.16, these structural morphisms are determined by the equalities in Figures 8.1 and 8.2, where $T = Z$ and X, Y run over all objects of $\mathcal{C}$. $\square$

9.2.4 Example

Let G be a finite group. Recall from Example 9.1.3 the centralizer Z of the additive pivotal fusion $\Bbbk$-category $\mathcal{C} = G\text{-}\mathrm{vect}^1_{\Bbbk}$. By Theorems 9.2 and 9.3, Z is an involutory quasitriangular $\Bbbk$-linear Hopf monad on $\mathcal{C}$ such that $\mathcal{Z}(\mathcal{C}) \simeq \mathcal{C}^Z$ as braided pivotal $\Bbbk$-categories.

If the group G is non-abelian, then the Hopf monad Z is not representable by a Hopf algebra (see Remark 8.1.6). In fact, the functor Z is not isomorphic to an endofunctor of $\mathcal{C}$ of the form $A\otimes ?$ with $A \in \mathrm{Ob}(\mathcal{C})$. To see it, recall from Example 9.1.3 the representative set $\{\Bbbk_h\}_{h\in G}$ of simple objects of $\mathcal{C}$. For $h \in G$,

$$Z(\Bbbk_h) = \bigoplus_{g\in G} Z(\Bbbk_h)_g = \bigoplus_{g\in G} \left(\bigoplus_{a\in G} (\Bbbk_h)_{aga^{-1}} \right) \simeq \bigoplus_{g\in G} N_{g,h} \Bbbk_g$$

where $N_{g,h} = \mathrm{card}\{a \in G \,|\, ag = ha\}$. Here, for $X \in \mathrm{Ob}(\mathcal{C})$ and a non-negative integer N, the symbol NX stands for the direct sum of N copies of X. Suppose now that $Z \simeq A\otimes ?$ for some $A \in \mathrm{Ob}(\mathcal{C})$. Then

$$A \simeq A \otimes \mathbb{1} \simeq Z(\mathbb{1}) = Z(\Bbbk_1) \simeq \bigoplus_{g\in G} N_{g,1} \Bbbk_g \simeq |G|\, \Bbbk_1$$

where $|G| \geq 1$ is the order of G. So, for all $h \in G$,

$$\bigoplus_{g\in G} N_{g,h} \Bbbk_g \simeq Z(\Bbbk_h) \simeq A \otimes \Bbbk_h \simeq (|G|\, \Bbbk_1) \otimes \Bbbk_h \simeq |G|(\Bbbk_1 \otimes \Bbbk_h) \simeq |G|\, \Bbbk_h.$$

Consequently, $N_{h,h} = |G|$ for all $h \in G$, that is, G is abelian.

9.3 Proof of Theorems 9.2 and 9.3

The key idea of the proof is to include the forgetful functor $\mathcal{U}\colon \mathcal{Z}(\mathcal{C}) \to \mathcal{C}$ into a Hopf adjunction $\mathcal{F}\colon \mathcal{C} \rightleftarrows \mathcal{Z}(\mathcal{C}) :\mathcal{U}$ whose associated Hopf monad is Z and whose comparison functor $\mathcal{Z}(\mathcal{C}) \to \mathcal{C}^Z$ is an isomorphism inverse to E.

For any $X, Y \in \mathrm{Ob}(\mathcal{C})$, Lemma 9.1(b) applied to $A = X$, $B = Y \otimes Z(X)$, $C = Y$, and $\xi_V = \partial_{X,V\otimes Y}$ yields a unique morphism

$$\sigma_Y^X : Z(X) \otimes Y \to Y \otimes Z(X)$$

such that for all $V \in \mathrm{Ob}(\mathcal{C})$,

$$(\mathrm{id}_V \otimes \sigma_Y^X)(\partial_{X,V} \otimes \mathrm{id}_Y) = \partial_{X,V\otimes Y}$$

that is

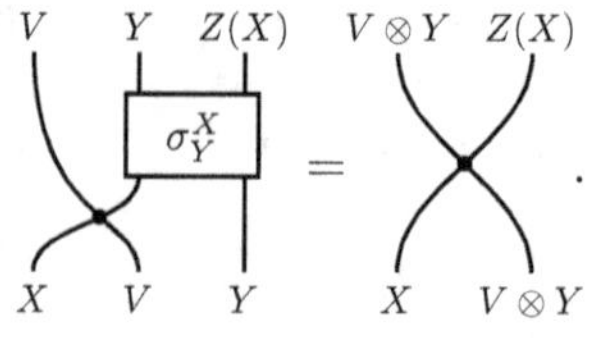

. (9.4)

We define a functor $\mathcal{F}\colon \mathcal{C} \to \mathcal{Z}(\mathcal{C})$ by

$$\mathcal{F}(X) = (Z(X), \sigma^X = \{\sigma_Y^X\}_{Y\in \mathrm{Ob}(\mathcal{C})}) \quad \text{and} \quad \mathcal{F}(f) = Z(f)$$

for all $X \in \mathrm{Ob}(\mathcal{C})$ and all morphisms f in $\mathcal{C}$.

Lemma 9.5. *$\mathcal{F}$ is a well-defined functor.*

Proof. We first prove that the family $\sigma = \{\sigma_Y^X\}_{X,Y \in \mathrm{Ob}(\mathcal{C})}$ is a natural transformation. Let $f\colon X \to X'$ and $g\colon Y \to Y'$ be morphisms in $\mathcal{C}$. For all $V \in \mathrm{Ob}(\mathcal{C})$,

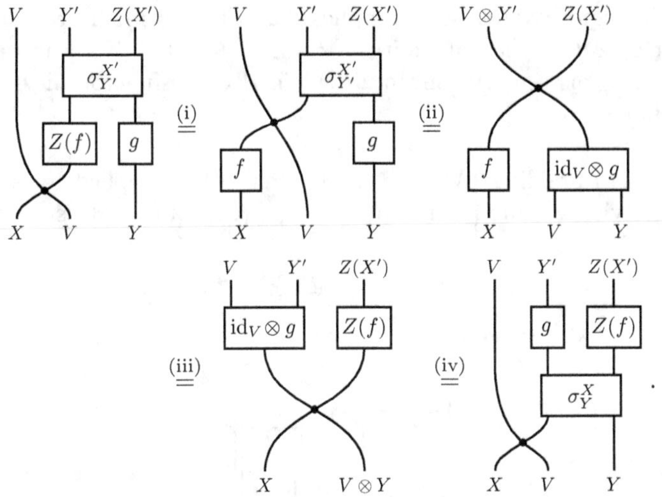

Here, (i) and (iii) follow from the naturality of ∂, and (ii) and (iv) from (9.4). The uniqueness in Lemma 9.1(b) implies that

$$\sigma_{Y'}^{X'}(Z(f) \otimes g) = (g \otimes Z(f))\sigma_Y^X.$$

Thus, σ is a natural transformation. We now prove that for any $X \in \mathrm{Ob}(\mathcal{C})$, the pair $\mathcal{F}(X) = (Z(X), \sigma^X)$ is a half-braiding of $\mathcal{C}$. Since σ is a natural transformation, so is σ^X. Applying (9.4) several times and using the unitality of the monoidal product and composition in $\mathcal{C}$, we obtain that for all $V, Y_1, Y_2 \in \mathrm{Ob}(\mathcal{C})$,

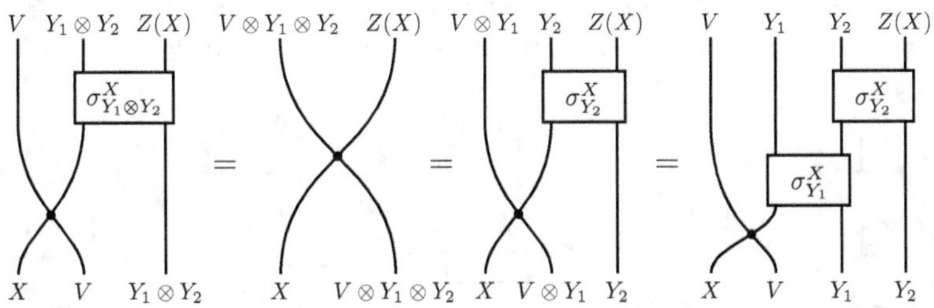

and

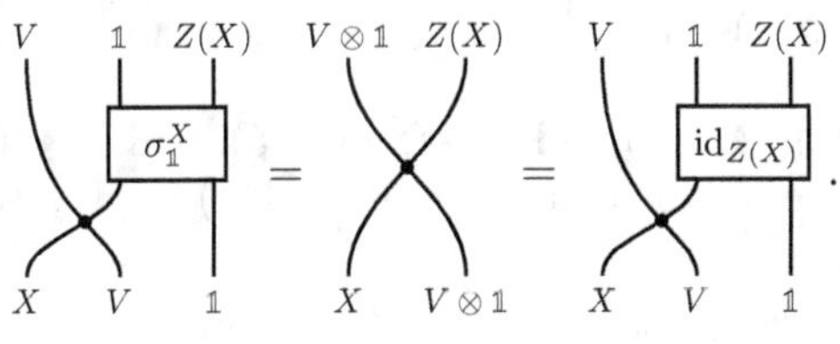

Then, the uniqueness in Lemma 9.1(b) implies that

$$\sigma^X_{Y_1\otimes Y_2} = (\mathrm{id}_{Y_1} \otimes \sigma^X_{Y_2})(\sigma^X_{Y_1} \otimes \mathrm{id}_{Y_2}) \quad \text{and} \quad \sigma^X_{\mathbb{1}} = \mathrm{id}_{Z(X)}.$$

Since $\mathcal{C}$ is right rigid, Exercise 5.1.5 implies that $\mathcal{F}(X)$ is a half-braiding of $\mathcal{C}$, that is, an object of $\mathcal{Z}(\mathcal{C})$. The naturality of σ implies that if f is a morphism in $\mathcal{C}$, then $\mathcal{F}(f) = Z(f)$ is a morphism in $\mathcal{Z}(\mathcal{C})$. Finally, the functoriality of Z implies that $\mathcal{F}$ is a functor. □

For any half-braiding (M, σ) of $\mathcal{C}$, Lemma 9.1(a) applied to $A = B = M$ and $\xi_V = \sigma_V$ yields a unique morphism $\varepsilon_{(M,\sigma)} \colon Z(M) \to M$ such that for all $V \in \mathrm{Ob}(\mathcal{C})$,

$$(\mathrm{id}_V \otimes \varepsilon_{(M,\sigma)})\partial_{M,V} = \sigma_V$$

that is

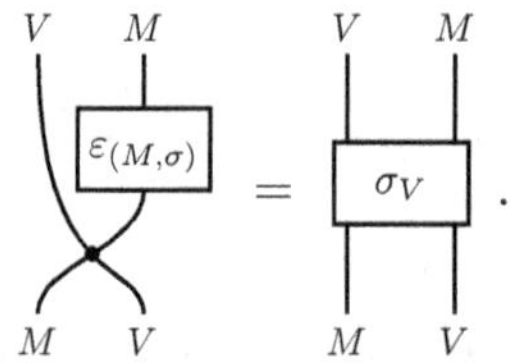

(9.5)

Consider the families

$$\eta = \{\eta_X = \partial_{X,\mathbb{1}}\}_{X\in\mathrm{Ob}(\mathcal{C})} \quad \text{and} \quad \varepsilon = \{\varepsilon_{(M,\sigma)}\}_{(M,\sigma)\in\mathrm{Ob}(\mathcal{Z}(\mathcal{C}))}.$$

Lemma 9.6. *The pair $(\mathcal{F}, \mathcal{U})$ is an adjunction with unit η and counit ε.*

Proof. The naturality of ∂ implies that η is a natural transformation from $1_{\mathcal{C}}$ to $Z = \mathcal{U}\mathcal{F}$. For any half-braiding (M, σ) of $\mathcal{C}$, $\varepsilon_{(M,\sigma)}$ is a morphism in $\mathcal{Z}(\mathcal{C})$ from $\mathcal{F}\mathcal{U}(M, \sigma) = \mathcal{F}(M) = (Z(M), \sigma^M)$ to (M, σ). Indeed, for all $V, Y \in \mathrm{Ob}(\mathcal{C})$,

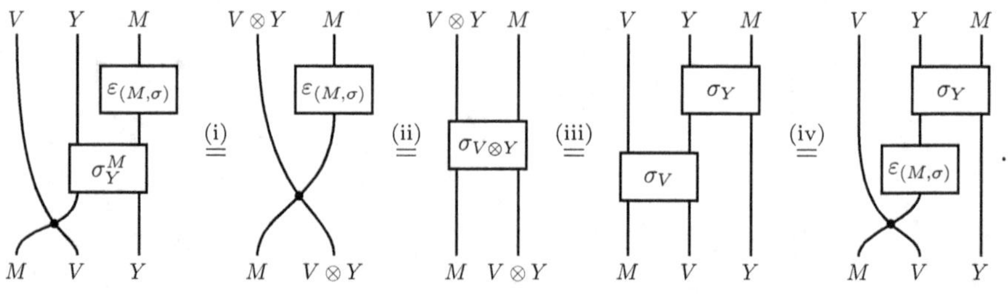

Here, (i) follows from (9.4), (ii) and (iv) from (9.5), and (iii) from the ⊗-multiplicativity (5.1) of σ. The uniqueness in Lemma 9.1(b) implies that

$$(\mathrm{id}_Y \otimes \varepsilon_{(M,\sigma)})\sigma^M_Y = \sigma_Y(\varepsilon_{(M,\sigma)} \otimes \mathrm{id}_Y)$$

for all $Y \in \mathrm{Ob}(\mathcal{C})$. This means that $\varepsilon_{(M,\sigma)} : \mathcal{FU}(M,\sigma) \to (M,\sigma)$ is a morphism in $\mathcal{Z}(\mathcal{C})$. Now, let $f : (M,\sigma) \to (N,\gamma)$ be a morphism in $\mathcal{Z}(\mathcal{C})$. For all $Y \in \mathrm{Ob}(\mathcal{C})$,

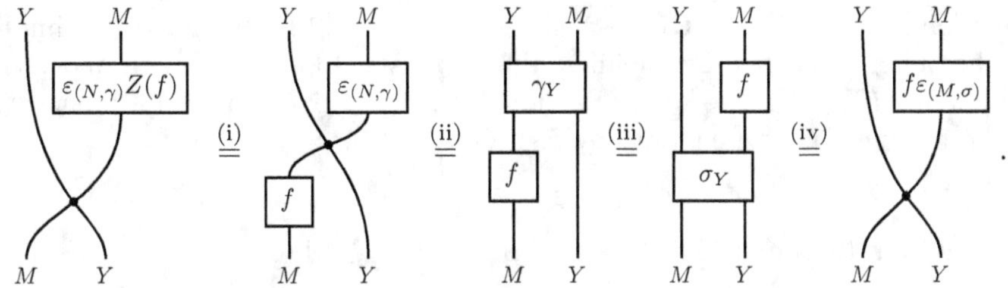

Here, (i) follows from the naturality of ∂, (ii) and (iv) from (9.5), and (iii) from the fact that f is a morphism in $\mathcal{Z}(\mathcal{C})$. The uniqueness in Lemma 9.1(a) implies that

$$\varepsilon_{(N,\gamma)} Z(f) = f\varepsilon_{(M,\sigma)}.$$

Consequently, ε is a natural transformation from $\mathcal{FU}$ to $1_{\mathcal{Z}(\mathcal{C})}$.

We now prove that η and ε satisfy (7.1). Pick $X \in \mathrm{Ob}(\mathcal{C})$. For all $Y \in \mathrm{Ob}(\mathcal{C})$,

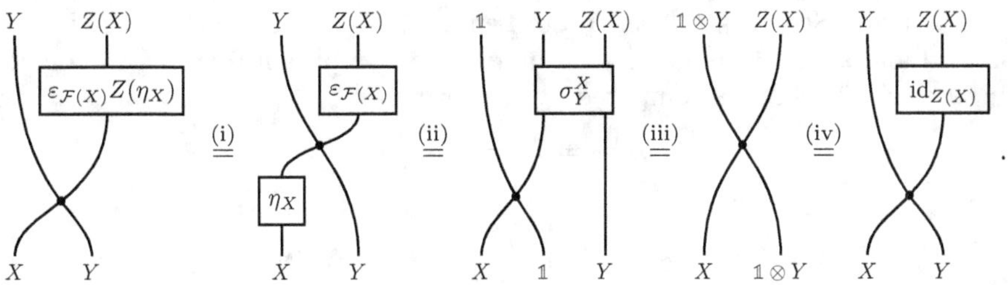

Here, (i) follows from the naturality of ∂, (ii) from (9.5) and the definition of η, (iii) from (9.4), and (iv) from the unitality of the monoidal product and composition in $\mathcal{C}$. The uniqueness in Lemma 9.1(a) implies that $\varepsilon_{\mathcal{F}(X)} Z(\eta_X) = \mathrm{id}_{Z(X)}$. Thus,

$$\varepsilon_{\mathcal{F}(X)} \mathcal{F}(\eta_X) = \mathrm{id}_{\mathcal{F}(X)}.$$

Also, for any $(M,\sigma) \in \mathrm{Ob}(\mathcal{Z}(\mathcal{C}))$,

$$\mathcal{U}(\varepsilon_{(M,\sigma)})\eta_{\mathcal{U}(M,\sigma)} \overset{(i)}{=} \varepsilon_{(M,\sigma)}\eta_M$$

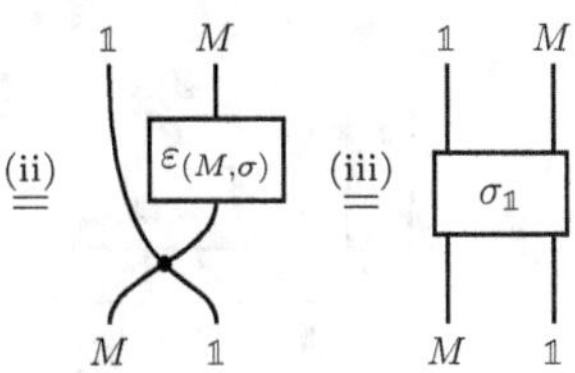

$$\overset{(iv)}{=} \mathrm{id}_M \overset{(v)}{=} \mathrm{id}_{\mathcal{U}(M,\sigma)}.$$

Here, (i) and (v) follow from the definition of $\mathcal{U}$, (ii) from the definition of η_M, (iii) from (9.5), and (iv) from Section 5.1.1. Thus, η and ε satisfy (7.1). By Section 7.2.1, the pair $(\mathcal{F},\mathcal{U})$ is an adjunction with unit η and counit ε. □

Since the forgetful functor $\mathcal{U}\colon \mathcal{Z}(\mathcal{C}) \to \mathcal{C}$ is strict monoidal, Lemma 7.10 implies that the functor $\mathcal{F}\colon \mathcal{C} \to \mathcal{Z}(\mathcal{C})$ has a unique structure of a comonoidal functor such that η and ε are comonoidal. Then the adjunction $\mathcal{F} \dashv \mathcal{U}$ is a comonoidal adjunction. Since $\mathcal{C}$ and $\mathcal{Z}(\mathcal{C})$ are rigid (see Section 5.2.1), Corollary 8.5 implies that the bimonad of this comonoidal adjunction is a Hopf monad with underlying endofunctor $\mathcal{U}\mathcal{F} = Z\colon \mathcal{C} \to \mathcal{C}$. By Section 7.2.2, the unit of this Hopf monad is η and the product μ is defined for any $X \in \mathrm{Ob}(\mathcal{C})$ by

$$\mu_X = \mathcal{U}(\varepsilon_{\mathcal{F}(X)}) = \varepsilon_{\mathcal{F}(X)}.$$

It follows from (9.4) and (9.5) that for all $X, Y_1, Y_2 \in \mathrm{Ob}(\mathcal{C})$,

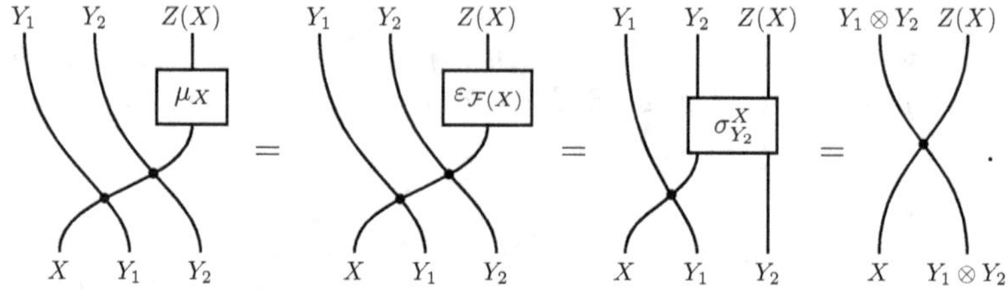

This is the first equality in Figure 9.1 and, by Lemma 9.1(c), it uniquely characterizes μ. Using Section 7.6.2, the proof Lemma 7.10, and the fact that $\mathcal{U}$ is strict monoidal and acts as the identity on morphisms, we obtain that the comonoidal structure (Z_2, Z_0) of Z is computed by the formulas

$$Z_0 = \mathcal{F}_0 = \varepsilon_{\mathbb{1}_{\mathcal{Z}(\mathcal{C})}}$$

and, for all $X_1, X_2 \in \mathrm{Ob}(\mathcal{C})$,

$$Z_2(X_1, X_2) = \mathcal{F}_2(X_1, X_2) = \varepsilon_{\mathcal{F}(X_1)\otimes\mathcal{F}(X_2)}\, Z(\eta_{X_1} \otimes \eta_{X_2}).$$

Then, for all $X_1, X_2, Y \in \mathrm{Ob}(\mathcal{C})$,

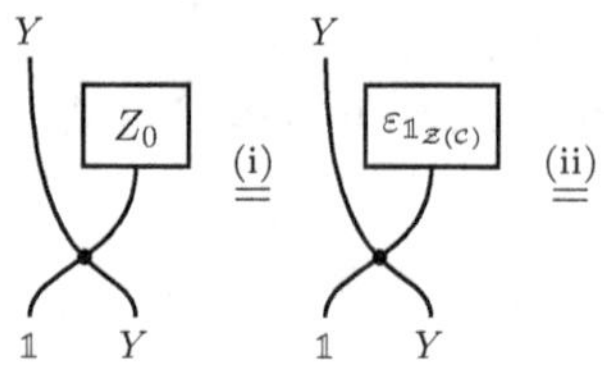

and

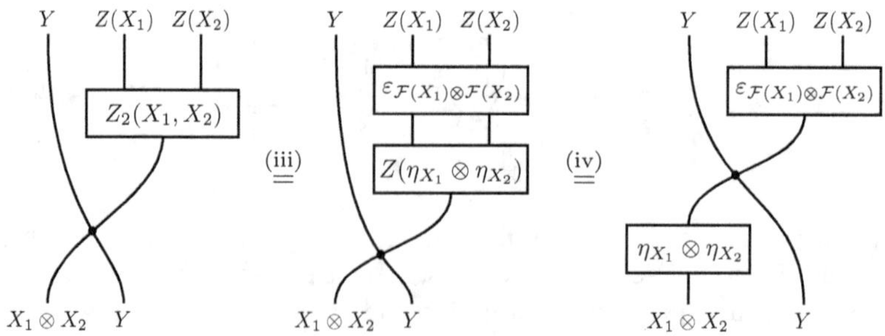

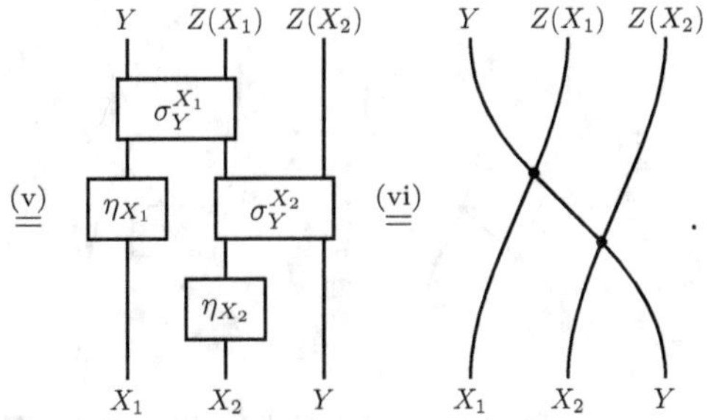

Here, (i) follows from the above computation of Z_0, (ii) from (9.5) and the definition of the monoidal unit of $\mathcal{Z}(\mathcal{C})$, (iii) from the above computation of $Z_2(X_1, X_2)$, (iv) from the naturality of ∂, (v) from (9.5) and the definition of the monoidal product in $\mathcal{Z}(\mathcal{C})$, and (vi) from the definition of η and (9.4). The resulting equalities coincide with the equalities in the second row of Figure 9.1 and, by Lemma 9.1, they uniquely characterize Z_0 and Z_2.

Let $K\colon \mathcal{Z}(\mathcal{C}) \to \mathcal{C}^Z$ be the comparison functor of the adjunction $\mathcal{F} \dashv \mathcal{U}$ (see Section 7.2.2). By definition, $K(M,\sigma) = (M, \varepsilon_{(M,\sigma)})$ for all $(M,\sigma) \in \mathrm{Ob}(\mathcal{Z}(\mathcal{C}))$. Since $\mathcal{U}$ acts as the identity on morphisms, $K(f) = f$ for all morphisms $f \in \mathcal{Z}(\mathcal{C})$. By Section 7.6.2, the functor K is strict monoidal and $U_Z K = \mathcal{U}$ as monoidal functors. We define a functor $E\colon \mathcal{C}^Z \to \mathcal{Z}(\mathcal{C})$ by

$$E(M,r) = (M, \sigma^{(M,r)} = \{\sigma_X^{(M,r)}\}_{X \in \mathrm{Ob}(\mathcal{C})}) \quad \text{and} \quad E(f) = f$$

for all $(M,r) \in \mathrm{Ob}(\mathcal{C}^Z)$ and all morphisms f in $\mathcal{C}^Z$, where $\sigma_X^{(M,r)}$ is as in (9.3).

Lemma 9.7. *The functor E is well defined and is inverse to K.*

Proof. Consider an arbitrary Z-module (M,r). We have:

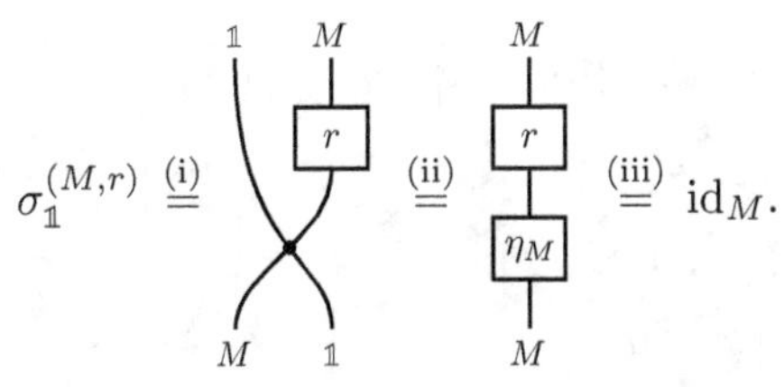

Here, (i) follows from (9.3), (ii) from the definition of η, and (iii) from the fact that r is a Z-action. Also, for all $X, Y \in \mathrm{Ob}(\mathcal{C})$,

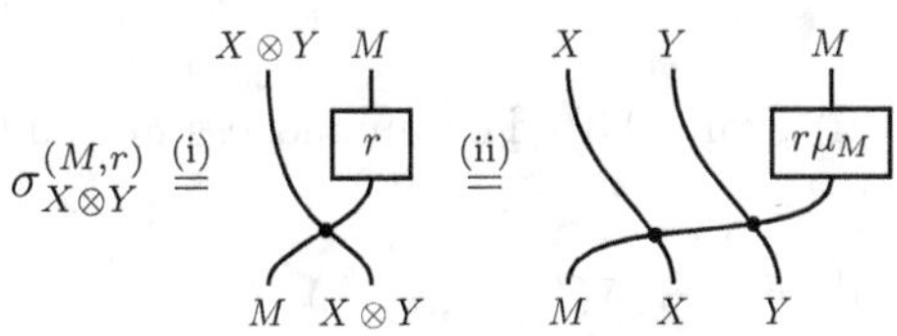

$\overset{\text{(iii)}}{=}$ 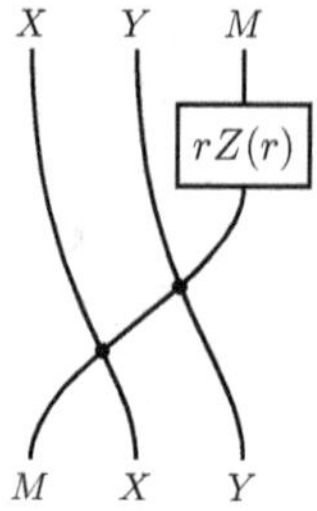 $\overset{\text{(iv)}}{=}$ 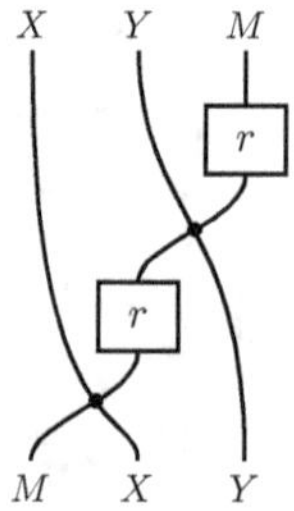 $\overset{\text{(v)}}{=}$ 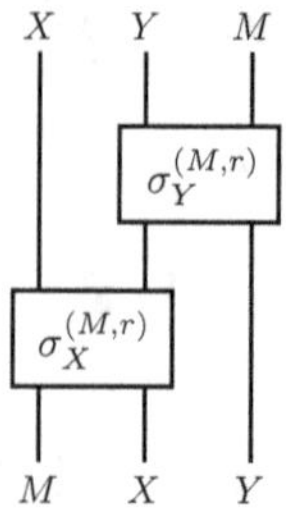 .

Here, (i) and (v) follow from (9.3), (ii) from the characterization of μ (see the first equality in Figure 9.1), (iii) from the fact that r is a Z-action, and (iv) from the naturality of ∂. Since $\mathcal{C}$ is right rigid, Exercise 5.1.5 implies that $E(M,r) = (M, \sigma^{(M,r)})$ is a half-braiding of $\mathcal{C}$, i.e., an object of $\mathcal{Z}(\mathcal{C})$. Let $f\colon (M,r) \to (N,s)$ be a morphism in $\mathcal{C}^Z$. For all $X \in \mathrm{Ob}(\mathcal{C})$,

$$\begin{aligned}\sigma_X^{(N,s)}(f \otimes \mathrm{id}_X) &\overset{\text{(i)}}{=} (\mathrm{id}_X \otimes s)\partial_{N,X}(f \otimes \mathrm{id}_X)\\ &\overset{\text{(ii)}}{=} (\mathrm{id}_X \otimes sZ(f))\partial_{M,X} \overset{\text{(iii)}}{=} (\mathrm{id}_X \otimes fr)\partial_{M,X} \overset{\text{(iv)}}{=} (\mathrm{id}_X \otimes f)\sigma_X^{(M,r)}.\end{aligned}$$

Here, (i) and (iv) follow from (9.3), (ii) from the naturality of ∂, and (iii) from the Z-linearity of f. Thus $E(f) = f$ is a morphism in $\mathcal{Z}(\mathcal{C})$. The functoriality of $E\colon \mathcal{C}^Z \to \mathcal{Z}(\mathcal{C})$ follows from the fact that E acts as the identity on morphisms. Hence, E is a well-defined functor.

Let us prove that K and E are inverse to each other. Let $(M,\sigma) \in \mathrm{Ob}(\mathcal{Z}(\mathcal{C}))$. For all $X \in \mathrm{Ob}(\mathcal{C})$,

$\sigma_X^{(M,\varepsilon_{(M,\sigma)})} \overset{\text{(i)}}{=}$ 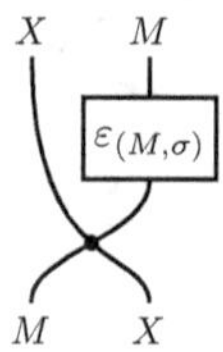 $\overset{\text{(ii)}}{=} \sigma_X.$

Here, (i) follows from (9.3) and (ii) from (9.5). Thus,

$$EK(M,\sigma) = E(M, \varepsilon_{(M,\sigma)}) = (M, \sigma^{(M,\varepsilon_{(M,\sigma)})}) = (M,\sigma).$$

Let $(M,r) \in \mathrm{Ob}(\mathcal{C}^Z)$. For all $X \in \mathrm{Ob}(\mathcal{C})$,

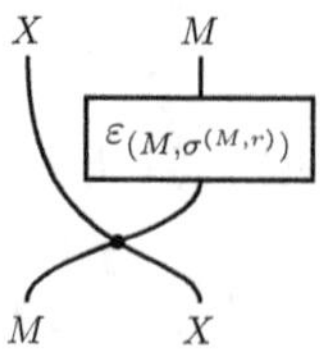 $\overset{\text{(i)}}{=} \sigma_X^{(M,r)} \overset{\text{(ii)}}{=}$ 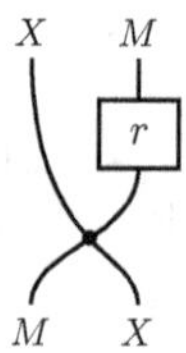 .

Here, (i) follows from (9.5) and (ii) from (9.3). The uniqueness in Lemma 9.1(a) implies that $\varepsilon_{(M,\sigma^{(M,r)})} = r$. Thus,

$$KE(M,r) = K(M, \sigma^{(M,r)}) = (M, \varepsilon_{(M,\sigma^{(M,r)})}) = (M,r).$$

Since both E and K act as the identity on morphisms, we conclude that they are isomorphisms inverse to each other. □

Since the functor K is strict monoidal, so is its inverse $E\colon \mathcal{C}^Z \to \mathcal{Z}(\mathcal{C})$. Therefore, we can use E to transport the left duality (5.2) and the right duality (5.3) in $\mathcal{Z}(\mathcal{C})$ into a left duality and a right duality in $\mathcal{C}^Z$. The latter dualities are denoted respectively by

$$\{({}^\vee(M,r),\, \mathrm{ev}_{(M,r)})\}_{(M,r)\in \mathrm{Ob}(\mathcal{C}^Z)} \quad \text{and} \quad \{((M,r)^\vee,\, \widetilde{\mathrm{ev}}_{(M,r)})\}_{(M,r)\in \mathrm{Ob}(\mathcal{C}^Z)}.$$

In the notation of Section 5.2.1, for any Z-module (M,r), we have

$$\begin{gathered}
{}^\vee(M,r) = E^{-1}({}^\vee E(M,r)) = E^{-1}({}^\vee M, {}^\dagger(\sigma^{(M,r)})) = ({}^\vee M, \varepsilon_{({}^\vee M, {}^\dagger(\sigma^{(M,r)}))}),\\
(M,r)^\vee = E^{-1}(E(M,r)^\vee) = E^{-1}(M^\vee, (\sigma^{(M,r)})^\dagger) = (M^\vee, \varepsilon_{(M^\vee,(\sigma^{(M,r)})^\dagger)}),\\
\mathrm{ev}_{(M,r)} = (\mathrm{ev}_{E(M,r)})^K = \mathrm{ev}_M,\\
\widetilde{\mathrm{ev}}_{(M,r)} = (\widetilde{\mathrm{ev}}_{E(M,r)})^K = \widetilde{\mathrm{ev}}_M.
\end{gathered}$$

It is clear from these formulas that our left and right dualities in $\mathcal{C}^Z$ are lifts along the forgetful functor $\mathcal{C}^Z \to \mathcal{C}$ of the distinguished left and right dualities in $\mathcal{C}$. The uniqueness in Theorem 8.7(b) implies that these left and right dualities in $\mathcal{C}^Z$ coincide with those induced by the left antipode s^l and the right antipode s^r of Z, see Section 8.3.1. We conclude that for all Z-modules (M,r),

$$\varepsilon_{({}^\vee M, {}^\dagger(\sigma^{(M,r)}))} = s^l_M Z({}^\vee r) \quad \text{and} \quad \varepsilon_{(M^\vee,(\sigma^{(M,r)})^\dagger)} = s^r_M Z(r^\vee). \tag{9.6}$$

These equalities give rise to the equalities in the third row of Figure 9.1. Indeed, for $X \in \mathrm{Ob}(\mathcal{C})$, we have:

$$s^l_X \overset{(i)}{=} s^l_X Z({}^\vee Z(\eta_X)^\vee \mu_X) \overset{(ii)}{=} {}^\vee\eta_X\, s^l_{Z(X)} Z({}^\vee \mu_X) \overset{(iii)}{=} {}^\vee\eta_X \varepsilon_{({}^\vee Z(X), {}^\dagger(\sigma^{F_Z(X)}))}.$$

Here, (i) follows from the unitality of μ, (ii) from the naturality of s^l, and (iii) from the first equality of (9.6) applied to the free Z-module $F_Z(X) = (Z(X), \mu_X)$. Thus, for all $Y \in \mathrm{Ob}(\mathcal{C})$, we have:

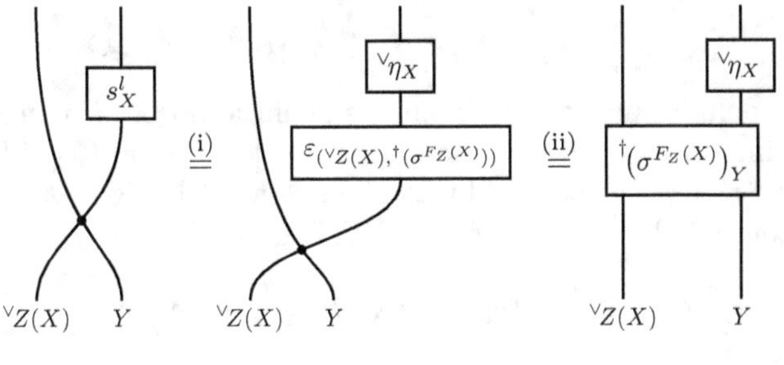

$\overset{\text{(iii)}}{=}$

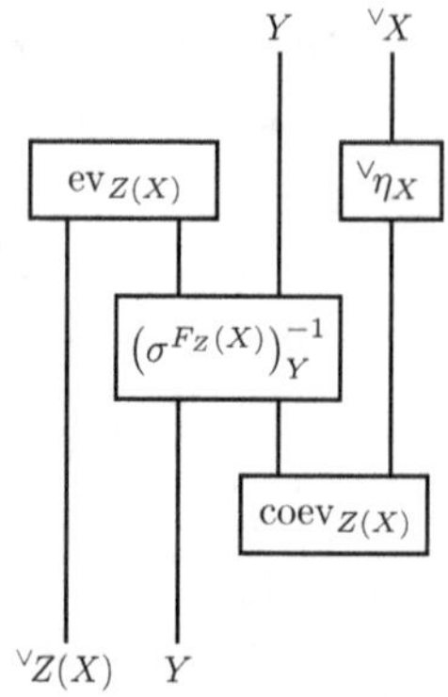

$\overset{\text{(iv)}}{=}$

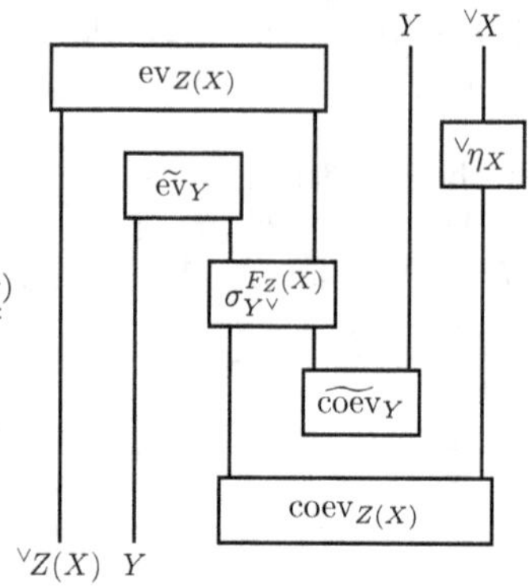

$\overset{\text{(v)}}{=}$

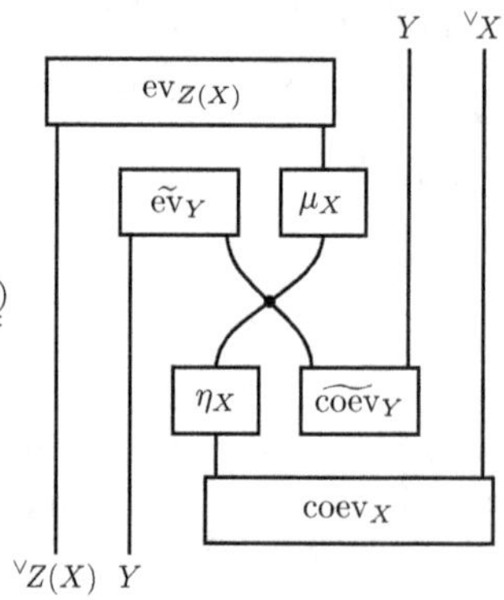

$\overset{\text{(vi)}}{=}$

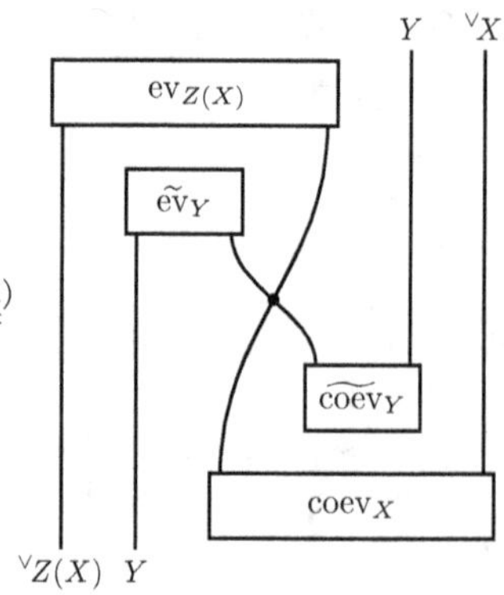

.

Here, (i) follows from the above computation of s^l_X, (ii) from (9.5), (iii) from the definition of the left dual of a half-braiding in Section 5.2.1, (iv) from the computation of the inverse of a half-braiding in Exercise 5.1.5, (v) from (2.2) and (9.3), and (vi) from the naturality of ∂ and the unitality of μ. The resulting equality is the first equality in the third row of Figure 9.1, and, by Lemma 9.1(a), it characterizes s^l uniquely. The equality in Figure 9.1 characterizing s^r is proved similarly.

The strict monoidal isomorphism $E\colon \mathcal{C}^Z \to \mathcal{Z}(\mathcal{C})$ can be similarly used to transport the braiding τ in $\mathcal{Z}(\mathcal{C})$ into a braiding τ' in $\mathcal{C}^Z$ so that E becomes braided. Explicitly, for any Z-modules (M,r) and (N,s),

$$\tau'_{(M,r),(N,s)} \overset{\text{(i)}}{=} E^{-1}\big(\tau_{E(M,r),E(N,s)}\big) \overset{\text{(ii)}}{=} \tau_{(M,\sigma^{(M,r)}),(N,\sigma^{(N,s)})} \overset{\text{(iii)}}{=} \sigma_N^{(M,r)} \overset{\text{(iv)}}{=} (\mathrm{id}_N \otimes r)\partial_{M,N}.$$

Here, (i) follows from the definition of τ', (ii) from the definition of E, (iii) from the definition of the braiding in $\mathcal{Z}(\mathcal{C})$ (see Section 5.1.2), and (iv) from (9.3). By Theorem 7.11, the braiding τ' of $\mathcal{C}^Z$ is induced by a unique $*$-invertible R-matrix R for Z. This R-matrix is computed as follows: for any $X, Y \in \mathrm{Ob}(\mathcal{C})$,

$$R_{X,Y} \overset{\text{(i)}}{=} \tau'_{F_Z(X),F_Z(Y)}(\eta_X \otimes \eta_Y) \overset{\text{(ii)}}{=} (\mathrm{id}_{Z(Y)} \otimes \mu_X)\partial_{Z(X),Z(Y)}(\eta_X \otimes \eta_Y)$$

$$\overset{\text{(iii)}}{=} (\eta_Y \otimes \mu_X Z(\eta_X))\partial_{X,Y} \overset{\text{(iv)}}{=} (\eta_Y \otimes \mathrm{id}_{Z(X)})\partial_{X,Y}.$$

Here, (i) follows from the proof of Theorem 7.11, (ii) from the above computation of τ', (iii) from the naturality of ∂, and (iv) from the unitality of μ. This gives the last equality in Figure 9.1.

The arguments above show that Z is a quasitriangular Hopf monad on $\mathcal{C}$ (so that $\mathcal{C}^Z$ is a braided rigid category), the structural morphisms of Z are characterized by the equalities in Figure 9.1, the functor $E\colon \mathcal{C}^Z \to \mathcal{Z}(\mathcal{C})$ is a braided strict monoidal isomorphism, and $\mathcal{U}E = U_Z K E = U_Z$ as monoidal functors. This proves the first assertion of Theorem 9.2 and the part (a) of Theorem 9.3.

Suppose now that $\mathcal{C}$ is $\Bbbk$-linear. By Section 5.3.1, both the category $\mathcal{Z}(\mathcal{C})$ and the forgetful functor $\mathcal{U}\colon \mathcal{Z}(\mathcal{C}) \to \mathcal{C}$ are $\Bbbk$-linear. Thus, by Lemma 7.7, the left adjoint $\mathcal{F}$ of $\mathcal{U}$ is $\Bbbk$-linear. Consequently, $Z = \mathcal{U}\mathcal{F}$ is $\Bbbk$-linear. Then, by Lemma 7.6, the category $\mathcal{C}^Z$ has a unique structure of a $\Bbbk$-linear category such that the forgetful functor $U_Z\colon \mathcal{C}^Z \to \mathcal{C}$ is $\Bbbk$-linear. Since $\mathcal{U}$ and U_Z are $\Bbbk$-linear and $\mathcal{U}$ acts as the identity on morphisms, the equality $\mathcal{U}E = U_Z$ implies that E is $\Bbbk$-linear. This proves the second assertion of Theorem 9.2 and the part (b) of Theorem 9.3.

Finally, suppose that $\mathcal{C}$ is pivotal. By Section 5.2.2, $\mathcal{Z}(\mathcal{C})$ is pivotal and the forgetful functor $\mathcal{U}\colon \mathcal{Z}(\mathcal{C}) \to \mathcal{C}$ is strictly pivotal. The strict monoidal isomorphism $E\colon \mathcal{C}^Z \to \mathcal{Z}(\mathcal{C})$ transports the pivotal duality (5.4) in $\mathcal{Z}(\mathcal{C})$ into a pivotal duality $\{((M,r)^*, \mathrm{ev}_{(M,r)}, \widetilde{\mathrm{ev}}_{(M,r)})\}_{(M,r)\in \mathrm{Ob}(\mathcal{C}^Z)}$ in $\mathcal{C}^Z$. Explicitly,

$$(M,r)^* = E^{-1}\big((E(M,r))^*\big) = E^{-1}\big(M^*, (\sigma^{(M,r)})^\dagger\big) = \big(M^*, \varepsilon_{(M^*,(\sigma^{(M,r)})^\dagger)}\big),$$
$$\mathrm{ev}_{(M,r)} = \big(\mathrm{ev}_{E(M,r)}\big)^K = \mathrm{ev}_M,$$
$$\widetilde{\mathrm{ev}}_{(M,r)} = \big(\widetilde{\mathrm{ev}}_{E(M,r)}\big)^K = \widetilde{\mathrm{ev}}_M.$$

This pivotal duality turns $\mathcal{C}^Z$ into a pivotal category such that U_Z and E are strictly pivotal (see Section 1.7.5). By Theorem 8.10, Z is involutory. This proves the third assertion of Theorem 9.2 and the part (c) of Theorem 9.3.

9.4 The case of pivotal fusion categories

Throughout this section, $\mathcal{C}$ is an additive pivotal fusion $\Bbbk$-category. We apply the theory of centralizers to $\mathcal{C}$ and prove, in particular, that the center $\mathcal{Z}(\mathcal{C})$ of $\mathcal{C}$ is modular in the sense of Lyubashenko.

Fix throughout this section a representative set I of simple objects of $\mathcal{C}$.

9.4.1 The centralizer

Since the centralizability of the category $\mathcal{C}$ can be verified object-wise (see Section 9.1.1), Example B.1.2 shows that $\mathcal{C}$ is centralizable. We construct a centralizer

(Z, ρ) of $\mathcal{C}$ as follows. For any $X \in \mathrm{Ob}(\mathcal{C})$, set

$$Z(X) = \bigoplus_{i \in I} i^* \otimes X \otimes i \in \mathrm{Ob}(\mathcal{C}). \tag{9.7}$$

For each $i \in I$, let

$$\rho_{X,i} \colon i^* \otimes X \otimes i \to Z(X) \quad \text{and} \quad \nu_{X,i} \colon Z(X) \to i^* \otimes X \otimes i$$

be respectively the inclusion and the projection associated with the direct sum decomposition (9.7). For any morphism $f \colon X \to X'$ in $\mathcal{C}$, set

$$Z(f) = \sum_{i \in I} \rho_{X',i}(\mathrm{id}_{i^*} \otimes f \otimes \mathrm{id}_i)\nu_{X,i} \colon Z(X) \to Z(X').$$

It follows from the definition of a direct sum that $Z \colon \mathcal{C} \to \mathcal{C}$ is a well-defined functor.

Next, for any $X, Y \in \mathrm{Ob}(\mathcal{C})$, pick an I-partition $(p_\alpha \colon Y \to i_\alpha,\, q_\alpha \colon i_\alpha \to Y)_\alpha$ of Y and set

$$\rho_{X,Y} = \sum_\alpha \rho_{X,i_\alpha}\big((q_\alpha)^* \otimes \mathrm{id}_X \otimes p_\alpha\big) \colon Y^* \otimes X \otimes Y \to Z(X).$$

Note that for $Y = i \in I$, we have $\rho_{X,Y} = \rho_{X,i}$ because the pair $(\mathrm{id}_i, \mathrm{id}_i)$ is an I-partition of i. It is easy to check that the family

$$\rho = \{\rho_{X,Y} \colon Y^* \otimes X \otimes Y \to Z(X)\}_{X,Y \in \mathrm{Ob}(\mathcal{C})}$$

is well defined, natural in X, and dinatural in Y. Also, by Example B.1.2, for any $X \in \mathrm{Ob}(\mathcal{C})$, the dinatural transformation $\{\rho_{X,Y} \colon Y^* \otimes X \otimes Y \to Z(X)\}_{Y \in \mathrm{Ob}(\mathcal{C})}$ is universal. Hence, the pair (Z, ρ) is a centralizer of $\mathcal{C}$. By Section 9.1.2 and (2.2), the universal coaction ∂ associated with ρ is computed for all $X, Y \in \mathrm{Ob}(\mathcal{C})$ by

$$\partial_{X,Y} = \sum_\alpha (q_\alpha \otimes \rho_{X,i_\alpha})(\mathrm{coev}_{i_\alpha} \otimes \mathrm{id}_X \otimes p_\alpha) \colon X \otimes Y \to Y \otimes Z(X),$$

where $(p_\alpha \colon Y \to i_\alpha,\, q_\alpha \colon i_\alpha \to Y)_\alpha$ is an I-partition of Y. Using the enriched graphical calculus from Section 4.6, we have:

$Z(f) = \sum_{i \in I}$ 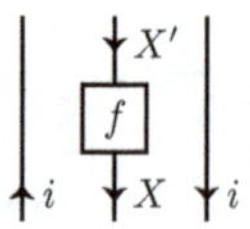 $\colon Z(X) \to Z(X')$,

$\rho_{X,Y} = \sum_{i \in I}$ 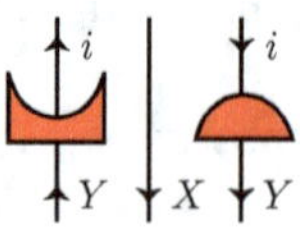 $\colon Y^* \otimes X \otimes Y \to Z(X)$,

$\partial_{X,Y} = \sum_{i \in I}$ 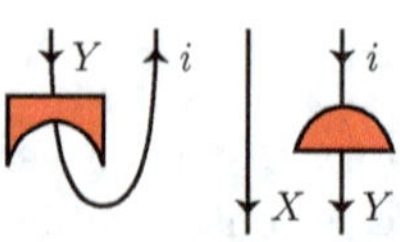 $\colon X \otimes Y \to Y \otimes Z(X)$.

In these figures and in the sequel, we drop the inclusions and projections associated with the direct sum (9.7).

The morphism $r\colon Z(A) \to B$ associated by Lemma 9.1(a) to a natural transformation $\xi = \{\xi_Y : A \otimes Y \to Y \otimes B\}_{Y \in \mathrm{Ob}(\mathcal{C})}$ is computed by

$$r = \sum_{i \in I} (\mathrm{ev}_i \otimes \mathrm{id}_B)(\mathrm{id}_{i^*} \otimes \xi_i)\nu_{A,i}. \tag{9.8}$$

Graphically:

$r = \sum_{i \in I}$ 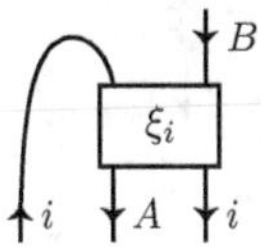 $: Z(A) \to B.$

Also, the morphism $r\colon Z^2(A) \to B$ associated by Lemma 9.1(c) to a natural transformation

$$\xi = \{\xi_{Y_1,Y_2} : A \otimes Y_1 \otimes Y_2 \to Y_1 \otimes Y_2 \otimes B\}_{Y_1,Y_2 \in \mathrm{Ob}(\mathcal{C})}$$

is computed by

$$r = \sum_{i,j \in I} \big(\mathrm{ev}_j(\mathrm{id}_{j^*} \otimes \mathrm{ev}_i \otimes \mathrm{id}_j) \otimes \mathrm{id}_B\big)(\mathrm{id}_{j^* \otimes i^*} \otimes \xi_{i,j})(\mathrm{id}_{j^*} \otimes \nu_{A,i} \otimes \mathrm{id}_j)\nu_{Z(A),j}. \tag{9.9}$$

Graphically:

$r = \sum_{i,j \in I}$ 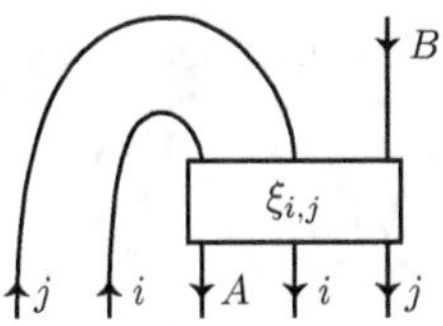 $: Z^2(A) \to B.$

Formulas (9.8) and (9.9) are deduced from Formula (B.1) in Example B.1.2 by following the proof of Lemma 9.1 line by line.

By Theorem 9.2, Z is a $\Bbbk$-linear involutory quasitriangular Hopf monad on $\mathcal{C}$. Its comonoidal structure (Z_2, Z_0), product μ, unit η, antipode s, and R-matrix R are computed in Figure 9.2, where $X, Y \in \mathrm{Ob}(\mathcal{C})$. These computations are derived from the formulas in Figure 9.1 as well as Formulas (9.8) and (9.9). In Figure 9.2, the dotted lines represent $\mathrm{id}_{\mathbb{1}}$ or $\mathrm{id}_{\mathbb{1}^*}$ and can be removed without changing the morphisms. We depict them to record the relevant direct summands of (9.7).

Lemma 9.8. *The Hopf monad Z is semisimple if and only if $\dim(\mathcal{C})$ is invertible in $\Bbbk$.*

Proof. By Theorem 8.11, Z is semisimple if and only if there exists a morphism $\Lambda\colon \mathbb{1} \to Z(\mathbb{1})$ in $\mathcal{C}$ such that

$$\Lambda Z_0 = \mu_{\mathbb{1}} Z(\Lambda) \quad \text{and} \quad Z_0 \Lambda = \mathrm{id}_{\mathbb{1}}. \tag{9.10}$$

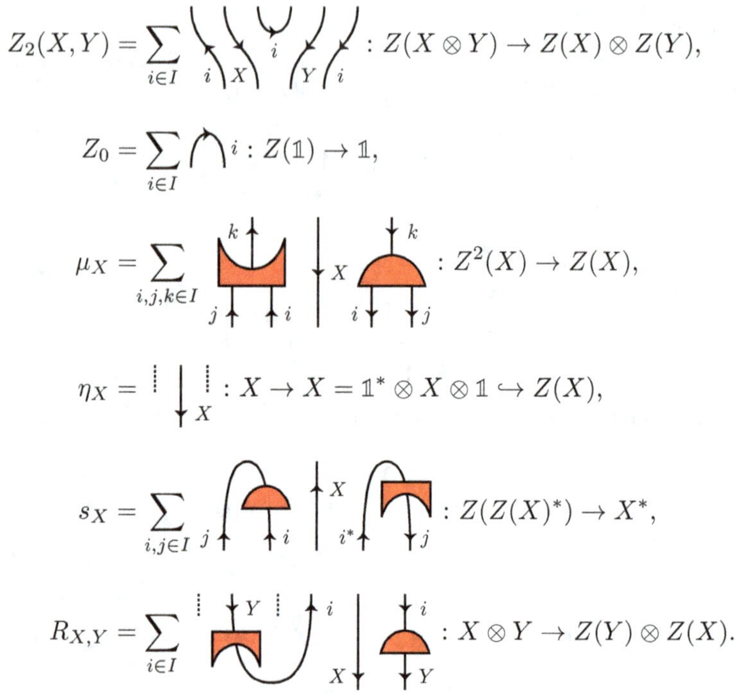

Figure 9.2: The structural morphisms of Z in the fusion case

As in the proof of Lemma 6.4, the family

$$\Big(e_i = \rho_{\mathbb{1},i}\, \widetilde{\mathrm{coev}}_i \colon \mathbb{1} \to Z(\mathbb{1}) = \bigoplus_{i\in I} i^* \otimes i\Big)_{i\in I}$$

is a basis of the free $\Bbbk$-module $\mathrm{Hom}_{\mathcal{C}}(\mathbb{1}, Z(\mathbb{1}))$. Any morphism $\Lambda\colon \mathbb{1} \to Z(\mathbb{1})$ expands as $\Lambda = \sum_{i\in I} a_i\, e_i$ with $a_i \in \Bbbk$. Using the above description of the structural morphisms of Z, we obtain

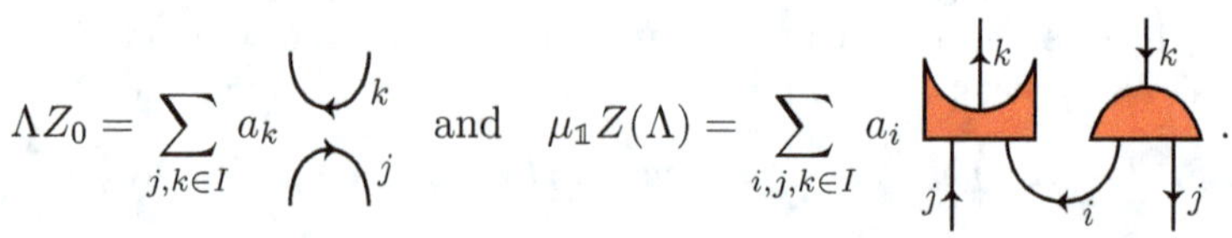

Now, Lemma 4.10 implies that Λ satisfies $\Lambda Z_0 = \mu_{\mathbb{1}} Z(\Lambda)$ if and only if $a_i = a_{\mathbb{1}} \dim_r(i)$ for all $i \in I$, that is, if and only if $\Lambda = a_{\mathbb{1}} \sum_{i\in I} \dim_r(i)\, e_i$. The equalities

$$Z_0\, e_i = Z_0\, \rho_{\mathbb{1},i}\, \widetilde{\mathrm{coev}}_i = \mathrm{ev}_i\, \widetilde{\mathrm{coev}}_i = \dim_l(i)$$

imply that for $\Lambda = a_{\mathbb{1}} \sum_{i\in I} \dim_r(i)\, e_i$, we have

$$Z_0\Lambda = a_{\mathbb{1}} \sum_{i\in I} \dim_r(i)\, Z_0 e_i = a_{\mathbb{1}} \sum_{i\in I} \dim_r(i) \dim_l(i) = a_{\mathbb{1}} \dim(\mathcal{C}).$$

Hence a morphism $\Lambda\colon \mathbb{1} \to Z(\mathbb{1})$ satisfying (9.10) exists if and only if $\dim(\mathcal{C})$ is invertible in $\Bbbk$. □

The following lemma will be instrumental in Section 16.2.

Lemma 9.9. *If $\dim(\mathcal{C})$ is invertible in $\Bbbk$, then for any objects $(A,\sigma), (B,\beta)$ of $\mathcal{Z}(\mathcal{C})$, the set $\mathrm{Hom}_{\mathcal{Z}(\mathcal{C})}((A,\sigma),(B,\beta))$ is the image of the $\Bbbk$-linear projector*

$$\pi\colon \mathrm{Hom}_{\mathcal{C}}(A,B) \to \mathrm{Hom}_{\mathcal{C}}(A,B)$$

carrying any $f \in \mathrm{Hom}_{\mathcal{C}}(A,B)$ to

$$\pi(f) = (\dim(\mathcal{C}))^{-1} \sum_{i\in I} \dim_r(i)$$ 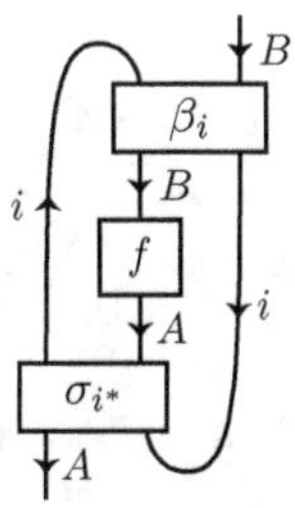 . (9.11)

Proof. By Theorem 9.3, the functor $E\colon \mathcal{C}^Z \to \mathcal{Z}(\mathcal{C})$ of Section 9.2.2 is an isomorphism acting as the identity on morphisms. By Section 9.3, for any half-braiding (M,σ) of $\mathcal{C}$, we have $E^{-1}(M,\sigma) = (M, \varepsilon_{(M,\sigma)})$ where $\varepsilon_{(M,\sigma)}\colon Z(M) \to M$ is defined by (9.5). Therefore

$$\mathrm{Hom}_{\mathcal{Z}(\mathcal{C})}\big((A,\sigma),(B,\beta)\big) = \mathrm{Hom}_{\mathcal{C}^Z}\big((A,\varepsilon_{(A,\sigma)}),(B,\varepsilon_{(B,\beta)})\big). \tag{9.12}$$

By Lemma 9.8, the Hopf monad Z is semisimple. Then Z is separable by Theorem 8.11. Consequently, Lemma 7.5 implies that the set (9.12) is the image of the $\Bbbk$-linear projector

$$\pi\colon \mathrm{Hom}_{\mathcal{C}}(A,B) \to \mathrm{Hom}_{\mathcal{C}}(A,B), \quad f \mapsto \pi(f) = \varepsilon_{(B,\beta)} Z(f\varepsilon_{(A,\sigma)})\gamma_A,$$

where $\gamma\colon 1_{\mathcal{C}} \to Z^2$ is a natural transformation as in Lemma 7.4. For any half-braiding (M,σ) of $\mathcal{C}$, the formulas $\mathrm{id}_{Z(M)} = \sum_{i\in I} \rho_{M,i}\,\nu_{M,i}$ and (9.5) imply that

$$\varepsilon_{(M,\sigma)} = \sum_{i\in I} \varepsilon_{(M,\sigma)}\rho_{M,i}\,\nu_{M,i} = \sum_{i\in I} (\mathrm{ev}_i \otimes \mathrm{id}_M)(\mathrm{id}_{i^*} \otimes \sigma_i)\nu_{M,i} = \sum_{i\in I}$$ 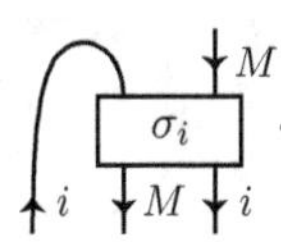 .

The proof of Lemma 9.8 shows that the morphism

$$\Lambda = (\dim(\mathcal{C}))^{-1} \sum_{i\in I} \dim_r(i)$$  $: \mathbb{1} \to Z(\mathbb{1})$

satisfies the conditions (9.10). Then, the proof of Theorem 8.11 yields that

$$\gamma_A = (H^l_{\mathbb{1},A})^{-1}(\Lambda \otimes \eta_A)\colon A \to Z^2(A),$$

where H^l is the left fusion operator of Z. By Exercise 8.3.7,

$$\begin{aligned}(H^l_{\mathbb{1},A})^{-1} = \big(&\mathrm{id}_{Z^2(A)} \otimes \mathrm{ev}_{Z(A)}(s^l_{Z(A)} Z({}^\vee\mu_A) \otimes \mathrm{id}_{Z(A)})\big)\\ &\circ \big(Z_2(Z(A), {}^\vee Z(A)) Z(\mathrm{coev}_{Z(A)}) \otimes \mathrm{id}_{Z(A)}\big).\end{aligned}$$

The naturality of s^l and unitality of μ imply that

$${}^\vee\eta_A s^l_{Z(A)} Z({}^\vee\mu_A) = s^l_A Z({}^\vee Z(\eta_A){}^\vee\mu_A) = s^l_A.$$

Therefore, using (2.2), we obtain that

$$\begin{aligned}(H^l_{\mathbb{1},A})^{-1}(\mathrm{id}_{Z(\mathbb{1})} \otimes \eta_A) = \big(&\mathrm{id}_{Z^2(A)} \otimes \mathrm{ev}_A(s^l_A \otimes \mathrm{id}_A)\big)\\ &\circ \big(Z_2(Z(A), {}^\vee Z(A)) Z(\mathrm{coev}_{Z(A)}) \otimes \mathrm{id}_A\big).\end{aligned}$$

Using the computations above and the expressions for the structural morphisms of Z in Figure 9.2, we obtain that

$$\gamma_A = (\dim(\mathcal{C}))^{-1} \sum_{i,j\in I} \dim_r(i)$$

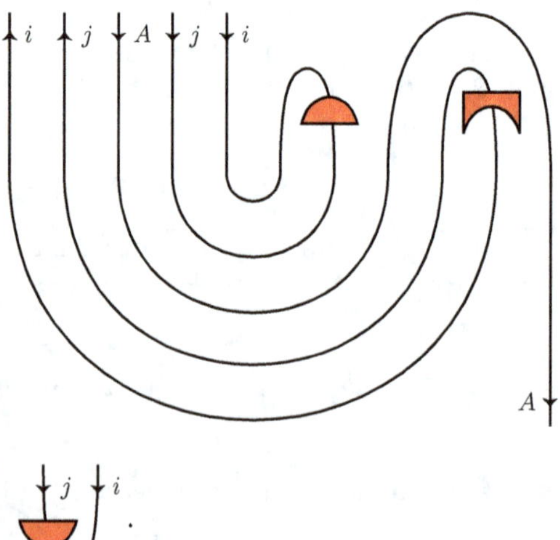

$$= \sum_{i,j\in I} \frac{\dim_r(i)}{\dim(\mathcal{C})}$$

.

Hence, for any $f \in \mathrm{Hom}_{\mathcal{C}}(A, B)$,

$$\pi(f) \overset{\text{(i)}}{=} \varepsilon_{(B,\beta)} Z(f \varepsilon_{(A,\sigma)}) \gamma_A \overset{\text{(ii)}}{=} \sum_{i,j \in I} \frac{\dim_r(i)}{\dim(\mathcal{C})}$$

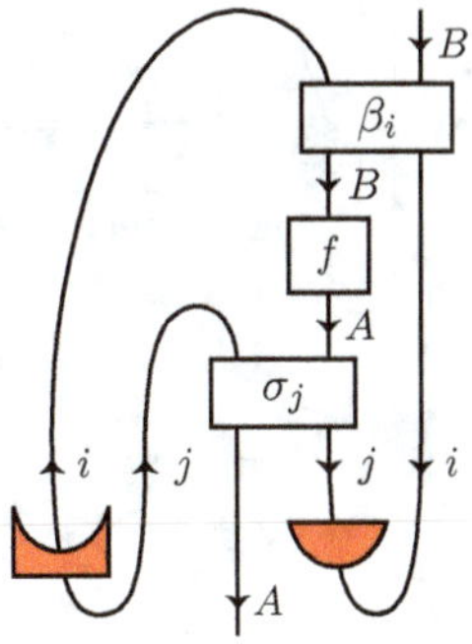

$$\overset{\text{(iii)}}{=} \sum_{i,j \in I} \frac{\dim_r(i)}{\dim(\mathcal{C})}$$

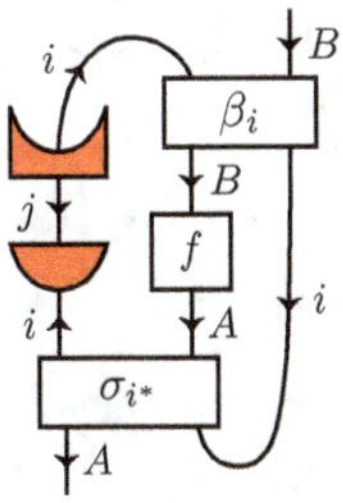

$$\overset{\text{(iv)}}{=} \sum_{i \in I} \frac{\dim_r(i)}{\dim(\mathcal{C})}$$

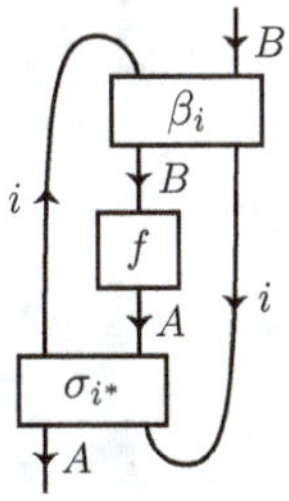

.

Here, (i) follows from the definition of π, (ii) from the above computations of ε and γ_A, (iii) from the naturality of σ, and (iv) from Lemma 4.9(a). □

9.4.2 The coend of the center

Set

$$C = \bigoplus_{i,j \in I} i^* \otimes j^* \otimes i \otimes j \in \mathrm{Ob}(\mathcal{C}) \tag{9.13}$$

and define a family $\sigma = \{\sigma_X\}_{X \in \mathrm{Ob}(\mathcal{C})}$ of morphisms in $\mathcal{C}$ by

$$\sigma_X = \sum_{i,j,k,\ell,n \in I}$$

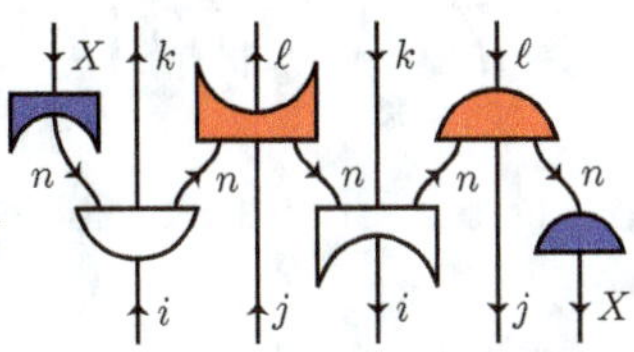

$$: C \otimes X \to X \otimes C,$$

Here, we drop the inclusions and projections associated with the direct sum decomposition (9.13).

Theorem 9.10. *The pair (C, σ) is a coend of $\mathcal{Z}(\mathcal{C})$. The product m, coproduct Δ, unit u, counit ε, antipode S, and canonical pairing ω of the Hopf algebra (C, σ) are computed in Figure* 9.3, *where the dotted lines represent* $\mathrm{id}_{\mathbb{1}}$ *or* $\mathrm{id}_{\mathbb{1}^*}$ *and serve to indicate which direct summand of C is concerned.*

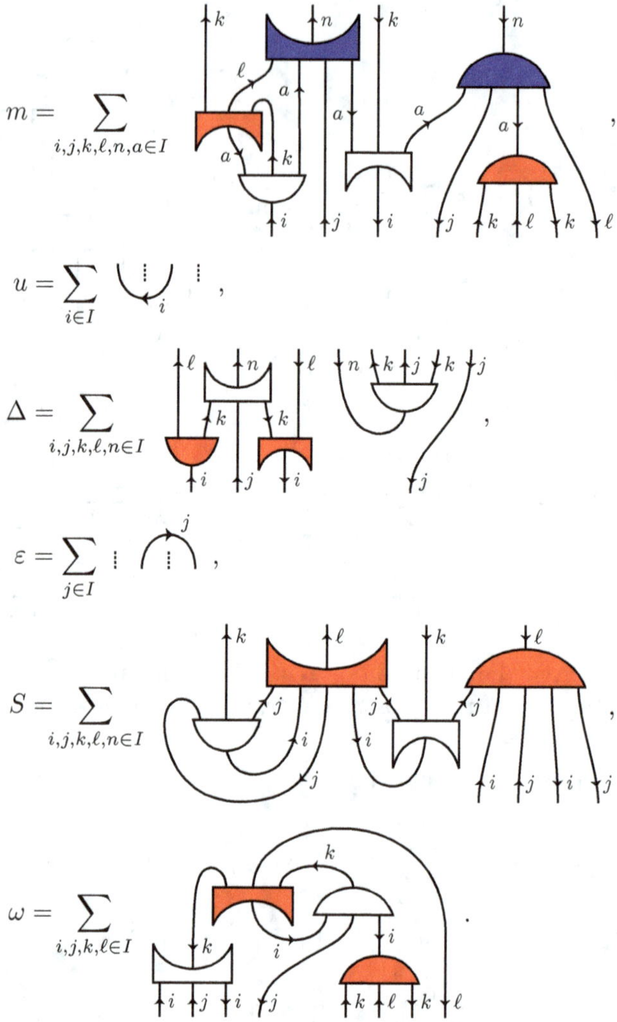

Figure 9.3: The Hopf algebra structure of the coend of $\mathcal{Z}(\mathcal{C})$

Proof. Consider the centralizer (Z, ρ) of $\mathcal{C}$ constructed in Section 9.4.1. Recall that $Z(X) = \bigoplus_{i \in I} i^* \otimes X \otimes i$ for any $X \in \mathrm{Ob}(\mathcal{C})$. For all $i \in I$, we have the associated inclusion and projection

$$\rho_{X,i} \colon i^* \otimes X \otimes i \to Z(X) \quad \text{and} \quad \nu_{X,i} \colon Z(X) \to i^* \otimes X \otimes i.$$

Also, for all $i, j \in I$, we have the inclusion and projection

$$q_{i,j} \colon i^* \otimes j^* \otimes i \otimes j \to C \quad \text{and} \quad p_{i,j} \colon C \to i^* \otimes j^* \otimes i \otimes j$$

associated with the direct sum decomposition (9.13). The monoidal constraints of the dual functor and the pivotal structure of $\mathcal{C}$ (see Section 2.5) induce for each $X \in \mathrm{Ob}(\mathcal{C})$ an isomorphism

$$Z(X)^* \simeq \bigoplus_{i \in I} (i^* \otimes X \otimes i)^* \simeq \bigoplus_{i \in I} i^* \otimes X^* \otimes i^{**} \simeq \bigoplus_{i \in I} i^* \otimes X^* \otimes i = Z(X^*).$$

Since Z is $\Bbbk$-linear, Example B.1.2 shows that the coend $\int^{X \in \mathcal{C}} Z(X)^* \otimes X$ exists and is isomorphic to the direct sum $\bigoplus_{j \in I} Z(j)^* \otimes j$. In view of the isomorphisms

$$\bigoplus_{j \in I} Z(j)^* \otimes j \simeq \bigoplus_{j \in I} Z(j^*) \otimes j \simeq \bigoplus_{i,j \in I} i^* \otimes j^* \otimes i \otimes j = C,$$

we can set $\int^{X \in \mathcal{C}} Z(X)^* \otimes X = C$ with dinatural transformation $\varrho = \{\varrho_X\}_{X \in \mathrm{Ob}(\mathcal{C})}$ given for all $X \in \mathrm{Ob}(\mathcal{C})$ by

$\varrho_X = \sum_{i,j \in I}$

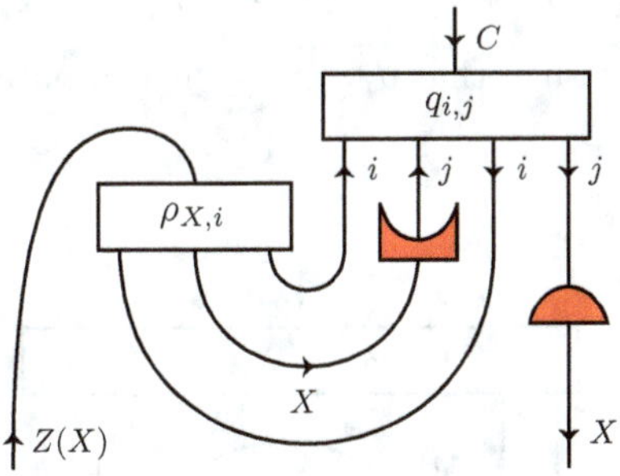

$: Z(X)^* \otimes X \to C.$

It follows from Formula (B.2) that the morphisms factorizing dinatural transformations are computed as follows. First, for any dinatural transformation

$$d = \{d_X \colon Z(X)^* \otimes X \to M\}_{X \in \mathrm{Ob}(\mathcal{C})},$$

the morphism $\varphi \colon C \to M$ such that $d = \varphi\varrho$ is computed by

$\varphi = \sum_{i,j \in I}$

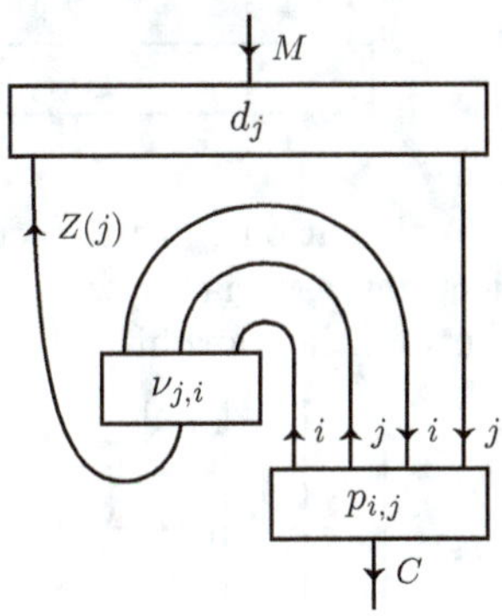

. (9.14)

Second, for any dinatural transformation

$$d = \{d_{X,Y} \colon Z(X)^* \otimes X \otimes Z(Y)^* \otimes Y \to M\}_{X,Y \in \mathrm{Ob}(\mathcal{C})},$$

the morphism $\varphi\colon C \otimes C \to M$ such that $d = \varphi(\varrho \otimes \varrho)$ is computed by

$$\varphi = \sum_{i,j,k,l \in I}$$ 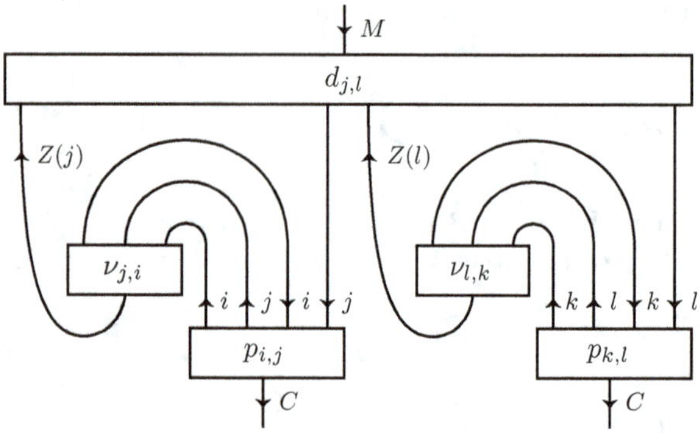 . (9.15)

Third, for any dinatural transformation

$$d = \{d_X \colon Z(Z(X)^* \otimes X) \to M\}_{X \in \mathrm{Ob}(\mathcal{C})},$$

the morphism $\varphi\colon Z(C) \to M$ such that $d = \varphi Z(\varrho)$ is computed by

$$\varphi = \sum_{i,j,k \in I}$$ 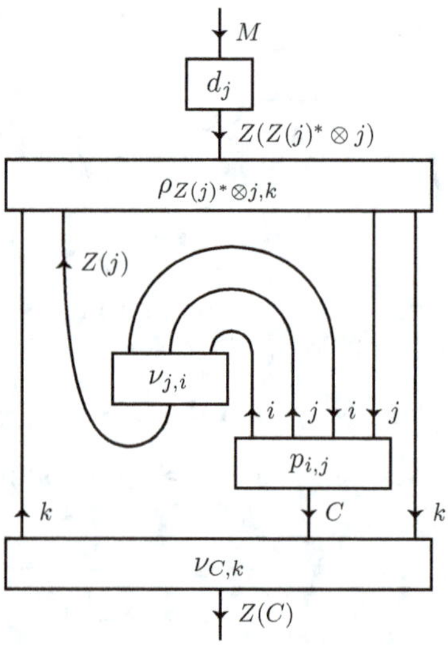 . (9.16)

To simplify the graphical calculations, in the pictures to follow we do not depict the inclusions and projections associated with direct sum decompositions. For example, with this convention, the morphism ϱ_X is graphically represented as

$$\varrho_X = \sum_{i,j \in I}$$ 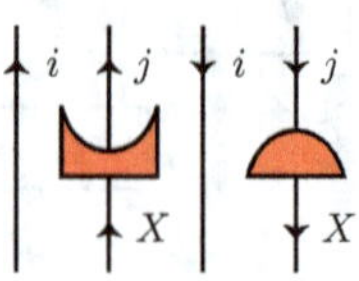 $: Z(X)^* \otimes X \to C.$

By Theorem 9.2, Z is a $\Bbbk$-linear involutory quasitriangular Hopf monad on $\mathcal{C}$. Let μ, η, s, R be respectively its product, unit, antipode, and R-matrix. We compute the coend (C, σ) of $\mathcal{Z}(\mathcal{C})$ and its structural morphisms. Recall from Section 9.2.3 that for all $X \in \mathrm{Ob}(\mathcal{C})$,

$$\sigma_X = (\mathrm{id}_X \otimes \alpha)\, \partial_{C,X},$$

where ∂ is the universal coaction associated with ρ (see Section 9.1.2) and the morphism $\alpha\colon Z(C) \to C$ is such that, for all $Y \in \mathrm{Ob}(\mathcal{C})$,

$$\alpha Z(\varrho_Y) = \varrho_{Z(Y)} \left({}^\vee\!\mu_Y\, s^l_{Z(Y)} Z({}^\vee\!\mu_Y) \otimes \mathrm{id}_{Z(Y)}\right) Z_2({}^\vee Z(Y), Y).$$

Observe that

$$\alpha \overset{\text{(i)}}{=} \sum_{i,j,k,\ell,a,b,n \in I}$$

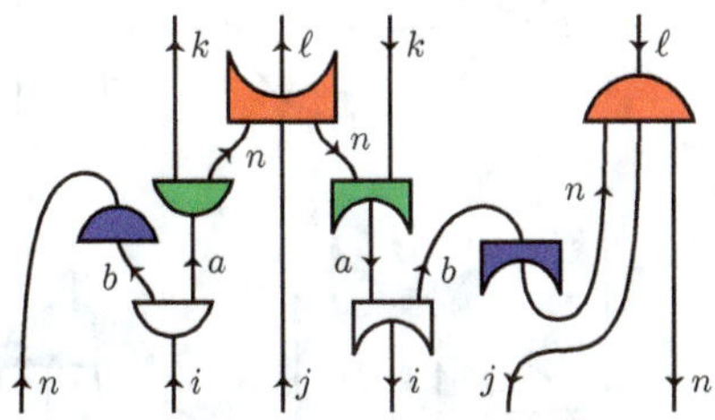

$$\overset{\text{(ii)}}{=} \sum_{i,j,k,\ell,n \in I}$$

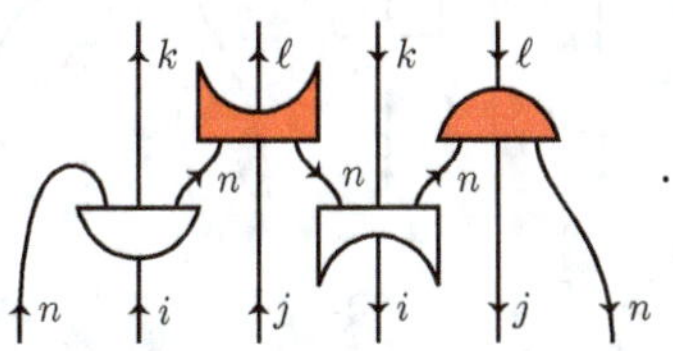

Here, (i) follows from Formula (9.16) and the computation of the structural morphisms of Z in Figure 9.2, and (ii) from Lemma 4.9(a), (g). For any $X \in \mathrm{Ob}(\mathcal{C})$,

$$\partial_{C,X} = \sum_{i,j,n \in I}$$

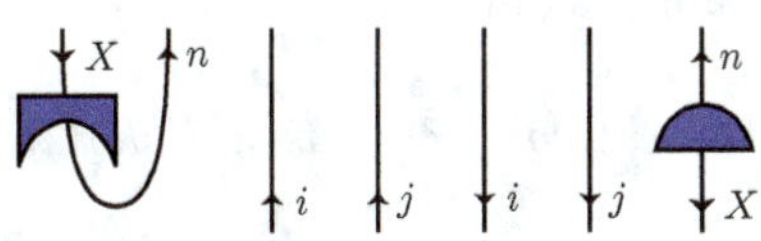

and consequently

$$\sigma_X = \sum_{i,j,k,\ell,n \in I}$$

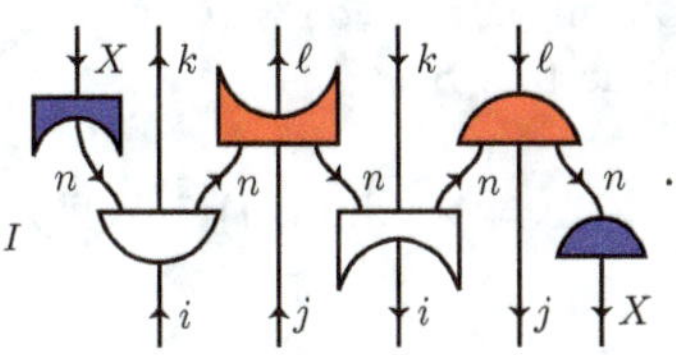

By Theorem 9.4, the structural morphisms of the Hopf algebra (C, σ) are characterized by the equalities in Figures 8.1 and 8.2, where $T = Z$ and X, Y run over all objects of $\mathcal{C}$. We compute these structural morphisms using Formulas (9.14)

and (9.15), the computation of the structural morphisms of the Hopf monad Z in Figure 9.2, and the properties of the enriched graphical calculus (see Section 4.6). In particular,

$$m \overset{\text{(i)}}{=} \sum_{\substack{i,j,k,\ell,n\in I\\ a,b,c,d,e\in I}}$$

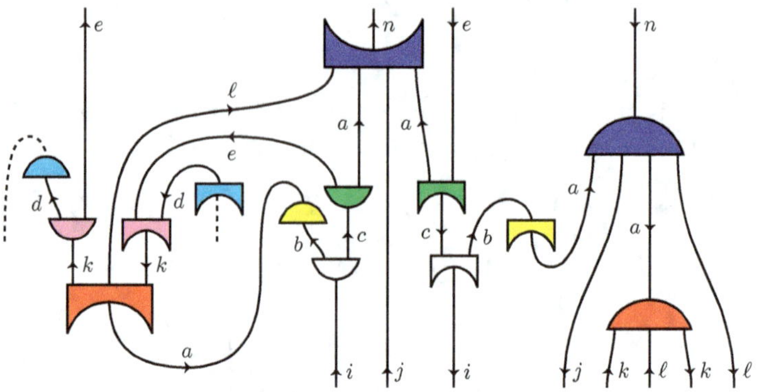

$$\overset{\text{(ii)}}{=} \sum_{i,j,k,\ell,n,a\in I}$$

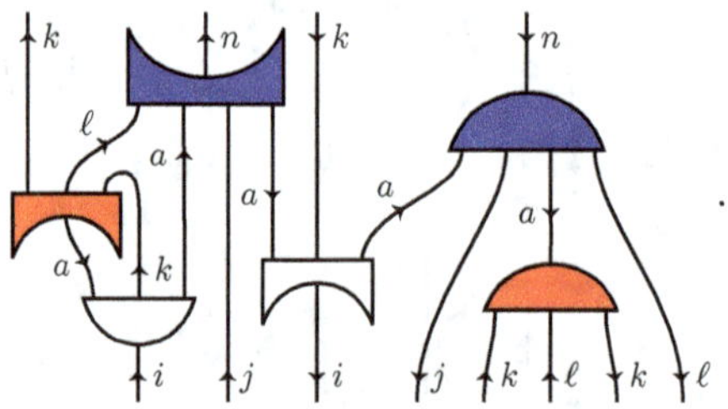

Here, (i) follows from (9.15), the equality characterizing m in Figure 8.2, and the computation of the structural morphisms of Z in Figure 9.2, and (ii) follows from Lemma 4.9(a),(g). This gives the expansion of m shown in Figure 9.3. The expansions of the other structural morphisms of the Hopf algebra (C,σ) shown in Figure 9.3 are obtained similarly. □

Theorem 9.11. *The category $\mathcal{Z}(\mathcal{C})$ is modular in the sense of Lyubashenko (see Section 6.7).*

Proof. Consider the coend (C,σ) of $\mathcal{Z}(\mathcal{C})$ from Theorem 9.10. We need to prove that the canonical pairing $\omega\colon (C,\sigma)\otimes(C,\sigma)\to \mathbb{1}_{\mathcal{Z}(\mathcal{C})}$ is non-degenerate. To this end, we will use Lemma 6.2. Set

$$\Lambda = \sum_{j\in I} \dim_r(j)$$  $$: \mathbb{1}\to C$$

and

$$\lambda = \sum_{j\in I} \dim_r(j)$$  $$: C\to \mathbb{1}.$$

For any object X of $\mathcal{C}$, we have

$$\sigma_X(\Lambda \otimes \mathrm{id}_X) \overset{\text{(i)}}{=} \sum_{j,k,\ell,n \in I} \dim_r(j)$$

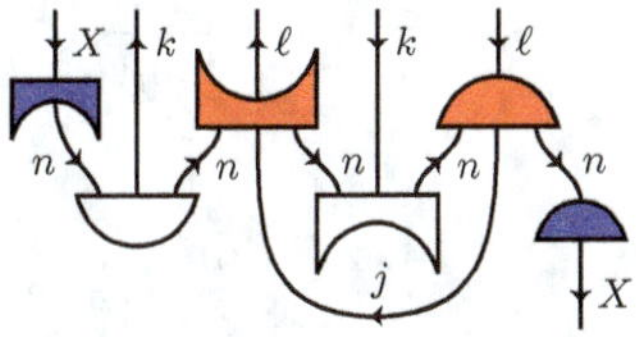

$$\overset{\text{(ii)}}{=} \sum_{j,k,\ell,n \in I} \frac{\dim_r(\ell)\dim_r(n)}{\dim_l(n)}$$

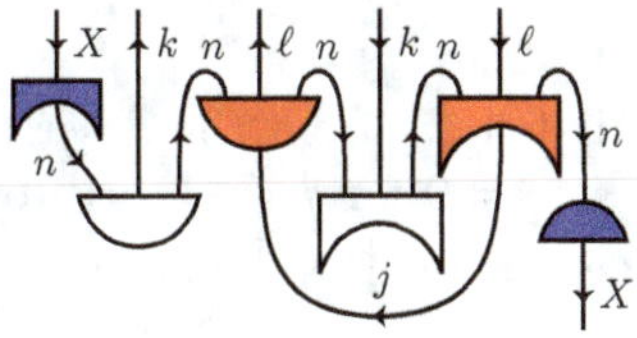

$$\overset{\text{(iii)}}{=} \sum_{k,\ell,n \in I} \frac{\dim_r(\ell)\dim_r(n)}{\dim_l(n)}$$

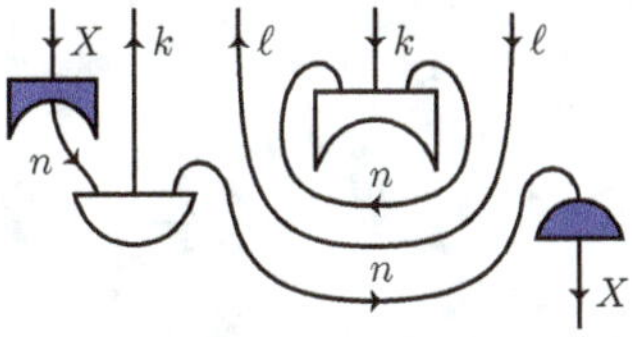

$$\overset{\text{(iv)}}{=} \sum_{\ell,n \in I} \frac{\dim_r(\ell)\dim_r(n)}{\dim_l(n)}$$

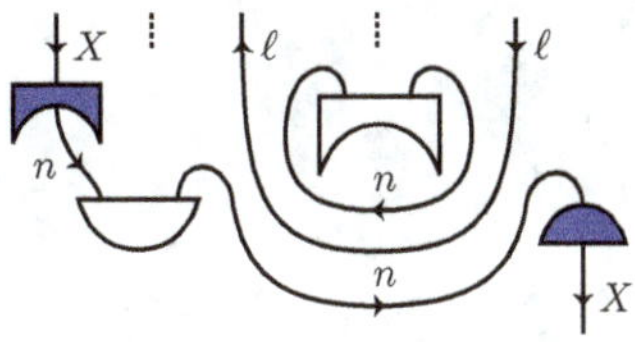

$$\overset{\text{(v)}}{=} \sum_{\ell,n \in I} \dim_r(\ell)$$

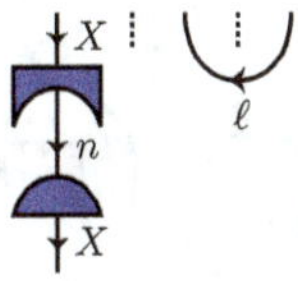

$$\overset{\text{(vi)}}{=} \mathrm{id}_X \otimes \Lambda.$$

Here, (i) follows from the definition of σ_X, (ii) from Lemma 4.9(e), (iii) and (vi) from (4.8), (iv) from the equality $\mathrm{Hom}_{\mathcal{C}}(\mathbb{1}, k) = 0$ for $k \neq \mathbb{1}$, and (v) from Lemma 4.9(c). Thus, Λ is a morphism in $\mathcal{Z}(\mathcal{C})$ from $\mathbb{1}_{\mathcal{Z}(\mathcal{C})} = (\mathbb{1}, \mathrm{id})$ to (C, σ). Next,

$$\omega(\mathrm{id}_C \otimes \Lambda) \overset{\text{(i)}}{=} \sum_{i,j,\ell \in I} \dim_r(\ell)$$

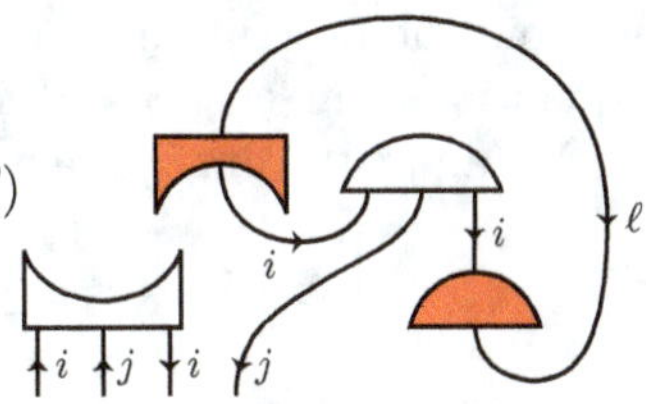

$$\overset{\text{(ii)}}{=} \sum_{i,j,\ell \in I} \dim_r(\ell)$$

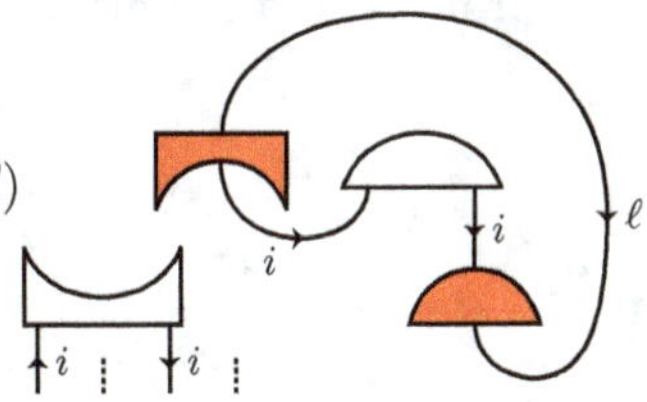

$$\overset{\text{(iii)}}{=} \sum_{i,\ell \in I} \frac{\dim_r(\ell)}{\dim_l(i)}$$

$$\overset{\text{(iv)}}{=} \sum_{i,\ell \in I} \frac{\dim_r(\ell)}{\dim_l(i)} \, \delta_{\ell,i^*}$$

$$\overset{\text{(v)}}{=} \lambda.$$

Here, (i) follows from the description of ω in Figure 9.3, (ii) from the equality $\mathrm{Hom}_{\mathcal{C}}(j, \mathbb{1}) = 0$ for $j \neq \mathbb{1}$, (iii) from Lemma 4.9(b), (iv) from Lemma 4.9(a), and (v) from the identity $\dim_r(i^*) = \dim_l(i)$. Likewise,

$$\omega(\Lambda \otimes \mathrm{id}_C) \overset{\text{(i)}}{=} \sum_{j,k,\ell \in I} \dim_r(j)$$

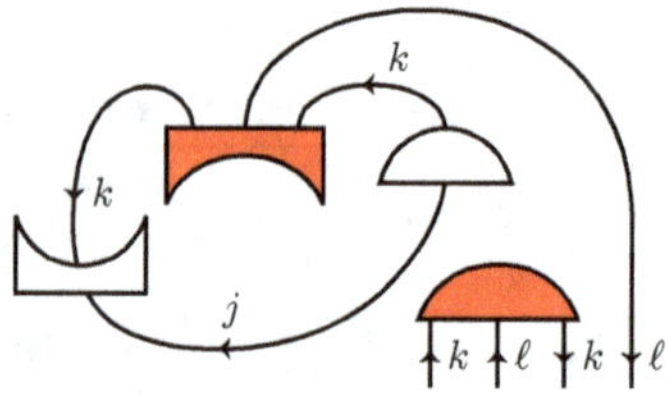

$$\overset{\text{(ii)}}{=} \sum_{j,k \in I} \dim_r(j) \, \delta_{j,k^*}$$

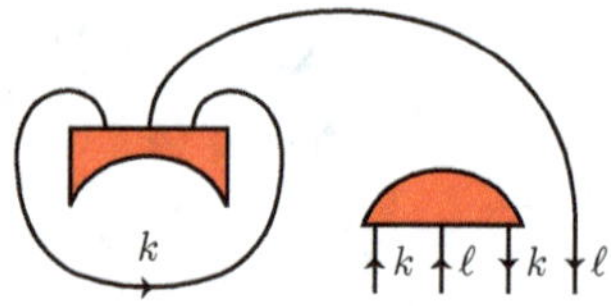

$$\overset{\text{(iii)}}{=} \sum_{k \in I} \dim_l(k)$$

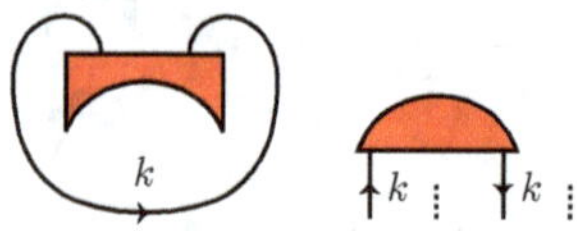

$$\overset{\text{(iv)}}{=} \sum_{k \in I}$$

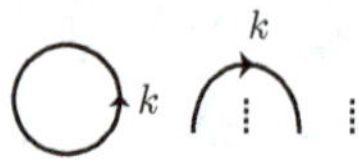

$$\overset{\text{(v)}}{=} \lambda.$$

Here, (i) follows from the description of ω in Figure 9.3, (ii) from Lemma 4.9(a), (iii) from the equalities $\mathrm{Hom}_{\mathcal{C}}(\mathbb{1}, \ell^*) = 0$ for $\ell \neq \mathbb{1}$ and $\dim_r(k^*) = \dim_l(k)$, (iv) from Lemma 4.9(b), and (v) from the definition of $\dim_r(k)$. Thus

$$\omega(\Lambda \otimes \mathrm{id}_C) = \omega(\mathrm{id}_C \otimes \Lambda) = \lambda.$$

Considering the coproduct Δ and unit u of (C, σ), we obtain

$$(\mathrm{id}_C \otimes \lambda)\Delta \overset{\text{(i)}}{=} \sum_{i,k,\ell,n \in I} \dim_r(k)$$

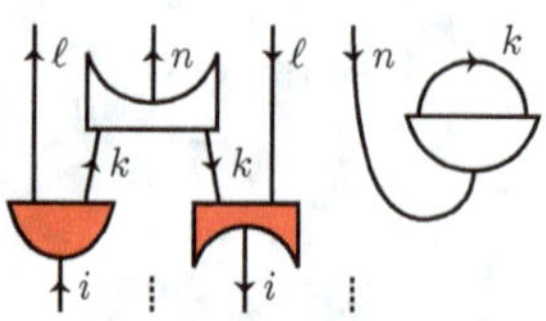

$$\overset{\text{(ii)}}{=} \sum_{i,k,\ell \in I} \dim_r(k)$$ 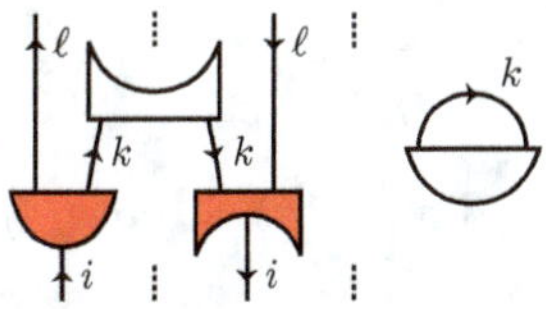

$$\overset{\text{(iii)}}{=} \sum_{i,k,\ell \in I} \dim_r(k)$$ 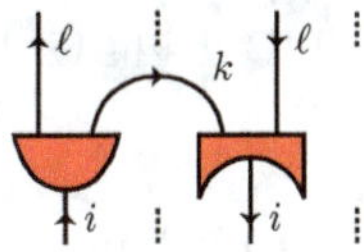

$$\overset{\text{(iv)}}{=} \sum_{i,k,\ell \in I} \dim_r(i)$$ 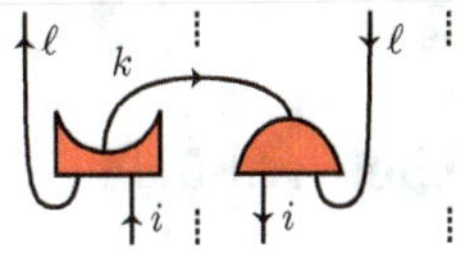

$$\overset{\text{(v)}}{=} \sum_{i,\ell \in I} \dim_r(i)$$ 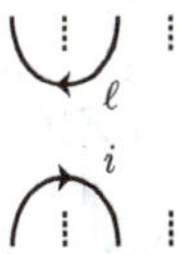 $$\overset{\text{(vi)}}{=} u\,\lambda.$$

Here, (i) and (vi) follow from the description of Δ and u in Figure 9.3, (ii) from the equality $\mathrm{Hom}_{\mathcal{C}}(\mathbb{1}, n) = 0$ for $n \neq \mathbb{1}$, (iii) from Lemma 4.9(b), (iv) from Lemma 4.9(e), and (v) from (4.8). Thus λ is a left cointegral of (C, σ). Finally,

$$\omega(\Lambda \otimes \Lambda) = \lambda\Lambda = \dim_r(\mathbb{1}) = \mathrm{id}_{\mathbb{1}}.$$

Thus, the morphisms Λ and $\Lambda' = \Lambda$ of $\mathcal{Z}(\mathcal{C})$ satisfy the conditions of Lemma 6.2. By this lemma, ω is non-degenerate. Hence, the category $\mathcal{Z}(\mathcal{C})$ is modular. □

9.5 Proof of Theorems 5.3–5.5

9.5.1 Proof of Theorem 5.3

Consider the centralizer Z of $\mathcal{C}$ constructed in Section 9.4.1. By Theorem 9.2, Z is a $\Bbbk$-linear involutory Hopf monad on $\mathcal{C}$. Consequently, the category $\mathcal{C}^Z$ of Z-modules is a pivotal $\Bbbk$-category. Since $\dim(\mathcal{C}) \neq 0$, Lemma 9.8 implies that the monad Z is semisimple. Since $\Bbbk$ is an algebraically closed field, Theorem 8.12 implies that $\mathcal{C}^Z$ is an abelian (and so additive) fusion $\Bbbk$-category. By Theorem 9.3, $\mathcal{C}^Z$ is isomorphic to $\mathcal{Z}(\mathcal{C})$ as a pivotal $\Bbbk$-category. Hence, $\mathcal{Z}(\mathcal{C})$ is an additive pivotal fusion $\Bbbk$-category.

It remains to compute $\dim(\mathcal{Z}(\mathcal{C}))$. To this end, pick a representative set I of simple objects of $\mathcal{C}$, and let (C, σ) be the coend of $\mathcal{Z}(\mathcal{C})$ from Theorem 9.10, where

$$C = \bigoplus_{i,j \in I} i^* \otimes j^* \otimes i \otimes j \in \mathrm{Ob}(\mathcal{C}).$$

By Lemma 6.3,

$$\begin{aligned}\dim(\mathcal{Z}(\mathcal{C})) &= \dim_l(C,\sigma) = \dim_l(C) = \dim_l\left(\bigoplus_{i,j\in I} i^*\otimes j^*\otimes i\otimes j\right)\\ &= \sum_{i,j\in I} \dim_l(i^*)\dim_l(j^*)\dim_l(i)\dim_l(j)\\ &= \left(\sum_{i\in I}\dim_r(i)\dim_l(i)\right)\left(\sum_{j\in I}\dim_r(j)\dim_l(j)\right) = (\dim(\mathcal{C}))^2.\end{aligned}$$

9.5.2 Proof of Theorem 5.4

By Theorem 5.3, $\mathcal{Z}(\mathcal{C})$ is an additive pivotal fusion $\Bbbk$-category and $\dim(\mathcal{Z}(\mathcal{C})) = (\dim(\mathcal{C}))^2$. By Lemma 5.2, $\mathcal{Z}(\mathcal{C})$ is ribbon. By Theorems 9.11 and 6.6, the S-matrix of $\mathcal{Z}(\mathcal{C})$ is invertible, and so $\mathcal{Z}(\mathcal{C})$ is modular. We need only to prove that $\Delta_+ = \Delta_- = \dim(\mathcal{C})$. Pick a representative set I of simple objects of $\mathcal{C}$. Consider the coend (C,σ) of $\mathcal{Z}(\mathcal{C})$ from Theorem 9.10. The proofs of Theorem 9.11 and Lemma 6.2 imply that the morphism

$$\Lambda = \sum_{j\in I} \dim(j)$$  $$: \mathbb{1}_{\mathcal{Z}(\mathcal{C})} \to (C,\sigma)$$

is a right integral of (C,σ). The description of the counit $\varepsilon\colon (C,\sigma)\to \mathbb{1}_{\mathcal{Z}(\mathcal{C})}$ of the Hopf algebra (C,σ) given in Figure 9.3 implies that

$$\varepsilon\Lambda = \sum_{j\in I}(\dim(j))^2 = \dim(\mathcal{C}).$$

Thus $\Lambda' = \dim(\mathcal{C})\Lambda$ is a right integral of (C,σ) such that

$$\varepsilon\Lambda' = \dim(\mathcal{C})\varepsilon\Lambda = (\dim(\mathcal{C}))^2 = \dim(\mathcal{Z}(\mathcal{C})).$$

Then, by Lemma 6.5,

$$\Delta_\pm = \theta_\pm\Lambda' = \dim(\mathcal{C})\,\theta_\pm\Lambda,$$

where θ_+ and θ_- are the twist forms of (C,σ). So, we need only to prove that $\theta_+\Lambda = \theta_-\Lambda = 1_\Bbbk$. Consider the centralizer Z of $\mathcal{C}$ constructed in Section 9.4.1. As in the proof of Theorem 9.10, the characterization of θ_+ and θ_- in Figure 8.4 (for $T = Z$), Formula (9.14), the computation of the structural morphisms of Z in Figure 9.2, and Lemma 4.9 yield that

$$\theta_+ = \sum_{i,j\in I}$$ 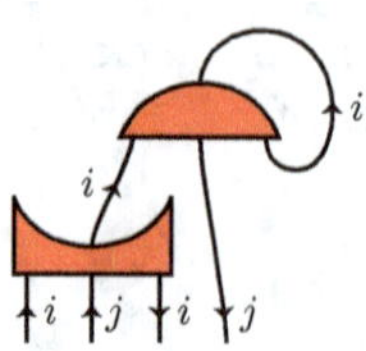 and $$\theta_- = \sum_{i,j\in I}$$ 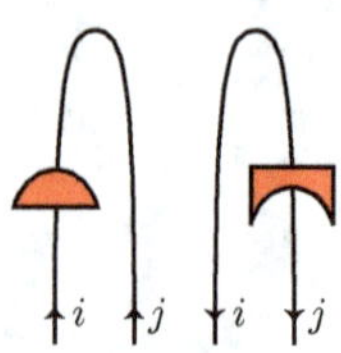 .

Therefore, using Lemma 4.9(a), we obtain that

$$\theta_+\Lambda = \sum_{j\in I} \dim(j)$$ 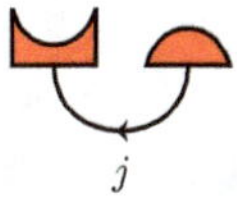 $$= \sum_{j\in I} \dim(j)\, \delta_{j,\mathbb{1}} = \dim(\mathbb{1}) = 1_{\mathbb{k}}$$

and

$$\theta_-\Lambda = \sum_{j\in I} \dim(j)$$ 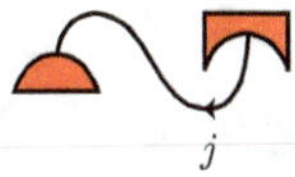 $$= \sum_{j\in I} \dim(j)\, \delta_{j,\mathbb{1}} = \dim(\mathbb{1}) = 1_{\mathbb{k}}.$$

9.5.3 Proof of Theorem 5.5

Suppose that $\mathcal{Z}(\mathcal{C})$ is pre-fusion. By Theorems 9.2 and 9.3, the centralizer Z of $\mathcal{C}$ from Section 9.4.1 is a $\mathbb{k}$-linear involutory Hopf monad and $\mathcal{C}^Z \simeq \mathcal{Z}(\mathcal{C})$ as pivotal $\mathbb{k}$-categories. Since $\mathcal{Z}(\mathcal{C})$ is additive and pre-fusion, so is $\mathcal{C}^Z$.

We prove now that the monad Z is semisimple. By Theorem 8.9, the functor $Z\colon \mathcal{C} \to \mathcal{C}$ has a right adjoint. By Lemma C.2(b), this functor preserves cokernels. Then Lemma 7.8(c) implies that the forgetful functor $U_Z\colon \mathcal{C}^Z \to \mathcal{C}$ creates cokernels. To prove that Z is semisimple, pick any Z-module (M,r) and recall that $r\colon F_Z(M) \to (M,r)$ is a morphism in $\mathcal{C}^Z$. Since r has a section in $\mathcal{C}$ (given by the unit $\eta_M\colon M \to Z(M)$ of Z), $r = U_Z(r)$ is an epimorphism in $\mathcal{C}$. By Lemma C.1(b), the morphism $0\colon M \to \mathbf{0}$ is a cokernel of r. Since U_Z creates cokernels, the morphism $0\colon (M,r) \to (\mathbf{0},0)$ is a cokernel of r in $\mathcal{C}^Z$. Then, by Lemma C.1(b), r is an epimorphism in $\mathcal{C}^Z$. Since $\mathcal{C}^Z \simeq \mathcal{Z}(\mathcal{C})$ is an additive pre-fusion $\mathbb{k}$-category and $\mathbb{k}$ is a field, Theorem C.6 implies that r has a section in $\mathcal{C}^Z$. Therefore, by Lemma 7.3, the Z-module (M,r) is projective. Hence, Z is semisimple.

The semisimplicity of Z and Lemma 9.8 imply that $\dim(\mathcal{C}) \neq 0$. This contradicts the assumptions of the theorem. We conclude that $\mathcal{Z}(\mathcal{C})$ is not pre-fusion.

9.5.4 Exercise*

Let $\mathcal{C}$ be an additive pivotal fusion $\mathbb{k}$-category. Prove that the forgetful functor $\mathcal{U}\colon \mathcal{Z}(\mathcal{C}) \to \mathcal{C}$ is Frobenius in the sense that there is a functor $\mathcal{C} \to \mathcal{Z}(\mathcal{C})$ which is both a left adjoint and a right adjoint to $\mathcal{U}$.

Part III

State Sum Topological Field Theory

Chapter 10

Topological Quantum Field Theory

We review foundations of Topological Quantum Field Theory. We start by discussing manifolds, cobordisms, and the category of cobordisms.

10.1 The category of cobordisms

10.1.1 Manifolds

We use the symbol $\mathbb{R}$ for the field of real numbers and set $\mathbb{R}_+ = \{x \in \mathbb{R} \,|\, x \geq 0\}$. For an integer $n \geq 0$, by an *n-manifold*, we mean an n-dimensional topological manifold, that is, a paracompact Hausdorff topological space whose each point has a neighborhood homeomorphic to $\mathbb{R}^n$ or $\mathbb{R}^n_+ = \mathbb{R}^{n-1} \times \mathbb{R}_+$ if $n \geq 1$ and to $\mathbb{R}^0 = \{0\}$ if $n = 0$. (We will often impose an additional condition of compactness which implies paracompactness.) The set of points of an n-manifold M having a neighborhood homeomorphic to $\mathbb{R}^n$ is denoted $\mathrm{Int}(M)$ and called the *interior* of M. The set $\partial M = M \setminus \mathrm{Int}(M)$ is called the *boundary* of M. For $n \geq 1$, ∂M is an $(n-1)$-manifold and for $n = 0$, $\partial M = \emptyset$.

For the definition of an oriented manifold, we refer to [Sp]. If a neighborhood of a point of an n-manifold is identified with $\mathbb{R}^n$ or $\mathbb{R}^n_+ \subset \mathbb{R}^n$, then an orientation in this neighborhood may be represented by a basis of $\mathbb{R}^n$. The boundary ∂M of an oriented manifold M is oriented so that at any point of ∂M, the orientation of M is given by a direction away from M followed by the orientation of ∂M. Given an oriented manifold M, we let $-M$ be the same manifold with opposite orientation. Clearly, $\partial(-M) = -\partial M$. The ordered disjoint union of oriented n-manifolds M_0, M_1 is the oriented n-manifold

$$M_0 \sqcup M_1 = (M_0 \times \{0\}) \cup (M_1 \times \{1\}).$$

Clearly, $\partial(M_0 \sqcup M_1) = \partial M_0 \sqcup \partial M_1$ and $-(M_0 \sqcup M_1) = (-M_0) \sqcup (-M_1)$.

A *closed manifold* is a compact manifold with empty boundary. The empty set $\emptyset$ is considered as a closed oriented manifold of arbitrary dimension.

10.1.2 Cobordisms

Let $n \geq 1$ be an integer. An *n-dimensional cobordism* is a quadruple $(M, \Sigma_0, \Sigma_1, h)$, where M is a compact oriented n-manifold, Σ_0, Σ_1 are closed oriented $(n-1)$-manifolds, and $h\colon (-\Sigma_0) \sqcup \Sigma_1 \to \partial M$ is an orientation preserving homeomorphism:

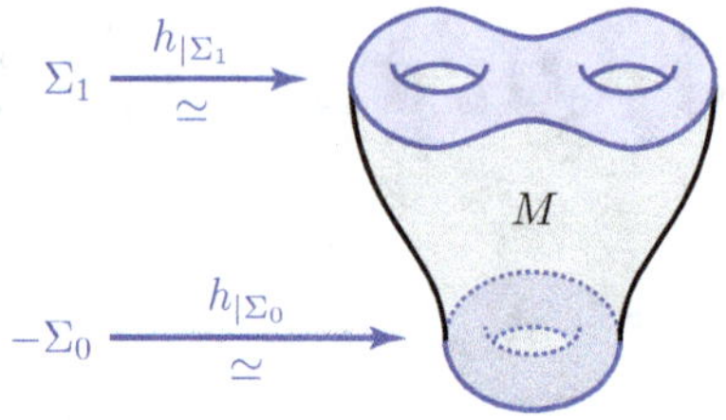

Instead of saying "a cobordism $(M, \Sigma_0, \Sigma_1, h)$" one often abusively says "a cobordism M between Σ_0 and Σ_1".

For example, for any closed oriented $(n-1)$-manifold Σ, the cylinder $C_\Sigma = \Sigma \times [0,1]$ with the product orientation and the tautological homeomorphism

$$(-\Sigma) \sqcup \Sigma = (-\Sigma \times \{0\}) \cup (\Sigma \times \{1\}) = \partial C_\Sigma$$

yields a cobordism between Σ and itself.

Two cobordisms $(M, \Sigma_0, \Sigma_1, h)$ and $(M', \Sigma_0, \Sigma_1, h')$ between Σ_0 and Σ_1 are *homeomorphic* if there is an orientation preserving homeomorphism $\alpha\colon M \to M'$ such that $h' = \alpha h$.

Given n-dimensional cobordisms $(M_0, \Sigma_0, \Sigma_1, h_0)$ and $(M_1, \Sigma_1, \Sigma_2, h_1)$, we form an n-dimensional cobordism $(M, \Sigma_0, \Sigma_2, h)$, where M is obtained by gluing M_0 to M_1 along $h_1 h_0^{-1}\colon h_0(\Sigma_1) \to h_1(\Sigma_1)$ and

$$h = h_0|_{\Sigma_0} \sqcup h_1|_{\Sigma_2}\colon (-\Sigma_0) \sqcup \Sigma_2 \to \partial M.$$

We say that the cobordism $(M, \Sigma_0, \Sigma_2, h)$ is obtained by *gluing* the cobordisms $(M_0, \Sigma_0, \Sigma_1, h_0)$ and $(M_1, \Sigma_1, \Sigma_2, h_1)$ along Σ_1.

10.1.3 The category Cob_n

For any integer $n \geq 1$, we define the category Cob_n of n-dimensional cobordisms. The objects of Cob_n are closed oriented $(n-1)$-manifolds. A morphism $f\colon \Sigma_0 \to \Sigma_1$ in Cob_n is a homeomorphism class of n-dimensional cobordisms between Σ_0 and Σ_1. We say that the cobordisms in this class *represent* f. The identity morphism of a closed oriented $(n-1)$-manifold Σ is represented by the cylinder C_Σ. The composition of morphisms $\Sigma_0 \to \Sigma_1$ and $\Sigma_1 \to \Sigma_2$ represented respectively by cobordisms M_0 and M_1 is represented by the cobordism obtained via gluing of M_0 and M_1 along Σ_1.

Each orientation preserving homeomorphism of closed oriented $(n-1)$-manifolds $\xi\colon \Sigma \to \Sigma'$ determines a morphism $\Sigma \to \Sigma'$ in Cob_n represented by the cobordism

$$(C_{\Sigma'} = \Sigma' \times [0,1],\, \Sigma,\, \Sigma',\, h\colon (-\Sigma) \sqcup \Sigma' \to \partial C_{\Sigma'}),$$

where $h(x) = (\xi(x), 0)$ for $x \in \Sigma$ and $h(x') = (x', 1)$ for $x' \in \Sigma'$. Composition of homeomorphisms corresponds under this construction to composition in Cob_n. Consequently, the morphisms in Cob_n derived in this way from homeomorphisms are isomorphisms.

We provide Cob_n with a structure of a symmetric monoidal category. The monoidal product in Cob_n is the ordered disjoint union of manifolds and cobordisms. The unit object of Cob_n is the empty set $\emptyset$ viewed as a closed oriented $(n-1)$-manifold. The associativity and unitality constraints are induced by the tautological homeomorphisms

$$(\Sigma \sqcup \Sigma') \sqcup \Sigma'' \simeq \Sigma \sqcup (\Sigma' \sqcup \Sigma'') \quad \text{and} \quad \emptyset \sqcup \Sigma \simeq \Sigma \simeq \Sigma \sqcup \emptyset,$$

where $\Sigma, \Sigma', \Sigma''$ run over all closed oriented $(n-1)$-manifolds. The symmetry in Cob_n is induced by the obvious permutation homeomorphisms

$$\Sigma \otimes \Sigma' = \Sigma \sqcup \Sigma' \simeq \Sigma' \sqcup \Sigma = \Sigma' \otimes \Sigma.$$

By definition, the monoid $\mathrm{End}_{\mathrm{Cob}_n}(\emptyset)$ consists of homeomorphism classes of closed oriented n-manifolds with monoid multiplication induced by the disjoint union. In particular, each closed oriented n-manifold represents a morphism $\emptyset \to \emptyset$ in Cob_n.

10.1.4 Pivotality of Cob_n

We now turn Cob_n into a pivotal category. Consider a closed oriented $(n-1)$-manifold Σ and the cobordisms

$$(\Sigma \times [0,1], \Sigma \sqcup (-\Sigma), \emptyset, h) \quad \text{and} \quad (\Sigma \times [0,1], \emptyset, (-\Sigma) \sqcup \Sigma, h)$$

where h is the tautological homeomorphism

$$(-\Sigma) \sqcup \Sigma \longrightarrow (-\Sigma \times \{0\}) \sqcup (\Sigma \times \{1\}) = \partial(\Sigma \times [0,1]).$$

These cobordisms represent morphisms in Cob_n, respectively,

$$\omega_\Sigma\colon \Sigma \otimes (-\Sigma) \to \emptyset \quad \text{and} \quad \Omega_\Sigma\colon \emptyset \to (-\Sigma) \otimes \Sigma.$$

It is easy to check that Ω_Σ is the inverse of the pairing ω_Σ. These morphisms determine a structure of a pivotal category on Cob_n: for any closed oriented $(n-1)$-manifold Σ, we set $\Sigma^* = -\Sigma$ and

$$\mathrm{ev}_\Sigma = \omega_{-\Sigma}, \quad \mathrm{coev}_\Sigma = \Omega_{-\Sigma}, \quad \widetilde{\mathrm{ev}}_\Sigma = \omega_\Sigma, \quad \widetilde{\mathrm{coev}}_\Sigma = \Omega_\Sigma.$$

This structure of a pivotal category on Cob_n arises as in Lemma 3.5 from the left duality $\{(-\Sigma, \mathrm{ev}_\Sigma)\}_\Sigma$ and the symmetry in Cob_n defined above. Therefore the category Cob_n is ribbon and, consequently, spherical. The trace of endomorphisms in Cob_n is computed as follows. Let Σ be a closed oriented $(n-1)$-manifold and let $f\colon \Sigma \to \Sigma$ be a morphism in Cob_n represented by an n-dimensional cobordism $(M, \Sigma, \Sigma, h\colon (-\Sigma) \sqcup \Sigma \to \partial M)$. The morphism

$$\mathrm{tr}(f) = \mathrm{ev}_\Sigma \circ (\mathrm{id}_{-\Sigma} \otimes f) \circ \widetilde{\mathrm{coev}}_\Sigma\colon \emptyset \to \emptyset$$

is represented by the closed oriented n-manifold obtained from M via the identification of $h(-\Sigma)$ with $h(\Sigma)$ induced by the identity map $-\Sigma \to \Sigma$. In particular, $\dim(\Sigma) = \mathrm{tr}(\mathrm{id}_\Sigma)$ is the homeomorphism class of the oriented n-manifold $\Sigma \times S^1$.

10.1.5 Exercise

Prove that if a morphism $\Sigma_0 \to \Sigma_1$ in Cob_n is represented by a cobordism $(M, \Sigma_0, \Sigma_1, h)$, then the dual morphism $-\Sigma_1 \to -\Sigma_0$ is represented by the cobordism $(M, -\Sigma_1, -\Sigma_0, h')$, where h' is the composition of the permutation homeomorphism $\Sigma_1 \sqcup (-\Sigma_0) \to (-\Sigma_0) \sqcup \Sigma_1$ with $h\colon (-\Sigma_0) \sqcup \Sigma_1 \to \partial M$.

10.2 Generalities on TQFTs

10.2.1 TQFTs

The following definition of a TQFT reformulates in the categorical language the axiomatic approach of Sir Michael Atiyah [At] inspired by the work of Edward Witten [Wi] and an earlier work of Albert Schwarz [Sc] in theoretical physics. Pick an integer $n \geq 1$ and consider the symmetric monoidal category Cob_n of n-dimensional cobordisms. Recall the symmetric monoidal category $\mathrm{Mod}_\Bbbk$ of (left) $\Bbbk$-modules. An *n-dimensional Topological Quantum Field Theory (TQFT)* is a symmetric strong monoidal functor

$$Z\colon \mathrm{Cob}_n \to \mathrm{Mod}_\Bbbk\,.$$

Such a functor Z includes isomorphisms

$$Z_0\colon \Bbbk \to Z(\emptyset) \quad \text{and} \quad Z_2(\Sigma, \Sigma')\colon Z(\Sigma) \otimes_\Bbbk Z(\Sigma') \to Z(\Sigma \sqcup \Sigma')$$

for any closed oriented $(n-1)$-manifolds Σ, Σ'. Applying Z to the morphism in Cob_n determined by an orientation preserving homeomorphism $\xi\colon \Sigma \to \Sigma'$, we obtain the induced action $Z(\xi)\colon Z(\Sigma) \to Z(\Sigma')$. As an exercise, the reader may verify that isotopic homeomorphisms act in the same way and that $Z(\xi\psi) = Z(\xi)\, Z(\psi)$ for all composable orientation preserving homeomorphisms ξ, ψ of closed oriented $(n-1)$-manifolds.

A TQFT allows us to derive $\Bbbk$-linear homomorphisms from manifolds. Consider a compact oriented n-dimensional manifold M and disjoint closed $(n-1)$-dimensional manifolds $\Sigma_0, \Sigma_1 \subset \partial M$ whose union is equal to ∂M. We orient Σ_0, Σ_1 so that $\partial M = (-\Sigma_0) \cup \Sigma_1$ and let $(M, \Sigma_0, \Sigma_1)\colon \Sigma_0 \to \Sigma_1$ be the morphism in Cob_n represented by the cobordism

$$(M, \Sigma_0, \Sigma_1, \mathrm{id}_{\partial M}\colon (-\Sigma_0) \sqcup \Sigma_1 \to \partial M).$$

Applying an n-dimensional TQFT Z to this morphism we obtain a $\Bbbk$-linear homomorphism $Z(M, \Sigma_0, \Sigma_1)\colon Z(\Sigma_0) \to Z(\Sigma_1)$. In particular, any closed oriented n-manifold M yields in this way a $\Bbbk$-linear homomorphism $Z(M, \emptyset, \emptyset)\colon Z(\emptyset) \to Z(\emptyset)$. Since the $\Bbbk$-module $Z(\emptyset)$ is isomorphic to $\Bbbk$, the latter homomorphism is multiplication by a unique element of $\Bbbk$. This element is denoted $Z(M)$ and is a topological invariant of M. Indeed, if there is an orientation preserving homeomorphism of closed oriented n-manifolds $M \to M'$, then M and M' represent the same morphism $\emptyset \to \emptyset$ in Cob_n and therefore $Z(M) = Z(M')$. It follows from the definitions that $Z(M_0 \sqcup M_1) = Z(M_0)\, Z(M_1)$ for any closed oriented n-manifolds M_0, M_1.

The following lemma shows that in the definition of a TQFT one can safely replace the category $\mathrm{Mod}_\Bbbk$ by its full subcategory $\mathrm{proj}_\Bbbk$ consisting of projective modules of finite type.

Lemma 10.1. *For any n-dimensional TQFT $Z\colon \mathrm{Cob}_n \to \mathrm{Mod}_\Bbbk$ and any closed oriented $(n-1)$-manifold Σ, the $\Bbbk$-module $Z(\Sigma)$ is projective of finite type and its dimension is equal to $Z(\Sigma \times S^1) \in \Bbbk$. The resulting strong monoidal functor $\mathrm{Cob}_n \to \mathrm{proj}_\Bbbk$ is pivotal with respect to the standard structures of a pivotal category on Cob_n and on $\mathrm{proj}_\Bbbk$ defined respectively in Section* 10.1.4 *and Example* 1.7.2.

Proof. As explained in Section 10.1.4, the manifold Σ viewed as an object of the category Cob_n has a left dual in Cob_n. By Section 1.6.4, the $\Bbbk$-module $Z(\Sigma)$ has a left dual in $\mathrm{Mod}_\Bbbk$. Then, by Lemma 1.6, the $\Bbbk$-module $Z(\Sigma)$ is projective of finite type. So, Z defines a strong monoidal functor $\mathrm{Cob}_n \to \mathrm{proj}_\Bbbk$. It is pivotal by Exercise 3.3.5. Applying Lemma 2.8, we obtain that

$$\dim(Z(\Sigma)) = Z_0^{-1} Z(\dim(\Sigma)) Z_0 = Z(\Sigma \times S^1) \in \Bbbk. \qquad \square$$

10.2.2 Isomorphisms of TQFTs

An *isomorphism* between two n-dimensional TQFTs $Z, Z'\colon \mathrm{Cob}_n \to \mathrm{Mod}_\Bbbk$ is a monoidal natural isomorphism $Z \to Z'$ in the sense of Section 1.4.2. By definition, such an isomorphism is a family $\varphi = \{\varphi_\Sigma\colon Z(\Sigma) \to Z'(\Sigma)\}_\Sigma$ of $\Bbbk$-linear isomorphisms, where Σ runs over all closed oriented $(n-1)$-manifolds. These isomorphisms should be multiplicative under disjoint unions of manifolds and satisfy $\varphi_\emptyset = Z_0' Z_0^{-1}$ and $Z'(M)\varphi_\Sigma = \varphi_{\Sigma'} Z(M)$ for all n-dimensional cobordisms M between Σ and Σ'. Two TQFTs are *isomorphic* if there is an isomorphism between them. If two n-dimensional TQFTs Z and Z' are isomorphic, then $Z(M) = Z'(M)$ for any closed oriented n-manifold M.

10.2.3 TQFTs and connected sums

The *connected sum* $M = M_0 \# M_1$ of compact connected oriented n-dimensional manifolds M_0, M_1 is obtained by taking embedded closed n-balls $B_0 \subset \mathrm{Int}(M_0)$, $B_1 \subset \mathrm{Int}(M_1)$ and gluing the manifolds $N_0 = M_0 \setminus \mathrm{Int}(B_0)$, $N_1 = M_1 \setminus \mathrm{Int}(B_1)$ along a homeomorphism $\varphi\colon \partial B_0 \to \partial B_1$ such that the orientations of N_0, N_1 induced by those of M_0, M_1 extend to an orientation of M. Then M is a compact connected oriented n-dimensional manifold independent of the choice of B_0, B_1, φ up to orientation preserving homeomorphisms constant on $\partial M = \partial M_0 \sqcup \partial M_1$. In the following lemma, S^m denotes an oriented m-dimensional sphere.

Lemma 10.2. *Let $Z\colon \mathrm{Cob}_n \to \mathrm{Mod}_{\Bbbk}$ be an n-dimensional TQFT such that*

$$Z(S^{n-1}) \simeq \Bbbk.$$

Then, for any connected sum $M = M_0 \# M_1$ as above, we have the following equality of homomorphisms $Z(-\partial M_0) \to Z(\partial M_1)$:

$$Z(S^n)\, Z(M, -\partial M_0, \partial M_1) = Z(M_1, \emptyset, \partial M_1) \circ Z(M_0, -\partial M_0, \emptyset).$$

In particular, if $\partial M_0 = \partial M_1 = \emptyset$, then $Z(S^n)\, Z(M) = Z(M_0)\, Z(M_1)$.

Proof. In the notation above, for $i = 0, 1$, we provide the ball $B_i \subset \mathrm{Int}(M_i)$ with the orientation induced by that of M_i and provide the sphere ∂B_i with the orientation induced by that of B_i (and opposite to the orientation induced by that of $N_i = M_i \setminus \mathrm{Int}(B_i)$). Pick orientation preserving homeomorphisms $\varphi_0\colon S^{n-1} \to -\partial B_0$ and $\varphi_1\colon S^{n-1} \to \partial B_1$. Let

$$b_0\colon S^{n-1} \to \emptyset, \quad b_1\colon \emptyset \to S^{n-1}, \quad n_0\colon -\partial M_0 \to S^{n-1}, \quad n_1\colon S^{n-1} \to \partial M_1$$

be the morphisms in Cob_n represented respectively by the cobordisms

$$\begin{gathered}
\big(B_0, S^{n-1}, \emptyset, \varphi_0\colon -S^{n-1} \to \partial B_0\big),\ \big(B_1, \emptyset, S^{n-1}, \varphi_1\colon S^{n-1} \to \partial B_1\big),\\
\big(N_0, -\partial M_0, S^{n-1}, \mathrm{id} \sqcup \varphi_0\colon \partial M_0 \sqcup S^{n-1} \to \partial N_0 = \partial M_0 \sqcup (-\partial B_0)\big),\\
\big(N_1, S^{n-1}, \partial M_1, \varphi_1 \sqcup \mathrm{id}\colon (-S^{n-1}) \sqcup \partial M_1 \to \partial N_1 = (-\partial B_1) \sqcup \partial M_1\big).
\end{gathered}$$

In the notation of Section 10.2.1 we have equalities of morphisms

$$b_0 \circ b_1 = (S^n, \emptyset, \emptyset), \quad b_0 \circ n_0 = (M_0, -\partial M_0, \emptyset), \quad n_1 \circ b_1 = (M_1, \emptyset, \partial M_1).$$

By assumption, $Z(S^{n-1}) \simeq \Bbbk \simeq Z(\emptyset)$. Pick a $\Bbbk$-linear isomorphism $z\colon Z(S^{n-1}) \to Z(\emptyset)$. Then $Z(b_0) = \lambda_0 z$ and $Z(b_1) = \lambda_1 z^{-1}$ for some $\lambda_0, \lambda_1 \in \Bbbk$. It follows from the definition of $Z(S^n) \in \Bbbk$ and the functoriality of Z that

$$Z(S^n)\, \mathrm{id}_{Z(\emptyset)} = Z(S^n, \emptyset, \emptyset) = Z(b_0 \circ b_1) = Z(b_0)\, Z(b_1) = \lambda_0 \lambda_1\, \mathrm{id}_{Z(\emptyset)}.$$

Therefore $Z(S^n) = \lambda_0\lambda_1$. To form $M = M_0\#M_1$, we glue N_0 and N_1 along $\varphi_1\varphi_0^{-1}\colon \partial B_0 \to \partial B_1$. Then $(M, -\partial M_0, \partial M_1) = n_1 \circ n_0$ and

$$\begin{aligned}
&Z(S^n)\, Z(M, -\partial M_0, \partial M_1)\\
&= \lambda_0\lambda_1\, Z(n_1 \circ n_0)\\
&= \lambda_0\lambda_1\, Z(n_1) \circ Z(n_0) = \lambda_0\lambda_1\, Z(n_1) \circ z^{-1} \circ z \circ Z(n_0)\\
&= Z(n_1) \circ (\lambda_1 z^{-1}) \circ (\lambda_0 z) \circ Z(n_0) = Z(n_1) \circ Z(b_1) \circ Z(b_0) \circ Z(n_0)\\
&= Z(n_1 \circ b_1) \circ Z(b_0 \circ n_0) = Z(M_1, \emptyset, \partial M_1) \circ Z(M_0, -\partial M_0, \emptyset).
\end{aligned}$$

The last claim of the lemma follows from the fact that when $\partial M_0 = \partial M_1 = \emptyset$, the endomorphisms

$$Z(M_0, -\partial M_0, \emptyset), \quad Z(M_1, \emptyset, \partial M_1), \quad Z(M, -\partial M_0, \partial M_1)$$

of $Z(\emptyset) \simeq \Bbbk$ are multiplications by $Z(M_0), Z(M_1), Z(M) \in \Bbbk$, respectively. □

10.2.4 Remark

Categories similar to Cob_n may be defined in terms of smooth (or piecewise-linear) manifolds. For $n \in \{1, 2, 3\}$, the resulting categories are equivalent to Cob_n and, consequently, there is no difference between n-dimensional TQFTs defined in terms of topological, smooth, or piecewise-linear manifolds. A 1-dimensional TQFT is easily seen to be determined by a projective $\Bbbk$-module of finite type (the value of the TQFT on the point). A 2-dimensional TQFT is determined by a commutative Frobenius $\Bbbk$-algebra (whose underlying $\Bbbk$-module is the value of the TQFT on the circle, see [Ko]). The 3-dimensional TQFTs are extensively discussed in the next chapters.

Chapter 11

Skeletons of 3-manifolds

We define skeletons of compact 3-manifolds and study their local transformations.

11.1 Stratified 2-polyhedra

We define (finite) graphs and introduce our key geometric tool, that of a stratified 2-polyhedron.

11.1.1 Graphs

By a *graph* we mean a topological space G obtained from a finite number of disjoint copies of the closed interval $[0,1]$ by identification of certain endpoints. The images of the copies of $[0,1]$ in G are called *edges* of G. The endpoints of the edges of G (that is, the images of $0,1 \in [0,1]$) are called *vertices* of G. Each edge of G connects two (possibly, coinciding) vertices, and each vertex of G is incident to at least one edge. By *half-edges* of G, we mean the images of the closed intervals $[0,1/2] \subset [0,1]$ and $[1/2,1] \subset [0,1]$ in G. The number of half-edges of G incident to a vertex v of G is greater than or equal to 1 and is called the *valence* of v.

A graph is *oriented* if all its edges are oriented. The empty set is viewed as an oriented graph with no vertices and no edges.

11.1.2 Stratified 2-polyhedra

By a 2-*polyhedron*, we mean a compact topological space that can be triangulated using a finite number of simplices of dimension ≤ 2 so that all 0-simplices and 1-simplices are faces of 2-simplices. For a 2-polyhedron P, denote by $\mathrm{Int}(P)$ the subspace of P consisting of all points having a neighborhood homeomorphic to $\mathbb{R}^2$. By the definition of a 2-polyhedron, the surface $\mathrm{Int}(P)$ is dense in P. (By *surfaces*, we mean 2-dimensional manifolds, possibly with boundary.) A *stratification* of a 2-polyhedron P is an (unoriented) graph G embedded in P so that $P \setminus \mathrm{Int}(P) \subset G$.

A *stratified polyhedron* is a 2-polyhedron P endowed with a stratification $G \subset P$. The graph G is then denoted by $P^{(1)}$. The vertices and edges of G are called respectively the *vertices* and *edges* of P. To specify a stratification of P, it suffices to specify the edges of P because the vertices of P are just the endpoints of the edges.

Any 2-polyhedron P can be stratified. For example, the 0-simplices and the 1-simplices of a triangulation of P form a graph which is a stratification of P. Any (possibly, empty) graph embedded in a compact surface without boundary is a stratification of this surface.

11.1.3 Regions and branches

Cutting a stratified 2-polyhedron P along the graph $P^{(1)} \subset P$, we obtain a compact surface $\widetilde{P}$ with interior $P \setminus P^{(1)}$. The polyhedron P can be recovered by gluing $\widetilde{P}$ to $P^{(1)}$ along a surjective map $p\colon \partial\widetilde{P} \to P^{(1)}$. The set

$$p^{-1}(\text{the set of vertices of} P) \subset \partial\widetilde{P}$$

is closed and discrete, and therefore is finite. The points of this set split $\partial\widetilde{P}$ into arcs whose interiors are mapped by p homeomorphically onto the interiors of edges of P. The (connected) components of $\widetilde{P}$ are called the *regions* of P. Each component of $P \setminus P^{(1)} \subset \widetilde{P}$ is the interior of a unique region. We let $\mathrm{Reg}(P)$ be the (finite) set of all regions of P. We say that a region r of P *embeds* in P if the map $p|_{\partial r}\colon \partial r \to P^{(1)}$ is injective.

A *branch* of a stratified 2-polyhedron P at a vertex x of P is a germ at x of an adjacent region. More formally, a branch of P at x is a homotopy class of paths $[0,1] \to P$ starting in x and carrying $(0,1]$ to $P \setminus P^{(1)}$. The number of branches of P at x is equal to $\mathrm{card}(p^{-1}(x))$, where $p\colon \partial\widetilde{P} \to P^{(1)}$ is the map above. Similarly, a *branch* of P at an edge e of P is a germ at e of an adjacent region. Formally, a branch of P at e is the homotopy class of paths $[0,1] \to P$ starting in the interior of e and carrying $(0,1]$ to $P \setminus P^{(1)}$. There is an obvious bijective correspondence between the branches of P at e and the connected components of $p^{-1}(\text{interior of} e)$. The set of branches of P at e is denoted P_e. This set is finite and non-empty. The number of elements of P_e is called the *valence* of e.

An *orientation* of a stratified 2-polyhedron P is an orientation of the surface $\widetilde{P}$ or, equivalently, an orientation of the surface $P \setminus P^{(1)} = \mathrm{Int}(\widetilde{P})$. To orient P, one must orient all its regions.

11.1.4 The boundary graph

The edges of a stratified 2-polyhedron P of valence 1 together with their vertices form a graph called the *boundary* of P and denoted ∂P. An orientation of P induces an orientation of ∂P: each edge e of ∂P is oriented so that the orientation of the unique region of P adjacent to e is determined by a pair (a vector directed outward at an interior point of e, a positive tangent vector of e).

A stratified 2-polyhedron P is *∂-cylindrical* if each vertex v of the graph ∂P is adjacent to a unique edge d_v of P not contained in ∂P, this edge is adjacent to all branches of P at v, and its second endpoint is distinct from v. Then the valence of v in ∂P is equal to the valence of d_v in P and is greater than or equal to 2 (because d_v is not contained in ∂P). If P is ∂-cylindrical, then ∂P has a neighborhood in P homeomorphic to $\partial P \times [0,1]$.

11.1.5 Two classes of polyhedra

We introduce two classes of stratified 2-polyhedra which will be instrumental in the sequel. A stratified 2-polyhedron P is *regular* if all regions of P are disks and embed in P. If P is regular, then the branches of P at any vertex lie in pairwise distinct regions (and the latter implies that the branches of P at any edge lie in pairwise distinct regions). For example, if the stratification of P is formed by the 0-simplices and 1-simplices of a triangulation of P, then the regions of P are the 2-simplices of this triangulation and P is regular.

Next, consider the set

$$\mathcal{B} = \{(x_1, x_2, x_3) \in \mathbb{R}^3 \mid x_3 = 0, \text{ or } x_1 = 0 \text{ and } x_3 > 0, \text{ or } x_2 = 0 \text{ and } x_3 < 0\}$$

shown in Figure 11.1. A point v of a topological space is said to be *special* if there is a homeomorphism of $\mathcal{B}$ onto a neighborhood of v carrying the origin $(0,0,0)$ to v. A 2-polyhedron P is said to be *special* if each point of P has a neighborhood homeomorphic to an open subset of $\mathcal{B}$, each connected component of P has at least one special point, and $P \setminus \mathrm{Int}(P)$ has no components homeomorphic to the circle S^1. Then $P \setminus \mathrm{Int}(P)$ is a graph whose vertices are the special points of P. This graph defines a canonical stratification of P. Unless explicitly stated to the contrary, special 2-polyhedra are always equipped with this stratification. A special 2-polyhedron has 6 branches at every vertex, 3 branches at every edge, and an empty boundary.

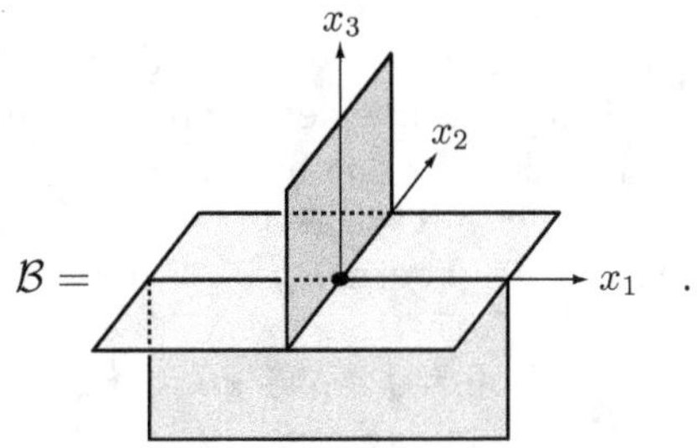

Figure 11.1: The set $\mathcal{B}$

11.1.6 Example

Let G be a graph whose all vertices have valence ≥ 2. The 2-polyhedron $P = G \times [0,1]$ carries a stratification with vertices $(v,0),(v,1)$ and edges $e \times \{0\}, e \times \{1\}, v \times [0,1]$ where v runs over vertices of G and e runs over edges of G. Clearly, ∂P is a disjoint union of two copies of G. The stratified polyhedron P is ∂-cylindrical and is not special. It is regular if and only if G has no edges with coinciding endpoints.

11.1.7 Example

Collapsing $G \times \{1\}$ to a point in the previous example, we obtain a stratified 2-polyhedron $P/(G \times \{1\})$ with boundary $G \times \{0\}$.

11.1.8 Exercise

Prove that if a stratified 2-polyhedron P has no edges of valence 2, then $P^{(1)} = P \setminus \mathrm{Int}(P)$.

11.1.9 Remark

Our definition of a special 2-polyhedron is less restrictive than the one in [Mat], where, additionally, all regions are required to be disks.

11.2 Skeletons of closed 3-manifolds

We define and study skeletons of closed 3-manifolds. For generalities on 3-manifolds, we refer to the monographs [Hem, Mat].

11.2.1 Skeletons

A *skeleton* of a closed (possibly, non-connected) 3-dimensional manifold M is an oriented stratified 2-polyhedron P embedded in M such that $\partial P = \emptyset$ and $M \setminus P$ is a disjoint union of open 3-balls. The number of these open 3-balls is denoted by $|M \setminus P|$. The condition $\partial P = \emptyset$ ensures that all edges of P have valence ≥ 2. Any vertex x of P has a closed ball neighborhood $B_x \subset M$ such that $\Gamma_x = P \cap \partial B_x$ is a non-empty graph and $P \cap B_x$ is the cone over Γ_x with summit x. The vertices of Γ_x are the intersection points of the 2-sphere ∂B_x with the edges of P incident to x. The edges of Γ_x are the intersections of ∂B_x with the branches of P at x. Since all edges of P have valence ≥ 2, so do all vertices of Γ_x. We call B_x a *P-cone neighborhood* of x and call $\Gamma_x \subset \partial B_x$ the *link graph* of x. The pair (B_x, Γ_x) is determined by the triple (M, P, x) uniquely up to homeomorphism.

11.2.2 Regular and special skeletons

A skeleton of a closed 3-manifold M is *regular* if the underlying stratified 2-polyhedron is regular. For example, the 2-skeleton $t^{(2)}$ of any triangulation t of M is a regular skeleton of M whose vertices, edges, and regions are respectively the 0-simplices, 1-simplices, and 2-simplices of t and whose orientation is chosen arbitrarily. The triangulation t of M also determines a cellular decomposition t^* of M dual to t. The 2-skeleton $(t^*)^{(2)}$ of t^* is a regular skeleton of M whose vertices, edges, and regions are respectively the 0-cells, 1-cells, and 2-cells of t^* and whose orientation is chosen arbitrarily. Clearly, $|M \setminus t^{(2)}|$ is the number of 3-simplices of t and $|M \setminus (t^*)^{(2)}|$ is the number of 0-simplices of t.

A skeleton of a closed 3-manifold M is *special* is its underlying 2-polyhedron is special and its stratification is the canonical one. For example, the skeleton $(t^*)^{(2)}$ of M derived from an arbitrary triangulation t of M is both regular and special. For brevity, special skeletons will be called s-skeletons.

11.2.3 Example

An oriented 2-sphere S^2 embedded in the 3-sphere S^3 is a regular skeleton of S^3 with no vertices and no edges. Clearly $|S^3 \setminus S^2| = 2$.

11.2.4 Example

Pick a point $a \in S^1$ and an embedded loop $\ell \subset S^2$. Then the 2-polyhedron $P = (\{a\} \times S^2) \cup (S^1 \times \ell)$ with a single edge $\{a\} \times \ell$ and an arbitrary orientation is a (non-regular) skeleton of $S^1 \times S^2$. Clearly $|(S^1 \times S^2) \setminus P| = 2$.

11.2.5 Example

Skeletons of 3-dimensional lens spaces may be constructed as follows. A lens space is determined by relatively prime integers $p \geq 2, q \geq 1$ and is obtained by gluing two solid tori. More precisely, let $D = \{z \in \mathbb{C} \,|\, |z| \leq 1\}$ be the unit disk in $\mathbb{C}$ with counterclockwise orientation. Let $S^1 = \partial D$ be the unit circle with counterclockwise orientation. Let U, V be two copies of the solid torus $S^1 \times D$. The boundaries of U and V are copies of $S^1 \times S^1$. The *lens space* $L(p,q) = U \cup_\varphi V$ is obtained by gluing U to V along any homeomorphism $\varphi\colon \partial U \to \partial V$ such that $\varphi(1,t) = (t^p, t^q)$ for all $t \in S^1$. The space $L(p,q)$ is a closed connected orientable 3-manifold. We provide it with orientation extending the product orientation in $V = S^1 \times D$. Considered up to orientation preserving homeomorphism, the oriented manifold $L(p,q)$ does not depend on the choice of φ as above. A skeleton P of $L(p,q)$ is obtained from the disk $\{1\} \times D \subset U$ and the circle $S^1 \times \{0\} \subset V$ by pulling the boundary of this disk towards this circle in V. More precisely,

$$P = (\{1\} \times D) \cup_\varphi \{(t^p, kt^q) \,|\, t \in S^1, k \in [0,1]\} \subset U \cup_\varphi V = L(p,q).$$

The polyhedron P contains the circle $S^1 \times \{0\} \subset V$ which, viewed as a loop based at the point $(1, 0)$, determines a stratification of P. The stratified polyhedron P has only one region which is a disk and which we orient arbitrarily. In this way, P becomes a skeleton of $L(p, q)$. It is neither regular nor special, and $|L(p, q) \setminus P| = 1$.

11.3 Moves on skeletons

We study local transformations (moves) on skeletons of closed 3-manifolds.

11.3.1 Primary moves

Let M be a closed 3-manifold. We define four local moves T_0, T_1, T_2, T_3 on a skeleton P of M transforming P into a new skeleton of M. Each of these moves modifies P inside a closed 3-ball in M as in Figure 11.2 where the shaded horizontal plane represents (a piece of) P and the upper half-space represents an adjacent component of $M \setminus P$. The edges shown in the horizontal plane before the moves may be adjacent to regions of P lying in the lower half-space (and not shown in the picture).

The "bubble move" T_0 adds to P an embedded disk $D_+ \subset M$ such that $\partial D_+ = D_+ \cap P \subset P \setminus P^{(1)}$, the circle ∂D_+ bounds a disk D_- in $P \setminus P^{(1)}$, and the 2-sphere $D_+ \cup D_-$ bounds a ball in M meeting P precisely along D_-. A point of the circle $\partial D_+ = \partial D_-$ is chosen as a vertex of the skeleton $T_0(P) = P \cup D_+$, and the circle itself is an edge of $T_0(P)$. The move T_0 increases the number $\#v$ of vertices of the skeleton by 1, the number $\#e$ of edges by 1, the number $\#r$ of regions by 2, and the number $|M \setminus P|$ of connected components of $M \setminus P$ by 1.

The "phantom edge move" T_1 keeps P as a polyhedron and adds a new 2-valent edge meeting $P^{(1)}$ solely at its endpoints which must be distinct vertices of P. This move preserves $|M \setminus P|$ and $\#v$, increases $\#e$ by 1, and either preserves $\#r$ or increases $\#r$ by 1.

The "contraction move" T_2 collapses into a point an edge of P with distinct endpoints. This move is allowed only when at least one endpoint of the collapsing edge is an endpoint of some other edge. (The valence of the collapsing edge may be arbitrary.) The move T_2 decreases $\#v$ and $\#e$ by 1 and preserves $\#r$ and $|M \setminus P|$.

The "percolation move" T_3 pushes a branch of P through a vertex x of P. The branch is pushed across a small disk D lying in another branch of P at x so that these branches are adjacent to the same component of $M \setminus P$ and

$$D \cap P^{(1)} = \partial D \cap P^{(1)} = \{x\}.$$

The loop ∂D based at x becomes an edge of the resulting 2-polyhedron. The move T_3 preserves $|M \setminus P|$ and $\#v$ and increases both $\#e$ and $\#r$ by 1.

Recall that all regions of a skeleton are oriented. The moves T_0, T_1, T_2, T_3 preserve orientation in the "big" regions in the pictures while orientation in the

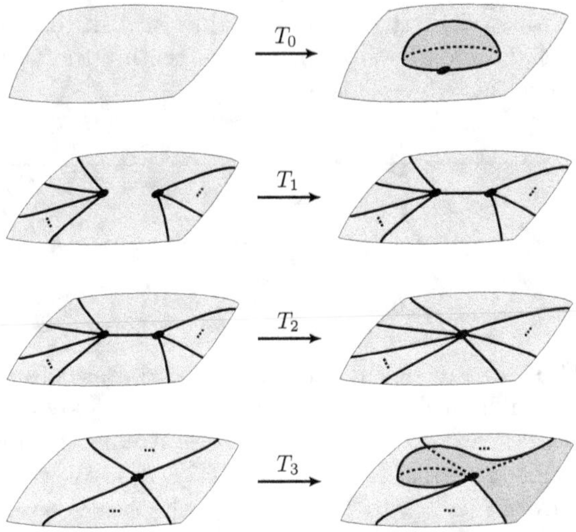

Figure 11.2: Local moves on skeletons

small disk regions created by T_0 and T_3 may be arbitrary. We will apply the same convention to all local moves on skeletons considered in the sequel.

The moves above have obvious inverses: a skeleton $Q \subset M$ is obtained from a skeleton $P \subset M$ by T_i^{-1} with $i \in \{0,1,2,3\}$ if P is obtained from Q by T_i. A move T_0^{-1} is obvious from the picture of T_0. A move T_1^{-1} deletes a 2-valent edge with distinct endpoints which are also endpoints of some other edges. This move is allowed only when the orientations of the two regions adjacent to that edge are compatible. A move T_2^{-1} stretches a vertex into an edge. In particular, using T_2^{-1}, one can add a new vertex in the interior of an edge keeping all other edges and all regions. A move T_3^{-1} collapses an embedded disk region whose boundary is formed by a single 3-valent edge with coinciding endpoints.

By *primary moves* on skeletons of M we mean orientation-preserving ambient isotopies of skeletons in M and the moves $T_0^{\pm 1}, T_1^{\pm 1}, T_2^{\pm 1}, T_3^{\pm 1}$. We now state the main result of this section.

Theorem 11.1. *Any two skeletons of a closed* 3*-manifold can be related by primary moves.*

We prove Theorem 11.1 in Section 11.4 using the results of Section 11.3.2.

11.3.2 Further moves

We will use several other local moves on skeletons of a 3-manifold. They all expand as compositions of primary moves. The move T_0' shown in Figure 11.3 acts on a skeleton P as the bubble move T_0 with the only difference that the circles ∂D_+ and ∂D_- meet $P^{(1)}$ in a vertex x of P. The move T_0' expands as a product of primary moves as follows. First, a move T_0 glues a disk to P along a circle in $\mathrm{Int}(D_-)$. After that, T_1 adds an edge in D_- connecting the vertex of that circle to x, and T_2 collapses this edge. This gives $T_0'(P)$.

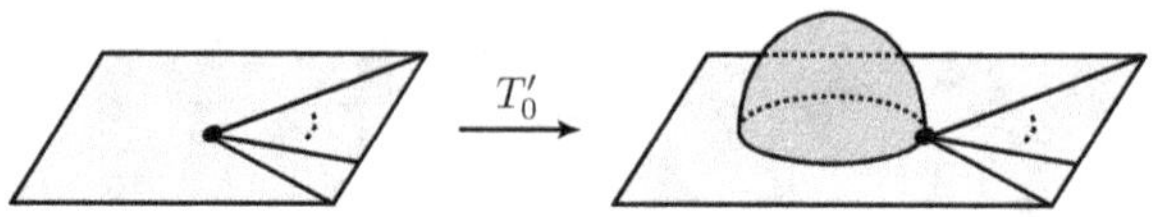

Figure 11.3: The move T_0'

The move T_0'' shown in Figure 11.4 attaches to a skeleton P an embedded disk D_+ in a small neighborhood of an interior point of an edge e of P such that $D_+ \cap P = \partial D_+$ meets e at two interior points of e. Under this move, e splits into three consecutive subedges e_1, e_2, e_3 so that $\partial e_2 = D_+ \cap e$. We construct a sequence of primary moves from $P \cup D_+$ to P. First, T_2 collapses e_2 to a point, then T_3^{-1} pushes ∂D_+ into a region adjacent to e to give $T_0'(P)$. The latter skeleton, as we know, can be turned into P by primary moves. Inverting the resulting sequence of moves $P \cup D_+ \rightsquigarrow P$ we expand T_0'' as a composition of primary moves.

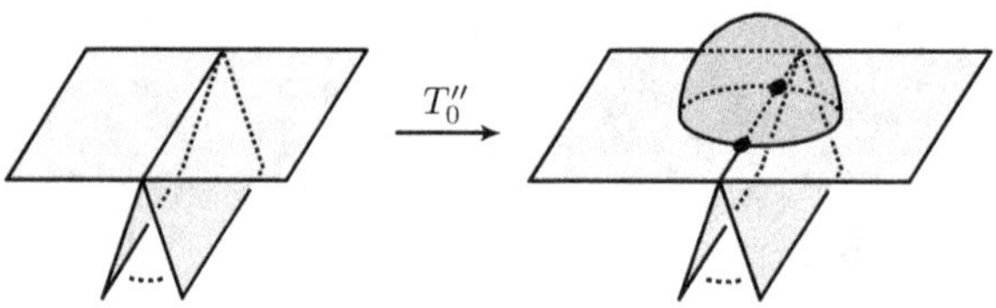

Figure 11.4: The move T_0''

For any integers $m, n \geq 0$, we define a move on skeletons $T^{m,n}$ pushing a region across a vertex, see Figure 11.5. This move is called an $m+1 \mapsto n+1$ move for the numbers of vertices in the picture before and after the move. For $n = 0$, this move is allowed only when the orientations of the two regions united under the move are compatible. The move inverse to $T^{m,n}$ is $T^{n,m}$. To see that $T^{m,n}$ is a composition of primary moves, we split $T^{m,n}$ as a composition of a move $T^{(m)}$ and the inverse of $T^{(n)}$ as shown in Figure 11.5. For $m > 0$, the move $T^{(m)}$ is obtained by applying m times T_2 to collapse the short edges going from the central vertex in the picture to the left and then applying $m-1$ times T_3^{-1} to remove the resulting $m-1$ loops. The move $T^{(0)}$ also expands as a composition of primary moves: first, use T_2^{-1} to create a vertex v on the boundary line of the vertical half-plane, then use T_1 to add an edge in the horizontal plane connecting v to the cental vertex, and use T_2 to collapse this new edge into a point.

Finally, we will need the *lune moves* $\mathcal{L}$ and $\mathcal{L}^{-1}$ shown in Figure 11.6. The move $\mathcal{L}^{-1}$ is allowed only when the orientations of two regions united under this move are compatible. The lune moves $\mathcal{L}^{\pm 1}$ may be obtained by combining $T^{0,2}$, $T^{2,0}$

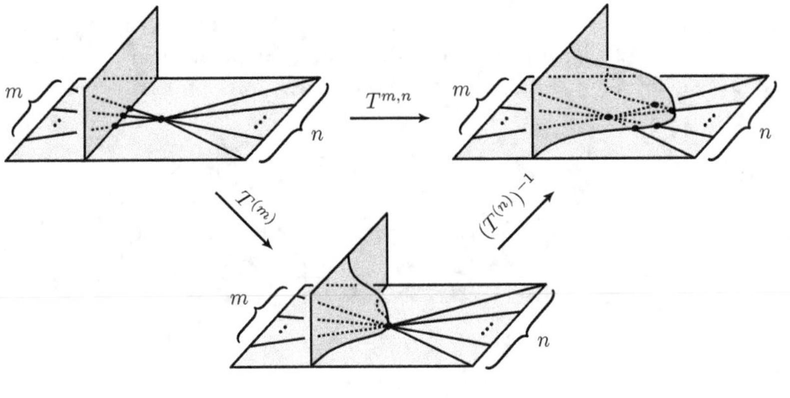

Figure 11.5: The move $T^{m,n}$

with the moves $T_2^{\pm 1}$ which allow to create or delete vertices in the interiors of edges. Therefore, the lune moves expand as products of primary moves.

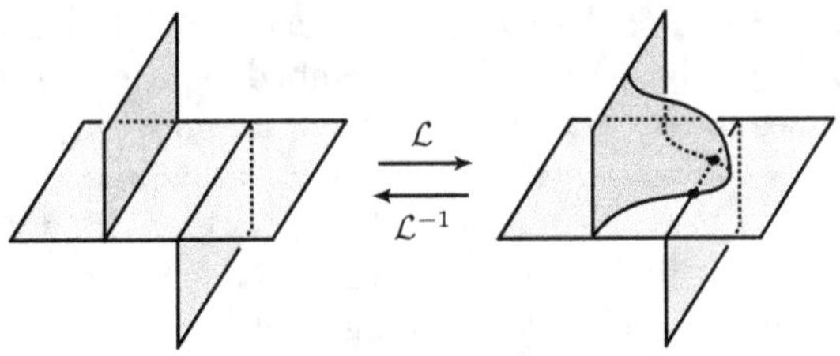

Figure 11.6: The lune moves $\mathcal{L}$ and $\mathcal{L}^{-1}$

11.4 Proof of Theorem 11.1

Theorem 11.1 is proved by a reduction to the theory of Matveev–Piergallini moves on special skeletons. We begin by defining these moves.

11.4.1 MP-moves

By *Matveev–Piergallini* moves or *MP-moves* on s-skeletons of a closed 3-manifold, we mean orientation-preserving ambient isotopies of s-skeletons in this manifold and the moves $(T_0'')^{\pm 1}$, $T^{1,2}$, $T^{2,1} = (T^{1,2})^{-1}$, $\mathcal{L}^{\pm 1}$, see Figures 11.4, 11.6, 11.7. These moves on special skeletons were first considered by Sergei Matveev and Riccardo Piergallini, see the monograph [Mat] for further references. All MP-moves transform s-skeletons into s-skeletons keeping orientations in the "big" regions in the pictures. The orientation of an s-skeleton can obstruct an application of

$\mathcal{L}^{-1}$ but cannot obstruct an application of other MP-moves. By Section 11.3.2, all MP-moves on s-skeletons expand as products of primary moves (though the intermediate skeletons produced by primary moves may be non-special).

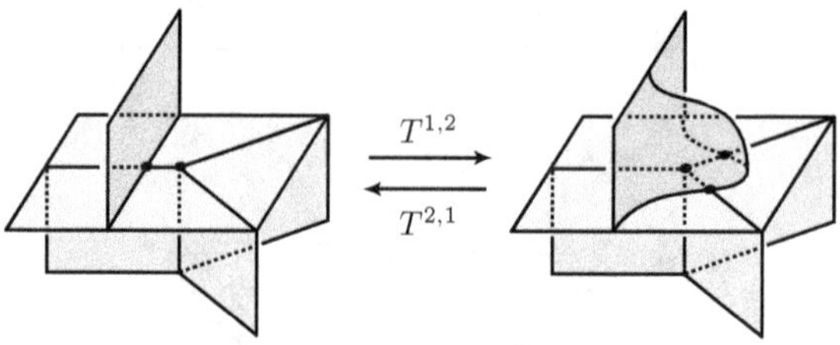

Figure 11.7: The moves $T^{1,2}$ and $T^{2,1}$

Lemma 11.2. *Any two s-skeletons of a closed 3-manifold can be related by MP-moves.*

Proof. It suffices to handle the case of a connected closed 3-manifold M. Pick a regular s-skeleton P of M. We will prove that any s-skeleton Q of M can be transformed into P by MP-moves. If α is an embedded arc in a region r of Q with endpoints inside edges of ∂r, then we can apply the move T_0'' near one endpoint, pull a piece of the newly attached disk along α, and further pull this piece by the lune move $\mathcal{L}$ into a branch of Q near the second endpoint of α. This creates several disk regions and splits r along α. Applying this procedure several times, we can transform Q into a new s-skeleton, again denoted by Q, whose all regions are disks. Furthermore, applying the move T_0'' to the skeletons P, Q several times, we can transform them into s-skeletons P', Q', respectively, such that both P' and Q' have at least two vertices and $|M \setminus P'| = |M \setminus Q'|$. Since all regions of P and Q are disks, so are the regions of P' and Q'. In terminology of [Mat], P' and Q' are special spines of the compact 3-manifold obtained from M by removing $|M \setminus P'| = |M \setminus Q'|$ disjoint open 3-balls. By [Mat, Theorem 1.2.30], there is a (finite) sequence of moves $T^{1,2}$, $T^{2,1}$ transforming Q' into P' up to orientation. Combining the moves T_0'' from Q to Q' with this sequence $Q' \rightsquigarrow P'$ and with the moves $(T_0'')^{-1}$ from P' to P, we obtain a sequence of MP-moves transforming Q into P up to orientation. It remains to show that for any region r of P, there is a sequence of MP-moves transforming P into the s-skeleton P_r obtained from P by reversing orientation in r. By the assumptions on P, the region r is a disk that embeds in P. Then the circle ∂r embeds in $P - \mathrm{Int}(r)$ onto a union of several consecutive edges of P. At each of these edges, one branch of P lies in r and two other branches of P lie in $P - \mathrm{Int}(r)$. The union of the latter branches over all edges forming ∂r yields an annulus neighborhood of ∂r in $P - \mathrm{Int}(r)$. Pushing r in a normal direction so that ∂r sweeps half of this annulus, we obtain a solid cylinder $r \times [0,1] \subset M$ meeting P at $\partial r \times [0,1]$ and at $r = r \times \{0\}$. Set $r_- = r \times \{1\}$. Then $P \cup r_-$ is an s-skeleton of M whose regions contained in P receive the induced orientation while the orientation of the region r_- is opposite to that of r. We can transform $P \cup r_-$ into an s-skeleton isotopic to P by MP-moves pushing subarcs of ∂r_- into r and, at the last step, eliminating r_-. This is achieved by a move $T^{1,2}$ followed by several $T^{2,1}$ and a single $(T_0'')^{-1}$ at the end. A similar elimination of r transforms $P \cup r_-$ into an s-skeleton isotopic to P_r. Hence P and P_r are related by MP-moves. □

11.4.2 Further lemmas

The aim of this section is to establish Lemma 11.4 which claims that all skeletons of a closed 3-manifold can be transformed by primary moves into s-skeletons. We begin with a preliminary lemma.

Lemma 11.3. *Let M be a closed 3-manifold, $B \subset M$ be a closed 3-ball, and $S = \partial B$. Let P be a skeleton of M such that $S \subset P \subset M \setminus \mathrm{Int}(B)$, all edges of P lying in S are 3-valent, and at least one edge of P has one endpoint in S and another endpoint in $M \setminus B$. Then the oriented stratified 2-polyhedron*

$$P/S \subset M/B \simeq M,$$

obtained from P by collapsing S, is a skeleton of M. Moreover, there is a sequence of primary moves transforming P into P/S.

Proof. In this lemma, we identify M/B with M in the obvious way. Our assumptions imply that S is a union of faces of P and the set $S \cap P^{(1)}$ is a union of vertices and edges of P. The projection $P \to P/S$ carries vertices, edges, and regions of P not contained in S to vertices, edges, and regions of the 2-polyhedron P/S, respectively. The point $S/S \in P/S$ is also a vertex of P/S. This gives a stratification of P/S. The orientation of P induces an orientation of P/S via the homeomorphism $\mathrm{Int}(P \setminus S) \to \mathrm{Int}(P/S)$ induced by the projection $P \to P/S$. The complement of P/S in M is a disjoint union of the open 3-balls forming $M \setminus P$ and distinct from $\mathrm{Int}(B)$. Thus, P/S is a skeleton of M.

We construct primary moves relating P to P/S. By assumption, P has at least one vertex lying in S. If P has just one vertex in S and no edges in S, then P/S is obtained from P by a move T_3 followed by $(T_0')^{-1}$. If P has one vertex and $n \geq 1$ edges in S, then these edges are loops and we can enumerate them inductively $e_1, \ldots, e_n$ so that each e_i bounds a 2-disk in S whose interior does not meet $e_{i+1}, \ldots, e_n$. Collapsing this disk, we transform P into a skeleton P_i of M. For $i = 1, \ldots, n-1$, the skeleton P_i can be obtained from P_{i-1} by a move T_3^{-1}; here $P_0 = P$ and we use the assumption that e_i is a trivalent edge of P and so is adjacent to a single branch of P lying in $M \setminus \mathrm{Int}(B)$. The skeleton P/S is obtained from P_{n-1} by $(T_0')^{-1}$. This gives a sequence of primary moves $P = P_0 \to P_1 \to \cdots \to P_{n-1} \to P/S$.

If two or more vertices of P lie in S, then we use the moves T_1, T_2 adding new edges to P in S and further collapsing them to ensure that the vertices and

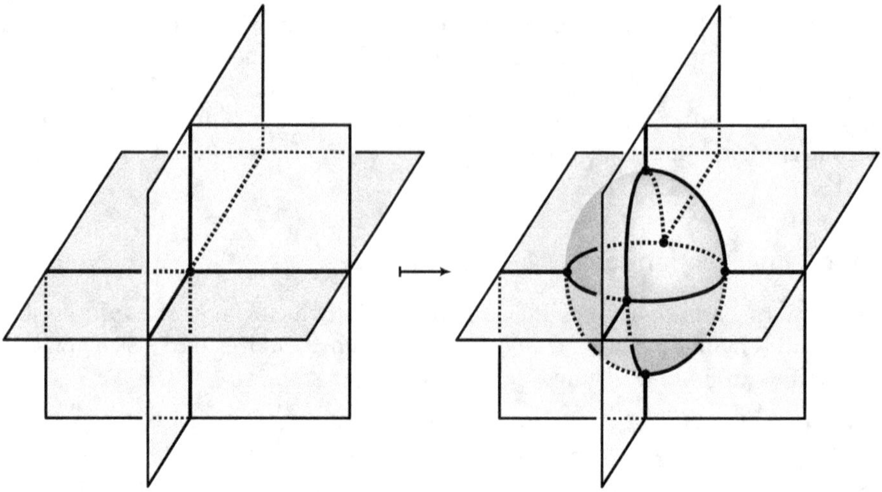

Figure 11.8: Blowing up a vertex

edges of P lying in S form a connected graph. Next, we apply the moves T_2 to P collapsing the edges of a maximal tree in this graph. This yields a new skeleton of M, again denoted by P, which gives the same quotient skeleton P/S as before and which has only one vertex in S. By the above, P and P/S are related by primary moves. □

Lemma 11.4. *Any skeleton of a closed 3-manifold can be transformed by primary moves into an s-skeleton.*

Proof. Let P be a skeleton of a closed 3-manifold M. We define a "blowing up" of P at a vertex x. Let $\Gamma_x \subset S_x$ be the link graph of x, where $S_x = \partial B_x$ and $B_x \subset M$ is a P-cone neighborhood of x, see Section 11.2.1. Set

$$P^x = (P \setminus \mathrm{Int}(B_x)) \cup S_x \subset M. \tag{11.1}$$

The 2-polyhedron P^x has the same edges and vertices as P except that the vertex x is deleted, all edges incident to x are cut near x, and the vertices and edges of Γ_x are added to the stratification, see Figure 11.8. This turns P^x into a stratified 2-polyhedron. The regions of P^x lying in P are provided with orientation induced by that of P and the regions of P^x lying on the 2-sphere S_x are oriented arbitrarily. Clearly, $M \setminus P^x$ is a disjoint union of the open 3-ball $\mathrm{Int}(B_x)$ and a 3-manifold homeomorphic to $M \setminus P$. Therefore, P^x is a skeleton of M.

It is clear that

$$P^x/S_x = (P \cup B_x)/B_x = P.$$

By Lemma 11.3, there is a sequence of primary moves

$$P \rightsquigarrow P^x.$$

Blowing up P recursively at all vertices, we obtain a new skeleton P_1 of M related to P by primary moves. The skeleton P_1 has edges of two types: the edges of the graphs $\{\Gamma_x\}_x$ above and the edges of P with small neighborhoods of their endpoints removed. We call these edges of P_1 respectively *short* and *long*. By construction, all short edges are trivalent. Since $\partial P = \emptyset$, all long edges have valence at least 2. Consider next a closed 2-disk D with center O, and for $n \geq 2$, let $Y_n \subset D$ be a union of n radii of D. A neighborhood in M of a long n-valent edge e of P_1 can be identified with $D \times [0,1]$ so that

$$e = \{O\} \times [0,1] \quad \text{and} \quad P_1 \cap (D \times [0,1]) = (D \times \{0,1\}) \cup (Y_n \times [0,1]).$$

We "blow up" e by deleting $Y_n \times (0,1)$ from P_1 and adding $\partial D \times (0,1)$ instead, see Figure 11.9. This gives a stratified 2-polyhedron embedded in M. Its regions lying in the sphere $S = \partial(D \times [0,1])$ are oriented arbitrarily while all other regions inherit their orientation from P_1. The resulting skeleton, P_1^e, of M contains S and meets the conditions of Lemma 11.3. By this lemma, there is a sequence of primary moves $P_1^e \rightsquigarrow P_1^e/S$. The skeleton P_1^e/S can also be obtained from P_1 by the move T_2 collapsing the edge e. Consequently, the skeleton P_1^e is related to P_1 and to P by primary moves.

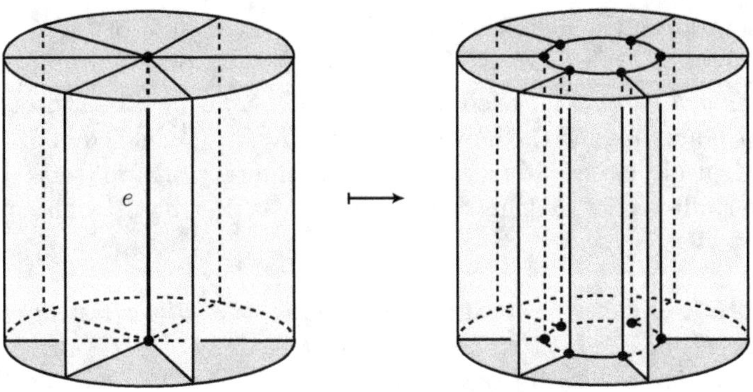

Figure 11.9: Blowing up an edge e

We recursively blow up in this way all long edges of P_1. This gives a special skeleton of M related to P by primary moves. □

11.4.3 Proof of Theorem 11.1

Theorem 11.1 follows from Lemmas 11.2, 11.4 and the fact that the MP-moves are compositions of primary moves.

11.5 Skeletons of 3-manifolds with boundary

We study skeletons of compact 3-manifolds with boundary and focus from now on oriented 3-manifolds. Fix throughout this section a compact oriented 3-manifold M. Recall that ∂M carries the orientation induced by that of M, see Section 10.1.1.

11.5.1 Skeletons of pairs

Consider an oriented graph $G \subset \partial M$ with all vertices of valence ≥ 2. A *skeleton* of the pair (M, G) is an oriented ∂-cylindrical stratified 2-polyhedron P embedded in M such that

(i) $\partial P = G$ as oriented graphs and $P \setminus \partial P \subset M \setminus \partial M$;

(ii) $M \setminus P$ is a disjoint union of open 3-balls and a 3-manifold homeomorphic to $(\partial M \setminus G) \times [0, 1)$ via a homeomorphism extending the identity map

$$\partial(M \setminus P) = \partial M \setminus G = (\partial M \setminus G) \times \{0\}.$$

It follows from the definitions that for any skeleton P of (M, G), there is a closed neighborhood N of ∂M in M such that $P \cap N$ is homeomorphic to $G \times [0, 1]$.

By *primary moves* on skeletons of (M, G), we mean ambient isotopies in M keeping ∂M pointwise and the moves $T_0^{\pm 1}, T_1^{\pm 1}, T_2^{\pm 1}, T_3^{\pm 1}$ defined in Section 11.3.1 and applied inside closed 3-balls lying in the interior $\mathrm{Int}(M) = M \backslash \partial M$ of M. In particular, all vertices/edges/regions created or destroyed by the moves lie in $\mathrm{Int}(M)$.

Theorem 11.5. *Let $G \subset \partial M$ be an oriented graph with all vertices of valence ≥ 2. The pair (M, G) has a skeleton, and any two skeletons of (M, G) can be related by primary moves in the class of skeletons of (M, G).*

Theorem 11.5 includes Theorem 11.1 (for oriented manifolds) as the case $G = \partial M = \emptyset$. If the graph G is trivalent (i.e., all its vertices have valence 3), then Theorem 11.5 can be proved following the lines of the proof of Theorem 11.1. Namely, one defines a special skeleton of (M, G) as a skeleton whose all components have special points and whose all vertices are the special points and the vertices of G. Theorem 1.2.30 of [Mat] extends to this setting and implies that any special skeletons of (M, G) can be related by the moves $(T_0'')^{\pm 1}$, $T^{1,2}, T^{2,1}$ up to the choice of orientation. The rest of the proof is modified accordingly. This approach does not cover the case where G has vertices of valence $\neq 3$. To handle this case, we introduce so-called frames of M and use them to study skeletons of (M, G) (Sections 11.5.2–11.5.4). Theorem 11.5 is proved in Section 11.5.5.

11.5.2 Frames

A *frame* of M is an oriented special 2-polyhedron F embedded in $\mathrm{Int}(M)$ and endowed with its canonical stratification such that there is an embedding

$$i\colon \partial M \times [0,1] \hookrightarrow M,$$

called an F-*collar*, with the following properties:

- $i(x,0) = x$ for all $x \in \partial M$ and $i^{-1}(F) = \partial M \times \{1\}$;
- the orientation of the regions of F lying in $i(\partial M \times \{1\}) \subset F$ is induced by that of ∂M;
- the set $M \setminus F$ is a disjoint union of $i(\partial M \times [0,1))$ and several open 3-balls.

Any two F-collars may be obtained from each other through composition with a self-homeomorphism of $\partial M \times [0,1]$ fixing $M \times \{0\}$ pointwise. Such a self-homeomorphism is isotopic to the identity map, and therefore any two F-collars $\partial M \times [0,1] \hookrightarrow M$ are isotopic in the class of F-collars.

Given a frame $F \subset \mathrm{Int}(M)$, we let $\widehat{F} \subset \mathrm{Int}(M)$ be the union of F with the ball components of $M \setminus F$. Clearly, $\widehat{F}$ is a compact 3-manifold whose boundary $\partial\widehat{F}$ is homeomorphic to ∂M. Set

$$\partial F = \partial\widehat{F} \subset F.$$

For any F-collar i, we have

$$\widehat{F} = M \setminus i(\partial M \times [0,1)) \quad \text{and} \quad \partial F = i(\partial M \times \{1\}).$$

Observe that each edge of F either lies in ∂F or meets ∂F at a subset (possibly, empty) of the set of its endpoints. The edges of F lying in ∂F and their endpoints form a trivalent graph $\Gamma_F = \partial F \cap F^{(1)}$.

Lemma 11.6. *The manifold M has a frame.*

Proof. Pick a triangulation t of M, and let ∂t be the induced triangulation of ∂M. Let t^* be the cellular decomposition of M dual to t. It is formed by the cells dual to the simplices of t in M and the cells of the cellular decomposition $(\partial t)^*$ of ∂M dual to ∂t. The 2-skeleton $F = (t^*)^{(2)} \subset M$ of t^* is a special 2-polyhedron such that $F \supset \partial M$ and $M \setminus F$ is a disjoint union of open 3-balls. All regions of F are embedded disks. The regions of F lying in ∂M inherit the orientation of ∂M. All other regions of F are oriented arbitrarily. Pushing F slightly into $\mathrm{Int}(M)$ we obtain a frame of M. □

11.5.3 From frames to skeletons

Consider an oriented graph $G \subset \partial M$ with all vertices of valence ≥ 2. We derive skeletons of (M, G) from a frame F of M. An F-collar $i\colon \partial M \times [0,1] \hookrightarrow M$ is said to be *G-generic* if the vertices of the graph $G_i = i(G \times \{1\})$ (which is a copy of G in ∂F) miss the graph $\Gamma_F = \partial F \cap F^{(1)}$ and the edges of G_i have only double transversal intersections with the edges of Γ_F. These intersections split the edges of G_i and Γ_F into smaller subedges. Consider the 2-polyhedron

$$P = P(F, i) = i(G \times [0,1]) \cup F \subset M.$$

We stratify P by taking as edges the above-mentioned subedges of G_i and Γ_F, the edges of F not lying in ∂F, the edges of G, and the arcs $i(v \times [0,1])$ where v runs over the vertices of G. The vertices of P are the endpoints of the edges. We orient P: the regions contained in F inherit their orientation from F, and the regions lying in $i(G \times [0,1])$ are oriented so that $\partial P = G$ as oriented graphs. Clearly, P is a skeleton of (M, G). We say that this skeleton is *associated* with F. Note that any F-collar $\partial M \times [0,1] \hookrightarrow M$ can be made G-generic by a small isotopy.

Lemma 11.7. *Any two skeletons of (M, G) associated with a frame F of M are related by primary moves.*

Proof. By Section 11.5.2, any two G-generic F-collars i_0, i_1 may be related by an isotopy $\{i_s\}_{s \in [0,1]}$ in the class of F-collars. We can assume the isotopy to be "generic" so that i_s is G-generic for all but a finite set of s where one of the G-genericity conditions does not hold in one point. When s varies in the set of non-exceptional values, the skeleton $P(F, i_s)$ is modified via an ambient isotopy in M. For an exceptional s, either a vertex of the graph $i_s(G \times \{1\})$ is also a vertex of the graph Γ_F, or a vertex of one of these graphs lies in the interior of an edge of the other graph, or these two graphs are tangent in a common interior point of edges. The first case does not occur in a generic isotopy. In the second case, the skeleton $P(F, i_s)$ is modified near the exceptional s via a move of type $T^{m,n}$. In the third case, $P(F, i_s)$ is modified near the exceptional s via a lune move $\mathcal{L}^{\pm 1}$. Orientations of the regions never obstruct these moves because all regions of $P(F, i_s)$ contained in ∂F are oriented coherently. We obtain thus a sequence of primary moves $P(F, i_0) \rightsquigarrow P(F, i_1)$. □

11.5.4 Frame moves

By *frame moves* on a frame F of M, we mean the MP-moves defined as in Section 11.4.1 and applied to F inside closed 3-balls in $\mathrm{Int}(M)$ which are either disjoint from the surface ∂F or meet this surface in a subset of a disk in ∂F. A frame move is allowed only when both the starting and the resulting 2-polyhedra are frames of M. As usual, all our moves must preserve orientation of the "big" regions in the pictures while the orientation of the small disk regions created or destroyed under the moves may be arbitrary. The move $\mathcal{L}^{-1}$ is allowed only when the orientations of two regions united under this move are compatible. The orientation of F can obstruct an application of $\mathcal{L}^{-1}$ but cannot obstruct an application of other frame moves.

Lemma 11.8. *Any two frames of M can be related by frame moves.*

Proof. Fix a triangulation t of M and the induced triangulation ∂t of ∂M. Let $Y \subset \partial M$ be the 1-skeleton of the cellular decomposition $(\partial t)^*$ of ∂M dual to ∂t. We provide the trivalent graph Y with an arbitrary orientation. By a Y-skeleton, we shall mean a skeleton P of the pair (M, Y) whose only vertices are the vertices of $\partial P = Y$ and the special points of P (cf. Section 11.1.5). We claim that any two Y-skeletons can be related by MP-moves. The proof repeats the proof of Lemma 11.2 with the following changes. Instead of [Mat, Theorem 1.2.30] we use a straightforward extension to Y-skeletons which implies that any Y-skeletons can be related by the moves $(T_0'')^{\pm 1}, T^{1,2}, T^{2,1}$ up to the choice of orientation. The role of a regular skeleton P in the proof of Lemma 11.2 is played by a skeleton of (M, Y) associated with the frame of M derived from t as in the proof of Lemma 11.6. This skeleton is a Y-skeleton, and all its regions are embedded disks. This allows us to apply the same orientation-reversing argument as in the proof of Lemma 11.2.

Any Y-skeleton P determines an oriented special 2-polyhedron $P \cup \partial M$. Pushing it slightly into $\mathrm{Int}(M)$, we obtain a frame of M denoted P_+. The frames obtained in this way from Y-skeletons are called Y-frames. Since any two Y-skeletons can be related by MP-moves in the class of Y-skeletons, any two Y-frames can be related by frame moves in the class of Y-frames.

To prove the lemma, it remains to show that any frame F of M can be obtained by frame moves from a Y-frame. Pick a Y-generic F-collar $i \colon \partial M \times [0,1] \hookrightarrow M$. The associated skeleton $P = P(F, i)$ of (M, Y) is a Y-skeleton. Consider the Y-frame

$$P_+ = i(\partial M \times \{1/2\}) \cup i(Y \times [1/2, 1]) \cup F.$$

The regions of P_+ contained in the surface $i(\partial M \times \{1/2\})$ are the 2-cells of the cellular decomposition of this surface induced by the cellular decomposition $(\partial t)^*$ of ∂M. The boundary circle of such a region of P_+ can be pushed by frame moves towards $\partial F \subset F$ (along $i(Y \times [1/2, 1])$) and further contracted in a position allowing us to apply the move $(T_0'')^{-1}$ eliminating this region. Eliminating consecutively all regions of P_+ contained in $i(\partial M \times \{1/2\})$, we obtain a sequence of frame moves $P_+ \rightsquigarrow F$. □

11.5.5 Proof of Theorem 11.5

Consider any frames F, F' of M related by a frame move modifying F in a 3-ball $B \subset M$. By the definition of a frame move, $B \cap \partial F$ is a subset of a disk in ∂F. We can pick a G-generic F-collar i so that the graph $G_i = i(G \times \{1\})$ does not meet this disk. Then i induces a G-generic F'-collar i' with $G_i = G_{i'}$, and the skeletons $P(F, i)$ and $P(F', i')$ are related by the same move. Since all MP-moves on skeletons are products of primary moves, the skeletons $P(F, i)$ and $P(F', i')$ are related by primary moves. From this and Lemmas 11.7, 11.8 we conclude that all skeletons of (M, G) associated with frames are related to each other by primary moves.

To finish the proof of the theorem, we need only to show that every skeleton P of (M, G) can be transformed by primary moves into a skeleton of (M, G) associated with a frame of M. We will say that a vertex/edge/region of P is *internal* if it lies in $\mathrm{Int}(M)$. Adding, if necessary, vertices to P by T_2^{-1}, we can ensure that all edges of P not lying in ∂P have at least one internal endpoint. Adding, if necessary, vertices and edges to P by T_2^{-1}, T_1, we can additionally ensure that all regions of P are disks. Now we blow up – in three steps – the internal part of P. First, we blow up all internal vertices x of P as in the proof of Lemma 11.4, see Figure 11.8. This gives a skeleton P_1 of (M, G) and can be achieved by primary moves. Here we need to be more specific about the orientation of the regions of P_1 lying in the 2-spheres $S_x = \partial B_x$: we endow them with orientation induced by that of M restricted to the 3-ball B_x. By the definition of a skeleton of (M, G), each vertex u of the graph $G = \partial P = \partial P_1$ is adjacent to a unique edge d_u of P_1 not lying in G. This edge connects u to an internal vertex of P_1 denoted by $\overline{u}$; by construction, $\overline{u} \in S_x$ for some internal vertex x of P. Besides the edges of G and the edges $\{d_u\}_u$, the skeleton P_1 has edges of two types:

- the "short" trivalent edges lying on the spheres $\{S_x\}_x$;
- the "long" edges obtained from the internal edges of P by removing small open neighborhoods of their endpoints.

Next, we blow up all long edges e of P_1 as in the proof of Lemma 11.4, see Figure 11.9. This gives a skeleton P_2 of (M, G) and can be achieved by primary moves. Here we equip the regions of P_2 lying in the 2-sphere $S_e = \partial(D_e \times [0, 1])$ with orientation induced by that of M restricted to the solid cylinder $D_e \times [0, 1]$. The edges $\{d_u\}_u$ of P_1 are disjoint from the long edges of P_1 and therefore survive as edges of P_2. All internal vertices of P_2 other than the vertices $\{\overline{u}\}_u$ are special. The skeleton P_2 has regions of three types:

- the "boundary" regions, i.e., the regions adjacent to edges of $G = \partial P_2$;
- the "small" regions created by the transformations $P \rightsquigarrow P_1 \rightsquigarrow P_2$ and lying in the spheres $\{S_x\}_x$ and $\{S_e\}_e$ contained in P_2;
- the "wide" regions obtained from the internal regions of P by cutting out small regular neighborhoods of their boundaries.

Since all regions of P are disks, so are all wide regions of P_2. Observe that a vertex or an edge of P_2 may have at most one branch lying in a wide region. This implies that all wide regions of P_2 embed in P_2 and are pairwise disjoint. A wide region r of P_2 has only special vertices and trivalent edges. As in the proof of Lemma 11.2, we can "double" r, i.e., add to P_2 a new region r_- parallel to r. This is done so that the orientations of r and r_- are induced by that of M restricted to the solid cylinder $r \times [0, 1]$ between r and r_-. This transformation can be achieved by primary moves, cf. the proof of Lemma 11.2. Doubling in this way all wide regions of P_2 we obtain a skeleton P_3 of (M, G). We claim that P_3 is associated with a frame of M. This will imply the theorem.

By the definition of a skeleton of (M, G), for each edge a of G, there is a unique boundary region D_a of P_2 such that $D_a \cap \partial P = a$. The region D_a is obtained from the region of P adjacent to a by cutting out a small regular neighborhood of internal vertices and internal edges in the boundary. Since all regions of P are disks, D_a is a disk. As above, a vertex or an edge of P_2 cannot be adjacent both to a branch lying in a wide region and to a branch lying in a boundary region. Consecutively, the disk D_a does not meet the wide regions of P_2 and survives as a region of P_3. The edges $\{d_u\}_u$ of P_2 also survive in P_3, as well as their internal endpoints $\{\overline{u}\}_u$. If v, w are the endpoints of a, then the loop ∂D_a runs from $\overline{v}$ to v along d_v, then from v to w along a, then from w to $\overline{w}$ along d_w, and then from $\overline{w}$ to $\overline{v}$ along a path $\overline{a}$ formed by internal edges of P_3. For a vertex u of G of valence $n \geq 2$, the vertex $\overline{u}$ of P_3 is adjacent to n boundary regions (counted with multiplicity). An internal vertex of P_3 not belonging to the set $\{\overline{u}\}_u$ may have at most one branch in P_3 lying in a boundary region. These observations imply that if an edge a of G is not a loop, then $\overline{a}$ is an injective path, and if a is a loop, then $\overline{a}$ is an injective loop. The edges of P_3 forming $\overline{a}$ are trivalent, and for each of them, the orientations of the two adjacent regions of P_3 other than D_a are compatible. Consider the 2-polyhedron $F \subset P_3 \cap \text{Int}(M)$ obtained from P_3 by removing $G = \partial P_3$, the interiors of the edges $\{d_u\}_u$, and the interiors of the regions $\{D_a\}_a$. The 2-polyhedron F can be stratified so that its edges are formed from the edges of P_3 not lying on the paths $\{p_a\}_a$. The regions of F are unions of regions of P_3 and inherit orientation from P_3. This turns F into an oriented stratified 2-polyhedron. Since P_3 is a skeleton of (M, G), the polyhedron F is a skeleton of M. Let $\widehat{F} \subset \text{Int}(M)$ be the union of F with the 3-ball components of $M \setminus F$. By construction, $\widehat{F}$ is nothing but the union of all 3-ball components of $M \setminus P$ and a regular neighborhood in M of the 2-polyhedron formed by the internal vertices, edges, and regions of P. Thus, $\widehat{F}$ is a 3-manifold with boundary. By our choice of orientations in the construction of P_1, P_2, P_3, the orientation of all regions of F contained in $\partial F = \partial \widehat{F}$ is induced by the orientation of M restricted to $\widehat{F}$. The 2-polyhedron F satisfies all requirements on a frame. It is clear that there is a G-generic F-collar $i \colon \partial M \times [0,1] \hookrightarrow M$ such that $i(\{u\} \times [0,1]) = d_u$ for all vertices u of G and $i(a \times [0,1]) = D_a$ for all edges a of G. Then $P_3 = P(F, i)$.

Chapter 12

Multiplicity modules and colored graphs

In this chapter, we associate with each pivotal $\Bbbk$-category a family of modules called multiplicity modules. Then we introduce an invariant of colored graphs which takes values in tensor products of multiplicity modules. This invariant will be our main tool in the construction of a 3-dimensional TQFT in the next chapter.

12.1 Permutation maps and multiplicity modules

In this section, $\mathcal{C}$ is a pivotal $\Bbbk$-category.

12.1.1 Permutation maps

For $X, Y \in \mathrm{Ob}(\mathcal{C})$, the *permutation map*

$$\pi_{X,Y} \colon \mathrm{Hom}_{\mathcal{C}}(\mathbb{1}, X \otimes Y) \to \mathrm{Hom}_{\mathcal{C}}(\mathbb{1}, Y \otimes X)$$

carries any $\alpha \in \mathrm{Hom}_{\mathcal{C}}(\mathbb{1}, X \otimes Y)$ to

$\pi_{X,Y}(\alpha) =$ 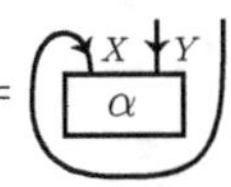

$= (\mathrm{ev}_X \otimes \mathrm{id}_{Y \otimes X})(\mathrm{id}_{X^*} \otimes \alpha \otimes \mathrm{id}_X)\, \widetilde{\mathrm{coev}}_X.$

Using the isotopy invariance of the graphical calculus (see Theorem 2.6), we get

$\pi_{X,Y}(\alpha) =$ 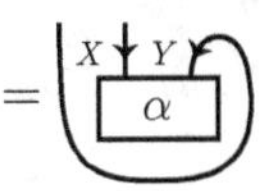

$= (\mathrm{id}_{Y \otimes X} \otimes \widetilde{\mathrm{ev}}_Y)(\mathrm{id}_Y \otimes \alpha \otimes \mathrm{id}_{Y^*})\, \mathrm{coev}_Y.$

It is easy to check that each map $\pi_{X,Y}$ is $\Bbbk$-linear and the family $\{\pi_{X,Y}\}_{X,Y \in \mathrm{Ob}(\mathcal{C})}$ is natural in the sense that

$$(g \otimes f)\pi_{X,Y}(\alpha) = \pi_{X',Y'}((f \otimes g)\alpha)$$

for any morphisms $f\colon X \to X'$, $g\colon Y \to Y'$, and $\alpha\colon \mathbb{1} \to X \otimes Y$ in $\mathcal{C}$.

The next lemma shows that the permutation maps are $\Bbbk$-linear isomorphisms and behave as permutations.

Lemma 12.1. *For any $X, Y, Z \in \mathrm{Ob}(\mathcal{C})$,*

(a) *$\pi_{X,Y}$ is a $\Bbbk$-linear isomorphism and $\pi_{X,Y}^{-1} = \pi_{Y,X}$;*

(b) *$\pi_{X,\mathbb{1}} = \pi_{\mathbb{1},X} = \mathrm{id}_{\mathrm{Hom}_{\mathcal{C}}(\mathbb{1},X)}$;*

(c) *$\pi_{X\otimes Y,Z} = \pi_{Y,Z\otimes X}\,\pi_{X,Y\otimes Z}$ and $\pi_{X,Y\otimes Z} = \pi_{Z\otimes X,Y}\,\pi_{X\otimes Y,Z}$.*

Proof. Claims (b) and (c) are direct consequences of the monoidality of the dual functor $?^*\colon \mathcal{C}^{\mathrm{rev}} \to \mathcal{C}$, see Section 1.7.1. From (b) and (c), we obtain that

$$\pi_{Y,X}\,\pi_{X,Y} = \pi_{X\otimes Y,\mathbb{1}} = \mathrm{id}_{\mathrm{Hom}_{\mathcal{C}}(\mathbb{1},X\otimes Y)},$$

hence (a). □

12.1.2 Cyclic sets

By Section 2.2.1, a tuple $S = ((X_1,\varepsilon_1),\ldots,(X_n,\varepsilon_n))$ of signed objects of $\mathcal{C}$ determines an object $X_S = X_1^{\varepsilon_1} \otimes \cdots \otimes X_n^{\varepsilon_n}$ of $\mathcal{C}$. The $\Bbbk$-module $\mathrm{Hom}_{\mathcal{C}}(\mathbb{1}, X_S)$ is preserved under cyclic permutations of S up to the permutation isomorphisms. To formalize this situation, we introduce the notion of a cyclic $\mathcal{C}$-set.

A *cyclic $\mathcal{C}$-set* is a triple (E, c, ε) consisting of a nonempty finite set E endowed with a cyclic order and two maps $c\colon E \to \mathrm{Ob}(\mathcal{C})$ and $\varepsilon\colon E \to \{+,-\}$. In other words, a cyclic $\mathcal{C}$-set is a nonempty cyclically ordered finite set whose elements are equipped with a signed object of $\mathcal{C}$. For shortness, we will often write E for (E, c, ε).

An *isomorphism* between two cyclic $\mathcal{C}$-sets E and E' is a bijection $E \to E'$ preserving the cyclic order and commuting with the maps to $\mathrm{Ob}(\mathcal{C})$ and $\{+,-\}$. More generally, a *weak isomorphism* between cyclic $\mathcal{C}$-sets (E, c, ε) and (E', c', ε') is a pair $\phi = (\rho, \varphi)$ consisting of a bijection $\rho\colon E \to E'$ preserving the cyclic order and a family of isomorphisms in $\mathcal{C}$

$$\varphi = \{\varphi_e\colon c(e)^{\varepsilon(e)} \to c'(\rho(e))^{\varepsilon'(\rho(e))}\}_{e\in E}.$$

12.1.3 Multiplicity modules

Each cyclic $\mathcal{C}$-set $E = (E, c, \varepsilon)$ determines a $\Bbbk$-module $H(E)$ called the *multiplicity module* of E. For $e \in E$, set

$$H_e = \mathrm{Hom}_{\mathcal{C}}(\mathbb{1}, c(e_1)^{\varepsilon(e_1)} \otimes \cdots \otimes c(e_n)^{\varepsilon(e_n)}),$$

where n is the number of elements of E and $e = e_1 < e_2 < \cdots < e_n$ are the elements of E ordered in the given cyclic order starting from e. If $f \in E \setminus \{e\}$, then $f = e_k$ for some $k \in \{2,\ldots,n\}$. Set

$$[ef) = c(e_1)^{\varepsilon(e_1)} \otimes \cdots \otimes c(e_{k-1})^{\varepsilon(e_{k-1})}$$

and

$$[fe) = c(e_k)^{\varepsilon(e_k)} \otimes \cdots \otimes c(e_n)^{\varepsilon(e_n)}.$$

Clearly,

$$H_e = \mathrm{Hom}_{\mathcal{C}}(\mathbb{1}, [ef) \otimes [fe)) \quad \text{and} \quad H_f = \mathrm{Hom}_{\mathcal{C}}(\mathbb{1}, [fe) \otimes [ef)).$$

Set $p_{e,e} = \mathrm{id}_{H_e} \colon H_e \to H_e$ and

$$p_{e,f} = \pi_{[ef),[fe)} \colon H_e \to H_f.$$

Lemma 12.1 implies that $p_{e,f}$ is a $\Bbbk$-linear isomorphism and $p_{f,g}\, p_{e,f} = p_{e,g}$ for all $e, f, g \in E$. Thus the family $(\{H_e\}_{e\in E}, \{p_{e,f}\}_{e,f\in E})$ is a projective system of $\Bbbk$-modules and $\Bbbk$-linear isomorphisms. The multiplicity module $H(E)$ is the projective limit of this system:

$$H(E) = \varprojlim H_e\,.$$

The module $H(E)$ depends only on $E = (E, c, \varepsilon)$ and is equipped with a family of $\Bbbk$-linear isomorphisms $\{\tau_e \colon H(E) \to H_e\}_{e\in E}$ such that $p_{e,f}\, \tau_e = \tau_f$ for all $e, f \in E$. We call τ_e the *cone isomorphism* and the family $\{\tau_e\}_{e\in E}$ the *universal cone*. Recall that $H(E)$ may be realized as the following submodule of the $\Bbbk$-module $\prod_{e\in E} H_e$:

$$H(E) = \Big\{ (x_e)_{e\in E} \in \prod_{e\in E} H_e \;\Big|\; p_{e,f}(x_e) = x_f \text{ for all } e, f \in E \Big\}.$$

The cone isomorphism $\tau_e \colon H(E) \to H_e$ associated with this realization is given by $\tau_e((x_f)_{f\in E}) = x_e$ and its inverse is computed by $\tau_e^{-1}(v) = (p_{e,f}(v))_{f\in E}$.

By Section 4.3.2, if the category $\mathcal{C}$ is non-degenerate, then the $\Bbbk$-module H_e is projective of finite type for all $e \in E$, and thus so is the multiplicity module $H(E)$. By Section 4.4.1, if $\mathcal{C}$ is pre-fusion, then all these $\Bbbk$-modules are free of finite type.

An isomorphism between two cyclic $\mathcal{C}$-sets E and E' induces a $\Bbbk$-linear isomorphism $H(E) \to H(E')$ in the obvious way. More generally, a weak isomorphism $\phi = (\rho, \varphi)$ between cyclic $\mathcal{C}$-sets E and E' induces $\Bbbk$-linear isomorphisms $\{H_e \to H_{\rho(e)}\}_{e\in E}$ which commute with the maps $p_{e,f}$ as above. These isomorphisms induce a $\Bbbk$-linear isomorphism $H(E) \to H(E')$ denoted $H(\phi)$.

12.1.4 Example

Any tuple $S = ((X_1, \varepsilon_1), \ldots, (X_n, \varepsilon_n))$ of $n \geq 1$ signed objects of $\mathcal{C}$ determines a cyclic $\mathcal{C}$-set $E_S = \{1, 2, \ldots, n\}$, where $1 < 2 < \cdots < n < 1$ and each $e \in E_S$ is labeled with the signed object (X_e, ε_e). In this case, the cone isomorphism τ_1 is a $\Bbbk$-linear isomorphism $H(E_S) \simeq \mathrm{Hom}_{\mathcal{C}}(\mathbb{1}, X_S)$.

12.2 An invariant of colored graphs

In this section, $\mathcal{C}$ is a pivotal $\Bbbk$-category. We define an invariant $\mathbb{F}_\mathcal{C}$ for $\mathcal{C}$-colored graphs in $\mathbb{R}^2$. When $\mathcal{C}$ is spherical, we extend $\mathbb{F}_\mathcal{C}$ to $\mathcal{C}$-colored graphs in the 2-sphere.

12.2.1 Colored graphs

Recall the notion of a graph from Section 11.1.1. A graph G is *$\mathcal{C}$-colored* if each edge of G is oriented and endowed with an object of $\mathcal{C}$ called the *color* of this edge. Given an oriented surface Σ, by a *$\mathcal{C}$-colored graph in* Σ, we mean a $\mathcal{C}$-colored graph embedded in Σ. A vertex v of a $\mathcal{C}$-colored graph G in Σ determines a cyclic $\mathcal{C}$-set $(E_v, c_v, \varepsilon_v)$ as follows: E_v is the set of half-edges of G incident to v with cyclic order induced by the opposite orientation of Σ, the map $c_v\colon E_v \to \mathrm{Ob}(\mathcal{C})$ assigns to a half-edge $e \in E_v$ the color of the edge of G containing e, and the map $\varepsilon_v\colon E_v \to \{+,-\}$ assigns to $e \in E_v$ the sign $+$ if e is oriented towards v and $-$ otherwise. Let $H_v(G) = H(E_v)$ be the multiplicity module of E_v. Set

$$H(G) = \otimes_v H_v(G),$$

where v runs over all vertices of G and $\otimes$ is the unordered tensor product of $\Bbbk$-modules (see Appendix E). To emphasize the role of Σ, we sometimes write $H_v(G;\Sigma)$ for $H_v(G)$ and $H(G;\Sigma)$ for $H(G)$. By definition, for $G = \emptyset$, we have $H(G) = \Bbbk$.

For a vertex v of G, the $\Bbbk$-module $H_v(G)$ can be described as follows. Let $n \geq 1$ be the valence of v and let $e_1 < e_2 < \cdots < e_n < e_1$ be the half-edges of G incident to v with cyclic order induced by the opposite orientation of Σ. Let $X_r = c_v(e_r)$ and $\varepsilon_r = \varepsilon_v(e_r)$ be the color and the sign of e_r. Then we have the cone isomorphism

$$\tau^v_{e_1}\colon H_v(G) \xrightarrow{\simeq} \mathrm{Hom}_\mathcal{C}(\mathbb{1}, X_1^{\varepsilon_1} \otimes \cdots \otimes X_n^{\varepsilon_n}).$$

By the definition of $H_v(G)$, the cone isomorphisms determined by different elements of E_v are related via composition with the permutation maps, see Section 12.1. For example, a trivalent vertex

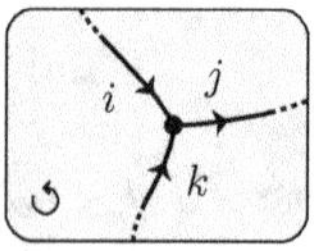

with $i,j,k \in \mathrm{Ob}(\mathcal{C})$ gives rise to the $\Bbbk$-module $H_v(G)$ isomorphic to the $\Bbbk$-modules

$$\mathrm{Hom}_\mathcal{C}(\mathbb{1}, i \otimes j^* \otimes k) \simeq \mathrm{Hom}_\mathcal{C}(\mathbb{1}, j^* \otimes k \otimes i) \simeq \mathrm{Hom}_\mathcal{C}(\mathbb{1}, k \otimes i \otimes j^*).$$

Clearly, for any disjoint $\mathcal{C}$-colored graphs G_1 and G_2 in Σ, there is a canonical $\Bbbk$-linear isomorphism

$$H(G_1 \sqcup G_2) \simeq H(G_1) \otimes H(G_2).$$

Any orientation preserving embedding f of Σ into an oriented surface Σ' carries a $\mathcal{C}$-colored graph $G \subset \Sigma$ into a $\mathcal{C}$-colored graph $G' = f(G) \subset \Sigma'$ preserving the vertices, the edges, and the orientations and colors of the edges. The map f induces a $\Bbbk$-linear isomorphism $H(f)\colon H(G;\Sigma) \to H(G';\Sigma')$ in the obvious way. This applies, in particular, when f is an orientation preserving self-homeomorphism of Σ.

Given $\mathcal{C}$-colored graphs G and G' in Σ, by an *isotopy* of G to G', we mean an ambient isotopy of G to G' in Σ preserving the vertices, the edges, and the orientations and colors of the edges. Such an isotopy ι induces an orientation preserving homeomorphism $\Sigma \to \Sigma$ carrying G to G'. This homeomorphism induces a $\Bbbk$-linear isomorphism $H(G) \to H(G')$ denoted $H(\iota)$.

12.2.2 The invariant $\mathbb{F}_\mathcal{C}$

We always orient the plane $\mathbb{R}^2$ counterclockwise. Let G be a $\mathcal{C}$-colored graph in $\mathbb{R}^2$. For each vertex v of G, pick a half-edge $e_v \in E_v$ and deform G near v so that the half-edges incident to v lie above v with respect to the second coordinate on $\mathbb{R}^2$ and e_v is the leftmost of them. Pick any $\alpha_v \in H_v(G)$ and replace v by a box colored with $\tau^v_{e_v}(\alpha_v)$, where τ^v is the universal cone of $H_v(G)$, see Figure 12.1. This transforms G into a $\mathcal{C}$-colored Penrose diagram without free ends. Let $\mathbb{F}_\mathcal{C}(G)(\otimes_v \alpha_v) \in \mathrm{End}_\mathcal{C}(\mathbb{1})$ be the associated morphism computed via the Penrose graphical calculus. This extends by linearity to a $\Bbbk$-linear homomorphism

$$\mathbb{F}_\mathcal{C}(G)\colon H(G) = \otimes_v H_v(G) \to \mathrm{End}_\mathcal{C}(\mathbb{1}).$$

By definition, for $G = \emptyset$, the map $\mathbb{F}_\mathcal{C}(G)\colon H(G) = \Bbbk \to \mathrm{End}_\mathcal{C}(\mathbb{1})$ is the $\Bbbk$-linear homomorphism carrying $1_\Bbbk$ to $\mathrm{id}_\mathbb{1}$.

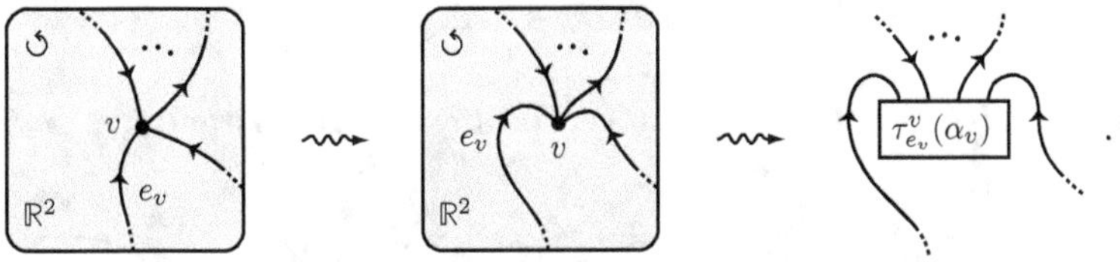

Figure 12.1: Transformation of a vertex

Lemma 12.2. *The homomorphism $\mathbb{F}_\mathcal{C}(G)\colon H(G) \to \mathrm{End}_\mathcal{C}(\mathbb{1})$ is a well-defined isotopy invariant of the $\mathcal{C}$-colored graph G in $\mathbb{R}^2$. More precisely, for any isotopy ι between $\mathcal{C}$-colored graphs G and G' in $\mathbb{R}^2$, we have $\mathbb{F}_\mathcal{C}(G')\,H(\iota) = \mathbb{F}_\mathcal{C}(G)$, where $H(\iota)\colon H(G) \to H(G')$ is the $\Bbbk$-linear isomorphism induced by ι.*

Proof. Independence of $\mathbb{F}_\mathcal{C}(G)$ of the choice of the half-edges e_v follows from the definition of $H_v(G)$. Invariance under isotopies follows from Theorem 2.6. □

12.2.3 Example

Consider the following $\mathcal{C}$-colored graph in $\mathbb{R}^2$ with two vertices u, v and two edges colored by $X, Y \in \mathrm{Ob}(\mathcal{C})$:

$G =$ 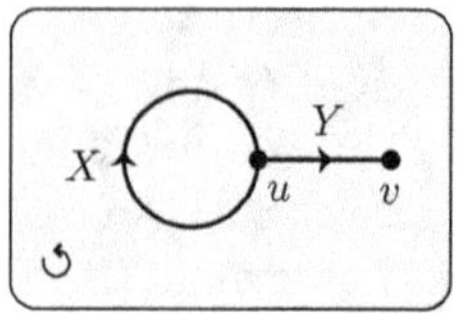 .

There are 3 half-edges e_1, e_2, e_3 incident to u and one half-edge e incident to v:

The cyclic order on $E_u = \{e_1, e_2, e_3\}$ is $e_1 < e_2 < e_3 < e_1$. The cone isomorphisms associated with the half-edges are:

$$\begin{aligned}
\tau^u_{e_1}&\colon H_u(G) \to \mathrm{Hom}_{\mathcal{C}}(\mathbb{1}, X^* \otimes X \otimes Y^*),\\
\tau^u_{e_2}&\colon H_u(G) \to \mathrm{Hom}_{\mathcal{C}}(\mathbb{1}, X \otimes Y^* \otimes X^*),\\
\tau^u_{e_3}&\colon H_u(G) \to \mathrm{Hom}_{\mathcal{C}}(\mathbb{1}, Y^* \otimes X^* \otimes X),\\
\tau^v_{e}&\colon H_v(G) \to \mathrm{Hom}_{\mathcal{C}}(\mathbb{1}, Y).
\end{aligned}$$

These isomorphisms are related to each other via composition with the permutation maps. For instance, for any $\alpha \in H_u(G)$,

$$\tau^u_{e_2}(\alpha) = \pi_{X^*, X\otimes Y^*}\, \tau^u_{e_1}(\alpha) =$$ 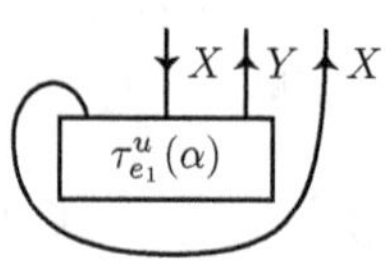 .

By definition, $H(G) = H_u(G) \otimes H_v(G)$. For any $\alpha \in H_u(G)$ and $\beta \in H_v(G)$,

$$\mathbb{F}_{\mathcal{C}}(G)(\alpha \otimes \beta) =$$ 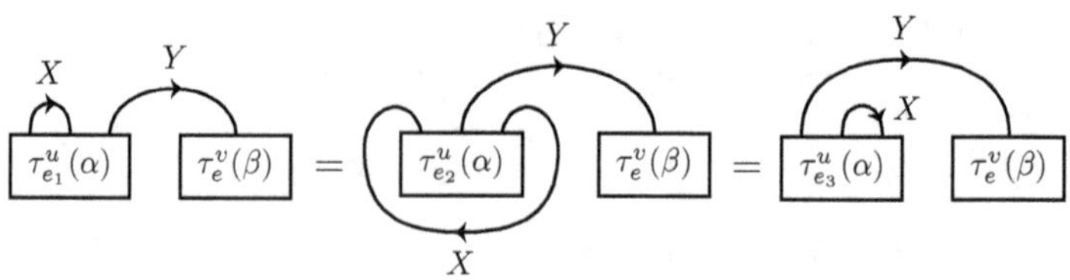 .

12.2.4 Properties of $\mathbb{F}_{\mathcal{C}}$

We state several properties of the invariant $\mathbb{F}_{\mathcal{C}}$ of $\mathcal{C}$-colored graphs in $\mathbb{R}^2$.

(i) Let G' be the $\mathcal{C}$-colored graph in $\mathbb{R}^2$ obtained from a $\mathcal{C}$-colored graph $G \subset \mathbb{R}^2$ by replacing the color X of an edge e by an isomorphic object X' of $\mathcal{C}$. Any isomorphism $X' \simeq X$ induces weak isomorphisms (see Section 12.1.2) between the cyclic $\mathcal{C}$-sets associated with the endpoints of e in G and G', and the latter induce a $\Bbbk$-linear isomorphism $\Phi\colon H(G') \to H(G)$. Then

$$\mathbb{F}_{\mathcal{C}}(G') = \mathbb{F}_{\mathcal{C}}(G)\,\Phi.$$

We call this property the *naturality* of $\mathbb{F}_{\mathcal{C}}$.

(ii) Let G' be the $\mathcal{C}$-colored graph in $\mathbb{R}^2$ obtained from a $\mathcal{C}$-colored graph G in $\mathbb{R}^2$ by reversing the orientation of an edge e and replacing the color of e by the dual object of $\mathcal{C}$. The pivotal structure of $\mathcal{C}$ (see Section 2.5) induces weak isomorphisms between the cyclic $\mathcal{C}$-sets associated with the endpoints of e in G and G'. These weak isomorphisms induce a $\Bbbk$-linear isomorphism $\Psi\colon H(G') \to H(G)$ and

$$\mathbb{F}_{\mathcal{C}}(G') = \mathbb{F}_{\mathcal{C}}(G)\,\Psi.$$

(iii) If an edge e of a $\mathcal{C}$-colored graph G in $\mathbb{R}^2$ is colored with $\mathbb{1}$ and the endpoints of e are also endpoints of other edges of G, then $G' = G \setminus \mathrm{Int}(e) \subset \mathbb{R}^2$ inherits from G the structure of a $\mathcal{C}$-colored graph, there is a canonical $\Bbbk$-linear isomorphism $\Delta\colon H(G') \to H(G)$, and

$$\mathbb{F}_{\mathcal{C}}(G') = \mathbb{F}_{\mathcal{C}}(G)\,\Delta.$$

Indeed, by the Penrose calculus, an edge colored with $\mathbb{1}$ can be deleted without changing the associated morphism.

(iv) If G, G' are disjoint $\mathcal{C}$-colored graphs in $\mathbb{R}^2$ lying on different sides of a straight line, then

$$\mathbb{F}_{\mathcal{C}}(G \amalg G') = \mu\big(\mathbb{F}_{\mathcal{C}}(G) \otimes \mathbb{F}_{\mathcal{C}}(G')\big)\Theta,$$

where $\Theta\colon H(G \amalg G') \to H(G) \otimes H(G')$ is the canonical isomorphism and μ is multiplication in $\mathrm{End}_{\mathcal{C}}(\mathbb{1})$. We call this property the $\otimes$-*multiplicativity* of $\mathbb{F}_{\mathcal{C}}$.

(v) If $\mathcal{C}$ is pure, then

$\mathbb{F}_{\mathcal{C}}$ 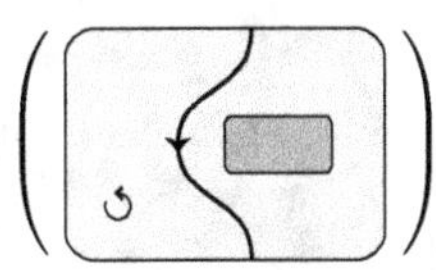 $= \mathbb{F}_{\mathcal{C}}$ 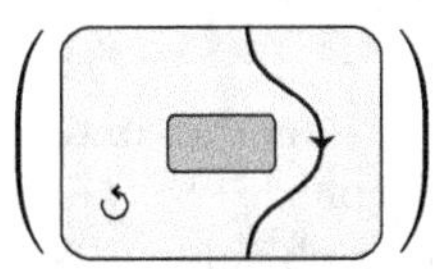

where the $\mathcal{C}$-colored graphs on the left and on the right coincide outside the big rectangles and the small rectangles on both sides stand for the same $\mathcal{C}$-colored graph.

(vi) Let $F\colon \mathcal{C} \to \mathcal{C}'$ be a pivotal $\Bbbk$-linear functor between pivotal $\Bbbk$-categories. Let G be a $\mathcal{C}$-colored graph in $\mathbb{R}^2$ and G' be the $\mathcal{C}'$-colored graph in $\mathbb{R}^2$ obtained from G by applying F to the colors of the edges. Lemma 2.8 implies that the following diagram of $\Bbbk$-linear homomorphisms commutes:

$$\begin{array}{ccc} H_{\mathcal{C}}(G) & \xrightarrow{\ \mathbb{F}_{\mathcal{C}}(G)\ } & \mathrm{End}_{\mathcal{C}}(\mathbb{1}) \\ \downarrow & & \downarrow \\ H_{\mathcal{C}'}(G') & \xrightarrow[\ \mathbb{F}_{\mathcal{C}'}(G')\ ]{} & \mathrm{End}_{\mathcal{C}'}(\mathbb{1}). \end{array}$$

Here the left vertical arrow is induced by F in the obvious way and the right vertical arrow represents the $\Bbbk$-algebra homomorphism carrying any $\alpha \in \mathrm{End}_{\mathcal{C}}(\mathbb{1})$ to $F_0^{-1}F(\alpha)F_0 \in \mathrm{End}_{\mathcal{C}'}(\mathbb{1})$. Note that if the functor F is fully faithful (e.g., if F is an equivalence), then both vertical arrows are isomorphisms.

12.2.5 The case of a spherical category

Suppose that the pivotal $\Bbbk$-category $\mathcal{C}$ is spherical. Then the invariant $\mathbb{F}_{\mathcal{C}}$ of $\mathcal{C}$-colored graphs in $\mathbb{R}^2$ extends uniquely to an isotopy invariant of $\mathcal{C}$-colored graphs in the 2-sphere $S^2 = \mathbb{R}^2 \cup \{\infty\}$ endowed with the orientation extending the counterclockwise orientation in $\mathbb{R}^2$. Indeed, consider a $\mathcal{C}$-colored graph G in S^2. Pushing G away from ∞ by an isotopy, we obtain a $\mathcal{C}$-colored graph G_0 in $\mathbb{R}^2$. The isotopy induces a $\Bbbk$-linear isomorphism $H(G; S^2) \simeq H(G_0; \mathbb{R}^2)$. Composing with $\mathbb{F}_{\mathcal{C}}(G_0) \colon H(G_0; \mathbb{R}^2) \to \mathrm{End}_{\mathcal{C}}(\mathbb{1})$ we obtain a $\Bbbk$-linear homomorphism

$$\mathbb{F}_{\mathcal{C}}(G) \colon H(G; S^2) \to \mathrm{End}_{\mathcal{C}}(\mathbb{1}).$$

Lemmas 2.9 and 12.2 imply that $\mathbb{F}_{\mathcal{C}}(G)$ does not depend on the way we push G in $\mathbb{R}^2$ and is an isotopy invariant of G. The properties of $\mathbb{F}_{\mathcal{C}}$ formulated in Section 12.2.4 extend to $\mathcal{C}$-colored graphs in S^2 in the obvious way. The condition in (iv) involving a straight line may be dropped here because for any pair of disjoint connected graphs in S^2, there is an isotopy of S^2 in itself carrying these graphs into new positions in $\mathbb{R}^2$ separated by a line. Similarly, the purity condition in (v) may be dropped. Applying inductively the property (iv), we obtain that for any $\mathcal{C}$-colored graph G in S^2 with connected components $\{G_i\}_i$,

$$\mathbb{F}_{\mathcal{C}}(G) = \prod_i \mathbb{F}_{\mathcal{C}}(G_i).$$

The invariant $\mathbb{F}_{\mathcal{C}}$ further extends to $\mathcal{C}$-colored graphs in an arbitrary oriented surface Σ homeomorphic to S^2. Given a $\mathcal{C}$-colored graph G in Σ, pick an orientation preserving homeomorphism $f \colon \Sigma \to S^2$ and set

$$\mathbb{F}_{\mathcal{C}}(G) = \mathbb{F}_{\mathcal{C}}(f(G))\, H(f) \colon H(G; \Sigma) \to \mathrm{End}_{\mathcal{C}}(\mathbb{1}).$$

Since all orientation preserving homeomorphisms $\Sigma \to S^2$ are isotopic, the homomorphism $\mathbb{F}_{\mathcal{C}}(G)$ does not depend on the choice of f.

12.2.6 The case of a pre-fusion category

Suppose that $\mathcal{C}$ is a pre-fusion spherical $\Bbbk$-category. Then $\mathrm{End}_{\mathcal{C}}(\mathbb{1}) = \Bbbk$. For any $\mathcal{C}$-colored graph G in S^2, the $\Bbbk$-module $H(G) = H(G; S^2)$ is free of finite rank and

$$\mathbb{F}_{\mathcal{C}}(G) \in H(G)^\star = \mathrm{Hom}_{\Bbbk}(H(G), \Bbbk).$$

Note two properties of $\mathbb{F}_\mathcal{C}$. First, for any non-isomorphic simple objects i, j of $\mathcal{C}$,

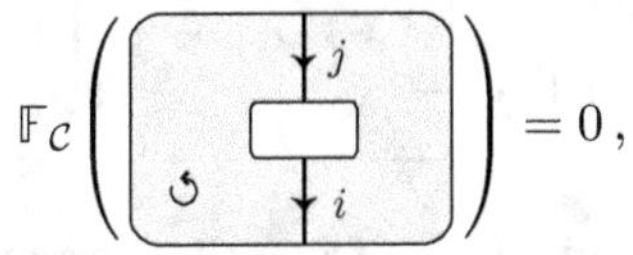

(12.1)

where the white box stands for any piece of a $\mathcal{C}$-colored graph with one input and one output as in the picture. Formula (12.1) holds because $\mathrm{Hom}_\mathcal{C}(i, j) = 0$. Second, for any simple object i of $\mathcal{C}$,

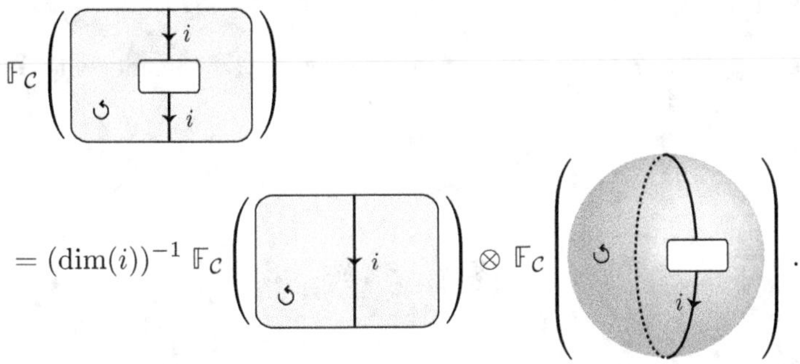

(12.2)

where the white box stands for a piece of a $\mathcal{C}$-colored graph (the same on the left-hand and right-hand sides) with one input and one output. Formula (12.2) follows from Lemma 4.2(b).

Lemma 12.3. *For any simple object i of $\mathcal{C}$,*

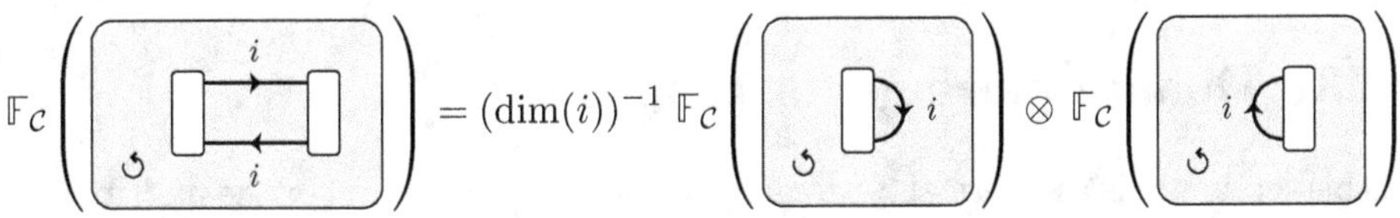

where the white boxes represent pieces of $\mathcal{C}$-colored graphs which are the same on the left-hand side and on the right-hand side.

Proof. We have:

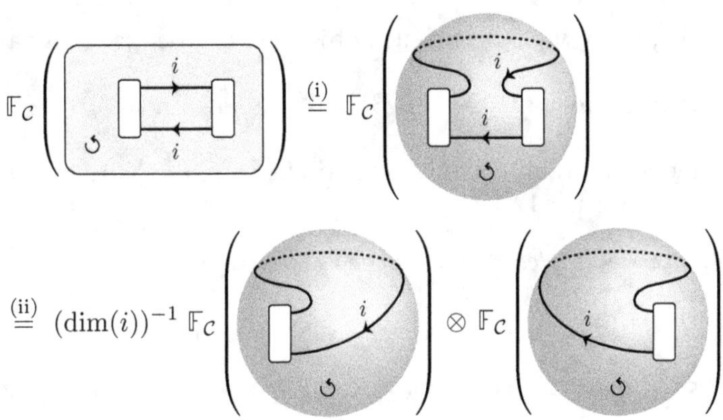

$\overset{\text{(iii)}}{=}$ $(\dim(i))^{-1}\, \mathbb{F}_{\mathcal{C}}$ (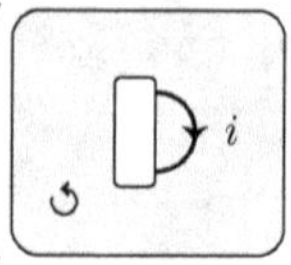) $\otimes\, \mathbb{F}_{\mathcal{C}}$ (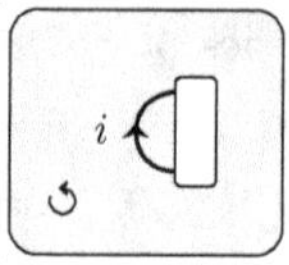) .

Here the equalities (i) and (iii) follow from the isotopy invariance of $\mathbb{F}_{\mathcal{C}}$ and (ii) from Formula (12.2). □

12.2.7 Example

For any objects i, j, k, l, m, n of a spherical $\Bbbk$-category $\mathcal{C}$, consider the following $\mathcal{C}$-colored graph $\Gamma \subset S^2$ with 4 vertices and 6 edges:

$\Gamma =$ 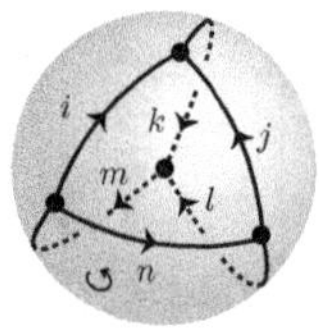 .

The $\Bbbk$-module $H(\Gamma)$ is the unordered tensor product of four multiplicity modules determined by i, j, k, l, m, n. The homomorphism $\mathbb{F}_{\mathcal{C}}(\Gamma)\colon H(\Gamma) \to \mathrm{End}_{\mathcal{C}}(\mathbb{1})$ and similar homomorphisms derived from other orientations of the edges of Γ are called *6j-symbols.* The 6j-symbols are further discussed in Appendix F.

12.3 Duality pairings for graphs

In this section, $\mathcal{C}$ is a spherical $\Bbbk$-category. We formulate duality for the multiplicity modules associated to dual vertices of $\mathcal{C}$-colored graphs.

12.3.1 Pairings and permutations

By Section 4.3.2, every tuple S of signed objects of $\mathcal{C}$ gives rise to a pairing

$$\omega_S\colon \mathrm{Hom}_{\mathcal{C}}(\mathbb{1}, X_{S^*}) \otimes_{\Bbbk} \mathrm{Hom}_{\mathcal{C}}(\mathbb{1}, X_S) \to \mathrm{End}_{\mathcal{C}}(\mathbb{1}).$$

The following lemma establishes the compatibility of these pairings with the permutation maps of Section 12.1.1.

Lemma 12.4. *For any tuples S, T of signed objects of $\mathcal{C}$,*

$$\omega_{TS}\big(\pi_{X_{T^*}, X_{S^*}} \otimes_{\Bbbk} \pi_{X_S, X_T}\big) = \omega_{ST},$$

where ST, TS are the concatenated tuples and S^*, T^* are the dual tuples.

Proof. Clearly, $X_{ST} = X_S \otimes X_T$ and $(ST)^* = T^*S^*$. Hence both sides of the required equality are $\Bbbk$-bilinear pairings

$$\operatorname{Hom}_{\mathcal{C}}(\mathbb{1}, X_{T^*} \otimes X_{S^*}) \otimes_{\Bbbk} \operatorname{Hom}_{\mathcal{C}}(\mathbb{1}, X_S \otimes X_T) \to \operatorname{End}_{\mathcal{C}}(\mathbb{1}).$$

Pick $\alpha \in \operatorname{Hom}_{\mathcal{C}}(\mathbb{1}, X_{T^*} \otimes X_{S^*})$ and $\beta \in \operatorname{Hom}_{\mathcal{C}}(\mathbb{1}, X_S \otimes X_T)$. The equality

$$\omega_{TS}\big(\pi_{X_{T^*}, X_{S^*}}(\alpha) \otimes_{\Bbbk} \pi_{X_S, X_T}(\beta)\big) = \omega_{ST}(\alpha \otimes_{\Bbbk} \beta)$$

is graphically represented by the following picture:

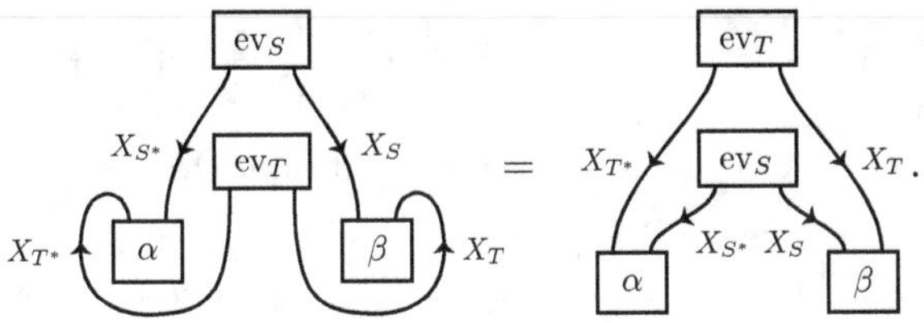

This equality follows from Lemma 2.9 since the two diagrams are isotopic in S^2. □

12.3.2 Dual cyclic sets

The *dual* of a cyclic $\mathcal{C}$-set (E, c, ε) is the cyclic $\mathcal{C}$-set $(E^{\mathrm{op}}, c, -\varepsilon)$, where E^{op} is the set E with opposite cyclic order. For each $e \in E$, we have the cone isomorphisms

$$H(E) \simeq \operatorname{Hom}_{\mathcal{C}}(\mathbb{1}, X_S) \quad \text{and} \quad H(E^{\mathrm{op}}) \simeq \operatorname{Hom}_{\mathcal{C}}(\mathbb{1}, X_{S^*})$$

where S is the tuple of signed objects of $\mathcal{C}$ obtained by enumerating the elements of E in the given cyclic order starting with e and recording the values of c and ε. Transporting the pairing ω_S along these isomorphisms, we obtain a $\Bbbk$-bilinear pairing

$$\widetilde{\omega}_E \colon H(E^{\mathrm{op}}) \otimes_{\Bbbk} H(E) \to \operatorname{End}_{\mathcal{C}}(\mathbb{1}).$$

Lemma 12.4 implies that $\widetilde{\omega}_E$ does not depend on the choice of $e \in E$. Moreover, the pairings $\widetilde{\omega}_E$ and $\widetilde{\omega}_{E^{\mathrm{op}}} \colon H(E) \otimes_{\Bbbk} H(E^{\mathrm{op}}) \to \operatorname{End}_{\mathcal{C}}(\mathbb{1})$ are equal up to permutation of the tensor factors. Consequently, they induce a $\Bbbk$-bilinear pairing

$$\omega_E \colon H(E^{\mathrm{op}}) \otimes H(E) \to \operatorname{End}_{\mathcal{C}}(\mathbb{1}),$$

where $\otimes$ is the unordered tensor product of $\Bbbk$-modules. By definition, $\omega_{E^{\mathrm{op}}} = \omega_E$. More generally, a *duality* between cyclic $\mathcal{C}$-sets E and E' is an isomorphism of cyclic $\mathcal{C}$-sets $\phi \colon E^{\mathrm{op}} \to E'$. Such a ϕ induces a $\Bbbk$-linear isomorphism $H(\phi) \colon H(E^{\mathrm{op}}) \to H(E')$ and a pairing

$$\omega_\phi = \omega_E(H(\phi)^{-1} \otimes \mathrm{id}_{H(E)}) \colon H(E') \otimes H(E) \to \operatorname{End}_{\mathcal{C}}(\mathbb{1}).$$

12.3.3 Dual vertices

We apply the definitions of the previous subsection to cyclic $\mathcal{C}$-sets associated with $\mathcal{C}$-colored graphs in surfaces. Let Σ be an oriented surface and let $-\Sigma$ be the same surface with opposite orientation. For a $\mathcal{C}$-colored graph G in Σ, we denote by G^{op} the $\mathcal{C}$-colored graph in $-\Sigma$ obtained from G by reversing orientation in all edges of G while keeping the colors of the edges. The cyclic $\mathcal{C}$-set E_v determined by a vertex v of $G \subset \Sigma$ is dual in the obvious way to the cyclic $\mathcal{C}$-set E_v^{op} determined by v as a vertex of $G^{\mathrm{op}} \subset -\Sigma$. Set

$$\omega_v = \omega_{E_v} : H_v(G^{\mathrm{op}}) \otimes H_v(G) = H(E_v^{\mathrm{op}}) \otimes H(E_v) \to \mathrm{End}_{\mathcal{C}}(\mathbb{1}).$$

The (unordered) tensor product of these pairings over all vertices of G yields a $\Bbbk$-bilinear pairing

$$\omega_G \colon H(G^{\mathrm{op}}) \otimes H(G) \to \mathrm{End}_{\mathcal{C}}(\mathbb{1}).$$

The pairing ω_v above can be generalized as follows. Besides $G \subset \Sigma$, consider a $\mathcal{C}$-colored graph G' in an oriented surface Σ'. By a *duality* between a vertex u of G' and a vertex v of G, we mean a duality between the cyclic $\mathcal{C}$-sets E_u and E_v, that is, an isomorphism of cyclic $\mathcal{C}$-sets $\phi\colon E_u^{\mathrm{op}} \to E_v$. Set

$$\omega_{u,v} = \omega_\phi \colon H_u(G') \otimes H_v(G) = H(E_u) \otimes H(E_v) \to \mathrm{End}_{\mathcal{C}}(\mathbb{1}).$$

Clearly, $\omega_{u,v} = \omega_{v,u}$, where $\omega_{v,u}$ is induced by the isomorphism $\phi^{-1}\colon E_v^{\mathrm{op}} \to E_u$. Also, $\omega_{v,v} = \omega_v$ for $\Sigma' = -\Sigma$, $G' = G^{\mathrm{op}}$, $u = v$, and $\phi = \mathrm{id}_{E_v}$.

12.3.4 The case of a non-degenerate category

Suppose that the spherical $\Bbbk$-category $\mathcal{C}$ is non-degenerate. Then all the pairings considered in Section 12.3.2 take values in $\mathrm{End}_{\mathcal{C}}(\mathbb{1}) = \Bbbk$ and are non-degenerate. Consequently, given a cyclic $\mathcal{C}$-set E, we can consider the contraction vectors

$$*_{\widetilde{\omega}_E} \in H(E) \otimes_\Bbbk H(E^{\mathrm{op}}) \quad \text{and} \quad *_{\widetilde{\omega}_{E^{\mathrm{op}}}} \in H(E^{\mathrm{op}}) \otimes_\Bbbk H(E)$$

associated with the pairings $\widetilde{\omega}_E$ and $\widetilde{\omega}_{E^{\mathrm{op}}}$ (see Section 1.5.2). These vectors are equal up to permutation of the tensor factors and determine a vector

$$*_E \in H(E) \otimes H(E^{\mathrm{op}}).$$

More generally, a duality $\phi\colon E^{\mathrm{op}} \to E'$ between cyclic $\mathcal{C}$-sets E and E' determines the *contraction vector*

$$*_\phi = (\mathrm{id}_{H(E)} \otimes H(\phi))(*_E) \in H(E) \otimes H(E').$$

For a $\mathcal{C}$-colored graph G in an oriented surface Σ and a vertex v of G, the pairings ω_v and ω_G introduced in Section 12.3.3 are non-degenerate and induce $\Bbbk$-linear isomorphisms

$$H_v(G^{\mathrm{op}}) \simeq H_v(G)^\star \quad \text{and} \quad H(G^{\mathrm{op}}) \simeq H(G)^\star.$$

For $\mathcal{C}$-colored graphs G, G' in oriented surfaces Σ, Σ' and for dual vertices v of G and u of G', the pairing $\omega_{u,v}$ introduced in Section 12.3.3 is non-degenerate and induces a $\Bbbk$-linear isomorphism

$$H_u(G') \simeq H_v(G)^\star.$$

The *contraction vector* of a duality $\phi\colon E_u^{\mathrm{op}} \to E_v$ between u and v is the vector

$$*_{u,v} = *_\phi \in H_u(G') \otimes H_v(G).$$

Clearly, $*_{u,v} = *_{v,u}$ where $*_{v,u}$ is determined by $\phi^{-1}\colon E_v^{\mathrm{op}} \to E_u$.

12.3.5 Example

Recall that $\mathcal{C}$ is a spherical $\Bbbk$-category. Let G be a $\mathcal{C}$-colored graph in an oriented surface and let v be a vertex of G incident to one incoming edge and one outgoing edge, both colored with the same object i of $\mathcal{C}$:

$G =$ 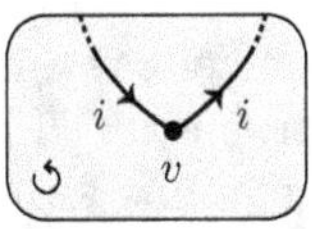

.

Let $a_i \in H_v(G)$ and $b_i \in H_v(G^{\mathrm{op}})$ be the pre-images of the coevaluation morphism $\mathrm{coev}_i\colon \mathbb{1} \to i \otimes i^*$ under the cone isomorphisms

$$H_v(G) \to \mathrm{Hom}_{\mathcal{C}}(\mathbb{1}, i \otimes i^*) \quad \text{and} \quad H_v(G^{\mathrm{op}}) \to \mathrm{Hom}_{\mathcal{C}}(\mathbb{1}, i \otimes i^*).$$

It follows from the definitions that

$\omega_v(a_i \otimes b_i) = \omega_{((i,+),(i,-))}(\mathrm{coev}_i \otimes_\Bbbk \mathrm{coev}_i) =$ 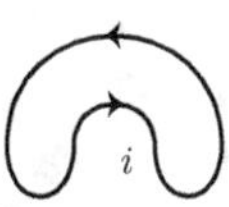

$=$ 

$= \dim(i).$

If the object i is simple, then this computes ω_v because $\mathrm{Hom}_{\mathcal{C}}(\mathbb{1}, i \otimes i^*) = \Bbbk\, \mathrm{coev}_i$ and so $H_v(G) = \Bbbk\, a_i$, $H_v(G^{\mathrm{op}}) = \Bbbk\, b_i$. If i is simple and $\mathcal{C}$ is non-degenerate, then the contraction vector $*_{v,v}$ determined by the obvious identification $E_v^{\mathrm{op}} \simeq E_v$ is computed by

$$*_{v,v} = (\dim(i))^{-1}\, b_i \otimes a_i \in H_v(G^{\mathrm{op}}) \otimes H_v(G).$$

12.3.6 Example

For any tuple $S = ((X_1, \varepsilon_1), \ldots, (X_n, \varepsilon_n))$ of $n \geq 1$ signed objects of $\mathcal{C}$, consider the following $\mathcal{C}$-colored graph γ_S in S^2:

$\gamma_S =$ 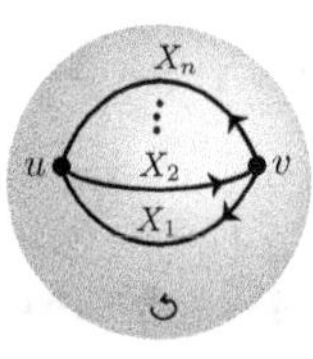

.

This graph has n edges connecting the vertices u and v, the i-th edge being colored with X_i and oriented towards v if $\varepsilon_i = +$ and towards u otherwise (in the picture $\varepsilon_1 = -$, $\varepsilon_2 = +$, $\varepsilon_n = -$). The cone isomorphisms induce an isomorphism

$$H(\gamma_S) = H_u(\gamma_S) \otimes H_v(\gamma_S) \simeq \operatorname{Hom}_{\mathcal{C}}(\mathbb{1}, X_{S^*}) \otimes \operatorname{Hom}_{\mathcal{C}}(\mathbb{1}, X_S)$$

where $X_S = X_1^{\varepsilon_1} \otimes \cdots \otimes X_n^{\varepsilon_n}$ and $S^* = ((X_n, -\varepsilon_n), \ldots, (X_1, -\varepsilon_1))$. The vertices u, v are in duality induced by the symmetry with respect to a vertical big circle and

$$\mathbb{F}_{\mathcal{C}}(\gamma_S) = \omega_{u,v} \colon H(\gamma_S) = H_u(\gamma_S) \otimes H_v(\gamma_S) \to \operatorname{End}_{\mathcal{C}}(\mathbb{1}).$$

If $\mathcal{C}$ is non-degenerate, then Lemma 3.6 implies that

$$\mathbb{F}_{\mathcal{C}}(\gamma_S)(*_{u,v}) = \omega_{u,v}(*_{u,v}) = \dim H_u(\gamma_S) = \dim H_v(\gamma_S) = \dim \operatorname{Hom}_{\mathcal{C}}(\mathbb{1}, X_S).$$

12.3.7 The case of a pre-fusion category

Suppose that $\mathcal{C}$ is a pre-fusion spherical $\Bbbk$-category. By Lemma 4.3, $\mathcal{C}$ is non-degenerate. In addition to the properties (12.1) and (12.2) of $\mathbb{F}_{\mathcal{C}}$ stated in Section 12.2.6, we formulate local relations for $\mathbb{F}_{\mathcal{C}}$ involving dual vertices and contraction vectors.

Lemma 12.5. *If I is a representative set of simple objects of $\mathcal{C}$, then*

$\mathbb{F}_{\mathcal{C}}\Big($ 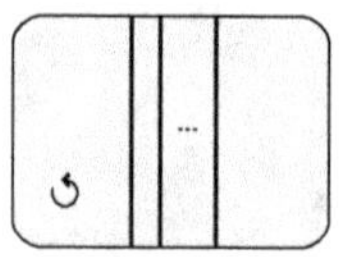 $\Big) = \sum_{i \in I} \dim(i)\, \mathbb{F}_{\mathcal{C}}\Big($ 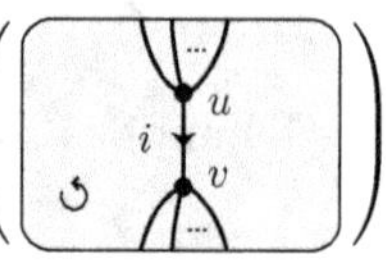 $\Big)(*_{u,v})$

where $*_{u,v}$ is the contraction vector provided by the duality between the vertices u and v induced by the symmetry with respect to a horizontal line (see Section 12.3.4).

Proof. We only need to compare the contributions to $\mathbb{F}_{\mathcal{C}}$ of the depicted pieces of $\mathcal{C}$-colored graphs. Let $S = ((X_1, \varepsilon_1), \ldots, (X_n, \varepsilon_n))$ be the tuple of signed objects of $\mathcal{C}$ determined by the vertical strands on the left-hand side of the equality in the lemma. Consider the dual tuple $S^* = ((X_n, -\varepsilon_1), \ldots, (X_1, -\varepsilon_n))$ and the morphisms

$$\operatorname{ev}_{S^*} \colon X_S \otimes X_{S^*} \to \mathbb{1}, \quad \operatorname{coev}_{S^*} \colon \mathbb{1} \to X_{S^*} \otimes X_S, \quad \Psi_{S^*} \colon X_{S^*} \to X_S^*$$

from Section 2.3.2. For $i \in I$, set $S_i = (i, -)S$. Then

$$S_i^* = S^*(i, +), \quad X_{S_i} = i^* \otimes X_S, \quad \text{and} \quad X_{S_i^*} = X_{S^*} \otimes i.$$

Since the category $\mathcal{C}$ is non-degenerate, the pairing

$$\omega_{S_i} \colon \operatorname{Hom}_{\mathcal{C}}(\mathbb{1}, X_{S_i^*}) \otimes_{\Bbbk} \operatorname{Hom}_{\mathcal{C}}(\mathbb{1}, X_{S_i}) \to \Bbbk$$

from Section 4.3.2 is non-degenerate. Let

$$*_i \in \mathrm{Hom}_{\mathcal{C}}(\mathbb{1}, X_{S_i}) \otimes_{\Bbbk} \mathrm{Hom}_{\mathcal{C}}(\mathbb{1}, X_{S_i^*})$$

be the associated contraction vector. Using the definitions of $\mathbb{F}_{\mathcal{C}}$ and $*_{u,v}$ and considering the isotopy

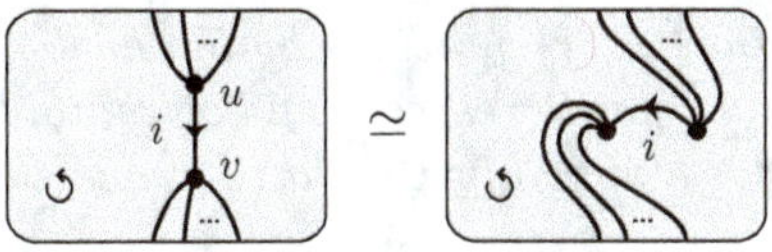

we reduce the lemma to the following claim: for some expansion $*_i = \sum_\alpha e_{i,\alpha} \otimes_{\Bbbk} f_{i,\alpha}$ with $e_{i,\alpha} \in \mathrm{Hom}_{\mathcal{C}}(\mathbb{1}, X_{S_i})$ and $f_{i,\alpha} \in \mathrm{Hom}_{\mathcal{C}}(\mathbb{1}, X_{S_i^*})$, we have

$$\sum_{i\in I}\sum_{\alpha} \dim(i)$$ 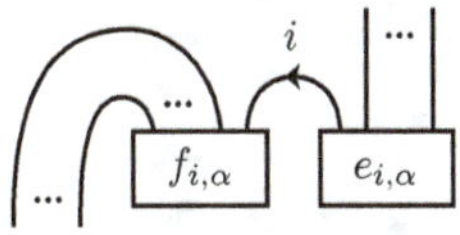 $$= \mathrm{id}_{X_S}, \tag{12.3}$$

where the top (respectively, bottom) free ends of the Penrose diagram are colored and oriented so that the corresponding tuple of signed objects (determined as in Section 2.2.2) is S. We verify (12.3) for the expansion

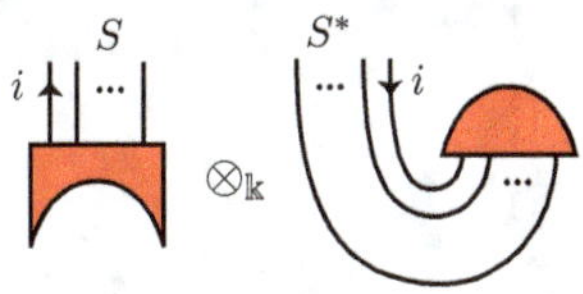

provided by Lemma 4.8. Then the left-hand side of (12.3) is equal to

$$\sum_{i\in I} \dim(i)$$ 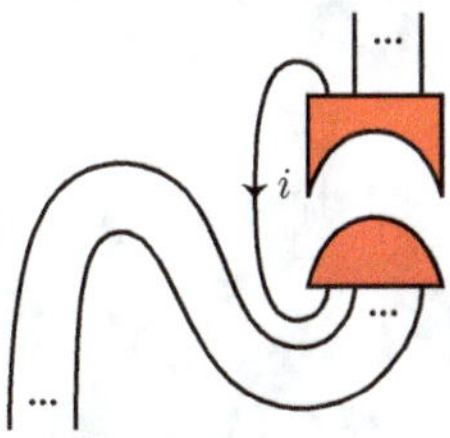

$$\overset{\text{(i)}}{=} \sum_{i\in I} \dim(i)$$ 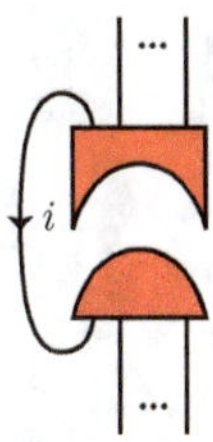 $$\overset{\text{(ii)}}{=} \sum_{i\in I}$$ 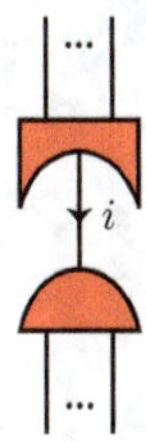 $$\overset{\text{(iii)}}{=} \mathrm{id}_{X_S}.$$

Here (i) follows from the isotopy invariance of the graphical calculus, (ii) from Lemma 4.9(d), and (iii) from (4.8). □

Lemma 12.6. *We have:*

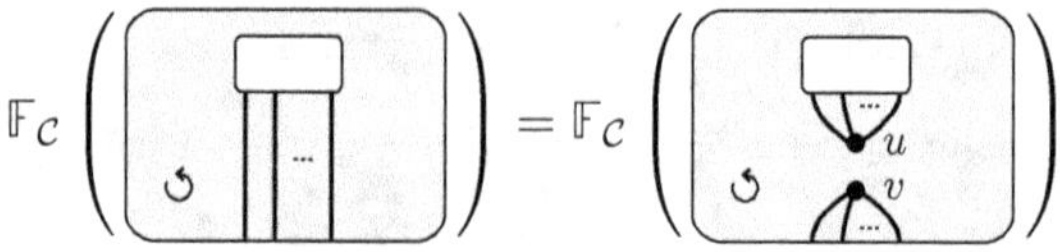

$$(*_{u,v})$$

where $*_{u,v}$ *is the contraction vector provided by the duality between the vertices* u *and* v *induced by the symmetry with respect to a horizontal line (see Section* 12.3.4) *and the white box stands for a piece of a* $\mathcal{C}$*-colored graph (the same on the left-hand and right-hand sides).*

Proof. The equality in this lemma follows from Lemma 12.5 since $\dim(\mathbb{1}) = 1_{\Bbbk}$ and $\mathrm{Hom}_{\mathcal{C}}(i, \mathbb{1}) = 0$ for all $i \in I$ distinct from $\mathbb{1}$. $\square$

Chapter 13

The state sum TQFT

Fix throughout this chapter a spherical fusion $\Bbbk$-category $\mathcal{C}$ and a representative set I of simple objects of $\mathcal{C}$. Unless explicitly stated to the contrary, we assume that $\dim(\mathcal{C})$ is invertible in $\Bbbk$. We derive from this data a 3-dimensional TQFT $|\cdot|_{\mathcal{C}}$. We start by defining a scalar topological invariant of closed oriented 3-manifolds.

13.1 An invariant of closed 3-manifolds

13.1.1 The state sum invariant

Let M be a closed oriented 3-manifold. Pick a skeleton P of M and recall from Section 11.1.3 the set $\mathrm{Reg}(P)$ of regions of P. A *coloring* of P is a map $c\colon \mathrm{Reg}(P) \to I$. The object $c(r) \in I$ assigned to a region r of P is called the *c-color* of r. We set

$$\dim(c) = \prod_{r\in \mathrm{Reg}(P)} (\dim c(r))^{\chi(r)} \in \Bbbk, \tag{13.1}$$

where χ is the Euler characteristic. Note that $\dim(c)$ is well defined since the dimension of a simple object is invertible, see Lemma 4.2.

Next, we define a scalar $|c| \in \Bbbk$ for any coloring c of P. If P has no vertices, then we set $|c| = 1_{\Bbbk}$. Consider now the case where P has at least one vertex. Note that such a P has at least one edge (stratified 2-polyhedra have no isolated vertices, see Section 11.1). By an *oriented edge* of P, we mean an edge of P endowed with an orientation. For each oriented edge e of P, the orientations of e and M determine a positive direction on a small loop in M encircling e. The resulting oriented loop determines a cyclic order on the set P_e of branches of P at e. To each branch $b \in P_e$ we assign the c-color of the region of P containing b and a sign equal to $+$ if the orientation of b induces the one of $e \subset \partial b$ (that is, the orientation of b is given by the orientation of e followed by a vector at a point of e directed inside b) and equal to $-$ otherwise. In this way, P_e becomes a cyclic $\mathcal{C}$-set. Let $H_c(e) = H(P_e)$

be its multiplicity module. Let

$$H_c = \bigotimes_e H_c(e)$$

be the unordered tensor product of the $\Bbbk$-modules $H_c(e)$ over all oriented edges e of P. By Appendix E, there is a canonical $\Bbbk$-linear isomorphism

$$H_c^\star \simeq \bigotimes_e H_c(e)^\star. \tag{13.2}$$

Each unoriented edge e of P gives rise to two opposite oriented edges e_1, e_2 of P. The cyclic $\mathcal{C}$-sets $E_1 = P_{e_1}$ and $E_2 = P_{e_2}$ are dual: $E_1 = E_2^{\mathrm{op}}$. By the non-degeneracy of $\mathcal{C}$ (Lemma 4.3) and Section 12.3.4, this yields a vector

$$*_e = *_{E_1} \in H_c(e_1) \otimes H_c(e_2)$$

independent of the numeration of e_1, e_2. Set

$$*_c = \otimes_e *_e \in H_c$$

where $\otimes_e$ is the unordered tensor product over all unoriented edges e of P.

For a vertex x of P, consider the link graph $\Gamma_x \subset \partial B_x$ where $B_x \subset M$ is a P-cone neighborhood of x, see Section 11.2.1. Here we endow ∂B_x with orientation induced by that of M restricted to $M \backslash \mathrm{Int}(B_x)$. Every edge a of Γ_x lies in a region r_a of P. We color a with $c(r_a) \in I$ and endow a with the orientation induced by that of $r_a \backslash \mathrm{Int}(B_x)$. In this way, Γ_x becomes a $\mathcal{C}$-colored graph in ∂B_x. It is denoted Γ_x^c. (An example is given in Figure 13.1, where ∂B_x is identified with $S^2 = \mathbb{R}^2 \cup \{\infty\}$ via the stereographic projection from a pole in the upper left quarter.) Section 12.2.5 yields a vector

$$\mathbb{F}_{\mathcal{C}}(\Gamma_x^c) \in H(\Gamma_x^c)^\star = \mathrm{Hom}_\Bbbk(H(\Gamma_x^c), \Bbbk). \tag{13.3}$$

Note that the cyclic $\mathcal{C}$-set associated with any vertex v of Γ_x^c (see Section 12.2.1) is canonically isomorphic to the cyclic $\mathcal{C}$-set P_e, where $e = e(v)$ is the edge of P containing v and oriented away from x. Therefore, there are canonical isomorphisms

$$H(\Gamma_x^c) \simeq \otimes_{e_x} H_c(e_x) \quad \text{and} \quad H(\Gamma_x^c)^\star \simeq \otimes_{e_x} H_c(e_x)^\star,$$

where e_x runs over all edges of P incident to x and oriented away from x. (An edge with both endpoints in x appears in each of these tensor products twice with opposite orientations.) The tensor product of the latter isomorphisms over all vertices x of P yields a $\Bbbk$-linear isomorphism

$$\bigotimes_x H(\Gamma_x^c)^\star \simeq \bigotimes_x \bigotimes_{e_x} H_c(e_x)^\star \simeq \bigotimes_e H_c(e)^\star \simeq H_c^\star$$

where e runs over all oriented edges of P. The image under this isomorphism of the unordered tensor product $\bigotimes_x \mathbb{F}_{\mathcal{C}}(\Gamma_x^c)$, where x runs over all vertices of P, is a vector $V_c \in H_c^\star$. We evaluate V_c on $*_c$ and set

$$|c| = V_c(*_c) \in \Bbbk.$$

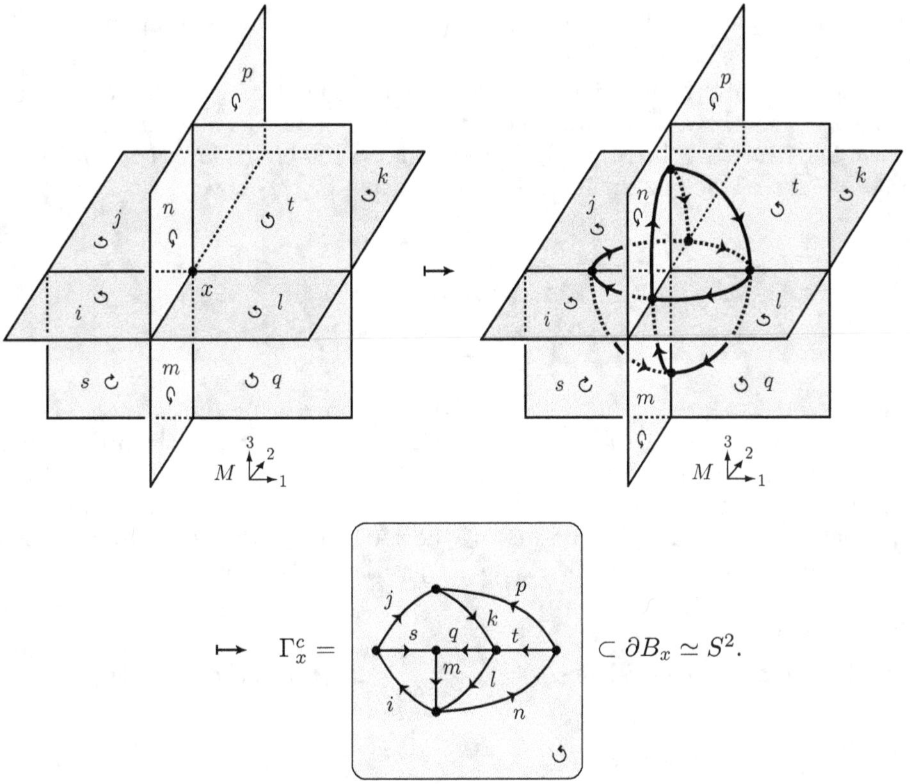

Figure 13.1: The colored graph Γ_x^c

Finally, set

$$|M|_{\mathcal{C}} = (\dim(\mathcal{C}))^{-|M \setminus P|} \sum_{c} \dim(c)\, |c| \in \Bbbk, \tag{13.4}$$

where $|M \setminus P|$ is the number of components of $M \setminus P$ and c runs over all colorings of P (i.e., over all maps $\mathrm{Reg}(P) \to I$). The right-hand side of (13.4) is well defined because both sets $\mathrm{Reg}(P)$ and I are finite.

Theorem 13.1. *$|M|_{\mathcal{C}}$ is a topological invariant of M independent of the choice of P and I.*

Proof. Denote the right-hand side of (13.4) by $|P|_{\mathcal{C}}$. Note that for any representative set I' of simple objects of $\mathcal{C}$, there is a unique bijection $\varphi\colon I \to I'$ such that the objects i and $\varphi(i)$ are isomorphic for all $i \in I$. Consequently, the naturality of $\mathbb{F}_{\mathcal{C}}$ (see Section 12.2.4) implies that $|P|_{\mathcal{C}}$ does not depend on the choice of the representative set I. An orientation preserving homeomorphism of closed oriented 3-manifolds $M \to M'$ carries a skeleton P of M to a skeleton P' of M' and, clearly,

$|P|_{\mathcal{C}} = |P'|_{\mathcal{C}}$. Thus, by Theorem 11.1, we need only to prove the invariance of $|P|_{\mathcal{C}}$ under the moves T_0, T_1, T_2, T_3 on P.

Let us verify the invariance of $|P|_{\mathcal{C}}$ under the bubble move T_0. Let x and e be respectively the vertex and the edge (with both endpoints in x) created by the move. Let $i, m, n \in I$ be respectively the colors of the big region where the move proceeds and of the two small disk regions created by the move:

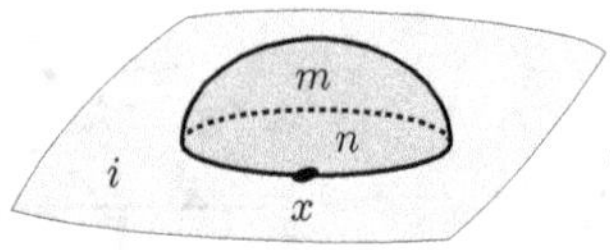

.

The $\mathcal{C}$-colored graph in S^2 associated with x is the graph γ_S in Example 12.3.6 where $S = ((i,\varepsilon),(m,\mu),(n,\nu))$. Here the signs ε, μ, ν depend on the orientations of the three regions involved in the move. Let u and v be the two vertices of the graph γ_S. Clearly $*_e = *_{u,v}$, where $*_{u,v}$ is the contraction vector determined by the duality between u and v induced by the symmetry with respect to a vertical great circle in γ_S (see Example 12.3.6). Then the contribution of x to the state sum is

$$\sum_{m,n\in I} \dim(m)\dim(n)\, \mathbb{F}_{\mathcal{C}}(\gamma_S)(*_{u,v}) \overset{\text{(i)}}{=} \sum_{m,n\in I} \dim(m)\dim(n)\, N^{\mathbb{1}}_{i^\varepsilon\otimes m^\mu\otimes n^\nu}$$

$$\overset{\text{(ii)}}{=} \sum_{m,n\in I} \dim(m)\dim(n)\, N^{\mathbb{1}}_{i^\varepsilon\otimes m\otimes n} \overset{\text{(iii)}}{=} \dim(i^\varepsilon)\dim(\mathcal{C}) \overset{\text{(iv)}}{=} \dim(i)\dim(\mathcal{C}).$$

Here (i) follows from Example 12.3.6, (ii) and (iv) from the equality $\dim(X^*) = \dim(X)$ for any $X \in \mathrm{Ob}(\mathcal{C})$, and (iii) from Lemma 4.7. The factor $\dim(i)\dim(\mathcal{C})$ is compensated by the change in the Euler characteristic of the big region and in the number of components of $M \setminus P$.

The move $P \xrightarrow{T_1} P'$ adds a new edge e connecting two distinct vertices of P. This modifies the link graphs of these vertices by adding a new vertex u (respectively, v) inside an edge. The colorings of P' assigning different colors to the regions of P' lying on the two sides of e contribute zero to the state sum by (12.1) (there are no such colorings if these regions coincide). The colorings of P' assigning the same color $i \in I$ to these two regions contribute the same as the colorings of P assigning i to the region of P containing e. Indeed,

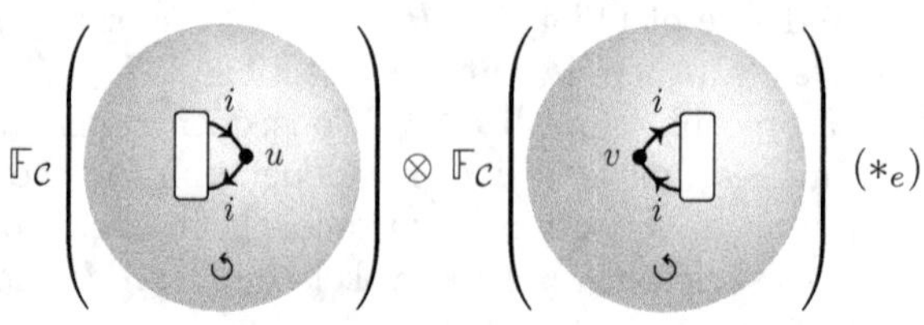

$\overset{\text{(i)}}{=} \mathbb{F}_{\mathcal{C}}\Bigg($ 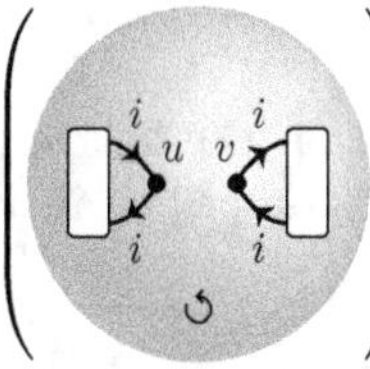

$\Bigg)(*_{u,v}) \overset{\text{(ii)}}{=} \mathbb{F}_{\mathcal{C}}\Bigg($ 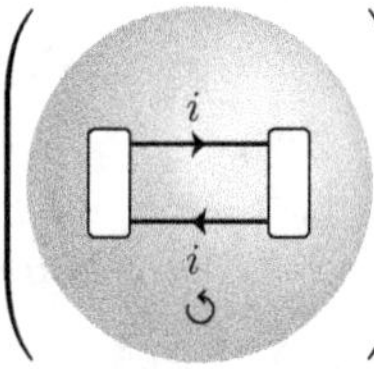

$\Bigg)$

$\overset{\text{(iii)}}{=} (\dim(i))^{-1}\, \mathbb{F}_{\mathcal{C}}\Bigg($ 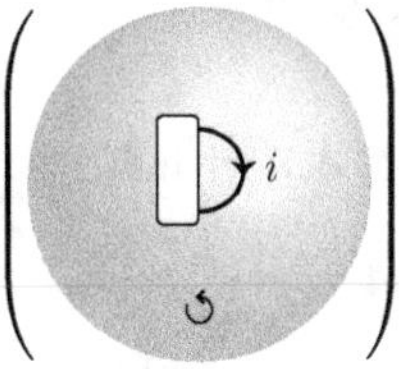

$\Bigg) \otimes \mathbb{F}_{\mathcal{C}}\Bigg($ 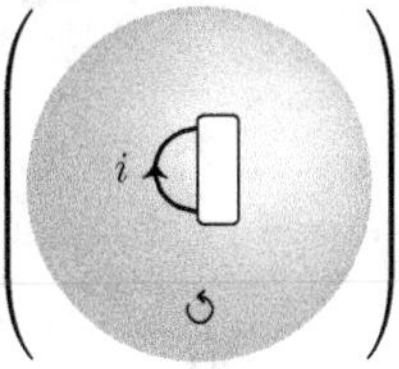

$\Bigg),$

where $*_{u,v}$ is the contraction vector determined by the duality between the vertices u and v induced by the symmetry with respect to a vertical great circle. Here (i) follows from the $\otimes$-multiplicativity of $\mathbb{F}_{\mathcal{C}}$, (ii) from Lemma 12.6, and (iii) from Lemma 12.3. The factor $(\dim(i))^{-1}$ is compensated by the change in the Euler characteristics of the regions.

The invariance under T_2 follows from Lemma 12.6. Finally, the invariance under T_3 follows from Lemma 12.5 with two vertical strands on the left-hand side. Here $i \in I$ is the color of the disk region created by the move. □

13.1.2 Properties

If P is a special skeleton of a closed oriented 3-manifold M, then the link graph of any vertex of P is the 1-skeleton of a tetrahedron and the vector (13.3) is a $6j$-symbol, see Section 12.2.7. Then the expression (13.4) is the Turaev-Viro-Barrett-Westbury invariant of M derived from $\mathcal{C}$, see [TV, BW1]. As any closed oriented 3-manifold M has a special skeleton, the invariant $|M|_{\mathcal{C}}$ is nothing but the TVBW invariant.

It follows from the definitions that for all closed oriented 3-manifolds M and N,

$$|M \amalg N|_{\mathcal{C}} = |M|_{\mathcal{C}}\, |N|_{\mathcal{C}} \quad \text{and} \quad |-M|_{\mathcal{C}} = |M|_{\mathcal{C}^{\otimes \mathrm{op}}},$$

where $-M$ is M with opposite orientation and $\mathcal{C}^{\otimes \mathrm{op}} = (\mathcal{C}, \otimes^{\mathrm{op}}, \mathbb{1})$.

The naturality of the invariant $\mathbb{F}_{\mathcal{C}}$ of $\mathcal{C}$-colored graphs (see Section 12.2.4) implies that

$$|M|_{\mathcal{C}'} = |M|_{\mathcal{C}}$$

for any closed oriented 3-manifold M and any spherical fusion $\Bbbk$-category $\mathcal{C}'$ equivalent to $\mathcal{C}$. In particular

$$|M|_{\mathcal{C}^{\mathrm{op}}} = |M|_{\mathcal{C}^{\otimes \mathrm{op}}} = |-M|_{\mathcal{C}} \quad \text{and} \quad |M|_{\mathcal{C}^{\mathrm{rev}}} = |M|_{\mathcal{C}},$$

since $\mathcal{C}^{\mathrm{op}} = (\mathcal{C}^{\mathrm{op}}, \otimes, \mathbb{1})$ is equivalent to $\mathcal{C}^{\otimes \mathrm{op}}$ and $\mathcal{C}^{\mathrm{rev}} = (\mathcal{C}^{\mathrm{op}}, \otimes^{\mathrm{op}}, \mathbb{1})$ is equivalent to $\mathcal{C}$, see Section 4.5.1.

13.1.3 Examples

We compute $|M|_{\mathcal{C}}$ for $M = S^3$, $M = S^1 \times S^2$ and for the lens spaces by applying Formula (13.4) to the skeletons of these manifolds described in Examples 11.2.3, 11.2.4, and 11.2.5. For the 3-sphere, this gives

$$|S^3|_{\mathcal{C}} = (\dim(\mathcal{C}))^{-2} \sum_{i \in I} (\dim(i))^2 = (\dim(\mathcal{C}))^{-1}.$$

For $S^1 \times S^2$, we obtain (using Lemma 4.7 with $X = Y^* \in I$) that $|S^1 \times S^2|_{\mathcal{C}} = 1$. For the lens space $L(p,q)$ with $p \geq 2$, we obtain

$$|L(p,q)|_{\mathcal{C}} = (\dim(\mathcal{C}))^{-1} \sum_{i \in I} \dim(i) \operatorname{Tr}(\sigma^q) \in \Bbbk,$$

where Tr is the standard trace of $\Bbbk$-linear endomorphisms of the free $\Bbbk$-module $K = \operatorname{Hom}_{\mathcal{C}}(\mathbb{1}, i^{\otimes p})$ and

$$\sigma = \pi_{i, i^{\otimes(p-1)}} : K = \operatorname{Hom}_{\mathcal{C}}(\mathbb{1}, i \otimes i^{\otimes(p-1)}) \to \operatorname{Hom}_{\mathcal{C}}(\mathbb{1}, i^{\otimes(p-1)} \otimes i) = K$$

is the permutation map from Section 12.1.1.

13.1.4 Remarks

1. As explained in Section 11.2.2, a triangulation t of M gives rise to skeletons $t^{(2)}$ and $(t^*)^{(2)}$ of M. For $t^{(2)}$, Formula (13.4) yields

$$|M|_{\mathcal{C}} = (\dim(\mathcal{C}))^{-|t|} \sum_{c} \left(\prod_{r} \dim c(r) \right) |c| \,,$$

where $|t|$ is the number of tetrahedra of t, r runs over all 2-faces of t, and c runs over all maps from the set of 2-faces of t to I. The skeleton $(t^*)^{(2)}$ is special and the associated expression (13.4) for $|M|_{\mathcal{C}}$ is the one in [TV, BW1].

2. The expression (13.4) can be viewed as a state sum (or a partition function) as follows. Provide all multiplicity modules in $\mathcal{C}$ with distinguished bases. By a state on a skeleton P, we mean a pair consisting of a coloring $c \colon \operatorname{Reg}(P) \to I$ and a choice of a basis vector $b_c(e) \in H_c(e)$ for every oriented edge e of P. Let $b_c^\star(e)$ be the corresponding vector in the basis of $H_c(e)^\star$ dual to the distinguished basis of $H_c(e)$. The Boltzmann weight associated with such a state is the product of the factors

$$\mathbb{F}_{\mathcal{C}}(\Gamma_x^c)(\otimes_e b_c(e)), \quad (b_c^\star(e^{\mathrm{op}}) \otimes b_c^\star(e))(*_e), \quad (\dim c(r))^{\chi(r)}, \quad (\dim(\mathcal{C}))^{-1}$$

contributed respectively by the vertices x of P, edges e of P, regions r of P, and connected components of $M \setminus P$. The invariant $|M|_{\mathcal{C}}$ is the sum of these Boltzmann weights over all states on P.

13.2 An invariant of 3-manifolds with boundary

We extend the state sum invariant $|\cdot|_{\mathcal{C}}$ of closed oriented 3-manifolds to compact oriented 3-manifolds endowed with a graph in the boundary.

13.2.1 The state sum

By an *I-colored graph*, we mean a $\mathcal{C}$-colored graph such that the colors of all edges lie in I and all vertices have valence ≥ 2. For a compact oriented 3-manifold M and an I-colored graph G embedded in ∂M, we define a vector $|M,G|_{\mathcal{C}} \in H(G)$, where $H(G) = H(G; \partial M)$ is the $\Bbbk$-module introduced in Section 12.2.1.

Pick a skeleton P of the pair (M,G), see Section 11.5.1. A *coloring* of P is a map $c\colon \mathrm{Reg}(P) \to I$ such that for every edge a of G, the map c carries the region of P adjacent to a to the color of a. Given a coloring c of P, we define $\dim(c) \in \Bbbk$ by (13.1). We also associate with c a vector $|c| \in H(G)$ as follows. If P has no vertices, then $G = \emptyset$, $H(G) = \Bbbk$, and we set $|c| = 1_{\Bbbk}$. We now define $|c|$ in the case where P has at least one vertex (and so has at least one edge). By an *oriented edge* of P, we mean an edge of P endowed with an orientation. Each oriented edge e of P originates at a vertex of P called the *tail* of e. Let $\mathcal{E}$ be the set of all oriented edges of P not lying in $\partial P = G$. For $e \in \mathcal{E}$, consider the $\Bbbk$-module $H_c(e) = H(P_e)$, where P_e is the set of branches of P at e turned into a cyclic $\mathcal{C}$-set as in Section 13.1.1. Set

$$H_c = \bigotimes_{e \in \mathcal{E}_0} H_c(e) \quad \text{and} \quad H_c^{\partial} = \bigotimes_{e \in \mathcal{E}_{\partial}} H_c(e),$$

where $\otimes$ is the unordered tensor product of $\Bbbk$-modules, $\mathcal{E}_0 \subset \mathcal{E}$ is the set of all oriented edges of P with tail in $\mathrm{Int}(M)$, and $\mathcal{E}_{\partial} = \mathcal{E} \setminus \mathcal{E}_0$. Observe that the tail $v(e)$ of any $e \in \mathcal{E}_{\partial}$ is a vertex of G and there is a canonical $\Bbbk$-linear isomorphism $H_c(e) \simeq H_{v(e)}(G)$. The formula $e \mapsto v(e)$ establishes a bijective correspondence between the edges $e \in \mathcal{E}_{\partial}$ and the vertices v of G. This gives a canonical $\Bbbk$-linear isomorphism

$$H_c^{\partial} \simeq \bigotimes_{v} H_v(G) = H(G).$$

As in Section 13.1.1, an unoriented edge e of P not lying in ∂P gives rise to two opposite oriented edges $e_1, e_2 \in \mathcal{E}$ and a vector $*_e \in H_c(e_1) \otimes H_c(e_2)$ independent of the numeration of e_1, e_2. The unordered tensor product of these vectors over all such e is a vector

$$\otimes_e *_e \in \bigotimes_{e \in \mathcal{E}} H_c(e).$$

We let $*_c \in H_c \otimes H(G)$ be the image of $\otimes_e *_e$ under the isomorphisms

$$\bigotimes_{e \in \mathcal{E}} H_c(e) \simeq H_c \otimes H_c^{\partial} \simeq H_c \otimes H(G).$$

As in Section 13.1.1, a vertex x of P lying in $\mathrm{Int}(M)$ determines a $\mathcal{C}$-colored graph Γ_x^c and a vector

$$\mathbb{F}_{\mathcal{C}}(\Gamma_x^c) \in H(\Gamma_x^c)^\star \simeq \otimes_{e_x} H_c(e_x)^\star,$$

where e_x runs over all edges of P incident to x and oriented away from x. Let $V_c \in H_c^\star$ be the image of the unordered tensor product $\otimes_x \mathbb{F}_{\mathcal{C}}(\Gamma_x^c)$ over all such x under the canonical $\Bbbk$-linear isomorphism

$$\bigotimes_x H(\Gamma_x^c)^\star \simeq \bigotimes_x \bigotimes_{e_x} H_c(e_x)^\star \simeq H_c^\star.$$

Set

$$|c| = (V_c \otimes \mathrm{id}_{H(G)})(*_c) \in \Bbbk \otimes H(G) \simeq H(G).$$

We now define $|M, G|_{\mathcal{C}} \in H(G)$ by

$$|M, G|_{\mathcal{C}} = (\dim(\mathcal{C}))^{-|M \setminus P|} \sum_c \dim(c)\, |c|,$$

where $|M \setminus P|$ is the number of connected components of $M \setminus P$ and c runs over all colorings of P.

Theorem 13.2. *The vector $|M, G|_{\mathcal{C}} \in H(G)$ is a topological invariant of the pair (M, G) independent of the choice of P.*

Proof. The topological invariance here means that for any orientation preserving homeomorphism f of M onto a compact oriented 3-manifold M', we have $|M', G'|_{\mathcal{C}} = H(f)(|M, G|_{\mathcal{C}})$ where $G' \subset \partial M'$ is the I-colored graph obtained by transporting G along f and $H(f)\colon H(G) \to H(G')$ is the isomorphism induced by f. This invariance follows from the independence of P, and the latter is verified exactly as in the proof of Theorem 13.1 using Theorem 11.5. □

If G and G' are isotopic I-colored graphs in ∂M, then $|M, G'|_{\mathcal{C}} = |M, G|_{\mathcal{C}}$ up to the isomorphism $H(G) \simeq H(G')$ induced by an isotopy of G to G'.

The invariant $|M, G|_{\mathcal{C}}$ is multiplicative: for any compact oriented 3-manifolds M_1, M_2 and any I-colored graphs $G_1 \subset M_1$, $G_2 \subset M_2$, we have

$$|M_1 \sqcup M_2, G_1 \sqcup G_2|_{\mathcal{C}} = |M_1, G_1|_{\mathcal{C}} \otimes |M_2, G_2|_{\mathcal{C}} \tag{13.5}$$

up to the canonical isomorphism

$$H(G_1 \sqcup G_2; \partial(M_1 \sqcup M_2)) \simeq H(G_1; \partial M_1) \otimes H(G_2; \partial M_2).$$

If a 3-manifold N is obtained from a compact oriented 3-manifold M by removing the interior of a closed 3-ball in $\mathrm{Int}(M)$, then $\partial M \subset \partial N$ and we can consider any I-colored graph $G \subset \partial M$ as an I-colored graph in ∂N. A skeleton of (N, G) yields a skeleton of (M, G) via the inclusion $N \subset M$. Consequently, $|N, G|_{\mathcal{C}} = |M, G|_{\mathcal{C}}$.

13.2.2 The scalar invariant

Taking $G = \emptyset$ in Theorem 13.2, we obtain a scalar topological invariant of a compact oriented 3-manifold M:

$$|M|_{\mathcal{C}} = |M, \emptyset|_{\mathcal{C}} \in H(\emptyset) = \Bbbk.$$

If M is closed, then this is the same invariant as in Theorem 13.1. If a 3-manifold N is obtained from a compact oriented 3-manifold M by removing the interior of a closed 3-ball in $\mathrm{Int}(M)$, then $|N|_{\mathcal{C}} = |M|_{\mathcal{C}}$. As a consequence,

$$|S^2 \times [0,1]|_{\mathcal{C}} = |B|_{\mathcal{C}} = |S^3|_{\mathcal{C}} = (\dim(\mathcal{C}))^{-1},$$

where B is a closed 3-ball. More generally, for any closed oriented surface Σ, we can compute $|\Sigma \times [0,1]|_{\mathcal{C}}$ on the skeleton $\Sigma \times \{1/2\}$ of $(\Sigma \times [0,1], \emptyset)$. This gives

$$|\Sigma \times [0,1]|_{\mathcal{C}} = (\dim(\mathcal{C}))^{-2} \sum_{i \in I} (\dim(i))^{\chi(\Sigma)}.$$

In particular,

$$|S^1 \times S^1 \times [0,1]|_{\mathcal{C}} = (\dim(\mathcal{C}))^{-2} \,\mathrm{card}(I).$$

Given a compact connected oriented 3-manifold M, a version of $|M|_{\mathcal{C}}$ can be defined for an arbitrary spherical fusion $\Bbbk$-category $\mathcal{C}$ without the assumption that $\dim(\mathcal{C})$ is invertible in $\Bbbk$. The definition uses the theory of spines. By a *spine* of M, we mean an oriented special 2-polyhedron $P \subset \mathrm{Int}(M)$ with at least one vertex such that either $\partial M = \emptyset$ and $M \setminus P$ is an open ball or $\partial M \neq \emptyset$ and $M \setminus P$ is homeomorphic to $\partial M \times [0,1)$. Any compact connected oriented 3-manifold M has a spine, and any two spines of M can be related by the moves $T^{1,2}$, $T^{2,1}$, and the lune moves (see Section 11.3.2). Up to the orientation of the regions, this follows from [Mat, Theorem 1.2.30] at least for spines with only disk regions and two or more vertices; the case of arbitrary spines easily follows using the lune moves. The orientation of any region may be inverted by the moves, as can be shown using [Mat, Proposition 1.2.8]. The sum

$$\|M\|_{\mathcal{C}} = \sum_{c} \dim(c)\, |c| \in \Bbbk$$

over all I-colorings c of a spine of M is preserved under the moves in question and so is a topological invariant of M. If $\dim(\mathcal{C})$ is invertible in $\Bbbk$, then

$$|M|_{\mathcal{C}} = (\dim(\mathcal{C}))^{-n}\, \|M\|_{\mathcal{C}}$$

where $n = 1$ if $\partial M = \emptyset$ and n is the number of components of ∂M if $\partial M \neq \emptyset$.

13.2.3 Exercise

Let G be a finite group and let $G_{\Bbbk} = G_{\Bbbk}^{1,1}$ be the spherical fusion $\Bbbk$-category from Example 4.5.4. Prove that for any compact connected oriented 3-manifold M, we have

$$\|M\|_{G_{\Bbbk}} = \mathrm{card}(\mathrm{Hom}(\pi, G)) \cdot 1_{\Bbbk},$$

where $\pi = \pi_1(M)$ if $\partial M = \emptyset$, $\pi = \pi_1(M/\partial M)$ if $\partial M \neq \emptyset$, and $\mathrm{Hom}(\pi, G)$ is the (finite) set of group homomorphisms from π to G.

13.3 Construction of the state sum TQFT

We construct a 3-dimensional TQFT $|\cdot|_{\mathcal{C}} \colon \mathrm{Cob}_3 \to \mathrm{Mod}_{\Bbbk}$ in three steps. First, we define operator invariants of morphisms in Cob_3. Second, we combine them into a "pseudo-TQFT". And third, we transform the latter into a genuine TQFT.

13.3.1 An operator invariant of morphisms

Let Σ_0, Σ_1 be closed oriented surfaces, and let $f \colon \Sigma_0 \to \Sigma_1$ be a morphism in Cob_3 represented by a pair (M, h) where M is a compact oriented 3-manifold and $h \colon (-\Sigma_0) \sqcup \Sigma_1 \to \partial M$ is an orientation preserving homeomorphism. Any I-colored graphs $G_0 \subset \Sigma_0$ and $G_1 \subset \Sigma_1$ yield an I-colored graph $h(G_0^{\mathrm{op}} \cup G_1)$ in ∂M, where G_0^{op} is the I-colored graph in $-\Sigma_0$ derived from G_0 as in Section 12.3.3. Theorem 13.2 gives a vector

$$\begin{aligned} |M, h(G_0^{\mathrm{op}} \cup G_1)|_{\mathcal{C}} &\in H(h(G_0^{\mathrm{op}} \cup G_1); \partial M) \simeq H(G_0^{\mathrm{op}} \cup G_1; (-\Sigma_0) \sqcup \Sigma_1) \\ &\simeq H(G_0^{\mathrm{op}}) \otimes H(G_1) \simeq H(G_0)^\star \otimes H(G_1) \simeq \mathrm{Hom}_{\Bbbk}\big(H(G_0), H(G_1)\big) \end{aligned}$$

where the first isomorphism is induced by h^{-1}, the second and fourth are obvious, and the third is induced by the isomorphism $H(G_0^{\mathrm{op}}) \simeq H(G_0)^\star$ discussed in Section 12.3.4. Set

$$|f, G_0, G_1| = \frac{(\dim(\mathcal{C}))^{|\Sigma_1 \setminus G_1|}}{\dim(G_1)} \, |M, h(G_0^{\mathrm{op}} \cup G_1)|_{\mathcal{C}} \colon H(G_0) \to H(G_1), \tag{13.6}$$

where $|\Sigma_1 \setminus G_1|$ is the number of components of $\Sigma_1 \setminus G_1$ and $\dim(G_1)$ is the product over all edges of G_1 of the dimensions of their colors. Theorem 13.2 implies that the homomorphism $|f, G_0, G_1|$ does not depend on the choice of the pair (M, h) representing f. To compute $|f, G_0, G_1|$ explicitly, we pick a basis $\{a_\alpha\}_\alpha$ of $H(G_0)$, consider the dual basis $\{b_\alpha\}_\alpha$ of $H(G_0^{\mathrm{op}})$, and expand

$$|M, h(G_0^{\mathrm{op}} \cup G_1)|_{\mathcal{C}} = \sum_\alpha b_\alpha \otimes z_\alpha \in H(G_0^{\mathrm{op}}) \otimes H(G_1)$$

with $z_\alpha \in H(G_1)$. Then, for all indices α,

$$|f, G_0, G_1|(a_\alpha) = \frac{(\dim(\mathcal{C}))^{|\Sigma_1 \setminus G_1|}}{\dim(G_1)} \, z_\alpha.$$

The normalization factor in the definition of $|f, G_0, G_1|$ is justified by the next lemma involving skeletons of surfaces. By a *skeleton* of a closed surface Σ, we mean an oriented graph G embedded in Σ such that all vertices of G have valence ≥ 2 and all components of $\Sigma \setminus G$ are open disks. For example, the vertices and edges of a triangulation of Σ (with any orientation of the edges) form a skeleton of Σ. For a skeleton $G \subset \Sigma$, we let $\mathrm{col}(G)$ be the set of all maps from the set of edges of G to I.

Lemma 13.3. *Let $f_0\colon \Sigma_0 \to \Sigma$ and $f_1\colon \Sigma \to \Sigma_1$ be morphisms in* Cob_3. *For any I-colored graphs $G_0 \subset \Sigma_0$, $G_1 \subset \Sigma_1$ and any skeleton G of Σ,*

$$|f_1 f_0, G_0, G_1| = \sum_{c \in \mathrm{col}(G)} |f_1, (G,c), G_1| \circ |f_0, G_0, (G,c)|\colon H(G_0) \to H(G_1).$$

Proof. Represent f_0 and f_1, respectively, by pairs

$$(M_0, h_0\colon (-\Sigma_0) \sqcup \Sigma \to \partial M_0) \quad \text{and} \quad (M_1, h_1\colon (-\Sigma) \sqcup \Sigma_1 \to \partial M_1)$$

as above. The morphism $f_1 f_0\colon \Sigma_0 \to \Sigma_1$ is represented by the pair (M, h), where M is obtained by gluing M_0 to M_1 along $h_1 h_0^{-1}\colon h_0(\Sigma) \to h_1(\Sigma)$ and

$$h = h_0|_{\Sigma_0} \sqcup h_1|_{\Sigma_2}\colon (-\Sigma_0) \sqcup \Sigma_2 \to \partial M.$$

To simplify notation, we identify $h_0(\Sigma) = h_1(\Sigma) \subset \mathrm{Int}(M)$ with Σ via h_0. Pick skeletons P_0 of $(M_0, h_0(G_0^{\mathrm{op}} \cup G))$ and P_1 of $(M_1, h_1(G^{\mathrm{op}} \cup G_1))$. For $i = 0, 1$, the ∂-cylindricity of P_i implies that every vertex v of $G = \Sigma \cap \partial P_i$ is adjacent to a single edge d_v^i of P_i not lying in ∂P_i. By the definition of ∂P_i, each edge a of G is adjacent to a single region D_a^i of P_i. Each component X of $\Sigma \backslash G$ is a disk adjacent to a single component B_X^i of $M_i \setminus P_i$, and, by the definition of a skeleton, B_X^i is homeomorphic to $X \times [0, 1)$. Clearly, $P = P_0 \cup P_1 \subset M$ is a 2-polyhedron which we stratify as follows. The vertices of P are the vertices of P_0 and P_1 not lying in G. The edges of P are the edges of P_0 and P_1 disjoint from G and the arcs $\{d_v^0 \cup d_v^1\}_v$ where v runs over all vertices of G. The regions of P are the regions of P_0 and P_1 disjoint from G and the disks $\{D_a = D_a^0 \cup D_a^1\}_a$ where a runs over all edges of G. The orientations of P_0 and P_1 are compatible at G and induce an orientation of the stratified 2-polyhedron P. Finally, the components of $M \setminus P$ are the components of $M_0 \setminus P_0$ and $M_1 \setminus P_1$ disjoint from Σ and the open 3-balls $\{B_X^0 \cup B_X^1\}_X$, where X runs over all components of $\Sigma \backslash G$. Consequently, P is a skeleton of $(M, h(G_0^{\mathrm{op}} \cup G_1))$ and

$$|M \setminus P| = |M_0 \setminus P_0| + |M_1 \setminus P_1| - |\Sigma \setminus G|.$$

A coloring c of P determines an element $\overline{c}$ of the set $\mathrm{col}(G)$ by $\overline{c}(a) = c(D_a)$. The colorings c of P inducing a given $\overline{c} \in \mathrm{col}(G)$ bijectively correspond via restriction to pairs (a coloring c_0 of P_0, a coloring c_1 of P_1) where $\partial P_0 = G_0^{\mathrm{op}} \cup (G, \overline{c})$ and $\partial P_1 = (G, \overline{c})^{\mathrm{op}} \cup G_1$. Under this correspondence,

$$\dim(c) = \dim(c_0) \dim(c_1) \, (\dim(G, \overline{c}))^{-1}.$$

The vector

$$|c| \in H(G_0^{\mathrm{op}} \cup G_1) \simeq H(G_0^{\mathrm{op}}) \otimes H(G_1)$$

is obtained from

$$\begin{aligned}|c_0| \otimes |c_1| &\in H(G_0^{\mathrm{op}} \cup (G,\overline{c})) \otimes H((G,\overline{c})^{\mathrm{op}} \cup G_1)\\ &\simeq H(G_0^{\mathrm{op}}) \otimes H(G,\overline{c}) \otimes H((G,\overline{c})^{\mathrm{op}}) \otimes H(G_1)\end{aligned}$$

by applying the evaluation pairing $\omega_{(G,\overline{c})}\colon H(G,\overline{c}) \otimes H((G,\overline{c})^{\mathrm{op}}) \to \Bbbk$ from Section 12.3.3. These computations directly imply the claim of the lemma. □

Lemma 13.4. *Let $f\colon \Sigma_0 \to \Sigma_1$ and $f'\colon \Sigma_0' \to \Sigma_1'$ be morphisms in* Cob_3. *For any I-colored graphs $G_i \subset \Sigma_i$, $G_i' \subset \Sigma_i'$ with $i \in \{0,1\}$,*

$$|f \otimes f', G_0 \sqcup G_0', G_1 \sqcup G_1'| = |f, G_0, G_1| \otimes |f', G_0', G_1'|$$

up to the canonical isomorphisms $H(G_i \sqcup G_i') \simeq H(G_i) \otimes H(G_i')$ with $i \in \{0,1\}$.

Proof. This follows from the multiplicativity of the invariant $|M,G|_{\mathcal{C}}$, see Formula (13.5). □

13.3.2 The TQFT

For a closed oriented surface Σ and a skeleton $G \subset \Sigma$, consider the $\Bbbk$-module

$$|\Sigma, G|^\circ = \bigoplus_{c \in \mathrm{col}(G)} H((G,c);\Sigma).$$

For a morphism $f\colon \Sigma_0 \to \Sigma_1$ in Cob_3 and skeletons $G_0 \subset \Sigma_0$ and $G_1 \subset \Sigma_1$, consider the $\Bbbk$-linear homomorphism

$$|f, G_0, G_1|^\circ\colon |\Sigma_0, G_0|^\circ \to |\Sigma_1, G_1|^\circ$$

whose restriction to every summand $H((G_0,c_0);\Sigma_0)$ of $|\Sigma_0, G_0|^\circ$ is equal to

$$\sum_{c_1 \in \mathrm{col}(G_1)} |f, (G_0,c_0), (G_1,c_1)|\colon H((G_0,c_0);\Sigma_0) \to |\Sigma_1, G_1|^\circ.$$

Lemma 13.3 implies that for any f_0, f_1 as in this lemma and for any skeletons $G_0 \subset \Sigma_0$, $G \subset \Sigma$, $G_1 \subset \Sigma_1$,

$$|f_1 f_0, G_0, G_1|^\circ = |f_1, G, G_1|^\circ \circ |f_0, G_0, G|^\circ. \tag{13.7}$$

This assigns a $\Bbbk$-module to each closed oriented surface endowed with a skeleton and assigns a homomorphism of these modules to each morphism in Cob_3 whose source and target are endowed with skeletons. This assignment satisfies all requirements on a functor from an appropriate version of Cob_3 to $\mathrm{Mod}_\Bbbk$ except one: this "functor" may carry the identity morphisms to non-identity homomorphisms.

Next, we transform the above "functor" into a TQFT and simultaneously get rid of the skeletons of surfaces (cf. [Tu1], Section VII.3). For any skeletons G, G' of a closed oriented surface Σ, set

$$p(G,G') = |\mathrm{id}_\Sigma, G, G'|^\circ \colon |\Sigma, G|^\circ \to |\Sigma, G'|^\circ,$$

where $\mathrm{id}_\Sigma \colon \Sigma \to \Sigma$ is the identity morphism in Cob_3. Formula (13.7) implies that

$$p(G,G'') = p(G',G'')\, p(G,G') \tag{13.8}$$

for any skeletons G, G', G'' of Σ. In particular, $(p(G,G))^2 = p(G,G)$. Set

$$|\Sigma, G| = \mathrm{Im}(p(G,G)) = \{x \in |\Sigma, G|^\circ \mid p(G,G)(x) = x\}.$$

Formula (13.8) implies that for any skeletons G, G' of Σ, the $\Bbbk$-linear homomorphism $p(G,G')$ restricts to an isomorphism $|\Sigma, G| \to |\Sigma, G'|$ and the family $(\{|\Sigma, G|\}_G, \{p(G,G')\}_{G,G'})$ is a projective system of $\Bbbk$-modules and $\Bbbk$-linear isomorphisms. The projective limit

$$|\Sigma| = \varprojlim |\Sigma, G|$$

is a $\Bbbk$-module depending only on Σ. For each skeleton G of Σ, we have a $\Bbbk$-linear cone isomorphism $\tau_G \colon |\Sigma| \to |\Sigma, G|$ and for any skeletons G, G' of Σ,

$$\tau_{G'} = p(G,G') \circ \tau_G. \tag{13.9}$$

We next associate with each morphism $f \colon \Sigma_0 \to \Sigma_1$ in Cob_3 a homomorphism $|f| \colon |\Sigma_0| \to |\Sigma_1|$. Pick any skeletons $G_0 \subset \Sigma_0$ and $G_1 \subset \Sigma_1$. Formulas (13.7) and $\mathrm{id}_{\Sigma_1} \circ f = f$ imply that

$$|f, G_0, G_1|^\circ = p(G_1, G_1)\, |f, G_0, G_1|^\circ \colon |\Sigma_0, G_0|^\circ \to |\Sigma_1, G_1|^\circ.$$

Consequently, the image of $|f, G_0, G_1|^\circ$ is contained in $|\Sigma_1, G_1|$. Set

$$|f| = \tau_{G_1}^{-1} \circ |f, G_0, G_1|^\circ \circ \tau_{G_0} \colon |\Sigma_0| \to |\Sigma_1|.$$

This homomorphism does not depend on the choice of G_0 and G_1. Indeed, for any skeletons $G_0' \subset \Sigma_0$ and $G_1' \subset \Sigma_1$, Formulas (13.7) and $f \circ \mathrm{id}_{\Sigma_0} = \mathrm{id}_{\Sigma_1} \circ f$ give

$$|f, G_0', G_1'|^\circ \circ p(G_0, G_0') = p(G_1, G_1') \circ |f, G_0, G_1|^\circ.$$

This and (13.9) imply that the following diagram commutes:

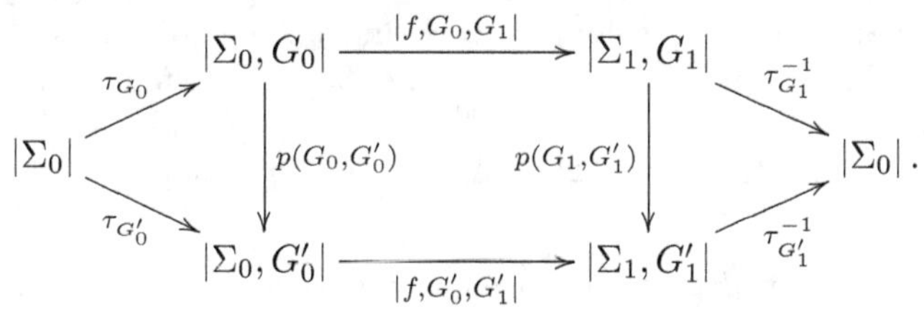

It is clear that the rule $\Sigma \mapsto |\Sigma|, f \mapsto |f|$ defines a functor $|\cdot|\colon \mathrm{Cob}_3 \to \mathrm{Mod}_{\Bbbk}$. We endow this functor with monoidal constraints as follows. Note that the empty surface $\emptyset$ has a unique (empty) skeleton $G_\emptyset$ and $p(G_\emptyset, G_\emptyset)$ is the identity automorphism of $|\emptyset, G_\emptyset|^\circ = H(G_\emptyset) = \Bbbk$. We let $|\cdot|_0\colon \Bbbk \to |\emptyset|$ be the $\Bbbk$-linear isomorphism inverse to the cone isomorphism

$$\tau_{G_\emptyset}\colon |\emptyset| \to |\emptyset, G_\emptyset| = |\emptyset, G_\emptyset|^\circ = \Bbbk.$$

Given two closed oriented surfaces Σ and Σ', we pick any skeletons G of Σ and G' of Σ' and define $|\cdot|_2(\Sigma, \Sigma')$ to be the composition of the $\Bbbk$-linear isomorphisms

$$|\Sigma| \otimes_{\Bbbk} |\Sigma'| \xrightarrow{\tau_G \otimes_{\Bbbk} \tau_{G'}} |\Sigma, G| \otimes_{\Bbbk} |\Sigma', G'| \simeq |\Sigma \sqcup \Sigma', G \sqcup G'| \xrightarrow{\tau^{-1}_{G \sqcup G'}} |\Sigma \sqcup \Sigma'|,$$

where the middle isomorphism is the composition of the canonical isomorphism $|\Sigma, G| \otimes_{\Bbbk} |\Sigma', G'| \simeq |\Sigma, G| \otimes |\Sigma', G'|$ with the isomorphism provided by Lemma 13.4 applied to $f = \mathrm{id}_\Sigma$ and $f' = \mathrm{id}_{\Sigma'}$. It is easy to check that $|\cdot|_2(\Sigma, \Sigma')$ does not depend on the choice of G and G'. These monoidal constraints turn $|\cdot|$ into a symmetric strong monoidal functor. We denote this functor by $|\cdot|_{\mathcal{C},I}$ or, shorter, by $|\cdot|_{\mathcal{C}}$. Combining the considerations above, we obtain the following theorem.

Theorem 13.5. *The functor $|\cdot|_{\mathcal{C}}$ is a 3-dimensional TQFT.*

For any closed oriented 3-manifold M, the scalar $|M|_{\mathcal{C}} \in \Bbbk$ produced by the TQFT $|\cdot|_{\mathcal{C}}$ is the invariant introduced in Section 13.1. The naturality of $\mathbb{F}_{\mathcal{C}}$ implies that the isomorphism type of the TQFT $|\cdot|_{\mathcal{C}}$ does not depend on the choice of the representative set I of simple objects of $\mathcal{C}$. Indeed, if I' is another representative set of simple objects of $\mathcal{C}$, then there is a unique bijection $\varphi\colon I \to I'$ such that the objects i and $\varphi(i)$ are isomorphic for all $i \in I$. Any family of isomorphisms $\{i \to \varphi(i)\}_{i \in I}$ canonically induces an isomorphism of TQFTs $|\cdot|_{\mathcal{C},I} \simeq |\cdot|_{\mathcal{C},I'}$.

13.4 Computations in the state sum TQFT

We prove that the $\Bbbk$-module $|S^2|_{\mathcal{C}}$ is isomorphic to $\Bbbk$ and use this result to describe the behaviour of $|\cdot|_{\mathcal{C}}$ under connected summation of manifolds.

13.4.1 The module associated with the 2-sphere

An explicit computation of the $\Bbbk$-modules associated by the TQFT $|\cdot|_{\mathcal{C}}$ with closed oriented surfaces requires considerable work and will be done in Section 16.2 (see Theorem 16.2). Here we compute the $\Bbbk$-module $|S^2|_{\mathcal{C}}$. By Lemma 10.1, the Hattori-Stallings dimension of $|S^2|_{\mathcal{C}}$ is equal to $|S^1 \times S^2|_{\mathcal{C}} = 1_{\Bbbk}$. Consequently, if $\Bbbk$ is a field of characteristic zero, then $|S^2|_{\mathcal{C}} \simeq \Bbbk$. We extend this result to any commutative ring $\Bbbk$.

Lemma 13.6. *The $\Bbbk$-module $|S^2|_{\mathcal{C}}$ is isomorphic to $\Bbbk$.*

Proof. Let $G \subset S^2$ be an oriented embedded circle viewed as a graph with one vertex x and one edge e. Clearly, G is a skeleton of S^2. Assigning $i \in I$ to e, we turn G into an I-colored graph G_i:

$G_i =$ 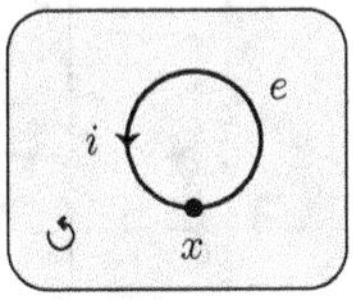 .

By definition, the $\Bbbk$-module $|S^2|_{\mathcal{C}}$ is isomorphic to the image of the $\Bbbk$-linear endomorphism $p(G,G) = \sum_{i,j \in I} p_i^j$ of

$$|G; S^2|^\circ = \bigoplus_{i \in I} H(G_i),$$

where

$$p_i^j = |\mathrm{id}_{S^2}, G_i, G_j| \colon H(G_i) \to H(G_j).$$

Each module $H(G_i) = H(G_i; S^2)$ is isomorphic to $\Bbbk$ via the cone isomorphism

$$H(G_i) \to \mathrm{Hom}_{\mathcal{C}}(\mathbb{1}, i \otimes i^*) \simeq \mathrm{Hom}_{\mathcal{C}}(i,i) \simeq \Bbbk.$$

To compute p_i^j, set $M = S^2 \times [0,1]$ and consider the 2-polyhedron

$$P = (G \times [0,1]) \cup \left(S^2 \times \left\{ \frac{1}{2} \right\} \right) \subset M.$$

We stratify P by taking the points $(x,0), (x,\frac{1}{2}), (x,1)$ as vertices and the arcs

$$\{x\} \times \left[0, \frac{1}{2}\right], \quad \{x\} \times \left[\frac{1}{2}, 1\right], \quad e_0 = e \times \{0\}, \quad e_{1/2} = e \times \left\{\frac{1}{2}\right\}, \quad e_1 = e \times \{1\}$$

as edges. We endow the two regions of P contained in $S^2 \times \{\frac{1}{2}\}$ with orientation induced by that of S^2 and orient the other two regions of P so that

$$\partial P = G^{\mathrm{op}} \sqcup G = (G^{\mathrm{op}} \times \{0\}) \cup (G \times \{1\}),$$

where G^{op} is obtained from G by reversing orientation of the edge e. Clearly, P is a skeleton of the pair $(M, G^{\mathrm{op}} \sqcup G)$. Consider the I-colored graph

$$G_i^j = G_i^{\mathrm{op}} \sqcup G_j \subset \partial M.$$

The maps $c \colon \mathrm{Reg}(P) \to I$ extending the coloring of G_i^j are numerated by the colors $z, w \in I$ of the two regions of P contained in $S^2 \times \{\frac{1}{2}\}$:

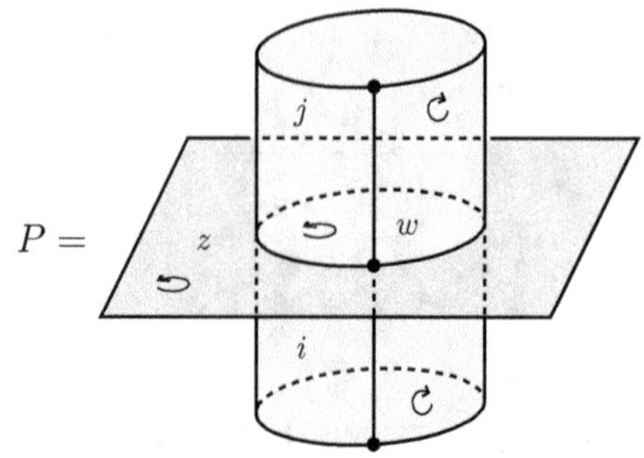

The link graph $\Gamma = \Gamma^c_{(x,\frac{1}{2})}$ of the vertex $(x, \frac{1}{2})$ of P is as follows:

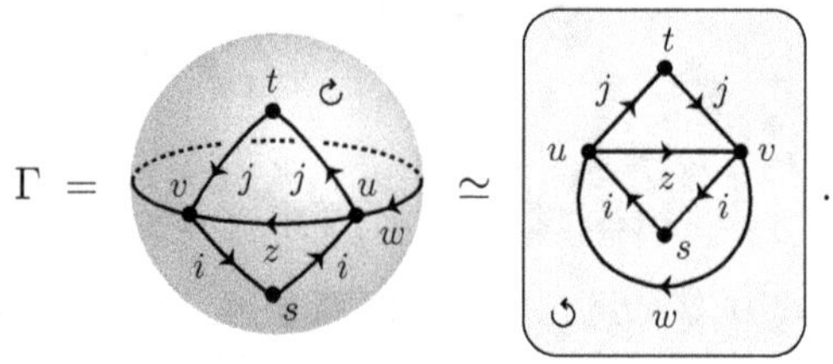

The oriented edges of P with tail in $\mathrm{Int}(M)$ are the edges of P starting at the vertex $(x, \frac{1}{2})$. Therefore

$$H_c = H(\Gamma) = H_s(\Gamma) \otimes H_t(\Gamma) \otimes H_u(\Gamma) \otimes H_v(\Gamma).$$

Consider the contraction vectors

$$*_{e_0} \in H(G_i^{\mathrm{op}}) \otimes H_s(\Gamma), \quad *_{e_1} \in H(G_j) \otimes H_t(\Gamma), \quad *_{e_{1/2}} \in H_u(\Gamma) \otimes H_v(\Gamma).$$

Note that $H(G_j) = \Bbbk\, a_j$, where $a_j \in H(G_j)$ is the pre-image of

$$\mathrm{coev}_j \in \mathrm{Hom}_{\mathcal{C}}(\mathbb{1}, j \otimes j^*) = \Bbbk\, \mathrm{coev}_j$$

under the cone isomorphism $H(G_j) \to \mathrm{Hom}_{\mathcal{C}}(\mathbb{1}, j \otimes j^*)$. Similarly,

$$H(G_i^{\mathrm{op}}) = \Bbbk\, b_i, \quad H_s(\Gamma) = \Bbbk\, g_i, \quad H_t(\Gamma) = \Bbbk\, h_j,$$

where b_i, g_i, h_j are the elements of these modules defined as the pre-images of coev_i and coev_j under the appropriate cone isomorphisms. By Example 12.3.5,

$$*_{e_0} = (\dim(i))^{-1}\, b_i \otimes g_i \quad \text{and} \quad *_{e_1} = (\dim(j))^{-1}\, a_j \otimes h_j.$$

By definition, the vector $*_c \in H_c \otimes H(G_i^j)$ is computed by

$$*_c = *_{e_0} \otimes *_{e_1} \otimes *_{e_{1/2}} = (\dim(i))^{-1}(\dim(j))^{-1}\, b_i \otimes g_i \otimes a_j \otimes h_j \otimes *_{e_{1/2}}$$

and the vector $|c| \in H(G_i^j) = H(G_i^{\mathrm{op}}) \otimes H(G_j)$ is computed by

$$|c| = (\mathbb{F}_{\mathcal{C}}(\Gamma) \otimes \mathrm{id}_{H(G_i^j)})(*_c) = (\dim(i))^{-1}(\dim(j))^{-1}\, \mathbb{F}_{\mathcal{C}}(\gamma)(*_{e_{1/2}})\, b_i \otimes a_j,$$

where

$\gamma =$ 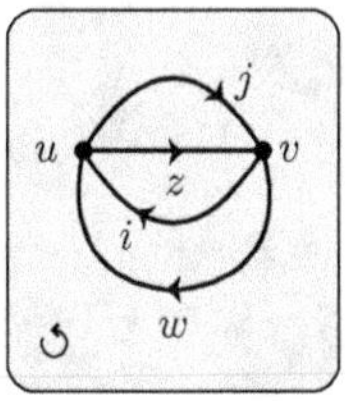

is the I-colored graph in the 2-sphere obtained from Γ by the transformation of Section 12.2.2 at the vertices s and t using the vectors g_i and h_j. Note that the vertices u, v of γ are in duality induced by reflection with respect to a vertical line and $*_{e_{1/2}} = *_{u,v}$. By Example 12.3.6,

$$\mathbb{F}_{\mathcal{C}}(\gamma)(*_{u,v}) = \dim H_u(\gamma) = \dim \mathrm{Hom}_{\mathcal{C}}(\mathbb{1}, j^* \otimes z^* \otimes i \otimes w) = N^{\mathbb{1}}_{j^*\otimes z^*\otimes i\otimes w}.$$

Therefore

$$|c| = (\dim(i))^{-1}(\dim(j))^{-1}\, N^{\mathbb{1}}_{j^*\otimes z^*\otimes i\otimes w}\, b_i \otimes a_j.$$

By definition, $\dim(c) = \dim(i)\dim(j)\dim(z)\dim(w)$ and

$$\begin{aligned} |M, G_i^j|_{\mathcal{C}} &= (\dim(\mathcal{C}))^{-|M\setminus P|} \sum_c \dim(c)|c| \\ &= (\dim(\mathcal{C}))^{-4} \sum_{z,w\in I} \dim(z)\dim(w) N^{\mathbb{1}}_{j^*\otimes z^*\otimes i\otimes w}\, b_i \otimes a_j \\ &= (\dim(\mathcal{C}))^{-3} \dim(i)\dim(j)\, b_i \otimes a_j \in H(G_i^{\mathrm{op}}) \otimes H(G_j) = H(G_i^j), \end{aligned}$$

where we use Lemma 4.7 and the identity $\dim(z^*) = \dim(z)$. The corresponding homomorphism $H(G_i) \to H(G_j)$ carries the generating vector $a_i \in H(G_i)$ to $(\dim(\mathcal{C}))^{-3}(\dim(i))^2 \dim(j)\, a_j$ because the evaluation of b_i on a_i is equal to $\dim(i)$. By (13.6) and the equality $|S^2 \setminus G| = 1$, we have

$$p_i^j(a_i) = (\dim(\mathcal{C}))^{-2}(\dim(i))^2\, a_j.$$

We conclude that

$$p(G,G)(a_i) = \sum_{j\in I} p_i^j(a_i) = (\dim(\mathcal{C}))^{-2}(\dim(i))^2 \sum_{j\in I} a_j.$$

Hence the image of $p(G,G)$ is generated by the non-zero vector $\sum_{j\in I} a_j$. Therefore $|S^2|_{\mathcal{C}} \simeq \Bbbk$. □

13.4.2 Connected sums

Since $|S^2|_{\mathcal{C}} \simeq \Bbbk$ by Lemma 13.6 and $|S^3|_{\mathcal{C}} = (\dim(\mathcal{C}))^{-1}$ by Section 13.1.3, we can apply Lemma 10.2 to the TQFT $|\cdot|_{\mathcal{C}}$ and obtain that for any compact connected oriented 3-manifolds M_0 and M_1,

$$\begin{aligned}&|M_0\#M_1, -\partial M_0, \partial M_1|_{\mathcal{C}}\\ &= \dim(\mathcal{C})\, |M_1, \emptyset, \partial M_1|_{\mathcal{C}} \circ |M_0, -\partial M_0, \emptyset|_{\mathcal{C}} \colon |-\partial M_0|_{\mathcal{C}} \to |\partial M_1|_{\mathcal{C}}.\end{aligned}$$

In particular, if $\partial M_0 = \partial M_1 = \emptyset$, then $|M_0\#M_1|_{\mathcal{C}} = \dim(\mathcal{C})\, |M_0|_{\mathcal{C}}\, |M_1|_{\mathcal{C}} \in \Bbbk$.

Part IV

Graph Topological Field Theory

Chapter 14

Ribbon graphs in 3-manifolds

Ribbon graphs in a 3-manifold M generalize framed knots and links in M by allowing free ends lying in ∂M and rectangular vertices lying in $\mathrm{Int}(M) = M \setminus \partial M$. We define ribbon graphs in terms of so-called plexuses. Then we explain how to present ribbon graphs by diagrams on skeletons of the manifold and introduce local moves on the diagrams preserving the associated ribbon graphs. The main results of the chapter (Theorems 14.3 and 14.4) claim that any diagrams of isotopic ribbon graphs may be related by our moves.

14.1 Plexuses and knotted plexuses

14.1.1 Plexuses

Plexuses are similar to graphs but differ in that their vertices are rectangles and some of their connected components may be circles without vertices. We proceed to precise definitions. A *circle* is a 1-manifold homeomorphic to S^1. An *arc* is a 1-manifold homeomorphic to the closed interval $[0, 1]$. The boundary points of an arc are called its *endpoints*. A *rectangle* is a 2-manifold with corners homeomorphic to $[0, 1] \times [0, 1]$. The four corner points of a rectangle split its boundary into four arcs called the *sides*. A *coupon* is an oriented rectangle with a distinguished side called the *bottom base*, the opposite side being the *top base*. In our pictures, the bottom bases of coupons are drawn boldface.

A *plexus* is a topological space obtained from a disjoint union of a finite number of oriented circles, oriented arcs, and coupons by gluing some endpoints of the arcs to the bases of the coupons. We require that different endpoints of the arcs are never glued to the same point of a (base of a) coupon. The endpoints of the arcs that are not glued to coupons are called *free ends*. The set of free ends of a plexus d is denoted by ∂d. The arcs and the circles of a plexus are collectively called *strands*. Figure 14.1 exhibits a plexus having 9 arcs, 2 circles, 5 coupons, and 6 free ends.

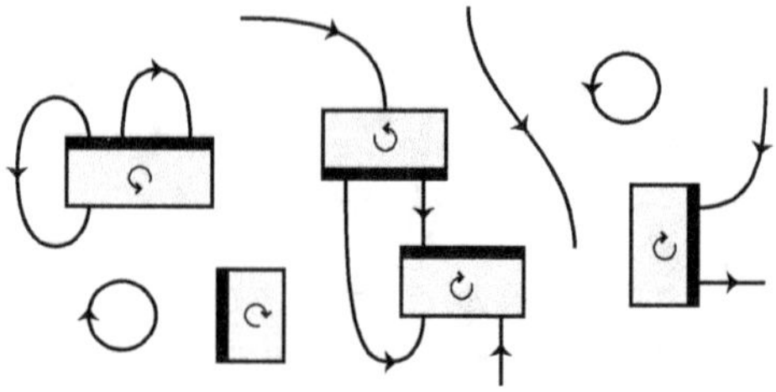

Figure 14.1: A plexus

14.1.2 Knotted plexuses

Let P be an oriented stratified 2-polyhedron. A knotted plexus in P is a plexus drawn in P possibly with double crossings of strands so that at every crossing, one of the two strands is distinguished. More precisely, we say that a plexus is *immersed* in P if it is endowed with a continuous map to P which is a local embedding. A *knotted plexus* d in P is a plexus immersed in P such that:

(i) all coupons of d are embedded in $P \setminus P^{(1)}$ preserving orientation;

(ii) all multiple points of d lie in $P \setminus P^{(1)}$ and are double transversal intersections of interiors of strands of d. At each double point, one of the two meeting strands is distinguished;

(iii) the plexus d is disjoint from the set of vertices of P and $d \cap \partial P = \partial d$;

(iv) the strands of d meet $P^{(1)}$ transversely as in Figure 14.2.

The double points of d are called *crossings*. They are finite in number and lie in $P \setminus P^{(1)}$ away from the coupons of d. Note that the coupons of d are pairwise disjoint in P and all free ends of d lie inside the edges of the graph ∂P.

The intersections of the strands of a knotted plexus d in P with the edges of P not lying in ∂P are called *switches*. Each switch w of d lies on an edge e_w of P. The valence k_w of the edge e_w is called the *valence* of w and is greater than or equal to 2. A neighborhood of w in P is formed by k_w half-planes adjacent to e_w so that the plexus d meets these half-planes along a segment contained in the union of two of them. We say that w is *positive* if the given orientations of the regions of P containing these two half-planes are compatible near w, i.e., if they are induced by an orientation of the horizontal plane in Figure 14.2. Otherwise, w is said to be *negative*. A switch w of d is negative if and only if the orientations of the two regions traversed by d near w induce the same direction on e_w.

14.2 Ribbon graphs and positive diagrams

Fix for the rest of the chapter an oriented compact 3-manifold M and an oriented graph $G \subset \partial M$ (possibly, empty) whose all vertices have valence ≥ 2. In this

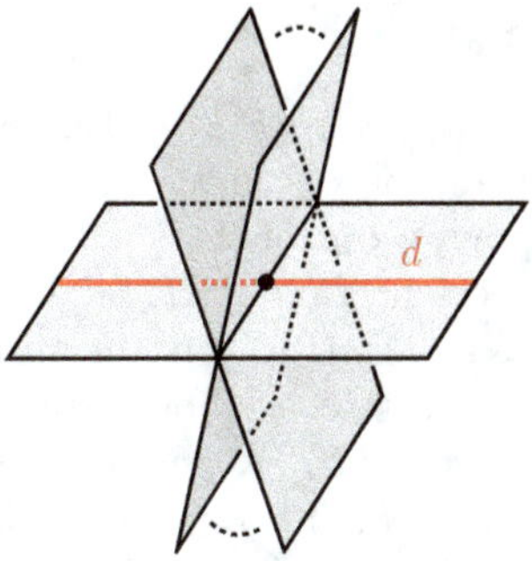

Figure 14.2: A neighborhood of a switch

section, we introduce ribbon graphs in (M, G) and show how to present them by diagrams on skeletons of M.

14.2.1 Ribbon graphs

A *ribbon graph* in M is a plexus embedded in M and equipped with a framing. Here, an *embedding* of a plexus d in M is an embedding $d \hookrightarrow M$ such that $\partial d = d \cap \partial M$ and the arcs of d are transverse to ∂M at all points of ∂d. A *framing* is a continuous map f assigning to every point $x \in d \subset M$ a direction in M at x such that

(i) if $x \in d$ lies in a strand/coupon of d, then the direction $f(x)$ is transverse to this strand/coupon;

(ii) at any point x of a coupon, the orientation of the coupon followed by $f(x)$ yields the given orientation of M;

(iii) if $x \in \partial d \subset \partial M$, then the direction $f(x)$ is tangent to ∂M.

For a ribbon graph $R = (d, f)$ in M, we set $\partial R = \partial d \subset \partial M$ and call the points of ∂R the *free ends* of R. Note that a ribbon graph in M has no free ends if and only if it lies in $\mathrm{Int}(M) = M \backslash \partial M$. Ribbon graphs in M without free ends and without coupons are nothing but framed oriented links in M.

Ribbon graphs in M with free ends in the graph $G \subset \partial M$ are called ribbon graphs in (M, G). More precisely, a *ribbon graph in* (M, G) is a ribbon graph $R = (d, f)$ in M such that each free end v of R lies in the interior of a certain edge e_v of G, the direction $f(v)$ is transverse to e_v in ∂M, and the pair (a positive tangent vector of e_v at v, $f(v)$) is positively oriented in ∂M. Two ribbon graphs in (M, G) are *isotopic* if one can be transformed into the other by an ambient isotopy of M fixing G pointwise.

14.2.2 Positive diagrams

A *ribbon diagram* is a pair (P, d) where P is a skeleton of (M, G) in the sense of Section 11.5.1 and d is a knotted plexus in P. Given a ribbon diagram (P, d), we pick a field of normal directions n on $\mathrm{Int}(P)$ such that the orientation of $\mathrm{Int}(P)$ followed by n yields the given orientation of M. Slightly pushing the distinguished strands at the crossings of d along n, we obtain an embedding $d \hookrightarrow M$ whose image is denoted d^n. To turn d^n into a ribbon graph, we must provide it with a framing. We do it for positive (P, d). A *positive diagram* is a ribbon diagram whose all switches are positive. If (P, d) is a positive diagram, then n continuously extends to all switches of d and determines a framing of d^n in the obvious way. This turns d^n into a ribbon graph in (M, G) well-defined up to isotopy. We say that a ribbon graph in (M, G) is *represented* by the diagram (P, d) if it is isotopic to d^n.

Lemma 14.1. *Any ribbon graph in (M, G) can be represented by a positive diagram.*

Proof. Consider first a ribbon graph $R \subset \mathrm{Int}(M)$ without free ends. Pick a closed regular neighborhood $U \subset \mathrm{Int}(M)$ of R and note that U is a handlebody or a disjoint union of several handlebodies. Pick a triangulation t of $M \setminus \mathrm{Int}(U)$ such that the graph $G \subset \partial M$ does not contain vertices of t and meets the edges of t in ∂M transversely. We extend the surface $\partial U \cup \partial M$ to a special 2-polyhedron $Q \subset M$ by adding the 2-skeleton of the cellular decomposition t^* of $M \setminus \mathrm{Int}(U)$ dual to t and adding meridional disks of U transverse to the edges of t^* lying in ∂U and to the strands of R. We equip the regions of Q contained in $\partial U \cup \partial M$ with the orientation induced by that of M restricted to $M \setminus \mathrm{Int}(U)$. All other regions of Q are oriented arbitrarily. Slightly pushing Q into $\mathrm{Int}(M)$ near ∂M we obtain a frame, Q', of M to which we attach a cylinder $G \times [0, 1]$ lying between ∂M and $\partial Q'$. This gives a skeleton P of (M, G). Pushing R along its framing into $\partial U \subset P$ we obtain a ribbon diagram (P, d) without crossings. This diagram has only positive switches because all regions of P lying in ∂U are oriented coherently. It is clear that (P, d) represents R up to isotopy. For a ribbon graph R in (M, G) with free ends, the construction is similar with a little extra care near the free ends. We first push a regular neighborhood U of R into $\mathrm{Int}(M)$, then proceed as above, and at the end add to the diagram the "legs" $\partial R \times [0, 1] \subset G \times [0, 1] \subset P$. $\square$

14.3 Moves on positive diagrams

We exhibit several local moves on positive diagrams and formulate the main results of the chapter (Theorems 14.3 and 14.4).

14.3.1 The moves

We define moves on positive diagrams preserving the associated ribbon graphs. We first fix our pictorial conventions. The orientation of M in all pictures is right-handed. We usually do not indicate in the pictures the orientations of the regions of skeletons and of the strands of plexuses but it is understood that they are the same before and after the moves. All crossings in the pictures will lie in the regions represented by horizontal planes or by planes parallel to the page. Such a region may be oriented counterclockwise or clockwise. In a region oriented counterclockwise, the distinguished strands at the crossings are represented by continuous (unbroken) lines. In a region oriented clockwise, the distinguished strands at the crossings are represented by broken lines, see Figure 14.3. These conventions ensure that our pictures visually embody the corresponding ribbon graphs: they look near the crossings exactly as the red strands.

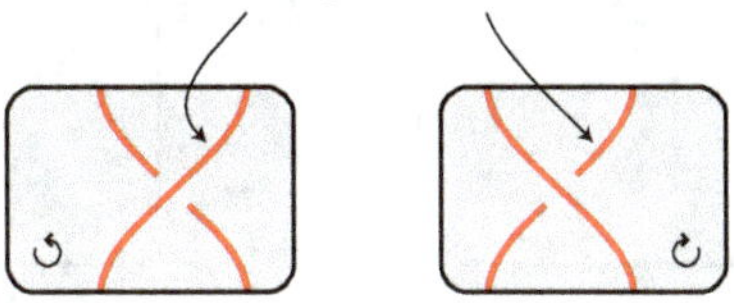

Figure 14.3: A crossing in a ribbon diagram

All our moves on a positive diagram (P, d) keep the graph $\partial P = G$ and the set $\partial d \subset G$. The first family of moves consists of ambient isotopies in M fixing G pointwise and the primary moves T_0, T_1, T_2, T_3 of Section 11.3.1 which modify P inside 3-balls in $\mathrm{Int}(M)$ disjoint from d, see Figure 11.2. These moves are collectively denoted ω_0. We now define further moves on (P, d) which keep P and modify d inside a 3-ball in $\mathrm{Int}(M)$. The moves $\omega_1, \omega_2, \omega_3$ in Figure 14.4 modify d inside a small 2-disk in $\mathrm{Int}(P)$ whose orientation is represented by the counterclockwise orientation of the plane of the page.

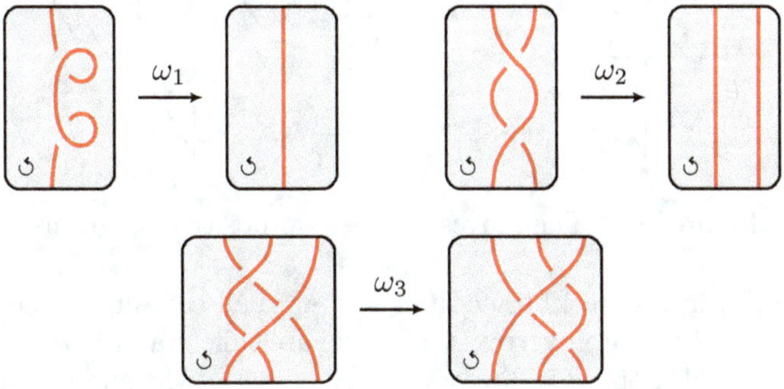

Figure 14.4: The moves $\omega_1 - \omega_3$ on positive diagrams

The moves $\omega_4, \omega_5, \omega_6, \omega_7, \omega_8$ in Figure 14.5 proceed near a trivalent edge of P. In particular, the move $\omega_{8,m,n}$ with $m, n \geq 0$ pushes a coupon with m inputs and n

outputs across a trivalent edge keeping the orientation of the coupon. Recall that the bottom bases of coupons are drawn boldface. The moves $\{\omega_{8,m,n}\}_{m,n}$ are collectively denoted ω_8.

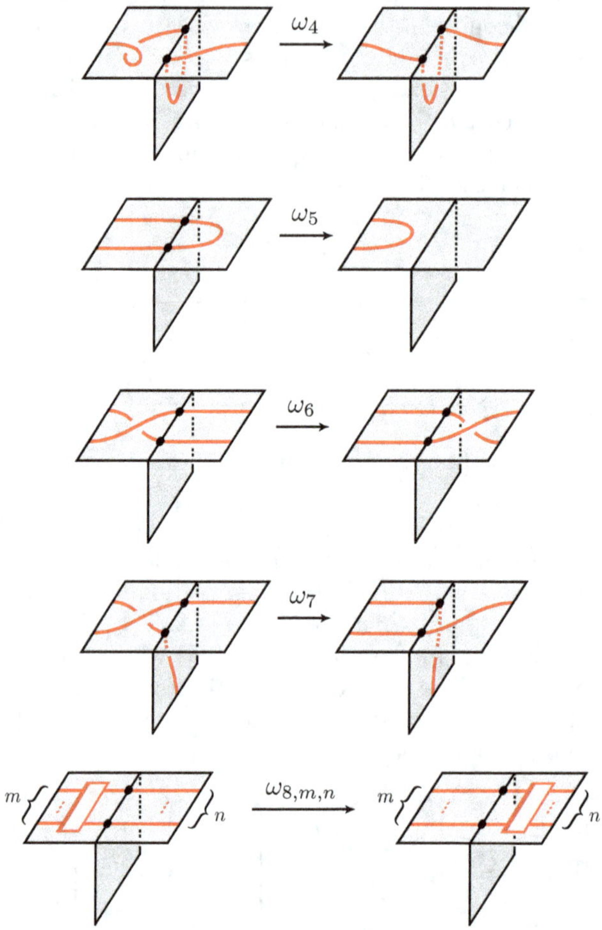

Figure 14.5: The moves $\omega_4 - \omega_8$ on positive diagrams

The move $\omega_{9,m,n}$ in Figure 14.6 modifies d in a 3-ball containing one switch w of valence $m+n+2$ and one vertex x of P whose link graph is a circle together with $m+n+2$ radii. Here $m \geq 0$ (respectively, $n \geq 0$) is the number of half-planes adjacent to the edge e_w and lying above (respectively, below) the horizontal plane in Figure 14.6. The move $\omega_{9,m,n}$ replaces the branch of d through w with a path in P encircling x from above and containing only trivalent switches. A similar move replacing the same branch of d with a path in P encircling x from below is nothing but $\omega_{9,n,m}$. Note that the move $\omega_{9,0,0}$ replaces a branch of d traversing a 2-valent edge of P by a branch traversing two 3-valent edges. The moves $\{\omega_{9,m,n}\}_{m,n}$ are collectively denoted ω_9. We will also consider the obvious inverse moves $\{\omega_*^{-1}\}_{*=0}^9$.

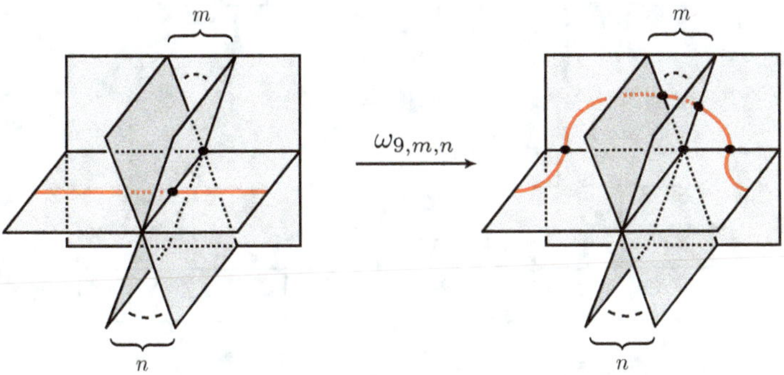

Figure 14.6: The move $\omega_{9,m,n}$ on positive diagrams

We stress that a move $\omega_*^{\pm 1}$ may be applied to a ribbon diagram only when both this diagram and the diagram produced by the move are positive. More precisely, the moves $\omega_5^{\pm 1}, \omega_6^{\pm 1}, \omega_8^{\pm 1}$ apply when the orientations of the two horizontal half-planes are compatible, i.e., are induced by an orientation of the horizontal plane. The move $\omega_4^{\pm 1}$ applies when the orientation of the vertical half-plane is compatible with the orientations of both horizontal half-planes. The move $\omega_7^{\pm 1}$ applies when the orientation of the left horizontal half-plane is compatible with orientations of the other two half-planes. The move $\omega_9^{\pm 1}$ applies when both horizontal half-planes and the regions lying in the vertical upper half-plane have compatible orientations.

14.3.2 Sufficiency of the moves

It is clear that positive diagrams related by the moves $\{\omega_*^{\pm 1}\}_{*=0}^9$ represent isotopic ribbon graphs in (M,G). We establish the converse, starting with the case of trivalent diagrams. A ribbon diagram is *trivalent* if all its switches have valence 3. The class of trivalent positive diagrams is closed under the moves

$$\{\omega_*^{\pm 1}\}_{*=0}^8, \quad \omega_{9,0,1}^{\pm 1}, \quad \omega_{9,1,0}^{\pm 1}. \tag{14.1}$$

The moves $\omega_{9,0,1}$ and $\omega_{9,1,0}$ are pictured in Figure 14.7.

Lemma 14.2. *If two trivalent positive diagrams represent isotopic ribbon graphs in (M,G), then they can be related by the moves* (14.1).

Lemma 14.2 is proved in Sections 14.4–14.7.

Theorem 14.3. *If two positive diagrams represent isotopic ribbon graphs in (M,G), then they can be related by the moves $\{\omega_*^{\pm 1}\}_{*=0}^9$.*

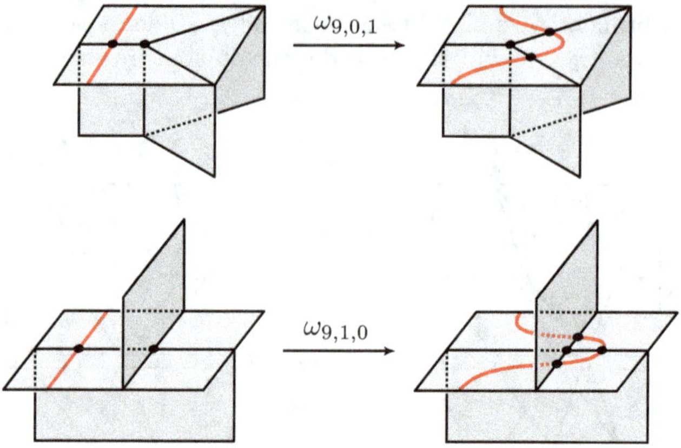

Figure 14.7: The moves $\omega_{9,0,1}$ and $\omega_{9,1,0}$ on positive diagrams

Proof. This follows from Lemma 14.2 and the following claim: any positive diagram (P, d) can be transformed by the moves ω_0 and ω_9 into a trivalent positive diagram. The transformation is done in three steps. First, we apply ω_0 to ensure that every edge of P of valence $\neq 1, 3$ has at least one endpoint in $\mathrm{Int}(M)$. Second, we blow up all vertices of P in $\mathrm{Int}(M)$ via ω_0 as in the proof of Lemma 11.4, see Figure 11.8. Third, for every edge e of P of valence $\neq 1, 3$, we pull along e the strands of d traversing e until they approach the 2-sphere S created by the blowing up at an endpoint of e. Furthermore, these strands are inductively pulled by ω_9 into S so that all newly created switches are positive. This can be done because the regions contained in S are oriented coherently. The described procedure transforms all switches of d on e into positive trivalent switches. This implies our claim and the theorem. □

The following theorem concerns ribbon diagrams without trivial components. A *trivial component* of a ribbon diagram (P, d) is a circle strand of d embedded in $\mathrm{Int}(P)$ and disjoint from the rest of $d \subset P$. If a trivial component bounds a disk in $\mathrm{Int}(P)$, then it represents a trivial knot, hence the name. A ribbon diagram is *neat* if it has no trivial components.

Theorem 14.4. *All ribbon graphs in (M, G) can be represented by neat positive diagrams. If two neat positive diagrams represent isotopic ribbon graphs, then they can be related by the moves $\{\omega_*^{\pm 1}\}_{*=0}^{9}$ in the class of neat positive diagrams.*

Proof. Each ribbon diagram (P, d) determines a neat diagram (P, d') obtained by the move ω_1^{-1} applied once to every trivial component of (P, d). If (P, d) is positive, then so is (P, d') and the associated ribbon graphs in (M, G) are isotopic. Thus, the first claim of the theorem directly follows from Lemma 14.1. By Theorem 14.3, to prove the second claim it suffices to show that for any positive diagrams (P_1, d_1) and $(P_2, d_2) = \omega_i(P_1, d_1)$ with $i = 0, \ldots, 9$, the neat diagrams (P_1, d_1') and (P_2, d_2')

are related by ω_i and the moves $\omega_1^{\pm 1}$ in the class of neat positive diagrams. This is obvious for $i = 0, 3, 4, 6, 7, 8, 9$ because ω_i does not modify the trivial components. For $i = 1, 2, 5$, trivial components of d_1 produce trivial components of $d_2 = \omega_i(d_1)$. The diagram d_2 may have additional trivial components arising when non-trivial circle strands of d_1 become trivial under ω_i. Then d_2' is obtained from d_1' by applying ω_1^{-1} to each of the corresponding non-trivial circle strands of d_1' and further applying ω_i. □

Applying Theorem 14.4 to the case where $\partial M = G = \emptyset$, we obtain the following corollary.

Corollary 14.5. *All ribbon graphs in a closed oriented 3-manifold can be represented by neat positive diagrams. If two neat positive diagrams in a closed oriented 3-manifold represent isotopic ribbon graphs, then they can be related by the moves* $\{\omega_*^{\pm 1}\}_{*=0}^{9}$ *in the class of neat positive diagrams.*

14.4 Moves on weighted diagrams

This is the first of four sections devoted to the proof of Lemma 14.2. We introduce weighted diagrams and use them to present ribbon graphs. Then we study moves on weighted diagrams and state our key lemma (Lemma 14.6).

14.4.1 Weighted diagrams

A *weight* on a ribbon diagram (P, d) is a map t from the set of strands of d to $\mathbb{Z}$ such that for every strand s of d the integer $t(s)$ is congruent modulo 2 to the number of negative switches on s. A *weighted diagram* is a trivalent ribbon diagram equipped with a weight. For example, each trivalent positive diagram (P, d) determines a weighted diagram $(P, d, t = 0)$ with zero weight. In this way, weighted diagrams generalize trivalent positive diagrams.

The construction of a ribbon graph in (M, G) from a positive diagram extends to weighted diagrams as follows. Let (P, d, t) be a weighted diagram. Let n and $d^n \subset M$ be as in Section 14.2.2. We now equip d^n with a framing. Note that the normal line bundle of $\mathrm{Int}(P)$ in M induces a normal line bundle on $d^n \subset M$ in the obvious way. Rotate the latter line bundle $t(s)/2$ times around every strand s of d^n. The positive direction of rotation is the one which together with the orientation of s determine the orientation of M. (For instance, $t(s) = 1$ gives rise to a positive half-rotation around s.) Denote by ν the resulting normal line bundle on d^n. The mod 2 condition on t ensures that if s is an arc strand of d^n, then the line bundle $\nu|_s$ has a non-zero section whose restriction to ∂s generates the normal direction $n|_{\partial s}$. If s is a circle strand of d^n, then the line bundle $\nu|_s$ is trivial and we take its arbitrary non-zero section. These sections determine a framing of d^n and turn d^n into a ribbon graph in (M, G) well-defined up to isotopy.

14.4.2 The moves

We define local moves on a weighted diagram (P, d, t) keeping $\partial P = G$ and $\partial d \subset G$ and preserving the isotopy type of the associated ribbon graph in (M, G). The first family of moves consists of ambient isotopies in M fixing G pointwise and the primary moves $T_0 - T_3$ modifying P inside 3-balls in $\mathrm{Int}(M)$ disjoint from d, see Figure 11.2. These moves keep d and t and are collectively denoted Ω_0. The next moves $\Omega_1 - \Omega_9$ are shown in Figures 14.8 and 14.9. They keep P and modify d in a small 3-ball in $\mathrm{Int}(M)$. Here we keep the pictorial conventions of Section 14.3.1 though the positivity requirement on the diagrams is dropped unless explicitly stated to the contrary. All the moves preserve the weights of the strands except the moves Ω_1 and Ω_4 which modify the weights by adding -2 and ± 1 as indicated in the picture by boxes with -2 and ± 1 inside. The move Ω_8 pushes a coupon across a trivalent edge of the skeleton keeping the orientation of the coupon and the weights of the strands. The move Ω_8 is defined only when the orientations of two horizontal half-planes in the picture are compatible; this condition follows from our convention that the orientation of the coupon must match the orientation of the ambient region. It is clear that the moves $\{\Omega_*^{\pm 1}\}_{*=0}^9$ preserve the isotopy type of the ribbon graph in (M, G) represented by the weighted diagram. The following lemma yields the converse.

Lemma 14.6. *If two weighted diagrams represent isotopic ribbon graphs in* (M, G), *then they can be related by the moves* $\{\Omega_*^{\pm 1}\}_{*=0}^9$.

The proof of Lemma 14.6 occupies the rest of this section and the next section. The proof uses the weight-preserving moves on weighted diagrams shown in Figure 14.10. Here $\Omega_{10,m,n}$ is defined for any integers $m, n \geq 0$ with $m + n \geq 1$. Figure 14.11 defines one more move X. Note that an inverse to Ω_{10} or to X is a move of the same type. Finally, for any $m, n \geq 0$, we need a version $\Omega'_{8,m,n}$ of $\Omega_{8,m,n}$ shown in Figure 14.12. The move $\Omega'_{8,m,n}$ applies when the orientations of the horizontal half-planes in the picture of $\Omega_{8,m,n}$ are incompatible. This move acts as $\Omega_{8,m,n}$ with an additional positive 180°-rotation of the bunch of incoming strands and a negative 180°-rotation of the bunch of outgoing strands together with the appropriate change of the weight. The orientation of the coupon is inverted under $\Omega'_{8,m,n}$ to match the orientation of the ambient region.

Lemma 14.7. *The moves* $X, \Omega'_6, \Omega'_7, \Omega'_8, \Omega_{10}, \Omega_{11}$ *expand as compositions of the moves* $\{\Omega_*^{\pm 1}\}_{*=0}^9$.

Proof. An expansion of X is shown in Figure 14.11. The move Ω'_6 expands as a composition of Ω_6^{-1} with $\Omega_2^{\pm 1}$: first apply Ω_2^{-1} to the strands in the right horizontal half-plane to create two new crossings, then use Ω_6^{-1} to push one of these crossing to the left horizontal half-plane, and finally apply Ω_2 in that half-plane. The move Ω'_7 expands as a composition of Ω_7^{-1} with Ω_2: first apply Ω_7^{-1} to create a second crossing in the left horizontal half-plane and then apply Ω_2 in that half-plane. The move $\Omega_{10,m,n}$ expands as follows: First, use the primary moves $\Omega_0^{\pm 1}$ to

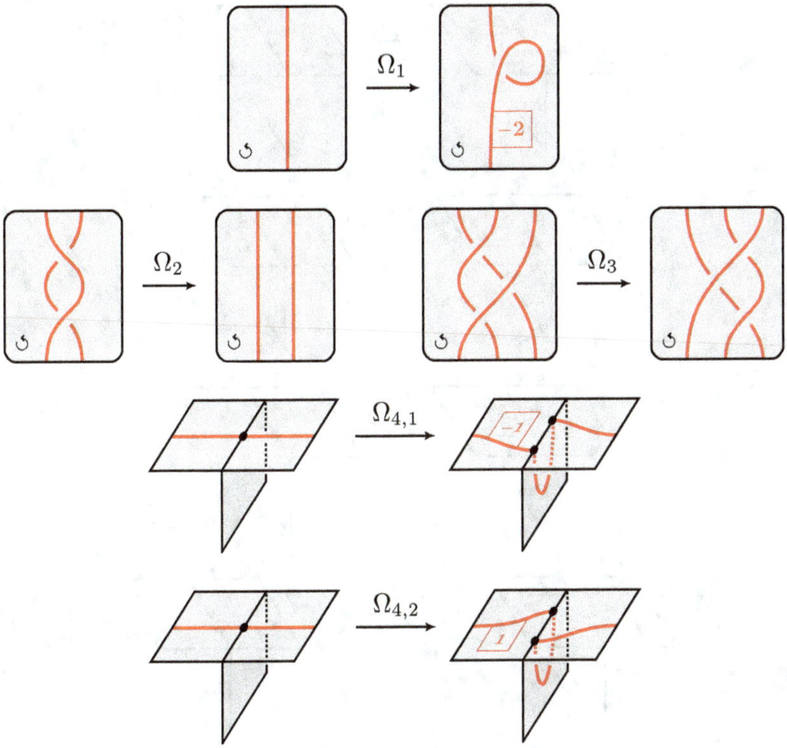

Figure 14.8: The moves $\Omega_1 - \Omega_4$ on weighted diagrams

attach a small cap in the upper half-space over the vertex of the skeleton, then apply $m+n$ times $\Omega_9^{\pm 1}$ to push the red strand across the cap, and finally use $\Omega_0^{\pm 1}$ again to remove the cap. The move Ω_{11} expands as follows: use Ω_0 to attach to the ambient region a small 2-disk D lying above the page, then push ∂D by Ω_5^{-1} so that it meets twice all incoming strands of the coupon near the bottom side, further push D across the coupon by a single Ω_8, then pull the moving strand into D by Ω_5^{-1}, Ω_7', and then repeat the same process in the inverse order near the top side of the coupon.

Consider the move $\Omega'_{8,m,n}$ pushing a coupon across a trivalent edge e of the skeleton. Denote the left horizontal half-plane in the picture of the move by L, the right horizontal half-plane by R, and the vertical half-plane by V. By assumption, the orientations of L and R induce the same orientation on e. Suppose first that the orientation of V induces the opposite orientation on e. Applying $\Omega_{4,2}$ at the n switches of the left diagram, we obtain a diagram meeting V along n small consecutive half-circles. We apply Ω_7^{-1} to the latter diagram $n(n-1)/2$ times so that the resulting diagram looks as before in L and R but meets V at n concentric half-circles with an n-strand braid woven on them. We push this braid in R us-

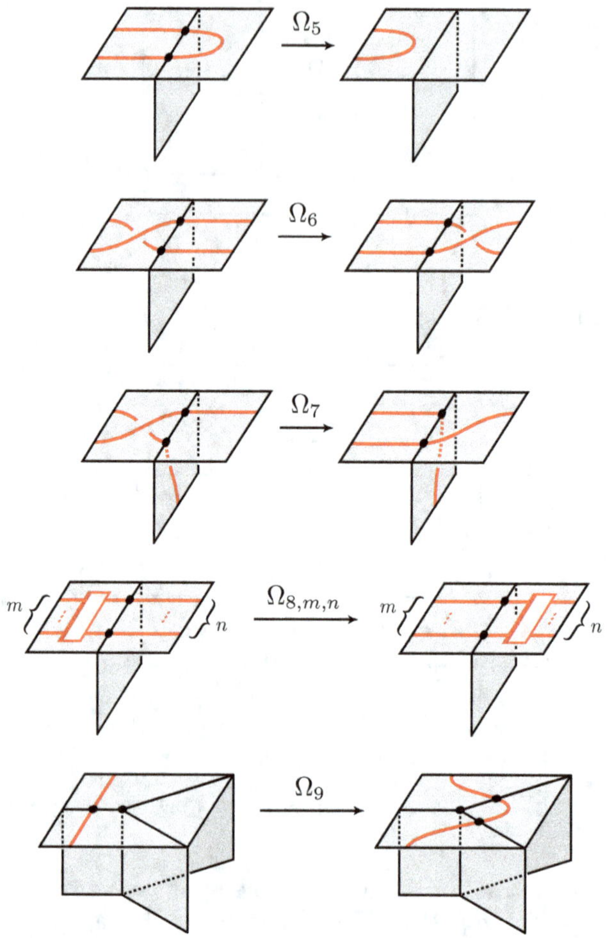

Figure 14.9: The moves $\Omega_5 - \Omega_9$ on weighted diagrams

ing Ω_6 and Ω_6'. Next, we push our coupon in V and then in R using, in both cases, $\Omega_{8,m,n}$. This leaves m concentric half-circles in V which are then pulled into R by the process inverse to that above. The resulting diagram, d', lies in $L \cup R$, meets L along m disjoint arcs, and meets R along a coupon with a braid on m strings attached to the bottom base and a braid on n strings attached to the top base. Using the isotopy invariance of the associated ribbon graph (or using the construction of d'), one easily identifies d' with the diagram produced by $\Omega_{8,m,n}'$, at least up to the moves $\Omega_2^{\pm 1}, \Omega_3^{\pm 1}$.

If the orientations of L, R, V induce the same orientation on e, then we apply $\Omega_0 = T_0''$ to add a small bubble at e as in Figure 14.13. This creates small disk regions $D \subset L$ and $D' \subset V$. We provide D with orientation induced by that

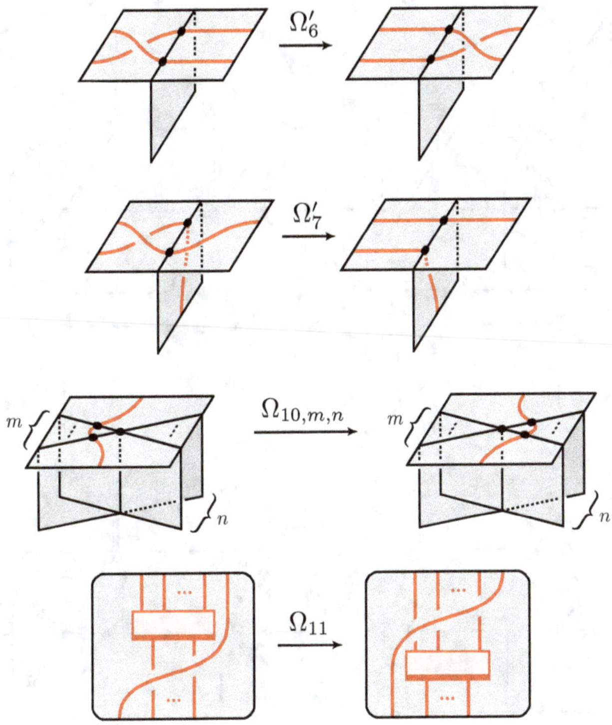

Figure 14.10: The moves Ω'_6, Ω'_7, Ω_{10}, Ω_{11} on weighted diagrams

of L and provide D' with orientation opposite to that of V. The third small disk region created by T''_0 is oriented arbitrarily. Next, we apply n times the move Ω_9 pushing n strands of our diagram in D. Now, each of these strands goes from $L \setminus D$ to D and then to R. We push our coupon into D by $\Omega_{8,m,n}$ and further into R by $\Omega'_{8,m,n}$. The latter move is as in the previous paragraph and expands as there. Then we use m times Ω_9^{-1} to pull the strands of the diagram away from D, and use $\Omega_0 = (T''_0)^{-1}$ to remove the bubble. This gives the desired expansion of $\Omega'_{8,m,n}$. □

14.5 Proof of Lemma 14.6

This is the second of four sections devoted to the proof of Lemma 14.2. Here we study weighted diagrams associated with frames and prove Lemma 14.6.

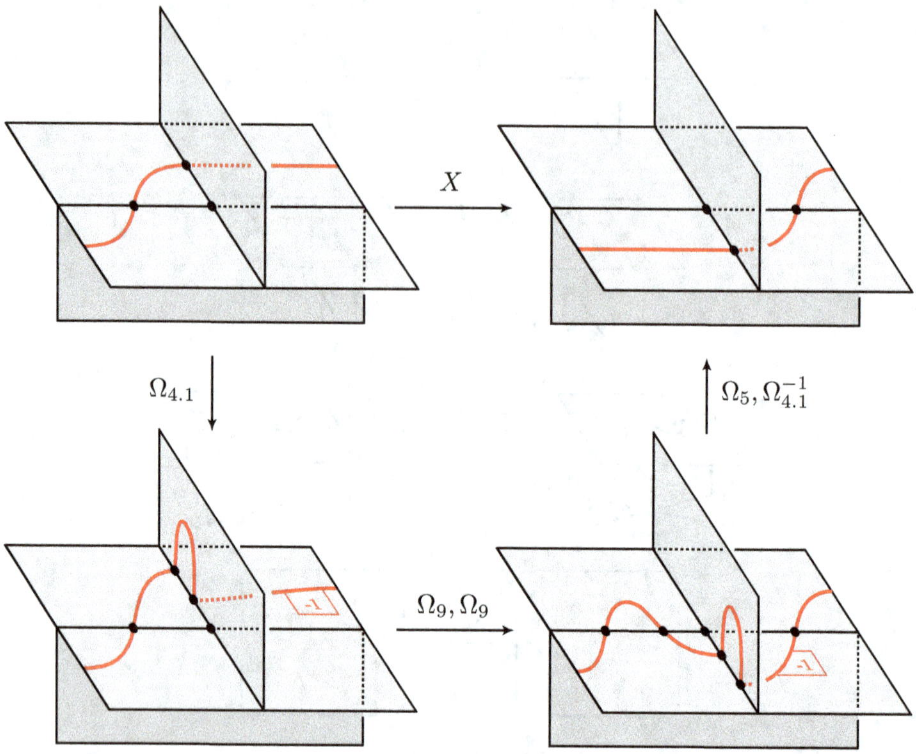

Figure 14.11: The move X on weighted diagrams

14.5.1 Diagrams associated with frames

Pick a frame $F \subset \mathrm{Int}(M)$ and a G-generic F-collar $i\colon \partial M \times [0,1] \hookrightarrow M$, see Sections 11.5.2 and 11.5.3. Recall the 3-manifold $\widehat{F} \subset \mathrm{Int}(M)$, the surface $\partial F = \partial\widehat{F}$, and the skeleton $P(F,i)$ of (M,G) associated with F and i. We say that a weighted diagram (P,d,t) is *associated with* (F,i) if $P = P(F,i)$, the knotted plexus d meets the regions of P adjacent to $\partial P = G$ along the segments $\{i(u \times [0,1])\}_{u\in\partial d}$, and the points $\{i(u,1)\}_{u\in\partial d}$ lie in the interiors of the regions of F contained in ∂F. Each of these points is then a switch of d at which d turns from a region of P adjacent to G to a region of P contained in ∂F. The knotted plexus d splits in three pieces: the union of disjoint "legs" $\{i(u \times [0,1])\}_{u\in\partial d}$, the part of d lying in $\mathrm{Int}(\widehat{F})$, and the part of d formed by the coupons and pieces of strands of d which lie (possibly, with crossings) in ∂F. The latter part of d is a knotted plexus in the surface ∂F and is denoted $d \cap \partial F$. Its free ends are the points $\{i(u,1)\}_{u\in\partial d}$ and the switches of d at which a strand of d turns from ∂F to $\mathrm{Int}(\widehat{F})$.

We show now that the class of weighted diagrams associated with F and i does not depend on i up to the moves $\{\Omega_*^{\pm 1}\}_{*=0}^9$. The weighted diagrams in this

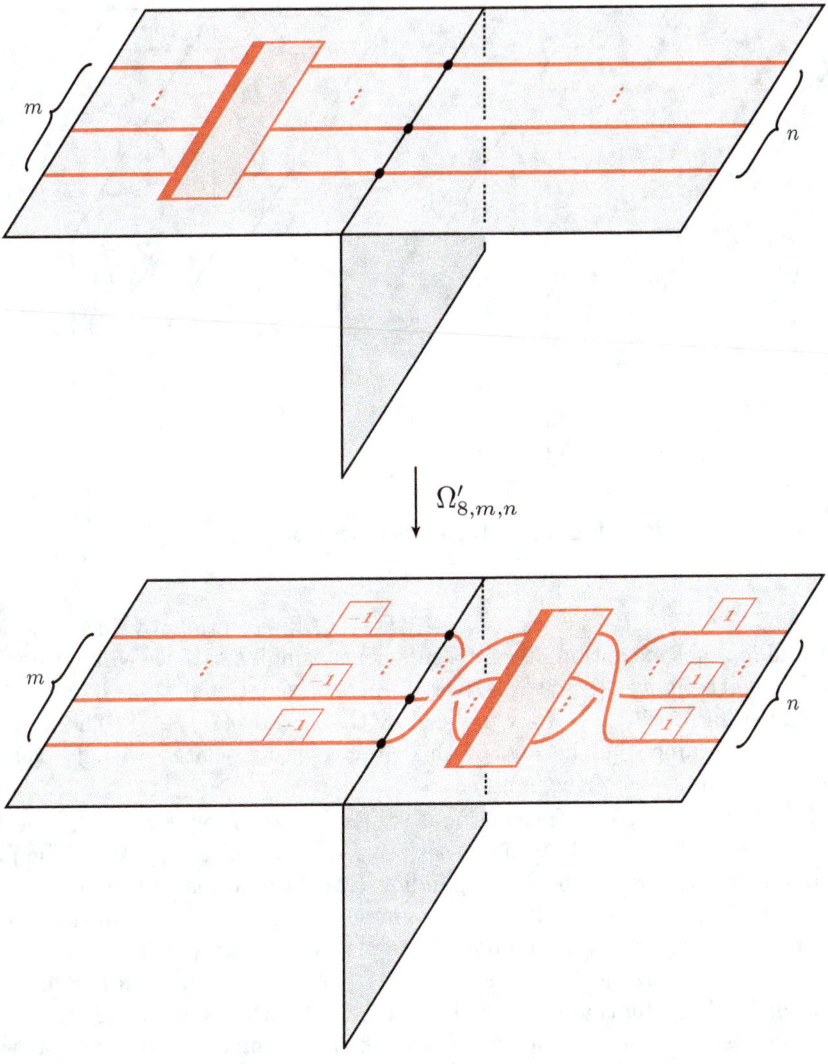

Figure 14.12: The move $\Omega'_{8,m,n}$ on weighted diagrams

class are said to be *associated with* F.

Lemma 14.8. *For any frame F of M, any G-generic F-collars i_0, i_1, and any weighted diagram $\mathcal{T}$ associated with (F, i_0), there is a weighted diagram associated with (F, i_1) and obtained from $\mathcal{T}$ by the moves $\{\Omega_*^{\pm 1}\}_{*=0}^{9}$.*

Proof. Pick a generic isotopy $\{i_s\}_{s\in[0,1]}$ from i_0 to i_1 in the class of F-collars. We start from $\mathcal{T}_0 = \mathcal{T}$ and construct for all but finitely many values of s a weighted

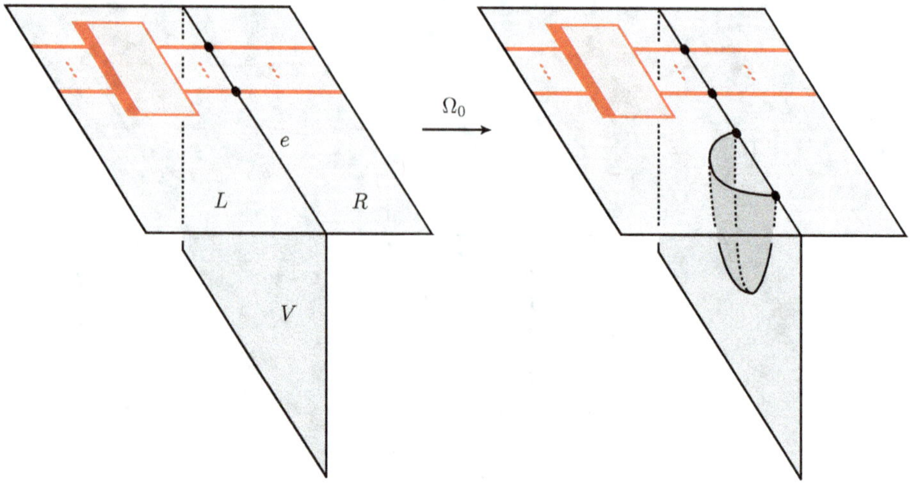

Figure 14.13: Attaching a bubble

diagram $\mathcal{T}_s = (P_s, d_s, t_s)$ associated with (F, i_s). All exceptional values of s lie in $(0, 1)$, and for each exceptional s and a sufficiently small $\varepsilon > 0$, we will obtain $\mathcal{T}_{s+\varepsilon}$ from $\mathcal{T}_{s-\varepsilon}$ by the moves $\{\Omega_*^{\pm 1}\}_{*=0}^{9}$. Away from the exceptional s, the diagram $\mathcal{T}_s$ is just isotoped in M via an isotopy induced by the isotopy $\{i_s\}_{s\in[0,1]}$. The weighted diagram $\mathcal{T}_1$ obtained at the end of this process is associated with (F, i_1) and satisfies the conditions of the lemma.

There are two types of exceptional values s. Exceptional values of the first type are the numbers $s \in (0, 1)$ such that the F-collar i_s is not generic. The proof of Lemma 11.7 describes the corresponding transformations of the skeleton P_s. Using an ambient isotopy of d_s in P_s, we may ensure that these transformations of P_s proceed away from d_s and modify $\mathcal{T}_s$ via Ω_0. Exceptional values of the second type are the numbers $s_0 \in (0, 1)$ such that i_{s_0} is generic but does not meet the conditions in the definition of a weighted diagram associated with (F, i_{s_0}). Since the isotopy $\{i_s\}_{s\in[0,1]}$ is generic, there is a sufficiently small $\varepsilon > 0$ such that when s increases from $s_0 - \varepsilon$ to $s_0 + \varepsilon$, one of the following transformations occurs:

(i) a vertex of the graph $G_s = i_s(G\times\{1\})$ is pulled across a strand of the knotted plexus $d_- = d_{s_0-\varepsilon} \cap \partial F$ in ∂F;

(ii) an edge of G_s is pulled across a coupon, a crossing, or a strand of d_-;

(iii) an edge of G_s is pulled across an intersection point of a strand of d_- and an edge of F contained in ∂F;

(iv) an edge of G_s is pulled across a free end of d_- at which a strand of d turns from ∂F to $\mathrm{Int}(\widehat{F})$;

(v) a point $i_s(u,1)$ with $u \in \partial d \subset G$ is pulled along an edge of G_s across an edge of F contained in ∂F.

The traces of these transformations in ∂F are depicted in Figure 14.14 where d_- is drawn in red, G_s in black, and the edges of F contained in ∂F are represented by dotted lines. Then $(P_{s_0+\varepsilon}, d_{s_0+\varepsilon})$ is obtained from $(P_{s_0-\varepsilon}, d_{s_0-\varepsilon})$ by Ω_{10} in the case (i), $\Omega_8^{\pm 1}$, $\Omega_6^{\pm 1}$, or $\Omega_5^{\pm 1}$ in the case (ii), X in the case (iii), Ω_9 in the case (iv), and Ω_9 in the case (v). It remains to observe that, by Lemma 14.7, the moves Ω_{10} and X expand as compositions of the moves $\{\Omega_*^{\pm 1}\}_{*=0}^9$. □

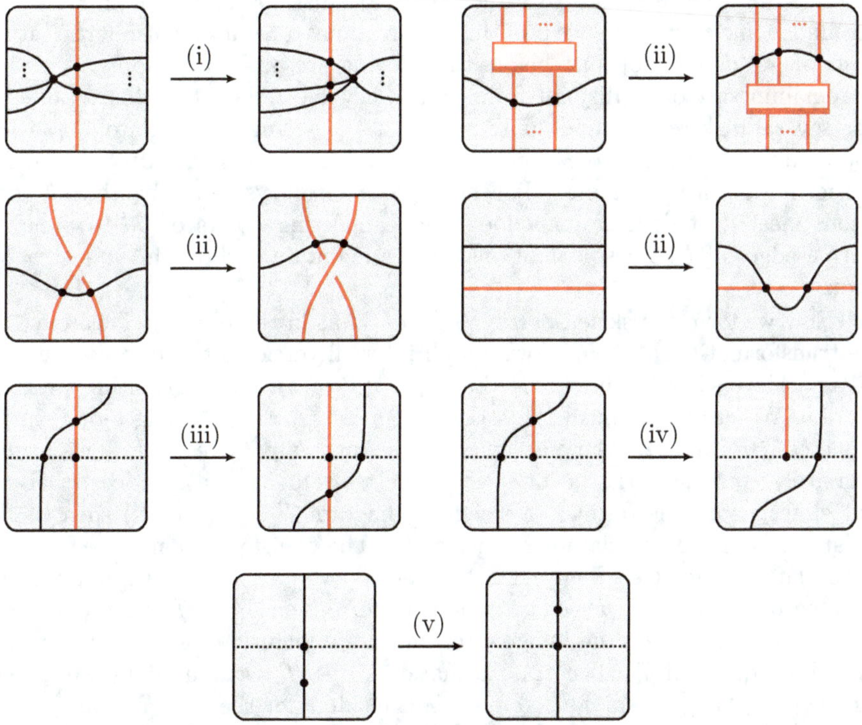

Figure 14.14: The transformations (i)–(v).

Lemma 14.9. *If two frames F, F' of M are related by a frame move (see Sections 11.4.1 and 11.5.4), then any weighted diagram associated with F can be transformed by the moves $\{\Omega_*^{\pm 1}\}_{*=0}^9$ into a weighted diagram associated with F'.*

Proof. For each frame move $(T_0'')^{\pm 1}$, $(T^{1,2})^{\pm 1}$, $\mathcal{L}^{\pm 1}$, we can use $\Omega_5^{-1}, \Omega_9^{\pm 1}$ and ambient isotopy to pull the knotted plexus in the skeleton away from the 3-ball where the frame move proceeds. Then the claim of the lemma becomes obvious. □

Lemma 14.10. *Any weighted diagram can be transformed by the moves $\{\Omega_*^{\pm 1}\}_{*=0}^{9}$ into a weighted diagram associated with a frame of M.*

Proof. The transformation of a weighted diagram $\mathcal{T} = (P, d, t)$ accompanies the transformation of P described in the proof of Lemma 11.4 and Theorem 11.5. First of all, we modify $\mathcal{T}$ to ensure that all regions of the underlying skeleton are disks. We cannot simply use T_1 to add new edges because they may meet d and then our moves T_1 are not of type Ω_0. Instead we show how to use the moves $\Omega_0^{\pm 1}, \Omega_5^{-1}$ to split a region r of P along a proper arc $\alpha \subset r$ with endpoints $x, y \in P^{(1)} \setminus d$. We isotop α so that it meets d transversely avoiding the crossings and the coupons. We use a Ω_0-type move T_0'' to attach to P a 2-disk D so that a subarc of ∂D surrounds x inside r. Then we pull this subarc along α towards y using Ω_5^{-1} at the points of $\alpha \cap d$. We stop just before this moving arc hits y and apply T_1 adding the remaining arc of α (disjoint from d) as a new edge. This transformation of P adds several disk regions and splits r along α. Choosing appropriate arcs α and proceeding in this way, we can ensure that all regions of P are disks. A similar transformation in the regions adjacent to $\partial P = G$ allows us to ensure that these regions meet d at disjoint embedded arcs connecting points of ∂d to points of internal edges of P. These conditions will be maintained through the rest of the proof.

Now we form the skeletons P_1, P_2, P_3 of M as in the proof of Theorem 11.5. The transformation $P \mapsto P_1$ blows up all internal vertices of P (see Figure 11.8) and is achieved by the moves Ω_0. Let $\mathcal{T}_1 = (P_1, d, t)$ be the resulting weighted diagram. We use Ω_9 to push the strands of d intersecting the edges of P in the 2-spheres $\{\partial B_x \subset P_1\}_x$ where x runs over the internal vertices of P. This is done recursively starting with the intersections closest to internal vertices, then the next closest, etc. The result is a weighted diagram $\mathcal{T}_1' = (P_1, d', t')$ such that d' is disjoint from the internal long edges of P_1. The transformation $P_1 \mapsto P_2$ blows up the latter edges (see Figure 11.9). Then d' survives in P_2 and determines a weighted diagram $\mathcal{T}_2 = (P_2, d', t')$. The transformation $\mathcal{T}_1' \mapsto \mathcal{T}_2$ can be achieved by the moves Ω_0. Next, the image of d' under the inclusion $P_2 \hookrightarrow P_3$ determines a weighted diagram $\mathcal{T}_3$. The transformation $\mathcal{T}_2 \mapsto \mathcal{T}_3$ is induced by a sequence of moves $\{\Omega_*^{\pm 1}\}_{*=0}^{8}$ pushing the circles ∂r_- inside r for the wide regions r of P_2. Namely, we apply the Ω_0-moves $T^{1,2}$, $T^{2,1}$ when ∂r_- crosses the vertices of r, the moves $X, \Omega_5^{\pm 1}, \Omega_9^{\pm 1}$ near the points of $d' \cap \partial r$ when ∂r_- crosses these points, and the moves $\Omega_5^{\pm 1}, \Omega_6^{\pm 1}, \Omega_8^{\pm 1}$ when ∂r_- contracts inside r. The rest of the proof of Theorem 11.5 extends to this setting and shows that the weighted diagram $\mathcal{T}_3$ is associated with a frame of M. □

14.5.2 Proof of Lemma 14.6

Consider two weighted diagrams representing isotopic ribbon graphs in (M, G). By Lemma 14.10, to relate these diagrams by the moves $\{\Omega_*^{\pm 1}\}_{*=0}^{9}$, it suffices to consider the case where both diagrams are associated with frames of M. By

Lemma 11.8, these frames may be related by frame moves. By Lemmas 14.8 and 14.9, it is enough to consider weighted diagrams associated with the same frame F and the same generic F-collar i. Since these diagrams represent isotopic ribbon graphs in (M, G), they have the same free ends. The ribbon graphs in question lie in the 3-manifold $\widehat{F}$ except their legs $i(u \times [0,1])$ where u runs over the free ends. Since M is obtained from $\widehat{F}$ by attaching the collar $\partial F \times [0,1] = \partial \widehat{F} \times [0,1]$, there is an ambient isotopy between our ribbon graphs constant on the legs. Since $\widehat{F} \setminus F$ is a disjoint union of balls, we can assume that throughout the isotopy the ribbon graph (minus the legs) is kept in a small regular neighborhood of F in M. The assumption that F is a special 2-polyhedron implies that such an isotopy may be induced by a sequence of moves $\{\Omega_*^{\pm 1}\}_{*=0}^{11}$ and $(\Omega_6')^{\pm 1}, (\Omega_7')^{\pm 1}, (\Omega_8')^{\pm 1}$, see [Tu1, Section IX.3]. Here an isotopy pushing a strand of the ribbon graph over $i(u, 1) \in \partial F$ with u as above is induced by $\Omega_7^{\pm 1}$ or $(\Omega_7')^{\pm 1}$, and an isotopy pushing a strand over $i(v, 1) \in \partial F$, where v is a vertex of G, is induced by $\Omega_{10}^{\pm 1}$. By Lemma 14.7, all these moves expand via $\{\Omega_*^{\pm 1}\}_{*=0}^{9}$. This proves Lemma 14.6.

14.6 A refinement of Lemma 14.6

This is the third of four sections devoted to the proof of Lemma 14.2. Here we establish a refined version of Lemma 14.6 as follows. For $j = 1, 2$, we let $\Omega_{4,j}^\circ$ be a move $\Omega_{4,j}$ as in Figure 14.8 such that both switches created by the move are positive (the switch replaced by the move is then negative). The moves $\Omega_{4,1}^\circ, \Omega_{4,2}^\circ$ are collectively denoted Ω_4°. A move Ω_7° is a move Ω_7 as in Figure 14.9 such that all switches in its picture are positive (they are positive before the move if and only if they are positive after the move). As we know, the moves on weighted diagrams

$$\{\Omega_*^{\pm 1} \,|\, * = 0, 1, 2, 3, 5, 6, 8, 9\} \cup \{(\Omega_4^\circ)^{\pm 1}, (\Omega_7^\circ)^{\pm 1}\}. \tag{14.2}$$

preserve the isotopy types of the associated ribbon graphs in (M, G).

Lemma 14.11. *If two weighted diagrams represent isotopic ribbon graphs in* (M, G), *then they can be related by the moves* (14.2).

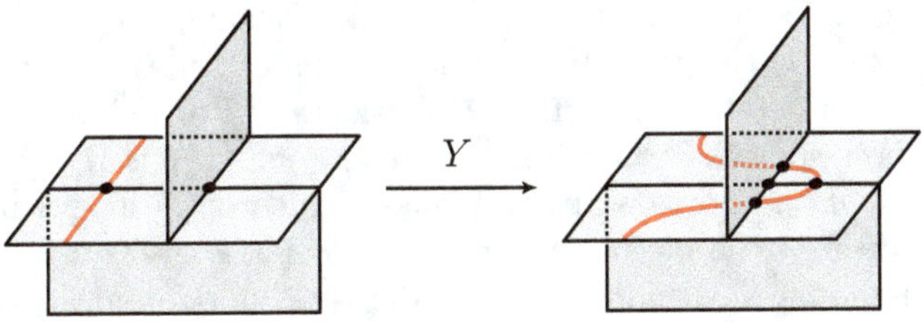

Figure 14.15: The move Y on weighted diagrams

Proof. Consider the generating set of moves $\{\Omega_*^{\pm 1}\}_{*=0}^9$ in Lemma 14.6. We show how to eliminate from this set all $\Omega_7^{\pm 1}$ other than $(\Omega_7^\circ)^{\pm 1}$. Let Y be the weight-preserving move shown in Figure 14.15. An arbitrary move Ω_7 is expanded in Figure 14.16 as follows. First, add a bubble via Ω_0 in such a way that the small horizontal half-disk is oriented as the right horizontal half-plane and the small vertical half-disk is oriented so that together these two half-disks form an oriented disk. Then pull the strands of the diagram into this bubble via Ω_9, Y, and Ω_6, apply Ω_7, pull the resulting strands away from the bubble via Y^{-1} and Ω_9^{-1}, and finally remove the bubble via Ω_0^{-1}. The choice of orientation of the half-disks above ensures that the move Ω_7 in this sequence is Ω_7°. It remains to observe that the move Y expands as a composition of a move Ω_5^{-1} pushing the red strand into the right horizontal half-plane and a move X. Consequently, Y is a composition of $\Omega_{4,1}^{\pm 1}, \Omega_5^{\pm 1}$, and Ω_9. This yields a required expansion of the original move Ω_7.

We next eliminate from the generating set the moves $\Omega_4^{\pm 1}$ other than $(\Omega_4^\circ)^{\pm 1}$. This requires a little preparation. We say that an edge e of a skeleton P is *branched* if among the branches of P at e there are two branches whose orientation induces opposite orientations of e. A move $\Omega_4^{\pm 1}$ at a branched edge is said to be *branched.* For example, all moves $(\Omega_4^\circ)^{\pm 1}$ are branched. We claim that any move $\Omega_{4,1}$ expands as a composition of moves $\Omega_0^{\pm 1}$, Ω_5, $\Omega_9^{\pm 1}$, and branched $\Omega_{4,1}^{\pm 1}$. Consider the expansion of $\Omega_{4,1}$ in Figure 14.17. The first move $\Omega_0 = T_0''$ creates three disk regions whose orientation is chosen so that the three short edges adjacent to them are branched. We apply Ω_9 and a branched move $\Omega_{4,1}^{-1}$ using that the short straight edge is branched. Next, we apply the move X defined in Figure 14.11 and expanded there as a composition of $\Omega_{4,1}^{\pm 1}, \Omega_5, \Omega_9$. Since the other two short edges mentioned above are branched, this expansion of X may be chosen to involve only branched $\Omega_{4,1}^{\pm 1}$. Finally, we apply Ω_9^{-1} to pull the strand away from the bubble and Ω_0^{-1} to remove the bubble. This gives a required expansion of the original move $\Omega_{4,1}$ and proves our claim. Similarly, any move $\Omega_{4,2}$ expands as a composition of moves $\Omega_0^{\pm 1}$, Ω_5, $\Omega_9^{\pm 1}$, and branched $\Omega_{4,2}^{\pm 1}$.

To complete the proof of Lemma 14.11, it remains to expand any branched move Ω_4 as a composition of the moves (14.2). Consider the move $\Omega_{4,1}$ in Figure 14.8 at a branched edge e. If the orientations of the horizontal half-planes induce the same orientation of e, then the orientation of the vertical half-plane must induce the opposite orientation of e. Then our $\Omega_{4,1}$ is $\Omega_{4,1}^\circ$. Suppose that the orientations of the horizontal half-planes induce opposite orientations of e. Rotating, if necessary, the picture around a vertical axis, we can assume that the orientations of the vertical half-plane and the right horizontal half-plane induce opposite orientations of e. Then the move $\Omega_{4,2}$ in Figure 14.18 is $\Omega_{4,2}^\circ$. This figure expands our $\Omega_{4,1}$ as a composition of Ω_5^{-1} and $(\Omega_{4,2}^\circ)^{-1}$. Similarly, a branched move $\Omega_{4,2}$ is $\Omega_{4,2}^\circ$ or expands as a composition of Ω_5^{-1} and $(\Omega_{4,1}^\circ)^{-1}$. □

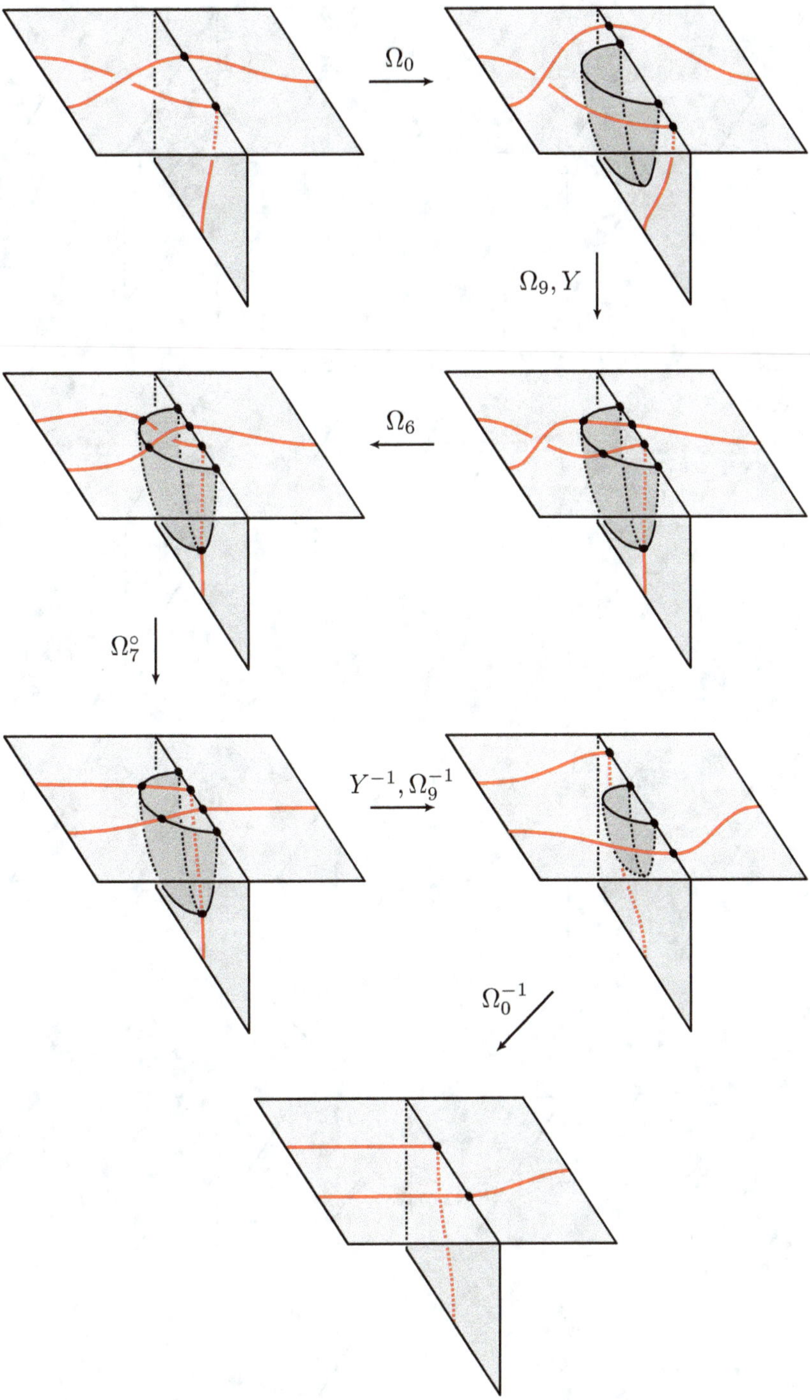

Figure 14.16: An expansion of Ω_7

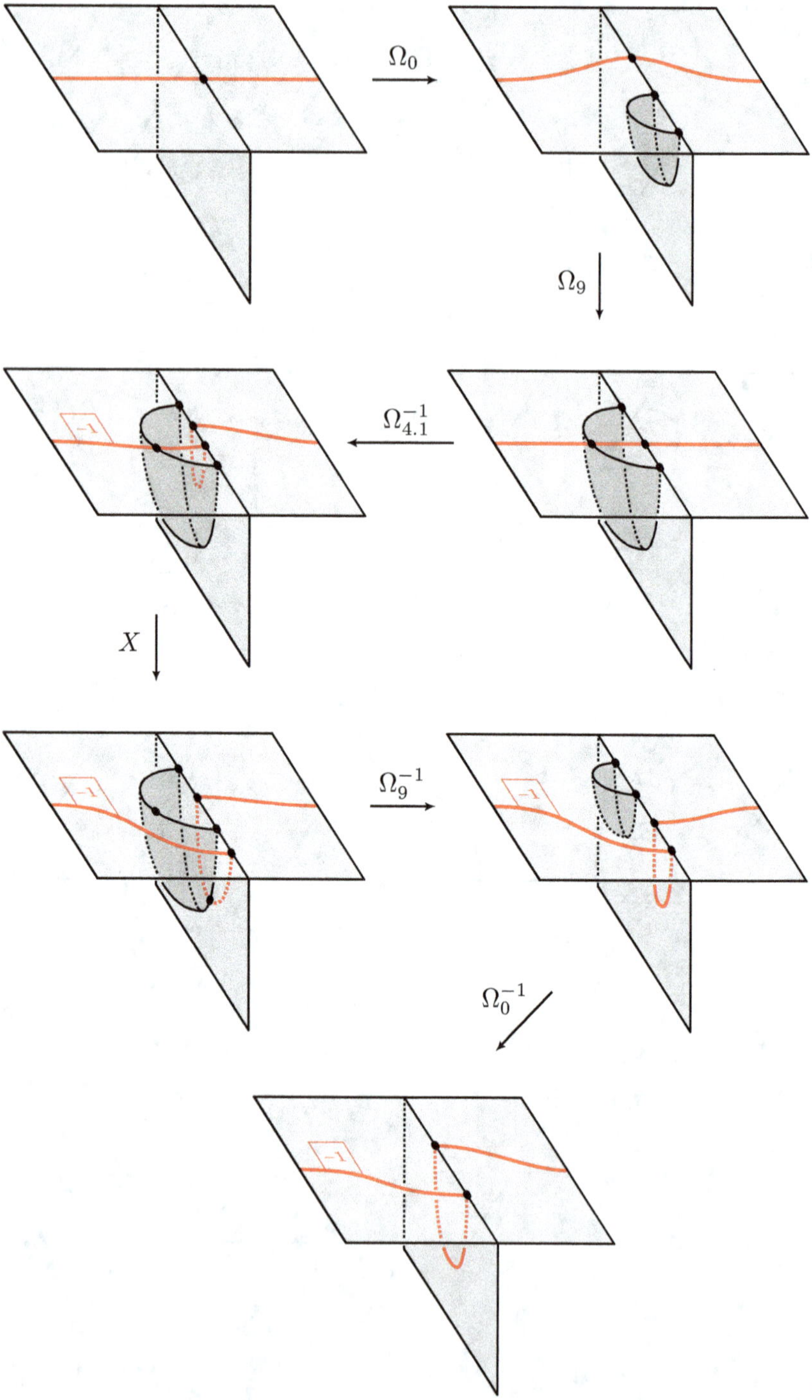

Figure 14.17: An expansion of $\Omega_{4,1}$

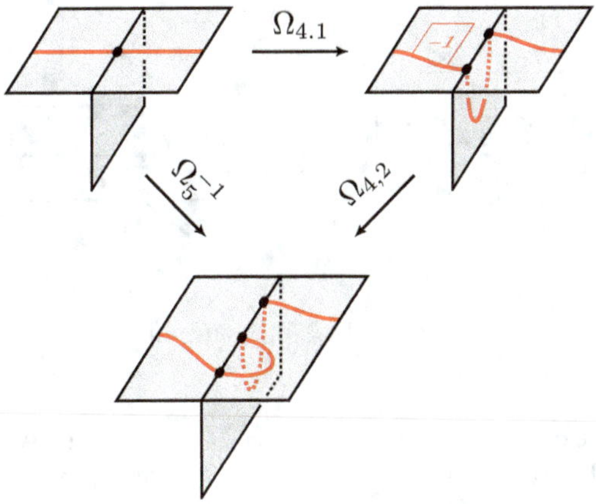

Figure 14.18: An expansion of a branched $\Omega_{4,1}$

14.7 Proof of Lemma 14.2

We introduce a tunneling transformation which turns weighted diagrams into positive diagrams. Then we complete the proof of Lemma 14.2.

14.7.1 Tunneling

Consider a stratified 2-polyhedron $P \subset M$ and an embedded arc $a \subset P$ whose interior lies in $\mathrm{Int}(P)$ and whose endpoints lie inside trivalent edges of P. We thicken a to a narrow rectangle $r \subset P$ meeting $P^{(1)}$ along opposite "short" sides. We next thicken r into a solid cylinder $C \subset M$ meeting P at r and at the disk bases of C, see the right picture in Figure 14.19. Set

$$P_a = (P \cup C) \setminus \mathrm{Int}(C).$$

The vertices/edges of P remain vertices/edges of the 2-polyhedron P_a except that the (open) short sides of r are removed from the edges of P containing ∂a. The remaining pieces of these edges become edges of P_a. We also add 6 new edges: the long sides of r and the half-circles bounding the disk bases of C. These new edges have 4 common endpoints which become vertices of P_a. This turns P_a into a stratified 2-polyhedron. We say that P_a is obtained from P by *drilling a tunnel* along a. Note that every region of P_a is contained in a region of P or in ∂C. The regions of P_a contained in ∂C are called *cylindrical regions*. In the pictures, we represent P_a by drawing the arc a in blue, see Figure 14.19.

We define a transformation, called *tunneling*, which turns any weighted diagram $\mathcal{T} = (P, d, t)$ into a trivalent positive diagram $\mathcal{T}^\dagger$. First, we construct a weighted diagram with only positive switches

$$\mathcal{T}^+ = (P^+, d^+, t^+).$$

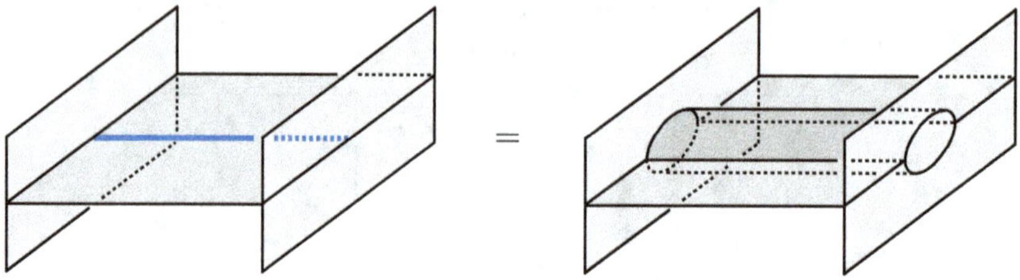

Figure 14.19: A tunnel

If all switches of $\mathcal{T}$ are positive, then $\mathcal{T}^+ = \mathcal{T}$. If $\mathcal{T}$ has negative switches, then P^+ is obtained from P by drilling a tunnel near each negative switch as in Figure 14.20, and d^+ is obtained from d as in this figure. The cylindrical regions of P^+ are oriented arbitrarily. Other regions of P^+ inherit their orientation from P except the small half-disk region in Figure 14.20 adjacent to the blue arc from above. This region is oriented so that both switches of d^+ in the picture are positive. The value of t^+ on a strand of d^+ is equal to the value of t on the corresponding strand of d minus the number of negative switches on that strand. This is indicated in Figure 14.20 by a box with -1 inside. It is clear from Figure 14.20 that the weighted diagrams $\mathcal{T}$ and $\mathcal{T}^+$ represent isotopic ribbon graphs in (M, G). The properties of t imply that all the values of t^+ are even.

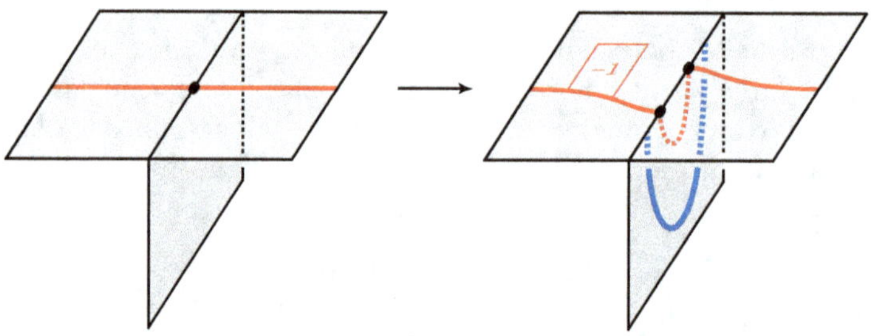

Figure 14.20: Tunneling near a negative switch

Next, we insert in each strand s of d^+ precisely $t^+(s)/2$ positive curls if $t^+(s) \geq 0$ and $-t^+(s)/2$ negative curls if $t^+(s) < 0$, see Figure 14.21. The curls are inserted at a point of $s \cap \mathrm{Int}(P^+)$. This turns (P^+, d^+) into a trivalent positive diagram denoted $\mathcal{T}^\dagger$. This diagram represents the same ribbon graph as $\mathcal{T}$ and $\mathcal{T}^+$. The construction of $\mathcal{T}^\dagger$ involves a choice of orientation on the cylindrical regions of P^+ and a choice of insertion places for the curls. Different orientations of cylindrical regions can be related by the moves $\omega_0^{\pm 1}$, see the proof of Lemma 11.2. Different insertion places for the curls can be related by ambient isotopy and the moves $\omega_2^{\pm 1}, \omega_3^{\pm 1}, \omega_5^{\pm 1}, \omega_6^{\pm 1}$. Thus, $\mathcal{T}^\dagger$ is well defined up to the moves (14.1).

Figure 14.21: Curls

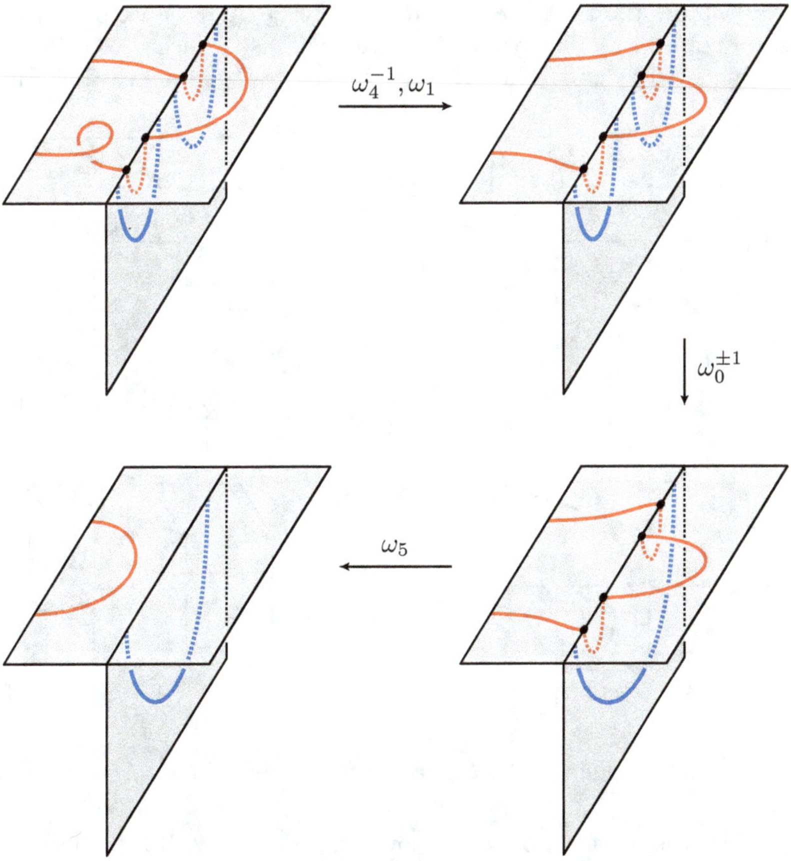

Figure 14.22: A transformation of $\mathcal{T}^\dagger$

Lemma 14.12. *If a weighted diagram $\mathcal{U}$ is obtained from a weighted diagram $\mathcal{T}$ by a move from the list* (14.2)*, then the positive diagrams $\mathcal{U}^\dagger$ and $\mathcal{T}^\dagger$ are related by the moves* (14.1).

Proof. For the moves Ω_* with $* = 0, 1, 2, 3, 8$, the claim is obvious: in this case

$\mathcal{U}^\dagger$ is obtained from $\mathcal{T}^\dagger$ by ω_*. The same argument works for Ω_7°. Consider the move Ω_5 in Figure 14.9. The two switches in the picture (before the move) are either both positive or both negative. In the former case $\mathcal{U}^\dagger$ is obtained from $\mathcal{T}^\dagger$ by ω_5. In the latter case, a sequence of transformations from $\mathcal{T}^\dagger$ to $\mathcal{U}^\dagger$ is shown in Figure 14.22. Here the first picture represents $\mathcal{T}^\dagger$, where the negative curl replaces the two weights -1 arising under tunneling at the two negative switches, see Figure 14.20. Applying ω_4^{-1} and using ω_1 to cancel the positive curl produced by ω_4^{-1} with the negative curl, we obtain the second picture. Modifying the blue arcs by $\omega_0^{\pm 1}$ we obtain the third picture. Now, applying ω_5 twice we obtain the last picture in Figure 14.22. The blue arc and the half-disk attached to it from above may be eliminated by $\omega_0^{\pm 1}$. This gives $\mathcal{U}^\dagger$. The move Ω_6 is treated similarly.

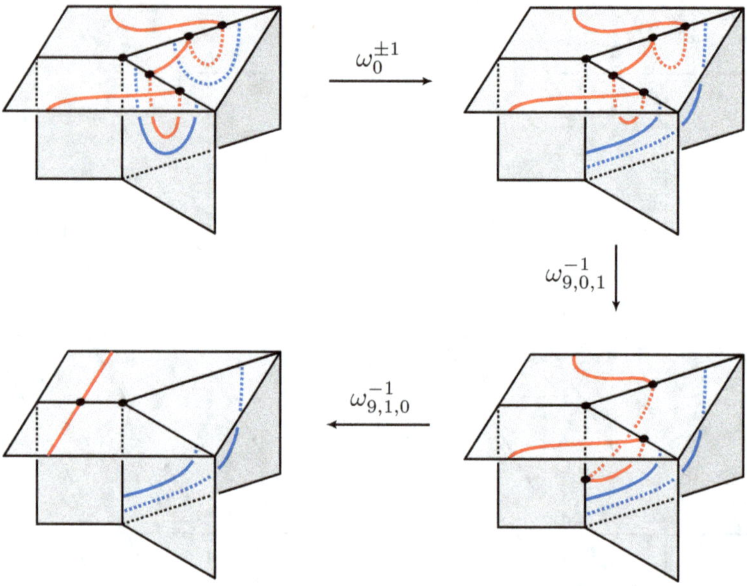

Figure 14.23: A transformation of $\mathcal{T}^\dagger$ bis

Consider the move Ω_9 on $\mathcal{T}$ as shown in Figure 14.9. If the orientations of the three horizontal regions are compatible (i.e., are induced by an orientation of the horizontal plane), then $\mathcal{U}^\dagger$ is obtained from $\mathcal{T}^\dagger$ by $\omega_{9,0,1}$. Suppose that these three orientations are incompatible. Then two of them must be compatible with each other. Applying $\Omega_5^{\pm 1}$ to $\mathcal{T}$ and using the results of the previous paragraph, we can reduce ourselves to the case where the switch before the move is positive. Then both after-move switches are negative. We have (locally) $\mathcal{T}^\dagger = \mathcal{T}$ while $\mathcal{U}^\dagger$ is shown in the first diagram in Figure 14.23 where we applied ω_4^{-1} in the upper part of $\mathcal{U}^\dagger$ and canceled the resulting curl with the two weights -1 arising under tunnelling at the two negative switches. The diagram $\mathcal{U}^\dagger$ can be transformed as in Figure 14.23: we pull both blue arcs through the vertex via a sequence of moves $\omega_0^{\pm 1}$, apply $\omega_{9,0,1}^{-1}$ and $\omega_{9,1,0}^{-1}$ (see Figure 14.7), and then delete the blue arcs via the moves $\omega_0^{\pm 1}$. This gives $\mathcal{T}^\dagger$.

Now we handle the move $\Omega_{4,1}^\circ$, see Figures 14.8 where, by the definition of $\Omega_{4,1}^\circ$, both after-move switches are positive and the before-move switch is negative. We have $\mathcal{U}^\dagger = \mathcal{U}$ while $\mathcal{T}^\dagger$ is as in the right picture in Figure 14.20. The diagrams $\mathcal{U}$ and $\mathcal{T}^\dagger$ differ only in the blue arc. The definition of the tunneling implies that the regions of $\mathcal{T}^\dagger$ adjacent to the blue arc from above and from below have compatible orientations. We can therefore eliminate the blue arc by a sequence of moves $\omega_0^{\pm 1}$. This transforms $\mathcal{T}^\dagger$ into $\mathcal{U}^\dagger$. The move $\Omega_{4,2}^\circ$ is treated similarly. □

14.7.2 Proof of Lemma 14.2

Consider trivalent positive diagrams $\mathcal{P}$ and $\mathcal{Q}$ representing isotopic ribbon graphs. Endowing $\mathcal{P}$ and $\mathcal{Q}$ with zero weights, we obtain weighted diagrams $\mathcal{P}'$ and $\mathcal{Q}'$ representing the same ribbon graphs. By Lemma 14.11, there is a finite sequence of weighted diagrams $\mathcal{P}' = \mathcal{T}_1, \mathcal{T}_2, \ldots, \mathcal{T}_n = \mathcal{Q}'$ with $n \geq 2$ in which every two consecutive diagrams are related by one of the moves (14.2). The tunneling yields positive diagrams $\mathcal{T}_1^\dagger, \mathcal{T}_2^\dagger, \ldots, \mathcal{T}_n^\dagger$, where, by Lemma 14.12, any consecutive diagrams are related by the moves (14.1). Clearly, $\mathcal{T}_1^\dagger = (\mathcal{P}')^\dagger = \mathcal{P}$ and $\mathcal{T}_n^\dagger = (\mathcal{Q}')^\dagger = \mathcal{Q}$. Thus, the diagrams $\mathcal{P}$ and $\mathcal{Q}$ are related by the moves (14.1).

Chapter 15

The state sum graph TQFT

We derive from a spherical fusion category with invertible dimension a graph TQFT which applies to colored ribbon graphs in 3-dimensional cobordisms. For cobordisms with empty graphs, we recover the TQFT of Chapter 13.

15.1 Colored surfaces and ribbon graphs

We introduce colored surfaces and colored ribbon graphs. They will be instrumental in the definition of a graph TQFT in Section 15.2. Throughout this section, $\mathcal{C}$ is a pivotal category.

15.1.1 Colored surfaces

A point of a surface Σ is *$\mathcal{C}$-colored* if it is endowed with a tangent direction in Σ, an object of $\mathcal{C}$ called the *color*, and a sign $\pm$. A *$\mathcal{C}$-colored surface* is a pair (Σ, A) where Σ is a closed oriented surface and A is a finite (possibly, empty) subset of Σ whose elements are $\mathcal{C}$-colored. A *homeomorphism* of colored surfaces $(\Sigma_0, A_0) \to (\Sigma_1, A_1)$ is an orientation preserving homeomorphism $\Sigma_0 \to \Sigma_1$ which carries A_0 onto A_1 keeping the distinguished tangent direction, the signs, and the colors of the points.

A $\mathcal{C}$-colored surface (Σ, A) determines an *opposite* $\mathcal{C}$-colored surface $(-\Sigma, -A)$ obtained by reversing the orientation of Σ and reversing the signs of the points of A while keeping the rest of the data. The disjoint union of $\mathcal{C}$-colored surfaces is defined in the obvious way.

For shortness, we will often write Σ for a $\mathcal{C}$-colored surface (Σ, A) specifying the set A whenever necessary.

15.1.2 Colorings of plexuses

Recall the notion of a plexus from Section 14.1.1. A $\mathcal{C}$*-coloring* of a plexus Γ is a function assigning to every strand of Γ an object of $\mathcal{C}$, called its $\mathcal{C}$*-color*, and assigning to every coupon Q of Γ a morphism $Q_\bullet \to Q^\bullet$ in $\mathcal{C}$, called its $\mathcal{C}$*-color*. Here $Q_\bullet$ and $Q^\bullet$ are objects of $\mathcal{C}$ defined as follows. Let us call the endpoints of the arcs of Γ lying on the bottom (respectively, top) base of Q the *inputs* (respectively, *outputs*) of Q. The orientation of the bottom base of Q induced by the orientation of Q determines an order in the set of the inputs. Let $X_i \in \mathrm{Ob}(\mathcal{C})$ be the color of the arc of Γ adjacent to the ith input. Set $\varepsilon_i = -$ if this arc is directed toward Q at the ith input and $\varepsilon_i = +$ otherwise. This determines a tuple $((X_1, \varepsilon_1), \ldots, (X_m, \varepsilon_m))$ of signed objects of $\mathcal{C}$, where m is the number of inputs of Q. The orientation of the top base of Q induced by the orientation of Q determines an order in the set of the outputs, and we take the opposite order. Let $Y_j \in \mathrm{Ob}(\mathcal{C})$ be the color of the arc of Γ adjacent to the jth output. Set $\nu_j = +$ if this arc is directed toward Q at the jth output and $\nu_j = -$ otherwise. This determines a tuple $((Y_1, \nu_1), \ldots, (Y_n, \nu_n))$ of signed objects of $\mathcal{C}$, where n is the number of outputs of Q. Then

$$Q_\bullet = X_1^{\varepsilon_1} \otimes \cdots \otimes X_m^{\varepsilon_m} \quad \text{and} \quad Q^\bullet = Y_1^{\nu_1} \otimes \cdots \otimes Y_n^{\nu_n},$$

where, as usual, $X^+ = X$ and $X^- = X^*$ for $X \in \mathrm{Ob}(\mathcal{C})$. For example, the coupon

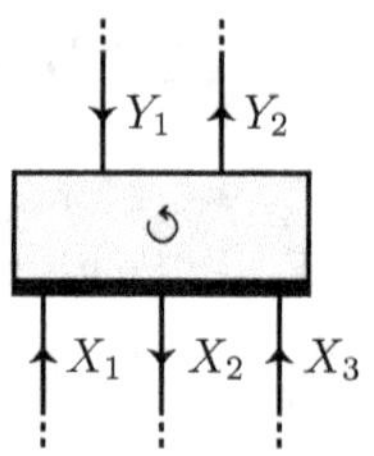

must be colored with a morphism $X_1^* \otimes X_2 \otimes X_3^* \to Y_1 \otimes Y_2^*$. As in Chapter 14, the bottom bases of coupons are drawn boldface.

15.1.3 Colored ribbon graphs

As in Section 14.2.1, a ribbon graph in a compact oriented 3-manifold M is a plexus embedded in M and equipped with a framing. A ribbon graph is $\mathcal{C}$*-colored* if its underlying plexus is endowed with a $\mathcal{C}$-coloring, see Section 15.1.2.

Given a $\mathcal{C}$-colored ribbon graph R in M, the pair $(\partial M, \partial R)$ acquires a structure of a $\mathcal{C}$-colored surface: the orientation of ∂M is induced by that of M and for any free end $a \in \partial R \subset \partial M$, the tangent direction of a is given by the framing of R, the sign of a is $+$ if the arc of R adjacent to a is directed away from M at a and is $-$ otherwise, and the color of a is the object of $\mathcal{C}$ coloring the arc of R adjacent to a.

Each $\mathcal{C}$-colored surface (Σ, A) determines a $\mathcal{C}$-colored ribbon graph $A \times [0,1]$ in $\Sigma \times [0,1]$. This ribbon graph has no coupons and no circle strands. The manifold

$\Sigma \times [0, 1]$ carries the product orientation, and $A \times [0, 1]$ carries the framing which projects to the given field of tangent directions on A in Σ and to zero on $[0, 1]$. For $a \in A$, the arc $\{a\} \times [0, 1]$ is colored with the same color as a and is directed toward $(a, 0)$ if the sign of a is $+$ and toward $(a, 1)$ otherwise. We call the pair $(\Sigma \times [0, 1], A \times [0, 1])$ the *cylinder* over (Σ, A) and denote it by C_Σ.

15.1.4 Moves on colored ribbon graphs

We define two moves on $\mathcal{C}$-colored ribbon graphs in a compact oriented 3-manifold M. These moves modify a $\mathcal{C}$-colored ribbon graph $R \subset M$ inside a 3-ball in $\mathrm{Int}(M)$ as shown in Figure 15.1. It is understood that the framing of the ribbon graph in this figure is everywhere orthogonal to the plane of the picture. The first move replaces two adjacent coupons colored by morphisms f, g in $\mathcal{C}$ with a coupon colored by $g \circ f$. The second move applies near a coupon Q having the same number of inputs and outputs $m \geq 0$ such that the corresponding inputs and outputs have the same direction and color (i.e., in the notation of Section 15.1.2 $X_i = Y_i$ and $\varepsilon_i = \nu_i$ for all $i = 1, \dots, m$) and the color of Q is the identity morphism $\mathrm{id}\colon Q_\bullet \to Q_\bullet = Q^\bullet$. The move deletes Q and connects the corresponding strands of the graph keeping their orientation and color. For $m = 0$, the move deletes a coupon having no inputs and no outputs and colored with $\mathrm{id}_{\mathbb{1}}$. All these moves on R keep the $\mathcal{C}$-colored surface $(\partial M, \partial R)$.

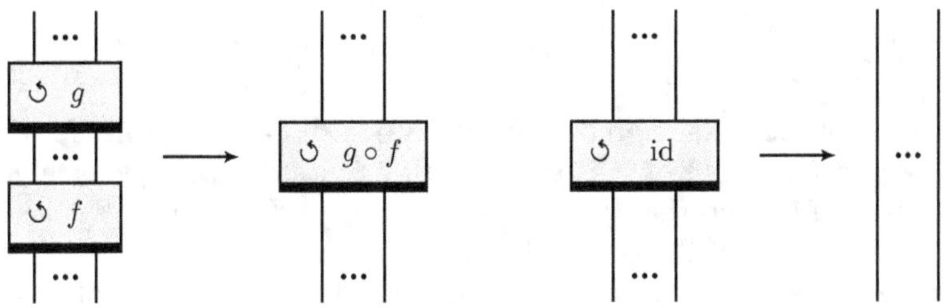

Figure 15.1: Moves on $\mathcal{C}$-colored ribbon graphs

15.1.5 Functoriality

Given a pivotal functor $F = (F, F_2, F_0)\colon \mathcal{C} \to \mathcal{D}$ between pivotal categories, there is a natural method transforming a $\mathcal{C}$-colored ribbon graph R in a compact oriented 3-manifold M into a $\mathcal{D}$-colored ribbon graph R^F in M. Namely, one keeps the underlying ribbon graph and replaces the color of each strand by its image under F. For every coupon of R, consider the tuples $((X_1, \varepsilon_1), \dots, (X_m, \varepsilon_m))$ and $((Y_1, \nu_1), \dots, (Y_n, \nu_n))$ of signed objects of $\mathcal{C}$ determined by the bottom and top bases of the coupon as in Section 15.1.2. We replace the color α of the coupon by the morphism

$$F(X_1)^{\varepsilon_1} \otimes \cdots \otimes F(X_m)^{\varepsilon_m} \simeq F(X_1^{\varepsilon_1} \otimes \cdots \otimes X_m^{\varepsilon_m})$$
$$\xrightarrow{F(\alpha)} F(Y_1^{\nu_1} \otimes \cdots \otimes Y_n^{\nu_n}) \simeq F(Y_1)^{\nu_1} \otimes \cdots \otimes F(Y_n)^{\nu_n}$$

in $\mathcal{D}$, where the isomorphisms are induced by the monoidal constraints F_2, F_0 and the natural isomorphism F^1 (see Section 1.7.5). This yields the $\mathcal{D}$-colored ribbon graph $R^F \subset M$.

15.2 Graph TQFTs

In this section, $\mathcal{C}$ is a pivotal category. We introduce a category $\mathrm{Cob}_3^{\mathcal{C}}$ of $\mathcal{C}$-colored ribbon graphs in 3-cobordisms and define graph TQFTs over $\mathcal{C}$.

15.2.1 The category $\mathrm{Cob}_3^{\mathcal{C}}$

We define a category $\mathrm{Cob}_3^{\mathcal{C}}$ as follows. The objects of $\mathrm{Cob}_3^{\mathcal{C}}$ are $\mathcal{C}$-colored surfaces. A morphism $\Sigma_0 \to \Sigma_1$ in $\mathrm{Cob}_3^{\mathcal{C}}$ between $\mathcal{C}$-colored surfaces Σ_0 and Σ_1 is a homeomorphism class of 3-cobordisms (with $\mathcal{C}$-colored ribbon graphs) between Σ_0 and Σ_1. Here, a *3-cobordism between* Σ_0 *and* Σ_1 is a triple (M, R, h), where M is a compact oriented 3-manifold, R is a $\mathcal{C}$-colored ribbon graph in M, and

$$h\colon (-\Sigma_0) \sqcup \Sigma_1 \to (\partial M, \partial R)$$

is a homeomorphism of $\mathcal{C}$-colored surfaces. Two such 3-cobordisms (M, R, h) and (M', R', h') are *homeomorphic* if there is an orientation preserving homeomorphism $f\colon M \to M'$ such that $fh = h'$ and $f(R) = R'$ where the latter equality stands for an equality of $\mathcal{C}$-colored ribbon graphs in M'.

Composition of morphisms in $\mathrm{Cob}_3^{\mathcal{C}}$ is defined through gluing of cobordisms: the composition of two morphisms $\Sigma_0 \to \Sigma_1$ and $\Sigma_1 \to \Sigma_2$, represented respectively by triples (M_0, R_0, h_0) and (M_1, R_1, h_1), is the morphism $\Sigma_0 \to \Sigma_2$ represented by the triple (M, R, h), where M is the result of gluing M_0 to M_1 along $h_1 h_0^{-1}\colon h_0(\Sigma_1) \to h_1(\Sigma_1)$, R is the union of the images of R_1 and R_2 under the embeddings $M_1 \hookrightarrow M$ and $M_2 \hookrightarrow M$, and

$$h = h_0|_{\Sigma_0} \sqcup h_1|_{\Sigma_2} \colon (-\Sigma_0) \sqcup \Sigma_2 \to (\partial M, \partial R).$$

The identity endomorphism of a $\mathcal{C}$-colored surface Σ is represented by the cylinder C_Σ over Σ (see Section 15.1.3) together with the homeomorphism of $\mathcal{C}$-colored surfaces $(-\Sigma) \sqcup \Sigma \simeq \partial C_\Sigma$ carrying any $x \in -\Sigma$ to $(x, 0)$ and any $y \in \Sigma$ to $(y, 1)$.

Each homeomorphism of $\mathcal{C}$-colored surfaces $\xi\colon \Sigma \to \Sigma'$ determines a morphism $\Sigma \to \Sigma'$ in $\mathrm{Cob}_3^{\mathcal{C}}$ represented by the cylinder $C_{\Sigma'}$ over Σ' endowed with the homeomorphism $(-\Sigma) \sqcup \Sigma' \to \partial C_{\Sigma'}$ carrying any $x \in \Sigma$ to $(\xi(x), 0)$ and any

$x' \in \Sigma'$ to $(x', 1)$. Composition of homeomorphisms corresponds under this construction to composition in $\mathrm{Cob}_3^{\mathcal{C}}$. Consequently, the morphisms in $\mathrm{Cob}_3^{\mathcal{C}}$ derived in this way from homeomorphisms of $\mathcal{C}$-colored surfaces are isomorphisms.

We provide $\mathrm{Cob}_3^{\mathcal{C}}$ with a structure of a symmetric monoidal category. The monoidal product in $\mathrm{Cob}_3^{\mathcal{C}}$ is given by the ordered disjoint union. The unit object of $\mathrm{Cob}_3^{\mathcal{C}}$ is an empty $\mathcal{C}$-colored surface $\emptyset$. The associativity and unitality constraints are induced by the tautological homeomorphisms of $\mathcal{C}$-colored surfaces

$$(\Sigma \sqcup \Sigma') \sqcup \Sigma'' \simeq \Sigma \sqcup (\Sigma' \sqcup \Sigma'') \quad \text{and} \quad \emptyset \sqcup \Sigma \simeq \Sigma \simeq \Sigma \sqcup \emptyset,$$

where $\Sigma, \Sigma', \Sigma''$ run over all $\mathcal{C}$-colored surfaces. The symmetry in $\mathrm{Cob}_3^{\mathcal{C}}$ is induced by the obvious permutation homeomorphisms of $\mathcal{C}$-colored surfaces

$$\Sigma \otimes \Sigma' = \Sigma \sqcup \Sigma' \simeq \Sigma' \sqcup \Sigma = \Sigma' \otimes \Sigma.$$

Note that any endomorphism of $\emptyset$ in $\mathrm{Cob}_3^{\mathcal{C}}$ is represented by a triple (M, R, h), where M is a closed oriented 3-manifold, R is a $\mathcal{C}$-colored ribbon graph in M, and $h = \mathrm{id}_\emptyset$. This endomorphism is fully determined by the pair (M, R). Two such pairs (M, R) and (M', R') represent the same endomorphism of $\emptyset$ if and only if they are *homeomorphic* in the sense that there is an orientation preserving homeomorphism $M \to M'$ carrying R into R' (and preserving the orientations and $\mathcal{C}$-colors of the strands and coupons).

The category $\mathrm{Cob}_3^{\mathcal{C}}$ has a distinguished left duality $\{(-\Sigma, \mathrm{ev}_\Sigma)\}_\Sigma$, where Σ runs over all $\mathcal{C}$-colored surface. Here $-\Sigma$ is the opposite $\mathcal{C}$-colored surface (see Section 15.1.1) and the morphism $\mathrm{ev}_\Sigma \colon (-\Sigma) \otimes \Sigma \to \emptyset$ is represented by the cylinder $C_{-\Sigma}$ together with the homeomorphism of $\mathcal{C}$-colored surfaces

$$-((-\Sigma) \otimes \Sigma) \sqcup \emptyset = \Sigma \sqcup (-\Sigma) \simeq (\Sigma \times \{0\}) \sqcup (-\Sigma \times \{1\}) = \partial C_{-\Sigma}\,.$$

By Lemma 3.5, the left duality $\{(-\Sigma, \mathrm{ev}_\Sigma)\}_\Sigma$ turns the symmetric monoidal category $\mathrm{Cob}_3^{\mathcal{C}}$ into a ribbon category. In particular, $\mathrm{Cob}_3^{\mathcal{C}}$ is spherical.

The trace of endomorphisms in $\mathrm{Cob}_3^{\mathcal{C}}$ is computed as follows. If Σ is a $\mathcal{C}$-colored surface and $f \colon \Sigma \to \Sigma$ is a morphism in $\mathrm{Cob}_3^{\mathcal{C}}$ represented by a triple $(M, R, h \colon (-\Sigma) \sqcup \Sigma \to (\partial M, \partial R))$, then the morphism

$$\mathrm{tr}(f) = \mathrm{ev}_\Sigma \circ (\mathrm{id}_{-\Sigma} \otimes f) \circ \widetilde{\mathrm{coev}}_\Sigma \colon \emptyset \to \emptyset$$

is represented by the pair obtained from (M, R) via the identification $h(-\Sigma) = h(\Sigma)$ induced by the identity map $-\Sigma \to \Sigma$. In particular, $\dim(\Sigma) = \mathrm{tr}(\mathrm{id}_\Sigma)$ is the homeomorphism class of the pair $(\Sigma \times S^1, A \times S^1)$, where A is the set of $\mathcal{C}$-colored points of Σ and $\Sigma \times S^1$ carries the product orientation. The ribbon graph $A \times S^1 \subset \Sigma \times S^1$ has only circle strands and its framing projects to the given field of tangent directions on A in Σ and to zero on S^1. For any $a \in A$, the circle $\{a\} \times S^1$ is colored with the same color as a and inherits the orientation of S^1 if the sign of a is $+$ and the opposite orientation otherwise.

Every closed oriented surface can be equipped with an empty set of $\mathcal{C}$-colored points, and every compact oriented 3-manifold can be equipped with an empty $\mathcal{C}$-colored ribbon graph. This allows us to view the category of 3-cobordisms Cob_3 defined in Section 10.1.3 as a (non-full) pivotal symmetric subcategory of $\mathrm{Cob}_3^{\mathcal{C}}$.

15.2.2 Graph TQFTs

Recall the symmetric monoidal category $\mathrm{Mod}_{\Bbbk}$ of $\Bbbk$-modules. A *graph TQFT* over $\mathcal{C}$ is a symmetric strong monoidal functor

$$Z = (Z, Z_2, Z_0)\colon \mathrm{Cob}_3^{\mathcal{C}} \to \mathrm{Mod}_{\Bbbk}.$$

Such a functor Z includes isomorphisms

$$Z_0\colon \Bbbk \to Z(\emptyset) \quad \text{and} \quad Z_2(\Sigma, \Sigma')\colon Z(\Sigma) \otimes_{\Bbbk} Z(\Sigma') \to Z(\Sigma \sqcup \Sigma')$$

for any $\mathcal{C}$-colored surfaces Σ, Σ'. Applying Z to morphisms in $\mathrm{Cob}_3^{\mathcal{C}}$ determined by homeomorphisms of $\mathcal{C}$-colored surfaces, we obtain an induced action of homeomorphisms as in Section 10.2.1. Also, we can apply Z to the morphism $\emptyset \to \emptyset$ in $\mathrm{Cob}_3^{\mathcal{C}}$ induced by a pair (M, R), where M is a closed oriented 3-manifold and R is a $\mathcal{C}$-colored ribbon graph in M. Since $Z(\emptyset) \simeq \Bbbk$, the resulting $\Bbbk$-linear homomorphism $Z(\emptyset) \to Z(\emptyset)$ is multiplication by an element of $\Bbbk$. This element is denoted $Z(M, R)$ and is a homeomorphism invariant of the pair (M, R).

The restriction of an arbitrary graph TQFT to Cob_3 is a 3-dimensional TQFT. Lemma 10.1 directly extends to a graph TQFT Z: for any $\mathcal{C}$-colored surface Σ, the $\Bbbk$-module $Z(\Sigma)$ is projective of finite type and its dimension is equal to $Z(\Sigma \times S^1, A \times S^1)$, where A is the set of $\mathcal{C}$-colored points of Σ and the $\mathcal{C}$-colored ribbon graph $A \times S^1 \subset \Sigma \times S^1$ is as in Section 15.2.1. Clearly, the isomorphism class of the $\Bbbk$-module $Z(\Sigma)$ is preserved under isotopy of A in Σ. Also, each connected component of Σ may be treated as a $\mathcal{C}$-colored surface whose $\mathcal{C}$-colored points are the $\mathcal{C}$-colored points of Σ belonging to this component. Since Σ is a disjoint union of its connected components, the strong monoidality and the symmetry of Z yield a $\Bbbk$-linear isomorphism

$$Z(\Sigma) \simeq \bigotimes_{\Gamma \in \pi_0(\Sigma)} Z(\Gamma). \tag{15.1}$$

Here $\pi_0(\Sigma)$ is the set of connected components of Σ and $\otimes$ is the unordered tensor product of $\Bbbk$-modules. We conclude that the modules assigned by Z to $\mathcal{C}$-colored surfaces are determined, at least up to isomorphism, by the modules assigned by Z to connected $\mathcal{C}$-colored surfaces.

An *isomorphism* between two graph TQFTs $Z, Z'\colon \mathrm{Cob}_3^{\mathcal{C}} \to \mathrm{Mod}_{\Bbbk}$ is a monoidal natural isomorphism $Z \to Z'$ in the sense of Section 1.4.2. By definition, such an isomorphism is a family $\varphi = \{\varphi_\Sigma\colon Z(\Sigma) \to Z'(\Sigma)\}_\Sigma$ of $\Bbbk$-linear isomorphisms, where Σ runs over all $\mathcal{C}$-colored surfaces. These isomorphisms should commute with the action of 3-cobordisms with $\mathcal{C}$-colored ribbon graphs, be multiplicative under disjoint unions of $\mathcal{C}$-colored surfaces, and satisfy $\varphi_\emptyset = Z'_0 Z_0^{-1}$.

As a consequence, if two graph TQFTs Z and Z' over $\mathcal{C}$ are isomorphic, then $Z(M,R) = Z'(M,R)$ for any $\mathcal{C}$-colored ribbon graph R in a closed oriented 3-manifold M.

15.2.3 Two classes of graph TQFTs

We introduce two important classes of graph TQFTs. A graph TQFT Z over $\mathcal{C}$ is *non-degenerate* if for any $\mathcal{C}$-colored surface Σ, the $\Bbbk$-module $Z(\Sigma)$ is spanned by the images of the $\Bbbk$-linear homomorphisms $Z(f)\colon Z(\emptyset) \to Z(\Sigma)$ determined by all morphisms $f\colon \emptyset \to \Sigma$ in the category $\mathrm{Cob}_3^{\mathcal{C}}$. Formula (15.1) and the strong monoidality of Z imply that if this condition holds for all connected $\mathcal{C}$-colored surfaces Σ, then it holds for all disconnected $\mathcal{C}$-colored surfaces Σ as well.

The moves on $\mathcal{C}$-colored ribbon graphs introduced in Section 15.1.4 determine transformations of morphisms in $\mathrm{Cob}_3^{\mathcal{C}}$ which keep the source and the target objects and modify the underlying $\mathcal{C}$-colored ribbon graphs as in Figure 15.1. A graph TQFT $Z\colon \mathrm{Cob}_3^{\mathcal{C}} \to \mathrm{Mod}_{\Bbbk}$ is *regular* if $Z(f) = Z(f')$ for any morphisms f, f' in $\mathrm{Cob}_3^{\mathcal{C}}$ related by these transformations. The $\Bbbk$-module associated with a connected $\mathcal{C}$-colored surface Σ by a regular graph TQFT can be computed as follows. Let A be the set of $\mathcal{C}$-colored points of Σ. For $a \in A$, let $V_a \in \mathrm{Ob}(\mathcal{C})$ and $\varepsilon_a \in \{+,-\}$ be respectively the color and sign of a. Set

$$V_\Sigma = \bigotimes_{a\in A} V_a^{\varepsilon_a} \in \mathrm{Ob}(\mathcal{C})$$

where the monoidal product is determined by a total order on A and, as usual, $V_a^+ = V_a$ and $V_a^- = V_a^*$ for any $a \in A$. By definition, if $A = \emptyset$, then $V_\Sigma = \mathbb{1}$. Pick an arbitrary point $s \in \Sigma$ and equip it with a tangent direction in Σ, the color $V_\Sigma \in \mathrm{Ob}(\mathcal{C})$, and the sign $+$. The underlying oriented surface of Σ together with the singleton $\{s\}$ is then a $\mathcal{C}$-colored surface denoted Σ_s.

Lemma 15.1. *For any regular graph TQFT Z over $\mathcal{C}$ and any connected $\mathcal{C}$-colored surface Σ, the $\Bbbk$-modules $Z(\Sigma)$ and $Z(\Sigma_s)$ are isomorphic.*

Proof. Pick an embedded oriented segment $J \subset \Sigma$ such that $A \cup \{s\} \subset J$ and the total order in A is induced by the orientation of J. Applying if necessary an ambient isotopy, we can assume that the distinguished tangent directions in the points of $A \cup \{s\}$ are transversal to J and directed to the left of J. Consider the coupon $Q = J \times [\frac{1}{3}, \frac{2}{3}]$ in $\Sigma \times [0,1]$ with product orientation and bottom base $J \times \{\frac{1}{3}\}$. Set

$$R_1 = \left(A \times \left[0, \frac{1}{3}\right]\right) \cup Q \cup \left(\{s\} \times \left[\frac{2}{3}, 1\right]\right) \subset \Sigma \times [0,1],$$

$$R_2 = \left(\{s\} \times \left[0, \frac{1}{3}\right]\right) \cup Q \cup \left(A \times \left[\frac{2}{3}, 1\right]\right) \subset \Sigma \times [0,1].$$

We turn R_1 and R_2 into $\mathcal{C}$-colored ribbon graphs with a single coupon Q and $\mathrm{card}(A) + 1$ edges. For $a \in A$, we endow the edge $\{a\} \times [0, \frac{1}{3}]$ of R_1 with the color

V_a and direct it toward $(a,0)$ if $\varepsilon_a = +$ and toward $(a, \frac{1}{3})$ otherwise. For $a \in A$, we endow the edge $\{a\} \times [\frac{2}{3}, 1]$ of R_2 with the color V_a and direct it toward $(a, \frac{2}{3})$ if $\varepsilon_a = +$ and toward $(a,1)$ otherwise. The edges $\{s\} \times [\frac{2}{3}, 1]$ of R_1 and $\{s\} \times [0, \frac{1}{3}]$ of R_2 are colored with V_Σ and directed toward $(s, \frac{2}{3})$ and $(s,0)$, respectively. We color Q with the morphism $\mathrm{id}_{V_\Sigma} \colon V_\Sigma \to V_\Sigma$. Let $f_1 \colon \Sigma \to \Sigma_s$ and $f_2 \colon \Sigma_s \to \Sigma$ be the morphisms in $\mathrm{Cob}_3^{\mathcal{C}}$ represented by $(\Sigma \times [0,1], R_1, h)$ and $(\Sigma \times [0,1], R_2, h)$, respectively, where the homeomorphism $h \colon (-\Sigma) \cup \Sigma \to \partial(\Sigma \times [0,1])$ carries any $x \in -\Sigma$ to $(x,0)$ and any $y \in \Sigma$ to $(y,1)$. Then

$$Z(f_2) \circ Z(f_1) \overset{\text{(i)}}{=} Z(f_2 \circ f_1) \overset{\text{(ii)}}{=} Z(\mathrm{id}_\Sigma) \overset{\text{(iii)}}{=} \mathrm{id}_{Z(\Sigma)}$$

where (i) and (iii) follow from the functoriality of Z and (ii) from the regularity of Z. A similar argument yields $Z(f_1) \circ Z(f_2) = \mathrm{id}_{Z(\Sigma_s)}$. Thus the $\Bbbk$-linear homomorphism $Z(f_1) \colon Z(\Sigma) \to Z(\Sigma_s)$ is an isomorphism with inverse $Z(f_2)$. □

15.2.4 State sum graph TQFT

In the remaining part of this chapter, we construct a regular graph TQFT over the center of a spherical fusion $\Bbbk$-category. Our construction extends the state sum method of Chapter 13 using so-called colored knotted nets introduced below in Section 15.3. The results of the chapter are summarized in Theorem 15.9 stated in Section 15.7.3. We note here the following consequence of this theorem.

Theorem 15.2. *For any spherical fusion $\Bbbk$-category $\mathcal{C}$ such that* $\dim(\mathcal{C})$ *is invertible in $\Bbbk$, there is a regular graph TQFT over the center $\mathcal{Z}(\mathcal{C})$ of $\mathcal{C}$ whose restriction to* Cob_3 *is the state sum TQFT of Theorem* 13.5 *derived from $\mathcal{C}$.*

The non-degeneracy of the state sum graph TQFTs constructed here will be discussed in Section 17.5.

15.2.5 Exercise

Let $F \colon \mathcal{C} \to \mathcal{D}$ be a pivotal functor between pivotal categories. Define a functor $\widetilde{F} \colon \mathrm{Cob}_3^{\mathcal{C}} \to \mathrm{Cob}_3^{\mathcal{D}}$ as follows:

- $\widetilde{F}$ carries a $\mathcal{C}$-colored surface (Σ, A) to the $\mathcal{D}$-colored surface (Σ, A^F), where A^F is obtained from A by applying F to the colors of all points while keeping the tangent directions and signs of all points;
- $\widetilde{F}$ carries any morphism f in $\mathrm{Cob}_3^{\mathcal{C}}$ represented by (M, R, h) into the morphism in $\mathrm{Cob}_3^{\mathcal{D}}$ represented by (M, R^F, h), where $R^F \subset M$ is the $\mathcal{D}$-colored ribbon graph defined in Section 15.1.5.

Prove that $\widetilde{F}$ is a symmetric strict monoidal functor. As a consequence, any graph TQFT Z over $\mathcal{D}$ induces a graph TQFT $Z \circ \widetilde{F}$ over $\mathcal{C}$. Prove that if F is a pivotal equivalence, then $\widetilde{F}$ is a symmetric equivalence and the formula $Z \mapsto Z \circ \widetilde{F}$ defines a bijective correspondence between the isomorphism classes of graph TQFTs over $\mathcal{D}$ and the isomorphism classes of graph TQFTs over $\mathcal{C}$.

15.3 Colored knotted nets

15.3.1 Nets

Nets generalize both graphs and plexuses. A *net* Γ is a topological space obtained from a plexus by gluing some of its free ends to each other. The images in Γ of the free ends of the plexus are called the *vertices* of Γ. The images in Γ of the arcs, half-arcs, circles, and coupons of the plexus are called respectively *edges*, *half-edges*, *circles*, and *coupons* of Γ. The edges and circles of Γ are collectively called *strands*. We endow the edges, circles, and coupons of Γ with orientation induced by the given orientation of the arcs, circles, and coupons of the plexus. The half-edges of Γ inherit their orientation from that of edges. The *valence* of a vertex of Γ is the number of half-edges incident to this vertex.

It is clear from the definitions that oriented graphs are nets without circles and without coupons. Plexuses are nets whose all vertices have valence 1. Figure 15.2 exhibits a net having 1 circle, 3 coupons, 11 vertices, and 15 edges. As usual, bottom bases of coupons are drawn boldface.

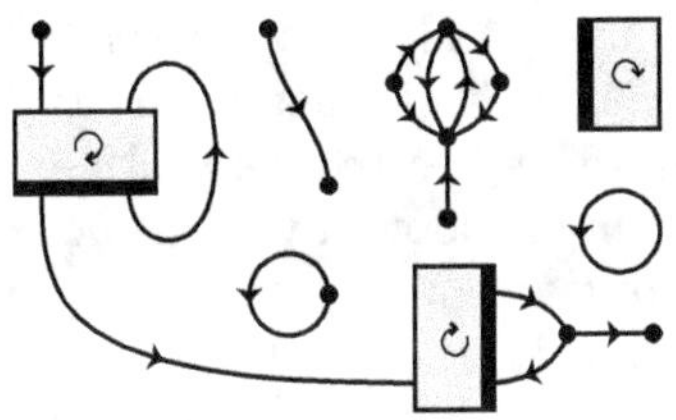

Figure 15.2: A net

15.3.2 Knotted nets

A knotted net in an oriented surface Σ is a net drawn in Σ, possibly with double crossings of strands such that at each crossing, one of the strands is distinguished. More precisely, a *knotted net* Γ in Σ is a net immersed in $\Sigma \setminus \partial\Sigma$ (i.e., endowed with a continuous map to $\Sigma \setminus \partial\Sigma$ which is a local embedding) such that:

(i) all coupons of Γ are embedded in Σ preserving orientation;

(ii) all multiple points of Γ are double transversal intersections of the interiors of strands. At every double point, one of the two meeting strands is distinguished.

Note that the coupons and vertices of a knotted net Γ are pairwise disjoint.

The double points of a knotted net are called *crossings*. They are finite in number and lie away from the coupons and the vertices. A crossing of a knotted net in Σ lies in an open disk in Σ represented in our pictures by a plane parallel to the page. The orientation of Σ is represented by an orientation of this plane. The distinguished strand at the crossing is represented by a red continuous (unbroken) line if the latter orientation is counterclockwise and by a red broken line otherwise:

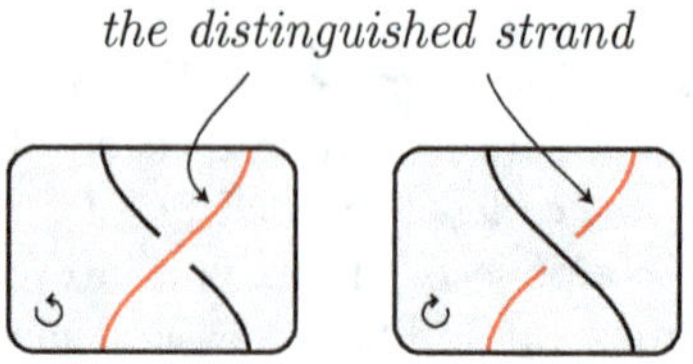

15.3.3 Colored knotted nets

Let $\mathcal{C}$ be a pivotal $\Bbbk$-category. Recall that the center $\mathcal{Z}(\mathcal{C})$ of $\mathcal{C}$ is a braided pivotal $\Bbbk$-category, see Sections 5.2.2 and 5.3.1.

A $\mathcal{C}$*-coloring* of a net Γ is a function assigning to every strand of Γ an object of $\mathcal{C}$, called its $\mathcal{C}$*-color*, and assigning to every coupon Q of Γ a morphism $Q_\bullet \to Q^\bullet$ in $\mathcal{C}$, called its $\mathcal{C}$*-color*. Here, $Q_\bullet$ and $Q^\bullet$ are the objects of $\mathcal{C}$ defined in the same way as for plexuses, see Section 15.1.2.

Consider an oriented surface Σ. By $\mathcal{C}$*-colorings* (respectively, $\mathcal{Z}(\mathcal{C})$*-colorings*) of a knotted net in Σ, we mean $\mathcal{C}$-colorings (respectively, $\mathcal{Z}(\mathcal{C})$-colorings) of the underlying net. We now introduce knotted nets with "mixed" colorings. A *knotted* $\mathcal{C}$*-net* in Σ is a knotted net in Σ endowed with a *coloring function* which assigns to every strand an object of $\mathcal{C}$ or an object of $\mathcal{Z}(\mathcal{C})$ and which assigns to every coupon a morphism in $\mathcal{Z}(\mathcal{C})$ so that:

(i) if a strand is incident to a coupon or is the distinguished strand of at least one crossing, then its color must be an object of $\mathcal{Z}(\mathcal{C})$;

(ii) the color of any coupon Q is a morphism $Q_\bullet \to Q^\bullet$ in $\mathcal{Z}(\mathcal{C})$, where $Q_\bullet$ (respectively, $Q^\bullet$) is the object of $\mathcal{Z}(\mathcal{C})$ determined by the colors and orientations of the inputs (respectively, outputs) of Q as in Section 15.1.2.

In the pictures, all strands as in (i) above will be drawn in red and all other strands will be drawn in black. Thus, the color of a red strand must belong to $\mathrm{Ob}(\mathcal{Z}(\mathcal{C}))$ while the color of a black strand belongs either to $\mathrm{Ob}(\mathcal{C})$ or to $\mathrm{Ob}(\mathcal{Z}(\mathcal{C}))$.

For example, every $\mathcal{Z}(\mathcal{C})$-coloring of a knotted net in Σ turns it into a knotted $\mathcal{C}$-net (with all colors in $\mathcal{Z}(\mathcal{C})$). Also, every $\mathcal{C}$-colored graph in Σ is a knotted $\mathcal{C}$-net without coupons and crossings (and with all colors in $\mathcal{C}$).

Given a knotted $\mathcal{C}$-net Γ in Σ, the *subcolor* of a strand colored with an object X is X itself if $X \in \mathrm{Ob}(\mathcal{C})$, and is the image of X under the forgetful functor $\mathcal{Z}(\mathcal{C}) \to \mathcal{C}$ if $X \in \mathrm{Ob}(\mathcal{Z}(\mathcal{C}))$. The *subcolor* of a coupon is the image of the color of this coupon under the forgetful functor $\mathcal{Z}(\mathcal{C}) \to \mathcal{C}$. The subcolors of strands and coupons of Γ form a $\mathcal{C}$-coloring of Γ.

The constructions of Section 12.2.1 extend to knotted $\mathcal{C}$-nets. In particular, a vertex v of a knotted $\mathcal{C}$-net Γ in Σ determines a cyclic $\mathcal{C}$-set $(E_v, c_v, \varepsilon_v)$ as there with the only difference that the map $c_v \colon E_v \to \mathrm{Ob}(\mathcal{C})$ carries each half-edge adjacent to v to its subcolor. We set $H_v(\Gamma) = H(E_v)$ and $H(\Gamma) = \otimes_v H_v(\Gamma)$, where v runs over all vertices of Γ and $\otimes$ is the unordered tensor product of $\Bbbk$-modules. To emphasize the role of Σ, we sometimes write $H_v(\Gamma;\Sigma)$ for $H_v(\Gamma)$ and $H(\Gamma;\Sigma)$ for $H(\Gamma)$. If Γ has no vertices, then by definition $H(\Gamma) = \Bbbk$. Note that for any disjoint knotted $\mathcal{C}$-nets Γ_1 and Γ_2 in Σ, there is a canonical $\Bbbk$-linear isomorphism

$$H(\Gamma_1 \sqcup \Gamma_2) \simeq H(\Gamma_1) \otimes H(\Gamma_2).$$

An *isotopy* of knotted $\mathcal{C}$-nets in Σ is an ambient isotopy in the class of knotted $\mathcal{C}$-nets in Σ preserving the vertices, the strands, the crossings (with their distinguished strand), the coupons (with their distinguished base), the orientations, and the colors. An isotopy between two knotted $\mathcal{C}$-nets Γ and Γ' in Σ induces a $\Bbbk$-linear isomorphism $H(\Gamma) \to H(\Gamma')$ in the obvious way.

Any orientation preserving embedding f of Σ into an oriented surface Σ' carries a knotted $\mathcal{C}$-net Γ in Σ into a knotted $\mathcal{C}$-net $\Gamma' = f(\Gamma)$ in Σ' and induces a $\Bbbk$-linear isomorphism $H(f)\colon H(\Gamma;\Sigma) \to H(\Gamma';\Sigma')$ in the obvious way. This applies, in particular, when f is an orientation preserving self-homeomorphism of Σ.

15.3.4 Example

Let Γ be the following knotted $\mathcal{C}$-net in $\mathbb{R}^2$:

$\Gamma =$ 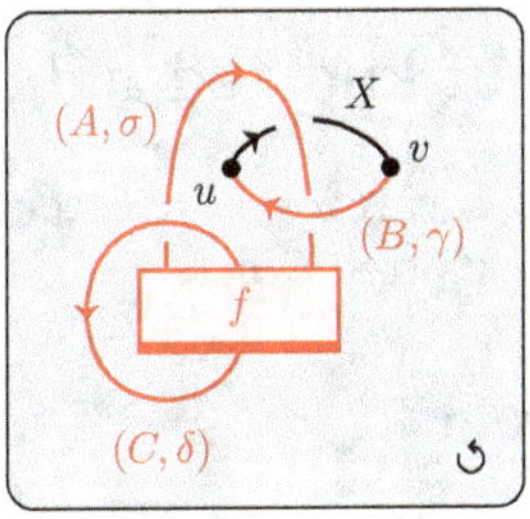

The net Γ has two vertices u and v, four edges colored by $(A,\sigma),(B,\gamma),(C,\delta) \in \mathrm{Ob}(\mathcal{Z}(\mathcal{C}))$ and $X \in \mathrm{Ob}(\mathcal{C})$, and one coupon colored by a morphism in $\mathcal{Z}(\mathcal{C})$

$$f\colon (C,\delta)^* \to (A,\sigma)^* \otimes (C,\delta)^* \otimes (A,\sigma).$$

There are two half-edges e_1, e_2 incident to u and two half-edges e_3, e_4 incident to v:

$e_1 =$  , $e_2 =$  , $e_3 =$  , $e_4 =$  .

Clearly, $H(\Gamma) = H_u(\Gamma) \otimes H_v(\Gamma)$ and the cone isomorphisms associated with the half-edges are:

$$\tau^u_{e_1}\colon H_u(\Gamma) \to \mathrm{Hom}_{\mathcal{C}}(\mathbb{1}, X^* \otimes B), \qquad \tau^v_{e_3}\colon H_v(\Gamma) \to \mathrm{Hom}_{\mathcal{C}}(\mathbb{1}, X \otimes B^*),$$
$$\tau^u_{e_2}\colon H_u(\Gamma) \to \mathrm{Hom}_{\mathcal{C}}(\mathbb{1}, B \otimes X^*), \qquad \tau^v_{e_4}\colon H_v(\Gamma) \to \mathrm{Hom}_{\mathcal{C}}(\mathbb{1}, B^* \otimes X).$$

15.3.5 Dual vertices

Let $\mathcal{C}$ be a spherical $\Bbbk$-category. The definition of dual vertices and the constructions of Sections 12.3.3 and 12.3.4 extend to knotted $\mathcal{C}$-nets as follows. Consider a knotted $\mathcal{C}$-net Γ in an oriented surface Σ and a knotted $\mathcal{C}$-net Γ' in an oriented surface Σ'. A *duality* between a vertex u of Γ' and a vertex v of Γ is a duality between the cyclic $\mathcal{C}$-sets E_u and E_v, that is, an isomorphism of cyclic $\mathcal{C}$-sets $\phi\colon E_u^{\mathrm{op}} \to E_v$. Set

$$\omega_{u,v} = \omega_\phi\colon H_u(\Gamma') \otimes H_v(\Gamma) = H(E_u) \otimes H(E_v) \to \mathrm{End}_{\mathcal{C}}(\mathbb{1}),$$

where ω_ϕ is the pairing associated with ϕ in Section 12.3.2. Clearly, $\omega_{u,v} = \omega_{v,u}$, where $\omega_{v,u}$ is induced by the isomorphism of cyclic $\mathcal{C}$-sets $\phi^{-1}\colon E_v^{\mathrm{op}} \to E_u$.

If the category $\mathcal{C}$ is non-degenerate, then the pairing $\omega_{u,v}$ takes values in $\mathrm{End}_{\mathcal{C}}(\mathbb{1}) = \Bbbk$ and is non-degenerate. The *contraction vector* of the duality ϕ between u and v is the vector

$$*_{u,v} = *_\phi \in H_u(\Gamma') \otimes H_v(\Gamma),$$

where $*_\phi$ is the contraction vector associated with ϕ (see Section 12.3.4). Clearly, $*_{u,v} = *_{v,u}$ where $*_{v,u}$ is determined by ϕ^{-1}.

15.4 An invariant of knotted nets

In this section, $\mathcal{C}$ is a pivotal $\Bbbk$-category. We extend the invariant $\mathbb{F}_{\mathcal{C}}$ of $\mathcal{C}$-colored graphs (see Section 12.2) to knotted $\mathcal{C}$-nets.

15.4.1 Knotted $\mathcal{C}$-nets in $\mathbb{R}^2$

Consider a knotted $\mathcal{C}$-net Γ in the plane $\mathbb{R}^2$ endowed with the counterclockwise orientation. Pick a vector $\alpha_v \in H_v(\Gamma)$ for every vertex v of Γ and transform Γ as follows. At each vertex v of Γ we proceed as in Figure 12.1: pick a half-edge $e_v \in E_v$, isotop Γ near v so that the half-edges incident to v lie above v with respect to the second coordinate on $\mathbb{R}^2$ and e_v is the leftmost of them, replace v by a box colored with $\tau^v_{e_v}(\alpha_v)$, where τ^v is the universal cone of $H_v(\Gamma)$. Next, isotoping Γ in $\mathbb{R}^2$ near the crossings, we ensure that the strands of Γ are oriented downward at all crossings (see, for instance, Figure 15.3). Then we replace each crossing of Γ by a box as in Figure 15.4, where $(A,\sigma) \in \mathrm{Ob}(\mathcal{Z}(\mathcal{C}))$ is the color of the distinguished strand and $X \in \mathrm{Ob}(\mathcal{C})$ is the subcolor of the second strand. At each coupon Q of Γ we proceed as in Figure 15.5: isotop Γ near Q to make the bases of Q horizontal and to ensure that the distinguished (bottom) base lies below the opposite base with respect to the second coordinate on $\mathbb{R}^2$, replace Q by a box with the same inputs and outputs as Q, and label this box with the subcolor of Q. Also, the colors of all edges of Γ are traded for the corresponding subcolors.

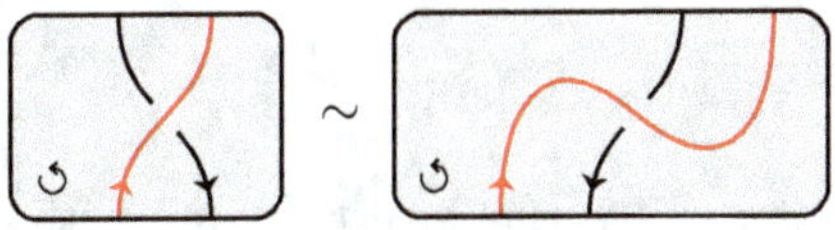

Figure 15.3: Isotopy near a crossing

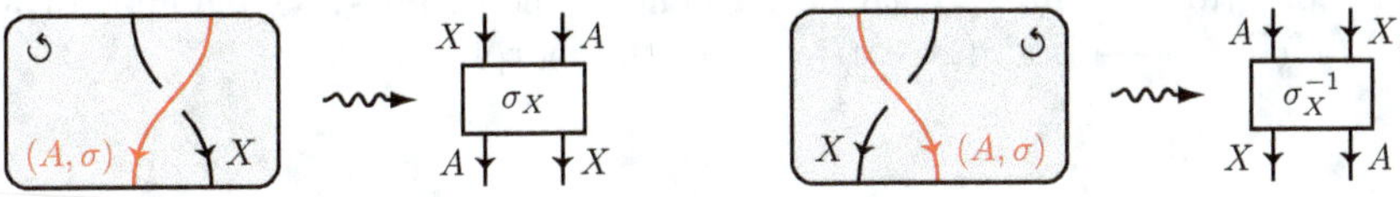

Figure 15.4: Transformation at a crossing

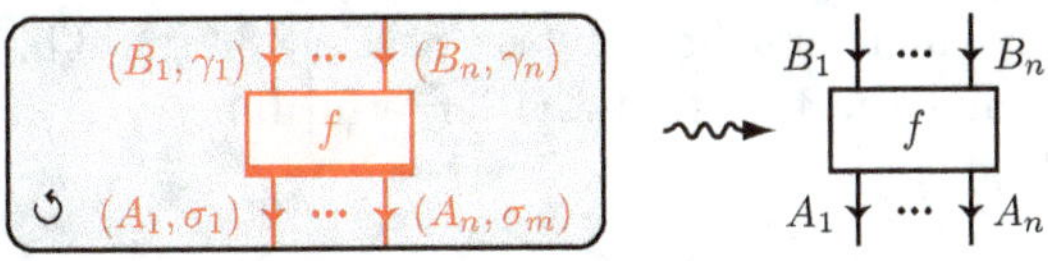

Figure 15.5: Transformation of a coupon

This turns Γ in a $\mathcal{C}$-colored Penrose diagram without free ends. The Penrose calculus associates with this diagram an element of $\mathrm{End}_{\mathcal{C}}(\mathbb{1})$ denoted $\mathbb{F}_{\mathcal{C}}(\Gamma)(\otimes_v \alpha_v)$. By linear extension, this procedure defines a $\Bbbk$-linear homomorphism

$$\mathbb{F}_{\mathcal{C}}(\Gamma)\colon H(\Gamma) = \otimes_v H_v(\Gamma) \to \mathrm{End}_{\mathcal{C}}(\mathbb{1}).$$

Lemma 15.3. *The homomorphism* $\mathbb{F}_{\mathcal{C}}(\Gamma)\colon H(\Gamma) \to \mathrm{End}_{\mathcal{C}}(\mathbb{1})$ *is a well-defined isotopy invariant of* Γ.

The isotopy invariance here is understood in the same sense as in Lemma 12.2. The proof of Lemma 15.3 repeats the proof of Lemma 12.2.

15.4.2 Example

For any object X of $\mathcal{C}$ or of $\mathcal{Z}(\mathcal{C})$ and for any endomorphism f of $Y \in \mathrm{Ob}(\mathcal{Z}(\mathcal{C}))$, the pictures

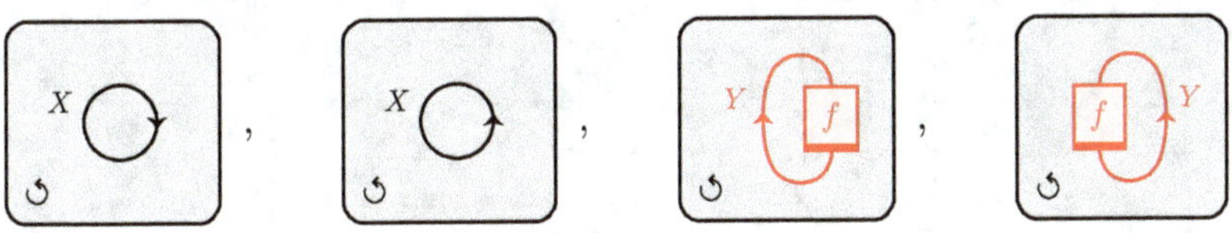

represent knotted $\mathcal{C}$-nets in $\mathbb{R}^2$ without vertices and without crossings. The corresponding $\Bbbk$-linear homomorphisms $\mathbb{F}_{\mathcal{C}}(\Gamma)\colon H(\Gamma) = \Bbbk \to \mathrm{End}_{\mathcal{C}}(\mathbb{1})$ carry $1_{\Bbbk}$ to $\dim_l(X), \dim_r(X), \mathrm{tr}_l(f), \mathrm{tr}_r(f)$, respectively.

15.4.3 Example

Let Γ be a knotted $\mathcal{C}$-net in $\mathbb{R}^2$ without vertices and with all colors of strands in $\mathrm{Ob}(\mathcal{Z}(\mathcal{C}))$. Isotoping the coupons of Γ so that their bases are horizontal and the distinguished bases lie below the opposite bases, we obtain a $\mathcal{Z}(\mathcal{C})$-colored Reidemeister diagram $\mathcal{R}_\Gamma$ in $\mathbb{R}^2$ with no free ends. Then the $\Bbbk$-linear homomorphism $\mathbb{F}_\mathcal{C}(\Gamma)\colon H(\Gamma) = \Bbbk \to \mathrm{End}_\mathcal{C}(\mathbb{1})$ carries $1_\Bbbk$ to the morphism

$$\langle \mathcal{R}_\Gamma \rangle \in \mathrm{End}_{\mathcal{Z}(\mathcal{C})}(\mathbb{1}_{\mathcal{Z}(\mathcal{C})}) \subset \mathrm{End}_\mathcal{C}(\mathbb{1})$$

defined by the Reidemeister calculus (see Section 3.2.1). This follows from the definition of $\mathcal{Z}(\mathcal{C})$ and the fact that the forgetful functor $\mathcal{Z}(\mathcal{C}) \to \mathcal{C}$ acts as the identity on morphisms.

15.4.4 Example

Consider the knotted $\mathcal{C}$-net Γ in $\mathbb{R}^2$ from Example 15.3.4. To compute $\mathbb{F}_\mathcal{C}(\Gamma)$, we first isotop Γ as follows:

$\Gamma \sim$ 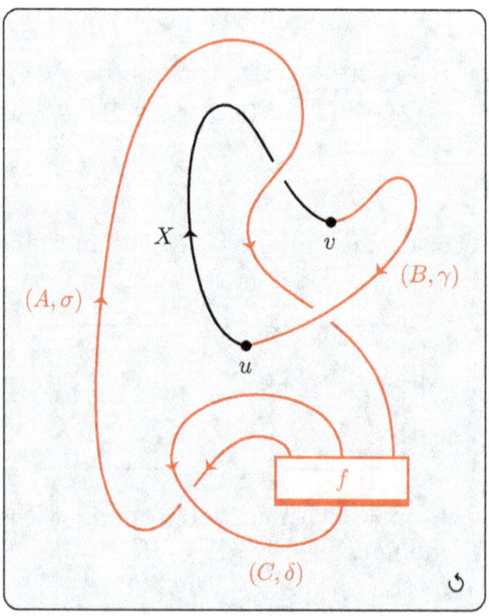 .

By definition, for any $\alpha \in H_u(\Gamma)$ and $\beta \in H_v(\Gamma)$, we have

$\mathbb{F}_{\mathcal{C}}(\Gamma)(\alpha \otimes \beta) =$

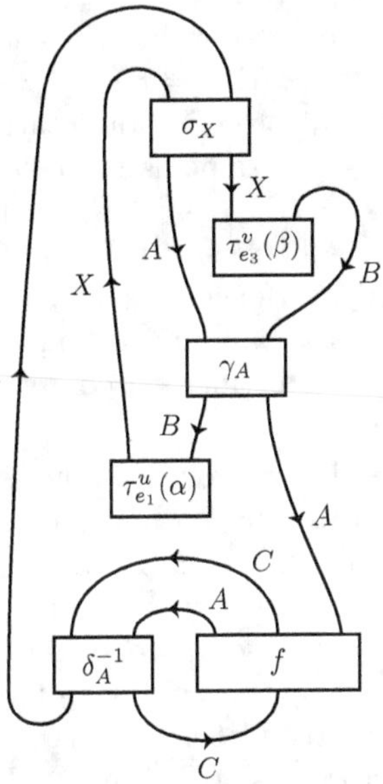

Here we use the half-edges e_1, e_3 defined in Example 15.3.4.

15.4.5 Properties of $\mathbb{F}_{\mathcal{C}}$

The properties of $\mathbb{F}_{\mathcal{C}}$ stated in Section 12.2.4 for $\mathcal{C}$-colored graphs in $\mathbb{R}^2$ generalize to knotted $\mathcal{C}$-nets in $\mathbb{R}^2$ with the obvious changes. We state such a generalization of the property (ii) used in the next subsection.

Lemma 15.4. *Let Γ' be a knotted $\mathcal{C}$-net obtained from a knotted $\mathcal{C}$-net Γ in $\mathbb{R}^2$ by reversing the orientation of a strand e not incident to a coupon and replacing the color of e by the dual object. If e is a circle, then $\mathbb{F}_{\mathcal{C}}(\Gamma') = \mathbb{F}_{\mathcal{C}}(\Gamma)$. If e is an edge, then the pivotal structure of $\mathcal{C}$ induces weak isomorphisms between the cyclic $\mathcal{C}$-sets associated with the endpoints of e in Γ and Γ'. These weak isomorphisms induce a $\Bbbk$-linear isomorphism $\Psi\colon H(\Gamma') \to H(\Gamma)$ and $\mathbb{F}_{\mathcal{C}}(\Gamma') = \mathbb{F}_{\mathcal{C}}(\Gamma)\Psi$.*

Proof. The case of a circle follows from Lemma 2.3. Suppose that e is an edge. Let $\Gamma_\bullet$ (respectively, $\Gamma'_\bullet$) be the knotted $\mathcal{C}$-net obtained from Γ (respectively, from Γ') by adding inside e a new vertex w which splits e into two subedges e_1, e_2:

$\Gamma =$ 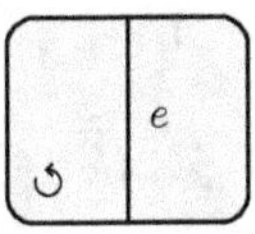 $\rightsquigarrow \Gamma_\bullet =$ 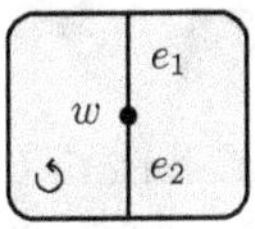 .

The strands and coupons of $\Gamma_\bullet$ (respectively, of $\Gamma'_\bullet$) inherit their orientations and colors from Γ (respectively, from Γ'). Clearly, $H(\Gamma_\bullet) \simeq H(\Gamma) \otimes H_w(\Gamma_\bullet)$. The

$\Bbbk$-linear map $\Bbbk \to H_w(\Gamma_\bullet)$ determined by the coevaluation as in Example 12.3.5 induces a linear map $\mu\colon H(\Gamma) \to H(\Gamma_\bullet)$ and $\mathbb{F}_{\mathcal{C}}(\Gamma) = \mathbb{F}_{\mathcal{C}}(\Gamma_\bullet)\,\mu$. Similarly, the $\Bbbk$-linear map $\Bbbk \to H_w(\Gamma'_\bullet)$ determined by the coevaluation induces a $\Bbbk$-linear map $\mu'\colon H(\Gamma') \to H(\Gamma'_\bullet)$ and $\mathbb{F}_{\mathcal{C}}(\Gamma') = \mathbb{F}_{\mathcal{C}}(\Gamma'_\bullet)\,\mu'$. Note that the knotted $\mathcal{C}$-net $\Gamma'_\bullet$ can be obtained from $\Gamma_\bullet$ by reversing orientation of the edges e_1, e_2 and replacing their colors (which are the same as the color of e) by the dual objects. We have $\mu\Psi = \Psi_\bullet \mu'$ where $\Psi_\bullet\colon H(\Gamma'_\bullet) \to H(\Gamma_\bullet)$ is the $\Bbbk$-linear isomorphism induced by the weak isomorphisms between the cyclic $\mathcal{C}$-sets associated with the endpoints of e_1 and e_2 in $\Gamma_\bullet$ and $\Gamma'_\bullet$. Hence, the claim of the lemma for e_1 and e_2 would imply the claim of the lemma for e. Subdividing e in this way, we reduce the lemma to the case where e has two distinct endpoints, has no self-crossings, and contains at most one crossing.

Let u and v be the endpoints of the edge e. The case where e contains no crossings follows from Lemma 2.2 and the definition of the $\Bbbk$-linear isomorphisms $H_u(\Gamma') \simeq H_u(\Gamma)$ and $H_v(\Gamma') \simeq H_v(\Gamma)$ induced by the weak isomorphisms between the cyclic $\mathcal{C}$-sets associated with u, v in Γ and Γ'. We focus now on the more difficult case where e traverses a crossing x and compare the contributions to $\mathbb{F}_{\mathcal{C}}$ of the vertices, crossings, and coupons of Γ and Γ'. Since the coupons are not incident to e, their contributions to $\mathbb{F}_{\mathcal{C}}(\Gamma)$ and $\mathbb{F}_{\mathcal{C}}(\Gamma')$ coincide. The same holds for all vertices and crossings of Γ except x. Isotoping Γ near x, we further reduce the lemma to one of the following four cases:

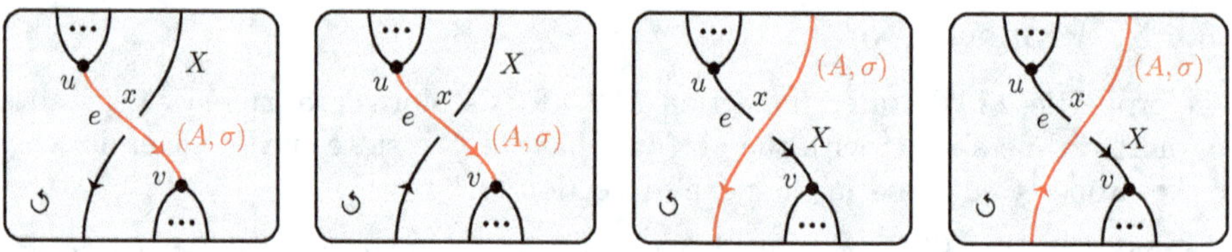

Case (a) Case (b) Case (c) Case (d)

where (A,σ) is the color of the distinguished strand at x and X is the subcolor of the other strand at x. We need to check that

$$\mathbb{F}_{\mathcal{C}}(\Gamma')(\alpha \otimes \beta) = \mathbb{F}_{\mathcal{C}}(\Gamma)(\Psi_u(\alpha) \otimes \Psi_v(\beta)) \tag{15.2}$$

for any $\alpha \in H_u(\Gamma')$ and $\beta \in H_v(\Gamma')$, where the $\Bbbk$-linear isomorphisms

$$\Psi_u\colon H_u(\Gamma') \to H_u(\Gamma) \quad \text{and} \quad \Psi_v\colon H_v(\Gamma') \to H_v(\Gamma)$$

are induced by the weak isomorphisms (determined by the pivotal structure of $\mathcal{C}$) between the cyclic $\mathcal{C}$-sets associated with the vertices u, v in Γ and Γ'. Let τ^u, τ^v, τ'^u, τ'^v be the cone isomorphisms in the definition of the modules $H_u(\Gamma)$, $H_v(\Gamma)$, $H_u(\Gamma')$, $H_v(\Gamma')$, respectively. Let us prove (15.2) in Case (a). By Section 5.2.2, we have $(A,\sigma)^* = (A^*, {}^\dagger\sigma)$. The isotopy

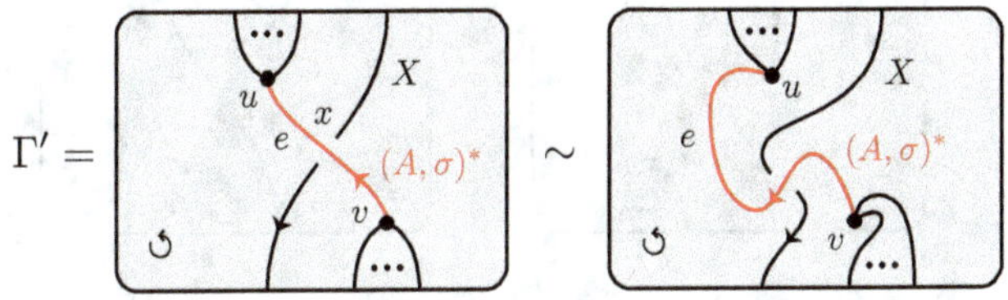

shows that the contribution of u, v, and x to $\mathbb{F}_\mathcal{C}(\Gamma')(\alpha \otimes \beta)$ is equal to

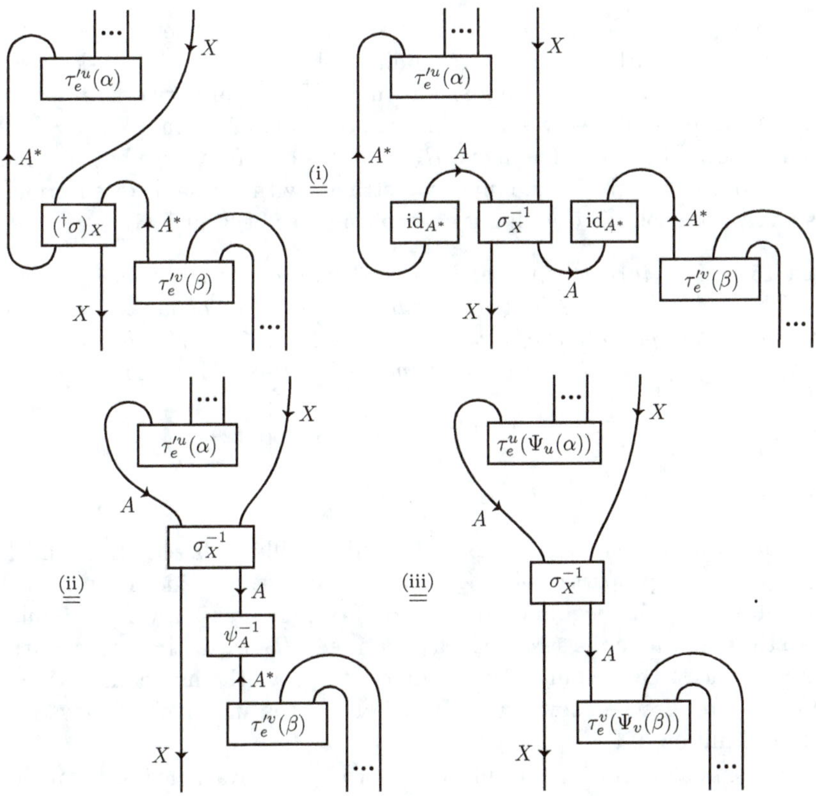

Here ψ is the pivotal structure of $\mathcal{C}$ (see Section 2.5), (i) follows from the definition of $({}^\dagger\sigma)_X$ (see Section 5.2.2), (ii) from the computation of ψ_A and ψ_A^{-1} in Lemma 2.2 and the equality $\mathrm{ev}_{A^*}(\psi_A \otimes \mathrm{id}_{A^*}) = \widetilde{\mathrm{ev}}_A$ (see Lemma 2.3), and (iii) from the definition of Ψ_u and Ψ_v. The latter graphical expression is precisely the contribution of u, v, and x to $\mathbb{F}_{\mathcal{C}}(\Gamma)(\Psi_u(\alpha) \otimes \Psi_v(\beta))$, as is clear from the isotopy

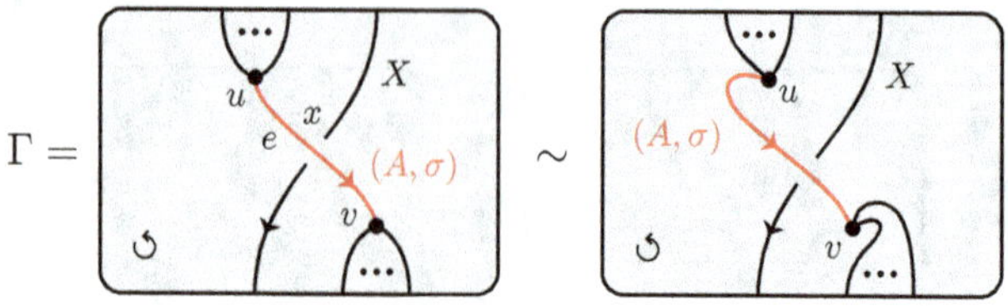

This proves (15.2) in Case (a). Cases (b), (c), (d) are treated similarly, using the equalities $(A,\sigma)^* = (A^*, {}^\dagger\sigma) = (A^*, \sigma^\dagger)$, see Lemma 5.1. □

15.4.6 Moves on knotted $\mathcal{C}$-nets

Figures 15.6 and 15.7 define eight local moves $R_1, \ldots, R_8$ in the class of knotted $\mathcal{C}$-nets in an oriented surface. Each move modifies a knotted $\mathcal{C}$-net in a disk preserving orientations and colors of all strands. The moves R_4, R_5 preserve the colors of the coupons while R_7 replaces two adjacent coupons colored by morphisms f, g with a coupon colored by $g \circ f$. The move R_8 deletes a coupon whose bottom and top bases are adjacent to the same number of strands with the same orientations and colors and whose color is the identity morphism (see Section 15.1.4).

Lemma 15.5. *The invariant $\mathbb{F}_{\mathcal{C}}$ of knotted $\mathcal{C}$-nets in $\mathbb{R}^2$ is preserved under the moves $R_2, \ldots, R_8$. More precisely, if Γ' is the knotted $\mathcal{C}$-net obtained from a knotted $\mathcal{C}$-net Γ in $\mathbb{R}^2$ by application of a move R_i with $2 \le i \le 8$, then the move induces in the obvious way a $\Bbbk$-linear isomorphism $\Phi\colon H(\Gamma) \to H(\Gamma')$ and*

$$\mathbb{F}_{\mathcal{C}}(\Gamma) = \mathbb{F}_{\mathcal{C}}(\Gamma')\,\Phi\colon H(\Gamma) \to \mathrm{End}_{\mathcal{C}}(\mathbb{1}).$$

If $\mathcal{Z}(\mathcal{C})$ is ribbon, then the same is true for $i = 1$.

Proof. The invariance under the moves R_7 and R_8 follows directly from the definition of $\mathbb{F}_{\mathcal{C}}$ and the properties of the graphical calculus. Consider now R_2. Adding new vertices as at the beginning of the proof of Lemma 15.4, we can reduce the proof to the case where the strands in the picture of R_2 are distinct edges with endpoints in vertices. By Lemma 15.4, it suffices to prove the invariance of $\mathbb{F}_{\mathcal{C}}$ when these edges are directed downward. In this case, the invariance follows directly from the definition of $\mathbb{F}_{\mathcal{C}}$.

By a similar argument, it suffices to prove the invariance of $\mathbb{F}_{\mathcal{C}}$ under the move R_3 in the case where the three strands in the picture of R_3 are directed downward. Let $(A,\sigma), (B,\gamma) \in \mathrm{Ob}(\mathcal{Z}(\mathcal{C}))$ be the colors of the first and second strands reading at the bottom from left to right. Let $X \in \mathrm{Ob}(\mathcal{C})$ be the subcolor of the third strand. The invariance of $\mathbb{F}_{\mathcal{C}}$ under R_3 would follow from the equality

$$(\gamma_X \otimes \mathrm{id}_A)(\mathrm{id}_B \otimes \sigma_X)(\sigma_B \otimes \mathrm{id}_X) = (\mathrm{id}_X \otimes \sigma_B)(\sigma_X \otimes \mathrm{id}_B)(\mathrm{id}_A \otimes \gamma_X).$$

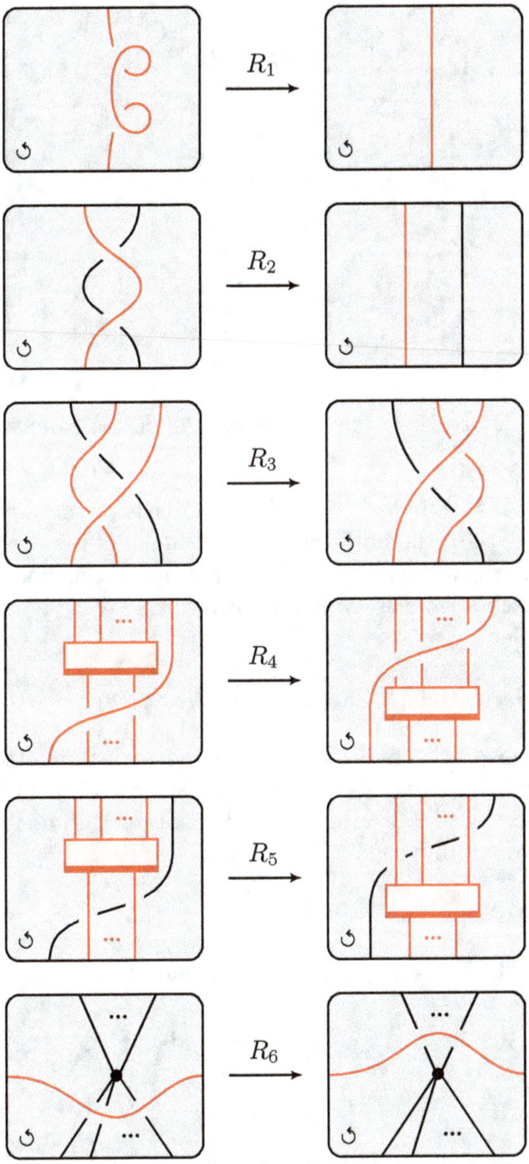

Figure 15.6: The moves R_1–R_6 on $\mathcal{C}$-nets

By the $\otimes$-multiplicativity (5.1) of σ, the left- and right-hand sides of this formula are equal respectively to $(\gamma_X \otimes \mathrm{id}_A)\sigma_{B\otimes X}$ and $\sigma_{X\otimes B}(\mathrm{id}_A \otimes \gamma_X)$. By the naturality of σ, the latter two expressions are equal. This implies the desired equality.

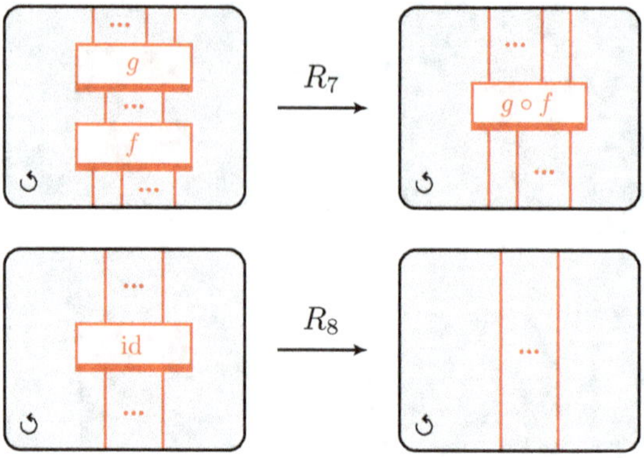

Figure 15.7: The moves R_7 and R_8 on $\mathcal{C}$-nets

The invariance of $\mathbb{F}_\mathcal{C}$ under R_4 and R_6 is similarly deduced from the naturality and $\otimes$-multiplicativity of the half-braiding which colors the over-passing strand.

To prove the invariance of $\mathbb{F}_\mathcal{C}$ under R_5, it suffices, as above, to consider the case where the black strand is directed downward. Let Q be the coupon in the picture of R_5 and let

$$(((A_1,\sigma_1),\varepsilon_1),\dots,((A_m,\sigma_m),\varepsilon_m)), \quad (((B_1,\gamma_1),\nu_1),\dots,((B_n,\gamma_n),\nu_n))$$

be the tuples of signed objects of $\mathcal{Z}(\mathcal{C})$ associated with the bottom and top bases of Q, respectively. For $(A,\sigma)\in \mathrm{Ob}(\mathcal{Z}(\mathcal{C}))$, we set $\sigma^+=\sigma$ and $\sigma^-={}^\dagger\sigma$, where ${}^\dagger\sigma$ is defined in Section 5.2.2. From the definition of the monoidal product and duality in $\mathcal{Z}(\mathcal{C})$, we obtain that

$$Q_\bullet=(A_1,\sigma_1)^{\varepsilon_1}\otimes\cdots\otimes(A_m,\sigma_m)^{\varepsilon_m}=(A_1^{\varepsilon_1}\otimes\cdots\otimes A_m^{\varepsilon_m},\alpha)$$

where the half-braiding $\alpha=\{\alpha_X\}_{X\in\mathrm{Ob}(\mathcal{C})}$ is given by

$$\alpha_X=\Big((\sigma_1^{\varepsilon_1})_X\otimes \mathrm{id}_{A_2^{\varepsilon_2}\otimes\cdots\otimes A_m^{\varepsilon_m}}\Big)\otimes\cdots\otimes\Big(\mathrm{id}_{A_1^{\varepsilon_1}\otimes\cdots\otimes A_{m-1}^{\varepsilon_{m-1}}}\otimes(\sigma_m^{\varepsilon_m})_X\Big).$$

Similarly,

$$Q^\bullet=(B_1,\gamma_1)^{\nu_1}\otimes\cdots\otimes(B_n,\gamma_n)^{\nu_n}=(B_1^{\nu_1}\otimes\cdots\otimes B_n^{\nu_n},\beta)$$

where the half-braiding $\beta=\{\beta_X\}_{X\in\mathrm{Ob}(\mathcal{C})}$ is given by

$$\beta_X=\Big((\gamma_1^{\nu_1})_X\otimes \mathrm{id}_{B_2^{\nu_2}\otimes\cdots\otimes B_n^{\nu_n}}\Big)\otimes\cdots\otimes\Big(\mathrm{id}_{B_1^{\nu_1}\otimes\cdots\otimes B_{n-1}^{\nu_{n-1}}}\otimes(\gamma_n^{\nu_n})_X\Big).$$

By the definition of a knotted $\mathcal{C}$-net, the coupon Q is colored with a morphism $f\colon Q_\bullet\to Q^\bullet$ in $\mathcal{Z}(\mathcal{C})$. Let $X\in\mathrm{Ob}(\mathcal{C})$ be the subcolor of the black strand in the picture of R_5. The contribution of a crossing between this strand and a red strand colored with (A,σ) is equal to $\sigma_X^+ = \sigma_X$ if the red strand is directed downward and to $\sigma_X^- = ({}^\dagger\sigma)_X$ otherwise. Indeed,

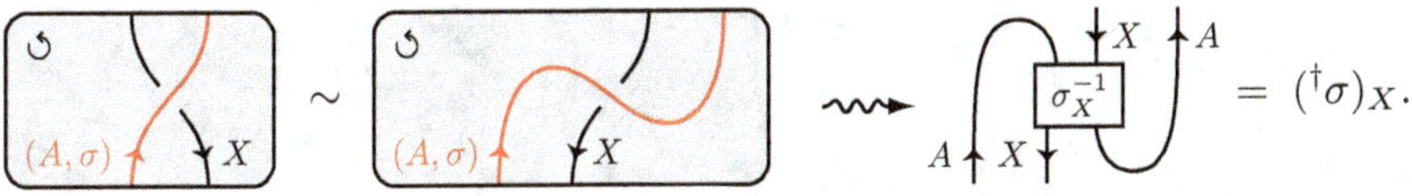

Therefore the invariance of $\mathbb{F}_\mathcal{C}$ under R_5 amounts to the equality

$$(\mathrm{id}_X \otimes f)\alpha_X = \beta_X(f \otimes \mathrm{id}_X).$$

This equality holds because f is a morphism in $\mathcal{Z}(\mathcal{C})$.

Suppose now that $\mathcal{Z}(\mathcal{C})$ is ribbon. If the strand in the picture of R_1 is oriented downward, then the invariance of $\mathbb{F}_\mathcal{C}$ under R_1 follows directly from the definition of the twist of $\mathcal{Z}(\mathcal{C})$ and its inverse, see Section 5.2.2 and Lemma 3.2. The case where the strand in the picture of R_1 is oriented upward is reduced to the previous case using Lemma 15.4. □

15.4.7 The case of a spherical category

Suppose that $\mathcal{C}$ is a spherical $\Bbbk$-category. Then the invariant $\mathbb{F}_\mathcal{C}$ of knotted $\mathcal{C}$-nets in $\mathbb{R}^2$ extends uniquely to an isotopy invariant of knotted $\mathcal{C}$-nets in the 2-sphere $S^2 = \mathbb{R}^2 \cup \{\infty\}$ endowed with the orientation extending the counterclockwise orientation in $\mathbb{R}^2$. Indeed, consider a knotted $\mathcal{C}$-net Γ in S^2. Isotoping Γ away from ∞, we obtain a knotted $\mathcal{C}$-net Γ_0 in $\mathbb{R}^2$. Composing $\mathbb{F}_\mathcal{C}(\Gamma_0)\colon H(\Gamma_0;\mathbb{R}^2) \to \mathrm{End}_\mathcal{C}(\mathbb{1})$ with the $\Bbbk$-linear isomorphism $H(\Gamma;S^2) \simeq H(\Gamma_0;\mathbb{R}^2)$ induced by the isotopy, we obtain a $\Bbbk$-linear homomorphism

$$\mathbb{F}_\mathcal{C}(\Gamma)\colon H(\Gamma;S^2) \to \mathrm{End}_\mathcal{C}(\mathbb{1}).$$

Lemmas 2.9 and 15.3 imply that $\mathbb{F}_\mathcal{C}(\Gamma)$ does not depend on the way we push Γ in $\mathbb{R}^2$ and is an isotopy invariant of Γ.

The invariant $\mathbb{F}_\mathcal{C}$ further extends to knotted $\mathcal{C}$-nets in an arbitrary oriented surface Σ homeomorphic to S^2. Namely, given a knotted $\mathcal{C}$-net Γ in Σ, pick an orientation preserving homeomorphism $f\colon \Sigma \to S^2$ and consider the $\Bbbk$-linear isomorphism $H(f)\colon H(\Gamma;\Sigma) \to H(f(\Gamma);S^2)$ induced by f (see Section 15.3.3). Since all such f are isotopic to each other, the homomorphism

$$\mathbb{F}_\mathcal{C}(\Gamma) = \mathbb{F}_\mathcal{C}(f(\Gamma))\, H(f)\colon H(\Gamma;\Sigma) \to \mathrm{End}_\mathcal{C}(\mathbb{1})$$

does not depend on the choice of f and is an isotopy invariant of Γ. Lemma 15.4 and the properties of $\mathbb{F}_\mathcal{C}$ stated for $\mathcal{C}$-colored graphs in Section 12.2.4 extend to knotted $\mathcal{C}$-nets in Σ with the obvious changes. As in Section 12.2.5, the condition in the property (iv) of Section 12.2.4 involving a straight line and the purity condition in (v) may be dropped. In particular, the property (v) becomes the following property of the invariant $\mathbb{F}_\mathcal{C}$ of knotted $\mathcal{C}$-nets in Σ:

$\mathbb{F}_\mathcal{C}$ 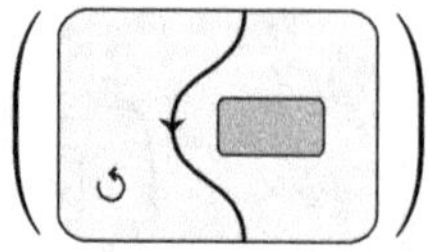 $= \mathbb{F}_\mathcal{C}$ 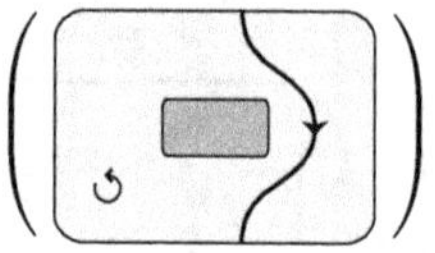 (15.3)

where the knotted $\mathcal{C}$-nets on the left and on the right coincide outside the big rectangles and in the small rectangles.

Lemma 15.6. *The invariant $\mathbb{F}_\mathcal{C}$ of knotted $\mathcal{C}$-nets in an oriented surface homeomorphic to S^2 is preserved under the moves $R_1, \ldots, R_8$.*

Proof. The invariance under the move R_i for $2 \leq i \leq 8$ directly follows from Lemma 15.5. The invariance under R_1 follows from the fact that the move R_1 on knotted $\mathcal{C}$-nets in S^2 expands as a composition of the moves $R_2^{\pm 1}, R_3^{\mp 1}, R_4^{\mp 1}, R_6^{\pm 1}$ by pushing the over-passing strand in the picture of R_1 to the left over the rest of the net and across the point $\infty \in S^2$. □

15.4.8 The case of a pre-fusion category

Suppose that $\mathcal{C}$ is a pre-fusion spherical $\Bbbk$-category. Then $\mathrm{End}_\mathcal{C}(\mathbb{1}) = \Bbbk$ and $\mathcal{C}$ is non-degenerate (see Lemma 4.3). For any knotted $\mathcal{C}$-net Γ in an oriented surface Σ homeomorphic to S^2, the $\Bbbk$-module $H(\Gamma) = H(\Gamma; \Sigma)$ is free of finite rank and

$$\mathbb{F}_\mathcal{C}(\Gamma) \in H(\Gamma)^\star = \mathrm{Hom}_\Bbbk(H(\Gamma), \Bbbk).$$

The local relations (12.1), (12.2) and Lemma 12.3 stated in Section 12.2.6 for the invariant $\mathbb{F}_\mathcal{C}$ of $\mathcal{C}$-colored graphs extend to knotted $\mathcal{C}$-nets in Σ in the obvious way. Likewise, Lemmas 12.5 and 12.6 extend to knotted $\mathcal{C}$-nets in Σ using the notions of dual vertices and contraction vectors introduced in Section 15.3.5.

15.5 An invariant of ribbon graphs

Throughout the rest of the chapter, we fix a spherical fusion $\Bbbk$-category $\mathcal{C}$ such that $\dim(\mathcal{C})$ is invertible in $\Bbbk$ and fix a representative set I of simple objects of $\mathcal{C}$.

This section is the first of three sections devoted to a construction of a graph TQFT over $\mathcal{Z}(\mathcal{C})$. We define here (via state sum techniques) an invariant of $\mathcal{Z}(\mathcal{C})$-colored ribbon graphs in closed oriented 3-manifolds.

15.5.1 The invariant

Consider a $\mathcal{Z}(\mathcal{C})$-colored ribbon graph R in a closed oriented 3-manifold M. By Corollary 14.5, the ribbon graph R can be represented by a neat positive diagram (P, d). Here P is a skeleton of M and d is a knotted plexus in P having only positive switches, no trivial components, and no free ends. The $\mathcal{Z}(\mathcal{C})$-coloring of R turns d into a $\mathcal{Z}(\mathcal{C})$-colored knotted plexus in P. Since $\partial d = \emptyset$, the strands of d are either circles or arcs with endpoints in the bases of the coupons. The vertices of P and the switches, crossings, and coupons of d are called the *nodes* of d. Since d has no trivial components, the complement of the nodes in $\widetilde{d} = d \cup P^{(1)} \subset P$ is a finite disjoint union of open intervals whose closures are called the *rims* of d. Each rim lies in $P^{(1)}$ or in d and connects two nodes (possibly, equal) called the *endpoints* of the rim. Cutting P along $\widetilde{d}$, we obtain a compact surface (with interior $P \setminus \widetilde{d}$) whose connected components are called the *faces* of d. Every face of d is contained in a unique region of P, and so the orientation of P induces an orientation of the faces of d. The (finite) set of faces of d is denoted by $\mathrm{Fac}(d)$.

A *coloring* of the pair (P, d) is a map $c\colon \mathrm{Fac}(d) \to I$. The object $c(r) \in I$ assigned to a face r of d is called the *c-color* of r. For each coloring c of (P, d), we define a scalar $|c| \in \Bbbk$ as follows. If $d = \emptyset$, then the nodes of d are the vertices of P, the rims of d are the edges of P, the faces of d are the regions of P, and $|c|$ is the same scalar as in Section 13.1.1. Suppose that $d \neq \emptyset$. By an *oriented rim* of d we mean a rim of d endowed with an orientation. For each oriented rim e of d, we define a cyclic $\mathcal{C}$-set P_e as follows. If $e \subset P^{(1)}$, then P_e is the set of branches of P at e turned into a cyclic $\mathcal{C}$-set as in Section 13.1.1. If $e \subset d$, then e is adjacent to two faces r_+, r_- of d such that the orientation of r_+ induces the given orientation of e and the orientation of r_- induces the opposite orientation of e:

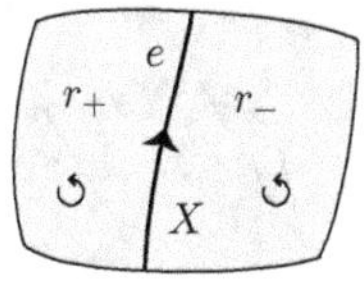

.

The arrow on e in this picture indicates the orientation of e (which may be compatible or not with the orientation of the strand of d containing e). Let $X \in \mathrm{Ob}(\mathcal{C})$ be the subcolor of the strand of d containing e. Set $\varepsilon = +$ if the orientation of this strand of d is compatible with that of e and set $\varepsilon = -$ otherwise. Then P_e is the cyclic $\mathcal{C}$-set $\{r_-, e, r_+\}$ where $r_- < e < r_+ < r_-$ and the map $\{r_-, e, r_+\} \to \mathrm{Ob}(\mathcal{C}) \times \{+, -\}$ carries $r_\pm$ to $(c(r_\pm), \pm)$ and e to (X, ε). We set

$$H_c(e) = H(P_e)$$

for each oriented rim e of d. In particular, if $e \subset d$, then in the notation above

$$H_c(e) \simeq \mathrm{Hom}_{\mathcal{C}}(\mathbb{1}, c(r_-)^* \otimes X^\varepsilon \otimes c(r_+)).$$

Let

$$H_c = \bigotimes_e H_c(e)$$

be the unordered tensor product of $\Bbbk$-modules $H_c(e)$ over all oriented rims of d. As for unoriented edges in Section 13.1.1, an unoriented rim e of d gives rise to two opposite oriented rims e_1, e_2 whose associated cyclic $\mathcal{C}$-sets P_{e_1} and P_{e_2} are dual to each other. This determines a contraction vector $*_e \in H_c(e_1) \otimes H_c(e_2)$. Set

$$*_c = \bigotimes_e *_e \in H_c \tag{15.4}$$

where e runs over all unoriented rims of d.

We next associate with each node x of d a knotted $\mathcal{C}$-net Γ_x^c in an oriented 2-sphere. There are four cases to consider depending on x.

(i) If x is a vertex of P, then Γ_x^c is defined as in Section 13.1.1.

(ii) If x is a switch of d, then a neighborhood of x in P looks as follows:

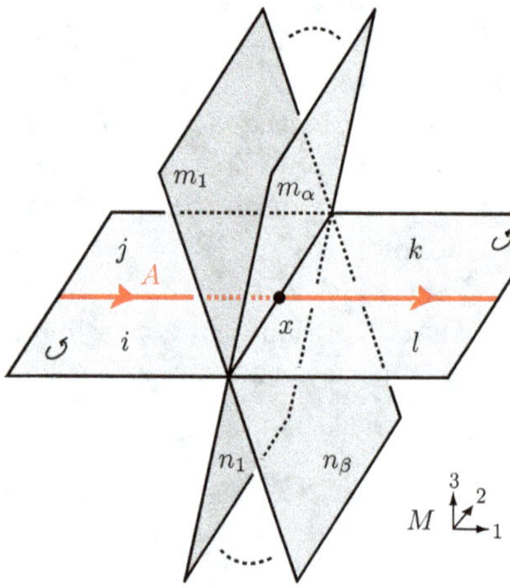

Here the orientation of M is right-handed (as always in our pictures), the object $A \in \mathrm{Ob}(\mathcal{Z}(\mathcal{C}))$ is the color of the strand of d passing through x, and

$$i, j, k, l, m_1, \ldots, m_\alpha, n_1, \ldots, n_\beta \in I$$

are the c-colors of the faces of d adjacent to x with $\alpha \geq 0$, $\beta \geq 0$. Set

$\Gamma_x^c =$ 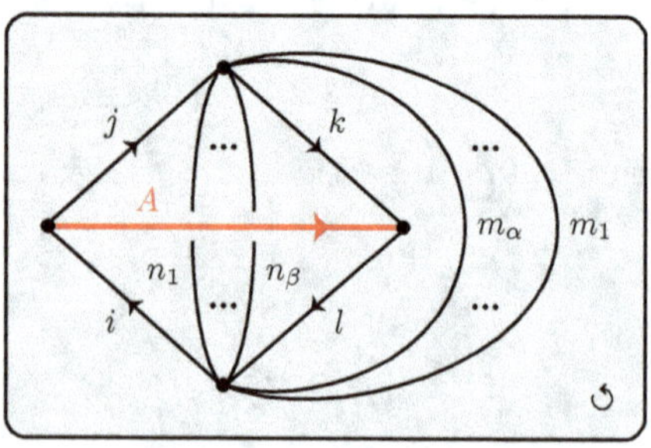 .

Here and below we draw Γ_x^c in the plane $\mathbb{R}^2$ (oriented counterclockwise) and consider Γ_x^c as a knotted $\mathcal{C}$-net in $S^2 = \mathbb{R}^2 \cup \{\infty\}$. The arc of Γ_x^c colored by m_p

with $1 \leq p \leq \alpha$ is directed upward if the orientation of the m_p-labeled face of d followed by that of d at x yields the positive orientation of M. Otherwise this arc is directed downward. The arc of Γ_x^c colored by n_q with $1 \leq q \leq \beta$ is directed downward if the orientation of the n_q-labeled face of d followed by that of d at x yields the positive orientation of M. Otherwise this arc is directed upward.

(iii) If x is a crossing of d, then a neighborhood of x in P looks as follows:

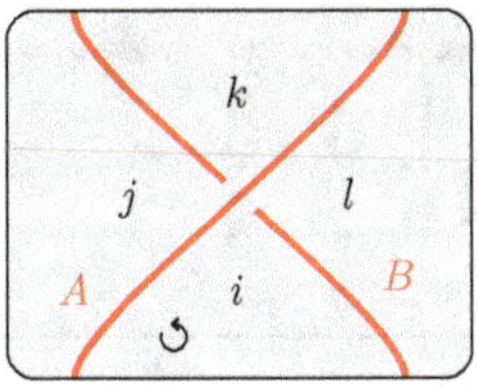

.

Here $A, B \in \mathrm{Ob}(\mathcal{Z}(\mathcal{C}))$ are the colors of the strands of d meeting at x, and $i, j, k, l \in I$ are the c-colors of the faces of d adjacent to x. Set

$\Gamma_x^c =$ 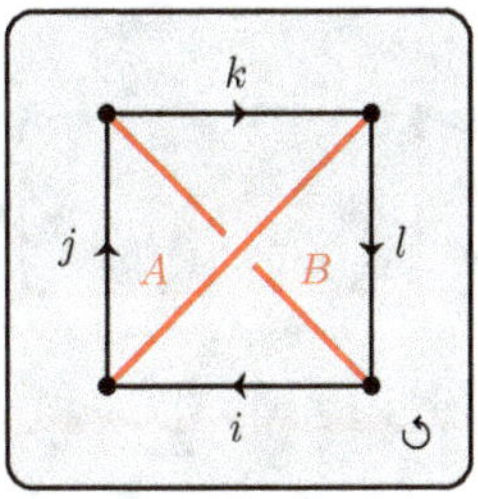 $\subset S^2,$

where the orientations of the diagonals (not shown in the picture) are induced by the orientations of the corresponding strands of d.

(iv) If x is a coupon of d with $m \geq 0$ inputs and $n \geq 0$ outputs, then a neighborhood of x in P looks as follows:

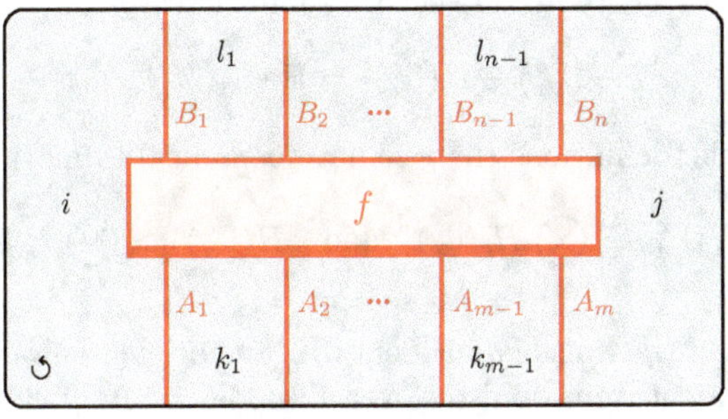

.

Here $i, j, k_1, \ldots, k_{m-1}, l_1, \ldots, l_{n-1} \in I$ are the c-colors of the faces of d adjacent to x, the objects $A_1, \ldots, A_m \in \mathrm{Ob}(\mathcal{Z}(\mathcal{C}))$ are the colors of the inputs of x, the objects $B_1, \dots, B_n \in \mathrm{Ob}(\mathcal{Z}(\mathcal{C}))$ are the colors of the outputs of x, and f is the morphism in $\mathcal{Z}(\mathcal{C})$ coloring x. For $m, n \geq 1$, set

$\Gamma_x^c =$ 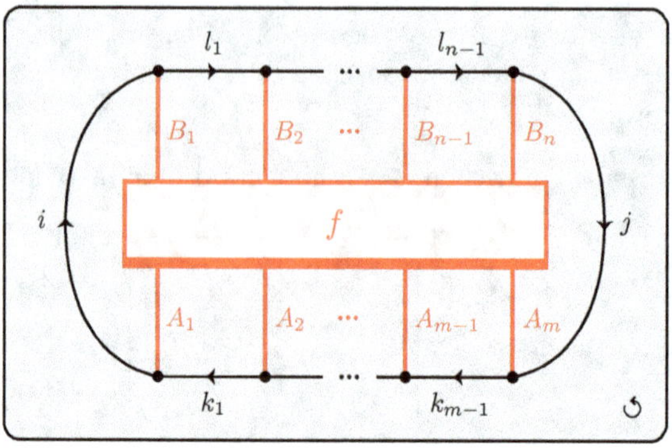 $\subset S^2,$

where the orientations of the vertical arcs (not shown in the picture) are induced by the orientations of the corresponding strands of d. For $m = 0$ and $n \geq 1$, set

$\Gamma_x^c =$ 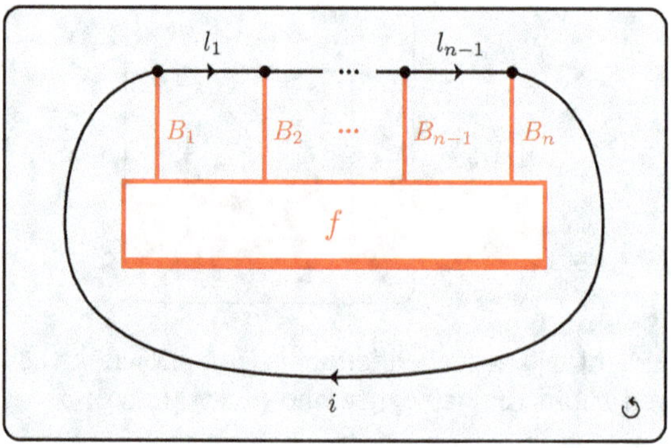 $\subset S^2,$

The case $n = 0$ is similar.

By Section 15.4.7, the knotted $\mathcal{C}$-net Γ_x^c yields a vector

$$\mathbb{F}_{\mathcal{C}}(\Gamma_x^c) \in H(\Gamma_x^c)^\star = \mathrm{Hom}_{\Bbbk}(H(\Gamma_x^c), \Bbbk).$$

It results from the definitions that we have canonical isomorphisms

$$H(\Gamma_x^c) \simeq \bigotimes_{e_x} H_c(e_x) \quad \text{and} \quad H(\Gamma_x^c)^\star \simeq \bigotimes_{e_x} H_c(e_x)^\star$$

where e_x runs over the rims of d incident to x and oriented away from x. The tensor product of the latter isomorphisms over all nodes x of d yields an isomorphism

$$\bigotimes_x H(\Gamma_x^c)^\star \simeq \bigotimes_x \bigotimes_{e_x} H_c(e_x)^\star \simeq H_c^\star.$$

The image under this isomorphism of the unordered tensor product $\bigotimes_x \mathbb{F}_\mathcal{C}(\Gamma_x)$ is a vector $V_c \in H_c^\star$. Recall the vector $*_c \in H_c$ defined in (15.4) and set

$$|c| = V_c(*_c) \in \Bbbk.$$

Set now

$$|M, R|_\mathcal{C} = (\dim(\mathcal{C}))^{-|M\setminus P|} \sum_c \dim(c)|c| \in \Bbbk, \tag{15.5}$$

where c runs over all colorings of (P, d), the positive integer $|M \setminus P|$ is the number of connected components of $M \setminus P$, and

$$\dim(c) = \prod_{r\in \mathrm{Fac}(d)} (\dim c(r))^{\chi(r)} \in \Bbbk. \tag{15.6}$$

As usual, χ is the Euler characteristic.

Theorem 15.7. *The scalar $|M, R|_\mathcal{C}$ is a topological invariant of the pair (M, R) and is invariant under the transformations of R shown in Figure* 15.1.

We prove this theorem in the next subsection. The naturality of $\mathbb{F}_\mathcal{C}$ implies that $|M, R|_\mathcal{C}$ is independent of the choice of the representative set I of simple objects of $\mathcal{C}$. For $R = \emptyset$, we obtain the invariant $|M, \emptyset|_\mathcal{C} = |M|_\mathcal{C}$ of M defined in Section 13.1.

15.5.2 Proof of Theorem 15.7

We will use the following properties of the invariant $\mathbb{F}_\mathcal{C}$ of knotted $\mathcal{C}$-nets in S^2:

(A) the invariance under (ambient) isotopies in S^2 (see Section 15.4.7);

(B) Property (15.3) (see Section 15.4.7);

(C) the invariance under the moves $\{R_i\}_{i=1}^8$ (see Lemma 15.6);

(D) Property (12.1) extended to knotted $\mathcal{C}$-nets (see Section 15.4.8);

(E) Lemma 12.3 extended to knotted $\mathcal{C}$-nets (see Section 15.4.8);

(F) Lemma 12.5 extended to knotted $\mathcal{C}$-nets (see Section 15.4.8);

(G) Lemma 12.6 extended to knotted $\mathcal{C}$-nets (see Section 15.4.8).

We will indicate equalities resulting from these properties by the corresponding symbols $(A), \ldots, (G)$ placed over the equality sign.

In the arguments to follow, a picture formed from a knotted $\mathcal{C}$-net Γ in S^2 and several dotted arcs connecting dual vertices of Γ represents the tensor

$$\mathbb{F}_\mathcal{C}(\Gamma)\,(\otimes_\ell *_\ell),$$

where ℓ runs over the dotted arcs and $*_\ell$ is the contraction vector associated with the duality between the endpoints of ℓ. This duality is the bijection between the sets of half-edges of Γ adjacent to the endpoints of ℓ which associates half-edges with the same color. We stress that the dotted arcs are not parts of Γ and serve solely to indicate that the tensor $\mathbb{F}_{\mathcal{C}}(\Gamma)$ is evaluated on the corresponding contraction vectors. For example,

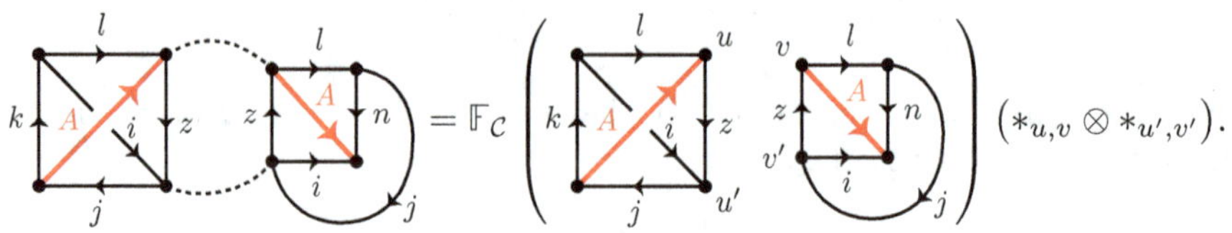

The topological invariance of $|M,R|_{\mathcal{C}}$ means that for any orientation preserving homeomorphism of M onto a closed oriented 3-manifold M', we have $|M',R'|_{\mathcal{C}} = |M,R|_{\mathcal{C}}$ where R' is the $\mathcal{Z}(\mathcal{C})$-colored ribbon graph in M' obtained by transporting R along the homeomorphism. The topological invariance of $|M,R|_{\mathcal{C}}$ directly follows from the fact that $|M,R|_{\mathcal{C}}$ is invariant under isotopy of R in M. By Corollary 14.5, to establish the isotopy invariance it suffices to verify the invariance of $|M,R|_{\mathcal{C}}$ under the moves $\{\omega_*\}_{*=0}^{9}$ in the class of neat positive diagrams of R. In our computations of the state sums before and after the move, we will specify only

- the labels $i,j,z,\ldots \in I$ of the faces attributed by a coloring map c;
- the contribution of the nodes involved in the move and the contribution $*_e$ of the rims e suppressed/added under the move. We depict such vectors $*_e$ by dotted arcs as above;
- the contribution $(\dim c(r))^{\chi(r)}$ of faces r whose Euler characteristic $\chi(r)$ changes under the move.

In our pictures, the neat positive diagram of R will be drawn in red. Recall that the strands of R are oriented and colored with objects of $\mathcal{Z}(\mathcal{C})$. The latter colors are preserved under the moves but play no role in the computations. Consequently, they will not be explicitly indicated in the pictures. For concreteness, we will prove the invariance of $|M,R|_{\mathcal{C}}$ for specific directions on the red strands. All other directions can be handled by exactly the same argument. Indeed, if the orientation of a red strand of a neat positive diagram is reversed, then the knotted $\mathcal{C}$-nets associated with the nodes change via the reversal of the direction of the corresponding red strands. This is clear for the nodes that are not switches. For a

switch, use the isotopies

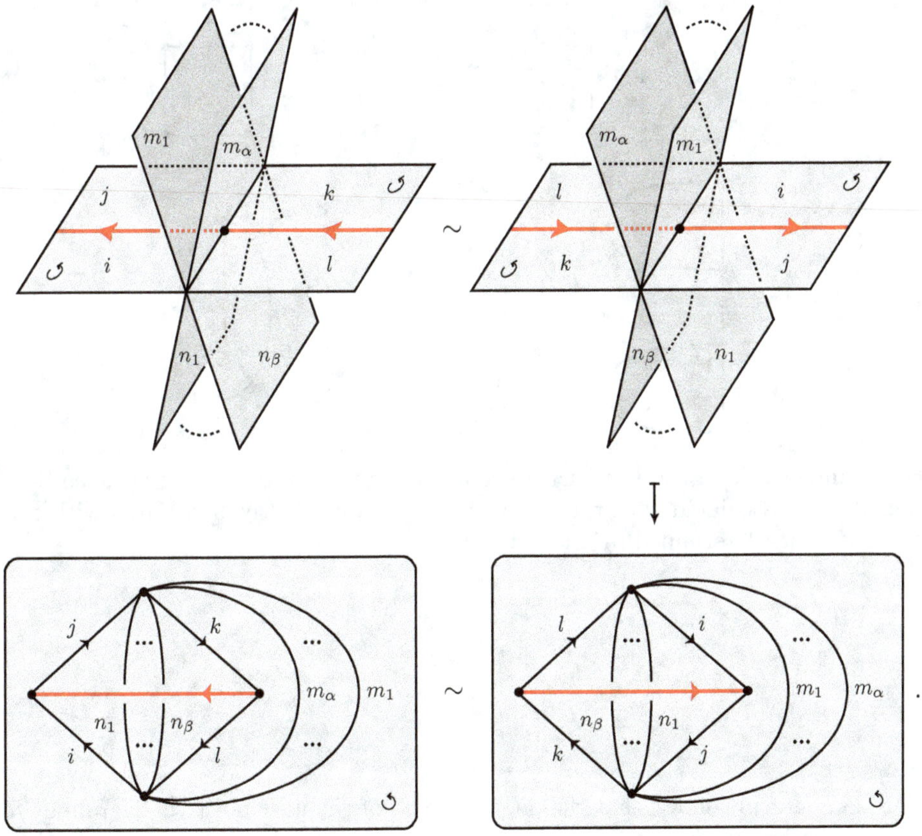

Note that the latter isotopy is compatible with our conventions concerning the orientations of the arcs.

The moves ω_0 do not involve red strands and are handled as in the proof of Theorem 13.1. Let us prove the invariance of the state sum $|M, R|_\mathcal{C}$ under ω_1. In the following computation, the arrow $\mapsto$ indicates the contribution of the picture to the state sum and the summation over $z, w \in I$ indicates that we sum up the contributions of all the colorings of the diagram attributing z and w to the corresponding faces while keeping the colors of the other faces. We have:

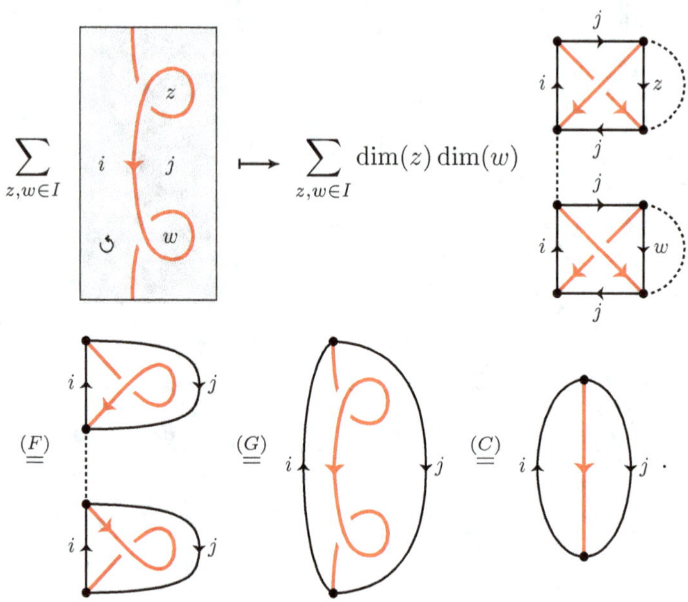

By Example 12.3.6, the last picture represents the duality pairing ω_E (see Section 12.3.2) associated with the two vertices or, equivalently, associated with the cyclic $\mathcal{C}$-set E determined by the red strand:

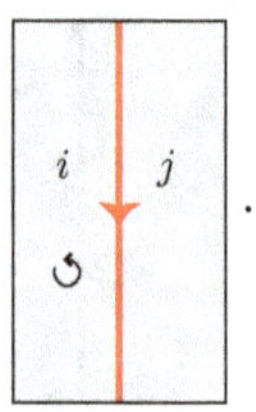

This computation shows that the depicted part of the neat positive diagram of R (before the application of the move) contributes ω_E to the state sum. Note that this part of the diagram is adjacent to two rims, one entering the picture from the top and one entering the picture from the bottom. These rims are distinct because the diagram produced by the move is neat. Each of these two rims contributes a contraction vector to the state sum. One of these vectors cancels with ω_E while the second vector is precisely the contribution to the state sum of the rim supporting the red strand after the move. This proves the invariance of the state sum under the move ω_1.

Let us prove the invariance of $|M, R|_{\mathcal{C}}$ under the move ω_2. Let r, r' be the faces colored by $k, l \in I$ in the picture below showing the neat positive diagram before the move. Set

$$\zeta_{k,l} = \begin{cases} \dim(k)^{\chi(r)} \dim(l)^{\chi(r')} & \text{if } r \neq r', \\ \dim(k)^{\chi(r)} & \text{if } r = r'. \end{cases}$$

We have:

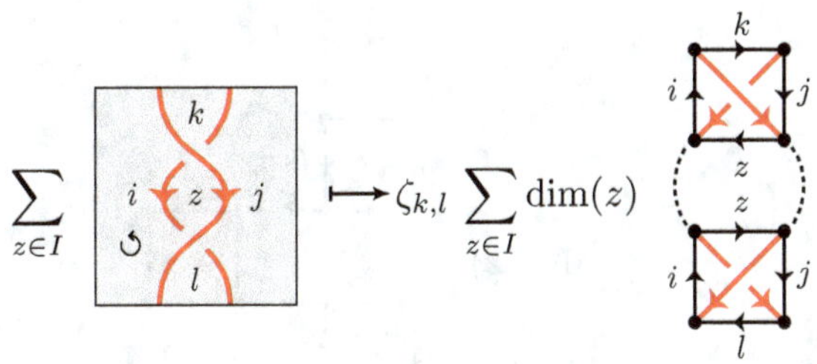

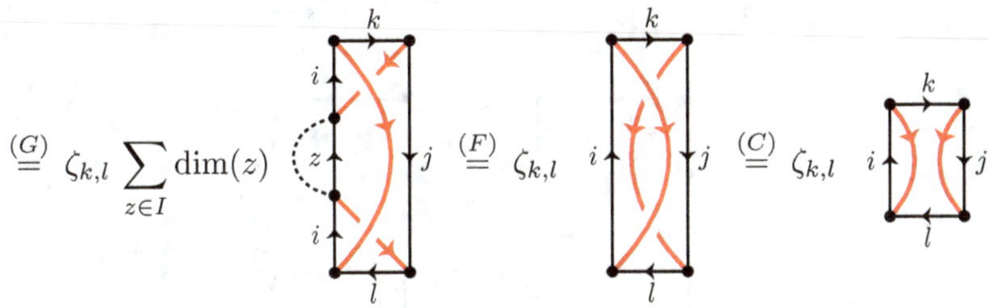

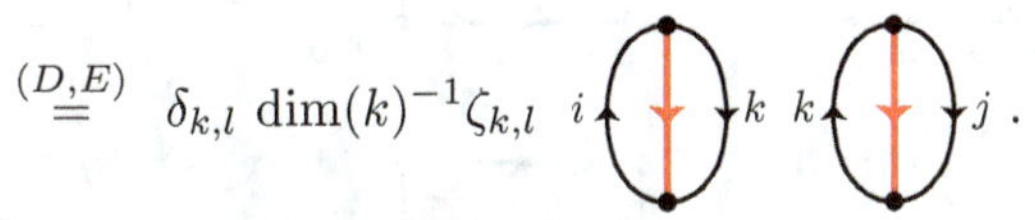

Therefore the total contribution to the state sum before the move of the colorings c with $c(r) \neq c(r')$ is equal to zero. The colorings c with $c(r) = c(r')$ induce colorings of the diagram after the move:

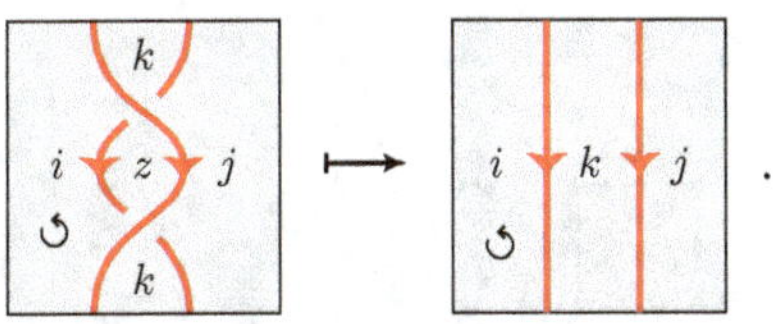

The computation above implies that, after summation over all $z \in I$, the contributions to the state sum of such colorings before the move is equal to the contribution to the state sum of the induced colorings after the move. Here we use the additivity of the Euler characteristic which yields the identity

$$\dim(k)^{-1}\zeta_{k,k} = \dim(k)^{\chi(r'')},$$

where r'' is the face colored by k after the move. These results imply the invariance of $|M, R|_{\mathcal{C}}$ under ω_2.

The invariance of $|M,R|_{\mathcal{C}}$ under the move ω_3 is verified as follows:

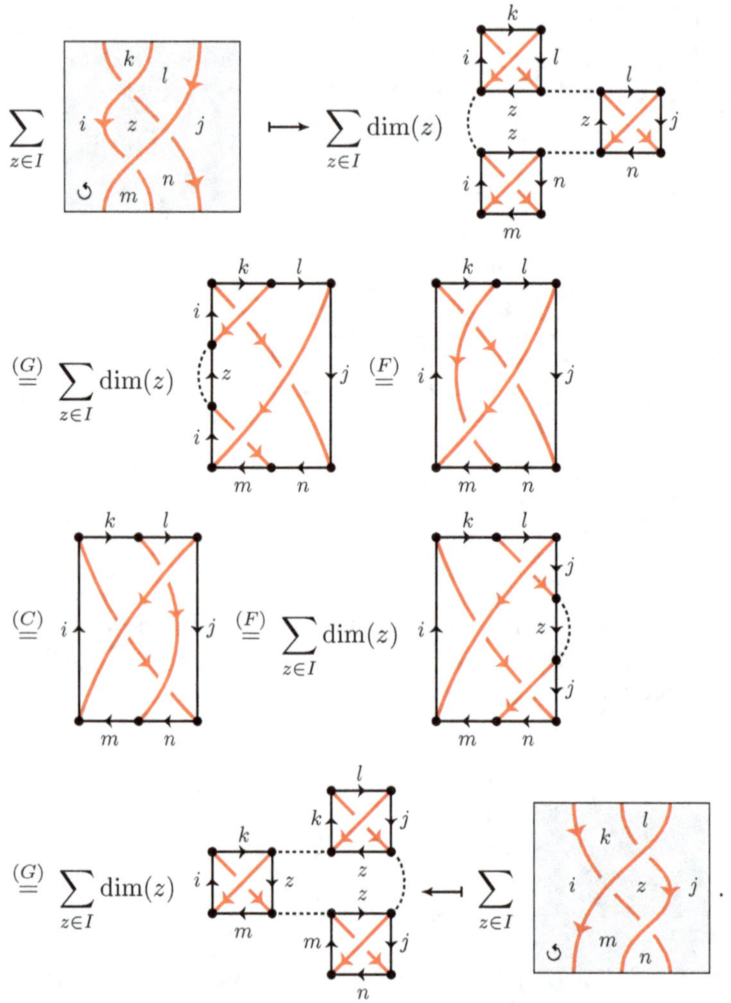

Let us prove the invariance of $|M,R|_{\mathcal{C}}$ under the move ω_4. There are two cases to consider, depending on the orientation of the left horizontal region in the picture of ω_4 which may be counterclockwise or clockwise. (The orientation of the other two regions is determined by the assumption that the switches in the picture are positive.) The first case:

$$\sum_{z,w\in I}$$

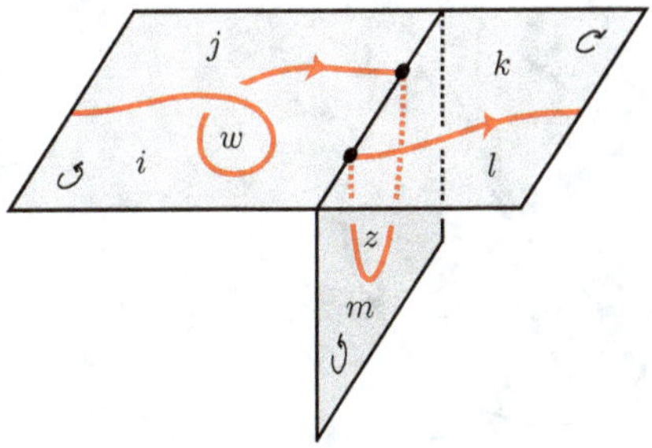

$$\longmapsto \sum_{z,w\in I} \dim(z)\dim(w)$$

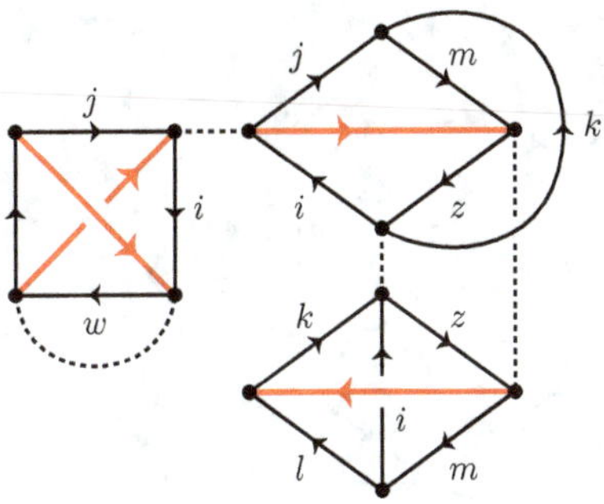

$$\overset{(F)}{=} \sum_{z\in I} \dim(z)$$

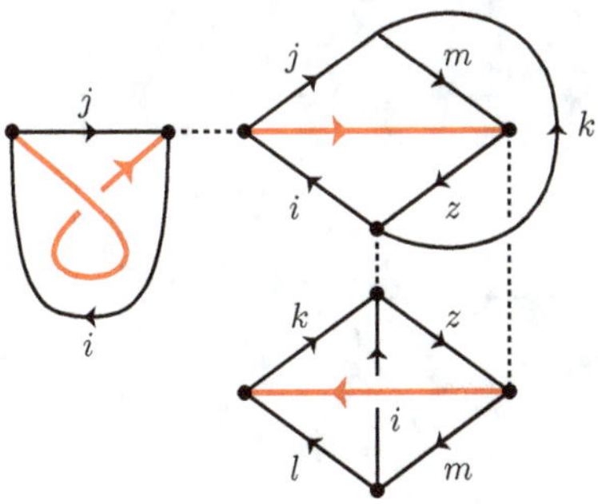

$$\overset{(G)}{=} \sum_{z\in I} \dim(z)$$

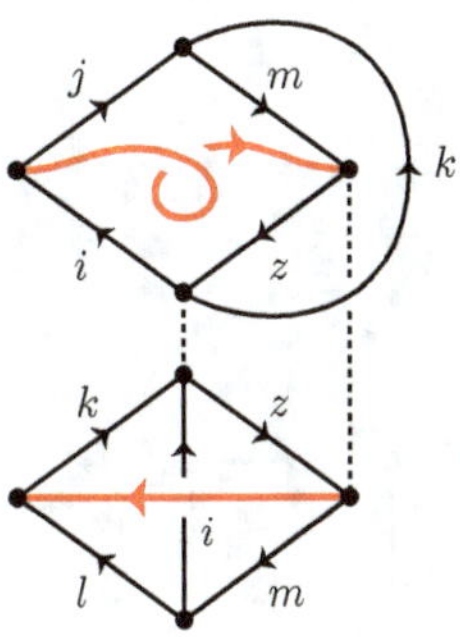

$$\overset{(B,A,C)}{=} \sum_{z\in I} \dim(z)$$

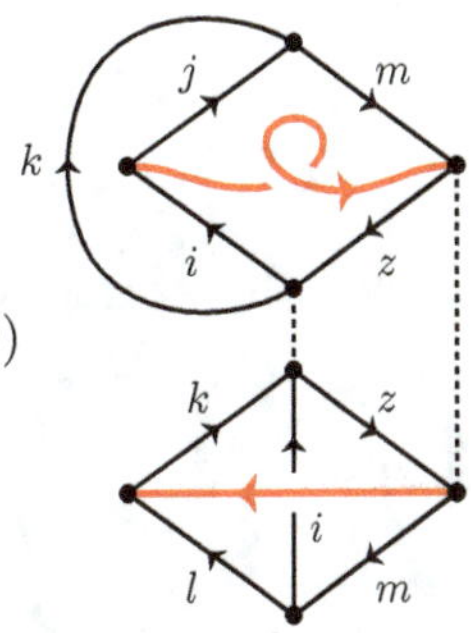

$\overset{(G,F)}{=}$ 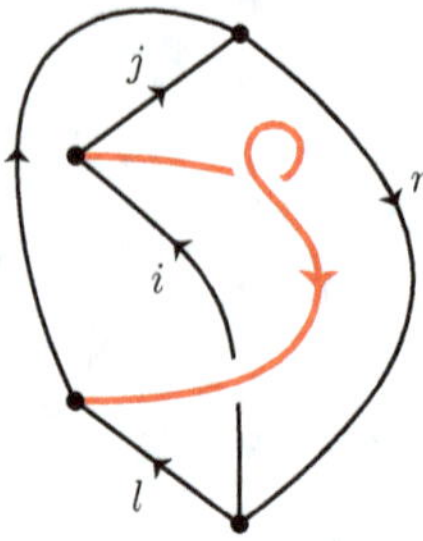

$\overset{(C)}{=}$ 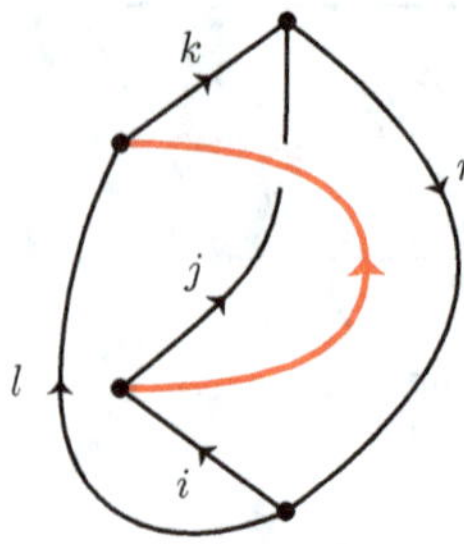

$\overset{(G,F)}{=} \sum_{z\in I} \dim(z)$ 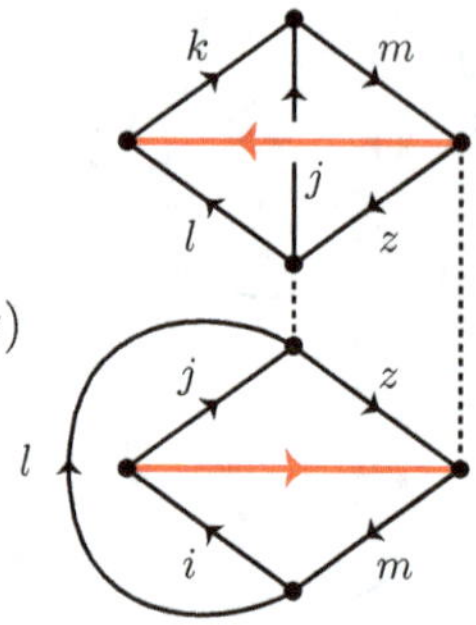

$\longleftarrow\!\!\!\!| \ \sum_{z\in I}$ 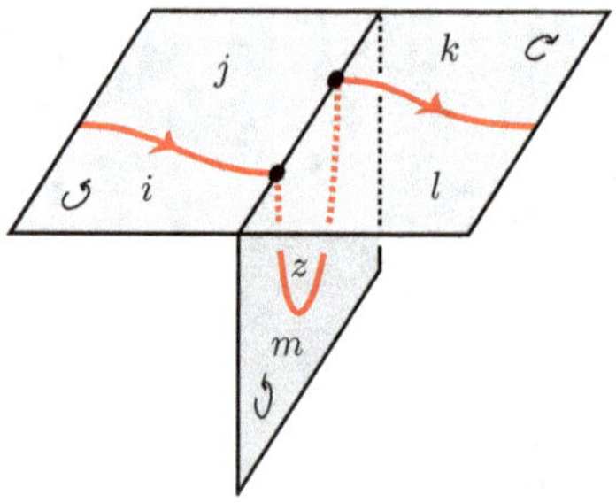 .

The second case:

$\sum_{z,w\in I}$ 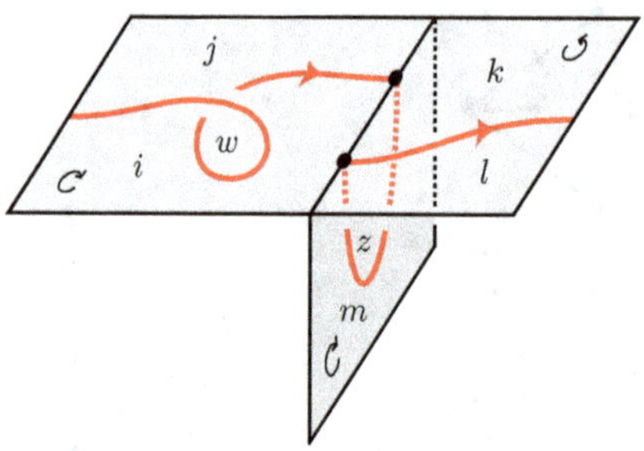

$\longmapsto \sum_{z,w\in I} \dim(z)\dim(w)$ 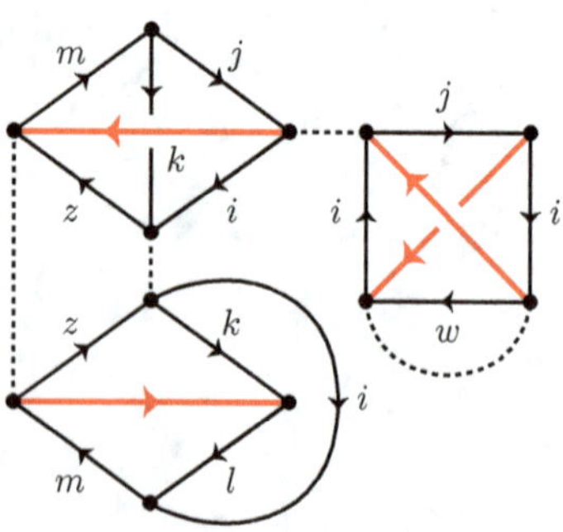

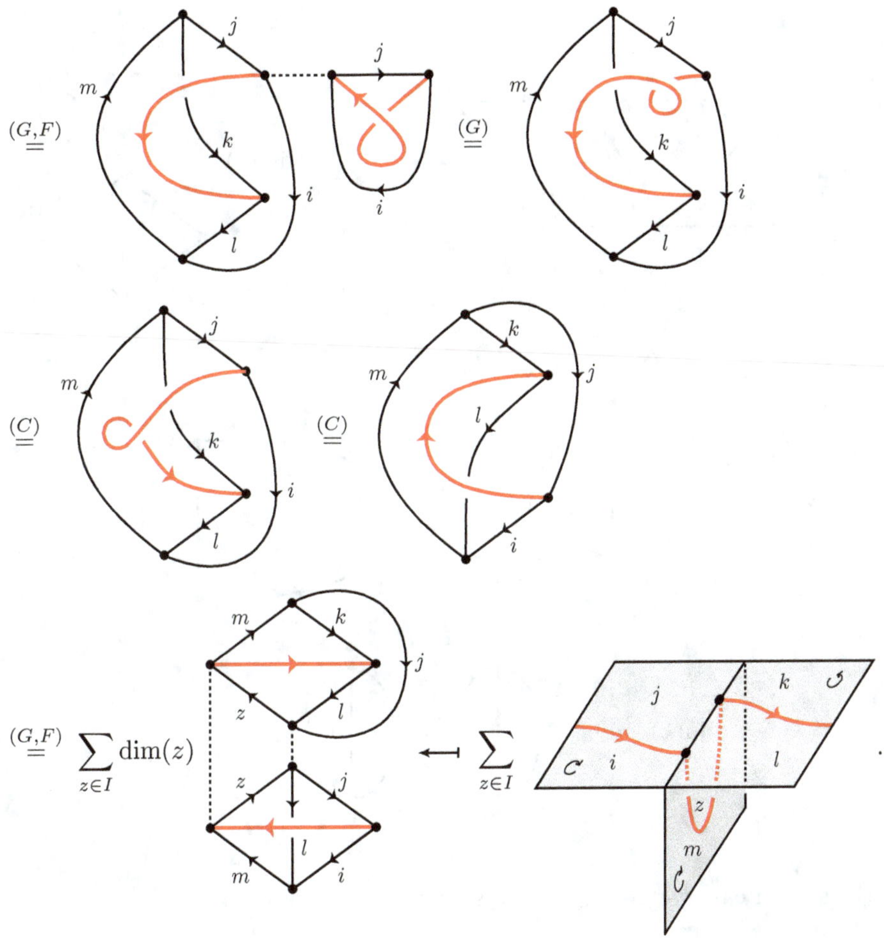

Let us prove the invariance of $|M,R|_{\mathcal{C}}$ under the move ω_5. Using the colors $k,l \in I$ of the faces shown in the pictures below, we introduce a scalar $\zeta_{k,l} \in \Bbbk^*$ by the same rule as in the proof of the invariance under ω_2. We consider two cases, depending on the (compatible) orientations of the horizontal regions. The orientation of the vertical region and the induced directions of the corresponding strands of knotted $\mathcal{C}$-graphs play no role in the argument and are not shown. The first case:

$\sum_{z\in I}$

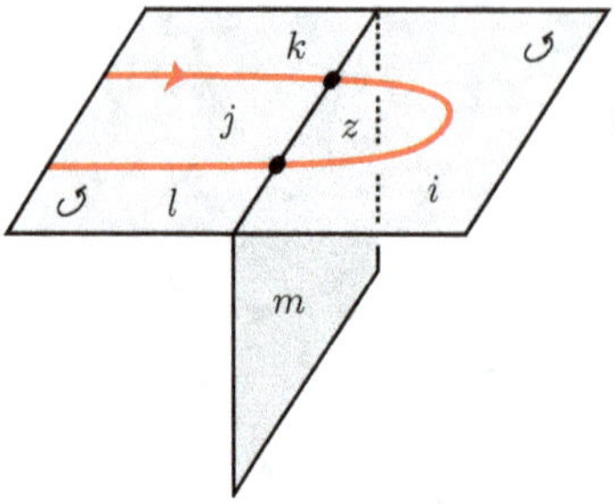

$\longmapsto \zeta_{k,l} \sum_{z\in I} \dim(z)$

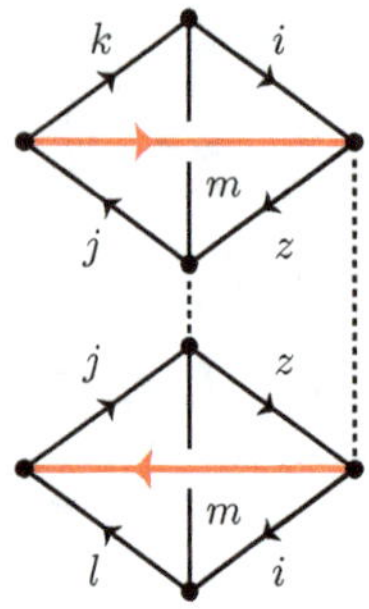

$\overset{(G,F)}{=} \zeta_{k,l}$

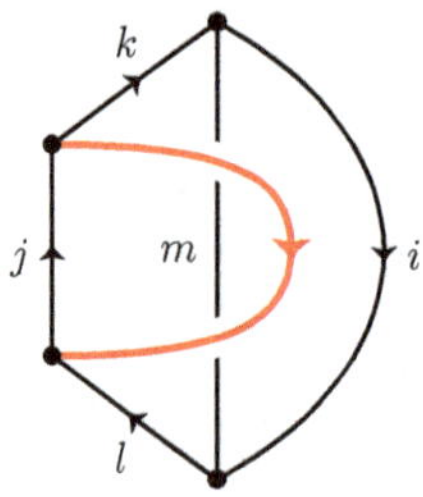

$\overset{(C)}{=} \zeta_{k,l}$

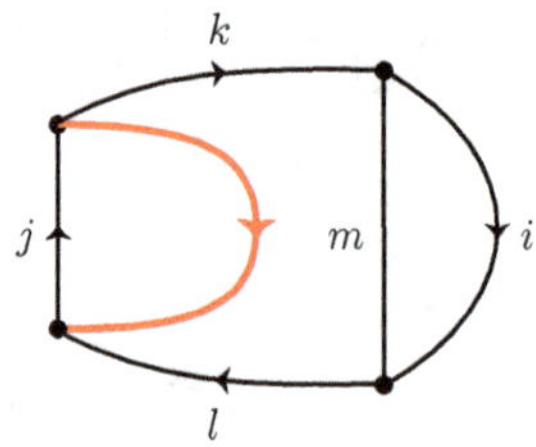

$\overset{(D,E)}{=} \delta_{k,l} \dim(k)^{-1} \zeta_{k,l}$

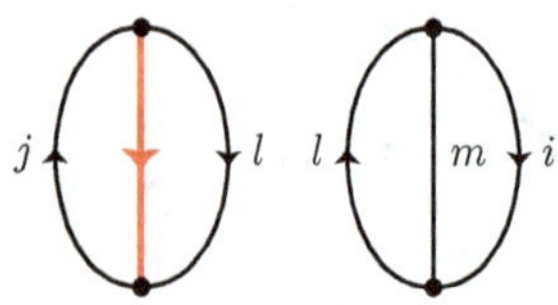

.

In the second case we first apply an isotopy $\sim$:

$\sum_{z\in I}$

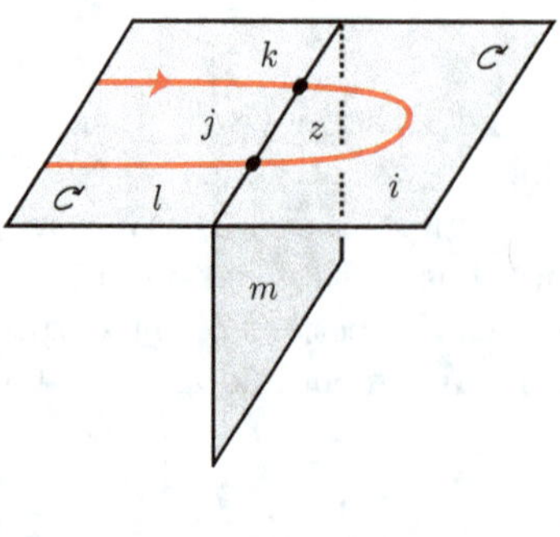

$\sim \sum_{z\in I}$

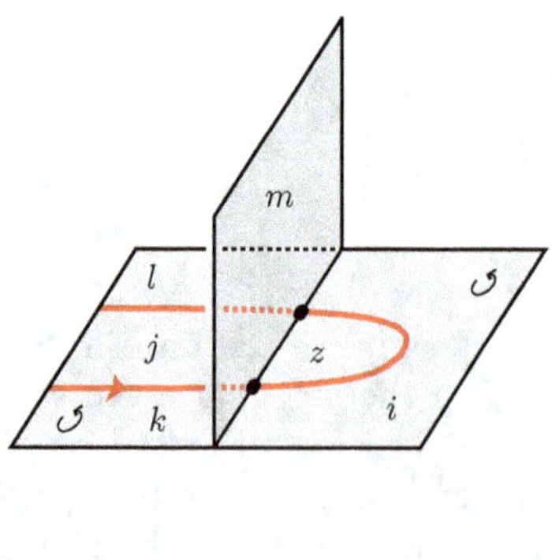

$\longmapsto \zeta_{k,l} \sum_{z\in I} \dim(z)$

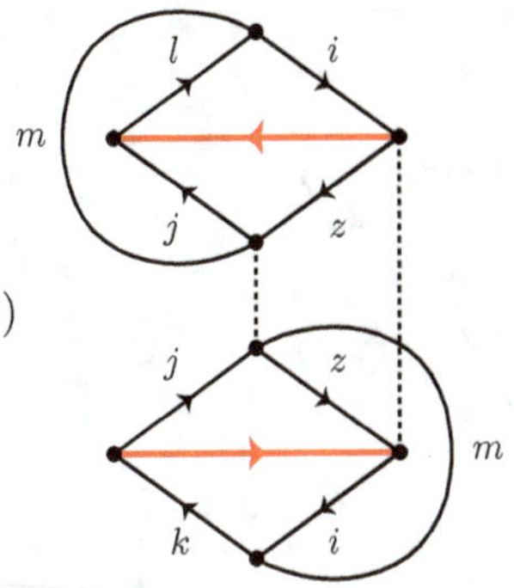

$\overset{(B,A)}{=} \zeta_{k,l} \sum_{z\in I} \dim(z)$

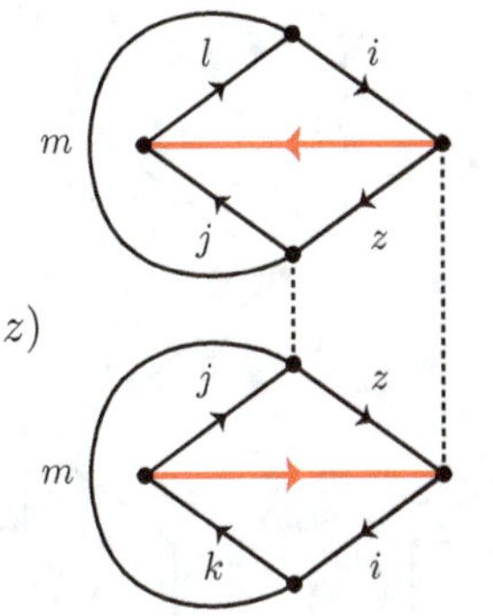

$\overset{(G,F)}{=} \zeta_{k,l}$

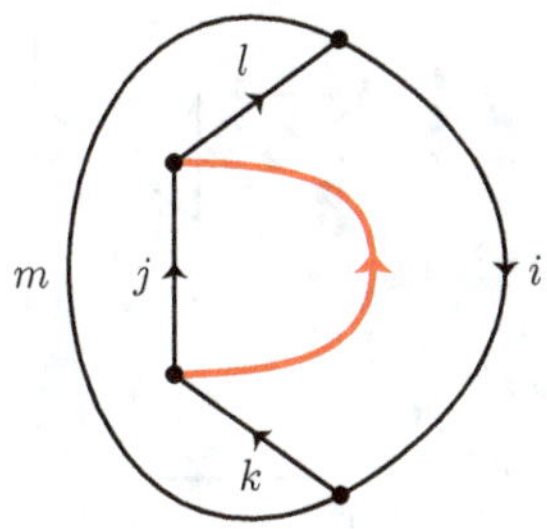

$\overset{(A)}{=} \zeta_{k,l}$

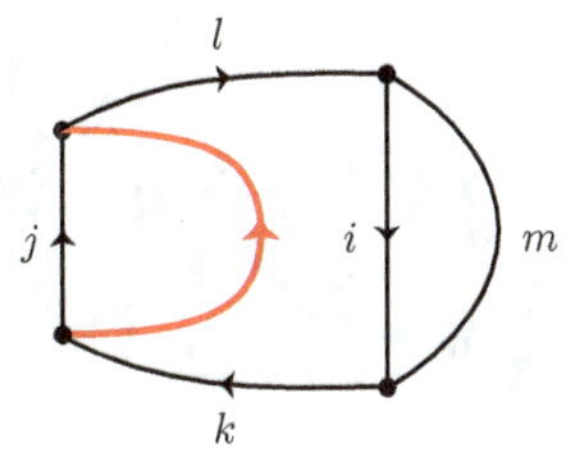

$\overset{(D,E)}{=} \delta_{k,l} \dim(k)^{-1}\zeta_{k,l}$

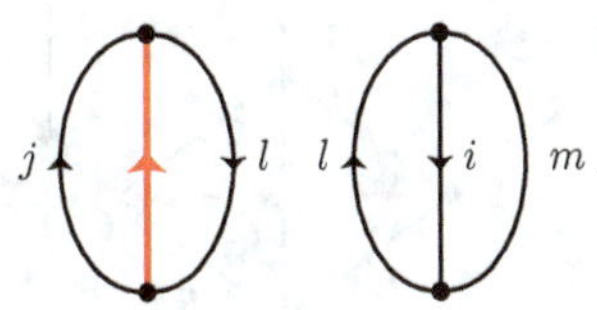

The end of the argument in both cases is similar to that in the proof of the invariance under ω_2.

To prove the invariance of $|M, R|_\mathcal{C}$ under the move ω_6, we consider two cases, depending on the (compatible) orientations of the horizontal regions. As above, the orientation of the vertical region and the induced direction of the corresponding strands of knotted $\mathcal{C}$-graphs play no role in the argument and are not shown. The first case:

$$\sum_{z\in I}$$

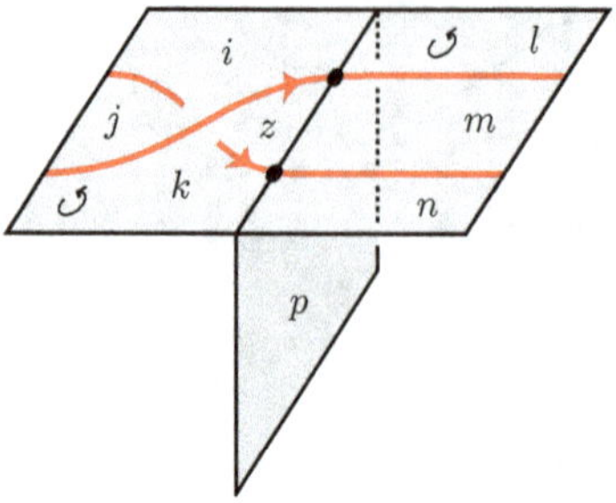

$$\longmapsto \sum_{z\in I} \dim(z)$$

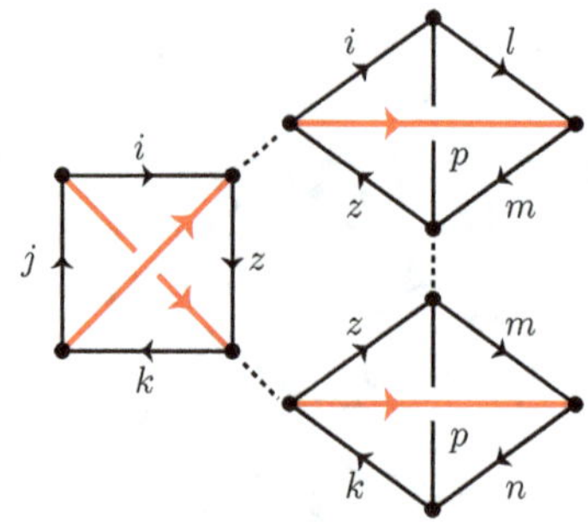

$$\overset{(G,F)}{=}$$

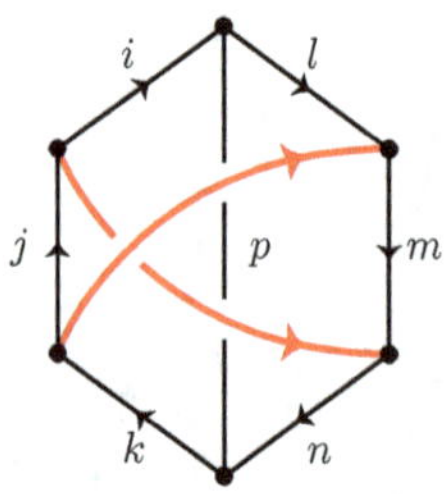

$$\overset{(C)}{=}$$

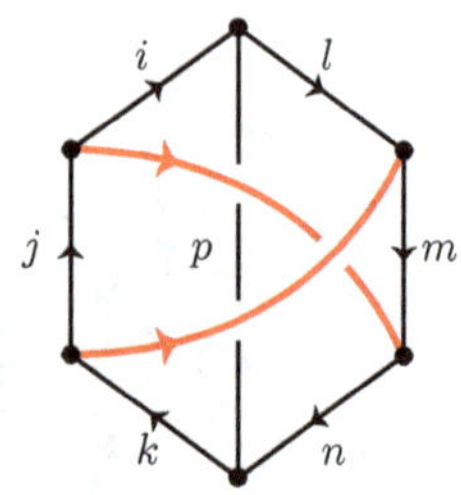

$$\overset{(G,F)}{=} \sum_{z\in I} \dim(z)$$

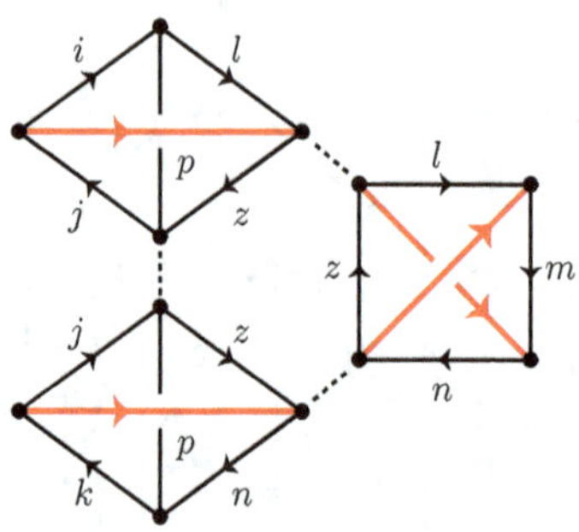

$\longmapsto \sum_{z\in I}$ 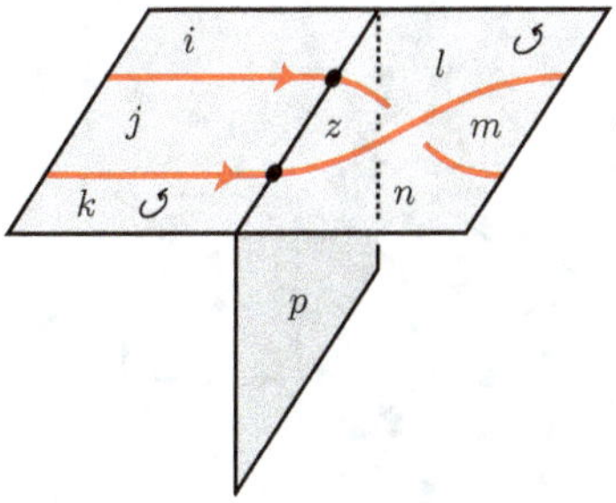 .

The second case:

$\sum_{z\in I}$ 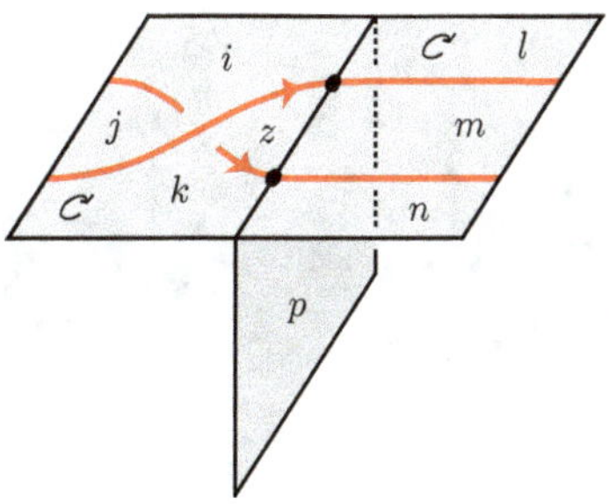 $\sim \sum_{z\in I}$ 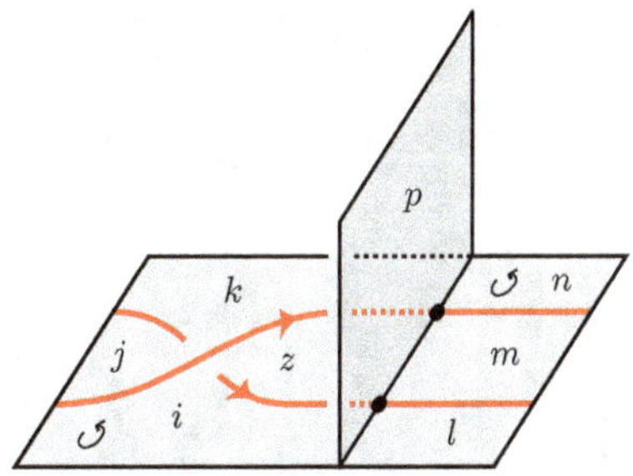

$\longmapsto \sum_{z\in I} \dim(z)$ 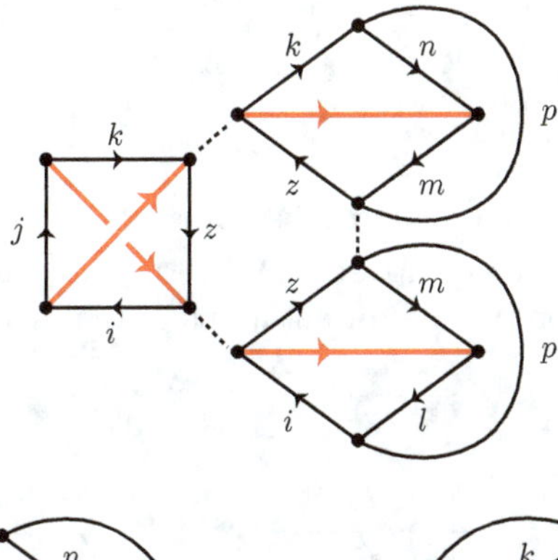

$\overset{(G,F)}{=}$ 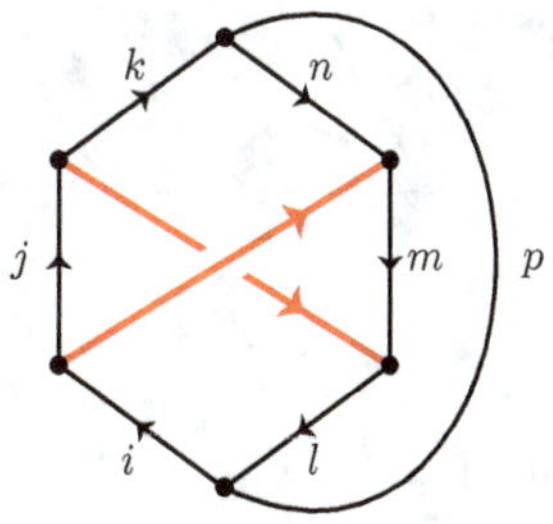 $\overset{(A)}{=}$ 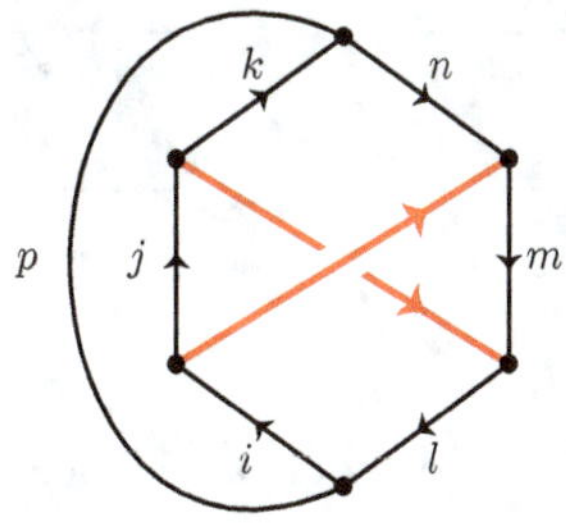

$$\stackrel{(G,F)}{=} \sum_{z\in I} \dim(z)$$

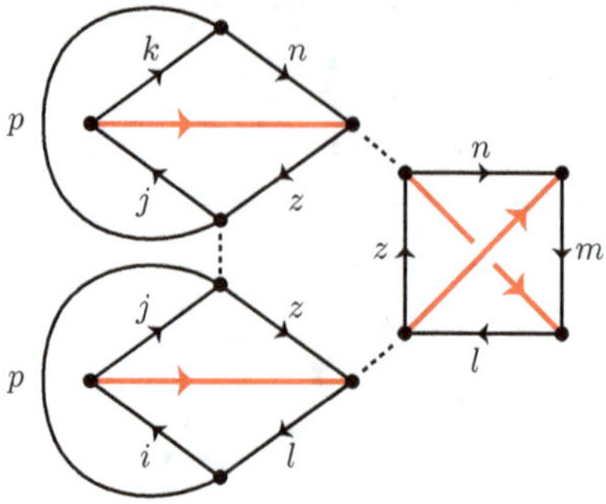

$$\longmapsto \sum_{z\in I}$$

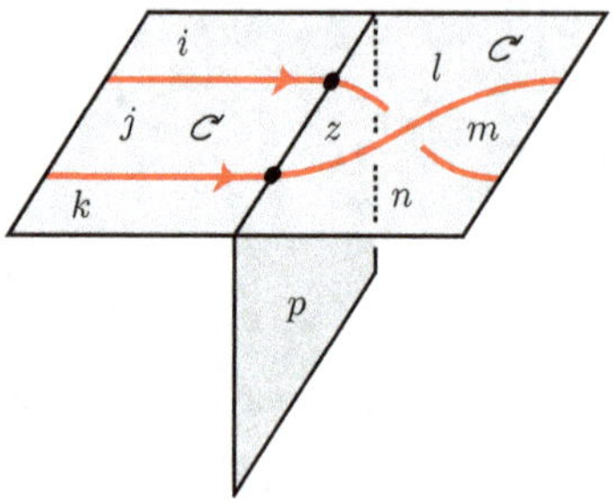

$$\sim \sum_{z\in I}$$

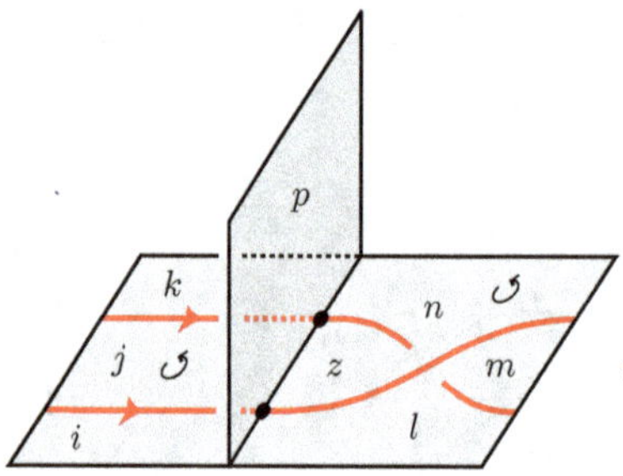

,

where, as above, the symbol $\sim$ stands for an isotopy.

Let us prove the invariance of $|M,R|_{\mathcal{C}}$ under the move ω_7. Up to isotopy, there is only one choice of (compatible) orientations of the regions. Indeed,

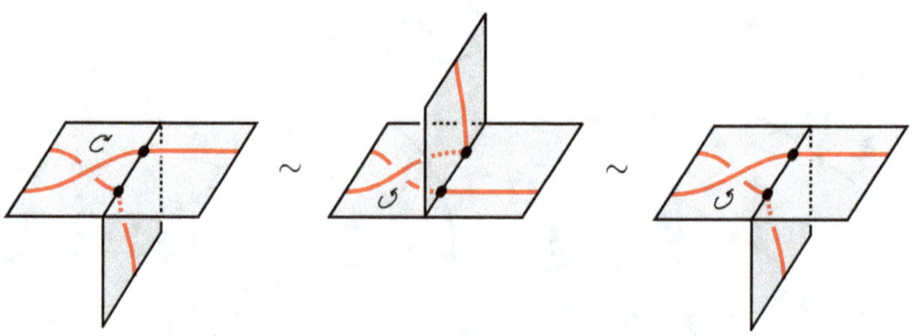

Thus, it is enough to consider one choice of orientations. We have:

$$\sum_{z\in I}$$

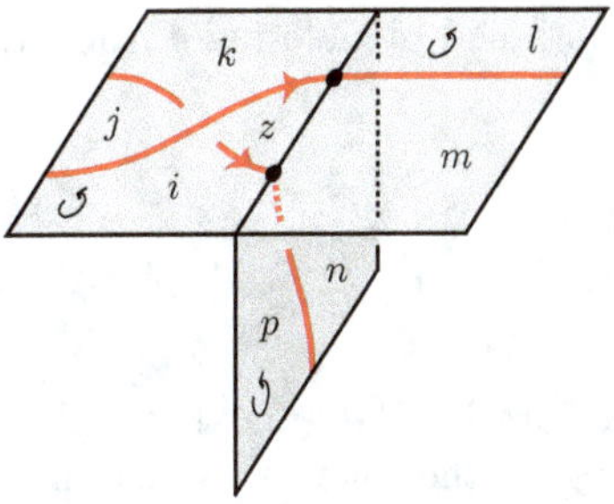

$$\longmapsto \sum_{z\in I} \dim(z)$$

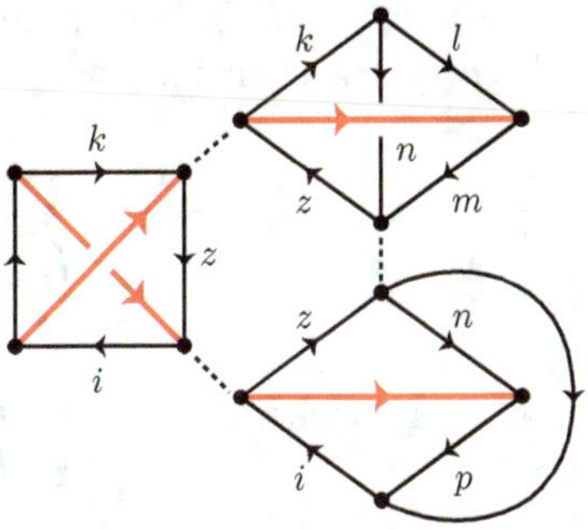

$$\overset{(G,F)}{=}$$

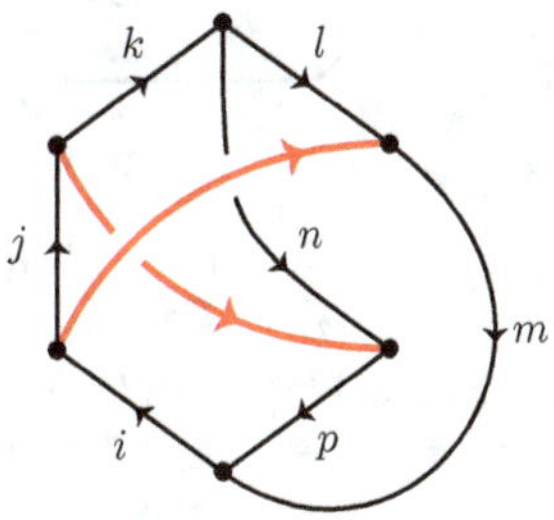

$$\overset{(C)}{=}$$

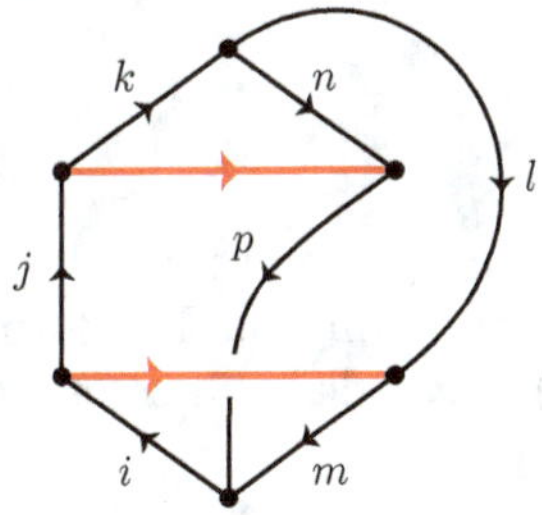

$$\overset{(G)}{=}$$

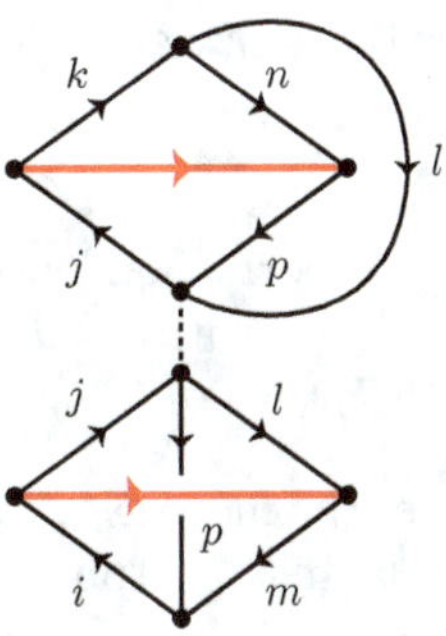

$$\longleftarrow$$

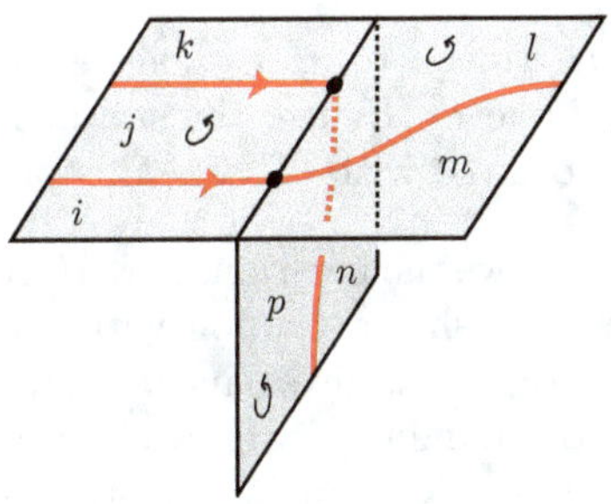

.

We now compute $|M, R|_{\mathcal{C}}$ in the case where R has a coupon with no inputs and no outputs. Let (P, d) be a neat positive diagram of R and let q be a coupon of d with no inputs and no outputs. Let $k \in \operatorname{End}_{\mathcal{Z}(\mathcal{C})}(\mathbb{1}_{\mathcal{Z}(\mathcal{C})}) = \Bbbk$ be the color of q. Deleting q from d, we obtain a neat positive diagram (P, d') representing the

$\mathcal{Z}(\mathcal{C})$-colored ribbon graph $R' \subset M$ obtained by deleting q from R. We claim that

$$|M, R|_{\mathcal{C}} = k\,|M, R'|_{\mathcal{C}} \tag{15.7}$$

where the left-hand side is computed from (P, d) and the right-hand side from (P, d'). To check this equality, let r_0 be the face of d encircling q. Consider a coloring c of d and set $i = c(r_0) \in I$. The contribution of q to the state sum $|M, R|_{\mathcal{C}}$ computed from (P, d) is the scalar factor

$$\mathbb{F}_{\mathcal{C}}(\Gamma_q^c) \in H(\Gamma_q^c)^\star = \Bbbk^\star \simeq \Bbbk, \quad \text{where} \quad \Gamma_q^c =$$ 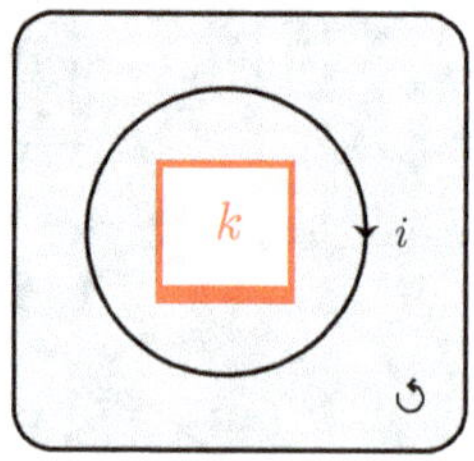

$$\subset S^2.$$

It follows from (B) and the definition of $\mathbb{F}_{\mathcal{C}}$ that $\mathbb{F}_{\mathcal{C}}(\Gamma_q^c) = k \dim(i)$. Let r_0' be the face of d' containing r_0. The formulas $c'(r_0') = c(r_0) = i$ and $c'(r) = c(r)$ for all $r \in \mathrm{Fac}(d) \setminus \{r_0\}$ define a coloring c' of d'. Since $\chi(r_0') = \chi(r_0) + 1$, we have

$$\dim(c(r_0))^{\chi(r_0)}\, \mathbb{F}_{\mathcal{C}}(\Gamma_q^c) = k \dim(i)^{\chi(r_0)+1} = k \dim(c'(r_0'))^{\chi(r_0')}.$$

Note that the rule $c \mapsto c'$ establishes a bijective correspondence between the colorings of d and d'. Hence, the state sum associated with (P, d) is k times the state sum associated with (P, d'). This proves (15.7).

Now, we prove the invariance of $|M, R|_{\mathcal{C}}$ under the move $\omega_{8,m,n}$. If $m = n = 0$, then this move pushes a coupon with no inputs and no outputs across an edge of the skeleton and the invariance follows from Formula (15.7). Suppose that at least one of the numbers m or n is non-zero. Then there is at least one switch in the picture of $\omega_{8,m,n}$ and therefore the orientations of two horizontal regions are compatible. As above, the orientation of the vertical region and the induced directions of the corresponding strands of knotted $\mathcal{C}$-graphs play no role in the argument and are not shown. Assume first that $m \neq 0$, $n \neq 0$, and both horizontal regions are oriented counterclockwise. Then

$$\sum_{z_2,\dots,z_{n-1}\in I}$$

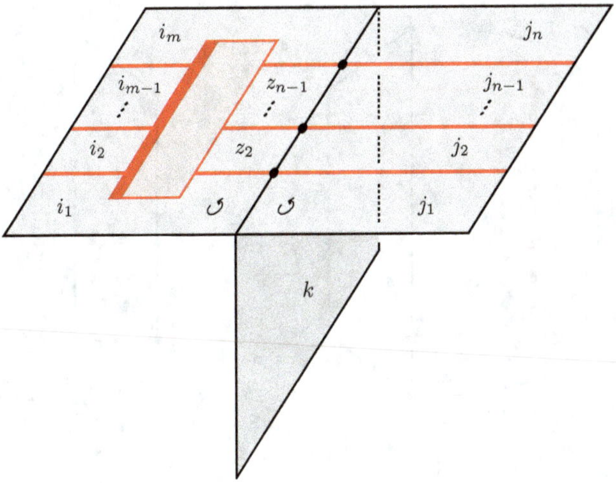

$$\longmapsto \sum_{z_2,\dots,z_{n-1}\in I}\ \prod_{\alpha=2}^{n-1}\dim(z_\alpha)$$

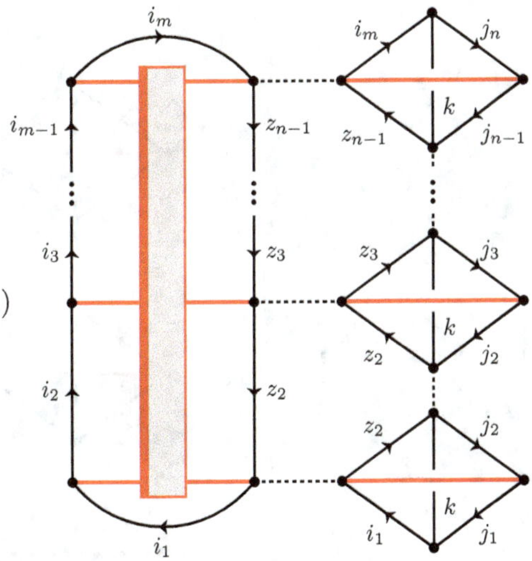

$\overset{(G,F)}{=}$

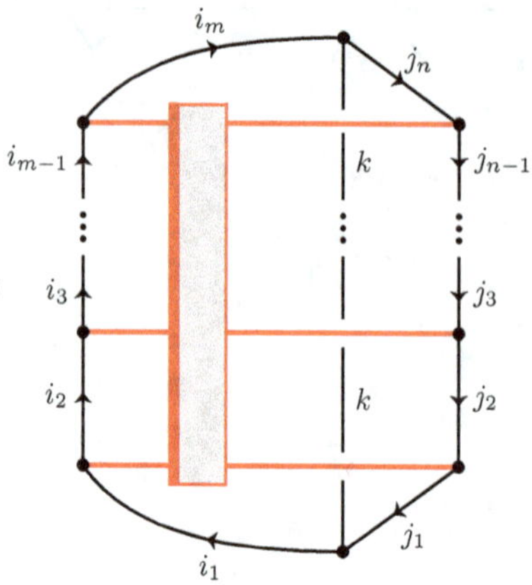

$\overset{(C)}{=}$

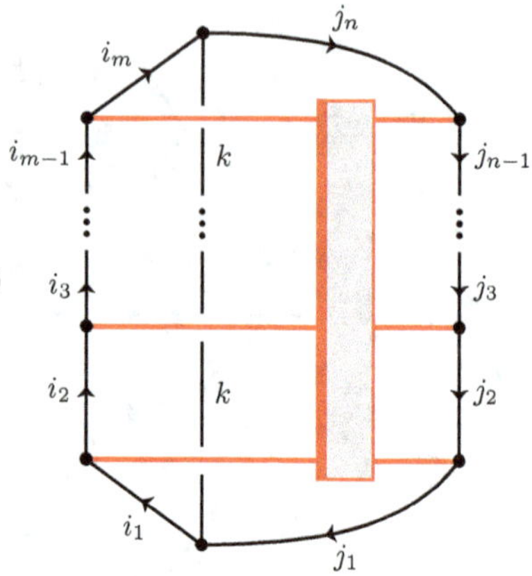

$$\overset{(G,F)}{=} \sum_{z_2,\ldots,z_{m-1}\in I} \prod_{\alpha=2}^{m-1} \dim(z_\alpha)$$

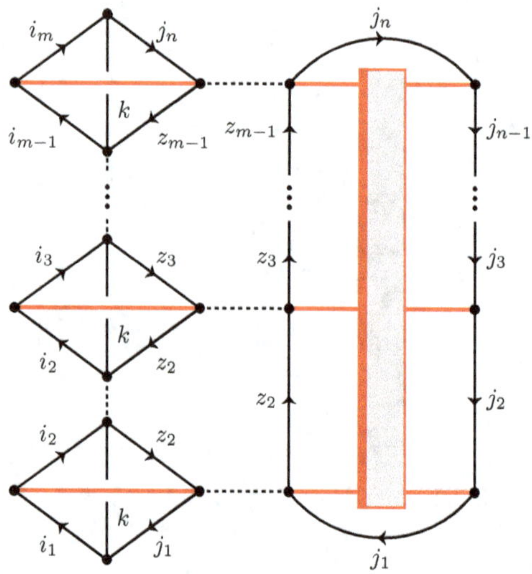

$$\longmapsto \sum_{z_2,\ldots,z_{m-1}\in I}$$

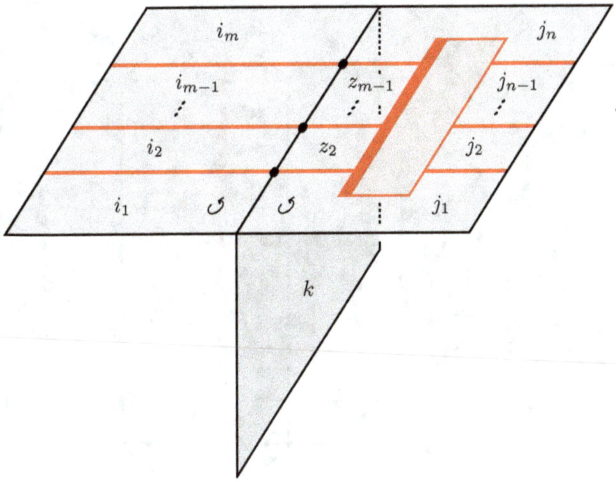

.

Assume now that $m \neq 0$, $n \neq 0$, and the horizontal regions are oriented clockwise. Then

$$\sum_{z_2,\ldots,z_{n-1}\in I}$$

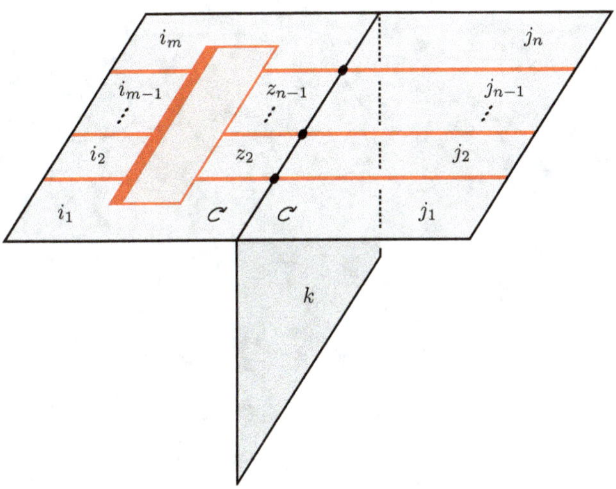

$$\mapsto \sum_{z_2,\ldots,z_{n-1}\in I} \prod_{\alpha=2}^{n-1} \dim(z_\alpha)$$

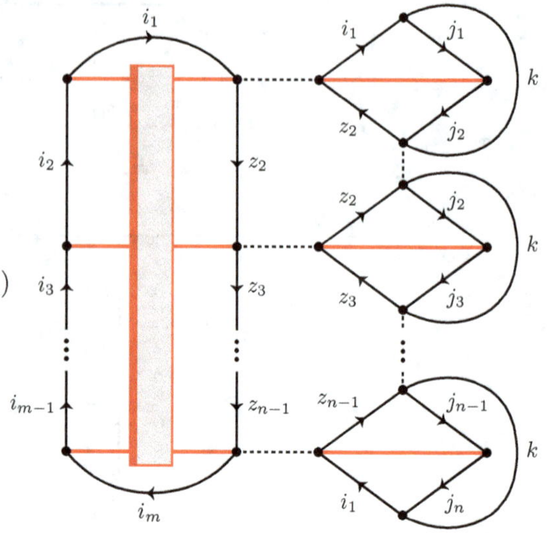

$\overset{(G,F)}{=}$

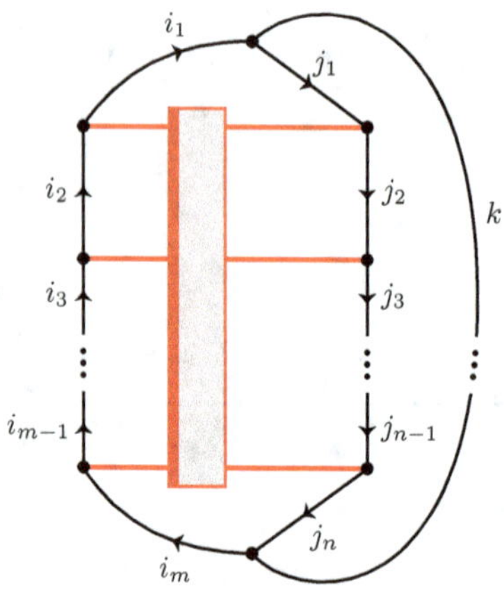

$\overset{(A)}{=}$

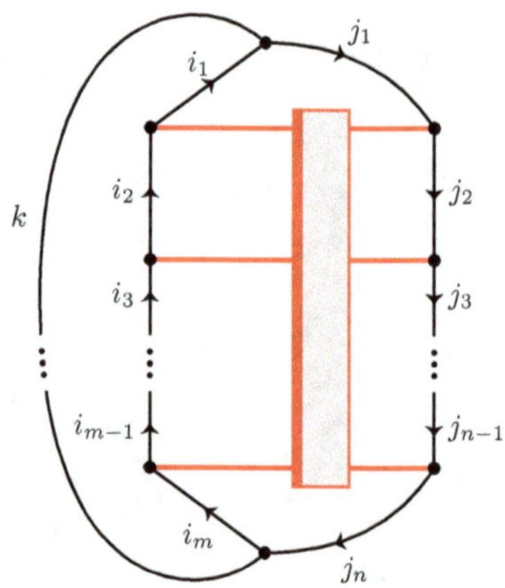

$$\overset{(G,F)}{=} \sum_{z_2,\dots,z_{m-1}\in I} \prod_{\alpha=2}^{m-1} \dim(z_\alpha)$$

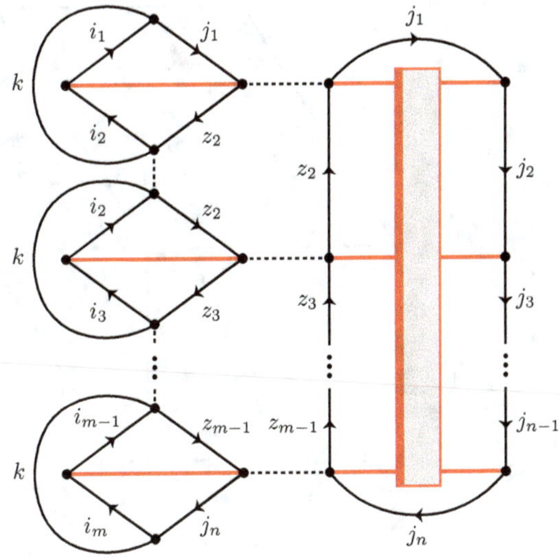

$$\longmapsto \sum_{z_2,\dots,z_{m-1}\in I}$$

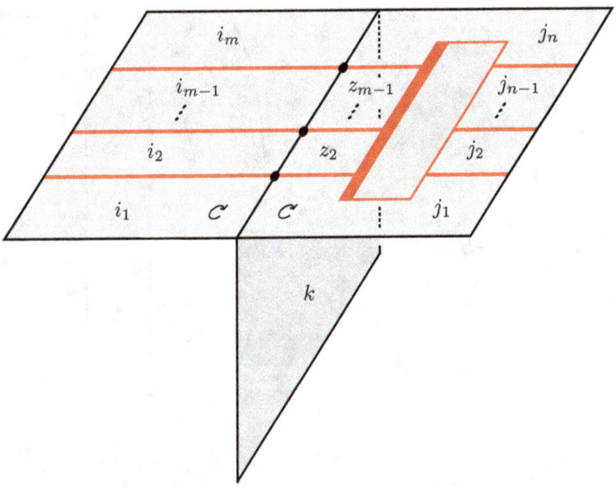

.

Assume next that $m = 0$, $n \neq 0$, and the horizontal regions are oriented counterclockwise. Let r, r' be the faces of the diagram colored by $j_1, j_n \in I$ in the picture below showing the neat positive diagram before the move. As in the proof of the invariance under ω_2, set

$$\zeta_{j_1,j_n} = \begin{cases} \dim(j_1)^{\chi(r)} \dim(j_n)^{\chi(r')} & \text{if } r \neq r', \\ \dim(j_1)^{\chi(r)} & \text{if } r = r'. \end{cases}$$

We have:

$$\sum_{z_2,\ldots,z_{n-1}\in I}$$

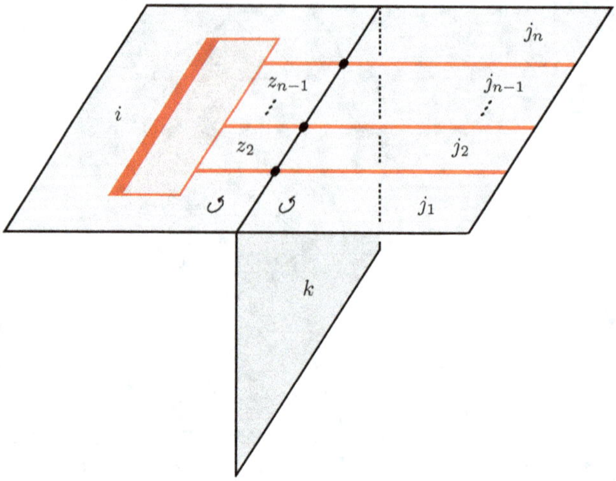

$$\longmapsto \zeta_{j_1,j_n} \sum_{z_2,\ldots,z_{n-1}\in I} \prod_{\alpha=2}^{n-1} \dim(z_\alpha)$$

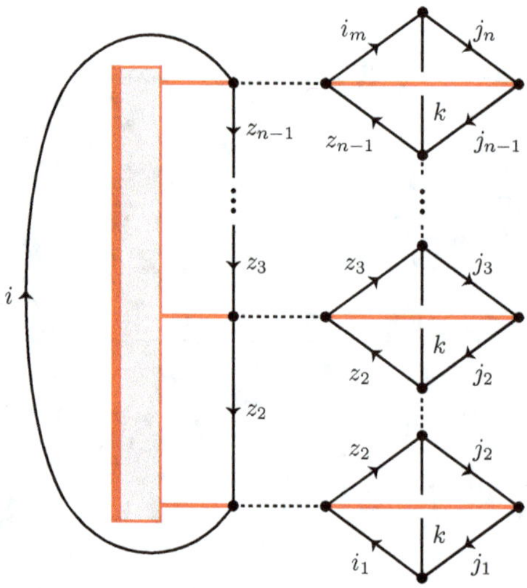

$\overset{(G,F)}{=} \zeta_{j_1,j_n}$ 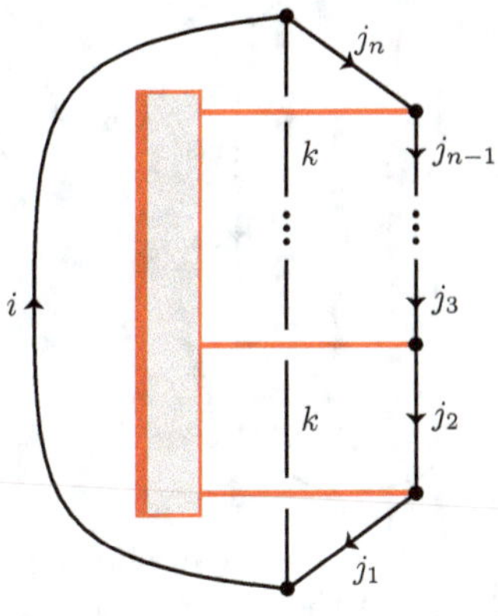

$\overset{(C)}{=} \zeta_{j_1,j_n}$ 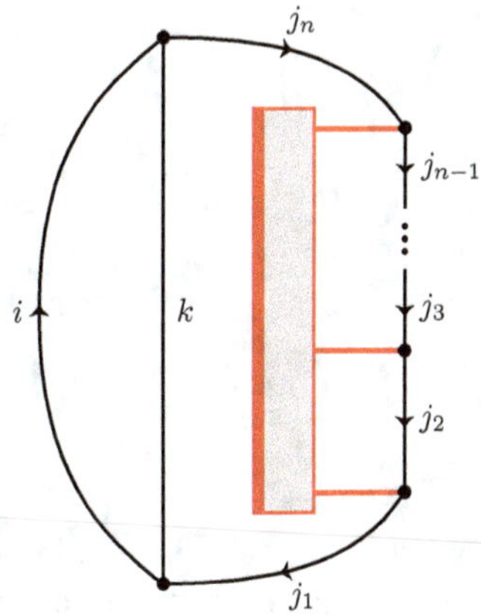

$\overset{(D,E)}{=} \delta_{j_1,j_n} \dim(j_1)^{-1} \zeta_{j_1,j_n}$ 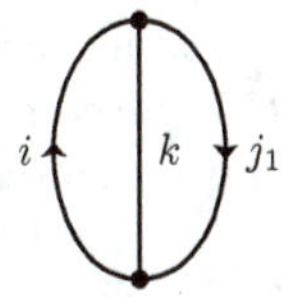

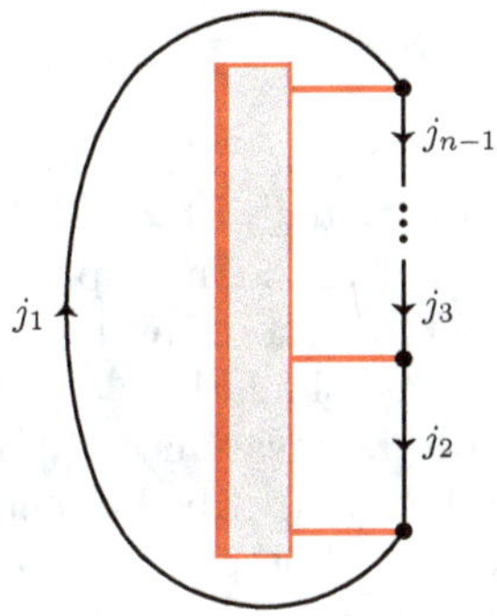

and

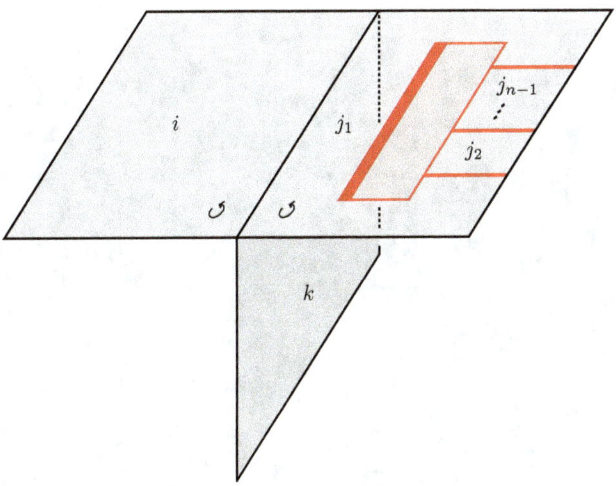

$\longmapsto \dim(j_1)^{\chi(r'')}$ 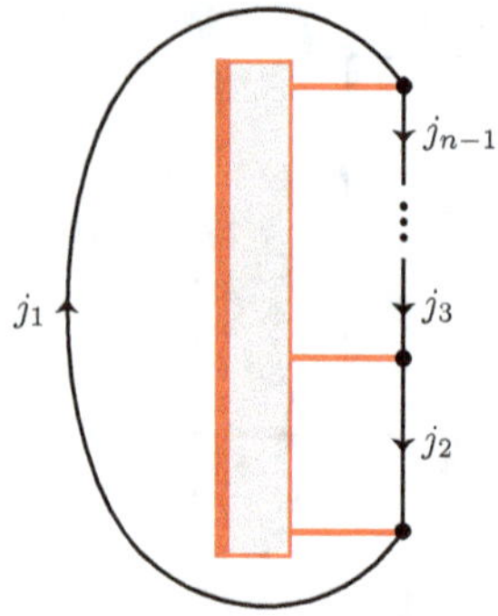

where r'' is the face colored by j_1 in the picture above showing the neat positive diagram after the move. The additivity of the Euler characteristic yields the identity

$$\dim(j_1)^{-1}\zeta_{j_1,j_1} = \dim(j_1)^{\chi(r'')}.$$

The end of the argument is similar to that in the proof of the invariance under ω_2: the pairing associated with the (i, k, j_1)-theta graph is cancelled by the contraction vector associated to the rim supported by the common edge of the involved regions. The case where the horizontal regions are oriented clockwise and the case $m \neq 0$, $n = 0$ are treated similarly.

To prove the invariance of $|M, R|_{\mathcal{C}}$ under $\omega_{9,m,n}$, we consider two cases, depending on the (compatible) orientations of the regions containing the red strand. The orientations of other regions and the induced directions of the corresponding strands of knotted $\mathcal{C}$-graphs play no role in the argument and are not shown. The first case:

$$\sum_{\substack{x_1 \in I \\ \vdots \\ x_{m+1} \in I}}$$

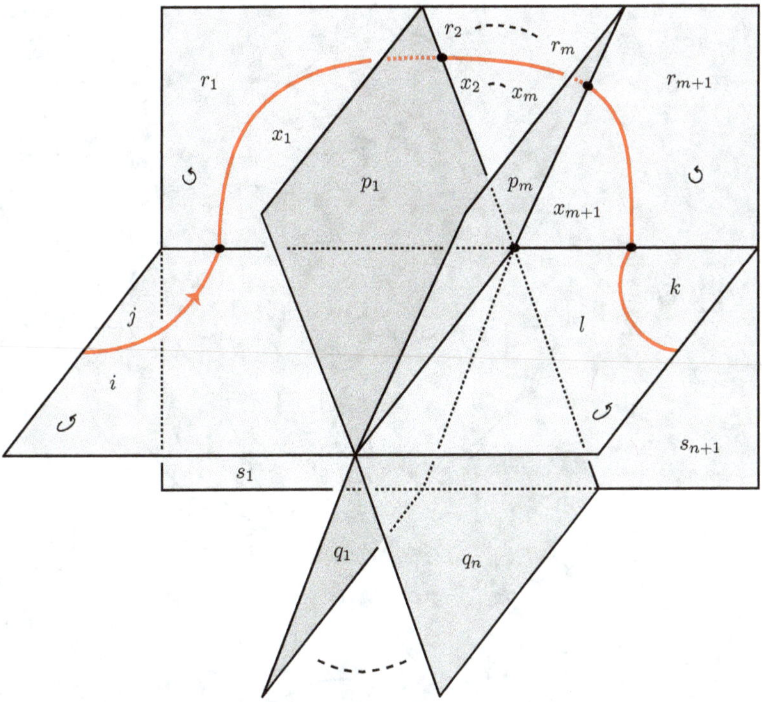

$$\mapsto$$

$$\sum_{\substack{x_1 \in I \\ \vdots \\ x_{m+1} \in I}} \prod_{\alpha=1}^{m+1} \dim(x_\alpha)$$

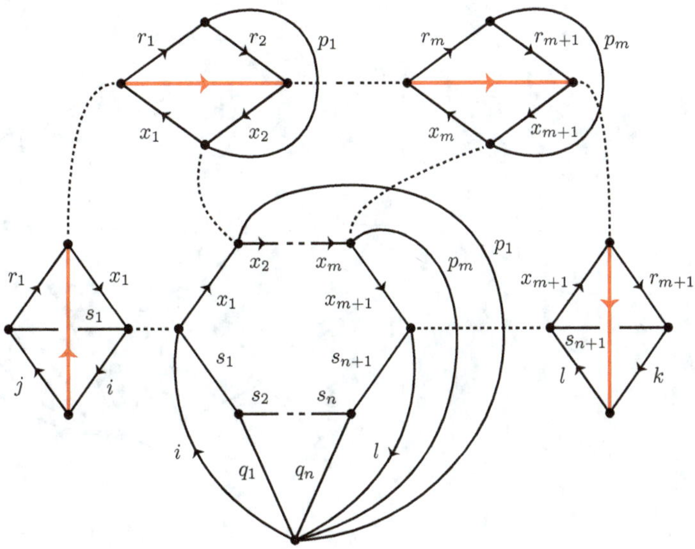

$$\overset{(G)}{=} \sum_{\substack{x_1 \in I \\ \vdots \\ x_{m+1} \in I}} \prod_{\alpha=1}^{m+1} \dim(x_\alpha)$$

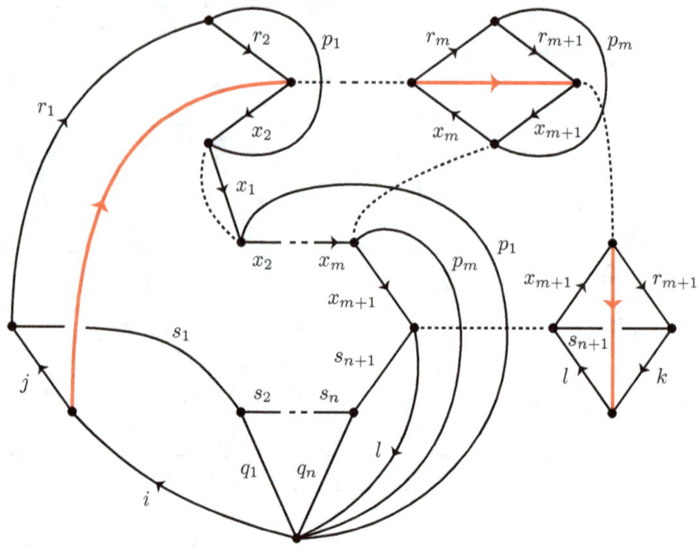

$$\overset{(F)}{=} \sum_{\substack{x_2 \in I \\ \vdots \\ x_{m+1} \in I}} \prod_{\alpha=2}^{m+1} \dim(x_\alpha)$$

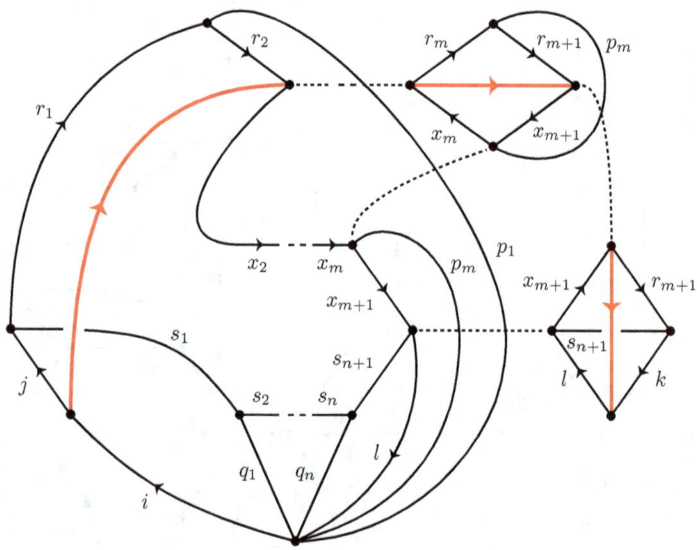

$$\stackrel{(B)}{=} \sum_{\substack{x_2 \in I \\ \vdots \\ x_{m+1} \in I}} \prod_{\alpha=2}^{m+1} \dim(x_\alpha)$$

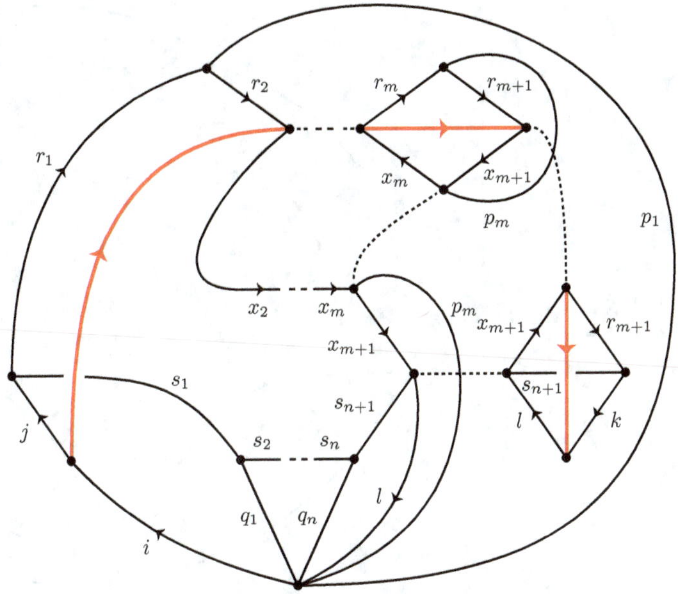

Repeating the last three steps $m-1$ times we transform the latter expression into

$$\sum_{x_{m+1} \in I} \dim(x_{m+1})$$

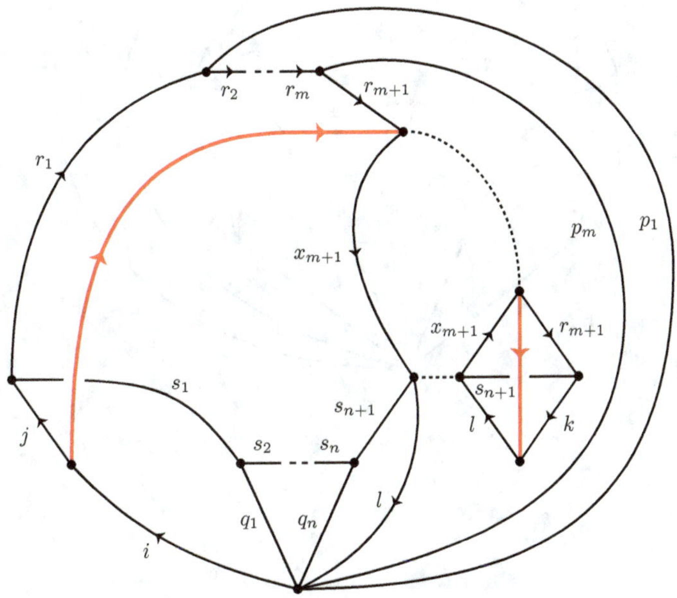

$$\overset{(G)}{=} \sum_{x_{m+1} \in I} \dim(x_{m+1})$$

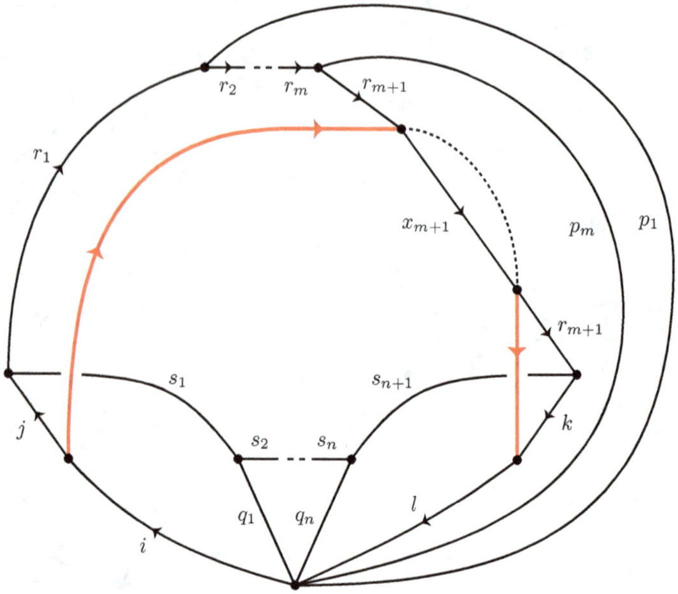

$$\overset{(F)}{=}$$

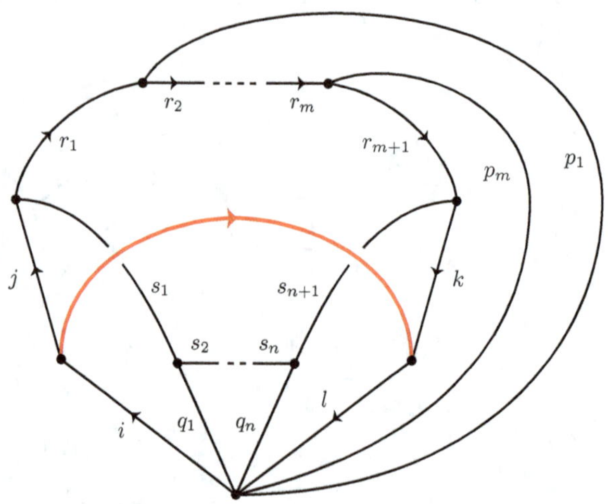

$\overset{(C)}{=}$

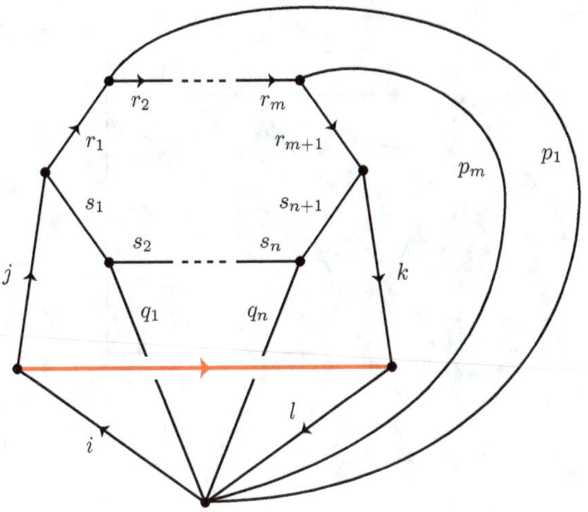

$\overset{(G)}{=}$

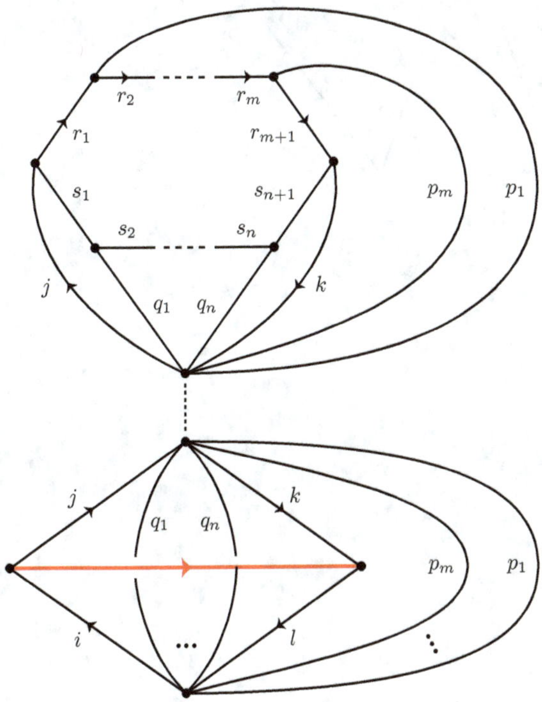

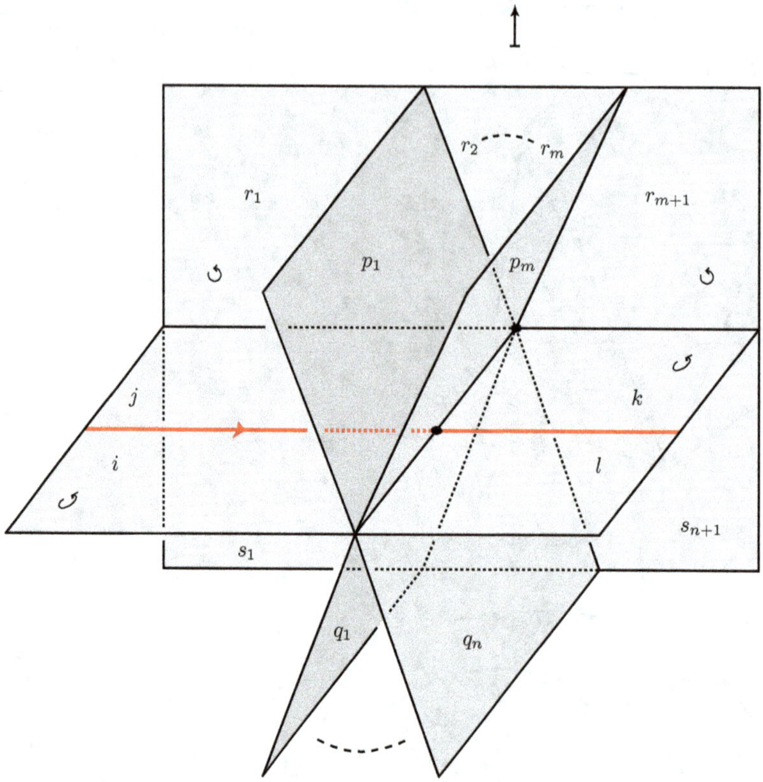
r_1
r_2
r_m
r_{m+1}
p_1
p_m
j
k
i
l
s_1
s_{n+1}
q_1
q_n

In the second case, the arguments proceed along similar lines. Note first that

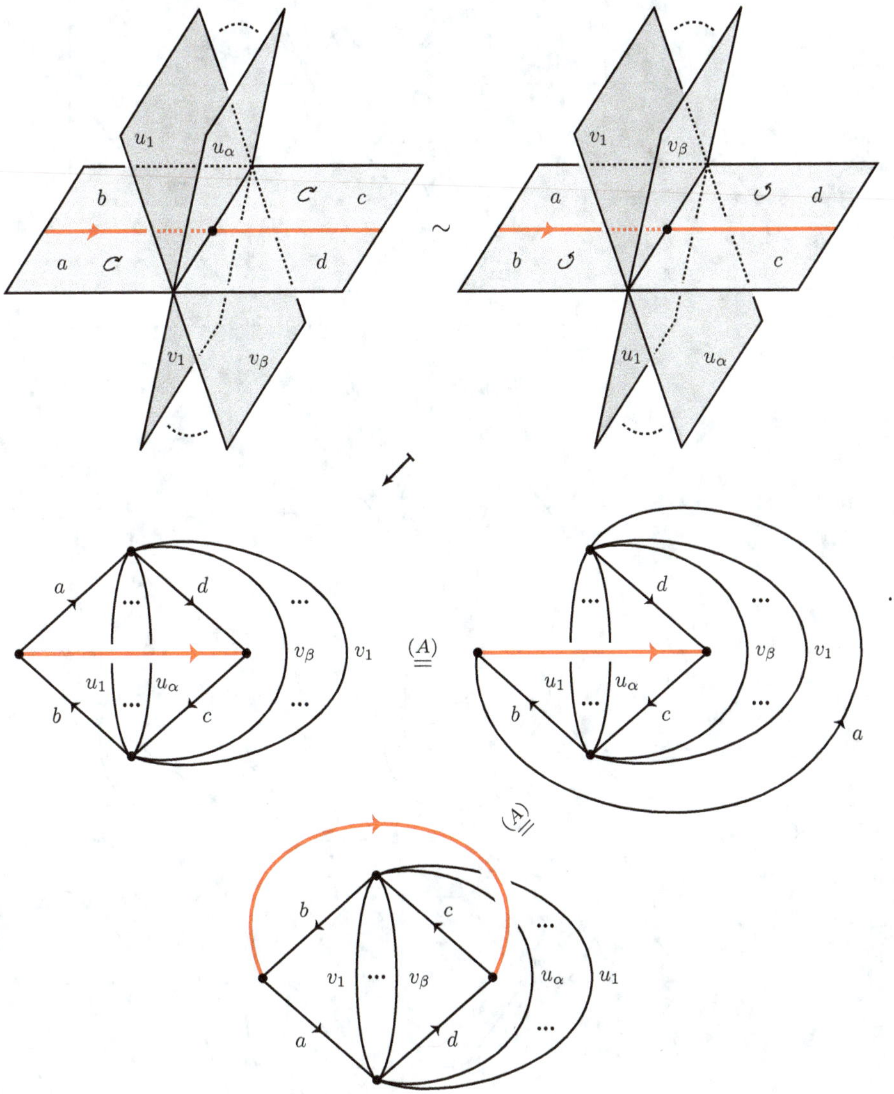

Therefore

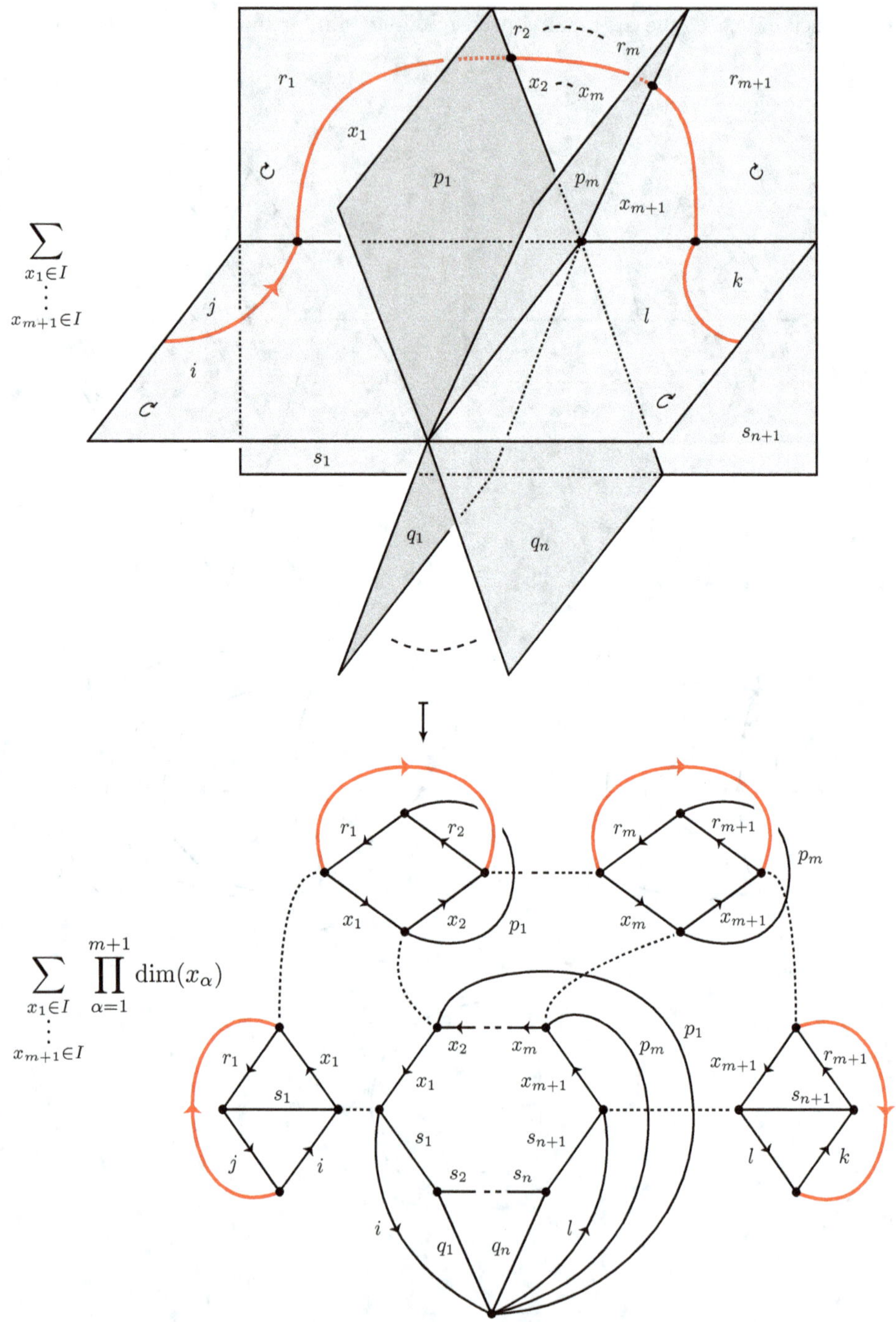
$\sum_{x_1 \in I, \ldots, x_{m+1} \in I}$
r_1 r_2 r_m r_{m+1}
x_1 x_2 x_m x_{m+1}
p_1 p_m
i j k l
s_1 s_{n+1}
q_1 q_n
$\sum_{x_1 \in I, \ldots, x_{m+1} \in I} \prod_{\alpha=1}^{m+1} \dim(x_\alpha)$
s_2 s_n

$$\overset{(G)}{=} \sum_{\substack{x_1\in I \\ \vdots \\ x_{m+1}\in I}} \prod_{\alpha=1}^{m+1} \dim(x_\alpha)$$

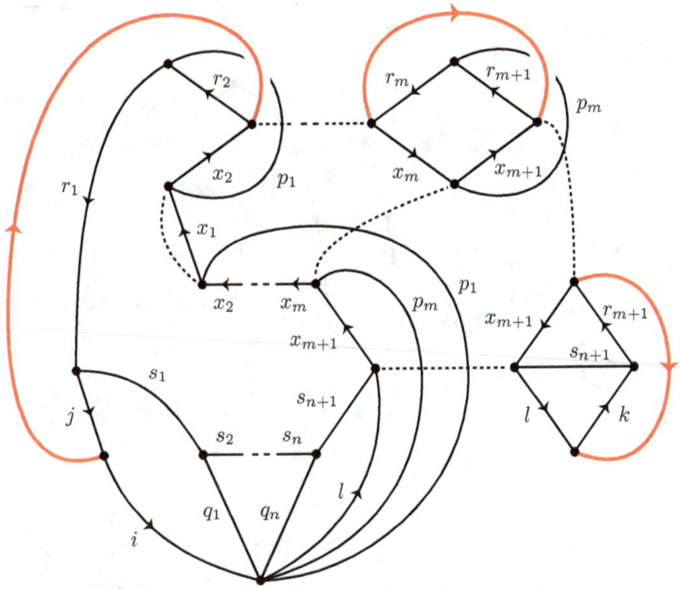

$$\overset{(F)}{=} \sum_{\substack{x_2\in I \\ \vdots \\ x_{m+1}\in I}} \prod_{\alpha=2}^{m+1} \dim(x_\alpha)$$

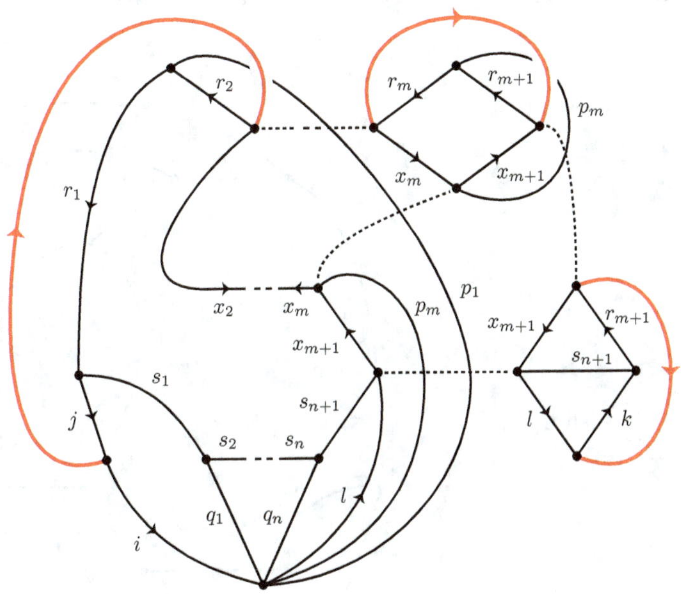

$$\overset{(B)}{=} \sum_{\substack{x_2 \in I \\ \vdots \\ x_{m+1} \in I}} \prod_{\alpha=2}^{m+1} \dim(x_\alpha)$$

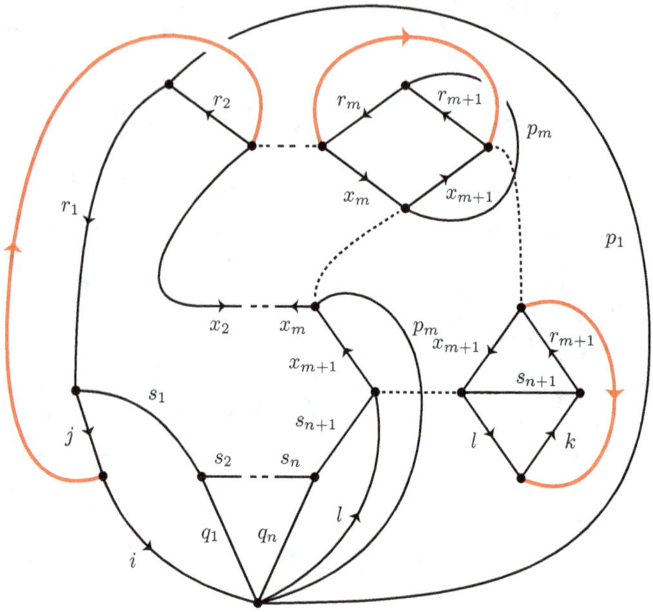

Repeating the last three steps $m-1$ times we transform the latter expression into

$$\sum_{x_{m+1} \in I} \dim(x_{m+1})$$

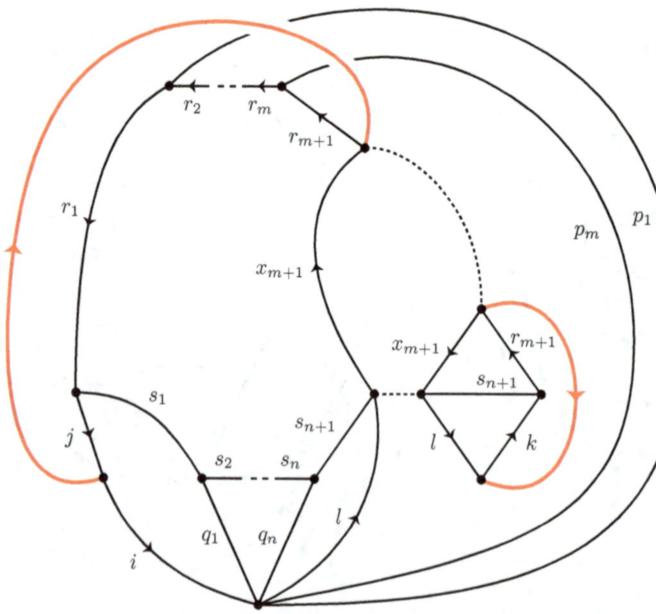

$$\stackrel{(G)}{=} \sum_{x_{m+1} \in I} \dim(x_{m+1})$$

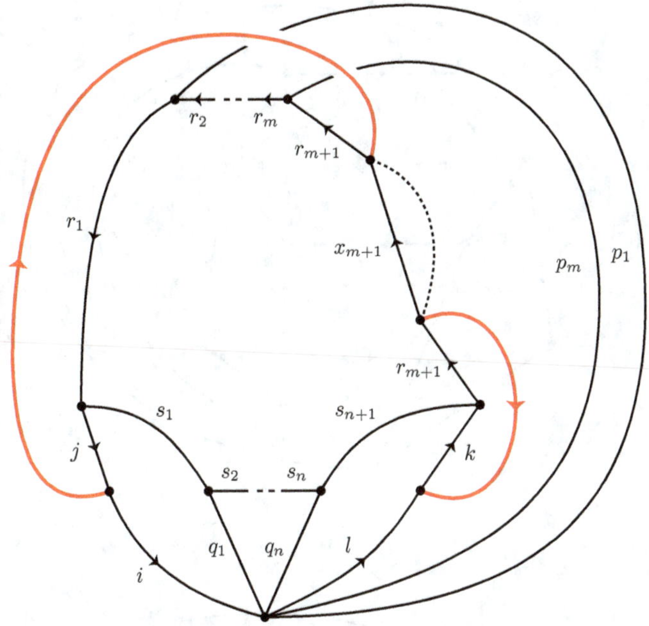

$$\stackrel{(F)}{=}$$

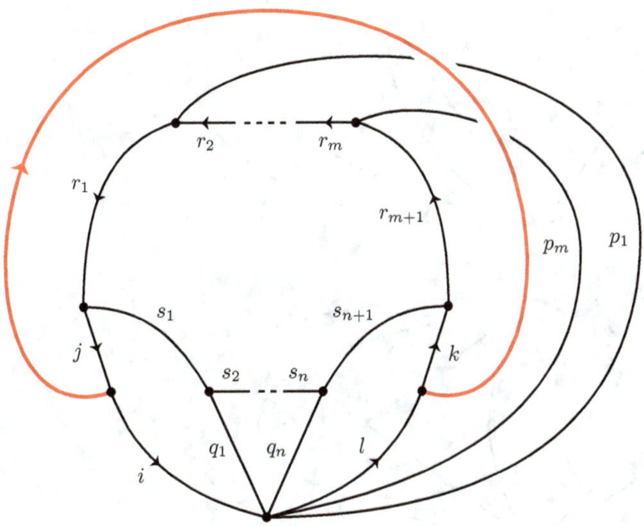

$\overset{(C)}{=}$

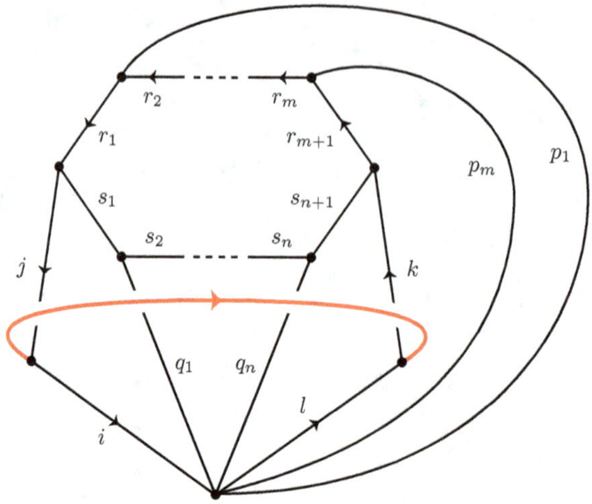

$\overset{(G)}{=}$

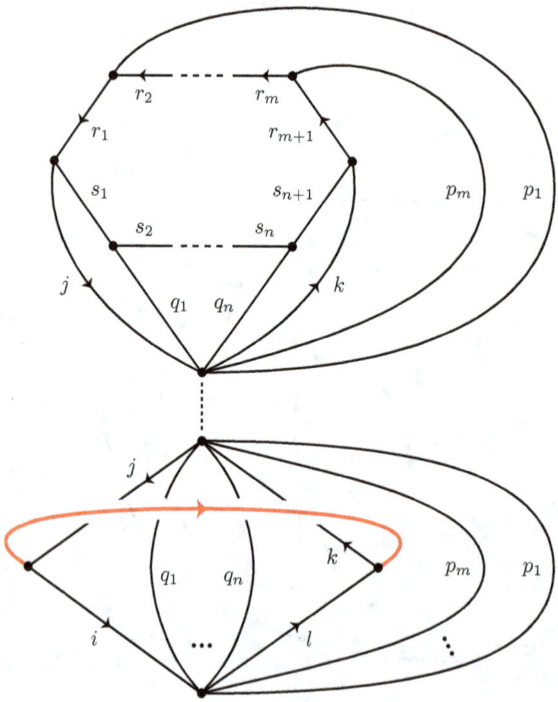

$\overset{(C)}{=}$

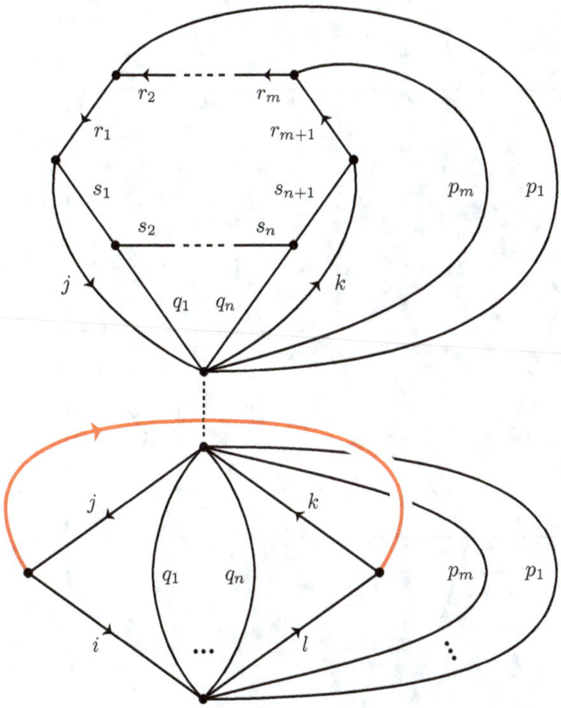

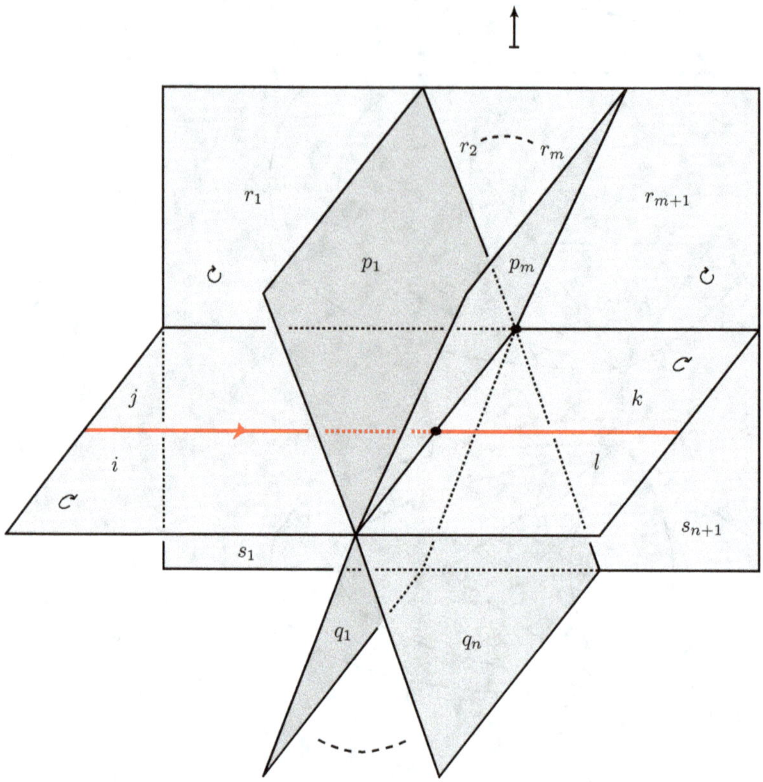

We have proved so far that the state sum $|M, R|_{\mathcal{C}}$ is invariant under the moves $\{\omega_*\}_{*=0}^{9}$ in the class of neat positive diagrams of R, and so is invariant under isotopy of R in M. It remains to verify that $|M, R|_{\mathcal{C}}$ is invariant under the moves in Figure 15.1. Let R' be a $\mathcal{Z}(\mathcal{C})$-colored ribbon graph in M obtained from R by one of these moves. We can find neat positive diagrams (P, d) of R and (P, d') of R' such that d' is obtained from d by one of the following transformations proceeding inside a disk in a region of P:

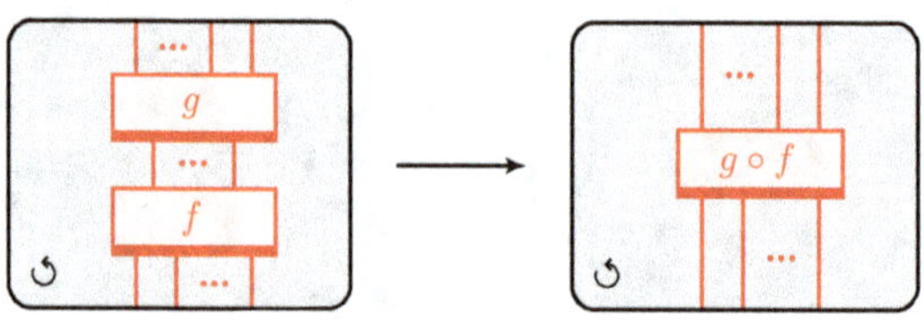

, (15.8)

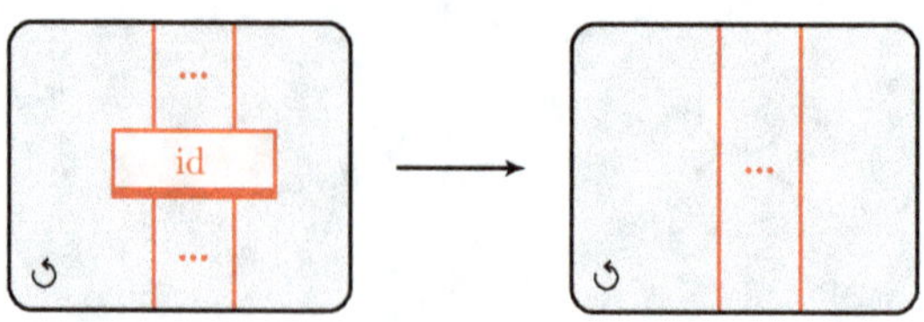

. (15.9)

Here, the orientations and colors of all strands are preserved under the transformations and are not indicated in the pictures. The morphisms $f, g, g \circ f$ and the identity morphism id are the colors of the coupons. We must prove that the state sums associated with (P, d) and (P, d') are equal. We start with (15.8) and let $m, n, s \geq 0$ be the integers such that the coupon colored by f has m inputs and n outputs and the coupon colored by g has n inputs and s outputs. Assume first that $n \geq 1$, that is, that there is at least one strand between the coupons. We have:

$$\sum_{z_1,\dots,z_{n-1}\in I}$$

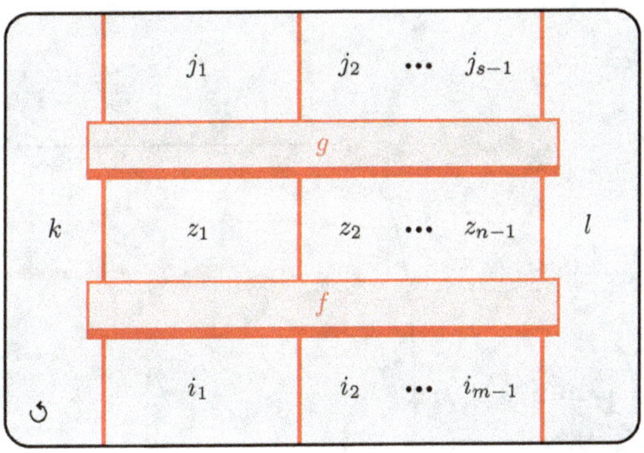

$$\longmapsto \sum_{z_1,\dots,z_{n-1}\in I} \prod_{\alpha=1}^{n-1} \dim(z_\alpha)$$

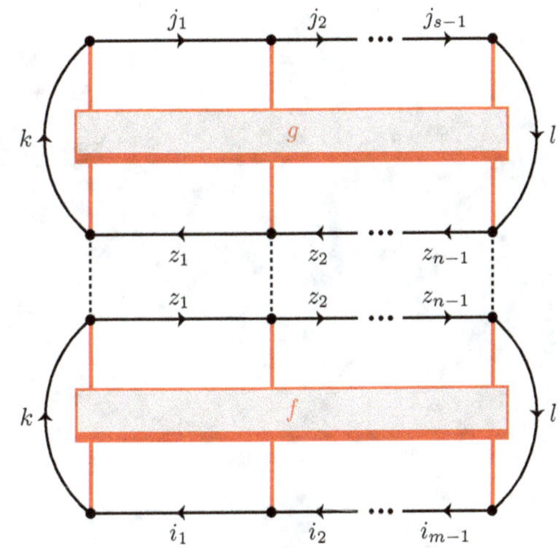

$$\overset{(G)}{=} \sum_{z_1,\dots,z_{n-1}\in I} \prod_{\alpha=1}^{n-1} \dim(z_\alpha)$$

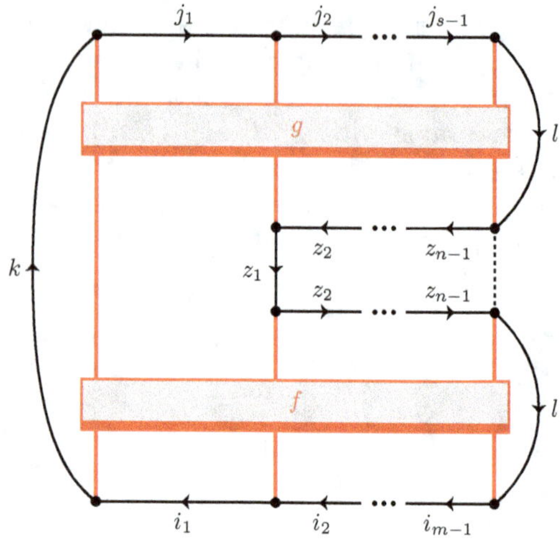

$$\overset{(F)}{=} \sum_{z_2,\dots,z_{n-1}\in I} \prod_{\alpha=2}^{n-1} \dim(z_\alpha)$$

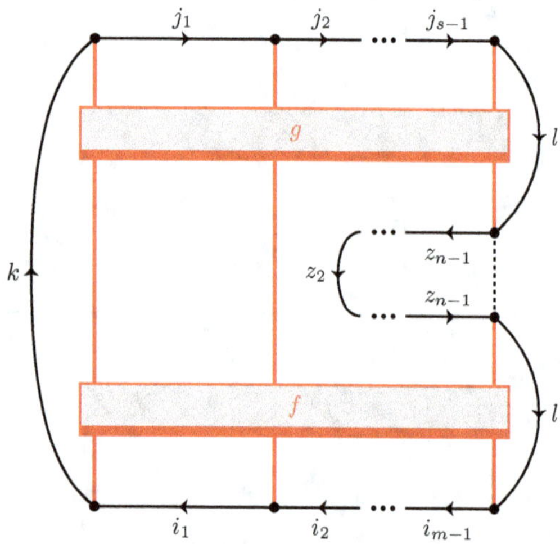

$$\cdots \overset{(F)}{=} \sum_{z_{n-1} \in I} \dim(z_{n-1})$$

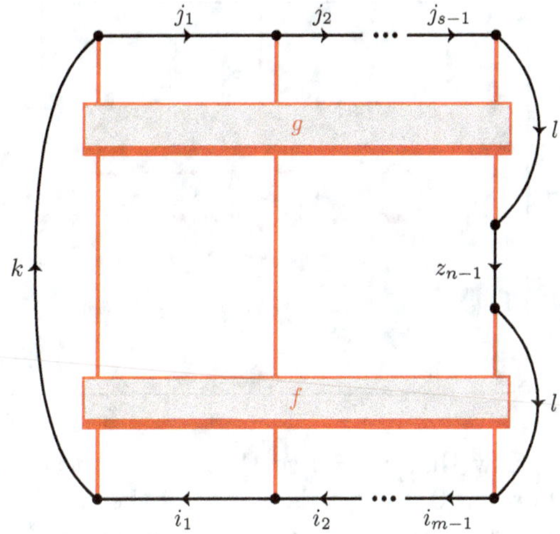

$$\overset{(F)}{=}$$

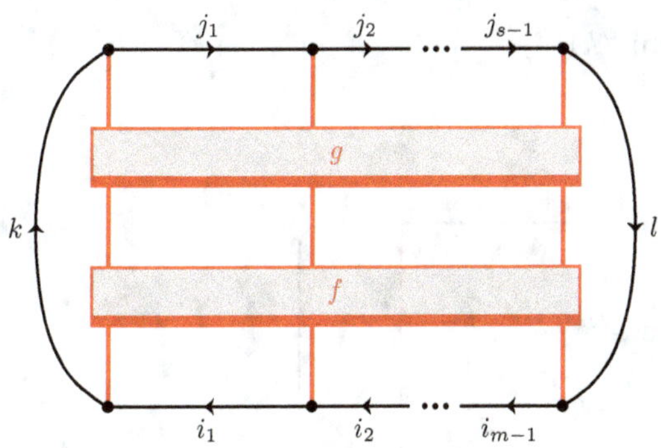

$$\overset{(C)}{=}$$

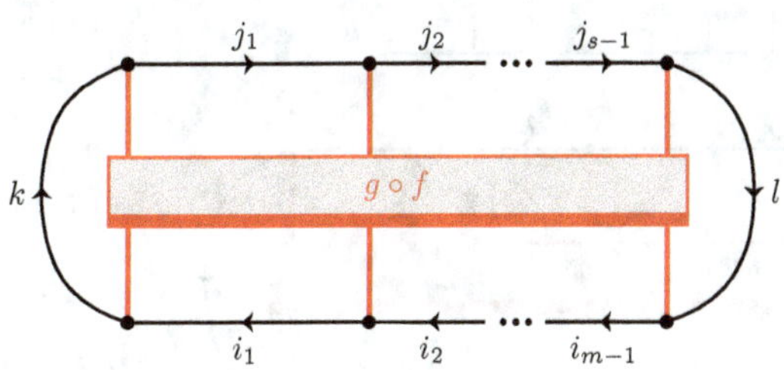

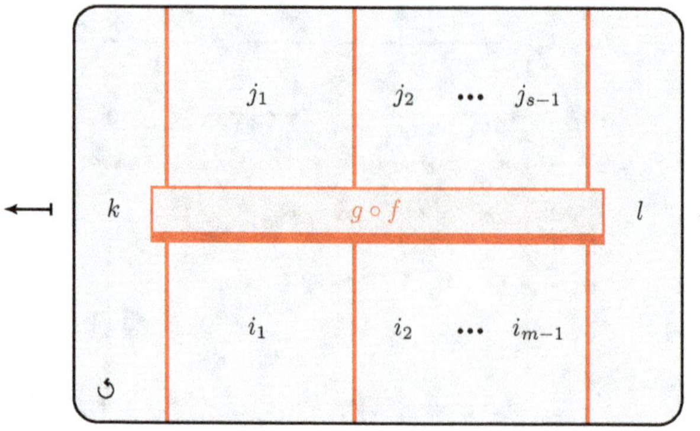

Assume now that $n = 0$ so that there are no strands between the coupons. If $m = 0$ or $s = 0$, then at least one of the coupons has no inputs and no outputs and the invariance follows from (15.7) using that $\mathrm{End}_{\mathcal{Z}(\mathcal{C})}(\mathbb{1}_{\mathcal{Z}(\mathcal{C})}) = \Bbbk$ as $\Bbbk$-algebras. Assume that $m \geq 1$ and $s \geq 1$. Let r be the face of d colored by $k \in I$ and let r', r'' be the faces of d' colored by $k, l \in I$ in the pictures below. Set

$$\zeta_{k,l} = \begin{cases} \dim(k)^{\chi(r')} \dim(l)^{\chi(r'')} & \text{if } r' \neq r'', \\ \dim(k)^{\chi(r')} & \text{if } r' = r''. \end{cases}$$

Then

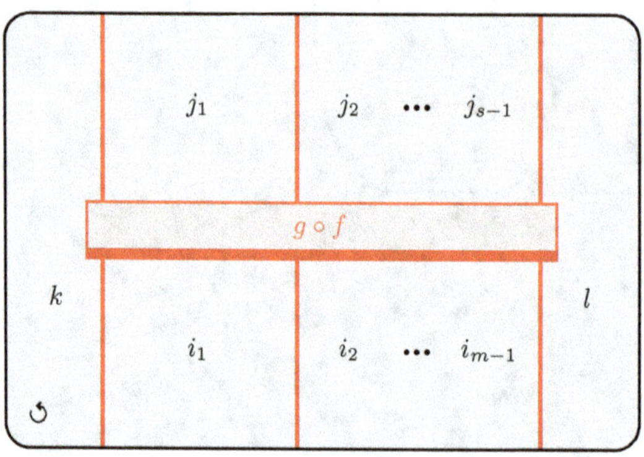

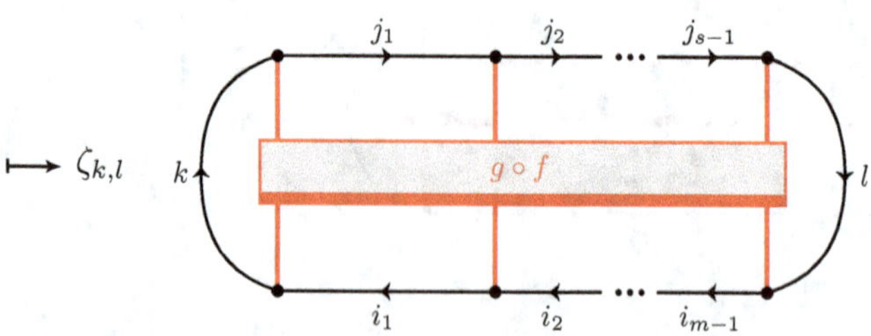

$\overset{(C)}{=} \zeta_{k,l}$

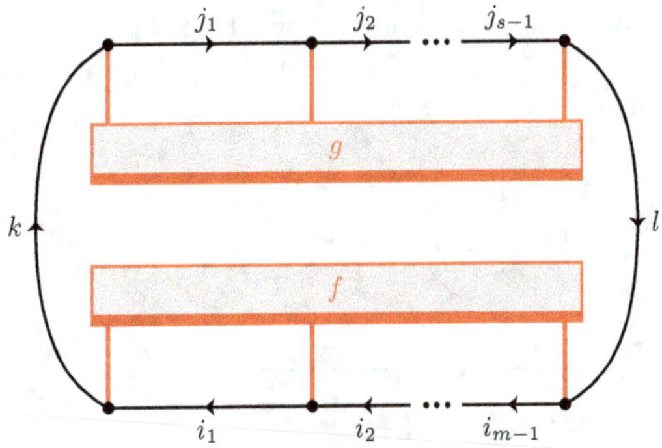

$\overset{(D,E)}{=} \delta_{k,l} \dim(k)^{-1} \zeta_{k,l}$

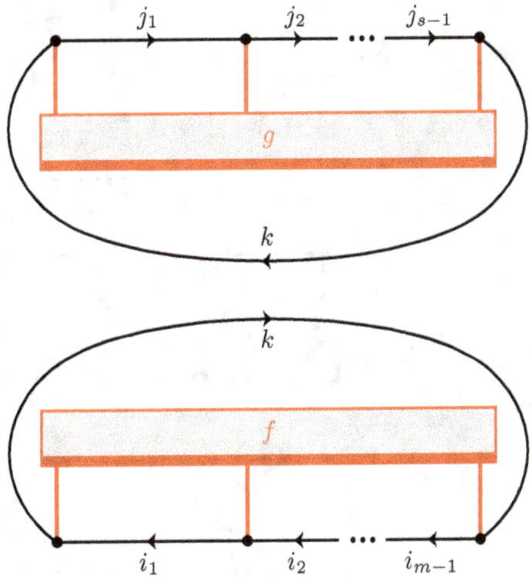

.

If $k \neq l$, then this contribution to the state sum is equal to zero. If $k = l$, then this contribution coincides with that of

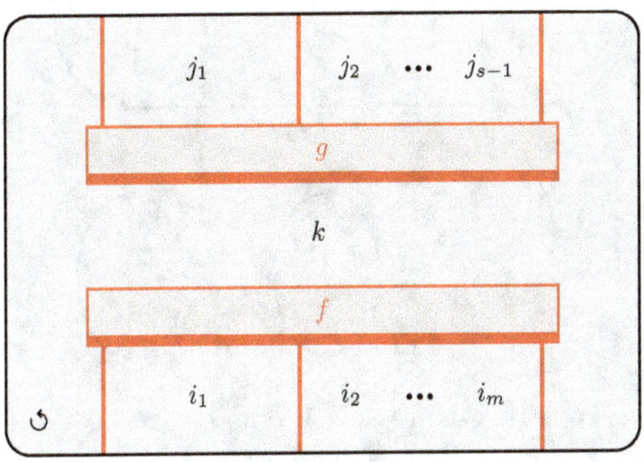

since $\dim(k)^{-1}\zeta_{k,k} = \dim(k)^{\chi(r)}$ by the additivity of the Euler characteristic.

Finally, we prove the invariance of the state sum under the transformation (15.9). Let m be the number of inputs (= the number of outputs) of the coupon at hand. If $m = 0$, then the invariance follows from (15.7). Assume that $m \geq 1$. For $s \in \{1, \ldots, m-1\}$, let r_s, r'_s be the faces of the diagram colored by $i_s, j_s \in I$, respectively, in the picture below. Set

$$\zeta_{i_s,j_s} = \begin{cases} \dim(i_s)^{\chi(r_s)} \dim(j_s)^{\chi(r'_s)} & \text{if } r_s \neq r'_s, \\ \dim(i_s)^{\chi(r_s)} & \text{if } r_s = r'_s. \end{cases}$$

Then

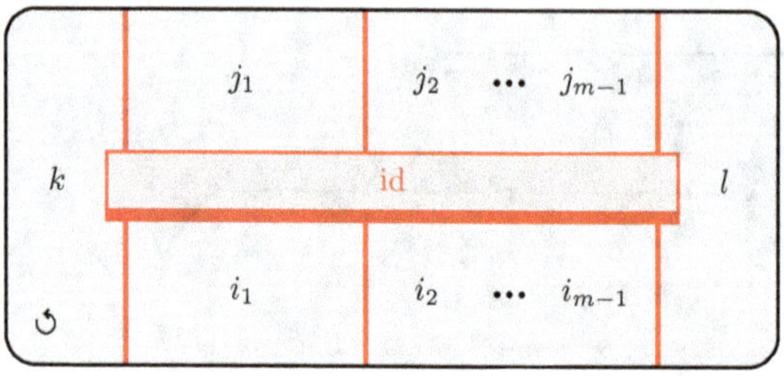

$$\longmapsto \prod_{s=1}^{m-1} \zeta_{i_s,j_s}$$

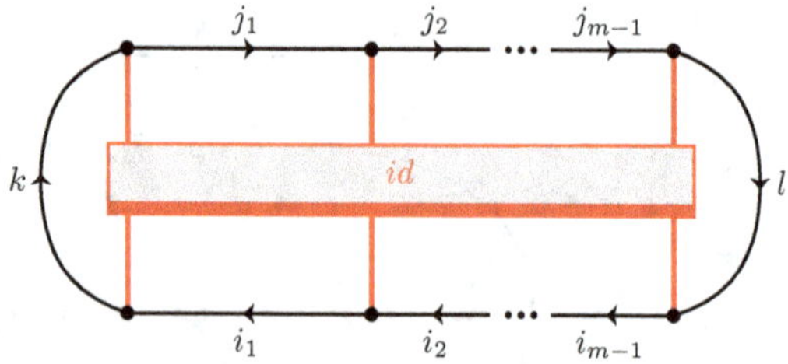

$$\overset{(C)}{=} \prod_{s=1}^{m-1} \zeta_{i_s,j_s}$$

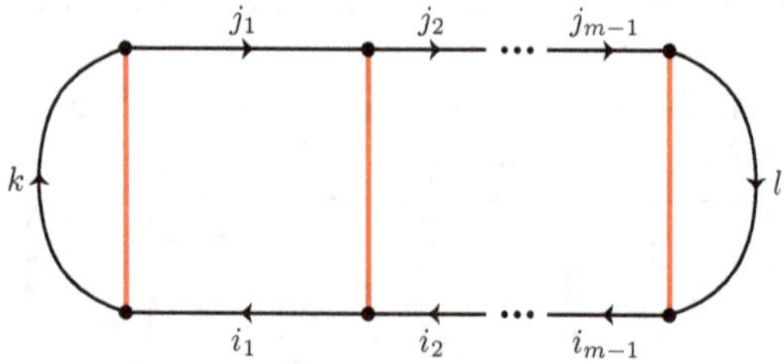

$$\overset{(D,E)}{=} \prod_{s=1}^{m-1} \delta_{i_s,j_s} \dim(i_s)^{-1} \zeta_{i_s,j_s}$$

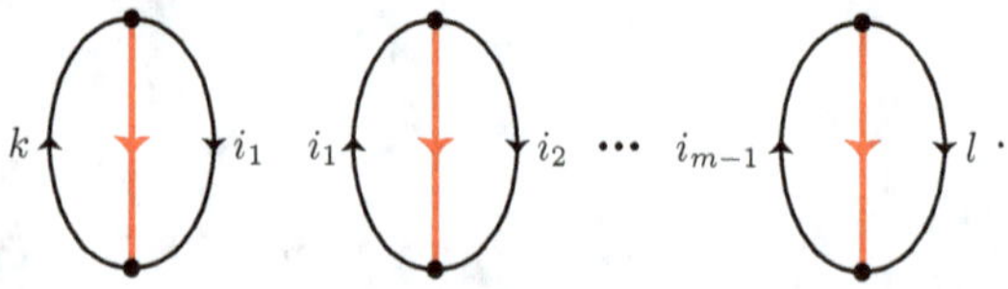

If $i_s \neq j_s$ for some s, then this expression is equal to zero. Assume that $i_s = j_s$ for all s. Then the contraction vector associated with the rim supported by the s-th input cancels the pairing associated with the corresponding theta graph. Also, we have $\dim(i_s)^{-1}\zeta_{i_s,i_s} = \dim(i_s)^{\chi(r''_s)}$, where r''_s is the face of the diagram (after the move) colored by i_s. Consequently, the contribution of the rims supported by the outputs of the coupon coincides with that of

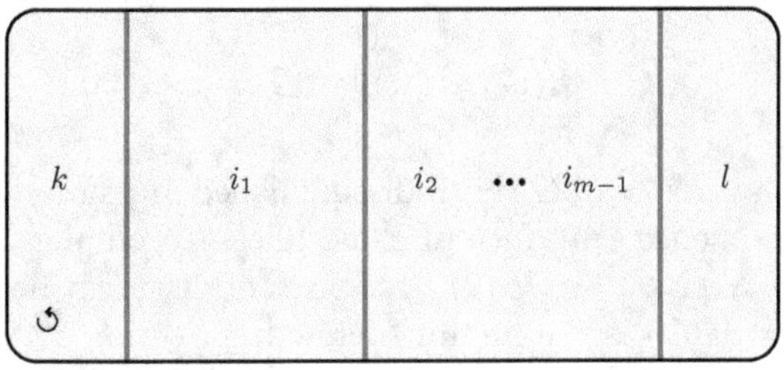

.

This implies the invariance of the state sum under the transformation (15.9) and concludes the proof of Theorem 15.7.

15.6 An invariant of ribbon triples

This is the second of three sections devoted to a construction of a graph TQFT over $\mathcal{Z}(\mathcal{C})$. Recall that $\mathcal{C}$ is a spherical fusion $\Bbbk$-category such that $\dim(\mathcal{C})$ is invertible in $\Bbbk$ and I is a representative set of simple objects of $\mathcal{C}$. We begin by generalizing the module associated to a $\mathcal{C}$-colored graph in a surface to so-called $\mathcal{C}$-enriched graphs. Then we introduce ribbon triples and their state sum invariants generalizing those of Theorem 15.7.

15.6.1 Enriched graphs

A vertex v of an oriented graph is *straight* if it is 2-valent and one of the two half-edges incident to v is oriented away from v while the other half-edge incident to v is oriented toward v. A *$\mathcal{C}$-enriched graph* G is a $\mathcal{C}$-colored graph such that some (possibly, none) straight vertices of G are decorated with a signed object of $\mathcal{C}$. These decorated vertices are called the *distinguished vertices* of G. Consider a distinguished vertex v of G decorated with (V,ε) where $V \in \mathrm{Ob}(\mathcal{C})$ and $\varepsilon \in \{+,-\}$. Let $e_{\mathrm{out}}, e_{\mathrm{in}}$ be the half-edges of G incident to v and oriented respectively away from v and toward v. Let $X_{\mathrm{out}}, X_{\mathrm{in}} \in \mathrm{Ob}(\mathcal{C})$ be their colors:

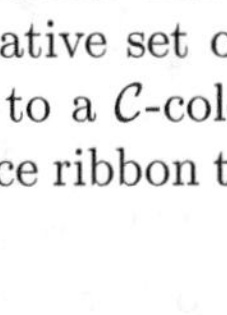

We turn the set $E_v = \{e_{\mathrm{out}}, v, e_{\mathrm{in}}\}$ into a cyclic $\mathcal{C}$-set with the cyclic order $e_{\mathrm{out}} < v < e_{\mathrm{in}} < e_{\mathrm{out}}$ and with the map $E_v \to \mathrm{Ob}(\mathcal{C}) \times \{+,-\}$ carrying e_{out} to $(X_{\mathrm{out}},-)$, v to (V,ε), and e_{in} to $(X_{\mathrm{in}},+)$. Set

$$\mathcal{H}_v(G) = H(E_v) \simeq \mathrm{Hom}_{\mathcal{C}}(\mathbb{1}, X_{\mathrm{out}}^* \otimes V^{\varepsilon} \otimes X_{\mathrm{in}}).$$

Given a $\mathcal{C}$-enriched graph G embedded in an oriented surface Σ, any non-distinguished vertex v of G determines a module $\mathcal{H}_v(G) = H_v(G)$ as in Section 12.2.1. Set

$$\mathcal{H}(G) = \bigotimes_v \mathcal{H}_v(G),$$

where v runs over all vertices of G (both distinguished and non-distinguished) and $\otimes$ is the non-ordered tensor product of $\Bbbk$-modules. To emphasize the role of Σ, we sometimes write $\mathcal{H}(G;\Sigma)$ for $\mathcal{H}(G)$. If the set of distinguished vertices of G is empty, then $\mathcal{H}(G) = H(G)$ is the module defined in Section 12.2.1. In particular, $\mathcal{H}(\emptyset) = H(\emptyset) = \Bbbk$. In general, the $\Bbbk$-module $\mathcal{H}(G)$ shares the properties of $H(G)$ stated in Section 12.2.1. For any disjoint $\mathcal{C}$-enriched graphs G, G' in Σ, there is a canonical $\Bbbk$-linear isomorphism

$$\mathcal{H}(G \sqcup G') \simeq \mathcal{H}(G) \otimes \mathcal{H}(G').$$

Any orientation preserving embedding f of Σ into an oriented surface Σ' carries a $\mathcal{C}$-enriched graph $G \subset \Sigma$ into a $\mathcal{C}$-enriched graph $G' = f(G) \subset \Sigma'$ preserving the vertices, the distinguished vertices and their decorations, as well as the orientations and the colors of the edges. The map f induces a $\Bbbk$-linear isomorphism $\mathcal{H}(f)\colon \mathcal{H}(G) \to \mathcal{H}(G')$ in the obvious way.

Given a $\mathcal{C}$-enriched graph G in an oriented surface Σ, we let G^{op} be the $\mathcal{C}$-enriched graph in $-\Sigma$ obtained from G by reversing orientation in all edges while keeping their colors and changing the decoration (V,ε) of each distinguished vertex to $(V,-\varepsilon)$. The cyclic $\mathcal{C}$-set determined by a vertex (distinguished or not) v of G is dual to the cyclic $\mathcal{C}$-set determined by v as a vertex of G^{op}. Section 12.3.2 yields a pairing in $\mathrm{Mod}_{\Bbbk}$

$$\mathcal{H}_v(G^{\mathrm{op}}) \otimes \mathcal{H}_v(G) \to \mathrm{End}_{\mathcal{C}}(\mathbb{1}) = \Bbbk.$$

Since $\mathcal{C}$ is non-degenerate (by Lemma 4.3), this pairing is non-degenerate and induces $\Bbbk$-linear isomorphisms

$$\mathcal{H}_v(G^{\mathrm{op}}) \simeq \mathcal{H}_v(G)^\star = \mathrm{Hom}_{\Bbbk}(\mathcal{H}_v(G),\Bbbk) \quad \text{and} \quad \mathcal{H}(G^{\mathrm{op}}) \simeq \mathcal{H}(G)^\star.$$

15.6.2 Ribbon triples

A *ribbon triple* (M,G,R) consists of

- a compact oriented 3-manifold M,
- an oriented graph $G \subset \partial M$ whose all vertices have valence ≥ 2,
- a $\mathcal{Z}(\mathcal{C})$-colored ribbon graph R in (M,G).

In particular, the set ∂R of free ends of R is finite and all its points lie inside the edges of G. These points viewed as new vertices subdivide G into a graph denoted G_R. The orientation of G induces an orientation of G_R, and the embedding $G \subset \partial M$ induces an embedding $G_R \subset \partial M$. The distinguished vertices of G_R are the points of ∂R. By construction, each distinguished vertex v of G_R is straight in the sense of Section 15.6.1 and is adjacent to a single strand e_v of R. We decorate v with the pair (V_v, ε_v) where $V_v \in \mathrm{Ob}(\mathcal{C})$ is the image of the $\mathcal{Z}(\mathcal{C})$-color of e_v under the forgetful functor $\mathcal{Z}(\mathcal{C}) \to \mathcal{C}$, while $\varepsilon_v = +$ if e_v is directed away from v and $\varepsilon_v = -$ if e_v is directed toward v.

A ribbon triple (M, G, R) is *I-colored* whenever the graph G_R is I-colored. Then G_R is a $\mathcal{C}$-enriched graph. We define a state sum generalizing those of Sections 13.2.1, 15.5 and yielding for any I-colored ribbon triple (M, G, R) a vector

$$|M, G, R|_{\mathcal{C}} \in \mathcal{H}(G_R) = \mathcal{H}(G_R; \partial M).$$

If $R = \emptyset$, then $G_R = G$, $\mathcal{H}(G_R) = H(G)$, and, by definition, $|M, G, R|_{\mathcal{C}} = |M, G|_{\mathcal{C}}$ is the vector from Theorem 13.2. If $\partial M = \emptyset$, then $G_R = G = \emptyset$, $\mathcal{H}(G_R) = \Bbbk$, and $|M, G, R|_{\mathcal{C}} = |M, R|_{\mathcal{C}}$ is the scalar from Theorem 15.7. Assume now that $R \neq \emptyset$ and $\partial M \neq \emptyset$. By Theorem 14.4, the ribbon graph R can be represented by a neat positive diagram (P, d) in (M, G). Here P is a skeleton of (M, G) and d is a knotted plexus in P having only positive switches and no trivial components. The $\mathcal{Z}(\mathcal{C})$-coloring of R turns d into a $\mathcal{Z}(\mathcal{C})$-colored knotted plexus with $\partial d = \partial R \subset \partial P = G$. The vertices of P and the switches, crossings, coupons, and free ends of d are called the *nodes* of d. A node of d is *internal* if it lies in $\mathrm{Int}(M)$. Internal nodes are the vertices of P not lying in ∂P and the switches, crossings, and coupons of d. Since d has no trivial components, the complement of the nodes in $\widetilde{d} = d \cup P^{(1)} \subset P$ is a finite disjoint union of open intervals. Among these intervals are the interiors of the edges of the graph G_R. All other open intervals in question lie in $\mathrm{Int}(M)$ and their closures are called the *rims* of d. Each rim connects two nodes (possibly, equal) called the *endpoints* of the rim. The interiors of rims are open intervals lying in $P^{(1)} \setminus G \subset \mathrm{Int}(M)$ or in $d \setminus \partial d \subset \mathrm{Int}(M)$. Cutting P along $\widetilde{d}$, we obtain a compact surface (with interior $P \setminus \widetilde{d}$) whose connected components are called the *faces* of d. Every face of d is contained in a unique region of P, and so the orientation of P induces an orientation of the faces of d. The (finite) set of faces of d is denoted by $\mathrm{Fac}(d)$.

A *coloring* of the diagram (P, d) is a map $c\colon \mathrm{Fac}(d) \to I$ carrying the faces of d adjacent to the edges of the graph G_R to the given I-colors of these edges. The object $c(r) \in I$ assigned to a face r of d is called the *c-color* of r. For each coloring c of (P, d), we define a scalar $|c| \in \Bbbk$ as follows. By an *oriented rim* of d, we mean a rim of d endowed with an orientation. Let $\mathcal{E}$ be the set of all oriented rims of d. For $e \in \mathcal{E}$, we consider the $\Bbbk$-module $H_c(e) = H(P_e)$, where P_e is the cyclic $\mathcal{C}$-set defined as in Section 15.5. Set

$$H_c = \bigotimes_{e \in \mathcal{E}_0} H_c(e) \quad \text{and} \quad H_c^{\partial} = \bigotimes_{e \in \mathcal{E}_\partial} H_c(e),$$

where $\otimes$ is the unordered tensor product of $\Bbbk$-modules, $\mathcal{E}_0 \subset \mathcal{E}$ is the set of all oriented rims of d with tail endpoint in $\mathrm{Int}(M)$, and $\mathcal{E}_\partial = \mathcal{E} \setminus \mathcal{E}_0$. Observe that the tail endpoint $v(e)$ of any $e \in \mathcal{E}_\partial$ is a vertex of G_R, and there is a canonical $\Bbbk$-linear isomorphism $H_c(e) \simeq \mathcal{H}_{v(e)}(G_R)$. The formula $e \mapsto v(e)$ establishes a bijective correspondence between the elements of $\mathcal{E}_\partial$ and the vertices of G_R. This gives a canonical $\Bbbk$-linear isomorphism

$$H_c^\partial = \bigotimes_{e \in \mathcal{E}_\partial} H_c(e) \simeq \bigotimes_v \mathcal{H}_v(G_R) = \mathcal{H}(G_R).$$

As in Section 13.1.1, any unoriented rim e of d gives rise to two opposite oriented rims e_1, e_2 whose associated cyclic $\mathcal{C}$-sets P_{e_1} and P_{e_2} are dual to each other. This determines a contraction vector $*_e \in H_c(e_1) \otimes H_c(e_2)$. The unordered tensor product of these vectors over all unoriented rims of d is a vector

$$\otimes_e *_e \in \bigotimes_{e \in \mathcal{E}} H_c(e).$$

We let $*_c \in H_c \otimes H(G_R)$ be the image of this vector under the isomorphisms

$$\bigotimes_{e \in \mathcal{E}} H_c(e) = H_c \otimes H_c^\partial \simeq H_c \otimes \mathcal{H}(G_R).$$

As in Section 15.5.1, each internal node x of d determines a knotted $\mathcal{C}$-net Γ_x^c in an oriented 2-sphere. By Section 15.4.7, this gives a vector $\mathbb{F}_\mathcal{C}(\Gamma_x^c) \in H(\Gamma_x^c)^\star$. It results from the definitions that we have canonical isomorphisms

$$H(\Gamma_x^c) \simeq \bigotimes_{e_x} H_c(e_x) \quad \text{and} \quad H(\Gamma_x^c)^\star \simeq \bigotimes_{e_x} H_c(e_x)^\star$$

where e_x runs over the rims of d incident to x and oriented away from x. The tensor product of the latter isomorphisms over all x is an isomorphism

$$\bigotimes_x H(\Gamma_x^c)^\star \simeq \bigotimes_x \bigotimes_{e_x} H_c(e_x)^\star \simeq H_c^\star.$$

Let $V_c \in H_c^\star$ be the image under this isomorphism of the vector $\bigotimes_x \mathbb{F}_\mathcal{C}(\Gamma_x)$. Set

$$|c| = (V_c \otimes \mathrm{id}_{\mathcal{H}(G_R)})(*_c) \in \Bbbk \otimes \mathcal{H}(G_R) \simeq \mathcal{H}(G_R).$$

Finally, set

$$|M, G, R|_\mathcal{C} = (\dim(\mathcal{C}))^{-|M \setminus P|} \sum_c \dim(c)\, |c| \in \mathcal{H}(G_R),$$

where c runs over all colorings of (P, d), the scalar $\dim(c) \in \Bbbk$ is defined by (15.6), and $|M \setminus P|$ is the number of connected components of $M \setminus P$.

Theorem 15.8. *The vector $|M, G, R|_\mathcal{C}$ is a topological invariant of the I-colored ribbon triple (M, G, R) independent of the choice of the diagram (P, d). Also, $|M, G, R|_\mathcal{C}$ is invariant under the transformations of R shown in Figure* 15.1.

Proof. The topological invariance of $|M, G, R|_{\mathcal{C}}$ means that for any orientation preserving homeomorphism f of M onto a compact oriented 3-manifold M', we have $|M', G', R'|_{\mathcal{C}} = \mathcal{H}(f)(|M, G, R|_{\mathcal{C}})$ where (M', G', R') is the I-colored ribbon triple obtained from (M, G, R) by transporting G, R along f and $\mathcal{H}(f)\colon \mathcal{H}(G_R) \to \mathcal{H}(G'_{R'})$ is the isomorphism induced by the homeomorphism $f|_{\partial M}\colon \partial M \to \partial M'$ (see Section 15.6.1). By Theorem 14.4, to prove the topological invariance of the state sum $|M, G, R|_{\mathcal{C}}$, it suffices to verify its invariance under the moves $\{\omega_*\}_{*=0}^{9}$ in the class of neat positive diagrams in (M, G) representing R. These moves involve only internal nodes and the invariance is verified in exactly the same way as in the proof of Theorem 15.7. The invariance under the transformations of R shown in Figure 15.1 is checked as in the proof of Theorem 15.7. □

The properties of the state sum invariant stated after Theorem 13.2 directly extend to ribbon triples. In particular, the vector $|M, G, R|_{\mathcal{C}}$ is multiplicative: the disjoint union $(M_1 \sqcup M_2, G_1 \sqcup G_2, R_1 \sqcup R_2)$ of I-colored ribbon triples (M_1, G_1, R_1) and (M_2, G_2, R_2) is an I-colored ribbon triple in the obvious way and

$$|M_1 \sqcup M_2, G_1 \sqcup G_2, R_1 \sqcup R_2|_{\mathcal{C}} = |M_1, G_1, R_1|_{\mathcal{C}} \otimes |M_2, G_2, R_2|_{\mathcal{C}}$$

up to the canonical isomorphism

$$\mathcal{H}\big((G_1 \sqcup G_2)_{R_1 \sqcup R_2}\big) = \mathcal{H}\big((G_1)_{R_1} \sqcup (G_2)_{R_2}\big) \simeq \mathcal{H}\big((G_1)_{R_1}\big) \otimes \mathcal{H}\big((G_2)_{R_2}\big).$$

If R is a $\mathcal{Z}(\mathcal{C})$-colored ribbon graph with no free ends in a compact oriented 3-manifold M, then $(M, \emptyset, R)$ is a ribbon triple and

$$|M, R|_{\mathcal{C}} = |M, \emptyset, R|_{\mathcal{C}} \in \mathcal{H}(\emptyset; \partial M) = \Bbbk$$

is a scalar topological invariant of R. In the case $\partial M = \emptyset$, this is the same scalar as in Theorem 15.7.

15.7 Construction of the state sum graph TQFT

We construct a regular graph TQFT $|\cdot|_{\mathcal{C}}\colon \mathrm{Cob}_3^{\mathcal{Z}(\mathcal{C})} \to \mathrm{Mod}_{\Bbbk}$. As above, $\mathcal{C}$ is a spherical fusion $\Bbbk$-category such that $\dim(\mathcal{C})$ is invertible in $\Bbbk$ and I is a representative set of simple objects of $\mathcal{C}$. Our construction follows the same lines as the construction of the state sum TQFT in Section 13.3 but uses the invariant of ribbon triples from Section 15.6.

15.7.1 The module $|G|^\circ$

Recall from Section 15.1.1 that a $\mathcal{Z}(\mathcal{C})$-colored surface (Σ, A) is formed by a closed oriented surface Σ and a finite set $A \subset \Sigma$ whose points are equipped with a tangent direction in Σ, an object of $\mathcal{Z}(\mathcal{C})$, and a sign. Let G be an oriented graph embedded in Σ such that

(i) each point a of A lies in the interior of an edge e_a of G transversal to the given tangent direction at a;

(ii) for all $a \in A$, the orientation of e_a followed by the tangent direction at a determine the positive orientation of Σ.

The points of A, viewed as new vertices, subdivide G into a graph G_A. The orientation of G induces an orientation of G_A, and the embedding $G \subset \Sigma$ induces an embedding $G_A \subset \Sigma$. Let $\mathrm{col}(G_A)$ be the set of all maps from the set of edges of G_A to I. Each such map c turns G_A into a $\mathcal{C}$-enriched graph G_A^c whose distinguished vertices are the points of A decorated with the images of their $\mathcal{Z}(\mathcal{C})$-colors under the forgetful functor $\mathcal{Z}(\mathcal{C}) \to \mathcal{C}$ and with their signs. Section 15.6.1 yields a $\Bbbk$-module $\mathcal{H}(G_A^c) = \mathcal{H}(G_A^c; \Sigma)$. Set

$$|G|^\circ = \bigoplus_{c \in \mathrm{col}(G_A)} \mathcal{H}(G_A^c).$$

We will apply this construction to skeletons of (Σ, A). A *skeleton* of (Σ, A) is an oriented graph G embedded in Σ, satisfying the conditions (i), (ii) above and such that all vertices of G have valence ≥ 2 and all components of $\Sigma \setminus G$ are open disks.

15.7.2 An operator invariant of morphisms

Let $\Sigma_0 = (\Sigma_0, A_0)$ and $\Sigma_1 = (\Sigma_1, A_1)$ be two $\mathcal{Z}(\mathcal{C})$-colored surfaces. Consider a morphism $f \colon \Sigma_0 \to \Sigma_1$ in $\mathrm{Cob}_3^{\mathcal{Z}(\mathcal{C})}$ represented by a triple (M, R, h), where M is a compact oriented 3-manifold, R is a $\mathcal{Z}(\mathcal{C})$-colored ribbon graph in M, and

$$h \colon (-\Sigma_0) \sqcup \Sigma_1 \to (\partial M, \partial R)$$

is a homeomorphism of $\mathcal{Z}(\mathcal{C})$-colored surfaces. For any skeletons G_0 of Σ_0 and G_1 of Σ_1, consider the $\Bbbk$-linear homomorphism

$$|f, G_0, G_1|^\circ \colon |G_0|^\circ = \bigoplus_{c_0 \in \mathrm{col}((G_0)_{A_0})} \mathcal{H}((G_0)_{A_0}^{c_0}) \to |G_1|^\circ = \bigoplus_{c_1 \in \mathrm{col}((G_1)_{A_1})} \mathcal{H}((G_1)_{A_1}^{c_1})$$

whose restriction to $\mathcal{H}((G_0)_{A_0}^{c_0})$ and corestriction to $\mathcal{H}((G_1)_{A_1}^{c_1})$ is the $\Bbbk$-linear homomorphism

$$|f, c_0, c_1|^\circ \colon \mathcal{H}((G_0)_{A_0}^{c_0}) \to \mathcal{H}((G_1)_{A_1}^{c_1})$$

defined as follows. Consider the $\mathcal{Z}(\mathcal{C})$-colored surface $-\Sigma_0$ opposite to Σ. The oriented graph G_0^{op} in $-\Sigma_0$, obtained from G_0 by reversing orientation of all edges, is a skeleton of $-\Sigma_0$, and the oriented graph $G = G_0^{\mathrm{op}} \sqcup G_1$ is a skeleton of $(-\Sigma_0) \sqcup \Sigma_1$. Then $h(G)$ is a skeleton of the $\mathcal{Z}(\mathcal{C})$-colored surface $(\partial M, \partial R)$ and

$(M, h(G), R)$ is a ribbon triple. The maps c_0, c_1 turn $(M, h(G), R)$ into an I-colored ribbon triple. Theorem 15.8 gives a vector

$$\begin{aligned}
&|M, h(G), R|_{\mathcal{C}} \in \mathcal{H}(h(G)_R; \partial M) \\
&\simeq \mathcal{H}(G^{c_0 \sqcup c_1}_{A_0 \sqcup A_1}; (-\Sigma_0) \sqcup \Sigma_1) \simeq \mathcal{H}((G_0^{\mathrm{op}})^{c_0}_{A_0}; -\Sigma_0) \otimes \mathcal{H}((G_1)^{c_1}_{A_1}; \Sigma_1) \\
&\simeq \mathcal{H}((G_0)^{c_0}_{A_0})^{\star} \otimes \mathcal{H}((G_1)^{c_1}_{A_1}) \simeq \mathrm{Hom}_{\Bbbk}\left(\mathcal{H}((G_0)^{c_0}_{A_0}), \mathcal{H}((G_1)^{c_1}_{A_1})\right).
\end{aligned}$$

Set

$$|f, c_0, c_1|^{\circ} = \frac{(\dim(\mathcal{C}))^{|\Sigma_1 \setminus G_1|}}{\dim(c_1)} \, |M, h(G)_R, R|_{\mathcal{C}} \colon \mathcal{H}((G_0)^{c_0}_{A_0}) \to \mathcal{H}((G_1)^{c_1}_{A_1}),$$

where $|\Sigma_1 \setminus G_1|$ is the number of components of $\Sigma_1 \setminus G_1$ and $\dim(c_1)$ is the product of the dimensions of the simple objects of $\mathcal{C}$ associated by c_1 with the edges of $(G_1)_{A_1}$. Theorem 15.8 implies that $|f, c_0, c_1|^{\circ}$ and

$$|f, G_0, G_1|^{\circ} = \sum_{c_0, c_1} |f, c_0, c_1|^{\circ},$$

do not depend on the choice of the triple (M, R, h) representing f and are invariant under the transformations of R shown in Figure 15.1.

Lemma 13.3 generalizes to this setting: for any morphisms $f_0 \colon \Sigma_0 \to \Sigma$ and $f_1 \colon \Sigma \to \Sigma_1$ in $\mathrm{Cob}_3^{\mathcal{Z}(\mathcal{C})}$ and any skeletons G_0, G, G_1 of the $\mathcal{Z}(\mathcal{C})$-colored surfaces $\Sigma_0, \Sigma, \Sigma_1$, respectively, we have

$$|f_1 f_0, G_0, G_1|^{\circ} = |f_1, G, G_1|^{\circ} \circ |f_0, G_0, G|^{\circ}. \tag{15.10}$$

Lemma 13.4 also generalizes to this setting: for any morphisms $f \colon \Sigma_0 \to \Sigma_1$ and $f' \colon \Sigma'_0 \to \Sigma'_1$ in $\mathrm{Cob}_3^{\mathcal{Z}(\mathcal{C})}$ and any skeletons G_i, G'_i of the $\mathcal{Z}(\mathcal{C})$-colored surfaces Σ_i, Σ'_i with $i \in \{0, 1\}$, we have

$$|f \otimes f', G_0 \sqcup G'_0, G_1 \sqcup G'_1|^{\circ} = |f, G_0, G_1|^{\circ} \otimes |f', G'_0, G'_1|^{\circ} \tag{15.11}$$

up to the canonical isomorphisms $|G_i \sqcup G'_i|^{\circ} \simeq |G_i|^{\circ} \otimes |G'_i|^{\circ}$ with $i \in \{0, 1\}$.

15.7.3 The graph TQFT

The constructions above assign a $\Bbbk$-module to every $\mathcal{Z}(\mathcal{C})$-colored surface endowed with a skeleton and assign a homomorphism of these modules to every morphism in $\mathrm{Cob}_3^{\mathcal{Z}(\mathcal{C})}$ whose source and target are endowed with skeletons. By (15.10), this assignment satisfies all requirements on a functor from an appropriate version of $\mathrm{Cob}_3^{\mathcal{Z}(\mathcal{C})}$ to $\mathrm{Mod}_{\Bbbk}$ except one: it may carry the identity morphisms to non-identity homomorphisms. We transform this assignment into a functor $|\cdot|_{\mathcal{C}} \colon \mathrm{Cob}_3^{\mathcal{Z}(\mathcal{C})} \to \mathrm{Mod}_{\Bbbk}$ by the procedure explained in Section 13.3.2 (this procedure gets rid of the skeletons of surfaces). Note that for any skeleton G of a $\mathcal{Z}(\mathcal{C})$-colored surface Σ, we have a cone isomorphism

$$|\Sigma|_{\mathcal{C}} \simeq \mathrm{Im}(|\mathrm{id}_\Sigma, G, G|^\circ \colon |G;\Sigma|^\circ \to |G;\Sigma|^\circ).$$

We endow $|\cdot|_{\mathcal{C}}$ with monoidal constraints as in Section 13.3.2. It follows from the property (15.11) that $|\cdot|_{\mathcal{C}}$ is a symmetric strong monoidal functor. It follows from the invariance of $|f, G_0, G_1|^\circ$ under the transformations in Figure 15.1 (see Section 15.7.2) that the graph TQFT $|\cdot|_{\mathcal{C}}$ is regular in the sense of Section 15.2.3. Also, it follows from the definitions that the restriction of $|\cdot|_{\mathcal{C}}$ to Cob_3 is the state sum TQFT defined in Chapter 13. We summarize these results in the following theorem.

Theorem 15.9. *The functor $|\cdot|_{\mathcal{C}}$ is a regular graph TQFT over $\mathcal{Z}(\mathcal{C})$ whose restriction to* Cob_3 *is the TQFT of Theorem* 13.5 *derived from $\mathcal{C}$.*

The naturality of $\mathbb{F}_{\mathcal{C}}$ implies that the isomorphism type of the state sum graph TQFT $|\cdot|_{\mathcal{C}}$ does not depend on the choice of the representative set I of simple objects of $\mathcal{C}$.

Theorem 15.9 directly implies Theorem 15.2.

15.7.4 Remark

Equivalent spherical fusion $\Bbbk$-categories with invertible dimension give rise to isomorphic graph TQFTs. More precisely, let $\mathcal{C}$ and $\mathcal{D}$ be equivalent spherical fusion $\Bbbk$-categories with invertible dimension. Pick any pivotal $\Bbbk$-linear equivalence $\mathcal{C} \to \mathcal{D}$. By Exercise 5.2.3, this equivalence induces a braided pivotal $\Bbbk$-linear equivalence $F\colon \mathcal{Z}(\mathcal{C}) \to \mathcal{Z}(\mathcal{D})$. By Exercise 15.2.5, the latter induces a symmetric strict monoidal equivalence $\widetilde{F}\colon \mathrm{Cob}_3^{\mathcal{Z}(\mathcal{C})} \to \mathrm{Cob}_3^{\mathcal{Z}(\mathcal{D})}$. It follows from the definitions that the graph TQFTs $|\cdot|_{\mathcal{C}}$ and $|\cdot|_{\mathcal{D}} \circ \widetilde{F}$ over $\mathcal{C}$ are isomorphic.

Chapter 16

Properties of the state sum graph TQFT

We establish here several properties of the state sum graph TQFT. Two of these properties (Theorems 16.1 and 16.2) will be used in the next chapter.

Throughout the chapter, $\mathcal{C}$ is a spherical fusion $\Bbbk$-category such that $\dim(\mathcal{C})$ is invertible in $\Bbbk$.

16.1 Ribbon graphs in the 3-sphere

We compute the state sum invariants of $\mathcal{Z}(\mathcal{C})$-colored ribbon graphs in the sphere S^3 in terms of the braiding in $\mathcal{Z}(\mathcal{C})$. We also compute the state sum invariants of $\mathcal{Z}(\mathcal{C})$-colored ribbon graphs in connected sums of 3-manifolds.

16.1.1 Ribbon graphs in S^3

A ribbon graph $R \subset S^3$ colored over a ribbon category $\mathcal{B}$ determines an isotopy invariant $\langle R \rangle_{\mathcal{B}} \in \mathrm{End}_{\mathcal{B}}(\mathbb{1})$ as follows. Recall that R is formed by a finite number of rectangular coupons and 1-dimensional strands, each strand being either a circle or an arc with both endpoints in the bases of the coupons. Pushing R away from the point $\infty \in S^3 = \mathbb{R}^3 \cup \{\infty\}$, we can deform R into $\mathbb{R}^3$ so that the following conditions hold:

- the image of the projection of R to the coordinate plane $\mathbb{R}^2 = \mathbb{R}^2 \times \{0\}$ lies in $\mathbb{R} \times (0,1) \subset \mathbb{R}^2$;
- the only multiple points of the projection of R are double transversal intersections of the interiors of the strands;
- the orientation of the coupons matches the counterclockwise orientation of $\mathbb{R}^2$;

- the distinguished bases of the projections of the coupons are horizontal and lie below the opposite bases;
- the framing of R is given by the vector $(0, 0, 1)$ in all points of R.

Recording the over/under-crossings and orientations of the strands of R, as well as the $\mathcal{B}$-colors of the strands and the coupons, we turn the projection of R into a $\mathcal{B}$-colored Reidemeister diagram $\mathcal{R}$ representing R. The Reidemeister calculus of Section 3.2 yields a morphism $\langle \mathcal{R} \rangle \colon \mathbb{1} \to \mathbb{1}$ in $\mathcal{B}$. Since any planar $\mathcal{B}$-colored Reidemeister diagrams representing R are related by ambient isotopies and the moves $\{R_i^{\pm 1}\}_{i=1}^5$ of Figure 3.2, Theorem 3.3 implies that $\langle R \rangle_{\mathcal{B}} = \langle \mathcal{R} \rangle \in \mathrm{End}_{\mathcal{B}}(\mathbb{1})$ is an isotopy invariant of R.

Consider now a $\mathcal{Z}(\mathcal{C})$-colored ribbon graph $R \subset S^3$ and the state sum invariant $|S^3, R|_{\mathcal{C}} \in \Bbbk$ from Theorem 15.7. Since $\mathcal{Z}(\mathcal{C})$ is a ribbon category (by Lemma 5.2), the constructions above yield the isotopy invariant

$$\langle R \rangle_{\mathcal{Z}(\mathcal{C})} \in \mathrm{End}_{\mathcal{Z}(\mathcal{C})}(\mathbb{1}_{\mathcal{Z}(\mathcal{C})}) = \mathrm{End}_{\mathcal{C}}(\mathbb{1}) = \Bbbk.$$

The following theorem relates these invariants.

Theorem 16.1. *We have:* $|S^3, R|_{\mathcal{C}} = (\dim(\mathcal{C}))^{-1} \langle R \rangle_{\mathcal{Z}(\mathcal{C})}$.

Proof. Set $\langle R \rangle = \langle R \rangle_{\mathcal{Z}(\mathcal{C})} \in \Bbbk$. If $R = \emptyset$, then $\langle R \rangle = 1_{\Bbbk}$ and the claim follows from the equality $|S^3|_{\mathcal{C}} = (\dim(\mathcal{C}))^{-1}$, see Section 13.1.3. If R has a coupon Q with no inputs and no outputs, then Q is colored with a scalar $k \in \mathrm{End}_{\mathcal{Z}(\mathcal{C})}(\mathbb{1}) = \Bbbk$. Deleting Q from R, we obtain a $\mathcal{Z}(\mathcal{C})$-colored ribbon graph $R' \subset S^3$. Formula (15.7) gives that $|S^3, R|_{\mathcal{C}} = k\, |S^3, R'|_{\mathcal{C}}$. Also, it follows from the definitions and the $\Bbbk$-linearity of $\mathcal{Z}(\mathcal{C})$ that $\langle R \rangle = k \, \langle R' \rangle$. Thus, if the claim holds for R', then it holds for R. Assume from now on that $R \neq \emptyset$ and that each coupon of R has at least one input or output.

Pick as in Section 16.1.1 a $\mathcal{Z}(\mathcal{C})$-colored Reidemeister diagram d representing R. Using the inclusions $\mathbb{R} \times (0,1) \subset \mathbb{R}^2 \subset S^2$, we can view d as a $\mathcal{Z}(\mathcal{C})$-colored knotted plexus (without free ends) in S^2. Recall that S^2 is a skeleton of S^3 without edges, see Example 11.2.3. The knotted plexus d has no switches and so the ribbon diagram (S^2, d) is positive. Since d represents R, so does the diagram (S^2, d).

Applying to d several moves R_2^{-1} as in Figure 3.2, we can ensure that d has no trivial components and is connected. Let $\underline{d} \subset \mathbb{R} \times (0,1)$ be the $\mathcal{Z}(\mathcal{C})$-colored Penrose diagram obtained from d by replacing all crossings by coupons as in Figure 3.1, where τ is the braiding in $\mathcal{Z}(\mathcal{C})$. Let $\underline{R}$ be the $\mathcal{Z}(\mathcal{C})$-colored ribbon graph in S^3 represented by $\underline{d}$. It follows from the definitions that $\langle R \rangle = \langle \underline{R} \rangle$. We can compute $|S^3, R|_{\mathcal{C}}$ and $|S^3, \underline{R}|_{\mathcal{C}}$ from the neat positive ribbon diagrams (S^2, d) and $(S^2, \underline{d})$. The definitions of the state sum (15.5) and of the invariant $\mathbb{F}_{\mathcal{C}}$ of knotted $\mathcal{C}$-nets (see Section 15.4) imply that, for any coloring $c \colon \mathrm{Fac}(d) = \mathrm{Fac}(\underline{d}) \to I$, the contribution $\mathbb{F}_{\mathcal{C}}(\Gamma_x^c)$ of any crossing x of d and the contribution $\mathbb{F}_{\mathcal{C}}(\Gamma_{\underline{x}}^c)$ of the corresponding coupon $\underline{x}$ of $\underline{d}$ are equal. So are the contributions of the faces and the rims. Consequently, $|S^3, R|_{\mathcal{C}} = |S^3, \underline{R}|_{\mathcal{C}}$. Thus, if the claim of the theorem holds for $\underline{R}$, then it holds for R. Hence, we can assume that d is a Penrose diagram (i.e., d has no crossings). Inserting, if necessary, in d and R several coupons with one input and one output, and colored with identity morphisms, we can additionally assume that d has no arcs connecting a coupon to itself. Indeed, such insertions preserve $\langle R \rangle$ as is clear from the Penrose calculus. They also preserve $|S^3, R|_{\mathcal{C}}$: the newly inserted coupons contribute duality pairings as in Example 12.3.6 and these pairings are cancelled by the contraction vectors associated with the rims adjacent to these coupons.

Pick a representative set I of simple objects of $\mathcal{C}$. Since all faces of the neat positive diagram (S^2, d) are disks, Formulas (15.5) and (15.6) with $M = S^3$ and $P = S^2$ give

$$|S^3, R|_{\mathcal{C}} = (\dim(\mathcal{C}))^{-2} \sum_{c\colon \mathrm{Fac}(d) \to I} \left(\prod_{r \in \mathrm{Fac}(d)} \dim c(r) \right) |c|.$$

To compute the right-hand side, pick a face $r_0 \in \mathrm{Fac}(d)$. For $i \in I$, set

$$S_i = \sum_{\substack{c\colon \mathrm{Fac}(d) \to I \\ c(r_0) = i}} \left(\prod_{r \in \mathrm{Fac}(d) \setminus \{r_0\}} \dim c(r) \right) |c|. \tag{16.1}$$

We claim that $S_i = \dim(i) \langle R \rangle$ for all $i \in I$. This will imply the theorem because

$$\begin{aligned} |S^3, R|_{\mathcal{C}} &= (\dim(\mathcal{C}))^{-2} \sum_{i \in I} \dim(i) S_i \\ &= (\dim(\mathcal{C}))^{-2} \sum_{i \in I} (\dim(i))^2 \langle R \rangle = (\dim(\mathcal{C}))^{-1} \langle R \rangle. \end{aligned}$$

To compute S_i, consider the Penrose diagram d in more detail. Recall that d is formed by coupons and disjoint embedded arcs with endpoints on the bases of the coupons. Let $x \geq 1$ be the number of coupons of d and $y \geq 1$ be the number of arcs of d. Let $z = \mathrm{card}(\mathrm{Fac(d)}) \geq 1$ be the number of connected components of $S^2 \setminus d$. Then $x - y + z = \chi(S^2) = 2$. As a consequence, $y = x + z - 2 \geq x - 1$.

We define a graph $d^* \subset S^2$ *dual* to $d \subset S^2$. Pick a point in the interior of each face of d and call it the *center* of the face. These z points are the vertices of d^*. Every arc e of d determines an *edge* e^* of d^* which connects the centers of the faces of d adjacent to e, meets the interior of e transversely in one point, and is disjoint from d otherwise. The edge e^* is either an embedded arc or an embedded loop in S^2. The edges of d^* are chosen so that they meet only in common vertices of d^*.

By a *subgraph* of a graph G we mean a graph formed by some vertices and edges of G. A subgraph of G is *full* if all vertices of G belong to this subgraph. A *tree* is a connected graph without cycles or, equivalently, a connected graph whose number of vertices exceeds the number of edges by 1. We apply the same terminology to d treating the coupons of d as vertices and the arcs of d as edges.

CLAIM. *The arcs of d can be enumerated $e_1, \ldots, e_y$ so that:*

(a) *the full subgraph of d with the edges $e_1, \ldots, e_{x-1}$ is a tree,*

(b) *for all $x \leq k \leq y$, the edges $e_k^*, \ldots, e_y^*$ and their vertices form a tree in d^*,*

(c) *the arc e_y of d is adjacent to the face r_0 of d.*

Let us prove this claim. Since d is connected, it contains a full subgraph F which is a tree. Since d and F have x vertices, F has $x-1$ edges which we enumerate $e_1, \ldots, e_{x-1}$ in an arbitrary way. This yields (a). If $y = x - 1$, then the condition (b) is void while (c) holds because $z = y - x + 2 = 1$ so that d has only one face r_0. Suppose that $y \geq x$. We must enumerate the remaining arcs of d. Removing (the interiors of) the edges $e_1^*, \ldots, e_{x-1}^*$ from d^* we obtain a full subgraph, H, of d^*. Clearly, $F \cap H = \emptyset$. Observe that every component of $S^2 \setminus H$ contains a vertex of d. Since any two vertices of d can be related by a path in $F \subset S^2 \setminus H$, the set $S^2 \setminus H$ is connected. Hence, H has no cycles and is a disjoint union of trees. Since H has z vertices and $y - (x-1) = z - 1$ edges, H is a single tree. In particular, the edges of H are not loops, i.e., have distinct vertices. Let O be the vertex of d^* represented by the center of the face r_0. We enumerate $h_1, \ldots, h_{z-1}$ the edges of H so that h_1 is incident to O and for all $t = 1, \ldots, z-1$, the edges $h_1, \ldots, h_t$ and their vertices form a tree. For $x \leq k \leq y$, we let e_k be the arc of d such that $e_k^* = h_{y+1-k}$. In particular, $e_x^* = h_{z-1}$ and $e_y^* = h_1$. It is clear that the conditions (b) and (c) above are met. This concludes the proof of the claim.

We now fix $i \in I$ and compute S_i. Recall that the term $|c|$ in the right-hand side of (16.1) is obtained by trading all coupons of d for knotted $\mathcal{C}$-nets in S^2, taking the tensor product of the associated invariants $\mathbb{F}_{\mathcal{C}}$, and tensor contracting it along the arcs of d. The colors of the arcs of these knotted $\mathcal{C}$-nets are determined by the map c and the given $\mathcal{Z}(\mathcal{C})$-coloring of d. To compute the right-hand side of (16.1), we shall perform the tensor contraction along one arc of d at a time in the order $e_1, \ldots, e_y$ as in Claim above. Condition (a) shows that for $k = 1, \ldots, x-1$, the tensor contraction at e_k involves two disjoint knotted $\mathcal{C}$-nets so that we can apply Lemma 12.6. At each of these $x-1$ steps, the tensor contraction of the tensor product of two $\mathbb{F}_{\mathcal{C}}$-invariants yields the $\mathbb{F}_{\mathcal{C}}$-invariant of the "fused" knotted $\mathcal{C}$-net in S^2. Condition (b) implies that for $k \geq x$, the edge e_k^* of d^* is not a loop and has a vertex which is not incident to the edges $e_{k+1}^*, \ldots, e_y^*$ and is distinct from the center O of r_0. This vertex is the center of a face, $r_k \in \mathrm{Fac}(d) \setminus \{r_0\}$, adjacent to e_k. All other arcs of d adjacent to r_k must belong to the list $e_1, \ldots, e_{k-1}$. Therefore the fusions along these arcs were effected before the k-th step. All knotted $\mathcal{C}$-nets associated with the vertices of r_k have an edge in r_k. Under the fusions in question, these edges are united together into a single $c(r_k)$-colored arc with the same endpoints u_k and v_k as e_k. Note that $u_k \neq v_k$ because all edges of d have distinct endpoints. We can now apply Lemma 12.5 where $u = u_k$, $v = v_k$, and i is replaced by $c(r_k) \in I$. This procedure eliminates the $c(r_k)$-colored arc and removes the factor $\dim c(r_k)$ on the right-hand side of (16.1). The faces $r_0, r_x, \dots, r_y$ are pairwise distinct and their number $y - x + 2 = z$ is equal to the number of faces of d. Thus, every face of d appears in this list precisely once. The final step ($k = y$) produces a "fused" knotted $\mathcal{C}$-net D in S^2 such that $S_i = \mathbb{F}_\mathcal{C}(D)$. This knotted $\mathcal{C}$-net is a union of d and an i-colored circle embedded in r_0. Therefore

$$S_i = \mathbb{F}_\mathcal{C}(D) = \dim(i)\, \mathbb{F}_\mathcal{C}(d) = \dim(i)\, \langle d \rangle = \dim(i)\, \langle R \rangle.$$

This concludes the proof of the theorem. □

16.1.2 Ribbon graphs in connected sums

Let R_0, R_1 be $\mathcal{Z}(\mathcal{C})$-colored ribbon graphs in closed connected oriented 3-manifolds M_0, M_1, respectively. Using 3-balls in $M_0 \setminus R_0$ and $M_1 \setminus R_1$ to construct the connected sum $M_0 \# M_1$, we obtain a $\mathcal{Z}(\mathcal{C})$-colored ribbon graph $R_0 \cup R_1$ in $M_0 \# M_1$. Then

$$|M_0 \# M_1, R_0 \cup R_1|_\mathcal{C} = \dim(\mathcal{C})\, |M_0, R_0|_\mathcal{C}\, |M_1, R_1|_\mathcal{C} \in \Bbbk. \tag{16.2}$$

The proof goes by extending Lemma 10.2 to graph TQFTs and applying it to the state sum graph TQFT $|\cdot|_\mathcal{C}$. Formula (16.2) easily generalizes to $\mathcal{Z}(\mathcal{C})$-colored ribbon graphs in connected sums of 3-cobordisms.

Taking $R_0 = \emptyset$, $M_1 = S^3$ in (16.2) and using Theorem 16.1, we obtain that for any closed connected oriented 3-manifold M and a $\mathcal{Z}(\mathcal{C})$-colored ribbon graph R in S^3,

$$|M, R_M|_\mathcal{C} = \dim(\mathcal{C})\, |M|_\mathcal{C}\, |S^3, R|_\mathcal{C} = |M|_\mathcal{C}\, \langle R \rangle_{\mathcal{Z}(\mathcal{C})}$$

where R_M is a copy of R in a ball in M.

16.2 Computation on surfaces

In this section we compute the $\Bbbk$-module $|\Sigma|_\mathcal{C}$ for a connected $\mathcal{Z}(\mathcal{C})$-colored surface Σ. (The case of non-connected surfaces is then deduced from Formula (15.1)). Let A be the set of $\mathcal{Z}(\mathcal{C})$-colored points of Σ. For $a \in A$, let $V_a \in \mathrm{Ob}(\mathcal{Z}(\mathcal{C}))$ and $\varepsilon_a \in \{+, -\}$ be the color and sign of a. Set

$$V_\Sigma = \bigotimes_{a \in A} V_a^{\varepsilon_a} \in \mathrm{Ob}(\mathcal{Z}(\mathcal{C})),$$

where the monoidal product is taken in accordance with a total order on A and, as usual, $V_a^+ = V_a$ and $V_a^- = V_a^*$. Note that since the category $\mathcal{Z}(\mathcal{C})$ is braided, the isomorphism class of V_Σ does not depend on the choice of the order. By definition, if $A = \emptyset$, then $V_\Sigma = \mathbb{1}_{\mathcal{Z}(\mathcal{C})}$.

The following theorem computes (up to isomorphism) the $\Bbbk$-module $|\Sigma|_\mathcal{C}$ when the category $\mathcal{C}$ is additive. In this theorem, $(C, \sigma) \in \mathrm{Ob}(\mathcal{Z}(\mathcal{C}))$ is a coend of $\mathcal{Z}(\mathcal{C})$, see Section 6.4.1 for the definition of a coend of a category and Theorem 9.10 for the existence of a coend of $\mathcal{Z}(\mathcal{C})$.

Theorem 16.2. *If $\mathcal{C}$ is additive, then*

$$|\Sigma|_{\mathcal{C}} \simeq \mathrm{Hom}_{\mathcal{Z}(\mathcal{C})}\big(\mathbb{1}_{\mathcal{Z}(\mathcal{C})}, (C,\sigma)^{\otimes g} \otimes V_\Sigma\big),$$

where $g \geq 0$ is the genus of Σ.

Proof. By Lemma 15.1, it suffices to handle the case where Σ has a single colored point a with color $V = (B, \beta) \in \mathrm{Ob}(\mathcal{Z}(\mathcal{C}))$ and sign $+$. Then $A = \{a\}$ and $V_\Sigma = V$. Below we exhibit a skeleton G of Σ and compute the $\Bbbk$-module $|G|^\circ$ defined in Section 15.7.1. By Section 15.7.3, the $\Bbbk$-module $|\Sigma|_{\mathcal{C}}$ is isomorphic to the image of the $\Bbbk$-linear homomorphism

$$p = |\mathrm{id}_\Sigma, G, G|^\circ \colon |G|^\circ \to |G|^\circ.$$

By Lemma 9.9, the $\Bbbk$-module $\mathrm{Hom}_{\mathcal{Z}(\mathcal{C})}(\mathbb{1}_{\mathcal{Z}(\mathcal{C})}, (C,\sigma)^{\otimes g} \otimes V)$ is the image of a $\Bbbk$-linear endomorphism π of $\mathrm{Hom}_{\mathcal{C}}(\mathbb{1}, C^{\otimes g} \otimes B)$. Thus, we need to prove that the images of p and π are isomorphic.

Fix for the rest of the proof a representative set I of simple objects of $\mathcal{C}$. We first treat the case $g = 0$. We can assume that $\Sigma = (S^2, A)$ with $A = \{a\}$. For a skeleton of Σ, we take an oriented embedded circle $G \subset S^2$ viewed as a graph with one vertex x and one edge whose interior contains the point a. As in Section 15.7.1, viewing a as a vertex, we turn G into an oriented graph $G_A \subset S^2$ with vertices x, a and two edges. An I-coloring of G_A (i.e., a map from the set of edges of G_A to I) is determined by the colors of the edges $i, j \in I$. We let $G_{i,j} \subset S^2$ be the resulting $\mathcal{C}$-enriched graph with distinguished vertex a decorated by $(B, +)$:

$G_{i,j} =$ 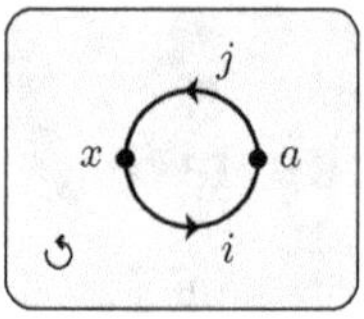 $\subset S^2.$

By definition,

$$\mathcal{H}(G_{i,j}) = \mathcal{H}_x(G_{i,j}) \otimes \mathcal{H}_a(G_{i,j}),$$

where $\otimes$ is the unordered tensor product of $\Bbbk$-modules. Note that

$$\mathcal{H}_x(G_{i,j}) \simeq \mathrm{Hom}_{\mathcal{C}}(\mathbb{1}, i^* \otimes j) \simeq \mathrm{Hom}_{\mathcal{C}}(i, j).$$

Consequently, if $i \neq j$, then $\mathcal{H}(G_{i,j}) = \mathcal{H}_x(G_{i,j}) = 0$. Therefore

$$|G|^\circ = \bigoplus_{i,j \in I} \mathcal{H}(G_{i,j}) = \bigoplus_{i \in I} \mathcal{H}(G_{i,i}).$$

For any $i \in I$, consider the $\Bbbk$-modules

$$U_i = \mathrm{Hom}_{\mathcal{C}}(\mathbb{1}, i^* \otimes i), \quad W_i = \mathrm{Hom}_{\mathcal{C}}(\mathbb{1}, i^* \otimes B \otimes i),$$

and the $\Bbbk$-linear isomorphism

$$\Psi_i \colon \mathcal{H}(G_{i,i}) \to U_i \otimes W_i \tag{16.3}$$

obtained as the tensor product of the cone isomorphisms $\mathcal{H}_x(G_{i,i}) \to U_i$ and $\mathcal{H}_a(G_{i,i}) \to W_i$ (see Sections 12.2.1 and 15.6.1). Set

$$F = \bigoplus_{i \in I} U_i \otimes W_i$$

and consider the $\Bbbk$-linear isomorphism

$$\Psi = \sum_{i \in I} \Psi_i \colon |G|^\circ = \bigoplus_{i \in I} \mathcal{H}(G_{i,i}) \to F.$$

Consider also the $\Bbbk$-module $E = \mathrm{Hom}_{\mathcal{C}}(\mathbb{1}, B)$ and the $\Bbbk$-linear homomorphism

$$\Phi = \sum_{i \in I} (\dim(i))^{-1}\, \Phi_i \colon E \to F,$$

where $\Phi_i \colon E \to U_i \otimes W_i$ carries any $f \in E$ to

$\Phi_i(f) =$ 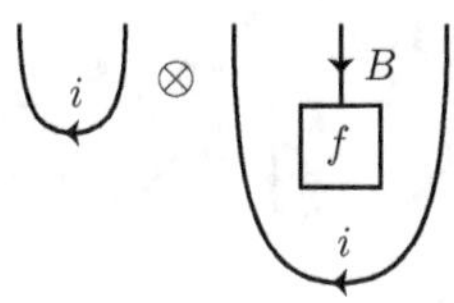 . (16.4)

The homomorphism Φ has a left inverse, and so is injective. To see this, consider for every $i \in I$ the $\Bbbk$-linear homomorphism $\Theta_i \colon U_i \otimes W_i \to E$ which carries $g \otimes h$ with any $g \in U_i$ and $h \in W_i$ to

$\Theta_i(g \otimes h) =$ 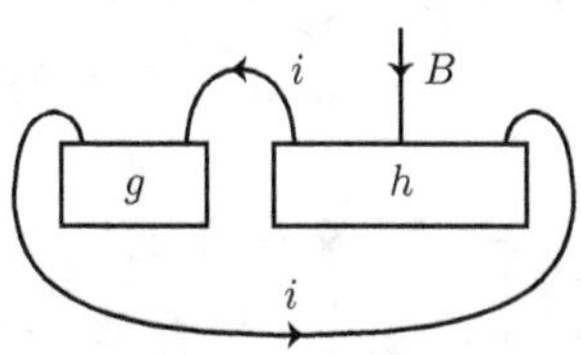 . (16.5)

Clearly, $\Theta_i \Phi_i(f) = \dim(i)\, f$ for any $f \in E$. Then the $\Bbbk$-homomorphism

$$\Theta = \sum_{i \in I} \frac{(\dim(i))^2}{\dim(\mathcal{C})}\, \Theta_i \colon F = \bigoplus_{i \in I} U_i \otimes W_i \to E$$

is a left inverse of Φ: for any $f \in E$,

$$\Theta\Phi(f) = \sum_{i \in I} \frac{\dim(i)}{\dim(\mathcal{C})}\, \Theta_i \Phi_i(f) = \sum_{i \in I} \frac{(\dim(i))^2}{\dim(\mathcal{C})}\, f = f.$$

To prove that the images of p and π are isomorphic, it suffices to prove that the following diagram commutes:

$$\begin{array}{ccccc} |G|^\circ & \xleftarrow{\Psi^{-1}} & F & \xrightarrow{\Theta} & E \\ {\scriptstyle p}\downarrow & & & & \downarrow{\scriptstyle \pi} \\ |G|^\circ & \xrightarrow{\Psi} & F & \xleftarrow{\Phi} & E. \end{array} \tag{16.6}$$

Indeed, since Ψ is an isomorphism, the images of p and $p' = \Psi p \Psi^{-1} \colon F \to F$ are isomorphic. The commutativity of (16.6) would imply that $p' = \Phi\pi\Theta$. Then $p'\Phi = \Phi\pi\Theta\Phi = \Phi\pi$, and so Φ restricts to a homomorphism $\mathrm{Im}(\pi) \to \mathrm{Im}(p')$. This homomorphism is surjective because $p' = \Phi\pi\Theta$ and injective because Φ is injective. Consequently, the module $\mathrm{Im}(\pi)$ is isomorphic to $\mathrm{Im}(p')$ and to $\mathrm{Im}(p)$.

We now compute π and p. By Lemma 9.9, π carries any $f \in E$ to

$$\pi(f) = \sum_{z\in I} \frac{\dim(z)}{\dim(\mathcal{C})} \Pi_z(f) \quad \text{where} \quad \Pi_z(f) =$$

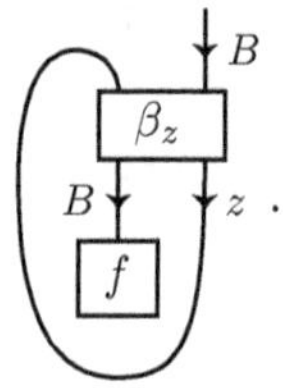

$.\qquad (16.7)$

The homomorphism $p\colon |G|^\circ \to |G|^\circ$ expands as

$$p = \sum_{i,j\in I} p_i^j \colon |G|^\circ = \bigoplus_{i\in I} \mathcal{H}(G_{i,i}) \to |G|^\circ = \bigoplus_{j\in I} \mathcal{H}(G_{j,j})$$

where the homomorphism $p_i^j \colon \mathcal{H}(G_{i,i}) \to \mathcal{H}(G_{j,j})$ is defined as follows. Recall from Section 15.1.3 the cylinder $(M = S^2 \times [0,1], R = A \times [0,1])$ over the $\mathcal{Z}(\mathcal{C})$-colored surface $\Sigma = (S^2, A = \{a\})$. The oriented graph

$$G' = (G^{\mathrm{op}} \times \{0\}) \sqcup (G \times \{1\}) \subset ((-S^2) \times \{0\}) \sqcup (S^2 \times \{1\}) = \partial M$$

is a skeleton of the $\mathcal{Z}(\mathcal{C})$-colored surface $(\partial M, \partial R)$. The elements $i, j \in I$ determine an I-coloring of the graph

$$G'_R = (G_A^{\mathrm{op}} \times \{0\}) \sqcup (G_A \times \{1\}) \subset \partial M.$$

and turns it into a $\mathcal{C}$-enriched graph

$$G_i^j = (G_{i,i}^{\mathrm{op}} \times \{0\}) \sqcup (G_{j,j} \times \{1\}) \subset \partial M.$$

Then (M, G', R) is an I-colored ribbon triple, and Theorem 15.8 yields a vector $|M, G', R|_{\mathcal{C}} \in \mathcal{H}(G'_R)$. Let Υ be the following composition of $\Bbbk$-linear isomorphisms:

$$\begin{aligned} \mathcal{H}(G'_R; \partial M) &\simeq \mathcal{H}(G_{i,i}^{\mathrm{op}} \sqcup G_{j,j}; -(S^2, A) \sqcup (S^2, A)) \\ &\simeq \mathcal{H}(G_{i,i}^{\mathrm{op}}; -(S^2, A)) \otimes \mathcal{H}(G_{j,j}; (S^2, A)) \\ &\simeq \mathcal{H}(G_{i,i})^\star \otimes \mathcal{H}(G_{j,j}) \simeq \mathrm{Hom}_\Bbbk(\mathcal{H}(G_{i,i}), \mathcal{H}(G_{j,j})). \end{aligned}$$

By definition,

$$p_i^j = \frac{(\dim(\mathcal{C}))^2}{(\dim(j))^2}\, \Upsilon(|M, G', R|_{\mathcal{C}}) \colon \mathcal{H}(G_{i,i}) \to \mathcal{H}(G_{j,j}).$$

To compute p_i^j, consider the 2-polyhedron

$$P = (G \times [0,1]) \cup \left(S^2 \times \left\{\frac{1}{2}\right\}\right) \subset M.$$

We stratify P by taking as its edges the following five arcs:

$$\{x\} \times \left[0, \frac{1}{2}\right], \quad \{x\} \times \left[\frac{1}{2}, 1\right], \quad e \times \{0\}, \quad e \times \left\{\frac{1}{2}\right\}, \quad e \times \{1\},$$

where e is the edge of G. The polyhedron P has three vertices:

$$x_{\mathrm{in}} = (x, 0), \quad x_{\mathrm{mid}} = \left(x, \frac{1}{2}\right), \quad x_{\mathrm{out}} = (x, 1),$$

and four regions: namely, $e \times [0, \frac{1}{2}]$, $e \times [\frac{1}{2}, 1]$, and two disk regions $r_0, r_1 \subset S^2 \times \{\frac{1}{2}\}$. We endow r_0 and r_1 with orientation induced by that of S^2 and orient the regions $e \times [0, \frac{1}{2}]$ and $e \times [\frac{1}{2}, 1]$ so that $\partial P = G'$ as oriented graphs. Then P becomes a skeleton of the pair (M, G') such that $|M \setminus P| = 4$. Consider the arc $d = \{a\} \times [0, 1]$ colored by $V = (B, \beta) \in \mathrm{Ob}(\mathcal{Z}(\mathcal{C}))$ and oriented toward $a_{\mathrm{in}} = (a, 0)$. This arc is a $\mathcal{Z}(\mathcal{C})$-colored knotted plexus in P with free ends a_{in} and $a_{\mathrm{out}} = (a, 1)$, a single (positive) switch $a_{\mathrm{mid}} = (a, \frac{1}{2})$, no crossings, and no coupons. The pair (P, d) is a neat positive diagram of R with six faces: r_0, r_1, $e \times [0, \frac{1}{2}]$, $e \times [\frac{1}{2}, 1]$, where e runs over both edges of the graph G_A. Colorings of the diagram (P, d) (i.e., maps $\mathrm{Fac}(d) \to I$ extending the coloring of the graph G_i^j) are numerated by the colors $z, w \in I$ of r_0, r_1:

$(P, d) =$ 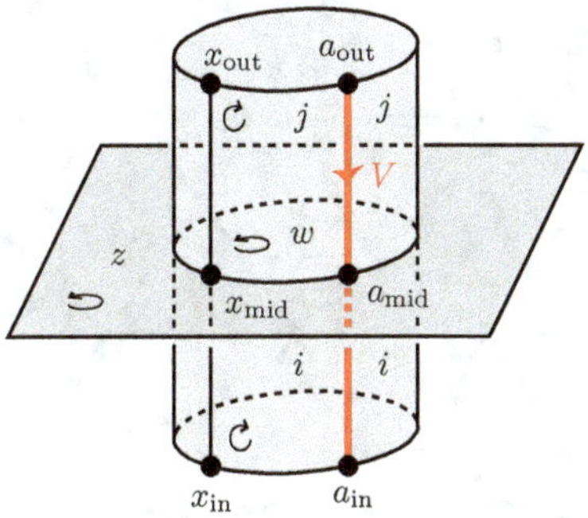

By Section 15.6.2, the internal nodes x_{mid} and a_{mid} of d give rise, respectively, to the following knotted $\mathcal{C}$-nets in an oriented 2-sphere:

$\Gamma_x^{z,w} =$ 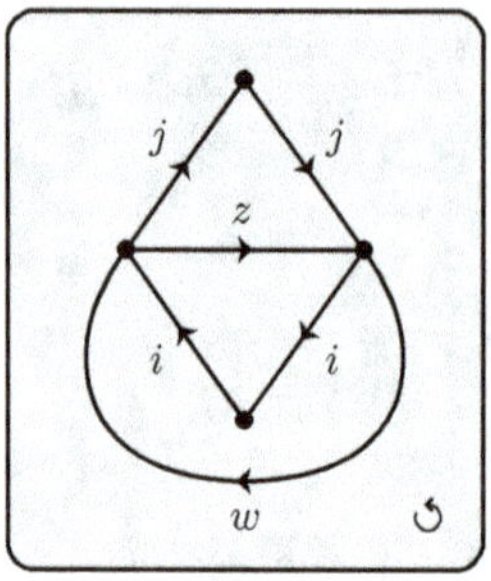 and $\Gamma_a^{z,w} =$ 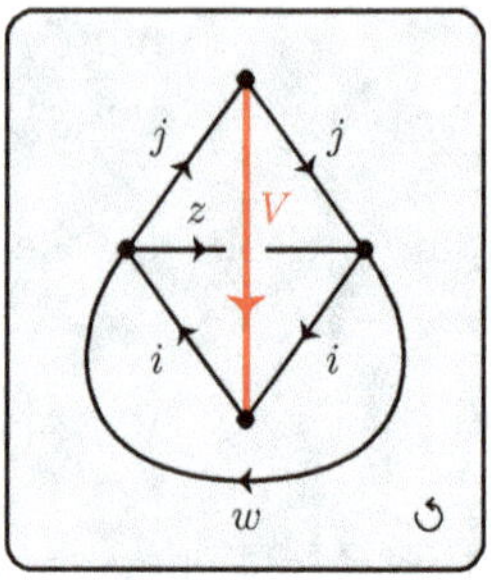 .

By Section 15.6.2,

$$|M, G', R|_{\mathcal{C}} = (\dim(i))^2 (\dim(j))^2 (\dim(\mathcal{C}))^{-4}$$
$$\times \sum_{z,w\in I} \dim(z)\dim(w)\left(\mathbb{F}_{\mathcal{C}}(\Gamma_x^{z,w}) \otimes \mathbb{F}_{\mathcal{C}}(\Gamma_a^{z,w}) \otimes \mathrm{id}_{\mathcal{H}(G'_R)}\right)(*_{z,w,d}) \in \mathcal{H}(G'_R),$$

where

$$*_{z,w,d} \in H(\Gamma_x^{z,w})^\star \otimes H(\Gamma_a^{z,w})^\star \otimes \mathcal{H}(G'_R)$$

is the unordered tensor product of the contraction vectors $*_e$ over all unoriented rims e of d. (Here and below we suppress the canonical isomorphisms of Section 15.6.2.) We expand $*_{z,w,d} = *_{z,w} \otimes *_d$, where $*_{z,w}$ is the unordered tensor product of the contraction vectors $*_e$ over the two unoriented rims e of d lying in $S^2 \times \{\frac{1}{2}\}$ and $*_d$ is the unordered tensor product of the contraction vectors $*_e$ over the four unoriented rims e of d not lying in $S^2 \times \{\frac{1}{2}\}$. Then

$$\left(\mathbb{F}_{\mathcal{C}}(\Gamma_x^{z,w}) \otimes \mathbb{F}_{\mathcal{C}}(\Gamma_a^{z,w}) \otimes \mathrm{id}_{\mathcal{H}(G'_R)}\right)(*_{z,w,d}) = \left(\mu_i^j(z,w) \otimes \mathrm{id}_{\mathcal{H}(G'_R)}\right)(*_d),$$

where

$$\mu_i^j(z,w) = \left(\mathbb{F}_{\mathcal{C}}(\Gamma_x^{z,w}) \otimes \mathbb{F}_{\mathcal{C}}(\Gamma_a^{z,w})\right)(*_{z,w}).$$

Using the graphical notation introduced at the beginning of Section 15.5.2, the vector $\mu_i^j(z,w)$ is computed by

$\mu_i^j(z,w) =$ 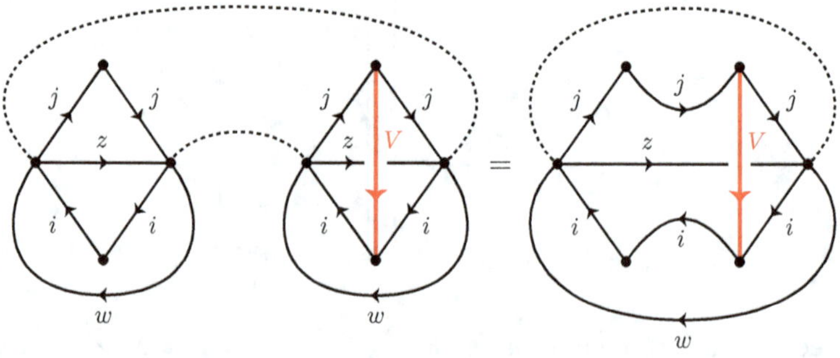 .

Here, the second equality follows from Lemma 12.6 extended to knotted $\mathcal{C}$-nets (see Section 15.4.8). Next we compute

$$M_i^j(z) = \sum_{w \in I} \dim(w)\, \mu_i^j(z,w) =$$

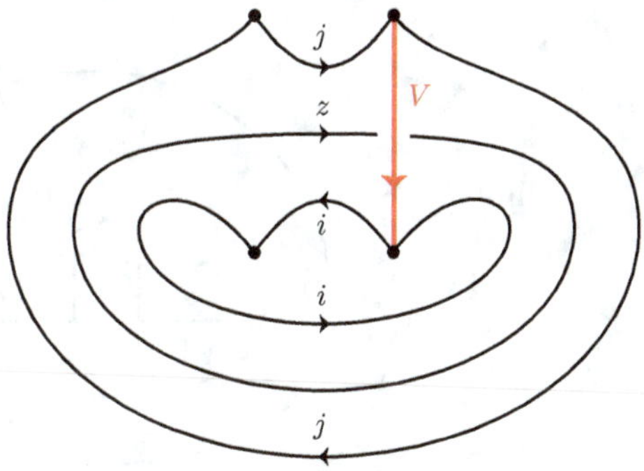

where we use Lemma 12.5 extended to knotted $\mathcal{C}$-nets. Clearly,

$$p_i^j = \frac{(\dim(i))^2}{(\dim(\mathcal{C}))^2} \sum_{z \in I} \dim(z)\, P_i^j(z) \tag{16.8}$$

where

$$P_i^j(z) = \Upsilon\Big(\big(M_i^j(z) \otimes \mathrm{id}_{\mathcal{H}(G'_R)}\big)(*_d)\Big) : \mathcal{H}(G_{i,i}) \to \mathcal{H}(G_{i,i}).$$

Consider now the isomorphisms Ψ_i and Ψ_j from (16.3). Using the above computation of $M_i^j(z)$, we obtain that for any $g \in U_i$ and $h \in W_i$,

$$\Psi_j P_i^j(z) \Psi_i^{-1}(g \otimes h) =$$

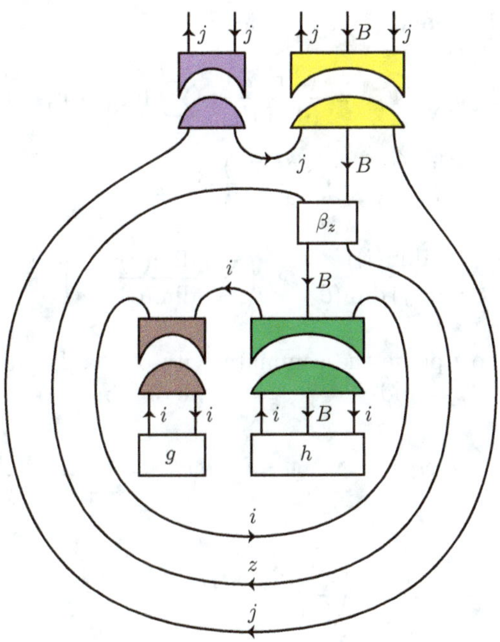

$= \dim(j)^{-1}$ 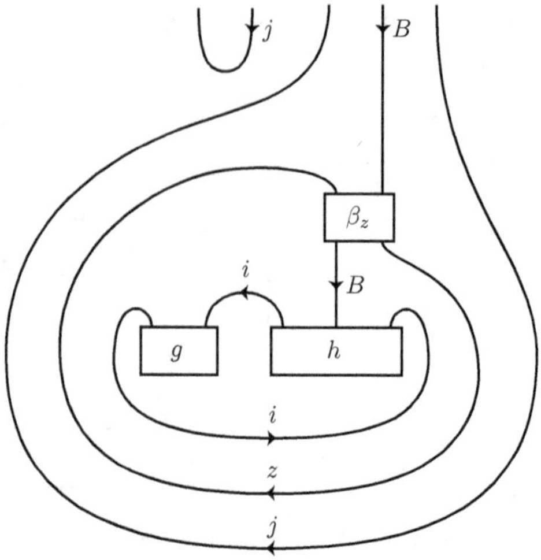

Here, in the first diagram, the four pairs of curvilinear boxes come from the contraction vectors associated to the four unoriented rims of d not lying in $S^2 \times \{\frac{1}{2}\}$ (see Lemma 4.8) while the three rectangular boxes come from the definition of $\mathbb{F}_{\mathcal{C}}$ (see Sections 15.4.1 and 15.4.7). The equality of the expressions in the first and second diagrams follows from Formula (4.9) and Lemma 4.9(b). This computation and the definitions of Φ_j, Θ_i, Π_z given in (16.4), (16.5), (16.7) imply that

$$\Psi_j P_i^j(z) \Psi_i^{-1} = \dim(j)^{-1}\, \Phi_j \Pi_z \Theta_i.$$

Using (16.8) and (16.7), we deduce that

$$\begin{aligned}\Psi_j\, p_i^j(z)\, \Psi_i^{-1} &= \frac{(\dim(i))^2}{(\dim(\mathcal{C}))^2} \sum_{z\in I} \dim(z)\, \Psi_j P_i^j(z) \Psi_i^{-1} \\ &= \frac{(\dim(i))^2}{(\dim(\mathcal{C}))^2} \sum_{z\in I} \frac{\dim(z)}{\dim(j)}\, \Phi_j \Pi_z \Theta_i \\ &= \frac{(\dim(i))^2}{\dim(j)\dim(\mathcal{C})}\, \Phi_j \left(\sum_{z\in I} \frac{\dim(z)}{\dim(\mathcal{C})}\, \Pi_z \right) \Theta_i = \frac{(\dim(i))^2}{\dim(j)\dim(\mathcal{C})}\, \Phi_j \pi \Theta_i.\end{aligned}$$

Let us now prove the commutativity of the diagram (16.6). Using the equalities $p = \sum_{i,j\in I} p_i^j$ and $\Psi = \sum_{i\in I} \Psi_i$, we obtain

$$\begin{aligned}\Psi p \Psi^{-1} &= \sum_{i,j\in I} \Psi_j\, p_i^j(z)\, \Psi_i^{-1} = \sum_{i,j\in I} \frac{(\dim(i))^2}{\dim(j)\dim(\mathcal{C})}\, \Phi_j \pi \Theta_i \\ &= \left(\sum_{j\in I} (\dim(j))^{-1}\, \Phi_j \right) \pi \left(\sum_{i\in I} \frac{(\dim(i))^2}{\dim(\mathcal{C})} \Theta_i \right) = \Phi \pi \Theta.\end{aligned}$$

This completes the proof of the theorem in the case $g = 0$.

Assume now that $g \geq 1$. We first compute the $\Bbbk$-module $\mathrm{Hom}_{\mathcal{C}}(\mathbb{1}, C^{\otimes g} \otimes B)$. By Theorem 9.10, we can assume that $C = \bigoplus_{i,j \in I} i^* \otimes j^* \otimes i \otimes j$. Then

$$C^{\otimes g} \otimes B = \bigoplus_{\mathbf{i},\mathbf{j} \in I^g} H_{\mathbf{i},\mathbf{j}} \otimes B$$

where

$$H_{\mathbf{i},\mathbf{j}} = i_1^* \otimes j_1^* \otimes i_1 \otimes j_1 \otimes \cdots \otimes i_g^* \otimes j_g^* \otimes i_g \otimes j_g \in \mathrm{Ob}(\mathcal{C})$$

for any tuples $\mathbf{i} = (i_1, \ldots, i_g) \in I^g$ and $\mathbf{j} = (j_1, \ldots, j_g) \in I^g$. Hence

$$\mathrm{Hom}_{\mathcal{C}}(\mathbb{1}, C^{\otimes g} \otimes B) = \bigoplus_{\mathbf{i},\mathbf{j} \in I^g} E_{\mathbf{i},\mathbf{j}} \quad \text{where} \quad E_{\mathbf{i},\mathbf{j}} = \mathrm{Hom}_{\mathcal{C}}(\mathbb{1}, H_{\mathbf{i},\mathbf{j}} \otimes B).$$

To compute the $\Bbbk$-module $E_{\mathbf{i},\mathbf{j}}$, consider the following object of $\mathcal{C}$:

$$D_{\mathbf{i},\mathbf{j}} = i_1^* \otimes j_1^* \otimes i_1 \otimes j_1 \otimes \cdots \otimes i_{g-1}^* \otimes j_{g-1}^* \otimes i_{g-1} \otimes j_{g-1} \otimes i_g^* \otimes j_g^* \otimes i_g.$$

For $s \in I$, we define $\Bbbk$-modules

$$U_{\mathbf{i},\mathbf{j},s} = \mathrm{Hom}_{\mathcal{C}}(\mathbb{1}, D_{\mathbf{i},\mathbf{j}} \otimes s) \quad \text{and} \quad W_{\mathbf{j},s} = \mathrm{Hom}_{\mathcal{C}}(\mathbb{1}, s^* \otimes B \otimes j_g).$$

Consider $\Bbbk$-linear homomorphisms

$$\Phi_{\mathbf{i},\mathbf{j},s} \colon U_{\mathbf{i},\mathbf{j},s} \otimes W_{\mathbf{j},s} \to E_{\mathbf{i},\mathbf{j}} \quad \text{and} \quad \widetilde{\Phi}_{\mathbf{i},\mathbf{j},s} \colon E_{\mathbf{i},\mathbf{j}} \to U_{\mathbf{i},\mathbf{j},s} \otimes W_{\mathbf{j},s}$$

defined by

$\Phi_{\mathbf{i},\mathbf{j},s}(g \otimes h) =$ 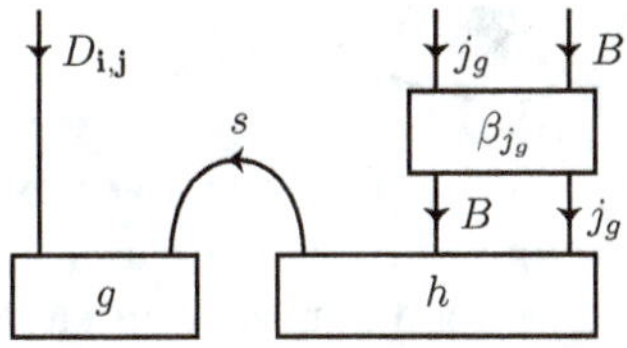 (16.9)

for all $g \in U_{\mathbf{i},\mathbf{j},s}$, $h \in W_{\mathbf{j},s}$, and

$\widetilde{\Phi}_{\mathbf{i},\mathbf{j},s}(f) =$ 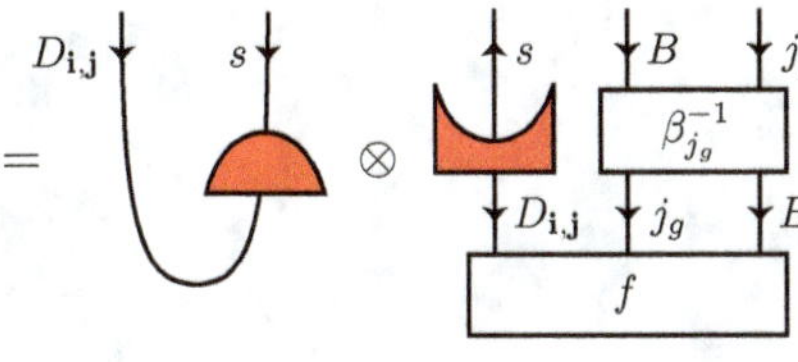 (16.10)

for all $f \in E_{\mathbf{i},\mathbf{j}}$. Formula (4.8) implies that for all such f, g, h,

$$\sum_{s \in I} \Phi_{\mathbf{i},\mathbf{j},s} \widetilde{\Phi}_{\mathbf{i},\mathbf{j},s}(f) = f \quad \text{and} \quad \widetilde{\Phi}_{\mathbf{i},\mathbf{j},s} \Phi_{\mathbf{i},\mathbf{j},s}(g \otimes h) = g \otimes h.$$

Thus, setting $\dim(\mathbf{j}) = \prod_{r=1}^{g} \dim(j_r)$, we obtain that the $\Bbbk$-linear homomorphism

$$\Phi_{\mathbf{i},\mathbf{j}} = \dim(\mathbf{j}) \sum_{s\in I} \dim(s)\, \Phi_{\mathbf{i},\mathbf{j},s} \colon \bigoplus_{s\in I} U_{\mathbf{i},\mathbf{j},s} \otimes W_{\mathbf{j},s} \to E_{\mathbf{i},\mathbf{j}} \tag{16.11}$$

is invertible with inverse

$$\Phi_{\mathbf{i},\mathbf{j}}^{-1} = \dim(\mathbf{j})^{-1} \sum_{s\in I} \dim(s)^{-1}\, \widetilde{\Phi}_{\mathbf{i},\mathbf{j},s} \colon E_{\mathbf{i},\mathbf{j}} \to \bigoplus_{s\in I} U_{\mathbf{i},\mathbf{j},s} \otimes W_{\mathbf{j},s}\,. \tag{16.12}$$

Next, we compute the $\Bbbk$-module $|G|^\circ$, where G is the skeleton of Σ formed by a single vertex x and $2g$ oriented edges such that the point a lies in the interior of one of these edges and $\Sigma \setminus G$ is a disk:

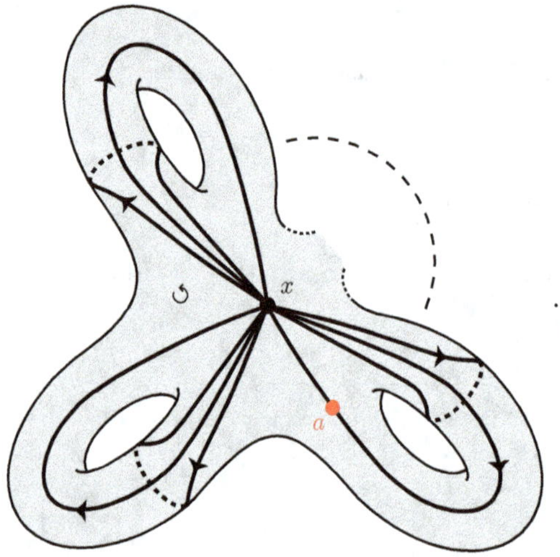

As in Section 15.7.1, viewing $a \in G$ as a vertex, we turn G into an oriented graph $G_A \subset \Sigma$ with $2g+1$ edges and two vertices x, a. Any tuples $\mathbf{i} = (i_1, \ldots, i_g) \in I^g$ and $\mathbf{j} = (j_1, \ldots, j_g) \in I^g$ together with an element $s \in I$ determine an I-coloring of G_A (i.e., a map from the set of edges of G_A to I) as follows:

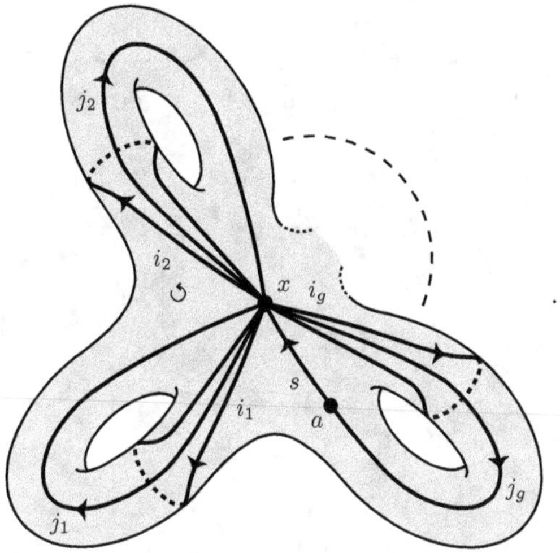

Denote by $G_{\mathbf{i},\mathbf{j},s}$ the resulting $\mathcal{C}$-enriched graph in Σ with the single distinguished vertex a decorated by $(B,+)$. By Section 15.7.1,

$$|G|^\circ = \bigoplus_{\mathbf{i},\mathbf{j}\in I^g, s\in I} \mathcal{H}(G_{\mathbf{i},\mathbf{j},s}).$$

Note that

$$\mathcal{H}(G_{\mathbf{i},\mathbf{j},s}) = \mathcal{H}_x(G_{\mathbf{i},\mathbf{j},s}) \otimes \mathcal{H}_a(G_{\mathbf{i},\mathbf{j},s}),$$

where $\otimes$ is the unordered tensor product of $\Bbbk$-modules. Consider the cone isomorphisms $\mathcal{H}_x(G_{\mathbf{i},\mathbf{j},s}) \to U_{\mathbf{i},\mathbf{j},s}$ and $\mathcal{H}_a(G_{\mathbf{i},\mathbf{j},s}) \to W_{\mathbf{j},s}$, see Sections 12.2.1 and 15.6.1. Their tensor product is a $\Bbbk$-linear isomorphism

$$\Psi_{\mathbf{i},\mathbf{j},s} \colon \mathcal{H}(G_{\mathbf{i},\mathbf{j},s}) \to U_{\mathbf{i},\mathbf{j},s} \otimes W_{\mathbf{j},s}. \tag{16.13}$$

Fixing $\mathbf{i},\mathbf{j} \in I^g$ and summing up over all $s \in I$, we obtain the $\Bbbk$-module

$$|G|^\circ_{\mathbf{i},\mathbf{j}} = \bigoplus_{s\in I} \mathcal{H}(G_{\mathbf{i},\mathbf{j},s})$$

and the $\Bbbk$-linear isomorphism

$$\Psi_{\mathbf{i},\mathbf{j}} = \sum_{s\in I} \Psi_{\mathbf{i},\mathbf{j},s} \colon |G|^\circ_{\mathbf{i},\mathbf{j}} \to \bigoplus_{s\in I} U_{\mathbf{i},\mathbf{j},s} \otimes W_{\mathbf{j},s}. \tag{16.14}$$

Composing with the isomorphism $\Phi_{\mathbf{i},\mathbf{j}}$ defined by (16.11) and summing up over all $\mathbf{i},\mathbf{j}$, we obtain the $\Bbbk$-linear isomorphism

$$\sum_{\mathbf{i},\mathbf{j}\in I^g} \Phi_{\mathbf{i},\mathbf{j}}\Psi_{\mathbf{i},\mathbf{j}} \colon |G|^\circ = \bigoplus_{\mathbf{i},\mathbf{j}\in I^g} |G|^\circ_{\mathbf{i},\mathbf{j}} \to \bigoplus_{\mathbf{i},\mathbf{j}\in I^g} E_{\mathbf{i},\mathbf{j}} = \mathrm{Hom}_{\mathcal{C}}(\mathbb{1}, C^{\otimes g} \otimes B).$$

We claim that this isomorphism conjugates the endomorphisms p of $|G|^\circ$ and π of $\mathrm{Hom}_{\mathcal{C}}(\mathbb{1}, C^{\otimes g} \otimes B)$ mentioned at the beginning of the proof. This implies that the images of p and π are isomorphic. It suffices to prove that for all $\mathbf{i}, \mathbf{j}, \mathbf{k}, \mathbf{l} \in I^g$, the following diagram commutes:

$$\begin{array}{ccc} |G|^\circ_{\mathbf{i},\mathbf{j}} & \xrightarrow{\Phi_{\mathbf{i},\mathbf{j}}\Psi_{\mathbf{i},\mathbf{j}}} & E_{\mathbf{i},\mathbf{j}} \\ {\scriptstyle p^{\mathbf{k},\mathbf{l}}_{\mathbf{i},\mathbf{j}}}\downarrow & & \downarrow{\scriptstyle \pi^{\mathbf{k},\mathbf{l}}_{\mathbf{i},\mathbf{j}}} \\ |G|^\circ_{\mathbf{k},\mathbf{l}} & \xrightarrow{\Phi_{\mathbf{k},\mathbf{l}}\Psi_{\mathbf{k},\mathbf{l}}} & E_{\mathbf{k},\mathbf{l}} \end{array} \tag{16.15}$$

where $p^{\mathbf{k},\mathbf{l}}_{\mathbf{i},\mathbf{j}} \colon |G|^\circ_{\mathbf{i},\mathbf{j}} \to |G|^\circ_{\mathbf{k},\mathbf{l}}$ and $\pi^{\mathbf{k},\mathbf{l}}_{\mathbf{i},\mathbf{j}} \colon E_{\mathbf{i},\mathbf{j}} \to E_{\mathbf{k},\mathbf{l}}$ are the summands in the direct sum decompositions

$$p = \sum_{\mathbf{i},\mathbf{j},\mathbf{k},\mathbf{l} \in I^g} p^{\mathbf{k},\mathbf{l}}_{\mathbf{i},\mathbf{j}} \colon \bigoplus_{\mathbf{i},\mathbf{j} \in I^g} |G|^\circ_{\mathbf{i},\mathbf{j}} \to \bigoplus_{\mathbf{k},\mathbf{l} \in I^g} |G|^\circ_{\mathbf{k},\mathbf{l}}$$

and

$$\pi = \sum_{\mathbf{i},\mathbf{j},\mathbf{k},\mathbf{l} \in I^g} \pi^{\mathbf{k},\mathbf{l}}_{\mathbf{i},\mathbf{j}} \colon \bigoplus_{\mathbf{i},\mathbf{j} \in I^g} E_{\mathbf{i},\mathbf{j}} \to \bigoplus_{\mathbf{k},\mathbf{l} \in I^g} E_{\mathbf{k},\mathbf{l}}.$$

By the definition of the monoidal product in $\mathcal{Z}(\mathcal{C})$,

$$(C,\sigma)^{\otimes g} \otimes V = (C,\sigma)^{\otimes g} \otimes (B,\beta) = (C^{\otimes g} \otimes B, \gamma)$$

where $\gamma = \{\gamma_X \colon C^{\otimes g} \otimes B \otimes X \to X \otimes C^{\otimes g} \otimes B\}_{X \in \mathrm{Ob}(\mathcal{C})}$ is defined by

$\gamma_X =$ 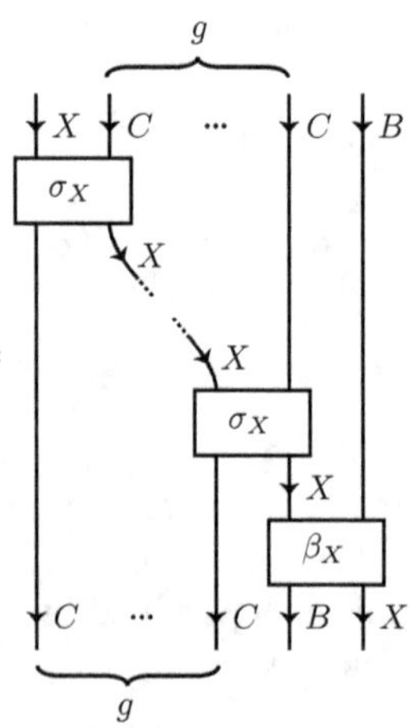

By Lemma 9.9, π is the endomorphism of the $\Bbbk$-module $\mathrm{Hom}_{\mathcal{C}}(\mathbb{1}, C^{\otimes g} \otimes B)$ carrying any element f of this module to

$$\pi(f) = (\dim(\mathcal{C}))^{-1} \sum_{z \in I} \dim(z)$$ 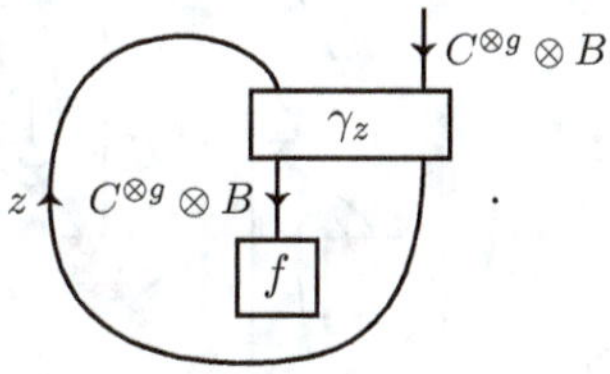 .

By Theorem 9.10, for any $z \in I$,

$$\sigma_z = \sum_{i,j,k,l \in I}$$ 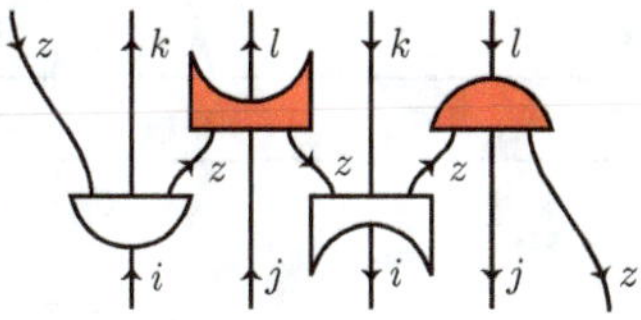 $: C \otimes z \to z \otimes C.$

Using Lemma 4.9(f) twice (once with $j = k = \mathbb{1}$ and $X = z \otimes k^* \otimes z^*$, and once with $i = k = \mathbb{1}$, $j = l$, and $X = z^* \otimes j \otimes z$), we rewrite σ_z as

$$\sigma_z = \sum_{i,j,k,l \in I} \dim(i) \dim(l)$$ 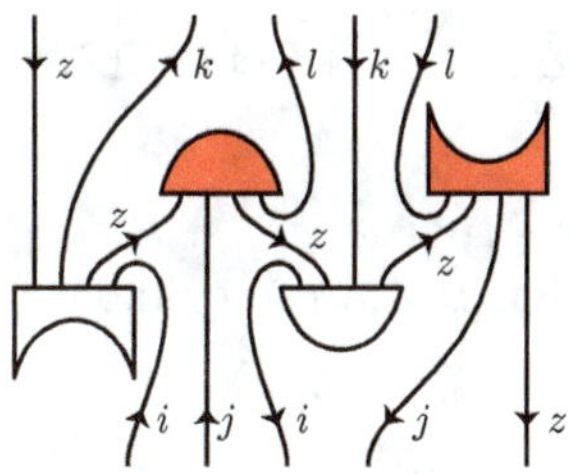 .

Hence, for all $\mathbf{i}, \mathbf{j}, \mathbf{k}, \mathbf{l} \in I^g$, the map $\pi_{\mathbf{i},\mathbf{j}}^{\mathbf{k},\mathbf{l}} \colon E_{\mathbf{i},\mathbf{j}} \to E_{\mathbf{k},\mathbf{l}}$ carries any $f \in E_{\mathbf{i},\mathbf{j}}$ to

$$\pi_{\mathbf{i},\mathbf{j}}^{\mathbf{k},\mathbf{l}}(f) = \frac{\dim(\mathbf{i}) \dim(\mathbf{l})}{\dim(\mathcal{C})} \sum_{z \in I} \dim(z)\, \Pi_{\mathbf{i},\mathbf{j}}^{\mathbf{k},\mathbf{l}}(z, f), \tag{16.16}$$

where $\Pi_{\mathbf{i},\mathbf{j}}^{\mathbf{k},\mathbf{l}}(z, f) \in E_{\mathbf{k},\mathbf{l}}$ is computed by

$$\Pi^{\mathbf{k},\mathbf{l}}_{\mathbf{i},\mathbf{j}}(z, f) =$$

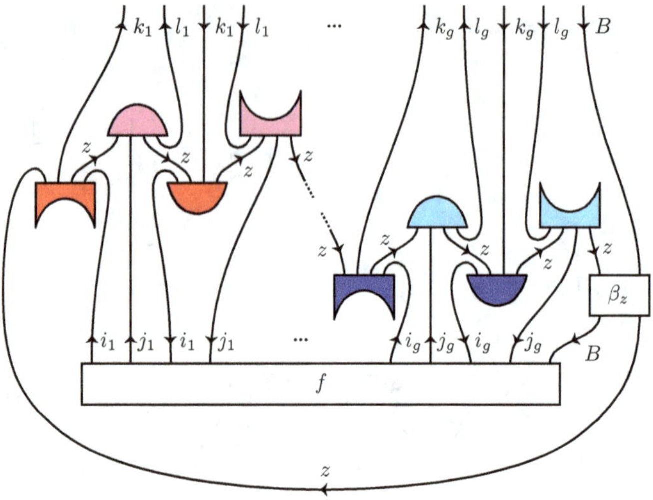

. (16.17)

In its turn, the homomorphism $p^{\mathbf{k},\mathbf{l}}_{\mathbf{i},\mathbf{j}} \colon |G|^{\circ}_{\mathbf{i},\mathbf{j}} \to |G|^{\circ}_{\mathbf{k},\mathbf{l}}$ expands as the sum

$$p^{\mathbf{k},\mathbf{l}}_{\mathbf{i},\mathbf{j}} = \sum_{s,t\in I} p^{\mathbf{k},\mathbf{l},t}_{\mathbf{i},\mathbf{j},s} \colon |G|^{\circ}_{\mathbf{i},\mathbf{j}} = \bigoplus_{s\in I} \mathcal{H}(G_{\mathbf{i},\mathbf{j},s}) \to |G|^{\circ}_{\mathbf{k},\mathbf{l}} = \bigoplus_{t\in I} \mathcal{H}(G_{\mathbf{k},\mathbf{l},t})$$

where the homomorphism $p^{\mathbf{k},\mathbf{l},t}_{\mathbf{i},\mathbf{j},s} \colon \mathcal{H}(G_{\mathbf{i},\mathbf{j},s}) \to \mathcal{H}(G_{\mathbf{k},\mathbf{l},t})$ is computed as follows. Recall from Section 15.1.3 the cylinder $(M = \Sigma \times [0,1], R = A \times [0,1])$ over the $\mathcal{Z}(\mathcal{C})$-colored surface $\Sigma = (\Sigma, A)$ where $A = \{a\}$. The oriented graph

$$G' = (G^{\mathrm{op}} \times \{0\}) \sqcup (G \times \{1\}) \subset ((-\Sigma) \times \{0\}) \sqcup (\Sigma \times \{1\}) = \partial M$$

is then a skeleton of the $\mathcal{Z}(\mathcal{C})$-colored surface $(\partial M, \partial R)$. The tuples $(\mathbf{i}, \mathbf{j}, s)$ and $(\mathbf{k}, \mathbf{l}, t)$ determine an I-coloring of the graph

$$G'_R = (G^{\mathrm{op}}_A \times \{0\}) \sqcup (G_A \times \{1\}) \subset \partial M$$

and turn it into a $\mathcal{C}$-enriched graph

$$G^{\mathbf{k},\mathbf{l},t}_{\mathbf{i},\mathbf{j},s} = (G^{\mathrm{op}}_{\mathbf{i},\mathbf{j},s} \times \{0\}) \sqcup (G_{\mathbf{k},\mathbf{l},t} \times \{1\}) \subset \partial M.$$

Then (M, G', R) is an I-colored ribbon triple and Theorem 15.8 gives a vector $|M, G', R|_{\mathcal{C}} \in \mathcal{H}(G'_R)$. Let Υ be the following composition of $\Bbbk$-linear isomorphisms:

$$\begin{aligned}
\mathcal{H}(G'_R) = \mathcal{H}(G'_R; \partial M) &\simeq \mathcal{H}\big(G^{\mathrm{op}}_{\mathbf{i},\mathbf{j},s} \sqcup G_{\mathbf{k},\mathbf{l},t}; -(\Sigma, A) \sqcup (\Sigma, A)\big) \\
&\simeq \mathcal{H}\big(G^{\mathrm{op}}_{\mathbf{i},\mathbf{j},s}; -(\Sigma, A)\big) \otimes \mathcal{H}\big(G_{\mathbf{k},\mathbf{l},t}; (\Sigma, A)\big) \\
&\simeq \mathcal{H}\big(G_{\mathbf{i},\mathbf{j},s}\big)^{\star} \otimes \mathcal{H}(G_{\mathbf{k},\mathbf{l},t}) \simeq \mathrm{Hom}_{\Bbbk}\big(\mathcal{H}(G_{\mathbf{i},\mathbf{j},s}), \mathcal{H}(G_{\mathbf{k},\mathbf{l},t})\big).
\end{aligned}$$

By definition,

$$p_{\mathbf{i},\mathbf{j},s}^{\mathbf{k},\mathbf{l},t} = \frac{\dim(\mathcal{C})}{\dim(\mathbf{k})\dim(\mathbf{l})\dim(t)} \Upsilon(|M, G', R|_{\mathcal{C}}) \colon \mathcal{H}(G_{\mathbf{i},\mathbf{j},s}) \to \mathcal{H}(G_{\mathbf{k},\mathbf{l},t}).$$

To compute the last expression, consider the 2-polyhedron

$$P = (G \times [0,1]) \cup \left(\Sigma \times \left\{\frac{1}{2}\right\}\right) \subset M.$$

We stratify P by taking as its edges the following $6g+2$ arcs:

$$\{x\} \times \left[0, \frac{1}{2}\right], \quad \{x\} \times \left[\frac{1}{2}, 1\right], \quad e \times \{0\}, \quad e \times \left\{\frac{1}{2}\right\}, \quad e \times \{1\},$$

where e runs over the $2g$ edges of G. The polyhedron P has three vertices

$$x_{\text{in}} = (x, 0), \quad x_{\text{mid}} = \left(x, \frac{1}{2}\right), \quad x_{\text{out}} = (x, 1),$$

and $4g+1$ regions, namely, the regions $e \times [0, \frac{1}{2}]$ and $e \times [\frac{1}{2}, 1]$, where e runs over the $2g$ edges of G, and the disk region r_Σ with interior $(\Sigma \setminus G) \times \{\frac{1}{2}\}$. We endow r_Σ with orientation induced by that of Σ and we orient all other regions of P so that $\partial P = G'$ as oriented graphs. Then P becomes a skeleton of the pair (M, G') such that $|M \setminus P| = 2$. Consider the arc $d = \{a\} \times [0,1]$ colored by $V = (B, \beta) \in \mathrm{Ob}(\mathcal{Z}(\mathcal{C}))$ and oriented toward $a_{\text{in}} = (a, 0)$. This arc is a $\mathcal{Z}(\mathcal{C})$-colored knotted plexus in P with free ends a_{in} and $a_{\text{out}} = (a, 1)$, a single (positive) switch $a_{\text{mid}} = (a, \frac{1}{2})$, no crossings, and no coupons. The pair (P, d) is a neat positive diagram of R with $4g+3$ faces r_Σ, $e \times [0, \frac{1}{2}]$, $e \times [\frac{1}{2}, 1]$, where e runs over the $2g+1$ edges of the graph G_A. Colorings of (P, d) (i.e., maps $\mathrm{Fac}(d) \to I$ extending the coloring of the graph $G_{\mathbf{i},\mathbf{j},s}^{\mathbf{k},\mathbf{l},t}$) are determined by the color $z \in I$ of r_Σ:

$(P, d) =$ 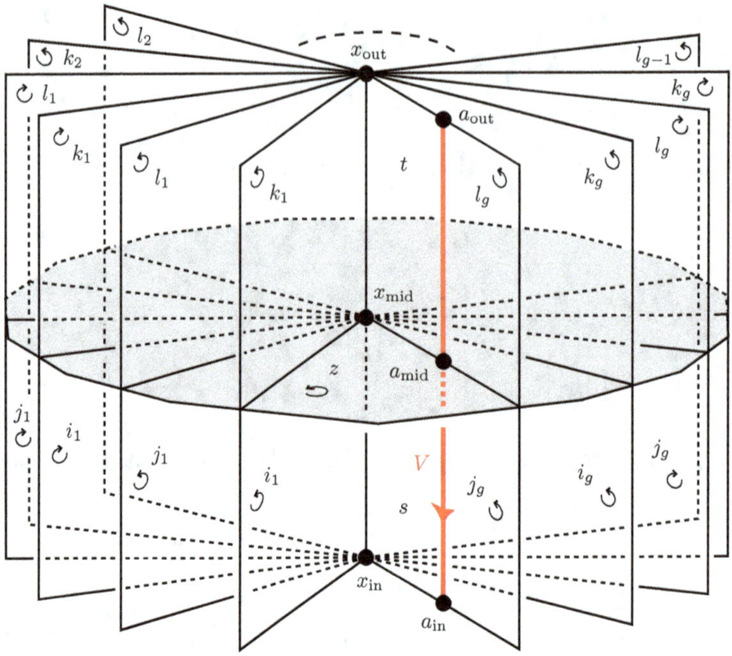 .

By Section 15.6.2, each internal node of d gives rise to a knotted $\mathcal{C}$-net in an oriented 2-sphere. The internal nodes of d are the points x_{mid} and a_{mid}. The knotted $\mathcal{C}$-net associated to x_{mid} is

$\Gamma_x^z =$ 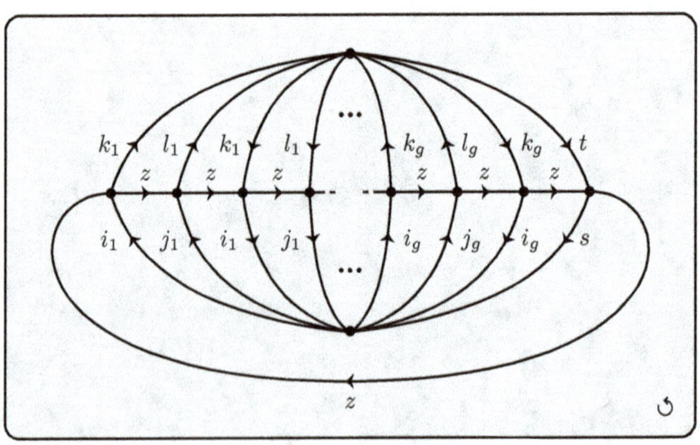 .

The knotted $\mathcal{C}$-net associated to a_{mid} is

$\Gamma_a^z =$ 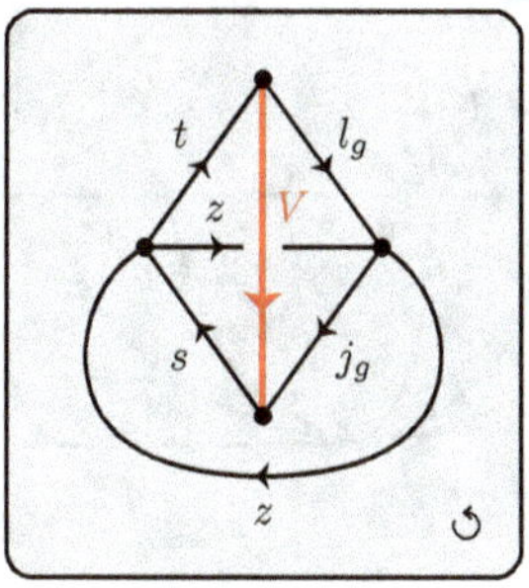

.

By Section 15.6.2,

$$|M, G', R|_{\mathcal{C}} = \dim(s)\dim(t)\dim(\mathbf{i})\dim(\mathbf{j})\dim(\mathbf{k})\dim(\mathbf{l})(\dim(\mathcal{C}))^{-2} \times \sum_{z \in I} \dim(z)\left(\mathbb{F}_{\mathcal{C}}(\Gamma_x^z) \otimes \mathbb{F}_{\mathcal{C}}(\Gamma_a^z) \otimes \mathrm{id}_{\mathcal{H}(G'_R)}\right)(*_{z,d}) \in \mathcal{H}(G'_R),$$

where

$$*_{z,d} \in H(\Gamma_x^z)^\star \otimes H(\Gamma_a^z)^\star \otimes \mathcal{H}(G'_R)$$

is the unordered tensor product of the contraction vectors $*_e$ over all unoriented rims e of d. We expand $*_{z,d} = *_z \otimes *_d$, where $*_z$ is the unordered tensor product of the vectors $*_e$ over the $2g+1$ unoriented rims e of d lying in $\Sigma \times \{\frac{1}{2}\}$ and $*_d$ is the unordered tensor product of the vectors $*_e$ over the four unoriented rims e of d not lying in $\Sigma \times \{\frac{1}{2}\}$. Then

$$\left(\mathbb{F}_{\mathcal{C}}(\Gamma_x^z) \otimes \mathbb{F}_{\mathcal{C}}(\Gamma_a^z) \otimes \mathrm{id}_{\mathcal{H}(G'_R)}\right)(*_{z,d}) = \left(\mu_z \otimes \mathrm{id}_{\mathcal{H}(G'_R)}\right)(*_d),$$

where

$$\mu_z = \left(\mathbb{F}_{\mathcal{C}}(\Gamma_x^z) \otimes \mathbb{F}_{\mathcal{C}}(\Gamma_a^z)\right)(*_z)$$

$=$ 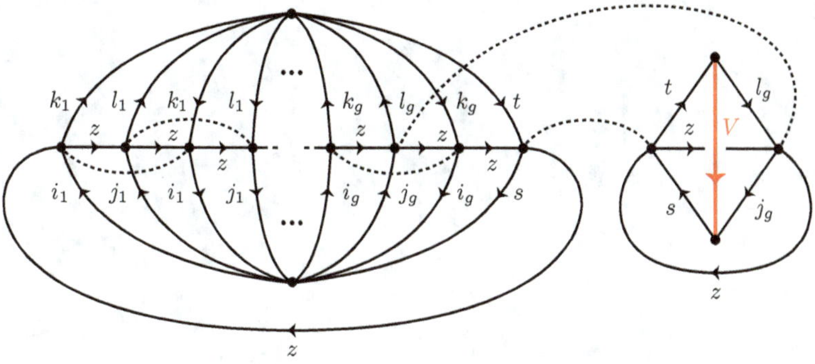

$=$ 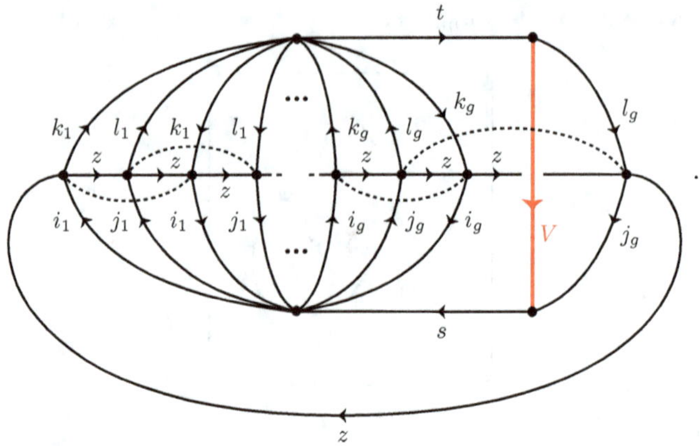

Here, the last equality follows from Lemma 12.6 extended to knotted $\mathcal{C}$-nets (see Section 15.4.8).

Combining the formulas above, we obtain that

$$p_{\mathbf{i},\mathbf{j},s}^{\mathbf{k},\mathbf{l},t} = \frac{\dim(\mathbf{i})\dim(\mathbf{j})\dim(s)}{\dim(\mathcal{C})} \sum_{z\in I} \dim(z)\, P_{\mathbf{i},\mathbf{j},s}^{\mathbf{k},\mathbf{l},t}(z) \tag{16.18}$$

where

$$P_{\mathbf{i},\mathbf{j},s}^{\mathbf{k},\mathbf{l},t}(z) = \Upsilon\Big(\big(\mu_z \otimes \mathrm{id}_{\mathcal{H}(G'_R)}\big)(*_d)\Big) : \mathcal{H}(G_{\mathbf{i},\mathbf{j},s}) \to \mathcal{H}(G_{\mathbf{k},\mathbf{l},t}).$$

Recall the isomorphisms $\Psi_{\mathbf{i},\mathbf{j},s}$ and $\Psi_{\mathbf{k},\mathbf{l},t}$, see (16.13). Using the above computation of μ_z, we obtain that for any $g \in U_{\mathbf{i},\mathbf{j},s}$ and $h \in W_{\mathbf{j},s}$, the vector

$$\Psi_{\mathbf{k},\mathbf{l},t}\, P_{\mathbf{i},\mathbf{j},s}^{\mathbf{k},\mathbf{l},t}(z)\, \Psi_{\mathbf{i},\mathbf{j},s}^{-1}(g\otimes h) \in U_{\mathbf{k},\mathbf{l},t} \otimes W_{\mathbf{l},t} \tag{16.19}$$

is computed by

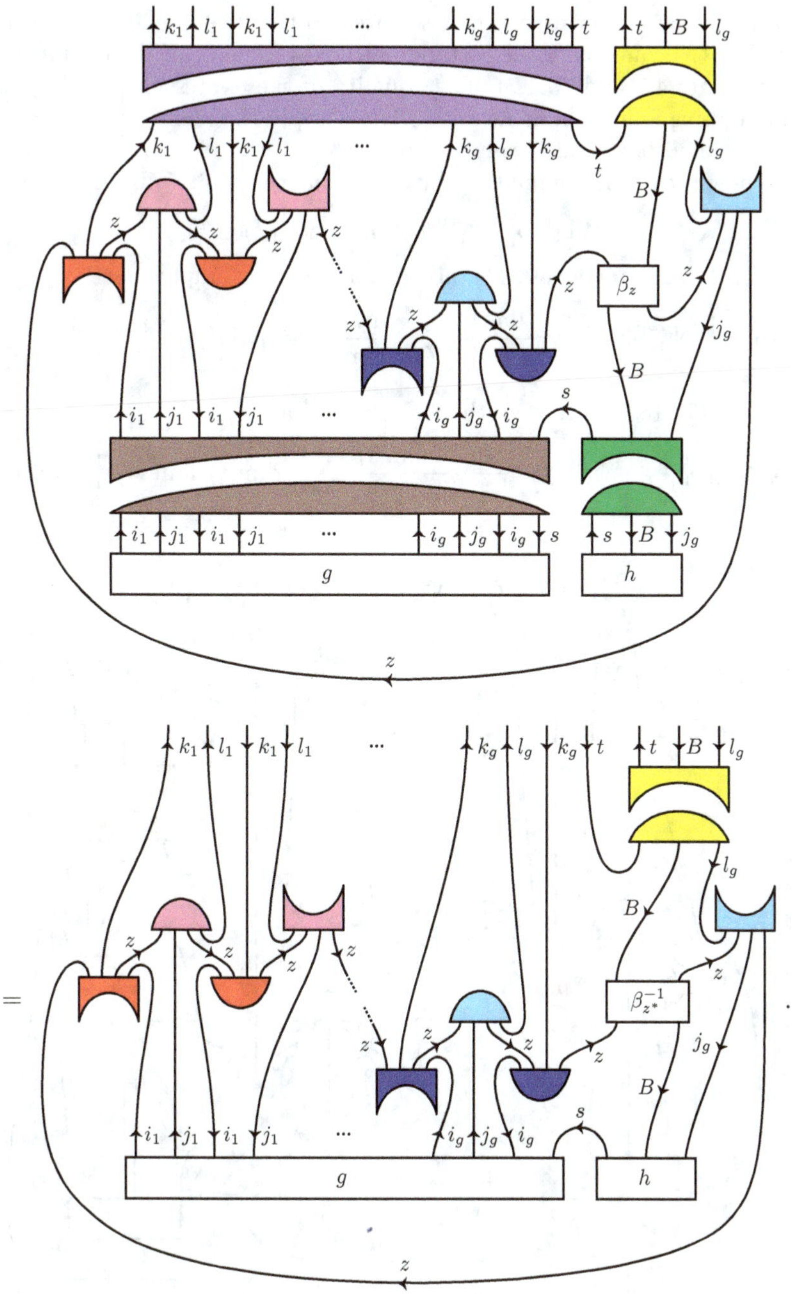

Here, in the first diagram, the pairs of purple, yellow, brown, and green curvilinear boxes come from the contraction vectors associated to the four unoriented rims of d not lying in $\Sigma \times \{\frac{1}{2}\}$ (see Lemma 4.8), the other pairs of curvilinear boxes similarly come from the contraction vectors associated to the $2g+1$ unoriented rims of d lying in $\Sigma \times \{\frac{1}{2}\}$, and the three rectangular boxes come from the definition of $\mathbb{F}_{\mathcal{C}}$ (see Sections 15.4.1 and 15.4.7). The equality of the expressions in the first and second diagrams follows from Lemma 5.1 (we apply the identity $({}^{\dagger}\beta)_z^{-1} = (\beta^{\dagger})_z^{-1}$ to the β_z-colored box) and Formula (4.9).

We now prove the commutativity of the diagram (16.15) by showing that

$$\Phi_{\mathbf{k},\mathbf{l}} \Psi_{\mathbf{k},\mathbf{l}}\, p_{\mathbf{i},\mathbf{j}}^{\mathbf{k},\mathbf{l}}\, \Psi_{\mathbf{i},\mathbf{j}}^{-1} \Phi_{\mathbf{i},\mathbf{j}}^{-1}(f) = \pi_{\mathbf{i},\mathbf{j}}^{\mathbf{k},\mathbf{l}}(f) \in E_{\mathbf{k},\mathbf{l}} \tag{16.20}$$

for any $f \in E_{\mathbf{i},\mathbf{j}}$. Formulas (16.11), (16.12), and (16.14) imply that

$$\Phi_{\mathbf{k},\mathbf{l}} \Psi_{\mathbf{k},\mathbf{l}}\, p_{\mathbf{i},\mathbf{j}}^{\mathbf{k},\mathbf{l}}\, \Psi_{\mathbf{i},\mathbf{j}}^{-1} \Phi_{\mathbf{i},\mathbf{j}}^{-1}(f) = \frac{\dim(\mathbf{l})}{\dim(\mathbf{j})} \sum_{s,t \in I} \frac{\dim(t)}{\dim(s)}\, \Phi_{\mathbf{k},\mathbf{l},t} \Psi_{\mathbf{k},\mathbf{l},t}\, p_{\mathbf{i},\mathbf{j},s}^{\mathbf{k},\mathbf{l},t}\, \Psi_{\mathbf{i},\mathbf{j},s}^{-1} \widetilde{\Phi}_{\mathbf{i},\mathbf{j},s}(f).$$

Substituting (16.18), we rewrite the last equality as

$$\Phi_{\mathbf{k},\mathbf{l}} \Psi_{\mathbf{k},\mathbf{l}}\, p_{\mathbf{i},\mathbf{j}}^{\mathbf{k},\mathbf{l}}\, \Psi_{\mathbf{i},\mathbf{j}}^{-1} \Phi_{\mathbf{i},\mathbf{j}}^{-1}(f) = \frac{\dim(\mathbf{i}) \dim(\mathbf{l})}{\dim(\mathcal{C})} \sum_{s,t,z \in I} \dim(t) \dim(z)\, \Delta_{\mathbf{i},\mathbf{j},s}^{\mathbf{k},\mathbf{l},t}(z,f),$$

where

$$\Delta_{\mathbf{i},\mathbf{j},s}^{\mathbf{k},\mathbf{l},t}(z,f) = \Phi_{\mathbf{k},\mathbf{l},t} \Psi_{\mathbf{k},\mathbf{l},t}\, P_{\mathbf{i},\mathbf{j},s}^{\mathbf{k},\mathbf{l},t}(z)\, \Psi_{\mathbf{i},\mathbf{j},s}^{-1} \widetilde{\Phi}_{\mathbf{i},\mathbf{j},s}(f)$$

$=$ 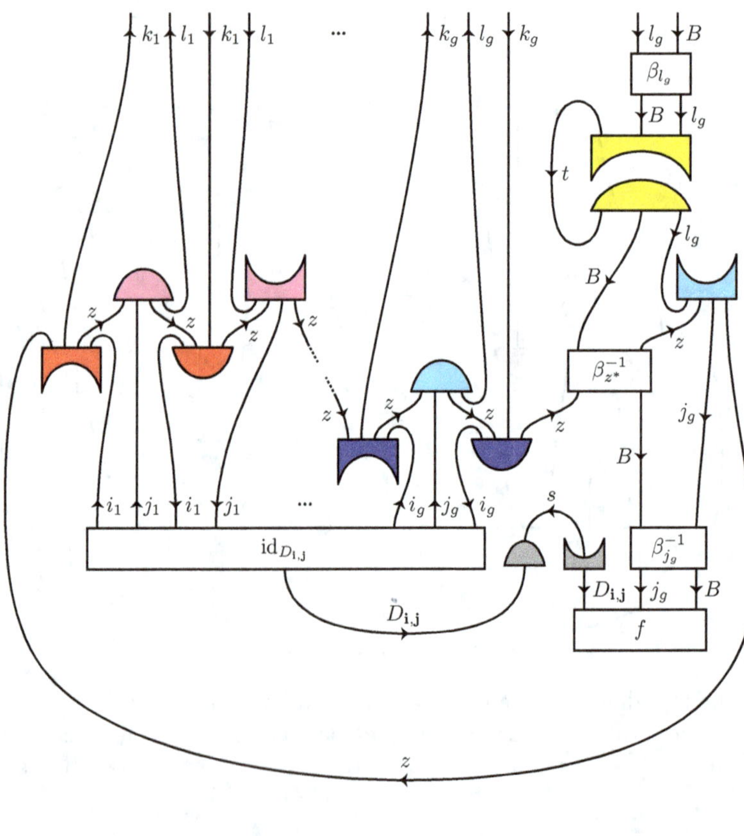

.

Here, the second equality follows from (16.9), (16.10), and the above computation of the vector (16.19). Thus

$$\Phi_{\mathbf{k},\mathbf{l}}\Psi_{\mathbf{k},\mathbf{l}}\, p^{\mathbf{k},\mathbf{l}}_{\mathbf{i},\mathbf{j}}\, \Psi^{-1}_{\mathbf{i},\mathbf{j}}\Phi^{-1}_{\mathbf{i},\mathbf{j}}(f) = \frac{\dim(\mathbf{i})\dim(\mathbf{l})}{\dim(\mathcal{C})} \sum_{z\in I} \dim(z)\, \Delta^{\mathbf{k},\mathbf{l}}_{\mathbf{i},\mathbf{j}}(z,f), \tag{16.21}$$

where

$$\Delta^{\mathbf{k},\mathbf{l}}_{\mathbf{i},\mathbf{j}}(z,f) = \sum_{s,t\in I} \dim(t)\, \Delta^{\mathbf{k},\mathbf{l},t}_{\mathbf{i},\mathbf{j},s}(z,f).$$

Now, by Lemma 4.9(d) and Formula (4.9), we have

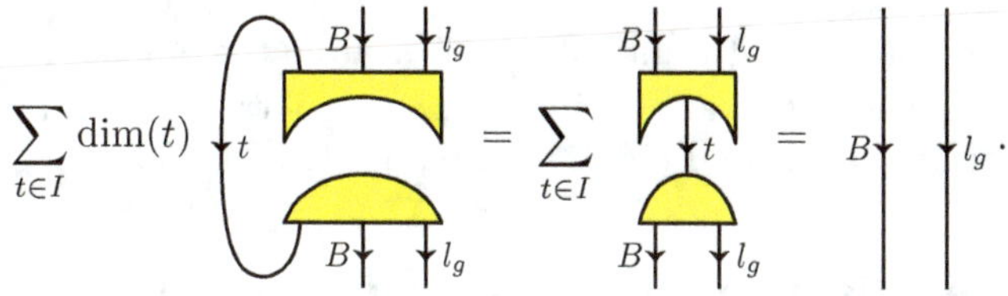

Also, Formula (4.9) gives that

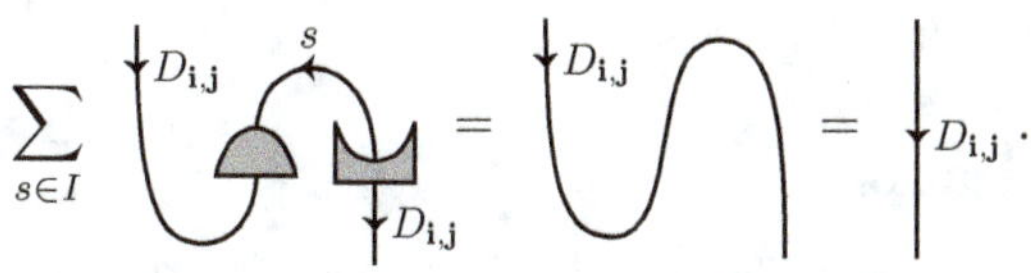

Using these formulas, the ⊗-multiplicativity (5.1) of β, and the above computation of the vector $\Delta^{\mathbf{k},\mathbf{l},t}_{\mathbf{i},\mathbf{j},s}(z,f)$, we obtain that

$\Delta^{\mathbf{k},\mathbf{l}}_{\mathbf{i},\mathbf{j}}(z,f) =$

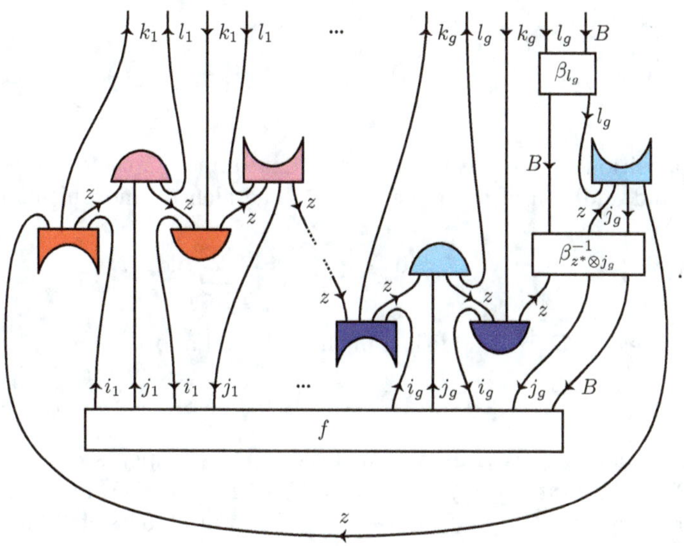

This expression can be simplified using that for any morphism $m\colon z^*\otimes j_g\otimes z\to l_g$,

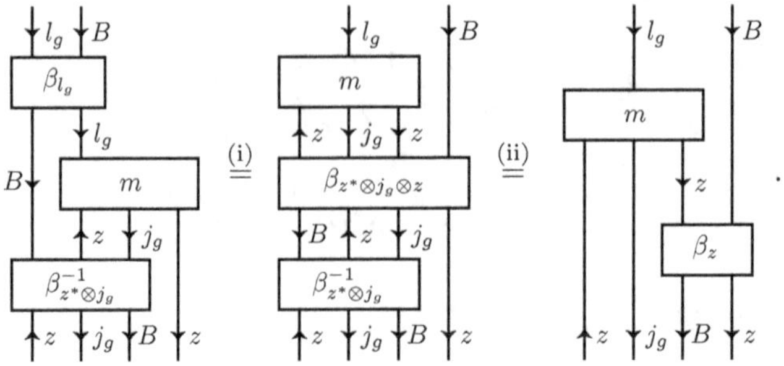

Here, (i) follows from the naturality of β and (ii) from the $\otimes$-multiplicativity (5.1) of β. As a consequence, comparing with (16.17), we obtain $\Delta^{\mathbf{k},\mathbf{l}}_{\mathbf{i},\mathbf{j}}(z,f)=\Pi^{\mathbf{k},\mathbf{l}}_{\mathbf{i},\mathbf{j}}(z,f)$. Finally, comparing (16.16) and (16.21), we obtain (16.20). □

16.3 Generalization to non-additive categories

We generalize Theorem 16.2 to non-additive categories. We start by explaining how to additivize linear categories.

16.3.1 Additivization

Let $\mathcal{D}$ be a $\Bbbk$-category. The *additivization* $\mathcal{D}^{\mathrm{add}}$ of $\mathcal{D}$ is the $\Bbbk$-category whose objects are finite sequences of objects of $\mathcal{D}$ and whose morphisms are matrices of morphisms in $\mathcal{D}$. In other words, the morphisms in $\mathcal{D}^{\mathrm{add}}$ from a sequence $(X_1,\dots,X_m)$ of $m\geq 0$ objects of $\mathcal{D}$ to a sequence $(Y_1,\dots,Y_n)$ of $n\geq 0$ objects of $\mathcal{D}$ are elements of the $\Bbbk$-module

$$\mathrm{Hom}_{\mathcal{D}^{\mathrm{add}}}\big((X_1,\dots,X_m),(Y_1,\dots,Y_n)\big)=\bigoplus_{1\leq i\leq m,\,1\leq j\leq n}\mathrm{Hom}_{\mathcal{D}}(X_i,Y_j).$$

Composition of morphisms in $\mathcal{D}^{\mathrm{add}}$ is determined in the obvious way by composition and addition of morphisms in $\mathcal{D}$. The identity morphisms in $\mathcal{D}^{\mathrm{add}}$ are given by

$$\mathrm{id}_{(X_1,\dots,X_m)}=\sum_{1\leq i\leq m}\mathrm{id}_{X_i}.$$

The category $\mathcal{D}^{\mathrm{add}}$ is additive: in the notation above,

$$(X_1,\dots,X_m)\oplus(Y_1,\dots,Y_n)=(X_1,\dots,X_m,Y_1,\dots,Y_n)$$

with obvious projections and inclusions. The empty sequence is a zero object of $\mathcal{D}^{\mathrm{add}}$. The *inclusion functor* $E_{\mathcal{D}}\colon\mathcal{D}\to\mathcal{D}^{\mathrm{add}}$ carries an object X of $\mathcal{D}$ to the 1-term sequence (X) and carries a morphism in $\mathcal{D}$ to itself. This functor is $\Bbbk$-linear and fully faithful. Any $\Bbbk$-linear functor from $\mathcal{D}$ to an additive $\Bbbk$-category $\mathcal{E}$ splits uniquely (up to unique natural isomorphism) as the composition of $E_\mathcal{D}$ with a $\Bbbk$-linear functor $\mathcal{D}^{\mathrm{add}} \to \mathcal{E}$. If $\mathcal{D}$ is additive, then $E_\mathcal{D}$ is an equivalence with inverse $(X_1, \ldots, X_m) \mapsto X_1 \oplus \cdots \oplus X_m$.

The properties of the $\Bbbk$-category $\mathcal{D}$ extend to $\mathcal{D}^{\mathrm{add}}$ as follows.

- If $\mathcal{D}$ is monoidal, then $\mathcal{D}^{\mathrm{add}}$ has a unique (up to monoidal equivalence) structure of a monoidal $\Bbbk$-category so that the inclusion functor $E_\mathcal{D}$ is strict monoidal. In particular $\mathbb{1}_{\mathcal{D}^{\mathrm{add}}} = (\mathbb{1})$ and $(X) \otimes (Y) = (X \otimes Y)$ for all $X, Y \in \mathrm{Ob}(\mathcal{D})$. A specific monoidal product in $\mathcal{D}^{\mathrm{add}}$ satisfying these conditions is defined using the formula
$$(X_1, \ldots, X_m) \otimes (Y_1, \ldots, Y_n) = (X_i \otimes Y_j)_{1 \le i \le m,\, 1 \le j \le n}$$
where the pairs (i, j) are ordered lexicographically.
- If $\mathcal{D}$ is pivotal, then $\mathcal{D}^{\mathrm{add}}$ has a structure of a pivotal $\Bbbk$-category such that $E_\mathcal{D}$ is strictly pivotal. If $\mathcal{D}$ is spherical, then so is $\mathcal{D}^{\mathrm{add}}$.
- If $\mathcal{D}$ is fusion, then so is $\mathcal{D}^{\mathrm{add}}$ and $E_\mathcal{D}$ carries any representative set of simple objects of $\mathcal{D}$ bijectively onto a representative set of simple objects of $\mathcal{D}^{\mathrm{add}}$. If $\mathcal{D}$ is pivotal fusion, then so is $\mathcal{D}^{\mathrm{add}}$ and $\dim(\mathcal{D}^{\mathrm{add}}) = \dim(\mathcal{D})$.

Lemma 16.3. *Let $\mathcal{D}$ be a monoidal $\Bbbk$-category and $E = E_\mathcal{D} \colon \mathcal{D} \to \mathcal{D}^{\mathrm{add}}$. There is a unique $\Bbbk$-linear functor $H = H_\mathcal{D} \colon \mathcal{Z}(\mathcal{D}) \to \mathcal{Z}(\mathcal{D}^{\mathrm{add}})$ such that*

1. *the diagram*
$$\begin{array}{ccc} \mathcal{Z}(\mathcal{D}) & \xrightarrow{\;H\;} & \mathcal{Z}(\mathcal{D}^{\mathrm{add}}) \\ \downarrow & & \downarrow \\ \mathcal{D} & \xrightarrow{\;E\;} & \mathcal{D}^{\mathrm{add}} \end{array}$$
commutes, where the vertical arrows stand for the forgetful functors;
2. *for any half-braiding (M, σ) of $\mathcal{D}$, the half-braiding $H(M, \sigma) = (E(M), \widetilde{\sigma})$ of $\mathcal{D}^{\mathrm{add}}$ satisfies $\widetilde{\sigma}_{E(X)} = \sigma_X$ for all $X \in \mathrm{Ob}(\mathcal{D})$.*

The functor H is fully faithful and braided strict monoidal. If $\mathcal{D}$ is pivotal, then the functor H is strictly pivotal.

Proof. For any half-braiding (M, σ) of $\mathcal{D}$, set $H(M, \sigma) = (E(M), \widetilde{\sigma})$ where
$$\widetilde{\sigma} = \{\widetilde{\sigma}_Y \colon E(M) \otimes Y \to Y \otimes E(M)\}_{Y \in \mathrm{Ob}(\mathcal{D}^{\mathrm{add}})}$$
is the half-braiding of $\mathcal{D}^{\mathrm{add}}$ defined as follows. For $Y \in \mathrm{Ob}(\mathcal{D}^{\mathrm{add}})$, pick a direct sum decomposition $Y = \bigoplus_\alpha E(X_\alpha)$ with $X_\alpha \in \mathrm{Ob}(\mathcal{D})$. Let $p_\alpha \colon Y \to E(X_\alpha)$ and $q_\alpha \colon E(X_\alpha) \to Y$ be the associated projection and inclusion. Set
$$\widetilde{\sigma}_Y = \sum_\alpha (q_\alpha \otimes \mathrm{id}_{E(M)}) \sigma_{X_\alpha} (\mathrm{id}_{E(M)} \otimes p_\alpha).$$
For any morphism f in $\mathcal{Z}(\mathcal{D})$, set $H(f) = E(f) = f$. It is straightforward to verify that the resulting functor H is well defined and has the required properties. □

16.3.2 Example

Given a group G, the additivization of the $\Bbbk$-category $G_\Bbbk$ from Example 4.2.5 is equivalent to the $\Bbbk$-category $G\text{-}\mathrm{vect}^1_\Bbbk$ of G-graded free $\Bbbk$-modules of finite rank (see Appendix A.2).

16.3.3 Generalization of Theorem 16.2

Recall that $\mathcal{C}$ is a spherical fusion $\Bbbk$-category such that $\dim(\mathcal{C})$ is invertible in $\Bbbk$. Then $\mathcal{C}^{\mathrm{add}}$ is an additive spherical fusion $\Bbbk$-category with $\dim(\mathcal{C}^{\mathrm{add}}) = \dim(\mathcal{C})$. These categories yield graph TQFTs $|\cdot|_\mathcal{C}$ over $\mathcal{Z}(\mathcal{C})$ and $|\cdot|_{\mathcal{C}^{\mathrm{add}}}$ over $\mathcal{Z}(\mathcal{C}^{\mathrm{add}})$. The former is computed from the latter as follows. Consider the pivotal functor

$$H = H_\mathcal{C} \colon \mathcal{Z}(\mathcal{C}) \to \mathcal{Z}(\mathcal{C}^{\mathrm{add}})$$

from Lemma 16.3. By Exercise 15.2.5, this functor induces a symmetric strict monoidal functor

$$\widetilde{H} \colon \mathrm{Cob}_3^{\mathcal{Z}(\mathcal{C})} \to \mathrm{Cob}_3^{\mathcal{Z}(\mathcal{C}^{\mathrm{add}})}.$$

Then the composition $|\cdot|_{\mathcal{C}^{\mathrm{add}}} \circ \widetilde{H} \colon \mathrm{Cob}_3^{\mathcal{Z}(\mathcal{C})} \to \mathrm{Mod}_\Bbbk$ is a graph TQFT over $\mathcal{Z}(\mathcal{C})$.

Lemma 16.4. *The graph TQFTs $|\cdot|_\mathcal{C}$ and $|\cdot|_{\mathcal{C}^{\mathrm{add}}} \circ \widetilde{H}$ are isomorphic. Consequently, the state sum TQFTs $|\cdot|_\mathcal{C}$ and $|\cdot|_{\mathcal{C}^{\mathrm{add}}}$ are isomorphic. In particular, $|M|_\mathcal{C} = |M|_{\mathcal{C}^{\mathrm{add}}}$ for any closed oriented 3-manifold M.*

Proof. The first claim follows directly from the definitions since the inclusion functor $\mathcal{C} \to \mathcal{C}^{\mathrm{add}}$ is a strictly pivotal strict monoidal functor which carries a representative set of simple objects of $\mathcal{C}$ bijectively onto a representative set of simple objects of $\mathcal{C}^{\mathrm{add}}$. Restricting the graph TQFTs $|\cdot|_\mathcal{C}$ and $|\cdot|_{\mathcal{C}^{\mathrm{add}}} \circ \widetilde{H}$ to Cob_3 and noticing that the composition of the inclusion $\mathrm{Cob}_3 \hookrightarrow \mathrm{Cob}_3^{\mathcal{Z}(\mathcal{C})}$ with $\widetilde{H}$ is the inclusion $\mathrm{Cob}_3 \hookrightarrow \mathrm{Cob}_3^{\mathcal{Z}(\mathcal{C}^{\mathrm{add}})}$, we obtain the second claim of the lemma. □

Theorem 16.5. *For any connected $\mathcal{Z}(\mathcal{C})$-colored surface Σ of genus $g \geq 0$,*

$$|\Sigma|_\mathcal{C} \simeq \mathrm{Hom}_{\mathcal{Z}(\mathcal{C}^{\mathrm{add}})}\big(\mathbb{1}_{\mathcal{Z}(\mathcal{C}^{\mathrm{add}})}, (C,\sigma)^{\otimes g} \otimes H(V_\Sigma)\big),$$

where $(C,\sigma) \in \mathrm{Ob}(\mathcal{Z}(\mathcal{C}^{\mathrm{add}}))$ is a coend of $\mathcal{Z}(\mathcal{C}^{\mathrm{add}})$ and $V_\Sigma \in \mathrm{Ob}(\mathcal{Z}(\mathcal{C}))$ is as in Section 16.2.

Proof. By Lemma 16.4 and Theorem 16.2, we have

$$|\Sigma|_\mathcal{C} \simeq |\widetilde{H}(\Sigma)|_{\mathcal{C}^{\mathrm{add}}} \simeq \mathrm{Hom}_{\mathcal{Z}(\mathcal{C}^{\mathrm{add}})}\big(\mathbb{1}_{\mathcal{Z}(\mathcal{C}^{\mathrm{add}})}, (C,\sigma)^{\otimes g} \otimes V_{\widetilde{H}(\Sigma)}\big).$$

We conclude by noticing that $V_{\widetilde{H}(\Sigma)} = H(V_\Sigma)$. □

To compute the $\Bbbk$-module $|\Sigma|_\mathcal{C}$ for a disconnected $\mathcal{Z}(\mathcal{C})$-colored surface Σ, one applies Theorem 16.5 to the connected components of Σ and takes the (non-ordered) tensor product of the resulting $\Bbbk$-modules, see Formula (15.1).

16.3.4 Exercise*

Let G be a finite non-abelian group and let $\mathcal{C} = G_{\Bbbk}^{1,1}$ be the monoidal $\Bbbk$-category from Example 4.2.5. Prove that the functor $H\colon \mathcal{Z}(\mathcal{C}) \to \mathcal{Z}(\mathcal{C}^{\mathrm{add}})$ from Lemma 16.3 and the induced functor $\mathcal{Z}(\mathcal{C})^{\mathrm{add}} \to \mathcal{Z}(\mathcal{C}^{\mathrm{add}})$ are not essentially surjective in the sense of Section 1.1.7. As a consequence, these functors are not equivalences.

16.3.5 Exercise

Prove that for any braided $\Bbbk$-category $\mathcal{D}$, its additivization $\mathcal{D}^{\mathrm{add}}$ has a structure of a braided $\Bbbk$-category such that the inclusion functor $\mathcal{D} \to \mathcal{D}^{\mathrm{add}}$ is braided.

16.4 Extensions of scalars

We briefly discuss the behavior of the state sum graph TQFTs under extensions of scalars. Fix throughout this section a ring homomorphism $\varphi\colon \Bbbk \to \mathbb{K}$, where $\Bbbk$ and $\mathbb{K}$ are commutative rings.

16.4.1 Preliminaries

The homomorphism $\varphi\colon \Bbbk \to \mathbb{K}$ induces a right $\Bbbk$-module structure on $\mathbb{K}$ defined by $\alpha \cdot k = \alpha\varphi(k)$ for all $\alpha \in \mathbb{K}$ and $k \in \Bbbk$. This structure determines a functor $\varphi_*\colon \mathrm{Mod}_{\Bbbk} \to \mathrm{Mod}_{\mathbb{K}}$ by

$$\varphi_*(M) = \mathbb{K} \otimes_{\Bbbk} M \quad \text{and} \quad \varphi_*(f) = \mathrm{id}_{\mathbb{K}} \otimes_{\Bbbk} f$$

for any (left) $\Bbbk$-module M and any $\Bbbk$-linear homomorphism f between $\Bbbk$-modules. Here the $\mathbb{K}$-module structure of $\varphi_*(M)$ is given by

$$\alpha \cdot (\beta \otimes_{\Bbbk} m) = (\alpha\beta) \otimes_{\Bbbk} m$$

for all $\alpha, \beta \in \mathbb{K}$ and $m \in M$. The functor φ_* is a symmetric strong monoidal functor with monoidal constraints

$$(\varphi_*)_0\colon \mathbb{K} \to \varphi_*(\Bbbk) \quad \text{and} \quad (\varphi_*)_2(M,N)\colon \varphi_*(M) \otimes_{\mathbb{K}} \varphi_*(N) \to \varphi_*(M \otimes_{\Bbbk} M)$$

given by $(\varphi_*)_0(\alpha) = \alpha \otimes_{\Bbbk} 1_{\Bbbk}$ and

$$(\varphi_*)_2(M,N)((\alpha \otimes_{\Bbbk} m) \otimes_{\mathbb{K}} (\beta \otimes_{\Bbbk} n)) = (\alpha\beta) \otimes_{\Bbbk} (m \otimes_{\Bbbk} n)$$

for all $\alpha, \beta \in \mathbb{K}$, $m \in M$, $n \in N$.

16.4.2 Transformation of $\Bbbk$-categories

Any $\Bbbk$-category $\mathcal{C}$ gives rise to a $\mathbb{K}$-category $\mathcal{C}_\varphi$ with the same objects as $\mathcal{C}$ and with

$$\mathrm{Hom}_{\mathcal{C}_\varphi}(X,Y) = \varphi_*(\mathrm{Hom}_{\mathcal{C}}(X,Y)) = \mathbb{K} \otimes_\Bbbk \mathrm{Hom}_{\mathcal{C}}(X,Y)$$

for any $X, Y \in \mathrm{Ob}(\mathcal{C})$. The composition in $\mathcal{C}_\varphi$ is the $\mathbb{K}$-bilinear extension of the rule

$$(\alpha \otimes_\Bbbk f) \circ (\beta \otimes_\Bbbk g) = \alpha\beta \otimes_\Bbbk fg$$

for any $\alpha, \beta \in \mathbb{K}$ and any composable morphisms f, g in $\mathcal{C}$. The identity morphism of an object $X \in \mathrm{Ob}(\mathcal{C}_\varphi) = \mathrm{Ob}(\mathcal{C})$ is $1_\mathbb{K} \otimes \mathrm{id}_X \in \mathrm{Hom}_{\mathcal{C}_\varphi}(X,X)$.

We define a functor $\Phi\colon \mathcal{C} \to \mathcal{C}_\varphi$ by

$$\Phi(X) = X \quad \text{and} \quad \Phi(f) = 1_\mathbb{K} \otimes_\Bbbk f$$

for any $X \in \mathrm{Ob}(\mathcal{C})$ and any morphism f in $\mathcal{C}$. Clearly,

$$\Phi(kf) = \varphi(k)\,\Phi(f) \quad \text{and} \quad \Phi(f+g) = \Phi(f) + \Phi(g)$$

for all $k \in \Bbbk$ and all morphisms f, g in $\mathcal{C}$ with the same source and target.

If $\mathcal{C}$ is a monoidal $\Bbbk$-category, then $\mathcal{C}_\varphi$ has a unique structure of a monoidal $\mathbb{K}$-category such that the functor Φ is strict monoidal. The monoidal unit of $\mathcal{C}_\varphi$ is the monoidal unit $\mathbb{1}$ of $\mathcal{C}$, and the monoidal product $\otimes_\varphi$ of $\mathcal{C}_\varphi$ is given by

$$X \otimes_\varphi Y = X \otimes Y \quad \text{and} \quad (\alpha \otimes_\Bbbk f) \otimes_\varphi (\beta \otimes_\Bbbk g) = (\alpha\beta) \otimes_\Bbbk (f \otimes g)$$

for any $X, Y \in \mathrm{Ob}(\mathcal{C})$, $\alpha, \beta \in \mathbb{K}$, and any morphisms f, g in $\mathcal{C}$. Moreover, if $\mathcal{C}$ is pivotal, then $\mathcal{C}_\varphi$ has a unique structure of a pivotal category such that the functor Φ is strictly pivotal. Then, for any endomorphism f in $\mathcal{C}$,

$$\mathrm{tr}_l(\Phi(f)) = \varphi(\mathrm{tr}_l(f)) \quad \text{and} \quad \mathrm{tr}_r(\Phi(f)) = \varphi(\mathrm{tr}_r(f)). \tag{16.22}$$

Consequently, for any object X of $\mathcal{C}$,

$$\dim_l(\Phi(X)) = \varphi(\dim_l(X)) \quad \text{and} \quad \dim_r(\Phi(X)) = \varphi(\dim_r(X)).$$

It also follows from (16.22) that if $\mathcal{C}$ is spherical, then so is $\mathcal{C}_\varphi$.

If $\mathcal{C}$ is a pre-fusion $\Bbbk$-category, then $\mathcal{C}_\varphi$ is a pre-fusion $\mathbb{K}$-category and any representative set I of simple objects of $\mathcal{C}$ is a representative set of simple objects of $\mathcal{C}_\varphi$. Indeed, all simple objects of $\mathcal{C}$ are simple objects of $\mathcal{C}_\varphi$ because $\mathbb{K} \otimes_\Bbbk \Bbbk \simeq \mathbb{K}$ as $\mathbb{K}$-modules. Hence, I is a set of simple objects of $\mathcal{C}_\varphi$. Since the functor Φ carries isomorphisms to isomorphisms, any objects $i, j \in I$ which are non-isomorphic in $\mathcal{C}_\varphi$ must be non-isomorphic in $\mathcal{C}$ and so

$$\mathrm{Hom}_{\mathcal{C}_\varphi}(i,j) = \mathbb{K} \otimes_\Bbbk \mathrm{Hom}_{\mathcal{C}}(i,j) = \mathbb{K} \otimes_\Bbbk 0 = 0.$$

Finally, since Φ is additive on morphisms, it preserves the direct sums, and so every object of $\mathcal{C}_\varphi$ is a direct sum (in $\mathcal{C}_\varphi$) of a finite family of elements of I.

As a consequence, if $\mathcal{C}$ is a pivotal fusion $\Bbbk$-category, then $\mathcal{C}_\varphi$ is a pivotal fusion $\mathbb{K}$-category and $\dim(\mathcal{C}_\varphi) = \varphi(\dim(\mathcal{C}))$.

16.4.3 Comparison of graph TQFTs

Let $\mathcal{C}$ be a spherical fusion $\Bbbk$-category such that $\dim(\mathcal{C})$ is an invertible element of $\Bbbk$. Then $\mathcal{C}_\varphi$ is a spherical fusion $\mathbb{K}$-category and $\dim(\mathcal{C}_\varphi) = \varphi(\dim(\mathcal{C}))$ is an invertible element of $\mathbb{K}$. We now compare the associated graph TQFTs $|\cdot|_\mathcal{C}$ over $\mathcal{Z}(\mathcal{C})$ and $|\cdot|_{\mathcal{C}_\varphi}$ over $\mathcal{Z}(\mathcal{C}_\varphi)$. To this end, we define a functor $F\colon \mathcal{Z}(\mathcal{C}) \to \mathcal{Z}(\mathcal{C}_\varphi)$ by

$$F(M,\sigma) = (M, \{\Phi(\sigma_X)\}_{X\in \mathrm{Ob}(\mathcal{C})}) \quad \text{and} \quad F(f) = \Phi(f)$$

for any $(M,\sigma) \in \mathrm{Ob}(\mathcal{Z}(\mathcal{C}))$ and any morphism f in $\mathcal{Z}(\mathcal{C})$, where $\Phi\colon \mathcal{C} \to \mathcal{C}_\varphi$ is the strictly pivotal strict monoidal functor of the previous section. The functor F is pivotal and so induces a symmetric strict monoidal functor

$$\widetilde{F}\colon \mathrm{Cob}_3^{\mathcal{Z}(\mathcal{C})} \to \mathrm{Cob}_3^{\mathcal{Z}(\mathcal{C}_\varphi)}$$

(see Exercise 15.2.5). It is a direct consequence of the definitions that the following diagram of symmetric monoidal functors commutes:

$$\begin{array}{ccc} \mathrm{Cob}_3^{\mathcal{Z}(\mathcal{C})} & \xrightarrow{\ \widetilde{F}\ } & \mathrm{Cob}_3^{\mathcal{Z}(\mathcal{C}_\varphi)} \\ {\scriptstyle |\cdot|_\mathcal{C}}\big\downarrow & & \big\downarrow{\scriptstyle |\cdot|_{\mathcal{C}_\varphi}} \\ \mathrm{Mod}_\Bbbk & \xrightarrow{\ \varphi_*\ } & \mathrm{Mod}_\mathbb{K}. \end{array}$$

In other words, we have the following equality of $\mathrm{Mod}_\mathbb{K}$-valued graph TQFTs over $\mathcal{Z}(\mathcal{C})$:

$$\varphi_* \circ |\cdot|_\mathcal{C} = |\cdot|_{\mathcal{C}_\varphi} \circ \widetilde{F}.$$

In particular, for any closed oriented 3-manifold M and any $\mathcal{Z}(\mathcal{C})$-colored ribbon graph R in M, we have

$$|M, R_\varphi|_{\mathcal{C}_\varphi} = \varphi(|M,R|_\mathcal{C}), \tag{16.23}$$

where R_φ is the $\mathcal{Z}(\mathcal{C}_\varphi)$-colored ribbon graph in M obtained from R by applying the functor F to all colors.

Chapter 17

Surgery computation

17.1 The main theorem

We formulate in this section the main result of the monograph (Theorem 17.1) which interprets the state sum graph TQFT as a surgery graph TQFT. We start by recalling the surgery graph TQFT due to Reshetikhin–Turaev [RT, Tu1].

17.1.1 The surgery graph TQFT

The Reshetikhin–Turaev construction starts with a modular $\Bbbk$-category $\mathcal{B}$ (see Section 4.5.2) equipped with a square root $\Delta \in \Bbbk$ of $\dim(\mathcal{B})$. The construction produces a symmetric strong monoidal functor $\tau_{\mathcal{B}}$ from a certain extension of the category $\mathrm{Cob}_3^{\mathcal{B}}$ of $\mathcal{B}$-colored ribbon graphs in 3-cobordisms to $\mathrm{Mod}_{\Bbbk}$. The extension in question is obtained by providing surfaces with Lagrangian subspaces in real 1-homology and by providing 3-cobordisms with integer weights. For detailed definitions, we refer to [Tu1]. Here we focus on the case where $\mathcal{B}$ is an anomaly free modular $\Bbbk$-category. Then the element $\Delta_+ = \Delta_-$ of $\Bbbk$ is a square root of $\dim(\mathcal{B})$, see Section 4.5.2. The functor $\tau_{\mathcal{B}}$ associated with $\mathcal{B}$ and $\Delta = \Delta_\pm$ does not depend on the choice of Lagrangian spaces in real homology of surfaces and of integer weights of cobordisms. This functor is a graph TQFT over $\mathcal{B}$ in the sense of Section 15.2.2. By [Tu1, Chapter IV], the graph TQFT $\tau_{\mathcal{B}}$ is non-degenerate and regular in the sense of Section 15.2.3.

We recall the definition of the scalar invariant $\tau_{\mathcal{B}}(M, R) \in \Bbbk$ of a pair (M, R) where M is a closed connected oriented 3-manifold and $R \subset M$ is a $\mathcal{B}$-colored ribbon graph. The Lickorish–Wallace theorem on surgery presentations of 3-manifolds implies that there are a framed link $L = L_1 \cup \cdots \cup L_n$ in S^3 and a $\mathcal{B}$-colored ribbon graph R' in $S^3 \setminus L$ such that the surgery on L turns the pair (S^3, R') into (M, R), at least up to homeomorphism. Let $\mathcal{I}$ be a representative set of simple objects of $\mathcal{B}$. We orient L in an arbitrary way and let $\mathrm{col}(L)$ be the set of maps $\{1, \ldots, n\} \to \mathcal{I}$. For each $\lambda \in \mathrm{col}(L)$, the pair $(L \cup R', \lambda)$ is a $\mathcal{B}$-colored ribbon graph in S^3. Section 16.1.1 yields an isotopy invariant $\langle L \cup R', \lambda \rangle_{\mathcal{B}} \in \Bbbk$. Then

$$\tau_{\mathcal{B}}(M, R) = \Delta^{-n-1} \sum_{\lambda \in \mathrm{col}(L)} \left(\prod_{q=1}^{n} \dim\big(\lambda(q)\big) \right) \langle L \cup R', \lambda \rangle_{\mathcal{B}} \in \Bbbk.$$

For a connected $\mathcal{B}$-colored surface Σ of genus $g \geq 0$, the $\Bbbk$-module $\tau_{\mathcal{B}}(\Sigma)$ is computed as follows. Let A be the set of $\mathcal{B}$-colored points of Σ (see Section 15.1.1). For $a \in A$, let $V_a \in \mathrm{Ob}(\mathcal{B})$ and $\varepsilon_a \in \{+, -\}$ be the color and sign of a. Set

$$V_\Sigma = \bigotimes_{a \in A} V_a^{\varepsilon_a} \in \mathrm{Ob}(\mathcal{B}),$$

where the monoidal product is determined by a total order on A and, as usual, $V_a^+ = V_a$ and $V_a^- = V_a^*$. Since $\mathcal{B}$ is braided, the isomorphism class of the object V_Σ does not depend on the order on A. If $A = \emptyset$, then $V_\Sigma = \mathbb{1}_{\mathcal{B}}$. We have

$$\tau_{\mathcal{B}}(\Sigma) \simeq \bigoplus_{i_1, \ldots, i_g \in \mathcal{I}} \mathrm{Hom}_{\mathcal{B}}(\mathbb{1}_{\mathcal{B}}, i_1^* \otimes i_1 \otimes \cdots \otimes i_g^* \otimes i_g \otimes V_\Sigma).$$

If $\mathcal{B}$ is additive (see Section 4.1.2), then

$$\tau_{\mathcal{B}}(\Sigma) \simeq \mathrm{Hom}_{\mathcal{B}}\Big(\mathbb{1}_{\mathcal{B}}, \bigoplus_{i_1, \ldots, i_g \in \mathcal{I}} i_1^* \otimes i_1 \otimes \cdots \otimes i_g^* \otimes i_g \otimes V_\Sigma\Big).$$

Setting $C = \bigoplus_{i \in \mathcal{I}} i^* \otimes i \in \mathrm{Ob}(\mathcal{B})$, we obtain that

$$\tau_{\mathcal{B}}(\Sigma) \simeq \mathrm{Hom}_{\mathcal{B}}(\mathbb{1}_{\mathcal{B}}, C^{\otimes g} \otimes V_\Sigma). \tag{17.1}$$

17.1.2 The comparison theorem

Let $\Bbbk$ be an algebraically closed field and let $\mathcal{C}$ be an additive spherical fusion $\Bbbk$-category such that $\dim(\mathcal{C}) \neq 0$. By Theorem 5.4, the center $\mathcal{Z}(\mathcal{C})$ of $\mathcal{C}$ is an anomaly free modular $\Bbbk$-category with $\Delta_+ = \Delta_- = \dim(\mathcal{C})$. Consider the state sum graph TQFT $|\cdot|_{\mathcal{C}}$ over $\mathcal{Z}(\mathcal{C})$ (see Theorem 15.9) and the Reshetikhin-Turaev surgery graph TQFT $\tau_{\mathcal{Z}(\mathcal{C})}$ over $\mathcal{Z}(\mathcal{C})$. Recall the notion of isomorphism for graph TQFTs, see Section 15.2.2.

Theorem 17.1. *The graph TQFTs $|\cdot|_{\mathcal{C}}$ and $\tau_{\mathcal{Z}(\mathcal{C})}$ are isomorphic.*

Theorem 17.1 yields a surgery computation of the state sum graph TQFT $|\cdot|_{\mathcal{C}}$ and exhibits it as a surgery TQFT. It also exhibits the surgery graph TQFT $\tau_{\mathcal{Z}(\mathcal{C})}$ as a state sum graph TQFT. Theorem 17.1 was first proved in our preprint [TVi]. In the case where the characteristic of $\Bbbk$ is equal to zero, Theorem 17.1 was independently proved by Kirillov Jr. and Balsam, see [KB, Ba1, Ba2].

Theorem 17.1 is proved in Section 17.4 using the results of Sections 17.2 and 17.3. In Section 17.5 we discuss corollaries of Theorem 17.1.

17.2 Isomorphisms of monoidal functors

We establish sufficient conditions for the existence of a monoidal isomorphism between $\mathrm{Mod}_{\Bbbk}$-valued strong monoidal functors. These conditions will be used to prove Theorem 17.1.

We say that a functor F from a monoidal category $\mathcal{C}$ to $\mathrm{Mod}_{\Bbbk}$ is *non-degenerate* if for any object X of $\mathcal{C}$, the $\Bbbk$-module $F(X)$ is spanned by the images of the $\Bbbk$-linear homomorphisms $F(g)\colon F(\mathbb{1}) \to F(X)$ determined by all $g \in \mathrm{Hom}_{\mathcal{C}}(\mathbb{1}, X)$. It is clear that a functor $\mathcal{C} \to \mathrm{Mod}_{\Bbbk}$ isomorphic to a non-degenerate functor is itself non-degenerate. Note also that this notion of a non-degenerate functor generalizes that of a non-degenerate graph TQFT, see Section 15.2.3.

Lemma 17.2. *Assume that $\Bbbk$ is a field. Let F and G be strong monoidal functors from a left rigid category $\mathcal{C}$ to $\mathrm{Mod}_{\Bbbk}$ such that:*

(a) *F is non-degenerate;*

(b) *for all $X \in \mathrm{Ob}(\mathcal{C})$, we have $\dim_{\Bbbk} F(X) \geq \dim_{\Bbbk} G(X)$;*

(c) *for all $\alpha \in \mathrm{End}_{\mathcal{C}}(\mathbb{1})$, we have*

$$F_0^{-1}F(\alpha)F_0 = G_0^{-1}G(\alpha)G_0 \in \mathrm{End}_{\mathrm{Mod}_{\Bbbk}}(\Bbbk) = \Bbbk.$$

Then the functors F and G are monoidally isomorphic.

Proof. Note first that since $\Bbbk$ is a field, the category $\mathrm{Mod}_{\Bbbk}$ is nothing but the category of $\Bbbk$-vector spaces and $\Bbbk$-linear homomorphisms. Consider an arbitrary strong monoidal functor $H\colon \mathcal{C} \to \mathrm{Mod}_{\Bbbk}$. For an object X of $\mathcal{C}$, we let $H^+(X)$ be the $\Bbbk$-vector subspace of the $\Bbbk$-vector space $H(X)$ spanned by the images of the $\Bbbk$-linear homomorphisms $H(g)\colon H(\mathbb{1}) \to H(X)$ determined by all $g \in \mathrm{Hom}_{\mathcal{C}}(\mathbb{1}, X)$. Since $H_0\colon \Bbbk \to H(\mathbb{1})$ is a $\Bbbk$-linear isomorphism, the vector space $H^+(X)$ is spanned by the vectors

$$X_g^H = H(g)\, H_0(1_{\Bbbk}) \in H(X)$$

where g runs over $\mathrm{Hom}_{\mathcal{C}}(\mathbb{1}, X)$ and $1_{\Bbbk} \in \Bbbk$ is the unit element of $\Bbbk$. Next, pick a left dual $({}^{\vee}X, \mathrm{ev}_X)$ of X and let

$$\omega_X^H = H_0^{-1} H(\mathrm{ev}_X)\, H_2({}^{\vee}X, X)\colon H({}^{\vee}X) \otimes_{\Bbbk} H(X) \to \Bbbk$$

be the non-degenerate pairing induced by H and ev_X, see Lemma 1.5. Consider the annihilator $H^-(X) \subset H(X)$ of $H^+({}^{\vee}X) \subset H({}^{\vee}X)$ with respect to ω_X^H. By the uniqueness of a left dual (up to isomorphism), the $\Bbbk$-vector space $H^-(X)$ does not depend on the choice of $({}^{\vee}X, \mathrm{ev}_X)$. Set

$$\bar{H}(X) = H^+(X)/(H^+(X) \cap H^-(X)).$$

The class of a vector $x \in H^+(X)$ in $\bar{H}(X)$ is denoted by $[x]$. By Lemma 1.6, the $\Bbbk$-vector spaces $H(X)$ and $\bar{H}(X)$ are finite-dimensional.

We apply these constructions to the functors F and G. For $X \in \mathrm{Ob}(\mathcal{C})$, we define a $\Bbbk$-linear homomorphism $\varphi_X \colon \bar{F}(X) \to \bar{G}(X)$ by $\varphi_X([X_g^F]) = [X_g^G]$ for all $g \in \mathrm{Hom}_{\mathcal{C}}(\mathbb{1}, X)$. To show that φ_X is well defined, it suffices to prove that for any finite family $\{g_r \in \mathrm{Hom}_{\mathcal{C}}(\mathbb{1}, X), \lambda_r \in \Bbbk\}_r$ satisfying $\sum_r \lambda_r X_{g_r}^F \in F^-(X)$, we have $\sum_r \lambda_r X_{g_r}^G \in G^-(X)$. For any $h \in \mathrm{Hom}_{\mathcal{C}}(\mathbb{1}, {}^\vee X)$, we have:

$$\begin{aligned}
&\omega_X^G\Big(({}^\vee X)_h^G \otimes_\Bbbk \sum_r \lambda_r X_{g_r}^G\Big)\\
&\overset{\text{(i)}}{=} \sum_r \lambda_r\, G_0^{-1} G(\mathrm{ev}_X) G_2({}^\vee X, X)(G(h)G_0(1_\Bbbk) \otimes_\Bbbk G(g_r)G_0(1_\Bbbk))\\
&\overset{\text{(ii)}}{=} \sum_r \lambda_r\, G_0^{-1} G(\mathrm{ev}_X(h \otimes g_r)) G_0(1_\Bbbk)\\
&\overset{\text{(iii)}}{=} \sum_r \lambda_r\, F_0^{-1} F(\mathrm{ev}_X(h \otimes g_r)) F_0(1_\Bbbk)\\
&\overset{\text{(iv)}}{=} \sum_r \lambda_r\, F_0^{-1} F(\mathrm{ev}_X) F_2({}^\vee X, X)(F(h)F_0(1_\Bbbk) \otimes_\Bbbk F(g_r)F_0(1_\Bbbk))\\
&\overset{\text{(v)}}{=} \omega_X^F\Big(({}^\vee X)_h^F \otimes_\Bbbk \sum_r \lambda_r X_{g_r}^F\Big) \overset{\text{(vi)}}{=} 0.
\end{aligned}$$

Here the equalities (i) and (v) follow from the definition of the pairings ω_X^F, ω_X^G and the vectors $X_{g_r}^F, X_{g_r}^G$, (ii) and (iv) from the monoidality of F and G, (iii) from the assumption (c), and (vi) from the assumption $\sum_r \lambda_r X_{g_r}^F \in F^-(X)$. Since the vectors $({}^\vee X)_h^G$ with $h \in \mathrm{Hom}_{\mathcal{C}}(\mathbb{1}, {}^\vee X)$ span the $\Bbbk$-vector space $G^+({}^\vee X)$, we conclude that $\sum_r \lambda_r X_{g_r}^G \in G^-(X)$. Hence, $\varphi_X \colon \bar{F}(X) \to \bar{G}(X)$ is a well defined $\Bbbk$-linear homomorphism. Exchanging F and G, we similarly construct a $\Bbbk$-linear homomorphism $\bar{G}(X) \to \bar{F}(X)$. It follows from the definitions that the latter homomorphism is inverse to φ_X. Hence φ_X is a $\Bbbk$-linear isomorphism.

By the assumption (a), $F^+(X) = F(X)$ and $F^+({}^\vee X) = F({}^\vee X)$. The non-degeneracy of ω_X^F implies that $F^-(X) = 0$, and so $\bar{F}(X) = F(X)$. Using the assumption (b), we obtain

$$\dim_\Bbbk \bar{G}(X) = \dim_\Bbbk \bar{F}(X) = \dim_\Bbbk F(X) \geq \dim_\Bbbk G(X).$$

Since $\bar{G}(X)$ is a quotient of the vector subspace $G^+(X)$ of $G(X)$, we must have $\bar{G}(X) = G^+(X) = G(X)$. Hence φ_X is a $\Bbbk$-linear isomorphism from $F(X)$ to $G(X)$ carrying X_g^F to X_g^G for all $g \in \mathrm{Hom}_{\mathcal{C}}(\mathbb{1}, X)$.

Let us prove that the family $\varphi = \{\varphi_X \colon F(X) \to G(X)\}_{X \in \mathrm{Ob}(\mathcal{C})}$ is a monoidal natural isomorphism. Let $f \colon X \to Y$ be a morphism in $\mathcal{C}$. For $g \in \mathrm{Hom}_{\mathcal{C}}(\mathbb{1}, X)$,

$$F(f)(X_g^F) = F(f)F(g)F_0(1_\Bbbk) = F(fg)F_0(1_\Bbbk) = Y_{fg}^F$$

and similarly $G(f)(X_g^G) = Y_{fg}^G$. Thus

$$G(f)\varphi_X(X_g^F) = G(f)(X_g^G) = Y_{fg}^G = \varphi_Y(Y_{fg}^F) = \varphi_Y F(f)(X_g^F).$$

Since the vectors X_g^F span $F(X)$, we deduce that $G(f)\varphi_X = \varphi_Y F(f)$. Hence φ is a natural transformation, which is invertible (since each φ_X is an isomorphism). It remains to check that φ is monoidal. By definition, $\varphi_{\mathbb{1}}$ carries $F_0(1_{\Bbbk})$ to $G_0(1_{\Bbbk})$, and so $\varphi_{\mathbb{1}} F_0 = G_0$. Finally, let $g \in \mathrm{Hom}_{\mathcal{C}}(\mathbb{1}, X)$ and $h \in \mathrm{Hom}_{\mathcal{C}}(\mathbb{1}, Y)$. Using the monoidality of F, we obtain

$$\begin{aligned} &F_2(X,Y)(X_g^F \otimes_{\Bbbk} Y_h^F) = F_2(X,Y)(F(g)F_0(1_{\Bbbk}) \otimes_{\Bbbk} F(h)F_0(1_{\Bbbk})) \\ &= F(g \otimes h)F_2(\mathbb{1},\mathbb{1})(F_0(1_{\Bbbk}) \otimes_{\Bbbk} F_0(1_{\Bbbk})) = F(g \otimes h)F_0(1_{\Bbbk}) = (X \otimes Y)_{g\otimes h}^F. \end{aligned}$$

Similarly, $G_2(X,Y)(X_g^G \otimes_{\Bbbk} Y_h^G) = (X \otimes Y)_{g\otimes h}^G$. Thus

$$\begin{aligned} &\varphi_{X\otimes Y} F_2(X,Y)(X_g^F \otimes_{\Bbbk} Y_h^F) = \varphi_{X\otimes Y}((X \otimes Y)_{g\otimes h}^F) = (X \otimes Y)_{g\otimes h}^G \\ &= G_2(X,Y)(X_g^G \otimes_{\Bbbk} Y_h^G) = G_2(X,Y)(\varphi_X \otimes_{\Bbbk} \varphi_Y)(X_g^F \otimes_{\Bbbk} Y_h^F). \end{aligned}$$

Since the vectors $X_g^F \otimes_{\Bbbk} Y_h^F$ span $F(X) \otimes_{\Bbbk} F(Y)$ as g runs over $\mathrm{Hom}_{\mathcal{C}}(\mathbb{1}, X)$ and h runs over $\mathrm{Hom}_{\mathcal{C}}(\mathbb{1}, Y)$, we deduce that

$$\varphi_{X\otimes Y} F_2(X,Y) = G_2(X,Y)(\varphi_X \otimes_{\Bbbk} \varphi_Y).$$

Hence $\varphi\colon F \to G$ is a monoidal natural isomorphism, and so F and G are isomorphic monoidal functors. □

17.3 Computations in a graph TQFT

In this section, $Z\colon \mathrm{Cob}_3^{\mathcal{D}} \to \mathrm{Mod}_{\Bbbk}$ is an arbitrary graph TQFT over a pivotal category $\mathcal{D}$. We introduce the so-called torus vector and use it to give a surgery formula for Z.

17.3.1 The torus vector

Any compact oriented 3-manifold M endowed with a $\mathcal{D}$-colored ribbon graph without free ends $R \subset \mathrm{Int}(M)$ represents in the obvious way two morphisms

$$(M,R)_+\colon \emptyset \to \partial M \quad \text{and} \quad (M,R)_-\colon -\partial M \to \emptyset$$

in the category $\mathrm{Cob}_3^{\mathcal{D}}$. Set

$$Z_+(M,R) = Z((M,R)_+)\, Z_0\colon \Bbbk \to Z(\partial M)$$

and

$$Z_-(M,R) = Z_0^{-1} Z((M,R)_-)\colon Z(-\partial M) \to \Bbbk.$$

If $\partial M = \emptyset$, then $(M,R)_+ = (M,R)_-$ and the $\Bbbk$-linear homomorphism

$$Z_0^{-1} Z_+(M,R) = Z_-(M,R) Z_0\colon \Bbbk \to \Bbbk$$

is multiplication by the scalar $Z(M,R) \in \Bbbk$ defined in Section 15.2.2.

We endow the unit disk $D^2 = \{z \in \mathbb{C}, |z| \leq 1\}$ and the unit circle $S^1 = \partial D^2$ with the counterclockwise orientation and endow the torus $S^1 \times S^1$ with the product orientations. Consider the solid torus $V = -(S^1 \times D^2)$ with orientation opposite to the product orientation. Then $\partial V = S^1 \times S^1$ in the category of oriented manifolds. The *torus vector* of Z is the vector

$$w(Z) = Z_+(V, \emptyset)(1_\Bbbk) \in Z(\partial V) = Z(S^1 \times S^1).$$

17.3.2 The surgery formula

Let $S^3 = \mathbb{R}^3 \cup \{\infty\}$ be the 3-sphere endowed with the right-handed orientation. Consider an n-component framed oriented link $L = L_1 \cup \cdots \cup L_n$ in S^3 with $n \geq 1$ and a $\mathcal{D}$-colored ribbon graph without free ends $R \subset S^3 \setminus L$. We establish a surgery formula for $Z(M, R_M)$, where M is the closed oriented 3-manifold obtained by the surgery on S^3 along L and $R_M \subset M$ is the $\mathcal{D}$-colored ribbon graph obtained from R via the surgery.

Pick a closed regular neighborhood $U \subset S^3 \setminus R$ of L and consider the exterior $E = S^3 \setminus \mathrm{Int}(U)$ of L. We endow both manifolds U and E with orientation induced by the right-handed orientation of S^3. Let $(S^1 \times S^1)^{\sqcup n}$ be a disjoint union of n copies of $S^1 \times S^1$. Pick a homeomorphism

$$f_L\colon (S^1 \times S^1)^{\sqcup n} \longrightarrow \partial U = -\partial E$$

which, for all points pt $\in S^1$ and all $q \in \{1, \ldots, n\}$, carries the q-th copy of $S^1 \times \{\mathrm{pt}\}$ onto a positively oriented meridian of L_q and carries the q-th copy of $\{\mathrm{pt}\} \times S^1$ onto a positively oriented longitude of L_q determined by the framing of L. The homeomorphism f_L is orientation preserving and induces a $\Bbbk$-linear isomorphism

$$Z(f_L)\colon Z\big((S^1 \times S^1)^{\sqcup n}\big) \to Z(-\partial E),$$

see Section 10.2.1. The pair $(E, R \subset \mathrm{Int}(E))$ determines a $\Bbbk$-linear homomorphism

$$Z_-(E,R)\colon Z(-\partial E) \to \Bbbk.$$

The monoidal constraints of Z induce a $\Bbbk$-linear isomorphism

$$\xi_n\colon \mathcal{A}^{\otimes n} \to Z\big((S^1 \times S^1)^{\sqcup n}\big).$$

Combining these three homomorphisms, we obtain a $\Bbbk$-linear form

$$Z_R^L = Z_-(E,R)\, Z(f_L)\, \xi_n\colon \mathcal{A}^{\otimes n} \to \Bbbk$$

independent of the choice of U.

Lemma 17.3. *We have:*

$$Z(M, R_M) = Z_R^L(w^{\otimes n}),$$

where $w = w(Z) \in \mathcal{A}$ is the torus vector of Z.

Proof. We have $(S^1 \times S^1)^{\sqcup n} = \partial V$ where V is a disjoint union of n copies of $-(S^1 \times D^2)$. The pair (M, R_M) is obtained by gluing V to the pair (E, R) along the homeomorphism $f_L \colon \partial V \to -\partial E$ above. The functoriality of Z implies that

$$Z_0^{-1} Z_+(M, R_M) = Z_-(E, R)\, Z(f_L)\, Z_+(V, \emptyset) \colon \Bbbk \to \Bbbk.$$

By the definition of the torus vector w and the monoidality of Z, the homomorphism $Z_+(V, \emptyset) \colon \Bbbk \to Z(\partial V)$ carries $1_\Bbbk$ to $\xi_n(w^{\otimes n})$. Therefore,

$$\begin{aligned} Z(M, R_M) = Z_0^{-1} Z_+(M, R_M)(1_\Bbbk) &= Z_-(E, R)\, Z(f_L)\, Z_+(V, \emptyset)(1_\Bbbk) \\ &= Z_-(E, R)\, Z(f_L)\, \xi_n(w^{\otimes n}) = Z_R^L(w^{\otimes n}). \end{aligned}$$

□

17.3.3 Computation of Z_R^L

We keep the notation of the previous subsection and compute the form Z_R^L on some elements of $\mathcal{A}$. For $X \in \mathrm{Ob}(\mathcal{D})$, let

$$K_X = \{0\} \times S^1 \subset \mathrm{Int}(D^2 \times S^1)$$

be the $\mathcal{D}$-colored knot with orientation induced by that of S^1, constant framing determined by a non-zero tangent vector of the disk D^2 at $0 \in D^2$, and color X. Set

$$[X] = Z_+(D^2 \times S^1, K_X)(1_\Bbbk) \in Z(\partial(D^2 \times S^1)) = Z(S^1 \times S^1) = \mathcal{A}. \qquad (17.2)$$

Let $T \subset S^3$ be the $\mathcal{D}$-colored ribbon graph obtained from the union $L \cup R$ by coloring the components $L_1, \ldots, L_n$ of L with some objects $X_1, \ldots, X_n$ of $\mathcal{D}$, respectively.

Lemma 17.4. $Z_R^L([X_1] \otimes_\Bbbk \cdots \otimes_\Bbbk [X_n]) = Z(S^3, T) \in \Bbbk$.

Proof. Let W be a disjoint union of n copies of $D^2 \times S^1$ so that $\partial W = (S^1 \times S^1)^{\sqcup n}$. For $q = 1, \ldots, n$, we view the knot K_{X_q} above as a $\mathcal{D}$-colored knot in the qth copy of $D^2 \times S^1$ in W. The monoidality of Z implies that

$$\xi_n([X_1] \otimes_\Bbbk \cdots \otimes_\Bbbk [X_n]) = Z_+(W, K_{X_1} \cup \cdots \cup K_{X_n})(1_\Bbbk).$$

Since the gluing of W to E along the homeomorphism $f_L \colon \partial W \to -\partial E$ yields the sphere S^3, the functoriality of Z implies that

$$\begin{aligned} Z_R^L([X_1] \otimes_\Bbbk \cdots \otimes_\Bbbk [X_n]) &= Z_-(E, R)\, Z(f_L)\, Z_+(W, K_{X_1} \cup \cdots \cup K_{X_n})(1_\Bbbk) \\ &= Z_0^{-1} Z((E, R)_-)\, Z(f_L)\, Z((W, K_{X_1} \cup \cdots \cup K_{X_n})_+)\, Z_0(1_\Bbbk) = Z(S^3, T). \end{aligned}$$

□

17.4 Proof of the main theorem

In this section, we prove Theorem 17.1. Throughout the proof, $Z = |\cdot|_\mathcal{C}$ is the state sum graph TQFT over $\mathcal{Z}(\mathcal{C})$, see Theorem 15.9. By Theorem 5.4, the $\Bbbk$-category

$\mathcal{Z}(\mathcal{C})$ is modular and in particular fusion. We fix a representative set $\mathcal{J}$ of simple objects of $\mathcal{Z}(\mathcal{C})$. The proof goes in four steps. In the first step, we find a basis of the $\Bbbk$-vector space

$$\mathcal{A} = Z(S^1 \times S^1) = |S^1 \times S^1|_{\mathcal{C}}.$$

In the second step, we compute the torus vector of Z as defined in Section 17.3. In the third step, we consider invariants of colored ribbon graphs in S^3. In the fourth step, we combine these results to accomplish the proof.

Step 1

Note that any $X \in \mathrm{Ob}(\mathcal{Z}(\mathcal{C}))$ determines a vector $[X] \in \mathcal{A}$ as in (17.2) where $\mathcal{D} = \mathcal{Z}(\mathcal{C})$. We claim that the vectors $\{[j]\}_{j\in\mathcal{J}}$ form a basis of $\mathcal{A}$. To see this, consider the Hopf link $H = K \cup L \subset S^3$ whose components K, L have zero framing and are oriented so that their linking number is equal to $+1$:

$H =$ 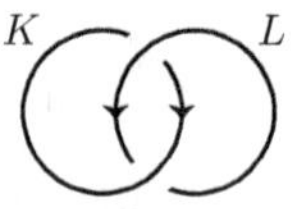 $\subset S^3.$

Coloring K with $i \in \mathcal{J}$, we obtain a $\mathcal{Z}(\mathcal{C})$-colored ribbon graph K_i in $S^3 \setminus L$. Section 17.3.2 yields a $\Bbbk$-linear homomorphism $Z^L_{K_i} : \mathcal{A} \to \Bbbk$. Coloring K with i and L with $j \in \mathcal{J}$, we turn H into a $\mathcal{Z}(\mathcal{C})$-colored ribbon graph $H_{i,j}$. Section 16.1.1 yields a scalar $S_{i,j} = \langle H_{i,j}\rangle_{\mathcal{Z}(\mathcal{C})} \in \Bbbk$. By Lemma 17.4 and Theorem 16.1,

$$Z^L_{K_i}([j]) = Z(S^3, H_{i,j}) = (\dim(\mathcal{C}))^{-1} S_{i,j}. \tag{17.3}$$

By Section 4.5.2, the matrix $[S_{i,j}]_{i,j\in\mathcal{J}}$ is the S-matrix of $\mathcal{Z}(\mathcal{C})$. Since the category $\mathcal{Z}(\mathcal{C})$ is modular, this matrix is invertible over $\Bbbk$. Therefore Formula (17.3) implies that the vectors $([j])_{j\in\mathcal{J}}$ are linearly independent.

By Section 6.6.1, the object $\bigoplus_{i\in\mathcal{J}} i^* \otimes i$ of $\mathcal{Z}(\mathcal{C})$ is a coend of $\mathcal{Z}(\mathcal{C})$. Theorem 16.2 yields $\Bbbk$-linear isomorphisms

$$\begin{aligned}\mathcal{A} = |S^1 \times S^1|_{\mathcal{C}} &\simeq \mathrm{Hom}_{\mathcal{Z}(\mathcal{C})}\Big(\mathbb{1}_{\mathcal{Z}(\mathcal{C})}, \bigoplus_{i\in\mathcal{J}} i^* \otimes i\Big)\\ &\simeq \bigoplus_{i\in\mathcal{J}} \mathrm{Hom}_{\mathcal{Z}(\mathcal{C})}(\mathbb{1}_{\mathcal{Z}(\mathcal{C})}, i^* \otimes i) \simeq \bigoplus_{i\in\mathcal{J}} \mathrm{Hom}_{\mathcal{Z}(\mathcal{C})}(i, i) \simeq \bigoplus_{i\in\mathcal{J}} \Bbbk \simeq \Bbbk^{\mathrm{card}(\mathcal{J})}.\end{aligned}$$

We conclude that $\dim_{\Bbbk}(\mathcal{A}) = \mathrm{card}(\mathcal{J})$. Hence, $([j])_{j\in\mathcal{J}}$ is a basis of $\mathcal{A}$.

Step 2

We prove now that the torus vector $w = w(Z) \in \mathcal{A}$ is computed by

$$w = (\dim(\mathcal{C}))^{-1} \sum_{j\in\mathcal{J}} \dim(j)\, [j]. \tag{17.4}$$

By Step 1, there is a unique expansion $w = \sum_{j \in \mathcal{J}} w_j[j]$ with $w_j \in \Bbbk$ for all j. We need only to compute the scalars $(w_j)_{j \in \mathcal{J}}$. For $i, j \in \mathcal{J}$, consider the $\mathcal{Z}(\mathcal{C})$-colored framed oriented link $H^+_{i,j} = K^+ \cup L^+$ in S^3 obtained from $H_{i,j}$ by adding a single positive twist to the framing of each component:

$$H^+_{i,j} = \quad \subset S^3.$$

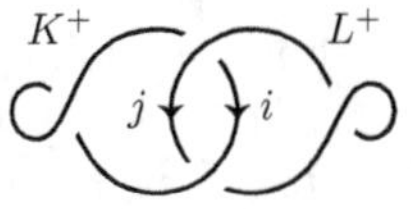

Since L^+ is an unknot with framing $+1$, the surgery on S^3 along L^+ gives again S^3. Under this surgery, K^+ turns into an oriented unknot in S^3 with zero framing and color i. Denote this $\mathcal{Z}(\mathcal{C})$-colored oriented framed unknot by K^i_0. By Theorem 16.1,

$$Z(S^3, K^i_0) = (\dim(\mathcal{C}))^{-1} \langle K^i_0 \rangle_{\mathcal{Z}(\mathcal{C})} = (\dim(\mathcal{C}))^{-1} \dim(i).$$

On the other hand, Lemmas 17.3 and 17.4 imply that

$$Z(S^3, K^i_0) = Z^{L^+}_{K^+_i}(w) = \sum_{j \in \mathcal{J}} w_j \, Z^{L^+}_{K^+_i}([j]) = \sum_{j \in \mathcal{J}} w_j \, Z(S^3, H^+_{i,j}),$$

where K^+_i is the framed oriented unknot K^+ colored with i and viewed as a $\mathcal{Z}(\mathcal{C})$-colored ribbon graph in $S^3 \setminus L^+$. Therefore

$$\dim(i) = \dim(\mathcal{C}) \sum_{j \in \mathcal{J}} w_j \, Z(S^3, H^+_{i,j}). \tag{17.5}$$

We now compute $Z(S^3, H^+_{i,j})$. Let θ be the twist of $\mathcal{Z}(\mathcal{C})$, see Section 3.3.2. Recall that $\theta_j = v_j \, \mathrm{id}_j$ for any $j \in \mathcal{J}$, where $v_j \in \Bbbk$ is the twist scalar, see Section 4.5.2. By Theorem 16.1 and Lemma 3.2,

$$Z(S^3, H^+_{i,j}) = (\dim(\mathcal{C}))^{-1} \langle H^+_{i,j} \rangle_{\mathcal{Z}(\mathcal{C})} = (\dim(\mathcal{C}))^{-1} v_i v_j S_{i,j}.$$

Substituting this expression in (17.5), we obtain that

$$\dim(i) = \sum_{j \in \mathcal{J}} w_j v_i v_j S_{i,j}$$

for all $i \in \mathcal{J}$. Since the S-matrix and the twist scalars are invertible, this system of linear equations in $(w_j)_{j \in \mathcal{J}}$ has a unique solution. Lemma 4.6(b) implies that

$$w_j = (\dim(\mathcal{C}))^{-1} \dim(j)$$

for all $j \in \mathcal{J}$.

Step 3

We prove now that for any closed oriented 3-manifold M and any $\mathcal{Z}(\mathcal{C})$-colored ribbon graph $R \subset M$,

$$|M,R|_{\mathcal{C}} = \tau_{\mathcal{Z}(\mathcal{C})}(M,R). \tag{17.6}$$

Since both sides are multiplicative under disjoint union, it suffices to consider the case where M is connected. Present M by surgery on S^3 along a framed oriented link $L = L_1 \cup \cdots \cup L_n \subset S^3$. Pushing R in the exterior of L in S^3, we obtain a $\mathcal{Z}(\mathcal{C})$-colored ribbon graph $R' \subset S^3 \setminus L$. By Lemma 17.3 and Formula (17.4),

$$|M,R|_{\mathcal{C}} = Z(M,R) = \sum_{j_1,\dots,j_n \in \mathcal{J}} \left(\prod_{q=1}^{n} \frac{\dim(j_q)}{\dim(\mathcal{C})} \right) Z_{R'}^{L}([j_1] \otimes_{\Bbbk} \cdots \otimes_{\Bbbk} [j_n])).$$

For $j_1,\dots,j_n \in \mathcal{J}$, let $T_{j_1,\dots,j_n}$ be the union of R' with the link L whose components $L_1,\dots,L_n$ are colored with $j_1,\dots,j_n$, respectively. Lemma 17.4 and Theorem 16.1 give

$$Z_{R'}^{L}([j_1] \otimes_{\Bbbk} \cdots \otimes_{\Bbbk} [j_n]) = Z(S^3, T_{j_1,\dots,j_n}) = (\dim(\mathcal{C}))^{-1} \langle T_{j_1,\dots,j_n} \rangle_{\mathcal{Z}(\mathcal{C})}.$$

Therefore

$$\begin{aligned} |M,R|_{\mathcal{C}} &= \sum_{j_1,\dots,j_n \in \mathcal{J}} \left(\prod_{q=1}^{n} \frac{\dim(j_q)}{\dim(\mathcal{C})} \right) (\dim(\mathcal{C}))^{-1} \langle T_{j_1,\dots,j_n} \rangle_{\mathcal{Z}(\mathcal{C})} \\ &= (\dim(\mathcal{C}))^{-n-1} \sum_{j_1,\dots,j_n \in \mathcal{J}} \left(\prod_{q=1}^{n} \dim(j_q) \right) \langle T_{j_1,\dots,j_n} \rangle_{\mathcal{Z}(\mathcal{C})} = \tau_{\mathcal{Z}(\mathcal{C})}(M,R), \end{aligned}$$

where the last equality is the definition of $\tau_{\mathcal{Z}(\mathcal{C})}(M,R)$, see Section 17.1.1.

Step 4

Observe that the strong monoidal functors $\tau_{\mathcal{Z}(\mathcal{C})}$ and $|\cdot|_{\mathcal{C}}$ from $\mathrm{Cob}_3^{\mathcal{Z}(\mathcal{C})}$ to $\mathrm{Mod}_{\Bbbk}$ satisfy Conditions (a)-(c) of Lemma 17.2. Indeed, Condition (a) holds because the graph TQFT $\tau_{\mathcal{Z}(\mathcal{C})}$ is non-degenerate, see Section 17.1.1. The inequality in Condition (b) is in fact an equality which holds because for any $\mathcal{Z}(\mathcal{C})$-colored surface Σ, the vector spaces $|\Sigma|_{\mathcal{C}}$ and $\tau_{\mathcal{Z}(\mathcal{C})}(\Sigma)$ are isomorphic. For connected Σ, this follows from Theorem 16.2, Formula (17.1) with $\mathcal{B} = \mathcal{Z}(\mathcal{C})$, and the fact that, by Section 6.6.1, the object $\bigoplus_{i \in \mathcal{J}} i^* \otimes i$ is a coend of $\mathcal{Z}(\mathcal{C})$. The case of disconnected Σ is deduced from the case of connected Σ using Formula (15.1) applied to the graph TQFTs $|\cdot|_{\mathcal{C}}$ and $\tau_{\mathcal{Z}(\mathcal{C})}$.

Finally, Condition (c) is a restatement of the identity (17.6). Now, Lemma 17.2 implies that there is a natural monoidal isomorphism between the functors $|\cdot|_{\mathcal{C}}$ and $\tau_{\mathcal{Z}(\mathcal{C})}$. Hence, the graph TQFTs $|\cdot|_{\mathcal{C}}$ and $\tau_{\mathcal{Z}(\mathcal{C})}$ are isomorphic.

17.5 Corollaries of the main theorem

We formulate several corollaries of Theorem 17.1.

17.5.1 Direct consequences

In this subsection, $\mathcal{C}$ is an additive spherical fusion $\Bbbk$-category of non-zero dimension, where $\Bbbk$ is an algebraically closed field.

Corollary 17.5. *The state sum TQFT $|\cdot|_{\mathcal{C}}$ from Theorem 13.5 is isomorphic to the 3-dimensional TQFT obtained by restricting the surgery graph TQFT $\tau_{\mathcal{Z}(\mathcal{C})}$ to* Cob_3.

Proof. The state sum TQFT $|\cdot|_{\mathcal{C}}$ is the restriction to Cob_3 of the state sum graph TQFT $|\cdot|_{\mathcal{C}}$. Therefore this corollary follows from Theorem 17.1 and the obvious fact that isomorphic graph TQFTs restrict to isomorphic TQFTs. □

Recall the notion of a non-degenerate graph TQFT, see Section 15.2.3.

Corollary 17.6. *The state sum graph TQFT $|\cdot|_{\mathcal{C}}$ is non-degenerate.*

Proof. This follows from Theorem 17.1, the non-degeneracy of the graph TQFT $\tau_{\mathcal{Z}(\mathcal{C})}$, and the obvious fact that a graph TQFT isomorphic to a non-degenerate graph TQFT is itself non-degenerate. □

The two claims in the next corollary concern the scalar state sum invariants of closed oriented 3-manifolds and ribbon graphs, see Theorems 13.1 and 15.7. Both claims directly follow from Theorem 17.1 since isomorphic graph TQFTs yield the same invariants of closed oriented 3-manifolds and ribbon graphs.

Corollary 17.7. (a) *For any closed oriented 3-manifold M,*

$$|M|_{\mathcal{C}} = \tau_{\mathcal{Z}(\mathcal{C})}(M).$$

(b) *For any $\mathcal{Z}(\mathcal{C})$-colored ribbon graph R in a closed oriented 3-manifold M,*

$$|M,R|_{\mathcal{C}} = \tau_{\mathcal{Z}(\mathcal{C})}(M,R).$$

Claim (a) of Corollary 17.7 was previously known in several special cases: when $\mathcal{C}$ is the category of representations of a finite group, when $\mathcal{C}$ is the category of bimodules associated with a subfactor [KSW], and when $\mathcal{C}$ is modular [Tu1, Wa]. In the latter case, it may be reformulated avoiding the notion of the center of a monoidal category as follows.

Corollary 17.8. *If $\mathcal{C}$ is a modular $\Bbbk$-category, then for any closed oriented 3-manifold M,*

$$|M|_{\mathcal{C}} = \tau_{\mathcal{C}}(M)\,\tau_{\mathcal{C}}(-M).$$

Proof. If $\mathcal{C}$ is modular, then its center $\mathcal{Z}(\mathcal{C})$ is braided equivalent to the Deligne tensor product $\mathcal{C} \boxtimes \overline{\mathcal{C}}$ (see [Mü, Theorem 7.10]), where $\overline{\mathcal{C}}$ is the mirror of $\mathcal{C}$ (see Example 3.1.7). Then

$$|M|_{\mathcal{C}} \overset{\text{(i)}}{=} \tau_{\mathcal{C}\boxtimes\overline{\mathcal{C}}}(M) \overset{\text{(ii)}}{=} \tau_{\mathcal{C}}(M)\, \tau_{\overline{\mathcal{C}}}(M) \overset{\text{(iii)}}{=} \tau_{\mathcal{C}}(M)\, \tau_{\mathcal{C}}(-M),$$

Here the equality (i) follows from Corollary 17.7(a) and the equalities (ii) and (iii) follow from the properties of the surgery graph TQFT (see [Tu1]). □

Note finally that if $\mathcal{C}$ is a unitary modular category (over $\Bbbk = \mathbb{C}$, see Section G.2.2), then for any closed oriented 3-manifold M,

$$|M|_{\mathcal{C}} = |\tau_{\mathcal{C}}(M)|^2 \in \mathbb{R}_+.$$

Indeed, the properties of the conjugation in $\mathcal{C}$ imply that $\tau_{\mathcal{C}}(-M) = \overline{\tau_{\mathcal{C}}(M)}$, and then the formula is just a rewriting of Corollary 17.8.

17.5.2 Weak equivalence

Two fusion $\Bbbk$-categories are said to be *weakly equivalent* if their centers are equivalent braided pivotal $\Bbbk$-categories. The next corollary relates the state sum TQFTs associated with weakly equivalent categories.

Corollary 17.9. *Let $\mathcal{C}$ and $\mathcal{D}$ be weakly equivalent additive spherical fusion $\Bbbk$-categories of non-zero dimension, where $\Bbbk$ is an algebraically closed field. Then the state sum 3-dimensional TQFTs $|\cdot|_{\mathcal{C}}$ and $|\cdot|_{\mathcal{D}}$ are isomorphic. In particular, $|M|_{\mathcal{C}} = |M|_{\mathcal{D}}$ for any closed oriented 3-manifold M.*

Proof. Since $\mathcal{C}$ and $\mathcal{D}$ are weakly equivalent, there is a braided pivotal $\Bbbk$-linear equivalence $\mathcal{Z}(\mathcal{C}) \to \mathcal{Z}(\mathcal{D})$. By Exercise 15.2.5, it induces a symmetric strict monoidal functor $\mathrm{Cob}_3^{\mathcal{Z}(\mathcal{C})} \to \mathrm{Cob}_3^{\mathcal{Z}(\mathcal{D})}$ which we denote by Ψ. We claim that the graph TQFTs

$$|\cdot|_{\mathcal{C}} \colon \mathrm{Cob}_3^{\mathcal{Z}(\mathcal{C})} \to \mathrm{Mod}_{\Bbbk} \quad \text{and} \quad |\cdot|_{\mathcal{D}} \circ \Psi \colon \mathrm{Cob}_3^{\mathcal{Z}(\mathcal{C})} \to \mathrm{Mod}_{\Bbbk}$$

are isomorphic. Restricting to Cob_3 and noting that the composition of the inclusion $\mathrm{Cob}_3 \hookrightarrow \mathrm{Cob}_3^{\mathcal{Z}(\mathcal{C})}$ with Ψ is the inclusion $\mathrm{Cob}_3 \hookrightarrow \mathrm{Cob}_3^{\mathcal{Z}(\mathcal{D})}$, we deduce the corollary. To prove the claim above, we consider the surgery graph TQFTs $\tau_{\mathcal{Z}(\mathcal{C})}$ over $\mathcal{Z}(\mathcal{C})$ and $\tau_{\mathcal{Z}(\mathcal{D})}$ over $\mathcal{Z}(\mathcal{D})$. Then $\tau_{\mathcal{Z}(\mathcal{D})} \circ \Psi$ is a graph TQFT over $\mathcal{Z}(\mathcal{C})$. The definitions of Section 17.1.1 imply that for any $\mathcal{Z}(\mathcal{C})$-colored surface Σ, the $\Bbbk$-vector spaces $\tau_{\mathcal{Z}(\mathcal{D})}(\Psi(\Sigma))$ and $\tau_{\mathcal{Z}(\mathcal{C})}(\Sigma)$ are isomorphic, and for any closed oriented 3-manifold M and any $\mathcal{Z}(\mathcal{C})$-colored ribbon graph $R \subset M$, we have

$$\tau_{\mathcal{Z}(\mathcal{D})}(\Psi(M,R)) = \tau_{\mathcal{Z}(\mathcal{C})}(M,R).$$

Since $\tau_{\mathcal{Z}(\mathcal{C})}$ is non-degenerate, Lemma 17.2 implies that the graph TQFTs $\tau_{\mathcal{Z}(\mathcal{C})}$ and $\tau_{\mathcal{Z}(\mathcal{D})} \circ \Psi$ are isomorphic. It follows from Theorem 17.1 that the graph TQFTs $|\cdot|_{\mathcal{C}}$ and $|\cdot|_{\mathcal{D}} \circ \Psi$ are isomorphic. □

17.5.3 The case of non-additive categories

To treat non-additive categories, we use the additivization construction from Section 16.3.1. Consider a (possibly, non-additive) spherical fusion $\Bbbk$-category $\mathcal{C}$ of non-zero dimension, where $\Bbbk$ is an algebraically closed field. The additivization $\mathcal{C}^{\mathrm{add}}$ of $\mathcal{C}$ is an additive spherical fusion $\Bbbk$-category such that $\dim(\mathcal{C}^{\mathrm{add}}) = \dim(\mathcal{C}) \neq 0$. By Theorem 5.4, the category $\mathcal{Z}(\mathcal{C}^{\mathrm{add}})$ is an anomaly free modular $\Bbbk$-category.

Corollary 17.10. *The 3-dimensional TQFTs $|\cdot|_{\mathcal{C}}$ and $\tau_{\mathcal{Z}(\mathcal{C}^{\mathrm{add}})}$ are isomorphic. In particular, for any closed oriented 3-manifold M, we have*

$$|M|_{\mathcal{C}} = \tau_{\mathcal{Z}(\mathcal{C}^{\mathrm{add}})}(M).$$

Proof. This follows from Corollary 17.5 and the fact that the state sum TQFTs $|\cdot|_{\mathcal{C}}$ and $|\cdot|_{\mathcal{C}^{\mathrm{add}}}$ are isomorphic, see Lemma 16.4. □

To extend the previous corollary to graph TQFTs, recall from Lemma 16.3 the functor $H = H_{\mathcal{C}} \colon \mathcal{Z}(\mathcal{C}) \to \mathcal{Z}(\mathcal{C}^{\mathrm{add}})$. By Exercise 15.2.5, this functor induces a symmetric strict monoidal functor

$$\widetilde{H} \colon \mathrm{Cob}_3^{\mathcal{Z}(\mathcal{C})} \to \mathrm{Cob}_3^{\mathcal{Z}(\mathcal{C}^{\mathrm{add}})}.$$

Corollary 17.11. *The graph TQFTs $|\cdot|_{\mathcal{C}}$ and $\tau_{\mathcal{Z}(\mathcal{C}^{\mathrm{add}})} \circ \widetilde{H}$ are isomorphic. In particular, for any closed oriented 3-manifold M and any $\mathcal{Z}(\mathcal{C})$-colored ribbon graph R in M, we have*

$$|M, R|_{\mathcal{C}} = \tau_{\mathcal{Z}(\mathcal{C}^{\mathrm{add}})}(M, R^{\mathrm{add}}),$$

where R^{add} is the $\mathcal{Z}(\mathcal{C}^{\mathrm{add}})$-colored ribbon graph in M obtained from R by applying to all colors the functor $H \colon \mathcal{Z}(\mathcal{C}) \to \mathcal{Z}(\mathcal{C}^{\mathrm{add}})$.

Proof. Lemma 16.4 and Theorem 17.1 yield the following isomorphisms of graph TQFTs over $\mathcal{Z}(\mathcal{C})$:

$$|\cdot|_{\mathcal{C}} \simeq |\cdot|_{\mathcal{C}^{\mathrm{add}}} \circ \widetilde{H} \simeq \tau_{\mathcal{Z}(\mathcal{C}^{\mathrm{add}})} \circ \widetilde{H}.$$

The second claim follows from the equality $\widetilde{H}(M, R) = (M, R^{\mathrm{add}})$. □

17.5.4 The case of rings

The assumption that the ground ring $\Bbbk$ is an algebraically closed field may be weakened as follows. Suppose just that $\Bbbk$ is a commutative ring and $\mathcal{C}$ is a spherical fusion $\Bbbk$-category whose dimension is invertible in $\Bbbk$. Consider a ring homomorphism φ from $\Bbbk$ into an algebraically closed field $\mathbb{K}$. Consider the spherical fusion $\mathbb{K}$-category $\mathcal{C}_\varphi$ obtained from $\mathcal{C}$ as in Section 16.4.2. Its additivization $\mathcal{C}_\varphi^{\mathrm{add}} = (\mathcal{C}_\varphi)^{\mathrm{add}}$ is an additive spherical fusion $\mathbb{K}$-category and

$$\dim(\mathcal{C}_\varphi^{\mathrm{add}}) = \dim(\mathcal{C}_\varphi) = \varphi(\dim(\mathcal{C})) \neq 0.$$

By Theorem 5.4, $\mathcal{Z}(\mathcal{C}_\varphi^{\mathrm{add}})$ is an additive anomaly free modular $\mathbb{K}$-category. Combining (16.23) with Corollary 17.11, we obtain that for any closed oriented 3-manifold M and any $\mathcal{Z}(\mathcal{C})$-colored ribbon graph R in M,

$$\varphi(|M,R|_{\mathcal{C}}) = \tau_{\mathcal{Z}(\mathcal{C}_\varphi^{\mathrm{add}})}(M, R_\varphi^{\mathrm{add}}), \tag{17.7}$$

where $R_\varphi^{\mathrm{add}} = (R_\varphi)^{\mathrm{add}}$.

In particular, given a prime ideal P of $\Bbbk$, we can take for $\mathbb{K}$ the algebraic closure of the fraction field of the integral domain $\Bbbk/P$ and for φ the composition of the projection $\Bbbk \to \Bbbk/P$ with the inclusion $\Bbbk/P \hookrightarrow \mathbb{K}$. Formula (17.7) gives then a surgery computation of $|M,R|_{\mathcal{C}} \,(\mathrm{mod}\, P) \in \Bbbk/P$. Varying P over all prime ideals of $\Bbbk$, we obtain a surgery computation of $|M,R|_{\mathcal{C}}$ modulo the nilradical of $\Bbbk$.

17.5.5 Invariants derived from involutory Hopf algebras

Theorem 17.1 allows us to relate several 3-manifold invariants derived from a finite-dimensional involutory Hopf algebra A over an algebraically closed field $\Bbbk$ such that $d = \dim_{\Bbbk}(A)1_{\Bbbk} \neq 0$. In terminology of Section 6.2, A is an involutory Hopf algebra in the symmetric category $\mathrm{vect}_{\Bbbk}$ of finite-dimensional $\Bbbk$-vector spaces. That A is involutory means that the antipode of A is an involutive map, see Example 6.2.6. By Exercise 6.1.8, the category mod_A of finite-dimensional left A-modules is an additive monoidal $\Bbbk$-category. By Section 6.2.3, since $\mathrm{vect}_{\Bbbk}$ is spherical, so is mod_A. Since A is semisimple by [LR, Corollary 2.6] and the ground field $\Bbbk$ is algebraically closed, mod_A is a pre-fusion $\Bbbk$-category. Moreover, since A is finite-dimensional, mod_A is a fusion $\Bbbk$-category. To compute its dimension, we use Lemma 6.3: by Exercise 6.4.7, $A^* = \mathrm{Hom}_{\Bbbk}(A, \Bbbk)$ is a coend of mod_A, and therefore

$$\dim(\mathrm{mod}_A) = \dim(A^*) = \dim_{\Bbbk}(A^*)1_{\Bbbk} = \dim_{\Bbbk}(A)1_{\Bbbk} = d \neq 0.$$

Let $D(A)$ be the Drinfeld double of A, see [Dri]. By [EG], the $\Bbbk$-category $\mathrm{mod}_{D(A)}$ of finite-dimensional left $D(A)$-modules is an anomaly free modular $\Bbbk$-category. Consequently, for any closed oriented 3-manifold M, we have the $\Bbbk$-valued invariants $|M|_{\mathrm{mod}_A}$ and $\tau_{\mathrm{mod}_{D(A)}}(M)$. Also, Kuperberg [Ku] derived from A a $\Bbbk$-valued invariant $\mathrm{Ku}_A(M)$ and Hennings, Kauffman, and Radford [He, KR] derived from $D(A)$ a $\Bbbk$-valued invariant $\mathrm{HKR}_{D(A)}(M)$.

Corollary 17.12. *For any closed oriented 3-manifold M, we have*

$$|M|_{\mathrm{mod}_A} = \tau_{\mathrm{mod}_{D(A)}}(M) \quad \text{and} \quad \mathrm{Ku}_A(M) = \mathrm{HKR}_{D(A)}(M) = d\,|M|_{\mathrm{mod}_A}.$$

Proof. Corollary 17.7(a) implies that

$$|M|_{\mathrm{mod}_A} = \tau_{\mathcal{Z}(\mathrm{mod}_A)}(M).$$

The braided pivotal $\Bbbk$-categories $\mathcal{Z}(\mathrm{mod}_A)$ and $\mathrm{mod}_{D(A)}$ are equivalent (see, for example, [Ka]). Therefore

$$\tau_{\mathcal{Z}(\mathrm{mod}_A)}(M) = \tau_{\mathrm{mod}_{D(A)}}(M).$$

These formulas imply the equality $|M|_{\mathrm{mod}_A} = \tau_{\mathrm{mod}_{D(A)}}(M)$. We conclude using that

$$\mathrm{Ku}_A(M) = d\,|M|_{\mathrm{mod}_A} \quad \text{and} \quad \mathrm{HKR}_{D(A)}(M) = d\,\tau_{\mathrm{mod}_{D(A)}}(M)$$

by [BW2] and [Ke, KL], respectively. □

17.5.6 Remark

Using the language of higher categories, one defines so-called extended TQFTs, see [La, Lu]. Any graph TQFT induces a 2-extended 3-dimensional TQFT with values in the 2-category of 2-vector spaces. Theorem 17.1 implies that (under the assumptions of this theorem) the 2-extended 3-dimensional TQFTs induced by the graph TQFTs $|\cdot|_{\mathcal{C}}$ and $\tau_{\mathcal{Z}(\mathcal{C})}$ are isomorphic.

Appendices

Appendix A

Examples of monoidal categories

We outline the standard constructions of monoidal categories from Hopf algebras and groups. As in the main text, the symbol $\Bbbk$ denotes a non-zero commutative ring.

A.1 The category of modules over a Hopf algebra

Consider the category $\mathrm{proj}_{\Bbbk}$ of projective $\Bbbk$-modules of finite type. Recall that $\mathrm{proj}_{\Bbbk}$ is a symmetric pivotal $\Bbbk$-category, see Examples 1.7.2, 3.1.4, and 4.2.4. Consider a Hopf $\Bbbk$-algebra A of finite type, i.e., a Hopf algebra in $\mathrm{proj}_{\Bbbk}$ in the sense of Section 6.2. Let mod_A be the $\Bbbk$-category of left A-modules in $\mathrm{proj}_{\Bbbk}$ (see Section 6.1.1), that is, the category of left A-modules, whose underlying $\Bbbk$-modules are projective of finite type, and of A-linear homomorphisms. By Exercise 6.1.8 and Lemma 6.1, mod_A is an additive rigid $\Bbbk$-category. If the square of the antipode S of A is the conjugation by a grouplike element of A, then mod_A is pivotal. Moreover, if A is involutory (that is, if $S^2 = \mathrm{id}_A$), then the forgetful functor $\mathrm{mod}_A \to \mathrm{proj}_{\Bbbk}$ is pivotal and the category mod_A is spherical, see Section 6.2.3.

Note next that if a Hopf $\Bbbk$-algebra A of finite type is quasitriangular in the sense of Drinfeld [Dri], then the monoidal category mod_A is braided. If $\Bbbk$ is a field, then for any finite-dimensional Hopf $\Bbbk$-algebra A, the center $\mathcal{Z}(\mathrm{mod}_A)$ of mod_A can be described in terms of a quasitriangular Hopf $\Bbbk$-algebra $D(A)$ called the Drinfeld double of A: the braided $\Bbbk$-categories $\mathcal{Z}(\mathrm{mod}_A)$ and $\mathrm{mod}_{D(A)}$ are equivalent. Note that the Drinfeld doubles of the dual Hopf algebra A^* and of the coopposite Hopf algebra A^{cop} are isomorphic. Consequently, the centers of the rigid $\Bbbk$-categories mod_{A^*} and $\mathrm{mod}_{A^{\mathrm{cop}}}$ are equivalent braided categories.

Suppose now that $\Bbbk$ is an algebraically closed field and A is an involutory Hopf $\Bbbk$-algebra of finite dimension relatively prime to $\mathrm{char}(\Bbbk)$. By Section 17.5.5, mod_A is an additive spherical fusion $\Bbbk$-category of non-zero dimension $\dim_{\Bbbk}(A)1_{\Bbbk}$. By Theorem 5.4, the center of mod_A is modular. Similar remarks hold for the Hopf algebras A^* and A^{cop}. By the above, the fusion $\Bbbk$-categories mod_{A^*} and $\mathrm{mod}_{A^{\mathrm{cop}}}$ are weakly equivalent in the sense of Section 17.5.2. By Corollary 17.9, they give rise to isomorphic 3-dimensional state sum TQFTs.

Finite-dimensional involutory Hopf algebras naturally arise in the study of finite groups. Consider a finite group G of order relatively prime to char($\Bbbk$). The group algebra $\Bbbk[G]$ and the algebra $\Bbbk^G$ of $\Bbbk$-valued functions on G are involutory Hopf algebras of dimension $|G|$. Since $\Bbbk[G]^{\mathrm{cop}} = \Bbbk[G]$ and $(\Bbbk[G])^* \simeq \Bbbk^G$ as Hopf algebras, the previous paragraph shows that $\mathrm{mod}_{\Bbbk[G]}$ and $\mathrm{mod}_{\Bbbk^G}$ are weakly equivalent additive spherical fusion $\Bbbk$-categories of non-zero dimension. They give rise to isomorphic 3-dimensional state sum TQFTs. We will discuss the corresponding 3-manifold invariants in Appendix H. Note that $\mathrm{mod}_{\Bbbk[G]}$ is just the category of finite-dimensional $\Bbbk$-linear representations of G while the category $\mathrm{mod}_{\Bbbk^G}$ is equivalent to the $\Bbbk$-category $G\text{-}\mathrm{vect}_{\Bbbk}^1$ discussed below in Section A.2.

There are various generalizations of Hopf algebras, and their categories of modules provide further examples of monoidal categories. In particular, weak Hopf algebras are defined similarly to Hopf algebras but with weaker compatibility conditions on multiplication and comultiplication. Any fusion category over a field is equivalent to the category of representations of a finite-dimensional weak Hopf algebra, see [Ha, Sz].

A.2 The category G - vect$_{\Bbbk}^{\alpha}$

Let G be a group with unit element 1 and let $\alpha\colon G^3 \to \Bbbk^*$ be a 3-cocycle. A free $\Bbbk$-module M of finite rank is *G-graded* if it is decomposed as a direct sum

$$M = \bigoplus_{g \in G} M_g$$

of free submodules labeled by $g \in G$. Each M_g necessarily has a finite rank, and $M_g = 0$ for all but a finite number of $g \in G$. For example, each $h \in G$ gives rise to a G-graded free $\Bbbk$-module $\Bbbk_h$ of finite rank where $(\Bbbk_h)_h = \Bbbk$ and $(\Bbbk_h)_g = 0$ for $g \in G \setminus \{h\}$.

Consider the category of G-graded free $\Bbbk$-modules of finite rank and $\Bbbk$-linear grading-preserving homomorphisms. We endow this category with the monoidal product defined on objects by

$$M \otimes N = \bigoplus_{g \in G} (M \otimes N)_g \quad \text{where} \quad (M \otimes N)_g = \bigoplus_{\substack{h,j \in G \\ hj = g}} M_h \otimes_{\Bbbk} N_j,$$

and on morphisms by $f \otimes g = f \otimes_{\Bbbk} g$. The associativity and unitality constraints

$$a^{\alpha}_{M,N,P}\colon (M \otimes N) \otimes P \to M \otimes (N \otimes P), \quad l^{\alpha}_M\colon \mathbb{1} \otimes M \to M, \quad r^{\alpha}_M\colon M \otimes \mathbb{1} \to M$$

are defined by the formulas

$$a^{\alpha}_{M,N,P}((m \otimes_{\Bbbk} n) \otimes_{\Bbbk} p) = \alpha(g,h,j)\, m \otimes_{\Bbbk} (n \otimes_{\Bbbk} p),$$
$$l^{\alpha}_{M}(1_{\Bbbk} \otimes_{\Bbbk} m) = \alpha(1,1,g)^{-1}\, m,$$
$$r^{\alpha}_{M}(m \otimes_{\Bbbk} 1_{\Bbbk}) = \alpha(g,1,1)\, m$$

for any $g,h,j \in G$, $m \in M_g$, $n \in N_h$, $p \in P_j$. This gives a monoidal $\Bbbk$-category with unit object $\mathbb{1} = \Bbbk_1$. It is additive: the direct sum of two G-graded free $\Bbbk$-modules M, N of finite rank is computed by

$$M \oplus N = \bigoplus_{g\in G} (M \oplus N)_g \quad \text{where} \quad (M \oplus N)_g = M_g \oplus N_g.$$

The resulting additive monoidal $\Bbbk$-category is denoted by $G\text{-}\mathrm{vect}_{\Bbbk}^{\alpha}$. It carries a canonical pivotal duality

$$\{(M^*, \mathrm{ev}_M, \widetilde{\mathrm{ev}}_M)\}_{M\in \mathrm{Ob}(G\text{-}\mathrm{vect}_{\Bbbk}^{\alpha})}$$

defined as follows. The dual of any object M is defined by

$$M^{\star} = \bigoplus_{g\in G} (M^{\star})_g \quad \text{where} \quad (M^{\star})_g = \mathrm{Hom}_{\Bbbk}(M_{g^{-1}}, \Bbbk).$$

The evaluations $\mathrm{ev}_M \colon M^{\star} \otimes M \to \mathbb{1}$ and $\widetilde{\mathrm{ev}}_M \colon M \otimes M^{\star} \to \mathbb{1}$ are defined by

$$\mathrm{ev}_M(f \otimes_{\Bbbk} m) = \delta_{h,g^{-1}}\, f(m) \quad \text{and} \quad \widetilde{\mathrm{ev}}_M(m \otimes_{\Bbbk} f) = \lambda_g\, \delta_{h,g^{-1}}\, f(m)$$

for any $g, h \in G$, $m \in M_g$, $f \in (M^{\star})_h$, where

$$\lambda_g = \alpha(1,1,g)\, \alpha(g,g^{-1},g)\, \alpha(g,1,1) \in \Bbbk^*.$$

The corresponding coevaluation morphisms are computed by the formulas

$$\mathrm{coev}_M(1_{\Bbbk}) = \sum_{g,i} \lambda_g^{-1}\, m_{g,i} \otimes_{\Bbbk} m^{\star}_{g,i} \quad \text{and} \quad \widetilde{\mathrm{coev}}_M(1_{\Bbbk}) = \sum_{g,i} m^{\star}_{g,i} \otimes_{\Bbbk} m_{g,i}$$

where $(m_{g,i})_i$ is any basis of M_g and $(m^{\star}_{g,i})_i$ is the dual basis of $(M^{\star})_{g^{-1}}$. It is easy to see that $G\text{-}\mathrm{vect}_{\Bbbk}^{\alpha}$ is an additive spherical pre-fusion $\Bbbk$-category with representative set of simple objects $\{\Bbbk_h\}_{h\in G}$. Note that $\dim(\Bbbk_h) = 1_{\Bbbk} \in \Bbbk$ for all $h \in G$ and there are isomorphisms $\Bbbk_g \otimes \Bbbk_h \simeq \Bbbk_{gh}$ and $\Bbbk_h^{\star} \simeq \Bbbk_{h^{-1}}$ for all $g, h \in G$.

We explain in Section A.3 (in a more general setting) that the category $G\text{-}\mathrm{vect}_{\Bbbk}^{\alpha}$ may be viewed as an additivization of the category $G_{\Bbbk}^{\alpha}$ from Example 1.2.6.

A.3 The category $G\text{-vect}_{\Bbbk}^{\alpha,\mathrm{d}}$

Let G and α be as in the previous subsection. The right evaluation morphisms in the category $G\text{-vect}_{\Bbbk}^{\alpha}$ may be modified using strong monoidal natural automorphisms of the identity functor, see Exercise 1.7.9. Each group homomorphism $d\colon G \to \Bbbk^*$ determines such an automorphism acting on a G-graded free $\Bbbk$-module $M = \bigoplus_{g\in G} M_g$ of finite rank via multiplication by $d(g)$ on every M_g. Using this automorphism, we modify the right duality as in Example 1.7.3 and obtain a new pivotal duality

$$\{(M^*, \mathrm{ev}_M, \widetilde{\mathrm{ev}}_M^d)\}_{M\in \mathrm{Ob}(G\text{-vect}_{\Bbbk}^{\alpha})}.$$

By definition, for any $g, h \in G$, $m \in M_g$, $f \in (M^*)_h$,

$$\widetilde{\mathrm{ev}}_M(m \otimes_{\Bbbk} f) = d(g)\, \lambda_g\, \delta_{h,g^{-1}}\, f(m).$$

The corresponding right coevaluation is computed by

$$\widetilde{\mathrm{coev}}_M^d(1_{\Bbbk}) = \sum_{g,i} d(g)^{-1} m_{g,i}^{\star} \otimes_{\Bbbk} m_{g,i}$$

where $(m_{g,i})_i$ is any basis of M_g and $(m_{g,i}^{\star})_i$ is the dual basis of $(M^{\star})_{g^{-1}}$. This new pivotal duality turns $G\text{-vect}_{\Bbbk}^{\alpha}$ into an additive pivotal pre-fusion $\Bbbk$-category denoted by $G\text{-vect}_{\Bbbk}^{\alpha,d}$. The left and right dimensions of the object $\Bbbk_h$ of $G\text{-vect}_{\Bbbk}^{\alpha,d}$ are computed by

$$\dim_l(\Bbbk_h) = d(h)^{-1} \quad \text{and} \quad \dim_r(\Bbbk_h) = d(h).$$

By Lemma 4.4, the category $G\text{-vect}_{\Bbbk}^{\alpha,d}$ is spherical if and only if $d(h)^2 = 1$ for all $h \in G$. If G is a finite group of order n, then $G\text{-vect}_{\Bbbk}^{\alpha,d}$ is a fusion $\Bbbk$-category and

$$\dim\big(G\text{-vect}_{\Bbbk}^{\alpha,d}\big) = n\, 1_{\Bbbk}.$$

Note that $G\text{-vect}_{\Bbbk}^{\alpha,1} = G\text{-vect}_{\Bbbk}^{\alpha}$ as pivotal $\Bbbk$-categories, where 1 is the trivial group homomorphism $G \to \Bbbk^*$.

We now compare the category $G\text{-vect}_{\Bbbk}^{\alpha,d}$ with the pivotal pre-fusion $\Bbbk$-category $G_{\Bbbk}^{\alpha,d}$ from Example 4.4.4. The formula $h \mapsto \Bbbk_h$ defines a pivotal $\Bbbk$-linear functor $G_{\Bbbk}^{\alpha,d} \to G\text{-vect}_{\Bbbk}^{\alpha,d}$. By Section 16.3.1, this functor induces a pivotal $\Bbbk$-linear functor

$$(G_{\Bbbk}^{\alpha,d})^{\mathrm{add}} \to G\text{-vect}_{\Bbbk}^{\alpha,d}$$

where $(G_{\Bbbk}^{\alpha,d})^{\mathrm{add}}$ is the additivization of $G_{\Bbbk}^{\alpha,d}$. It is easy to check that the latter functor is an equivalence. Hence, $G\text{-vect}_{\Bbbk}^{\alpha,d}$ may be seen as the additivization of $G_{\Bbbk}^{\alpha,d}$. Since two pivotal $\Bbbk$-categories are equivalent if and only if their additivizations are equivalent (as pivotal $\Bbbk$-categories), we deduce from Exercise 4.2.7 the following classification. For $i \in \{1, 2\}$, let G_i be a group, let α_i be a $\Bbbk^*$-valued

3-cocycle on G_i, and let $d_i\colon G_i \to \Bbbk^*$ be a group homomorphism. Then the pivotal $\Bbbk$-categories

$$G_1\text{-}\mathrm{vect}_{\Bbbk}^{\alpha_1,d_1} \quad \text{and} \quad G_2\text{-}\mathrm{vect}_{\Bbbk}^{\alpha_2,d_2}$$

are equivalent if and only if there is a group isomorphism $f\colon G_1 \to G_2$ such that $d_1 = d_2 f$ and $[\alpha_1] = [f^*(\alpha_2)]$ where the square brackets stand for the cohomology class of a cocycle. Consequently, the equivalence class of the pivotal $\Bbbk$-category $G\text{-}\mathrm{vect}_{\Bbbk}^{\alpha,d}$ depends only on the class of (α, d) in the set

$$\big(H^3(G,\Bbbk^*) \times H^1(G,\Bbbk^*)\big)/\mathrm{Aut}(G)$$

where $\mathrm{Aut}(G)$ is the group of automorphisms of G.

Appendix B

Coends

We review foundations of the theory of coends following [ML2].

B.1 Coends of functors

B.1.1 Dinatural transformations and coends

Let $\mathcal{C}$ be a category and let F be a functor from $\mathcal{C}^{\mathrm{op}} \times \mathcal{C}$ to a category $\mathcal{D}$. A *dinatural transformation* from F to an object D of $\mathcal{D}$ is a family

$$d = \{d_X : F(X, X) \to D\}_{X \in \mathrm{Ob}(\mathcal{C})}$$

of morphisms in $\mathcal{D}$ such that, for every morphism $f: X \to Y$ in $\mathcal{C}$, the diagram

$$\begin{array}{ccc} F(Y,X) & \xrightarrow{F(\mathrm{id}_Y, f)} & F(Y,Y) \\ {\scriptstyle F(f,\mathrm{id}_X)}\downarrow & & \downarrow{\scriptstyle d_Y} \\ F(X,X) & \xrightarrow[d_X]{} & D \end{array}$$

commutes. Here in the expression $F(f, \mathrm{id}_X)$ we view f as a morphism $Y \to X$ in $\mathcal{C}^{\mathrm{op}}$. We denote by $\mathrm{D{\small INAT}}(F, D)$ the class of dinatural transformations from F to D. The *composition* of $d \in \mathrm{D{\small INAT}}(F, D)$ with a morphism $\varphi: D \to D'$ in $\mathcal{D}$ is the dinatural transformation $\varphi d \in \mathrm{D{\small INAT}}(F, D')$ defined by

$$\varphi d = \{\varphi d_X : F(X, X) \to D'\}_{X \in \mathrm{Ob}(\mathcal{C})}.$$

A dinatural transformation ρ from a functor $F: \mathcal{C}^{\mathrm{op}} \times \mathcal{C} \to \mathcal{D}$ to an object $C \in \mathrm{Ob}(\mathcal{D})$ is *universal* if for any dinatural transformation d from F to any object

$D \in \mathrm{Ob}(\mathcal{D})$, there is a unique morphism $\varphi\colon C \to D$ such that $d = \varphi\rho$. In other words, ρ is universal if for every $D \in \mathrm{Ob}(\mathcal{D})$, the map

$$\varphi \in \mathrm{Hom}_{\mathcal{D}}(C, D) \mapsto \varphi\rho \in \mathrm{DINAT}(F, D)$$

is a bijection. A *coend* of a functor $F\colon \mathcal{C}^{\mathrm{op}} \times \mathcal{C} \to \mathcal{D}$ is a pair (C, ρ), where $C \in \mathrm{Ob}(\mathcal{D})$ and ρ is a universal dinatural transformation from F to C. The universality of a coend implies that a coend of F, if it exists, is unique up to isomorphism. More precisely, if (C, ρ) and (C', ρ') are coends of F, then there is a unique isomorphism $\varphi\colon C \to C'$ in $\mathcal{D}$ such that $\rho' = \varphi\rho$. We often write C for (C, ρ), specifying ρ whenever necessary. Following [ML2], we write $C = \int^{X \in \mathcal{C}} F(X, X)$.

Lemma B.1 (Naturality)**.** *Let $\mathcal{C}$ and $\mathcal{D}$ be categories. Let α be a natural transformation from a functor $F\colon \mathcal{C}^{\mathrm{op}} \times \mathcal{C} \to \mathcal{D}$ to a functor $G\colon \mathcal{C}^{\mathrm{op}} \times \mathcal{C} \to \mathcal{D}$. If F has a coend (C, ρ) and G has a coend (D, ϱ), then there exists a unique morphism $\varphi\colon C \to D$ in $\mathcal{D}$ such that the following diagram commutes for all $X \in \mathrm{Ob}(\mathcal{C})$:*

$$\begin{array}{ccc} F(X,X) & \xrightarrow{\alpha_{X,X}} & G(X,X) \\ {\scriptstyle \rho_X}\downarrow & & \downarrow{\scriptstyle \varrho_X} \\ C & \xrightarrow[\varphi]{} & D\,. \end{array}$$

If α is a natural isomorphism, then φ is an isomorphism.

Proof. The family $\varrho\alpha = \{\varrho_X \alpha_{X,X} \colon F(X, X) \to D\}_{X \in \mathrm{Ob}(\mathcal{C})}$ is a dinatural transformation from F to D and so $\varrho\alpha = \varphi\rho$ for a unique morphism $\varphi\colon C \to D$. If α is an isomorphism, then there is a unique morphism $\phi\colon D \to C$ such that $\rho\alpha^{-1} = \phi\varrho$. Consequently, $\phi\varphi\rho = \phi\varrho\alpha = \rho\alpha^{-1}\alpha = \rho$ and, similarly, $\varphi\phi\varrho = \varrho$. The uniqueness of the factorisation through a universal dinatural transformation implies that $\phi\varphi = \mathrm{id}_C$ and $\varphi\phi = \mathrm{id}_D$. Thus, φ is an isomorphism. □

B.1.2 Example

Let $\mathcal{C}$ be a fusion $\Bbbk$-category and let $\mathcal{D}$ be an additive $\Bbbk$-category. An arbitrary $\Bbbk$-linear functor $F\colon \mathcal{C}^{\mathrm{op}} \times \mathcal{C} \to \mathcal{D}$ has a coend (C, ρ) which can be constructed as follows. Pick a representative set I of simple objects of $\mathcal{C}$. Set

$$C = \bigoplus_{i \in I} F(i, i). \tag{B.1}$$

For each $i \in I$, let $\rho_i\colon F(i, i) \to C$ be the inclusion associated with the direct sum decomposition (B.1). For any $X \in \mathrm{Ob}(\mathcal{C})$, pick an I-partition $(p_\alpha\colon X \to i_\alpha,\, q_\alpha\colon i_\alpha \to X)_\alpha$ of X and set

$$\rho_X = \sum_\alpha \rho_{i_\alpha} F(q_\alpha, p_\alpha)\colon F(X, X) \to C.$$

Note that for $X = i \in I$, we have that $\rho_X = \rho_i$ because the pair $(\mathrm{id}_i, \mathrm{id}_i)$ is an I-partition of i and $F(\mathrm{id}_i, \mathrm{id}_i) = \mathrm{id}_{F(i,i)}$. It is easy to check that the family

$$\rho = \{\rho_X : F(X,X) \to C\}_{X \in \mathrm{Ob}(\mathcal{C})}$$

is a well-defined universal dinatural transformation from F to C. For any dinatural transformation d from F to an object $D \in \mathrm{Ob}(\mathcal{D})$, the unique morphism $\varphi : C \to D$ in $\mathcal{D}$ satisfying $d = \varphi\rho$ is computed by

$$\varphi = \sum_{i \in I} d_i \nu_i : C \to D \tag{B.2}$$

where $\nu_i : C \to F(i,i)$ is the projection associated with the decomposition (B.1).

B.1.3 Remark

Let $\mathcal{C}$ be a left rigid category with distinguished left duality $\{({}^\vee X, \mathrm{ev}_X)\}_{X \in \mathrm{Ob}(\mathcal{C})}$. Define a functor $F : \mathcal{C}^{\mathrm{op}} \times \mathcal{C} \to \mathcal{C}$ by

$$F(X,Y) = {}^\vee X \otimes Y \quad \text{and} \quad F(f,g) = {}^\vee f \otimes g$$

for all $X, Y \in \mathrm{Ob}(\mathcal{C})$ and all morphisms f, g in $\mathcal{C}$. A dinatural transformation in $\mathcal{C}$ to an object $D \in \mathrm{Ob}(\mathcal{C})$ in the sense of Section 6.4.1 is nothing but a dinatural transformation from F to D. Consequently, the category $\mathcal{C}$ has a coend in the sense of Section 6.4.1 if and only if the functor F has a coend.

B.2 Properties of coends

We state properties of coends used in Chapters 6 and 9.

B.2.1 Coends with parameters and Fubini theorem

For the proof of the following two theorems, see [ML2].

Theorem B.2 (Coends with parameters). *Let $\mathcal{P}, \mathcal{C}, \mathcal{D}$ be three categories and let $F : \mathcal{P} \times \mathcal{C}^{\mathrm{op}} \times \mathcal{C} \to \mathcal{D}$ be a functor such that for any $P \in \mathrm{Ob}(\mathcal{P})$, the functor $F(P, \cdot, \cdot) : \mathcal{C}^{\mathrm{op}} \times \mathcal{C} \to \mathcal{D}$ has a coend $(H(P), \rho^P)$. Then the correspondence $P \mapsto H(P)$ extends uniquely to a functor $H : \mathcal{P} \to \mathcal{D}$ such that, for any $X \in \mathrm{Ob}(\mathcal{C})$, the family $\rho_X = \{\rho_X^P : F(P,X,X) \to H(P)\}_{P \in \mathrm{Ob}(\mathcal{P})}$ is a natural transformation from $F(\cdot, X, X)$ to H.*

Theorem B.3 (Fubini theorem for coends). *Let $\mathcal{P}, \mathcal{C}, \mathcal{D}$ be three categories and let $F : \mathcal{P}^{\mathrm{op}} \times \mathcal{P} \times \mathcal{C}^{\mathrm{op}} \times \mathcal{C} \to \mathcal{D}$ be a functor such that for any $P, Q \in \mathrm{Ob}(\mathcal{P})$, the coend $\int^{X \in \mathcal{C}} F(P,Q,X,X)$ exists. Let $\mathcal{P}^{\mathrm{op}} \times \mathcal{P} \to \mathcal{D}$ be the functor obtained by* *applying Theorem* B.2 *to the correspondence* $(P,Q) \mapsto \int^{X\in\mathcal{C}} F(P,Q,X,X)$. *Then the coend of this functor is isomorphic to the coend of the functor*

$$(\mathcal{P}\times\mathcal{C})^{\mathrm{op}}\times\mathcal{P}\times\mathcal{C} = \mathcal{P}^{\mathrm{op}}\times\mathcal{C}^{\mathrm{op}}\times\mathcal{P}\times\mathcal{C}\to\mathcal{D}$$

obtained from F *via the obvious permutation of factors. More precisely, if one of these two coends exists, then they both exist and are isomorphic via a unique isomorphism commuting with the universal dinatural transformations.*

Theorem B.3 is usually expressed by the formula

$$\int^{P\in\mathcal{P}}\left(\int^{X\in\mathcal{C}} F(P,P,X,X)\right) \simeq \int^{(P,X)\in\mathcal{P}\times\mathcal{C}} F(P,P,X,X).$$

B.2.2 Coends and adjunctions

The following lemma concerns the behaviour of coends under adjunctions.

Lemma B.4. *For any categories* $\mathcal{C},\mathcal{D},\mathcal{E}$, *any functor* $G\colon \mathcal{D}^{\mathrm{op}}\times\mathcal{C}\to\mathcal{E}$, *and an adjunction* $L\colon\mathcal{C}\rightleftarrows\mathcal{D}:R$, *there is a canonical isomorphism*

$$\int^{X\in\mathcal{C}} G(L(X),X) \simeq \int^{Y\in\mathcal{D}} G(Y,R(Y)).$$

More precisely, if one of these two coends exists, then they both exist and are isomorphic via a unique isomorphism commuting with the universal dinatural transformations.

Proof. Let η be the unit and ε be the counit of the adjunction $L\dashv R$. For an object $E\in\mathrm{Ob}(\mathcal{E})$, define a map

$$\psi_E\colon \mathrm{D{\scriptsize INAT}}\big(G(L\times 1_{\mathcal{C}}),E\big)\to \mathrm{D{\scriptsize INAT}}\big(G(1_{\mathcal{D}^{\mathrm{op}}}\times R),E\big)$$

by

$$\psi_E(d)=\{(\psi_E(d))_Y = d_{R(Y)}G(\varepsilon_Y,\mathrm{id}_{R(Y)})\}_{Y\in\mathrm{Ob}(\mathcal{D})}.$$

This map is bijective with inverse

$$\psi_E^{-1}(t)=\{(\psi_E^{-1}(t))_X = t_{L(X)}G(\mathrm{id}_{L(X)},\eta_X)\}_{X\in\mathrm{Ob}(\mathcal{C})}.$$

Note that $\psi_{E'}(fd)=f\psi_E(d)$ for any morphism $f\colon E\to E'$ in $\mathcal{E}$. Thus, (C,ρ) is a coend of $G(L\times 1_{\mathcal{C}})$ if and only if $(C,\psi_C(\rho))$ is a coend of $G(1_{\mathcal{D}^{\mathrm{op}}}\times R)$. The lemma easily follows. □

B.2.3 Preservation and creation of coends

A functor $H\colon \mathcal{D} \to \mathcal{E}$ between categories $\mathcal{D}$ and $\mathcal{E}$ *preserves coends* if for any category $\mathcal{C}$ and any functor $F\colon \mathcal{C}^{\mathrm{op}} \times \mathcal{C} \to \mathcal{D}$ having a coend (C,ρ), the pair $(H(C), H(\rho))$ is a coend of the composed functor $HF\colon \mathcal{C}^{\mathrm{op}} \times \mathcal{C} \to \mathcal{E}$. Here $H(\rho)$ is the dinatural transformation from HF to $H(C)$ defined by $H(\rho)_X = H(\rho_X)$ for any $X \in \mathrm{Ob}(\mathcal{C})$.

Lemma B.5. *Any functor having a right adjoint preserves coends.*

Proof. Let $R\colon \mathcal{E} \to \mathcal{D}$ be a right adjoint of a functor $H\colon \mathcal{D} \to \mathcal{E}$. Let

$$\eta = \{\eta_D\colon D \to RH(D)\}_{D\in \mathrm{Ob}(\mathcal{D})} \quad \text{and} \quad \varepsilon = \{\varepsilon_E\colon HR(E) \to E\}_{E\in \mathrm{Ob}(\mathcal{E})}$$

be the unit and the counit of the adjunction $H \dashv R$, respectively. Pick any functor $F\colon \mathcal{C}^{\mathrm{op}} \times \mathcal{C} \to \mathcal{D}$ having a coend (C,ρ). We must prove that $(H(C), H(\rho))$ is a coend of the functor HF. Let d be a dinatural transformation from HF to an object $E \in \mathrm{Ob}(\mathcal{E})$. Then

$$\{R(d_X)\eta_{F(X,X)}\colon F(X,X) \to R(E)\}_{X\in \mathrm{Ob}(\mathcal{C})}$$

is a dinatural transformation from F to $R(E)$ and so there exists a unique morphism $\beta\colon C \to R(E)$ such that $R(d_X)\eta_{F(X,X)} = \beta\rho_X$ for all $X \in \mathrm{Ob}(\mathcal{C})$. Set

$$\alpha = \varepsilon_E H(\beta)\colon H(C) \to E.$$

The naturality of ε and (7.1) imply that for any $X \in \mathrm{Ob}(\mathcal{C})$,

$$\alpha H(\rho_X) = \varepsilon_E H(\beta\rho_X) = \varepsilon_E HR(d_X) H(\eta_{F(X,X)}) = d_X \varepsilon_{F(X,X)} H(\eta_{F(X,X)}) = d_X.$$

If there is another morphism $\alpha'\colon H(C) \to E$ with $\alpha' H(\rho) = d$, then the naturality of η implies that for any $X \in \mathrm{Ob}(\mathcal{C})$,

$$R(\alpha')\eta_C \rho_X = R(\alpha' H(\rho_X))\eta_{F(X,X)} = R(d_X)\eta_{F(X,X)} = \beta\rho_X,$$

Hence, $R(\alpha')\eta_C = \beta$. Using (7.1) and naturality of ε, we obtain

$$\alpha' = \alpha' \mathrm{id}_C = \alpha' \varepsilon_{H(C)} H(\eta_C) = \varepsilon_E H(R(\alpha')\eta_C) = \varepsilon_E H(\beta) = \alpha.$$

Thus there exists a unique morphism $\alpha\colon H(C) \to E$ such that $\alpha H(\rho) = d$, and so $(H(C), H(\rho))$ is a coend of the functor HF. □

A functor $H\colon \mathcal{D} \to \mathcal{E}$ between categories $\mathcal{D}$ and $\mathcal{E}$ *creates coends* if for any category $\mathcal{C}$ and any functor $F\colon \mathcal{C}^{\mathrm{op}} \times \mathcal{C} \to \mathcal{D}$ such that the composed functor $HF\colon \mathcal{C}^{\mathrm{op}} \times \mathcal{C} \to \mathcal{E}$ has a coend (C', ρ'), we have:

(i) there is a unique pair (C,ρ), where $C \in \mathrm{Ob}(\mathcal{D})$ and ρ is a dinatural transformation from F to C, such that $H(C) = C'$ and $H(\rho) = \rho'$;

(ii) the pair (C,ρ) from (i) is a coend of F.

Appendix C

Abelian categories

In this appendix we first summarize the basics of the theory of abelian categories (for further details, see, e.g., [Fr]). Next, we prove that additive pre-fusion categories over a field (in the sense of Chapter 4) are abelian.

C.1 Kernels and cokernels

C.1.1 Zero objects

An object of a category $\mathcal{C}$ is called a *zero object* and is denoted $\mathbf{0}$ if it is both initial and terminal, meaning that for any $X \in \mathrm{Ob}(\mathcal{C})$, there is a unique morphism $\mathbf{0} \to X$ and a unique morphism $X \to \mathbf{0}$. A zero object, if it exists, is unique up to isomorphism. Note that an object of a $\Bbbk$-linear category is a zero object if and only if has only the zero endomorphism.

If a category $\mathcal{C}$ has a zero object, then for any $X, Y \in \mathrm{Ob}(\mathcal{C})$, the composition of the unique morphisms $X \to \mathbf{0}$ and $\mathbf{0} \to Y$ is a morphism $X \to Y$ called the *zero morphism* and denoted by 0. This morphism does not depend on the choice of $\mathbf{0}$. The composition of a zero morphism with an arbitrary morphism in $\mathcal{C}$ is a zero morphism.

C.1.2 (Co)kernels of morphisms

Let $\mathcal{C}$ be a category with a zero object. A *kernel* of a morphism $f\colon X \to Y$ in $\mathcal{C}$ is a morphism $q\colon K \to X$ in $\mathcal{C}$ such that $fq = 0$ and any morphism $g\colon Z \to X$ with $fg = 0$ factors uniquely through q, i.e., $g = qh$ for a unique morphism $h\colon Z \to K$. Clearly, such a morphism q is a monomorphism in the sense of Section 7.3.1. The source of a kernel of f is unique up to isomorphism and is denoted $\mathrm{Ker}(f)$. One says that *kernels exist in* $\mathcal{C}$ if every morphism in $\mathcal{C}$ has a kernel.

A *cokernel* of a morphism $f\colon X \to Y$ in $\mathcal{C}$ is a morphism $p\colon Y \to C$ in $\mathcal{C}$ such that $pf = 0$ and any morphism $g\colon Y \to Z$ with $gf = 0$ factors uniquely through p, i.e., $g = hp$ for a unique morphism $h\colon C \to Z$. Then p is an epimorphism in the sense of Section 7.3.1. The target of a cokernel of f is unique up to isomorphism and is denoted $\mathrm{Coker}(f)$. One says that *cokernels exist in* $\mathcal{C}$ if every morphism in $\mathcal{C}$ has a cokernel. Note that a cokernel of a morphism in $\mathcal{C}$ is a kernel of the same morphism in $\mathcal{C}^{\mathrm{op}}$.

Lemma C.1. *Let $f\colon X \to Y$ be a morphism in a $\Bbbk$-category with a zero object. Then the following claims hold:*

(a) *f is a monomorphism if and only if the morphism $\mathbf{0} \to X$ is a kernel of f;*

(b) *f is an epimorphism if and only if the morphism $Y \to \mathbf{0}$ is a cokernel of f.*

Proof. Let us prove (a). Assume that the (unique) morphism $0\colon \mathbf{0} \to X$ is a kernel of f and let $g, h\colon Z \to X$ be any morphisms such that $fg = fh$. Then $f(g-h) = 0$ and so $g - h$ factors uniquely through $\mathbf{0}$. In particular $g - h = 0$, and so $g = h$. Hence f is a monomorphism. Conversely, assume that f is a monomorphism and let $g\colon Z \to X$ be a morphism such that $fg = 0$. Then $fg = f0$, and so $g = 0$ because f is a monomorphism. Consequently, g factors through the morphism $\mathbf{0} \to X$, necessarily in a unique way. Thus, the morphism $\mathbf{0} \to X$ is a kernel of f. Part (b) follows from (a) because a morphism in a category $\mathcal{C}$ is an epimorphism if and only if it is a monomorphism in $\mathcal{C}^{\mathrm{op}}$. □

C.1.3 Functors and (co)kernels

A functor *preserves monomorphisms* if it carries monomorphisms to monomorphisms, and *preserves epimorphisms* if it carries epimorphisms to epimorphisms. Clearly, a functor $U\colon \mathcal{D} \to \mathcal{C}$ preserves monomorphisms if and only if $U^{\mathrm{op}}\colon \mathcal{D}^{\mathrm{op}} \to \mathcal{C}^{\mathrm{op}}$ preserves epimorphisms.

Let $\mathcal{C}$ and $\mathcal{D}$ be categories having a zero object. A functor $U\colon \mathcal{D} \to \mathcal{C}$ *preserves kernels* if whenever a morphism q in $\mathcal{D}$ is a kernel of a morphism f in $\mathcal{D}$, the morphism $U(q)$ is a kernel of the morphism $U(f)$. A functor $U\colon \mathcal{D} \to \mathcal{C}$ *preserves cokernels* if whenever a morphism p in $\mathcal{D}$ is a cokernel of a morphism f in $\mathcal{D}$, the morphism $U(q)$ is a cokernel of the morphism $U(f)$. Note that a functor $U\colon \mathcal{D} \to \mathcal{C}$ preserves kernels if and only if $U^{\mathrm{op}}\colon \mathcal{D}^{\mathrm{op}} \to \mathcal{C}^{\mathrm{op}}$ preserves cokernels.

A functor $U\colon \mathcal{D} \to \mathcal{C}$ *creates kernels* if for any morphism f in $\mathcal{D}$ such that $U(f)$ has a kernel q, there is a unique morphism h in $\mathcal{D}$ such that $fh = 0$ and $U(h) = q$, and then h is a kernel of f in $\mathcal{D}$. A functor $U\colon \mathcal{D} \to \mathcal{C}$ *creates cokernels* if for any morphism f in $\mathcal{D}$ such that $U(f)$ has a cokernel p, there is a unique morphism h in $\mathcal{D}$ such that $hf = 0$ and $U(h) = p$, and then h is a cokernel of f in $\mathcal{D}$. Note that a functor $U\colon \mathcal{D} \to \mathcal{C}$ creates kernels if and only if $U^{\mathrm{op}}\colon \mathcal{D}^{\mathrm{op}} \to \mathcal{C}^{\mathrm{op}}$ creates cokernels.

Lemma C.2. *Let $U\colon \mathcal{D} \to \mathcal{C}$ be a $\Bbbk$-linear functor between $\Bbbk$-categories with a zero object. Then the following properties hold:*

(a) *If U has a left adjoint, then U preserves kernels;*

(b) *If U has a right adjoint, then U preserves cokernels;*

(c) *If U preserves kernels, then U preserves monomorphisms;*

(d) *If U preserves cokernels, then U preserves epimorphisms;*

(e) *If kernels exist in $\mathcal{C}$ and U creates kernels, then kernels exist in $\mathcal{D}$ and are preserved by U;*

(f) *If cokernels exist in $\mathcal{C}$ and U creates cokernels, then cokernels exist in $\mathcal{D}$ and are preserved by U.*

Proof. Note first that the $\Bbbk$-linearity of the functor U implies that it carries zero objects/morphisms of $\mathcal{D}$ to zero objects/morphisms of $\mathcal{C}$ and that the functor $U^{\mathrm{op}}: \mathcal{D}^{\mathrm{op}} \to \mathcal{C}^{\mathrm{op}}$ is $\Bbbk$-linear.

Let us prove (a). Let $F: \mathcal{C} \to \mathcal{D}$ be a left adjoint of U. By Lemma 7.7, the functor F is $\Bbbk$-linear. Let η and ε be the unit and counit of the adjunction $F \dashv U$. Recall from Lemma 7.1 that $\varepsilon_{F(C)} F(\eta_C) = \mathrm{id}_{F(C)}$ and $U(\varepsilon_D)\, \eta_{U(D)} = \mathrm{id}_{U(D)}$ for all $C \in \mathrm{Ob}(\mathcal{C})$ and $D \in \mathrm{Ob}(\mathcal{D})$. Let $f: X \to Y$ be a morphism in $\mathcal{D}$ with kernel $q: K \to X$. We must prove that the morphism $U(q): U(K) \to U(X)$ is a kernel of $U(f): U(X) \to U(Y)$. Firstly, $U(f)U(q) = U(fq) = U(0) = 0$. Secondly, let $g: Z \to U(X)$ be a morphism in $\mathcal{C}$ such that $U(f)g = 0$. Then

$$f\varepsilon_X F(g) = \varepsilon_Y F(U(f)g) = \varepsilon_Y F(0) = \varepsilon_Y 0 = 0.$$

Therefore there is a unique morphism $k: F(Z) \to K$ in $\mathcal{D}$ such that $\varepsilon_X F(g) = qk$. Set $h = U(k)\eta_Z : Z \to U(K)$. Then

$$U(q)h = U(qk)\eta_Z = U(\varepsilon_X F(g))\eta_Z = U(\varepsilon_X)\eta_{U(X)} g = \mathrm{id}_{U(X)} g = g.$$

Consider an arbitrary morphism $h': Z \to U(K)$ in $\mathcal{C}$ with $U(q)h' = g$. Then

$$q\varepsilon_K F(h') = \varepsilon_X F(U(q)h') = \varepsilon_X F(g) = qk.$$

Thus, $\varepsilon_K F(h') = k$ and

$$h = U(k)\eta_Z = U(\varepsilon_K F(h'))\eta_Z = U(\varepsilon_K)\eta_{U(K)} h' = \mathrm{id}_{U(K)} h' = h'.$$

We conclude that $U(q)$ is a kernel of the morphism $U(f)$.

Let us prove (b). If U has a right adjoint, then U^{op} has a left adjoint and preserves kernels by (a). Hence U preserves cokernels.

Let us prove (c). Let f be a monomorphism in $\mathcal{C}$. Then $\mathrm{Ker}(f) = \mathbf{0}$ by Lemma C.1(a). Since U preserves zero objects (because it is $\Bbbk$-linear) and U preserves kernels, we have

$$\mathrm{Ker}(U(f)) = U(\mathrm{Ker}(f)) = U(\mathbf{0}) = \mathbf{0}.$$

Thus, by Lemma C.1(a), the morphism $U(f)$ is a monomorphism.

Claim (d) is obtained by applying (c) to U^{op}. Let us prove (e). Pick any morphism $f\colon X \to Y$ in $\mathcal{D}$. Let $k\colon C \to U(X)$ be a kernel of $U(f)\colon U(X) \to U(Y)$. Since U creates kernels, there is a unique morphism $h\colon D \to X$ in $\mathcal{D}$ such that $fh = 0$ and $U(h) = k$, and this morphism h is a kernel of f in $\mathcal{D}$. Now, let $q\colon K \to X$ be a kernel of f. The uniqueness of kernels implies the existence of an isomorphism $\phi\colon K \to D$ in $\mathcal{D}$ such that $q = h\phi$. Then

$$U(q) = U(h\phi) = U(h)U(\phi) = kU(\phi).$$

Since k is a kernel of $U(f)$ and $U(\phi)$ is an isomorphism, we conclude that $U(q)$ is a kernel of $U(f)$. Hence U preserves kernels.

Finally, Claim (f) is obtained by applying (e) to U^{op}. □

C.1.4 Remark

Lemma C.2 implies that a $\Bbbk$-linear functor having a left adjoint preserves monomorphisms and a $\Bbbk$-linear functor having a right adjoint preserves epimorphisms. These claims actually hold without the assumption of $\Bbbk$-linearity, see Exercise 7.3.6.

C.2 Abelian categories

C.2.1 Definition and properties

A $\Bbbk$-category is *abelian* if it is additive, every morphism has a kernel and a cokernel, every monomorphism is a kernel, and every epimorphism is a cokernel. For example, the category $\mathrm{Mod}_{\Bbbk}$ of (left) $\Bbbk$-modules (see Example 1.1.3) is an abelian $\Bbbk$-category.

Any abelian $\Bbbk$-category has a zero object, see Section 4.1.2. To state further properties of abelian categories, recall the notions of retract, retraction, and section from Section 7.3.1. In a $\Bbbk$-category, an object Y is said to be a *direct summand* of an object X if $X \simeq Y \oplus Z$ for some object Z of the category.

Lemma C.3. *In an abelian $\Bbbk$-category, the following assertions hold:*

(a) *Any monomorphism is the kernel of its cokernel;*

(b) *Any epimorphism is the cokernel of its kernel;*

(c) *A morphism which is both a monomorphism and an epimorphism is an isomorphism;*

(d) *Any retract of an object is a direct summand of that object;*

(e) *Any monomorphism has a retraction if and only if any epimorphism has a section.*

Proof. Consider an abelian $\Bbbk$-category $\mathcal{C}$. We start by proving (a). Let $f\colon X \to Y$ be a monomorphism in $\mathcal{C}$. Let $p\colon Y \to C$ be a cokernel of f. Pick a morphism $g\colon Z \to Y$ in $\mathcal{C}$ such that $pg = 0$. We need to show that g factors uniquely through f. The monomorphism f is the kernel of some morphism $h\colon Y \to T$. Since $hf = 0$ and p is a cokernel of f, there is a unique morphism $\alpha\colon C \to T$ such that $h = \alpha p$. Now,

$$hg = \alpha pg = \alpha 0 = 0.$$

Thus, since f is the kernel of h, there is a unique morphism $\beta\colon Z \to X$ such that $g = f\beta$.

Claim (b) is obtained by applying (a) to the abelian $\Bbbk$-category $\mathcal{C}^{\mathrm{op}}$. Let us prove (c). Let $f\colon X \to Y$ be a morphism in $\mathcal{C}$ which is a monomorphism and an epimorphism. By Lemma C.1(b), the zero morphism $0\colon Y \to \mathbf{0}$ is a cokernel of f. By Claim (a), f is then a kernel of $0\colon Y \to \mathbf{0}$. Since $0\,\mathrm{id}_Y = 0$, there is a morphism $g\colon Y \to X$ such that $fg = \mathrm{id}_Y$. Since $fgf = \mathrm{id}_Y f = f\mathrm{id}_X$ and f is a monomorphism, we obtain that $gf = \mathrm{id}_X$. Hence f is invertible with inverse g.

Let us prove (d). Let $Y \in \mathrm{Ob}(\mathcal{C})$ be a retract of $X \in \mathrm{Ob}(\mathcal{C})$. There are morphisms $r\colon X \to Y$ and $s\colon Y \to X$ in $\mathcal{C}$ such that $rs = \mathrm{id}_Y$. Let $a\colon X \to Z$ be a cokernel of s. Set

$$f = \mathrm{id}_X - sr\colon X \to X.$$

Since $fs = s - srs = s - s\mathrm{id}_Y = 0$, there is a morphism $b\colon Z \to X$ such that $f = ba$, that is, $sr + ba = \mathrm{id}_X$. We have $as = 0$ because a is a cokernel of s. Since

$$rba = rf = r - rsr = r - \mathrm{id}_Y r = 0 = 0a,$$
$$aba = af = a - asr = a - 0r = a = \mathrm{id}_Z a,$$

and a is an epimorphism (as a cokernel), we obtain that $rb = 0$ and $ab = \mathrm{id}_Z$. Consequently, $X = Y \oplus Z$ and Y is a direct summand of X.

Finally, let us prove (e). Assume that any monomorphism in $\mathcal{C}$ has a retraction. Let us prove that any epimorphism $f\colon X \to Y$ in $\mathcal{C}$ has a section. Let $s\colon K \to X$ be a kernel of f. Then s is a monomorphism and so has a retraction, i.e., there is a morphism $r\colon X \to K$ such that $rs = \mathrm{id}_K$. Then, as in the proof of Claim (d) above, there are morphisms $a\colon X \to Z$ and $b\colon Z \to X$ such that

$$sr + ba = \mathrm{id}_X, \quad ab = \mathrm{id}_Z, \quad rb = 0, \quad as = 0.$$

Now, by Claim (b), f is a cokernel of s. Since $as = 0$, there is a morphism $t\colon Y \to Z$ such that $a = tf$. Set $g = bt\colon Y \to X$. Then

$$\mathrm{id}_Y f = f\mathrm{id}_X = f(sr + ba) = fsr + fba = 0r + fbtf = fgf.$$

Since f is an epimorphism, we conclude that $fg = \mathrm{id}_Y$, that is, g is a section of f. This proves the 'only if' part of (e). The 'if' part is obtained by applying the 'only if' part to $\mathcal{C}^{\mathrm{op}}$. □

C.2.2 Indecomposable objects

An object of $\Bbbk$-category is *indecomposable* if it is non-zero and is not isomorphic to a direct sum of two non-zero objects. For example, if the only idempotent elements of $\Bbbk$ are 0 and $1_\Bbbk$ (e.g., if $\Bbbk$ is an integral domain), then all simple objects of any $\Bbbk$-category are indecomposable.

Lemma C.4. *Let $\mathcal{A}$ be an abelian $\Bbbk$-category in which all monomorphisms have a retraction. Then any morphism between indecomposable objects of $\mathcal{A}$ is either an isomorphism or equal to zero.*

Proof. Let $f\colon X \to Y$ be a morphism in $\mathcal{A}$ between indecomposable objects. Let $q\colon \mathrm{Ker}(f) \to X$ be a kernel of f. We claim that $\mathrm{Ker}(f) = \mathbf{0}$ or q is an isomorphism. Indeed, q is a monomorphism in $\mathcal{A}$ and, by assumption, q has a retraction, i.e., there is a morphism $r\colon X \to \mathrm{Ker}(f)$ such that $rq = \mathrm{id}_{\mathrm{Ker}(f)}$. As in the proof of Lemma C.3(d), there are $Z \in \mathrm{Ob}(\mathcal{C})$ and morphisms $a\colon X \to Z$ and $b\colon Z \to X$ such that

$$X = \mathrm{Ker}(f) \oplus Z \quad \text{and} \quad qr + ba = \mathrm{id}_X.$$

Since X is indecomposable, $\mathrm{Ker}(f) = \mathbf{0}$ or $Z = \mathbf{0}$. In the latter case, $qr = \mathrm{id}_X$ (since $ba = 0$) and so q is an isomorphism. Similarly, let $p\colon Y \to \mathrm{Coker}(f)$ be a cokernel of f. Then p is an epimorphism, and the assumptions of the lemma and Lemma C.3(e) imply that p has a section. Using that Y is indecomposable, we conclude as above that $\mathrm{Coker}(f) = \mathbf{0}$ or p is an isomorphism. If $\mathrm{Ker}(f) = \mathbf{0}$ and $\mathrm{Coker}(f) = \mathbf{0}$, then f is an isomorphism by Lemmas C.1 and C.3(c). If q is an isomorphism, then $f = fqq^{-1} = 0q^{-1} = 0$. If p is an isomorphism, then $f = p^{-1}pf = p^{-1}0 = 0$. □

Lemma C.5. *Let $\Bbbk$ be a field and let X be an object of a $\Bbbk$-category $\mathcal{C}$ such that the $\Bbbk$-vector space $\mathrm{End}_\mathcal{C}(X)$ is finite-dimensional. Then X is a direct sum of finitely many indecomposable objects of $\mathcal{C}$.*

Proof. If X is a zero object, then X is a direct sum of an empty family of indecomposable objects of $\mathcal{C}$. Assume that X is non-zero and let $n \geq 1$ be the dimension of the $\Bbbk$-vector space $\mathrm{End}_\mathcal{C}(X)$. It follows from the results of Section 4.1.2 that if X is a direct sum of a finite family $(X_\alpha)_\alpha$ of objects of $\mathcal{C}$, then

$$\sum_\alpha \dim_\Bbbk(\mathrm{End}_\mathcal{C}(X_\alpha)) \leq \dim_\Bbbk(\mathrm{End}_\mathcal{C}(X)) = n.$$

Thus, X cannot be a direct sum of more than n non-zero objects of $\mathcal{C}$. Consequently, there is the largest integer $m \geq 1$ such that X is a direct sum of m non-zero objects of $\mathcal{C}$. All these non-zero objects must be indecomposable (otherwise X would be a direct sum of $m+1$ non-zero objects). □

C.2.3 Additive pre-fusion categories over a field are abelian

Theorem C.6. *If $\Bbbk$ is a field, then any additive pre-fusion $\Bbbk$-category $\mathcal{C}$ is abelian, any monomorphism in $\mathcal{C}$ has a retraction, and any epimorphism in $\mathcal{C}$ has a section.*

Proof. Fix a representative set I of simple objects of $\mathcal{C}$. Recall from Section 4.4.1 that the Hom-sets in $\mathcal{C}$ are finite-dimensional $\Bbbk$-vector spaces. Our first aim is to define for any $X, Y \in \mathrm{Ob}(\mathcal{C})$, an isomorphism between $\mathrm{Hom}_{\mathcal{C}}(X,Y)$ and the $\Bbbk$-vector space

$$V_{X,Y} = \prod_{i \in I} \mathrm{Hom}_{\Bbbk}\big(\mathrm{Hom}_{\mathcal{C}}(i,X), \mathrm{Hom}_{\mathcal{C}}(i,Y)\big).$$

To this end, we fix an i-partition $(p_\alpha, q_\alpha)_{\alpha \in A_i}$ of X for all $i \in I$. For any $\ell = (\ell_i)_{i \in I} \in V_{X,Y}$, set

$$\Gamma_{X,Y}(\ell) = \sum_{i \in I} \sum_{\alpha \in A_i} \ell_i(q_\alpha) p_\alpha \colon X \to Y.$$

It is clear that the resulting map $\Gamma_{X,Y} \colon V_{X,Y} \to \mathrm{Hom}_{\mathcal{C}}(X,Y)$ is a $\Bbbk$-linear homomorphism. Next, for all $f \in \mathrm{Hom}_{\mathcal{C}}(X,Y)$ and $i \in I$, consider the $\Bbbk$-linear homomorphism

$$\Delta_i^f \colon \mathrm{Hom}_{\mathcal{C}}(i,X) \to \mathrm{Hom}_{\mathcal{C}}(i,Y), \quad u \mapsto fu.$$

Consider the $\Bbbk$-linear homomorphism

$$\Delta_{X,Y} \colon \mathrm{Hom}_{\mathcal{C}}(X,Y) \to V_{X,Y}, \quad f \mapsto \Delta_{X,Y}(f) = (\Delta_i^f)_{i \in I}.$$

We claim that the homomorphisms $\Gamma_{X,Y}$ and $\Delta_{X,Y}$ are inverse to each other. (As a consequence, $\Gamma_{X,Y}$ does not depend on the choice of the i-partitions of X used in its definition.) To prove our claim we must show the identities

$$\Gamma_{X,Y}\big((\Delta_i^f)_{i \in I}\big) = f \quad \text{and} \quad \Delta_i^{\Gamma_{X,Y}(\ell)} = \ell_i \quad \text{for all } i \in I. \tag{C.1}$$

For all $f \in \mathrm{Hom}_{\mathcal{C}}(X,Y)$, we have:

$$\begin{aligned}\Gamma_{X,Y}\big((\Delta_i^f)_{i \in I}\big) &\overset{\text{(i)}}{=} \sum_{i \in I} \sum_{\alpha \in A_i} \Delta_i^f(q_\alpha) p_\alpha \\ &\overset{\text{(ii)}}{=} \sum_{i \in I} \sum_{\alpha \in A_i} f q_\alpha p_\alpha \overset{\text{(iii)}}{=} f\left(\sum_{i \in I} \sum_{\alpha \in A_i} q_\alpha p_\alpha\right) \overset{\text{(iv)}}{=} f.\end{aligned}$$

Here, (i) follows from the definition of $\Gamma_{X,Y}$, (ii) from the definition of Δ_i^f, (iii) from the $\Bbbk$-linearity of the composition in $\mathcal{C}$, and (iv) from the fact that a union of i-partitions of X over all $i \in I$ is an I-partition of X. Also, for all $i \in I$ and

$u \in \mathrm{Hom}_{\mathcal{C}}(i, X)$, we have:

$$\begin{aligned}\Delta_i^{\Gamma_{X,Y}(\ell)}(u) &\overset{\text{(i)}}{=} \Gamma_{X,Y}(\ell)u \overset{\text{(ii)}}{=} \sum_{j\in I}\sum_{\alpha\in A_j} \ell_j(q_\alpha)p_\alpha u \\ &\overset{\text{(iii)}}{=} \sum_{\alpha\in A_i} \ell_j(q_\alpha)p_\alpha u \overset{\text{(iv)}}{=} \sum_{\alpha\in A_i} \ell_i(q_\alpha p_\alpha u) \\ &\overset{\text{(v)}}{=} \ell_i\left(\Big(\sum_{\alpha\in A_i} q_\alpha p_\alpha\Big)u\right) \overset{\text{(vi)}}{=} \ell_i\left(\Big(\sum_{j\in I}\sum_{\alpha\in A_j} q_\alpha p_\alpha\Big)u\right) \overset{\text{(vii)}}{=} \ell_i(u).\end{aligned}$$

Here, (i) follows from the definition of $\Delta_i^{\Gamma_{X,Y}(\ell)}$, (ii) from the definition of $\Gamma_{X,Y}(\ell)$, (iii) and (vi) from the equality $\mathrm{Hom}_{\mathcal{C}}(i,j) = 0$ for all $j \in I$ with $j \neq i$, (iv) from the fact that $p_\alpha u \in \mathrm{End}_{\mathcal{C}}(i) = \Bbbk\,\mathrm{id}_i$ and the $\Bbbk$-linearity of ℓ_i, (v) from the $\Bbbk$-linearity of ℓ_i, and (vii) from the fact that a union of j-partitions of X over all $j \in I$ is an I-partition of X. Thus, $\Delta_i^{\Gamma_{X,Y}(\ell)} = \ell_i$. This completes the proof of (C.1).

We now verify the following two properties of the isomorphisms Γ. For all $X, Y, Z \in \mathrm{Ob}(\mathcal{C})$, $(\ell_i)_{i\in I} \in V_{X,Y}$, and $(\xi_i)_{i\in I} \in V_{Y,Z}$, we have:

$$\Gamma_{Y,Z}\big((\xi_i)_{i\in I}\big) \circ \Gamma_{X,Y}\big((\ell_i)_{i\in I}\big) = \Gamma_{X,Z}\big((\xi_i\ell_i)_{i\in I}\big), \tag{C.2}$$

$$\Gamma_{X,X}\big((\mathrm{id}_{\mathrm{Hom}_{\mathcal{C}}(i,X)})_{i\in I}\big) = \mathrm{id}_X. \tag{C.3}$$

To see it, pick i-partitions $(p_\alpha, q_\alpha)_{\alpha\in A_i}$ of X and $(a_\beta, b_\beta)_{\beta\in B_i}$ of Y for all $i \in I$. Then

$$\begin{aligned}\Gamma_{Y,Z}\big((\xi_i)_{i\in I}\big) \circ \Gamma_{X,Y}\big((\ell_i)_{i\in I}\big) &\overset{\text{(i)}}{=} \sum_{i,j\in I}\sum_{\substack{\alpha\in A_i\\ \beta\in B_j}} \xi_j(b_\beta)a_\beta\ell_i(q_\alpha)p_\alpha \\ &\overset{\text{(ii)}}{=} \sum_{i\in I}\sum_{\substack{\alpha\in A_i\\ \beta\in B_i}} \xi_i(b_\beta)a_\beta\ell_i(q_\alpha)p_\alpha \overset{\text{(iii)}}{=} \sum_{i\in I}\sum_{\substack{\alpha\in A_i\\ \beta\in B_i}} \xi_i\big(b_\beta a_\beta\ell_i(q_\alpha)\big)p_\alpha \\ &\overset{\text{(iv)}}{=} \sum_{i\in I}\sum_{\alpha\in A_i} \xi_i\left(\Big(\sum_{\beta\in B_i} b_\beta a_\beta\Big)\ell_i(q_\alpha)\right)p_\alpha \\ &\overset{\text{(v)}}{=} \sum_{i\in I}\sum_{\alpha\in A_i} \xi_i\left(\Big(\sum_{j\in I}\sum_{\beta\in B_j} b_\beta a_\beta\Big)\ell_i(q_\alpha)\right)p_\alpha \\ &\overset{\text{(vi)}}{=} \sum_{i\in I}\sum_{\alpha\in A_i} (\xi_i\ell_i)(q_\alpha)p_\alpha \overset{\text{(vii)}}{=} \Gamma_{X,Z}\big((\xi_i\ell_i)_{i\in I}\big).\end{aligned}$$

Here, (i) follows from the definition of $\Gamma_{X,Y}$ and $\Gamma_{Y,Z}$, (ii) and (v) from the equality $\mathrm{Hom}_{\mathcal{C}}(i,j) = 0$ for all $i,j \in I$ with $i \neq j$, (iii) from the fact that $a_\beta\ell_i(q_\alpha) \in \mathrm{End}_{\mathcal{C}}(i) = \Bbbk\,\mathrm{id}_i$ and the $\Bbbk$-linearity of ξ_i, (iv) from the $\Bbbk$-linearity of ξ_i, (vi) from the fact that a union of j-partitions of Y over all $j \in I$ is an I-partition of Y, and (vii) from the definition of $\Gamma_{X,Z}$. Also,

$$\Gamma_{X,X}\big((\mathrm{id}_{\mathrm{Hom}_{\mathcal{C}}(i,X)})_{i\in I}\big) \overset{\text{(i)}}{=} \sum_{i\in I}\sum_{\alpha\in A_i} q_\alpha p_\alpha \overset{\text{(ii)}}{=} \mathrm{id}_X.$$

Here, (i) follows from the definition of $\Gamma_{X,X}$ and (ii) from the fact that a union of i-partitions of X over all $i \in I$ is an I-partition of X.

We can now prove the theorem. We first prove that each monomorphism $f\colon X \to Y$ in $\mathcal{C}$ has a retraction. For any $i \in I$, the $\Bbbk$-linear homomorphism Δ_i^f is injective (because f is a monomorphism) and so has a $\Bbbk$-linear retraction (since $\Bbbk$ is a field). Thus, there is a $\Bbbk$-linear homomorphism

$$r_i\colon \mathrm{Hom}_{\mathcal{C}}(i,Y) \to \mathrm{Hom}_{\mathcal{C}}(i,X)$$

such that $r_i\Delta_i^f = \mathrm{id}_{\mathrm{Hom}_{\mathcal{C}}(i,X)}$. The morphism

$$g = \Gamma_{Y,X}\big((r_i)_{i\in I}\big)\colon Y \to X$$

is a retraction of f since

$$\begin{aligned} gf &\overset{\text{(i)}}{=} \Gamma_{Y,X}\big((r_i)_{i\in I}\big)\Gamma_{X,Y}\big((\Delta_i^f)_{i\in I}\big) \\ &\overset{\text{(ii)}}{=} \Gamma_{X,X}\big((r_i\Delta_i^f)_{i\in I}\big) \overset{\text{(iii)}}{=} \Gamma_{X,X}\big((\mathrm{id}_{\mathrm{Hom}_{\mathcal{C}}(i,X)})_{i\in I}\big) \overset{\text{(iv)}}{=} \mathrm{id}_X. \end{aligned}$$

Here, (i) follows from the definitions of g and (C.1), (ii) from (C.2), (iii) from the fact that r_i is a retraction of Δ_i^f, and (iv) from (C.3). Hence, any monomorphism in $\mathcal{C}$ has a retraction. Applying this result to the additive pre-fusion $\Bbbk$-category $\mathcal{C}^{\mathrm{op}}$, we obtain that any epimorphism in $\mathcal{C}$ has a section.

Let us prove that any morphism $f\colon X \to Y$ in $\mathcal{C}$ has a kernel. For any $i \in I$, let n_i be the dimension of the kernel

$$\mathrm{Ker}(\Delta_i^f) = \{u \in \mathrm{Hom}_{\mathcal{C}}(i,X) \,|\, \Delta_i^f(u) = 0\}$$

of the $\Bbbk$-linear homomorphism Δ_i^f. Set

$$K = \bigoplus_{i\in I} K_i,$$

where K_i is a direct sum of n_i copies of i. For $i \in I$, the $\Bbbk$-vector space $\mathrm{Hom}_{\mathcal{C}}(i,K)$ has dimension n_i and so is isomorphic to $\mathrm{Ker}(\Delta_i^f)$. Pick any $\Bbbk$-linear isomorphism

$$\theta_i\colon \mathrm{Hom}_{\mathcal{C}}(i,K) \to \mathrm{Ker}(\Delta_i^f)$$

and consider the inclusion

$$\rho_i\colon \mathrm{Ker}(\Delta_i^f) \hookrightarrow \mathrm{Hom}_{\mathcal{C}}(i,X).$$

Since $\Bbbk$ is a field, for any $i \in I$, the $\Bbbk$-linear injection ρ_i has a $\Bbbk$-linear retraction: there is a $\Bbbk$-linear homomorphism

$$r_i\colon \mathrm{Hom}_{\mathcal{C}}(i, X) \to \mathrm{Ker}(\Delta_i^f)$$

such that $r_i\rho_i = \mathrm{id}_{\mathrm{Ker}(\Delta_i^f)}$. Set

$$q = \Gamma_{K,X}\big((\rho_i\theta_i)_{i\in I}\big)\colon K \to X.$$

We claim that the morphism q is a kernel of f. First,

$$\begin{aligned} fq &\overset{\text{(i)}}{=} \Gamma_{X,Y}\big((\Delta_i^f)_{i\in I}\big)\Gamma_{K,X}\big((\rho_i\theta_i)_{i\in I}\big) \\ &\overset{\text{(ii)}}{=} \Gamma_{K,Y}\big((\Delta_i^f\rho_i\theta_i)_{i\in I}\big) \overset{\text{(iii)}}{=} \Gamma_{K,Y}\big((0)_{i\in I}\big) \overset{\text{(iv)}}{=} 0. \end{aligned}$$

Here, (i) follows from the definition of q and (C.1), (ii) from (C.2), (iii) from the fact that $\Delta_i^f\rho_i = 0$, and (iv) from the definition of $\Gamma_{K,Y}$. Second, let $g\colon Z \to X$ be a morphism in $\mathcal{C}$ such that $fg = 0$. Set

$$h = \Gamma_{Z,K}\big((\theta_i^{-1}r_i\Delta_i^g)_{i\in I}\big)\colon Z \to K.$$

Then

$$\begin{aligned} qh &\overset{\text{(i)}}{=} \Gamma_{K,X}\big((\rho_i\theta_i)_{i\in I}\big)\Gamma_{Z,K}\big((\theta_i^{-1}r_i\Delta_i^g)_{i\in I}\big) \\ &\overset{\text{(ii)}}{=} \Gamma_{Z,X}\big((\rho_i r_i\Delta_i^g)_{i\in I}\big) \overset{\text{(iii)}}{=} \Gamma_{Z,X}\big((\Delta_i^g)_{i\in I}\big) \overset{\text{(iv)}}{=} g. \end{aligned}$$

Here, (i) follows from the definitions of q and h, (ii) from (C.2), (iii) from the equality $\rho_i r_i\Delta_i^g = \Delta_i^g$ which follows from the fact that the image of Δ_i^g is included in $\mathrm{Ker}(\Delta_i^f)$ because $fg = 0$, and (iv) from (C.1). If $h'\colon Z \to K$ is a morphism in $\mathcal{C}$ such that $g = qh'$, then

$$\begin{aligned} h' &\overset{\text{(i)}}{=} \Gamma_{K,K}\big((\mathrm{id}_{\mathrm{Hom}_{\mathcal{C}}(i,K)})_{i\in I}\big)h' \overset{\text{(ii)}}{=} \Gamma_{K,K}\big((\theta_i^{-1}r_i\rho_i\theta_i)_{i\in I}\big)h' \\ &\overset{\text{(iii)}}{=} \Gamma_{X,K}\big((\theta_i^{-1}r_i)_{i\in I}\big)\Gamma_{K,X}\big((\rho_i\theta_i)_{i\in I}\big)h' \\ &\overset{\text{(iv)}}{=} \Gamma_{X,K}\big((\theta_i^{-1}r_i)_{i\in I}\big)qh' \overset{\text{(v)}}{=} \Gamma_{X,K}\big((\theta_i^{-1}r_i)_{i\in I}\big)qh \\ &\overset{\text{(vi)}}{=} \Gamma_{X,K}\big((\theta_i^{-1}r_i)_{i\in I}\big)\Gamma_{K,X}\big((\rho_i\theta_i)_{i\in I}\big)h \\ &\overset{\text{(vii)}}{=} \Gamma_{K,K}\big((\theta_i^{-1}r_i\rho_i\theta_i)_{i\in I}\big)h \overset{\text{(viii)}}{=} \Gamma_{K,K}\big((\mathrm{id}_{\mathrm{Hom}_{\mathcal{C}}(i,K)})_{i\in I}\big)h \overset{\text{(ix)}}{=} h. \end{aligned}$$

Here, (i) and (ix) follow from (C.3), (ii) and (viii) from the invertibility of θ_i and the fact that r_i is a retraction of ρ_i, (iii) and (vii) from (C.2), (iv) and (vi) from the definition of q, and (v) from the fact that $qh' = g = qh$. Hence, q is a kernel of f. Furthermore, if f is an epimorphism, then f is a cokernel of q. Indeed, for any $i \in I$, the $\Bbbk$-linear homomorphism Δ_i^f is surjective (because f, being an epimorphism, has a section by the above). Since $\Bbbk$ is a field, Δ_i^f has a $\Bbbk$-linear section: there is a $\Bbbk$-linear homomorphism

$$s_i\colon \mathrm{Hom}_{\mathcal{C}}(i,Y) \to \mathrm{Hom}_{\mathcal{C}}(i,X)$$

such that $\Delta_i^f s_i = \mathrm{id}_{\mathrm{Hom}_{\mathcal{C}}(i,Y)}$. For all $i \in I$, we have:

$$\Delta_i^g s_i \Delta_i^f = \Delta_i^g. \tag{C.4}$$

Indeed,

$$\mathrm{Ker}(\Delta_i^f) \overset{\text{(i)}}{=} \mathrm{Im}(\rho_i) \overset{\text{(ii)}}{=} \mathrm{Im}(\rho_i\theta_i) \overset{\text{(iii)}}{=} \mathrm{Im}(\Delta_i^q) \overset{\text{(iv)}}{\subset} \mathrm{Ker}(\Delta_i^g).$$

Here, (i) follows from the definition of ρ_i, (ii) from the fact that θ_i is a $\Bbbk$-linear isomorphism, (iii) from (C.1) and the definition of q, and (iv) from the fact that $\Delta_i^g\Delta_i^q = 0$ (because $gq = 0$). Then, the image of $s_i\Delta_i^f - \mathrm{id}_{\mathrm{Hom}_{\mathcal{C}}(i,X)}$ lies in $\mathrm{Ker}(\Delta_i^f)$ (because s_i is a section of Δ_i^f) and so lies in $\mathrm{Ker}(\Delta_i^g)$. This proves (C.4). Now, let $g\colon X \to Z$ be a morphism in $\mathcal{C}$ such that $gq = 0$. Set

$$h = \Gamma_{Y,Z}\big((\Delta_i^g s_i)_{i\in I}\big)\colon Y \to Z.$$

Then

$$\begin{aligned} hf &\overset{\text{(i)}}{=} \Gamma_{Y,Z}\big((\Delta_i^g s_i)_{i\in I}\big)\Gamma_{X,Y}\big((\Delta_i^f)_{i\in I}\big) \\ &\overset{\text{(ii)}}{=} \Gamma_{X,Z}\big((\Delta_i^g s_i \Delta_i^f)_{i\in I}\big) \overset{\text{(iii)}}{=} \Gamma_{X,Z}\big((\Delta_i^g)_{i\in I}\big) \overset{\text{(iv)}}{=} g. \end{aligned}$$

Here, (i) follows from the definitions of h and the fact that $\Gamma_{X,Y}$ and $\Delta_{X,Y}$ are inverse to each other, (ii) from (C.2), (iii) from (C.4), and (iv) from (C.1). The uniqueness of the morphism $h\colon Y \to Z$ satisfying $g = hf$ follows from the assumption that f is an epimorphism. Thus, f is a cokernel of q.

We have shown that every morphism in $\mathcal{C}$ has a kernel and every epimorphism in $\mathcal{C}$ is a cokernel (of its kernel). Applying this result to the additive pre-fusion $\Bbbk$-category $\mathcal{C}^{\mathrm{op}}$, we obtain that every morphism in $\mathcal{C}$ has a cokernel and every monomorphism in $\mathcal{C}$ is a kernel. This completes the proof of the theorem. □

Appendix D

Hopf monads vs Hopf algebras

We give an internal description of Hopf monads arising from Hopf algebras. In this appendix, $\mathcal{C}$ is a monoidal category.

D.1 Augmented bimonads

A *bimonad morphism* between bimonads (T, μ, η) and (T', μ', η') on $\mathcal{C}$ is a comonoidal natural transformation $f = \{f_X : T(X) \to T'(X)\}_{X \in \mathcal{C}}$ such that for all $X \in \mathrm{Ob}(\mathcal{C})$,

$$f_X \mu_X = \mu'_X f_{T'(X)} T(f_X) \quad \text{and} \quad f_X \eta_X = \eta'_X.$$

(See Exercise 7.5.8 for the relationships with the categories of modules.)

An *augmentation* of a bimonad T on $\mathcal{C}$ is a bimonad morphism from T to the trivial bimonad $1_{\mathcal{C}}$ (see Example 7.5.3). A bimonad is *augmented* if it is endowed with an augmentation.

For example, if $((A, \sigma), m, u, \Delta, \varepsilon)$ is a bialgebra in $\mathcal{Z}(\mathcal{C})$, then the bimonad $A \otimes_\sigma ?$ of Example 7.5.5 has the augmentation $\varepsilon \otimes ?$ given by

$(\varepsilon \otimes ?)_X = \varepsilon \otimes \mathrm{id}_X =$  $: A \otimes_\sigma X \to X$

for any $X \in \mathrm{Ob}(\mathcal{C})$.

D.2 Hopf monads represented by Hopf algebras

Consider the category $\mathcal{A} = \mathrm{HopfAlg}(\mathcal{Z}(\mathcal{C}))$ whose objects are Hopf algebras in $\mathcal{Z}(\mathcal{C})$ and whose morphisms are Hopf algebra morphisms. Consider also the category $\mathcal{B} = \mathrm{HopfMon}(\mathcal{C})/1_{\mathcal{C}}$ whose objects are augmented Hopf monads on $\mathcal{C}$ and whose morphisms are bimonad morphisms commuting with the augmentation.

Recall from Example 8.1.5 that any Hopf algebra (A,σ) in $\mathcal{Z}(\mathcal{C})$ induces a Hopf monad $A\otimes_\sigma ?$ on $\mathcal{C}$. By Section D.1, it is augmented with augmentation $\varepsilon\otimes ?$ induced by the counit ε of (A,σ). We use this construction to define a functor $\mathfrak{R}\colon \mathcal{A}\to\mathcal{B}$ by

$$\mathfrak{R}(A,\sigma)=(A\otimes_\sigma ?, \varepsilon\otimes ?) \quad \text{and} \quad \mathfrak{R}(f)=f\otimes ? = \{f\otimes \mathrm{id}_X\}_{X\in \mathrm{Ob}(\mathcal{C})}$$

for any Hopf algebra (A,σ) in $\mathcal{Z}(\mathcal{C})$ and any Hopf algebra morphism f.

Theorem D.1. *The functor $\mathfrak{R}$ is an equivalence of categories.*

Proof (sketch). We construct a quasi-inverse $\mathfrak{J}\colon \mathcal{B}\to\mathcal{A}$ of $\mathfrak{R}$ as follows. Let (T,μ,η) be a Hopf monad on $\mathcal{C}$ with augmentation e and fusion operators H^l and H^r. Then the natural transformation

$$\gamma_{(T,e)}=\{(\gamma_{(T,e)})_X=(\mathrm{id}_{T(\mathbb{1})}\otimes e_X)T_2(\mathbb{1},X)\colon T(X)\to T(\mathbb{1})\otimes X\}_{X\in\mathrm{Ob}(\mathcal{C})}$$

is an isomorphism with inverse

$$\big(\gamma^{-1}_{(T,e)}\big)_X = T(e_X)(H^l_{\mathbb{1},X})^{-1}(\mathrm{id}_{T(\mathbb{1})}\otimes \eta_X)\colon T(\mathbb{1})\otimes X\to T(X).$$

The natural transformation

$$\sigma=\{\sigma_X=(e_X\otimes \mathrm{id}_{T(\mathbb{1})})T_2(X,\mathbb{1})\big(\gamma^{-1}_{(T,e)}\big)_X\colon T(\mathbb{1})\otimes X\to X\otimes T(\mathbb{1})\}_{X\in\mathrm{Ob}(\mathcal{C})}$$

is a half-braiding, and $(T(\mathbb{1}),\sigma)$ is a Hopf algebra in $\mathcal{Z}(\mathcal{C})$ with product m, unit u, coproduct Δ, counit ε, antipode S, and the inverse S^{-1} given by

$$m=\mu_{\mathbb{1}}\big(\gamma^{-1}_{(T,e)}\big)_{T(\mathbb{1})}, \quad u=\eta_{\mathbb{1}}, \quad \Delta=T_2(\mathbb{1},\mathbb{1}), \quad \varepsilon=T_0,$$
$$S=e_{T(\mathbb{1})}(H^l_{\mathbb{1},\mathbb{1}})^{-1}(\mathrm{id}_{T(\mathbb{1})}\otimes\eta_{\mathbb{1}}), \quad S^{-1}=e_{T(\mathbb{1})}(H^r_{\mathbb{1},\mathbb{1}})^{-1}(\eta_{\mathbb{1}}\otimes \mathrm{id}_{T(\mathbb{1})}).$$

Set $\mathfrak{J}(T,e)=(T(\mathbb{1}),\sigma)$. For any morphism $f\colon (T,e)\to(T',e')$ in $\mathcal{B}$, set $\mathfrak{J}(f)=f_{\mathbb{1}}\colon T(\mathbb{1})\to T'(\mathbb{1})$. The functor $\mathfrak{J}$ is well defined, $\mathfrak{J}\mathfrak{R}=1_{\mathcal{A}}$, and the family

$$\gamma=\{\gamma_{(T,e)}\}_{(T,e)\in\mathrm{Ob}(\mathcal{B})}$$

is a natural isomorphism from $\mathfrak{R}\mathfrak{J}$ to $1_{\mathcal{B}}$. Hence $\mathfrak{R}$ is an equivalence. For details, see [BLV, Section 5.6]. □

Theorem D.1 shows that the Hopf monads on $\mathcal{C}$ representable by Hopf algebras in $\mathcal{Z}(\mathcal{C})$ are exactly the augmented ones. Note that there are Hopf monads which are not representable by Hopf algebras and so cannot be augmented, see Example 9.2.4.

The following assertion characterizes Hopf monads arising from Hopf algebras in braided categories.

Theorem D.2. *A Hopf monad T on a braided category $\mathcal{C}$ is isomorphic to the Hopf monad $A\otimes ?$ for some Hopf algebra A in $\mathcal{C}$ if and only if T has an augmentation $e\colon T \to 1_{\mathcal{C}}$ which is compatible with the braiding τ of $\mathcal{C}$ in the following sense:*

$$(e_X \otimes \mathrm{id}_{T(\mathbb{1})})\tau_{T(\mathbb{1}),T(X)}T_2(\mathbb{1}, X) = (e_X \otimes \mathrm{id}_{T(\mathbb{1})})T_2(X, \mathbb{1})$$

for all $X \in \mathrm{Ob}(\mathcal{C})$.

Proof. Consider first a Hopf algebra A in $\mathcal{C}$ and its associated Hopf monad $A\otimes ?$ on $\mathcal{C}$ (see Example 8.1.4). Then the augmentation $\varepsilon\otimes ? = \{\varepsilon \otimes \mathrm{id}_X\}_{X\in\mathrm{Ob}(\mathcal{C})}$ of $A\otimes ?$, where ε is the counit of A, is compatible with τ since for all $X \in \mathrm{Ob}(\mathcal{C})$,

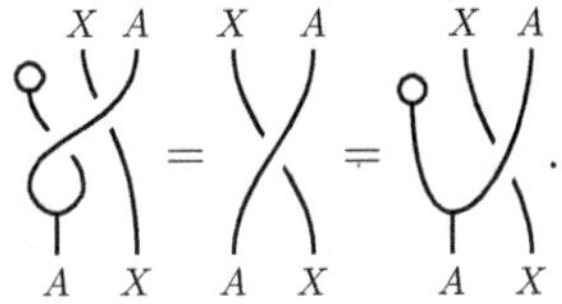

Consequently, any Hopf monad on $\mathcal{C}$ isomorphic to $A\otimes ?$ has an augmentation compatible with τ.

Conversely, let T be a Hopf monad on $\mathcal{C}$ with an augmentation e compatible with τ. Let $\gamma_{(T,e)}$ and σ be as is the proof of Theorem D.1, so that $(T(\mathbb{1}), \sigma)$ is a Hopf algebra in $\mathcal{Z}(\mathcal{C})$ such that $T \simeq T(\mathbb{1})\otimes_\sigma ?$ as Hopf monads. It follows from the definition of σ, the compatibility of e with τ, the naturality of τ, and the definition of $\gamma_{(T,e)}$ that for any $X \in \mathrm{Ob}(\mathcal{C})$,

$$\begin{aligned}\sigma_X &= (e_X \otimes \mathrm{id}_{T(\mathbb{1})})T_2(X, \mathbb{1})\big(\gamma^{-1}_{(T,e)}\big)_X \\ &= (e_X \otimes \mathrm{id}_{T(\mathbb{1})})\tau_{T(\mathbb{1}),T(X)}T_2(\mathbb{1}, X)\big(\gamma^{-1}_{(T,e)}\big)_X \\ &= \tau_{T(\mathbb{1}),X}(\mathrm{id}_{T(\mathbb{1})} \otimes e_X)T_2(\mathbb{1}, X)\big(\gamma^{-1}_{(T,e)}\big)_X = \tau_{T(\mathbb{1}),X}.\end{aligned}$$

Then $(T(\mathbb{1}), \tau_{T(\mathbb{1}),-})$ is a Hopf algebra in $\mathcal{Z}(\mathcal{C})$. Consequently, $T(\mathbb{1})$ is a Hopf algebra in $\mathcal{C}$ and $T(\mathbb{1})\otimes ? = T(\mathbb{1})\otimes_\sigma ?$ as Hopf monads on $\mathcal{C}$. Hence $T \simeq T(\mathbb{1})\otimes ?$. $\square$

Appendix E

Unordered tensor products of modules

By a module we mean a left module over a commutative ring $\Bbbk$. Given a finite family E of modules, we define the *unordered tensor product* $\otimes_{M \in E} M$ as follows. Let $n = \#E$ be the number of elements of E, and let $\mathcal{S} = \mathcal{S}(E)$ be the set of bijections $\{1, \ldots, n\} \to E$. For any bijection $\sigma \in \mathcal{S}$, consider the module

$$E_\sigma = \sigma(1) \otimes_\Bbbk \cdots \otimes_\Bbbk \sigma(n).$$

For $\sigma, \mu \in \mathcal{S}$, let $p_{\sigma,\mu} \colon E_\sigma \to E_\mu$ be the $\Bbbk$-linear isomorphism induced by the permutations of modules: given any vectors $m_i \in \sigma(i)$ with $i = 1, \ldots, n$,

$$p_{\sigma,\mu}(m_1 \otimes_\Bbbk \cdots \otimes_\Bbbk m_n) = m_{\sigma^{-1}\mu(1)} \otimes_\Bbbk \cdots \otimes_\Bbbk m_{\sigma^{-1}\mu(n)}.$$

It follows from the definitions that for arbitrary $\sigma, \mu, \nu \in \mathcal{S}$,

$$p_{\mu,\nu} p_{\sigma,\mu} = p_{\sigma,\nu} \colon E_\sigma \to E_\nu \quad \text{and} \quad p_{\sigma,\sigma} = \mathrm{id}_{E_\sigma}.$$

The *unordered tensor product* of the modules $M \in E$ is the projective limit of the system $(E_\sigma, p_{\sigma,\mu})_{\sigma,\mu \in \mathcal{S}}$:

$$\bigotimes_{M \in E} M = \varprojlim E_\sigma.$$

This is a module (over $\Bbbk$) equipped with an isomorphism $\otimes_{M \in E} M \simeq E_\sigma$ for each $\sigma \in \mathcal{S}$. The latter isomorphisms are called the *cone isomorphisms*. They commute with $p_{\sigma,\mu}$ for all $\sigma, \mu \in \mathcal{S}$. If all modules $M \in E$ are projective of finite type, then so is $\otimes_{M \in E} M$ and there is a canonical isomorphism

$$\left(\bigotimes_{M \in E} M\right)^\star \simeq \bigotimes_{M \in E} M^\star.$$

The unordered tensor product of an empty set of modules is the ground ring $\Bbbk$.

Given a bijection $\varphi\colon E \to F$ between two finite families of modules, an arbitrary family $\{f_M\colon M \to \varphi(M)\}_{M\in E}$ of $\Bbbk$-linear homomorphisms induces a $\Bbbk$-linear homomorphism $\bigotimes_{M\in E} f_M\colon \bigotimes_{M\in E} M \to \bigotimes_{N\in F} N$. It is uniquely determined by the property that for all $\sigma \in \mathcal{S}(E)$, the following diagram commutes:

$$\begin{array}{ccc} \bigotimes_{M\in E} M & \xrightarrow{\bigotimes_{M\in E} f_M} & \bigotimes_{N\in F} N \\ \wr\| & & \|\wr \\ E_\sigma & \xrightarrow[f_{\sigma(1)}\otimes_\Bbbk\cdots\otimes_\Bbbk f_{\sigma(n)}]{} & F_{\varphi\sigma}\,. \end{array}$$

Here the vertical isomorphisms are the cone isomorphisms. If all f_M are isomorphisms, then so is $\bigotimes_{M\in E} f_M$.

Appendix F

The $6j$-symbols

The $6j$-symbols defined in Example 12.2.7 were first introduced by the physicists E. P. Wigner and G. Racah in the theory of representations of $SU_2(\mathbb{C})$. The $6j$-symbols have been extensively used in the theory of angular momentum in quantum mechanics and in the Ponzano-Regge approach to quantum gravity in dimension three. The $6j$-symbols play a special role in 3-dimensional state sum TQFTs: they are the only tensors appearing in the state sums on special skeletons of 3-manifolds. Since all compact 3-manifolds have special skeletons, the $6j$-symbols are sufficient to fully compute the state sum TQFTs. However, such computations may be rather involved because special skeletons typically have more vertices, edges, and regions than non-special skeletons.

The $6j$-symbols satisfy beautiful algebraic identities including the tetrahedral symmetry, the Biedenharn-Elliott identity, and the orthonormality relation. We formulate these identities here for the $6j$-symbols associated with an arbitrary spherical pre-fusion $\Bbbk$-category $\mathcal{C}$. Recall from Example 12.1.4 that each tuple S of signed objects of $\mathcal{C}$ determines a module $H(E_S)$ which is preserved under cyclic permutations of S and is canonically isomorphic to the module $\mathrm{Hom}_{\mathcal{C}}(\mathbb{1}, X_S)$. Given a triple $S = ((i,\varepsilon),(j,\mu),(k,\nu))$ of signed objects of $\mathcal{C}$, we write $H(i\varepsilon, j\mu, k\nu)$ for $H(E_S)$.

Pick a tuple T of six signed objects of $\mathcal{C}$:

$$T = ((i,\varepsilon_i),(j,\varepsilon_j),(k,\varepsilon_k),(l,\varepsilon_l),(m,\varepsilon_m),(n,\varepsilon_n)).$$

Recall the $\mathcal{C}$-colored graph $\Gamma \subset S^2$ from Example 12.2.7:

$\Gamma =$ 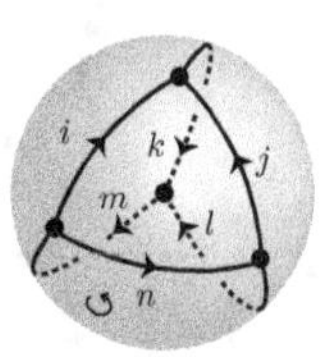 .

Let $\Gamma_T \subset S^2$ be the $\mathcal{C}$-colored graph obtained from Γ by reversing orientation on all edges colored by a letter $x \in \{i,j,k,l,m,n\}$ with $\varepsilon_x = -$. Set

$$\begin{vmatrix} i\,\varepsilon_i & j\,\varepsilon_j & k\,\varepsilon_k \\ l\,\varepsilon_l & m\,\varepsilon_m & n\,\varepsilon_n \end{vmatrix} = \mathbb{F}_{\mathcal{C}}(\Gamma_T) \in H(\Gamma_T)^*, \tag{F.1}$$

where $H(\Gamma_T)$ is the unordered tensor product of the modules

$$H(m\varepsilon_m, i(-\varepsilon_i), n(-\varepsilon_n)), \qquad H(j\varepsilon_j, i\varepsilon_i, k(-\varepsilon_k)),$$
$$H(n\varepsilon_n, j(-\varepsilon_j), l(-\varepsilon_l)), \qquad H(l\varepsilon_l, k\varepsilon_k, m(-\varepsilon_m)).$$

The tensor (F.1) is the *6j-symbol* determined by T. In particular, when all signs in T are $+$ we obtain

$$\begin{vmatrix} i+ & j+ & k+ \\ l+ & m+ & n+ \end{vmatrix} = \mathbb{F}_{\mathcal{C}}(\Gamma) \in H(\Gamma)^*.$$

The isotopy invariance of $\mathbb{F}_{\mathcal{C}}$ implies that the 6j-symbols have the symmetries of an oriented tetrahedron. These symmetries are generated by the following two:

$$\begin{vmatrix} i\,\varepsilon_i & j\,\varepsilon_j & k\,\varepsilon_k \\ l\,\varepsilon_l & m\,\varepsilon_m & n\,\varepsilon_n \end{vmatrix} = \begin{vmatrix} j\,\varepsilon_j & k(-\varepsilon_k) & i(-\varepsilon_i) \\ m\,\varepsilon_m & n\,\varepsilon_n & l\,\varepsilon_l \end{vmatrix} = \begin{vmatrix} k\,\varepsilon_k & l\,\varepsilon_l & m\,\varepsilon_m \\ n(-\varepsilon_n) & i\,\varepsilon_i & j(-\varepsilon_j) \end{vmatrix}.$$

For example,

$$\begin{vmatrix} i+ & j+ & k+ \\ l+ & m+ & n+ \end{vmatrix} = \begin{vmatrix} j+ & k- & i- \\ m+ & n+ & l+ \end{vmatrix} = \begin{vmatrix} k+ & l+ & m+ \\ n- & i+ & j- \end{vmatrix}.$$

For any triple $S = ((i,\varepsilon),(j,\mu),(k,\nu))$ of signed objects of $\mathcal{C}$, Section 12.3.2 yields the pairing

$$\omega_{i\varepsilon,j\mu,k\nu} = \omega_{E_S} \colon H(i\varepsilon, j\mu, k\nu) \otimes H(k(-\nu), j(-\mu), i(-\varepsilon)) \to \Bbbk.$$

By Section 12.3.4, this pairing is non-degenerate and determines the contraction vector

$$*_{i\varepsilon,j\mu,k\nu} \in H(i\varepsilon, j\mu, k\nu) \otimes H(k(-\nu), j(-\mu), i(-\varepsilon)).$$

Theorem F.1 (The Biedenharn–Elliott identity). *Let I be a representative set of simple objects of $\mathcal{C}$. For any objects a,b,c,i,j,k,l,m,n of $\mathcal{C}$,*

$$\sum_{z \in I} \dim(z) \left(\begin{vmatrix} i\pm & z\mp & b\pm \\ m\pm & n\pm & k\pm \end{vmatrix} \otimes \begin{vmatrix} z\pm & j\mp & c\pm \\ l\pm & m\pm & k\pm \end{vmatrix} \otimes \begin{vmatrix} i\pm & j\mp & a\mp \\ c\mp & b\pm & z\mp \end{vmatrix} \right) (*_{m\mp,k\pm,z\pm}$$
$$\otimes *_{c\mp,j\mp,z\pm} \otimes *_{b\pm,i\mp,z\pm}) = \left(\begin{vmatrix} i\pm & j\mp & a\mp \\ l\pm & n\pm & k\pm \end{vmatrix} \otimes \begin{vmatrix} n\pm & l\mp & a\mp \\ c\mp & b\pm & m\mp \end{vmatrix} \right) (*_{a\mp,n\mp,l\pm})$$

where the signs of a,b,c,i,j,k,l,m,n may be chosen independently from each other.

Proof. The summation on the left-hand side may involve only a finite number of non-zero terms so the sum is well defined. When all signs are $+$, a graphical proof is given in Figure F.1; to handle negative signs, reverse the direction of the corresponding edges. The equalities in Figure F.1 stand for the equalities of the $\mathbb{F}_{\mathcal{C}}$-invariants of the depicted colored $\mathcal{C}$-graphs in $S^2 = \mathbb{R}^2 \cup \{\infty\}$. The first equality is obtained by applying Lemma 12.6 twice (along the dotted lines) and then Lemma 12.5. The second equality follows from the isotopy invariance of $\mathbb{F}_{\mathcal{C}}$. The last equality follows from Lemma 12.6. □

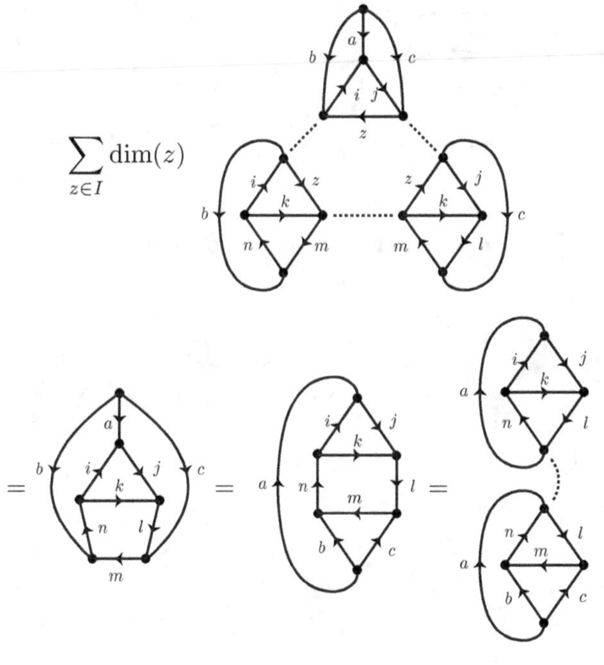

Figure F.1: Proof of the Biedenharn–Elliot identity

Theorem F.2 (The orthonormality relation)**.** *For any objects i, k, l, n of $\mathcal{C}$, any representative set I of simple objects of $\mathcal{C}$, and any $m, m' \in I$,*

$$\dim(m) \sum_{z\in I} \dim(z) \left(\begin{vmatrix} i\pm & z\pm & k\pm \\ l\pm & m\pm & n\pm \end{vmatrix} \otimes \begin{vmatrix} z\mp & i\mp & k\mp \\ m'\pm & l\pm & n\pm \end{vmatrix} \right) (*_{z\pm,i\pm,k\mp} \otimes *_{z\pm,n\mp,l\pm})$$

$$= \delta_{m,m'}\, \omega_{m\mp,n\pm,i\pm} \otimes \omega_{m\pm,k\mp,l\mp}$$

where the signs of i, j, k, l, m, m', n may be chosen independently from each other.

Proof. A graphical proof is given in Figure F.2 in the case when all signs are $+$. The first equality is obtained by applying Lemma 12.6 and then Lemma 12.5. The second equality follows from Lemma 12.3. We conclude using the computation in Example 12.3.6. □

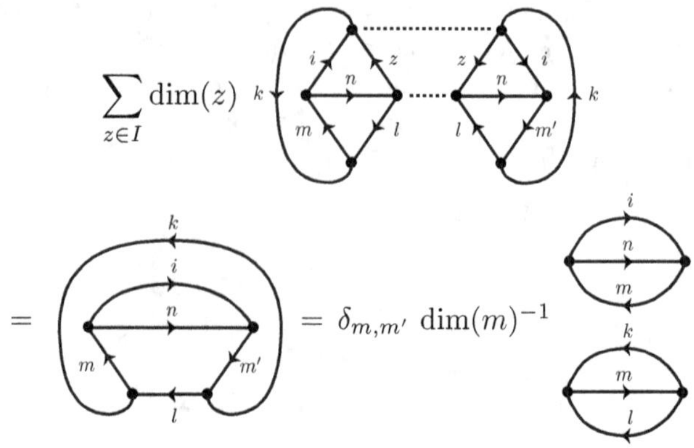

Figure F.2: Proof of the orthonormality relation

Appendix G

Unitary TQFTs

G.1 Conjugations and the unitary center

G.1.1 Conjugations in monoidal categories

By a *conjugation* in a monoidal category $\mathcal{C}$, we mean a strict monoidal functor

$$\overline{?} \colon \mathcal{C}^{\mathrm{op}} \to \mathcal{C}$$

which acts as the identity on objects and is involutory in the sense that

$$(\overline{?})^{\mathrm{op}} \circ \overline{?} = 1_{\mathcal{C}},$$

where $1_{\mathcal{C}} \colon \mathcal{C} \to \mathcal{C}$ is the identity functor, $\mathcal{C}^{\mathrm{op}} = (\mathcal{C}^{\mathrm{op}}, \otimes, \mathbb{1})$ is the monoidal opposite of $\mathcal{C}$ (see Section 1.2.2), and $(\overline{?})^{\mathrm{op}}$ is defined as in Section 1.4.1. In other words, a conjugation in $\mathcal{C}$ is a family of maps

$$\{f \in \mathrm{Hom}_{\mathcal{C}}(X,Y) \to \overline{f} \in \mathrm{Hom}_{\mathcal{C}}(Y,X)\}_{X,Y \in \mathrm{Ob}(\mathcal{C})},$$

such that for all morphisms f, g in $\mathcal{C}$ and all $X, Y, Z \in \mathrm{Ob}(\mathcal{C})$

$$\overline{gf} = \overline{f}\,\overline{g}, \qquad \overline{\mathrm{id}_X} = \mathrm{id}_X, \qquad \overline{\overline{f}} = f, \qquad \overline{f \otimes g} = \overline{f} \otimes \overline{g},$$
$$\overline{\mathbb{1}} = \mathbb{1}, \qquad \overline{a_{X,Y,Z}} = a^{-1}_{X,Y,Z}, \qquad \overline{l_X} = l_X^{-1}, \qquad \overline{r_X} = r_X^{-1},$$

where a, l, r are the associativity and unitality constraints of $\mathcal{C}$.

A conjugation $\overline{?} \colon \mathcal{C}^{\mathrm{op}} \to \mathcal{C}$ in a braided category $\mathcal{C}$ is *braided* if it is braided as a functor (with $\mathcal{C}^{\mathrm{op}}$ braided as in Example 3.1.8), that is, if

$$\overline{\tau_{X,Y}} = \tau^{-1}_{X,Y}$$

for all $X, Y \in \mathrm{Ob}(\mathcal{C})$, where τ is the braiding in $\mathcal{C}$. A conjugation in a symmetric category is *symmetric* if it is braided.

A conjugation $\overline{?}: \mathcal{C}^{\text{op}} \to \mathcal{C}$ in a pivotal category $\mathcal{C}$ is *pivotal* if it is strictly pivotal as a functor, that is, if

$$\overline{\text{ev}_X} = \widetilde{\text{coev}}_X \quad \text{and} \quad \overline{\text{coev}_X} = \widetilde{\text{ev}}_X$$

for all $X \in \text{Ob}(\mathcal{C})$, where $\{(X^*, \text{ev}_X, \widetilde{\text{ev}}_X)\}_{X \in \text{Ob}(\mathcal{C})}$ is the pivotal duality in $\mathcal{C}$. Note that then, for any endomorphism f of an object X of $\mathcal{C}$, we have

$$\text{tr}_l(\overline{f}) = \overline{\text{tr}_l(f)} \quad \text{and} \quad \text{tr}_r(\overline{f}) = \overline{\text{tr}_r(f)}. \tag{G.1}$$

Indeed, the axioms of a pivotal conjugation imply that

$$\begin{aligned}\text{tr}_l(\overline{f}) &= \text{ev}_X(\text{id}_{X^*} \otimes \overline{f})\, \widetilde{\text{coev}}_X = \overline{\widetilde{\text{coev}}_X}\,(\overline{\text{id}_{X^*}} \otimes \overline{f})\, \overline{\text{ev}_X} \\ &= \overline{\text{ev}_X(\text{id}_{X^*} \otimes f)\, \widetilde{\text{coev}}_X} = \overline{\text{tr}_l(f)},\end{aligned}$$

and similarly for the right trace. Consequently, the left and right dimensions of objects of $\mathcal{C}$ are invariant under the involution $\alpha \mapsto \bar{\alpha}$ in $\text{End}_{\mathcal{C}}(\mathbb{1})$. This is verified by applying (G.1) to $f = \text{id}_X$ and using the identity $\overline{\text{id}_X} = \text{id}_X$.

G.1.2 Example

For any $n \geq 1$, the symmetric pivotal category Cob_n of n-dimensional cobordisms (see Section 10.1) has a symmetric pivotal conjugation carrying the homeomorphism class of an n-dimensional cobordism

$$(M, \Sigma_0, \Sigma_1, h: (-\Sigma_0) \sqcup \Sigma_1 \to \partial M)$$

to the homeomorphism class of the cobordism

$$(-M, \Sigma_1, \Sigma_0, hP: (-\Sigma_1) \sqcup \Sigma_0 \to -\partial M = \partial(-M)),$$

where P is the orientation reversing permutation homeomorphism

$$(-\Sigma_1) \sqcup \Sigma_0 \simeq (-\Sigma_0) \sqcup \Sigma_1.$$

G.1.3 The case of $\mathbb{C}$-categories

Recall that $\mathbb{C}$ denotes the field of complex numbers. A conjugation $\overline{?}: \mathcal{C}^{\text{op}} \to \mathcal{C}$ in a monoidal $\mathbb{C}$-category $\mathcal{C}$ is *antilinear* if

$$\overline{f+g} = \bar{f} + \bar{g} \quad \text{and} \quad \overline{\lambda f} = \bar{\lambda}\, \bar{f}$$

for all morphisms f, g in $\mathcal{C}$ (with the same source and target) and all $\lambda \in \mathbb{C}$, where $\bar{\lambda}$ is the complex conjugate of λ.

Consider now a pivotal $\mathbb{C}$-category $\mathcal{C}$ whose unit object $\mathbb{1}$ is simple. Any antilinear pivotal conjugation $\overline{?}: \mathcal{C}^{\text{op}} \to \mathcal{C}$ in $\mathcal{C}$ determines, for all $X, Y \in \text{Ob}(\mathcal{C})$, two Hermitian forms $\langle , \rangle_l$ and $\langle , \rangle_r$ on $\text{Hom}_{\mathcal{C}}(X, Y)$ defined by

$$\langle f, g \rangle_l = \text{tr}_l(f\bar{g}) \quad \text{and} \quad \langle f, g \rangle_r = \text{tr}_r(f\bar{g}) \tag{G.2}$$

for all $f, g \in \mathrm{Hom}_{\mathcal{C}}(X, Y)$. Note that the forms $\langle , \rangle_l$ and $\langle , \rangle_r$ coincide whenever $\mathcal{C}$ is spherical. By a *unitary conjugation* in $\mathcal{C}$, we mean an antilinear pivotal conjugation in $\mathcal{C}$ such that the associated Hermitian forms (G.2) are positive definite for all $X, Y \in \mathrm{Ob}(\mathcal{C})$. If $\mathcal{C}$ has a unitray conjugation, then the left and right dimensions of any non-zero object of $\mathcal{C}$ are positive real numbers. This follows from the fact that $\dim_{l/r}(X) = \langle \mathrm{id}_X, \mathrm{id}_X \rangle_{l/r}$ for all $X \in \mathrm{Ob}(\mathcal{C})$.

G.1.4 Example

Let Hilb be the category of finite-dimensional Hilbert spaces over $\mathbb{C}$ and $\mathbb{C}$-linear homomorphisms between them. This is a symmetric monoidal $\mathbb{C}$-category with the usual tensor product of Hilbert spaces as monoidal product and with symmetry defined as in Example 3.1.4. The category Hilb is left rigid with distinguished left duality defined by the usual dual of a finite-dimensional Hilbert space. By Lemma 3.5, this left duality turns Hilb into a pivotal $\mathbb{C}$-category. Also, the symmetric pivotal $\mathbb{C}$-category Hilb is ribbon, and so spherical (the trace of endomorphisms in H coincides with the standard trace of $\mathbb{C}$-linear endomorphisms). Assigning to every $\mathbb{C}$-linear homomorphism between finite-dimensional Hilbert spaces its hermitian adjoint, we obtain a symmetric unitary conjugation in Hilb.

G.1.5 The unitary center

Let $\mathcal{C}$ be a monoidal category endowed with a conjugation. A half-braiding (A, σ) of $\mathcal{C}$ is *unitary* if $\overline{\sigma_X} = \sigma_X^{-1}$ for all $X \in \mathrm{Ob}(\mathcal{C})$. The full subcategory of the center $\mathcal{Z}(\mathcal{C})$ of $\mathcal{C}$ formed by the unitary half-braidings of $\mathcal{C}$ is a monoidal subcategory of $\mathcal{Z}(\mathcal{C})$ called the *unitary center* of $\mathcal{C}$ and denoted $\mathcal{Z}^u(\mathcal{C})$. The braiding in $\mathcal{Z}(\mathcal{C})$ restricted to $\mathcal{Z}^u(\mathcal{C})$ turns $\mathcal{Z}^u(\mathcal{C})$ into a braided category. The conjugation in $\mathcal{C}$ induces a conjugation in $\mathcal{Z}^u(\mathcal{C})$ in the obvious way.

If both $\mathcal{C}$ and its conjugation are pivotal, then the pivotal duality in $\mathcal{C}$ lifts uniquely along the forgetful functor $\mathcal{Z}^u(\mathcal{C}) \to \mathcal{C}$ into a pivotal duality in $\mathcal{Z}^u(\mathcal{C})$ and the conjugation in $\mathcal{Z}^u(\mathcal{C})$ is pivotal.

If $\mathcal{C}$ is $\mathbb{C}$-linear and its conjugation is antilinear, then the $\mathbb{C}$-linear structure in $\mathcal{Z}(\mathcal{C})$ (see Section 5.3.1) restricts to $\mathcal{Z}^u(\mathcal{C})$ and turns $\mathcal{Z}^u(\mathcal{C})$ into a monoidal $\mathbb{C}$-category with antilinear conjugation. If $\mathcal{C}$ is additive, then so is $\mathcal{Z}^u(\mathcal{C})$.

G.2 Unitary TQFTs and fusion categories

G.2.1 Unitary TQFTs

For an integer $n \geq 1$, a *unitary n-dimensional TQFT* is a symmetric strong monoidal functor $Z\colon \mathrm{Cob}_n \to \mathrm{Hilb}$ commuting with the conjugations in Cob_n and in Hilb defined in Examples G.1.2 and G.1.4. This definition is equivalent to the one in [Tu1] and implies, in particular, that $Z(-M) = \overline{Z(M)} \in \mathbb{C}$ for any closed oriented n-dimensional manifold M. Composing such a functor Z with the obvious forgetful functor Hilb $\to$ $\mathrm{Mod}_{\mathbb{C}}$, we obtain an n-dimensional TQFT in the sense of Section 10.2.1. We call Z a *unitary lift* of the latter TQFT.

G.2.2 Unitary fusion categories

A *unitary fusion category* is an additive spherical fusion $\mathbb{C}$-category $\mathcal{C}$ endowed with a unitary conjugation. Then the dimensions of all simple objects of $\mathcal{C}$ are positive real numbers and $\dim(\mathcal{C}) \neq 0$.

A *unitary modular category* is a ribbon unitary fusion category which is modular (in the sense of Section 4.5.2) and whose conjugation is braided.

G.2.3 Unitary lift of the state sum TQFT

Let $\mathcal{C}$ be a unitary fusion category. Since $\dim(\mathcal{C}) \neq 0$, Theorem 13.5 yields the state sum 3-dimensional TQFT $|\cdot|_{\mathcal{C}}$.

Theorem G.1. *The state sum TQFT $|\cdot|_{\mathcal{C}}$ has a unitary lift.*

Proof. The center $\mathcal{Z}(\mathcal{C})$ of $\mathcal{C}$ is additive $\mathbb{C}$-linear (by Section 5.3.1) and ribbon (by Lemma 5.2). Consequently, the unitary center $\mathcal{Z}^u(\mathcal{C})$ of $\mathcal{C}$ (see Section G.1.5) is an additive ribbon $\mathbb{C}$-category with an antilinear braided pivotal conjugation. By definition, the inclusion $\mathcal{Z}^u(\mathcal{C}) \hookrightarrow \mathcal{Z}(\mathcal{C})$ is a strictly pivotal strict monoidal (and so ribbon) fully faithful $\mathbb{C}$-linear functor. By [Mü, Theorem 6.4], the inclusion $\mathcal{Z}^u(\mathcal{C}) \hookrightarrow \mathcal{Z}(\mathcal{C})$ is an equivalence of categories. Since $\mathcal{Z}(\mathcal{C})$ is an anomaly free modular fusion $\mathbb{C}$-category (by Theorem 5.4), we conclude that $\mathcal{Z}^u(\mathcal{C})$ is an anomaly free unitary modular category.

Theorem 17.1 and the fact that $\mathcal{Z}(\mathcal{C})$ and $\mathcal{Z}^u(\mathcal{C})$ are equivalent modular $\mathbb{C}$-categories yield the isomorphisms of 3-dimensional TQFTs

$$|\cdot|_{\mathcal{C}} \simeq \tau_{\mathcal{Z}(\mathcal{C})} \simeq \tau_{\mathcal{Z}^u(\mathcal{C})}.$$

Now, since $\mathcal{Z}^u(\mathcal{C})$ is an anomaly free unitary modular category, it follows from [Tu1] that the TQFT $\tau_{\mathcal{Z}^u(\mathcal{C})}$ has a unitary lift. Hence, $|\cdot|_{\mathcal{C}}$ has a unitary lift. □

A unitary lift of $|\cdot|_{\mathcal{C}}$ can be constructed directly by turning all multiplicity modules into Hilbert spaces over $\mathbb{C}$. This is achieved by using that for any $X \in \mathrm{Ob}(\mathcal{C})$, the finite-dimensional $\mathbb{C}$-vector space $\mathrm{Hom}_{\mathcal{C}}(\mathbb{1}, X)$ is a Hilbert space with inner product

$$\langle f, g\rangle = \mathrm{tr}(f\bar{g}) = f\bar{g} \in \mathrm{End}_{\mathcal{C}}(\mathbb{1}) \simeq \mathbb{C}$$

for all $f, g \in \mathrm{Hom}_{\mathcal{C}}(\mathbb{1}, X)$, see Section G.1.3. We leave the details to the reader.

G.2.4 Unitary lift of the state sum graph TQFT

We briefly discuss a version of the results of Section G.2.3 for graph TQFTs. By Theorem 15.9, each unitary fusion category $\mathcal{C}$ gives rise to a state sum graph TQFT $|\cdot|_{\mathcal{C}}$ over $\mathcal{Z}(\mathcal{C})$ which we restrict to a graph TQFT over $\mathcal{Z}^u(\mathcal{C})$. We claim that this restriction has a unitary lift, i.e., it lifts to a symmetric strong monoidal functor

$$\mathrm{Cob}_3^{\mathcal{Z}^u(\mathcal{C})} \to \mathrm{Hilb}$$

commuting with the conjugations. Here, the conjugation in Hilb is defined in Example G.1.4 and the conjugation in $\mathrm{Cob}_3^{\mathcal{Z}^u(\mathcal{C})}$ carries the homeomorphism class of a 3-cobordism

$$(M, h\colon (-\Sigma_0) \sqcup \Sigma_1 \to \partial M, R)$$

between $\mathcal{Z}^u(\mathcal{C})$-colored surfaces Σ_0, Σ_1 to the homeomorphism class of the 3-cobordism

$$(-M, hP\colon (-\Sigma_1) \sqcup \Sigma_0 \to -\partial M = \partial(-M), \bar{R}),$$

where P is the obvious orientation reversing permutation homeomorphism

$$(-\Sigma_1) \sqcup \Sigma_0 \simeq (-\Sigma_0) \sqcup \Sigma_1$$

and $\bar{R}$ is the $\mathcal{Z}^u(\mathcal{C})$-colored ribbon graph in $-M$ obtained from the $\mathcal{Z}^u(\mathcal{C})$-colored ribbon graph R in M by reversing orientation in all strands while keeping their colors, reversing orientation in all coupons and exchanging their bottom/top bases, and replacing the colors of the coupons by the conjugate morphisms. The existence of a unitary lift is proved similarly to Theorem G.1 using that the surgery graph TQFT $\tau_{\mathcal{Z}^u(\mathcal{C})}$ over $\mathcal{Z}^u(\mathcal{C})$ has a unitary lift (see [Tu1]).

Appendix H

The Dijkgraaf–Witten invariants

Let G be a finite group whose order $|G| = \mathrm{card}(G)$ is invertible in the commutative ring $\Bbbk$. A 3-cocycle $\alpha\colon G^3 \to \Bbbk^*$ determines a spherical fusion $\Bbbk$-category $\mathcal{C} = G^{\alpha}_{\Bbbk} = G^{\alpha,1}_{\Bbbk}$ of dimension $|G| \cdot 1_{\Bbbk} \in \Bbbk$, see Sections 1.7.3, 2.7.2 and 4.5.4. We show here that the corresponding state sum $\Bbbk$-valued invariant $|\cdot|_{\mathcal{C}}$ of closed oriented 3-manifolds (see Theorem 13.1) coincides with the well-known Dijkgraaf–Witten [DW] invariant. We first recall the definition of the latter invariant. Let

$$[\alpha] \in H^3(G;\Bbbk^*) \cong H^3(BG;\Bbbk^*)$$

be the cohomology class of α, where $BG = K(G,1)$ is the (pointed) classifying space of the group G. The Dijkgraaf–Witten invariant of a closed connected oriented 3-manifold M is defined by the formula

$$\mathrm{DW}_{\alpha}(M) = |G|^{-1} \sum_{\varphi \in \mathrm{Hom}(\pi_1(M,*),G)} \langle \widetilde{\varphi}^*([\alpha]), [M] \rangle \in \Bbbk.$$

Here: $*$ is a base point in M, $[M] \in H_3(M;\mathbb{Z})$ is the fundamental class of M, the bracket $\langle \cdot,\cdot \rangle$ stands for the evaluation pairing $H^3(M;\Bbbk^*) \times H_3(M;\mathbb{Z}) \to \Bbbk^*$, the map $\widetilde{\varphi}^*\colon H^3(BG;\Bbbk^*) \to H^3(M;\Bbbk^*)$ is induced by a pointed map $\widetilde{\varphi}\colon M \to BG$ inducing φ in π_1. (Such a map $\widetilde{\varphi}$ exists and is unique up to homotopy in the class of pointed maps $M \to BG$.) The invariant $\mathrm{DW}_{\alpha}(M)$ does not depend on the choice of the base point $* \in M$.

Theorem H.1. *For any closed connected oriented 3-manifold M,*

$$|M|_{\mathcal{C}} = \mathrm{DW}_{\alpha^{-1}}(M) = \mathrm{DW}_{\alpha}(-M).$$

Proof. A 3-cocycle $a\colon G^3 \to \Bbbk^*$ is *normalized* if $a(g,1,h) = 1_{\Bbbk}$ for all $g,h \in G$. Then the cocycle condition (1.3) implies that $a(g,h,1) = a(1,g,h) = 1_{\Bbbk}$ for all $g, h \in G$. The given 3-cocyle α is cohomologous to a normalized 3-cocycle constructed as follows. Define the maps $\omega\colon G^2 \to \Bbbk^*$ and $\partial_2(\omega)\colon G^3 \to \Bbbk^*$ by

$$\omega(g,h) = \alpha(g,1,1)\,\alpha(1,1,h)^{-1},$$
$$\partial_2(\omega)(g,h,j) = \omega(h,j)\,\omega(gh,j)^{-1}\,\omega(g,hj)\,\omega(g,h)^{-1}$$

for all $g,h,j \in G$. Formula (1.3) applied to the quadruples $(g_1,g_2,g_3,g_4) = (1,1,1,1)$ and $(g_1,g_2,g_3,g_4) = (g,1,1,h)$ gives that

$$\alpha(1,1,1) = 1 \quad \text{and} \quad \alpha(g,1,h) = \alpha(g,1,1)\,\alpha(1,1,h).$$

Thus

$$\begin{aligned}\partial_2(\omega)(g,1,h) &= \omega(1,h)\,\omega(g,h)^{-1}\,\omega(g,h)\,\omega(g,1)^{-1} = \omega(1,h)\,\omega(g,1)^{-1}\\ &= \alpha(1,1,1)\,\alpha(1,1,h)^{-1}\,\alpha(g,1,1)^{-1}\,\alpha(1,1,1) = \alpha(g,1,h)^{-1}.\end{aligned}$$

Then $\alpha' = \alpha\,\partial_2(\omega)\colon G^3 \to \Bbbk^*$ is a normalized 3-cocycle cohomologous to α. Since the spherical fusion $\Bbbk$-categories $\mathcal{C} = G^\alpha_\Bbbk$ and $\mathcal{C}' = G^{\alpha'}_\Bbbk$ are equivalent (see Example 1.7.6), we have: $|M|_\mathcal{C} = |M|_{\mathcal{C}'}$ (see Section 13.1.2). Consequently, we can assume without loss of generality that α is normalized.

For convenience of the reader, we summarize here the definition of the category $\mathcal{C} = G^\alpha_\Bbbk$ (for normalized α). The objects of $\mathcal{C}$ are the elements of G. The sets of morphisms are defined by $\mathrm{Hom}_\mathcal{C}(g,g) = \Bbbk$ for all $g \in G$ and $\mathrm{Hom}_\mathcal{C}(g,h) = \{0_\Bbbk\}$ for distinct $g,h \in G$, with composition induced by multiplication in $\Bbbk$. The monoidal product is given by $g \otimes h = gh$ and $k \otimes l = kl$ for all $g,h \in G$ and $k,l \in \Bbbk$. The associativity and unitality constraints a, l, r in $\mathcal{C}$ are defined by

$$a_{g,h,j} = \alpha(g,h,j) \in \Bbbk^* \subset \Bbbk = \mathrm{End}_\mathcal{C}(ghj), \qquad l_g = r_g = 1_\Bbbk \in \Bbbk = \mathrm{End}_\mathcal{C}(g)$$

for all $g,h,j \in G$. The dual of any $g \in G$ is $g^{-1} \in G$ with (co)evaluation morphisms

$$\mathrm{ev}_g = 1_\Bbbk, \quad \widetilde{\mathrm{ev}}_g = \lambda_g, \quad \mathrm{coev}_g = \lambda_g^{-1}, \quad \widetilde{\mathrm{coev}}_g = 1_\Bbbk,$$

where

$$\lambda_g = \alpha(g, g^{-1}, g).$$

Since the objects of $\mathcal{C}$ are simple and pairwise non-isomorphic, the set G is a representative set of simple objects of $\mathcal{C}$. We have $\dim(g) = 1_\Bbbk$ for all $g \in G$ and $\dim(\mathcal{C}) = |G| \cdot 1_\Bbbk \in \Bbbk^*$.

Let M be a closed connected oriented 3-manifold. Pick a triangulation t of M. The simplices of t of dimension $0,1,2,3$ are called respectively vertices, edges, triangles, and tetrahedra of t. Consider the cellular decomposition t^* of M dual to t. The 2-skeleton $P = (t^*)^{(2)}$ of t^* is a skeleton of M whose vertices, edges, and regions are respectively the 0-cells, 1-cells, and 2-cells of t^*. We fix a strict total order $<$ in the set of vertices of t and provide P with induced orientation: the region of P dual to an edge of t with vertices $\beta < \gamma$ is oriented so that its

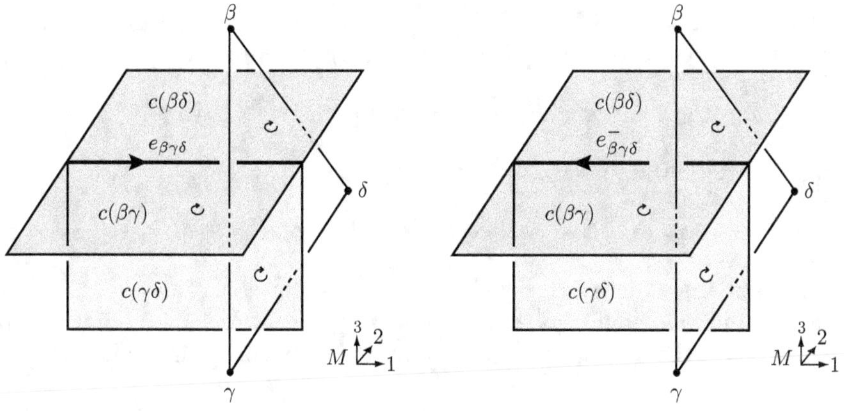

Figure H.1: Edges of P

orientation followed by the direction of that edge from β to γ yields the orientation of M. Since all objects of $\mathcal{C}$ have dimension $1_{\Bbbk}$, Formula (13.4) gives

$$|M|_{\mathcal{C}} = |G|^{-N} \sum_{c} |c|, \tag{H.1}$$

where $N = |M \setminus P|$ is the number of vertices of t and c runs over all maps from $\mathrm{Reg}(P)$ to G. We pick such a map c and compute $|c|$. We use the following notation. The value of c on the region of P dual to an edge of t with vertices $\beta < \gamma$ is denoted $c(\beta\gamma)$. The edge of P dual to the triangle of t with vertices $\beta < \gamma < \delta$ is denoted $e_{\beta\gamma\delta}$ and is oriented so that its orientation followed by the orientation of this triangle determined by the order $\beta < \gamma < \delta$ yields the orientation of M. The same edge with opposite orientation is denoted $e^-_{\beta\gamma\delta}$, see Figure H.1. For the edges $e = e_{\beta\gamma\delta}$ and $e^- = e^-_{\beta\gamma\delta}$, we have the cone isomorphisms

$$H_c(e) = H(P_e) \simeq \mathrm{Hom}_{\mathcal{C}}\big(1, \big(c(\beta\gamma) \otimes c(\gamma\delta)\big) \otimes c(\beta\delta)^{-1}\big),$$
$$H_c(e^-) = H(P_{e^-}) \simeq \mathrm{Hom}_{\mathcal{C}}\big(1, \big(c(\beta\delta) \otimes c(\gamma\delta)^{-1}\big) \otimes c(\beta\gamma)^{-1}\big).$$

Clearly, $H_c(e) = H_c(e^-) = 0$ and $|c| = 0$ unless

$$c(\beta\gamma)\, c(\gamma\delta) = c(\beta\delta). \tag{H.2}$$

Suppose that (H.2) holds and set $g = c(\beta\gamma)$ and $h = c(\gamma\delta)$. Then $c(\beta\delta) = gh$. We use the cone isomorphisms above to identify

$$H_c(e) = \mathrm{Hom}_{\mathcal{C}}\big(1, (g \otimes h) \otimes (gh)^{-1}\big) = \mathrm{Hom}_{\mathcal{C}}\big(1, 1\big) = \Bbbk$$

and

$$H_c(e^-) = \mathrm{Hom}_{\mathcal{C}}\big(1, (gh \otimes h^{-1}) \otimes g^{-1}\big) = \mathrm{Hom}_{\mathcal{C}}\big(1, 1\big) = \Bbbk.$$

Consider the duality pairing $\omega_{P_e}\colon H_c(e)\otimes H_c(e^-)\to \Bbbk$ from Section 12.3.2. The associated contraction vector (see Section 12.3.4) is computed by

$$*_e=\chi_e^{-1}\,1_\Bbbk\otimes 1_\Bbbk\in \Bbbk\otimes\Bbbk = H_c(e)\otimes H_c(e^-) \quad \text{where} \quad \chi_e=\omega_{P_e}(1_\Bbbk\otimes 1_\Bbbk)\in\Bbbk.$$

The computation in Figure H.2 below yields

$$\chi_e=\lambda_{gh}\,\lambda_g\,\lambda_h\,\alpha(g,h,h^{-1}).$$

In Figure H.2 the equality (i) follows from the definition of ω_{P_e}, (ii) from our conventions and Mac Lane's coherence (see Sections 1.3.4 and 1.4.5), and (iii) from the definitions of the associativity constraint, unitality constraints, and evaluation morphisms in $\mathcal{C}$.

Assume now that the map $c\colon \mathrm{Reg}(P)\to G$ satisfies (H.2) for all triangles of t (otherwise, $|c|=0$). We say that a tetrahedron of t is *positive* if its orientation induced by the order of the vertices is compatible with the orientation of M, and is *negative* otherwise. The vertex of P dual to a positive (respectively, negative) tetrahedron is said to be *positive* (respectively, *negative*). Let x be the vertex of P dual to a tetrahedron of t with vertices $\beta<\gamma<\delta<\varepsilon$. It follows from (H.2) that the colors of the six regions of P adjacent to x are determined by the colors $g=c(\beta\gamma)$, $h=c(\gamma\delta)$, $j=c(\delta\varepsilon)$. Let Γ_x^c be the $\mathcal{C}$-colored graph in the 2-sphere associated with x as in Section 13.1.1, see Figures H.3 and H.4. By definition,

$$H(\Gamma_x^c)=\begin{cases} H_c(e^-_{\beta\gamma\delta})\otimes H_c(e_{\beta\gamma\varepsilon})\otimes H_c(e^-_{\beta\delta\gamma})\otimes H_c(e_{\gamma\delta\varepsilon}) & \text{if } x \text{ is positive,}\\ H_c(e_{\beta\gamma\delta})\otimes H_c(e^-_{\beta\gamma\varepsilon})\otimes H_c(e_{\beta\delta\gamma})\otimes H_c(e^-_{\gamma\delta\varepsilon}) & \text{if } x \text{ is negative.}\end{cases}$$

Since c satisfies (H.2) for all triangles of t, we can identify all the modules H_c on the right-hand side with $\Bbbk$. Recall that $\mathbb{F}_\mathcal{C}(\Gamma_x^c)\in H(\Gamma_x^c)^\star=\mathrm{Hom}_\Bbbk(H(\Gamma_x^c),\Bbbk)$. Set

$$\Lambda_x=\mathbb{F}_\mathcal{C}(\Gamma_x^c)(1_\Bbbk\otimes 1_\Bbbk\otimes 1_\Bbbk\otimes 1_\Bbbk)\in\Bbbk.$$

For positive x, it follows from the definition of $\mathbb{F}_\mathcal{C}$ and Mac Lane's coherence that the Penrose diagram in Figure H.5 (including the associativity constraints) represents $\mathbb{F}_\mathcal{C}(\Gamma_x^c)$. Then, using the definition of $\mathcal{C}$, we obtain that

$$\Lambda_x=(\alpha(g,h,j))^{-1}\,\lambda_{ghj}\,\lambda_{gh}\,\lambda_{hj}\,\lambda_g\,\lambda_h\,\lambda_j\,\alpha(gh,j,j^{-1})\,\alpha(g,h,h^{-1}).$$

For negative x, a Penrose diagram (including the associativity constraints) representing $\mathbb{F}_\mathcal{C}(\Gamma_x^c)$ is given in Figure H.6. Then

$$\begin{aligned}\Lambda_x&=\alpha(g,h,j)\,\lambda_{ghj}\,\lambda_{gh}\,\lambda_{hj}\,\lambda_g\,\lambda_h\,\alpha(hj,j^{-1},j)\,\alpha(g,hj,(hj)^{-1})\\ &=\alpha(g,h,j)\,\lambda_{ghj}\,\lambda_{gh}\,\lambda_{hj}\,\lambda_g\,\lambda_h\,\lambda_j\,\alpha(h,j,j^{-1})\,\alpha(g,hj,(hj)^{-1})\end{aligned}$$

where we use the equality $\alpha(hj,j^{-1},j)=\lambda_j\,\alpha(h,j,j^{-1})$ which follows from the cocycle condition (1.3) applied to $(g_1,g_2,g_3,g_4)=(h,j,j^{-1},j)$.

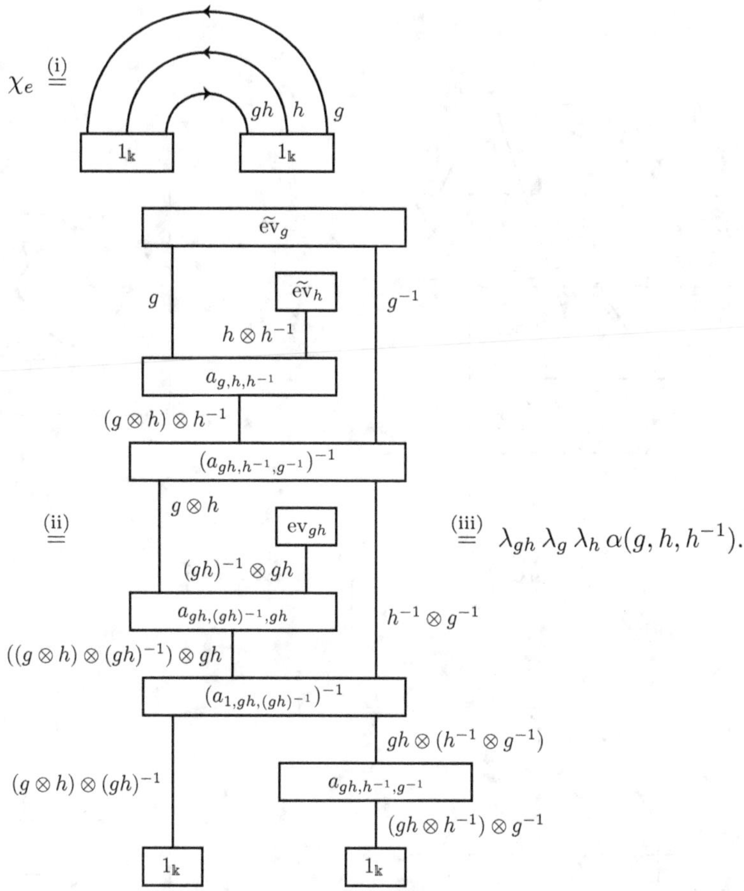

Figure H.2: χ_e for an edge e of P

The above computations and the definition of $|c|$ imply that

$$|c| = \prod_x \Lambda_x \prod_e \chi_e^{-1} \tag{H.3}$$

where x runs over all vertices of P and e runs over all edges of P. We next rewrite this expression for $|c|$. For a tetrahedron Δ of t with vertices $\beta < \gamma < \delta < \varepsilon$, we let $T^1_\Delta = (\gamma\delta\varepsilon), T^2_\Delta = (\beta\delta\varepsilon), T^3_\Delta = (\beta\gamma\varepsilon), T^4_\Delta = (\beta\gamma\delta)$ be its triangular faces. Set $\mu(\Delta) = 1$ if Δ is positive, $\mu(\Delta) = -1$ if Δ is negative, and

$$|\Delta|_c = \alpha\big(c(\beta\gamma), c(\gamma\delta), c(\delta\varepsilon)\big)^{-\mu(\Delta)} \in \Bbbk^*.$$

We call a triangular face T of Δ positive if the orientation of T induced by the

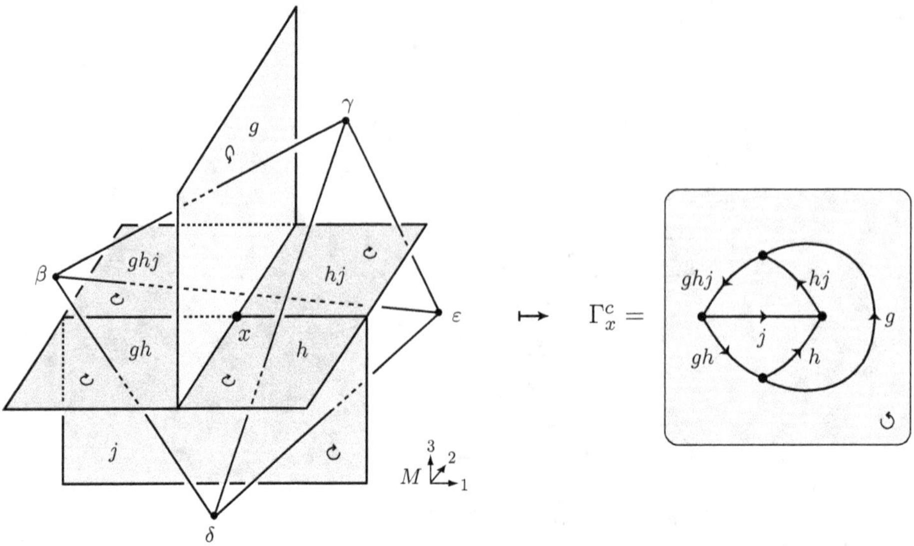

Figure H.3: A positive vertex x of P

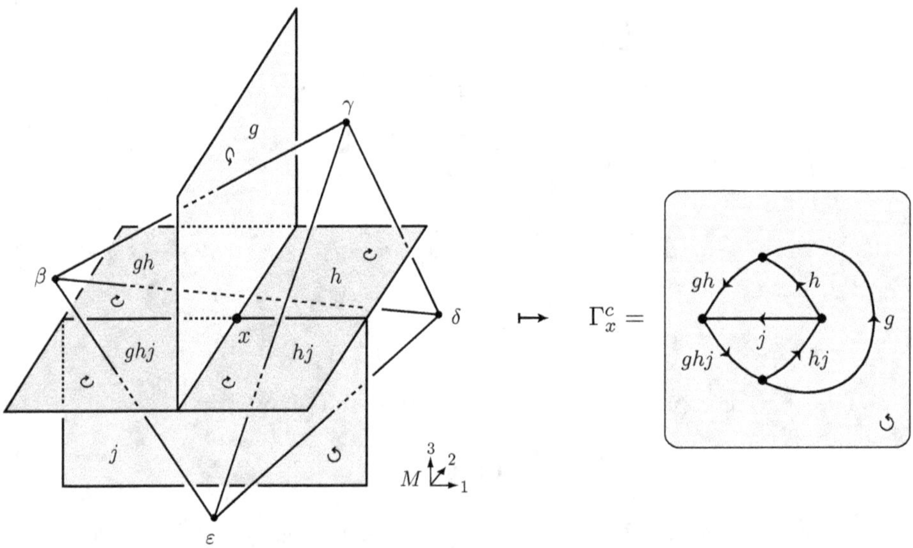

Figure H.4: A negative vertex x of P

order of its vertices is compatible with that of M restricted to Δ, and negative otherwise. The positive faces of Δ are T^1_Δ, T^3_Δ if Δ is positive, and T^2_Δ, T^4_Δ if Δ is

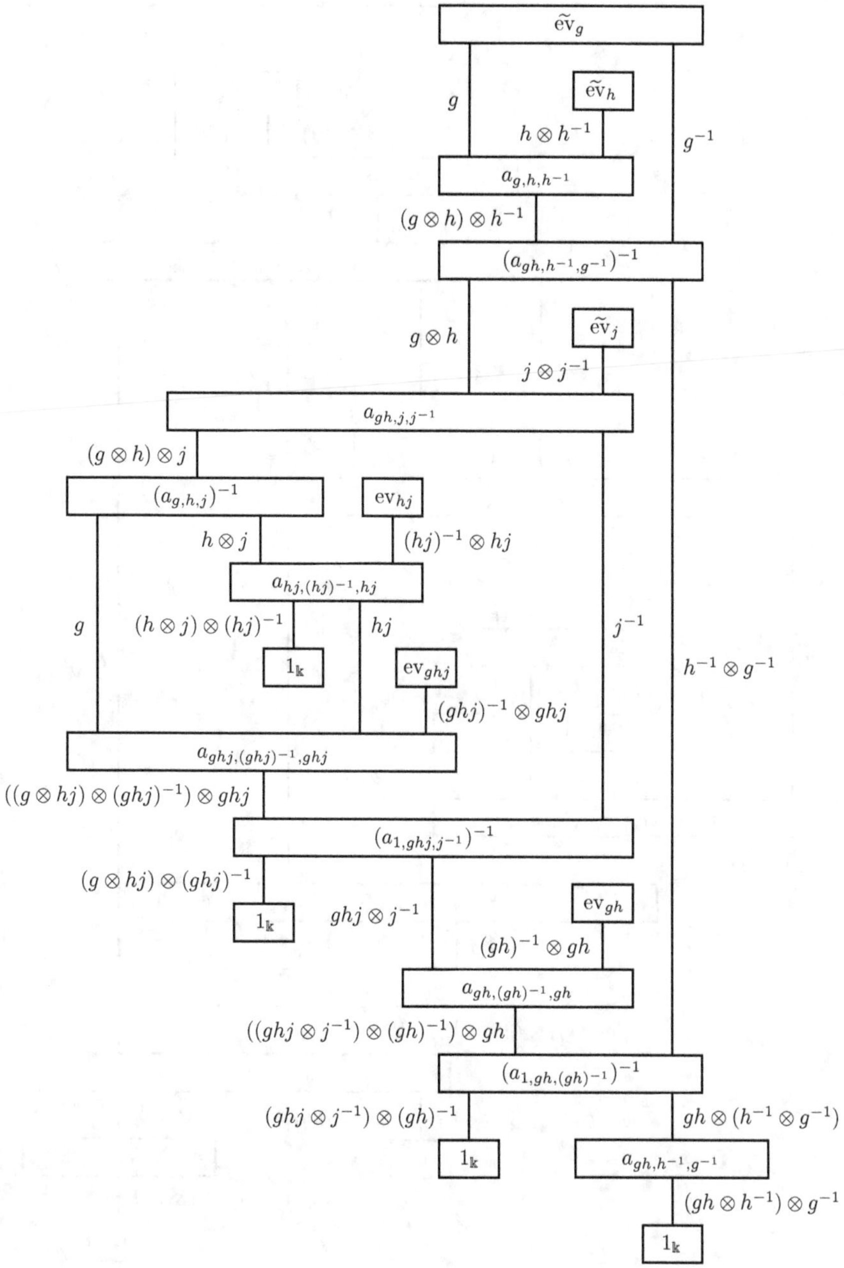

Figure H.5: $\mathbb{F}_{\mathcal{C}}(\Gamma_x^c)(1_{\Bbbk} \otimes 1_{\Bbbk} \otimes 1_{\Bbbk} \otimes 1_{\Bbbk})$ for a positive vertex x of P

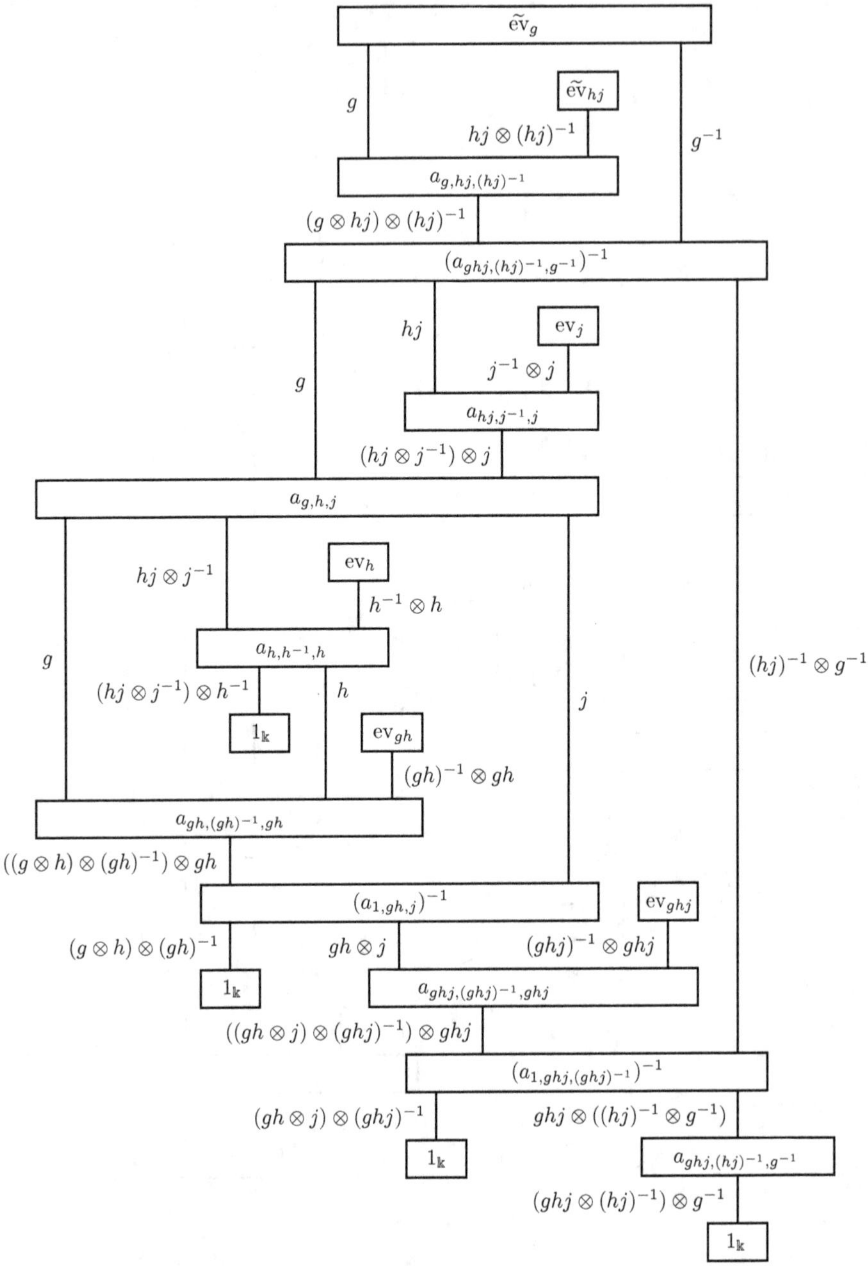

Figure H.6: $\mathbb{F}_{\mathcal{C}}(\Gamma_x^c)(1_{\mathbb{k}} \otimes 1_{\mathbb{k}} \otimes 1_{\mathbb{k}} \otimes 1_{\mathbb{k}})$ for a negative vertex x of P

negative. For a triangular face T of Δ with vertices $1<2<3$, set

$$[T/\Delta]=\begin{cases}\lambda_{c(12)}\,\lambda_{c(23)} & \text{if } T \text{ is positive in } \Delta,\\ \lambda_{c(13)}\,\alpha\big(c(12),c(23),c(23)^{-1}\big) & \text{if } T \text{ is negative in } \Delta.\end{cases}$$

For the vertex x of P dual to Δ, the above computation of Λ_x may be rewritten as

$$\Lambda_x=|\Delta|_c\prod_{i=1}^{4}[T^i_\Delta/\Delta]. \tag{H.4}$$

Any edge e of P is dual to a triangle T of t which is adjacent to two tetrahedra Δ^1_T and Δ^2_T of t. Clearly, T is a positive face of one these tetrahedra and a negative face of the other tetrahedron. The above computation of χ_e gives that

$$\chi_e=[T/\Delta^1_T]\,[T/\Delta^2_T]. \tag{H.5}$$

Since the vertices and edges of P bijectively correspond to the tetrahedra and triangles of t, respectively, Formulas (H.3), (H.4), (H.5) yield

$$|c|=\left(\prod_{\Delta}|\Delta|_c\prod_{i=1}^{4}[T^i_\Delta/\Delta]\right)\left(\prod_{T}([T/\Delta^1_T]\,[T/\Delta^2_T])^{-1}\right)$$

where Δ runs over all tetrahedra of t and T runs over all triangles of t. Now

$$\prod_{\Delta}\prod_{i=1}^{4}[T^i_\Delta/\Delta]=\prod_{T}([T/\Delta^1_T]\,[T/\Delta^2_T])$$

since any triangle T of t is adjacent to the two tetrahedra Δ^1_T and Δ^2_T. Thus

$$|c|=\prod_{\Delta}|\Delta|_c \tag{H.6}$$

where Δ runs over all tetrahedra of t. This expression has the following homological interpretation. Let E be the set of edges of t. Under the identification $E=\mathrm{Reg}(P)$ carrying an edge to the dual triangle, the coloring c corresponds to a map $E\to G$, again denoted by c, such that $c(\beta\gamma)\,c(\gamma\delta)=c(\beta\delta)$ for any triangle of t with vertices $\beta<\gamma<\delta$. For a path p in the 1-skeleton of t, set

$$c(p)=c(e_1)^{\nu_1}\cdots c(e_n)^{\nu_n}\in G,$$

where $e_1,\ldots,e_n$ are the consecutive edges of t traversed by p, $\nu_i=1$ if the orientation of e_i induced by the order of its vertices is compatible with that of p, and $\nu_i=-1$ otherwise. Now, pick a vertex $*$ of t as the base point of M. Any $a\in\pi_1(M,*)$ can be represented by a loop p_a in the 1-skeleton of t based at $*$.

The formula $\varphi_c(a) = c(p_a)$ defines a group homomorphism $\varphi_c\colon \pi_1(M,*) \to G$. Consider a pointed map $\widetilde{\varphi} = \widetilde{\varphi}_c\colon M \to BG$ inducing φ_c in π_1. Then

$$\prod_{\Delta} |\Delta|_c = \langle \widetilde{\varphi}^*([\alpha^{-1}]), [M]\rangle \in \Bbbk \tag{H.7}$$

where $\widetilde{\varphi}^*\colon H^3(BG;\Bbbk^*) \to H^3(M;\Bbbk^*)$ is the homomorphism induced by $\widetilde{\varphi}$. Indeed, the chain $\sum_\Delta \mu(\Delta)\Delta$ is a 3-cycle in M representing the fundamental class of M. Therefore

$$\begin{aligned}\langle \widetilde{\varphi}^*([\alpha^{-1}]), [M]\rangle &= \langle [\alpha^{-1}], \widetilde{\varphi}_*([M])\rangle \\ &= \big\langle [\alpha]^{-1}, \widetilde{\varphi}_*\big(\sum_\Delta \mu(\Delta)\,[\Delta]\big)\big\rangle = \prod_\Delta \widetilde{\alpha}(\widetilde{\varphi}_*(\Delta))^{-\mu(\Delta)}\end{aligned}$$

where $\widetilde{\alpha}$ is a $\Bbbk^*$-valued 3-cocycle on BG representing $[\alpha] \in H^3(BG;\Bbbk^*)$. Using an appropriate simplicial model of BG we can choose $\widetilde{\alpha}$ so that

$$\widetilde{\alpha}(\widetilde{\varphi}_*(\Delta)) = \alpha\big(c(\beta\gamma), c(\gamma\delta), c(\delta\varepsilon)\big)$$

where $\beta < \gamma < \delta < \varepsilon$ are the vertices of Δ. We have then $\widetilde{\alpha}(\widetilde{\varphi}_*(\Delta))^{-\mu(\Delta)} = |\Delta|_c$ and the equality (H.7) follows.

Let $\mathcal{A}$ be the set of maps $c\colon E \to G$ such that $c(\beta\gamma)\,c(\gamma\delta) = c(\beta\delta)$ for any triangle of t with vertices $\beta < \gamma < \delta$. Combining Formulas (H.1), (H.6), (H.7), we obtain that

$$\begin{aligned}|M|_{\mathcal{C}} &= |G|^{-N} \sum_{c\in\mathcal{A}} \langle \widetilde{\varphi}_c^*([\alpha^{-1}]), [M]\rangle \\ &= |G|^{-N} \sum_{\varphi\in \mathrm{Hom}(\pi_1(M,*),G)} n_\varphi \,\langle \widetilde{\varphi}^*([\alpha^{-1}]), [M]\rangle\end{aligned}$$

where $n_\varphi = \mathrm{card}\{c \in \mathcal{A} \,|\, \varphi_c = \varphi\}$. To accomplish the proof of the theorem, we need only to show that $n_\varphi = |G|^{N-1}$ for all φ.

Let $\mathcal{G}$ be the set of maps from the set of vertices of t to G which carry the vertex $*$ to $1 \in G$. Clearly, $\mathcal{G}$ is a group with pointwise multiplication. We define a left action $\cdot$ of $\mathcal{G}$ on $\mathcal{A}$ by

$$(\eta\cdot c)(\beta\gamma) = \eta(\beta)\, c(\beta\gamma)\, \eta(\gamma)^{-1}$$

for any $\eta \in \mathcal{G}$, $c \in \mathcal{A}$, and any edge $(\beta\gamma)$ of t with vertices $\beta < \gamma$. This action is easily seen to be free, and $\varphi_{\eta\cdot c} = \varphi_c$ for all $\eta \in \mathcal{G}, c \in \mathcal{A}$. Thus, the formula $c \mapsto \varphi_c$ defines a map

$$\mathcal{A}/\mathcal{G} \to \mathrm{Hom}(\pi_1(M,*), G). \tag{H.8}$$

This map is bijective, and its inverse is constructed as follows. For every vertex v of t, fix a path p_v from $*$ to v in the 1-skeleton of t so that p_* is the constant path at $*$. Given $\varphi \in \mathrm{Hom}(\pi_1(M,*), G)$, set

$$c^\varphi(\beta\gamma) = \varphi([p_\beta(\beta\gamma)p_\gamma^{-1}]) \in G$$

for any edge $(\beta\gamma)$ of t with $\beta < \gamma$. Here the square brackets stand for the homotopy class of a loop. Clearly, $c^\varphi \in \mathcal{A}$ and $\varphi_{c^\varphi} = \varphi$. For each $c \in \mathcal{A}$, we have $c^{\varphi_c} = \eta \cdot c$ where $\eta \in \mathcal{G}$ is defined by $\eta(v) = c(p_v) \in G$ for $v \in E$. Thus, the map

$$\mathrm{Hom}(\pi_1(M,*), G) \to \mathcal{A}/\mathcal{G}, \quad \varphi \mapsto \mathcal{G} \cdot c^\varphi$$

is inverse to the map (H.8). As a consequence, $n_\varphi = |\mathcal{G}| = |G|^{N-1}$ for all φ. This completes the proof of the theorem. □

Theorem H.1 and the properties of the state sum invariants of 3-manifolds imply that the Dijkgraaf–Witten invariants of 3-manifolds extend to 3-dimensional TQFTs. This was first established by Wakui [Wak].

Appendix I

Hints and solutions to exercises

In this appendix we give hints and solutions to the exercises marked by the asterisk in the main text.

Exercise 1.1.9 (Hint)

Pick a natural isomorphism $\gamma\colon FG \to 1_{\mathcal{D}}$ and, for each $Y \in \mathrm{Ob}(\mathcal{D})$, set

$$\beta_Y = \gamma_Y F(\alpha^{-1}_{G(Y)}) FG(\gamma_Y^{-1})\colon FG(Y) \to Y.$$

Use the invertibility and naturality of α and γ to prove that $GF(\alpha_X) = \alpha_{GF(X)}$ for all $X \in \mathrm{Ob}(\mathcal{C})$ and $FG(\gamma_Y) = \gamma_{FG(Y)}$ for all $Y \in \mathrm{Ob}(\mathcal{D})$. Deduce that the family $\beta = \{\beta_Y\}_{Y\in \mathrm{Ob}(\mathcal{D})}$ is a natural isomorphism satisfying (1.1). To prove the uniqueness, show that a natural isomorphism β satisfying (1.1) fulfills

$$\beta_Y = f F(\alpha_X^{-1}) FG(f^{-1})$$

for any isomorphism $f\colon F(X) \to Y$ in $\mathcal{D}$ with $X \in \mathrm{Ob}(\mathcal{C})$ and $Y \in \mathrm{Ob}(\mathcal{D})$.

Exercise 1.1.11 (Solution)

The implication (i) $\Rightarrow$ (ii) is clear: if f is an isomorphism, then each h_Z is bijective with inverse computed by $(h_Z)^{-1}(\beta) = \beta f^{-1}$ for all $\beta \in \mathrm{Hom}_{\mathcal{C}}(X, Z)$. The implication (ii) $\Rightarrow$ (iii) is straightforward. Let us prove the implication (iii) $\Rightarrow$ (i). Since h_X is surjective, there is $g \in \mathrm{Hom}_{\mathcal{C}}(Y, X)$ such that $h_X(g) = \mathrm{id}_X$, that is, $gf = \mathrm{id}_X$. Note that

$$h_Y(fg) = fgf = f\mathrm{id}_X = f = \mathrm{id}_Y f = h_Y(\mathrm{id}_Y).$$

Since h_Y is injective, $fg = \mathrm{id}_Y$. Hence, f is an isomorphism with inverse g.

Exercise 1.4.6 (Hint)

For any $A, B \in \mathrm{Ob}(\mathcal{D})$, define $G_2(A,B)\colon G(A) \otimes G(B) \to G(A \otimes B)$ by

$$G_2(A,B) = G(\beta_A \otimes \beta_B) G(F_2(G(A), G(B))^{-1}) \alpha_{G(A)\otimes G(B)}.$$

Set $G_0 = G(F_0^{-1})\alpha_{\mathbb{1}}\colon \mathbb{1} \to G(\mathbb{1})$. Prove that (G, G_2, G_0) is a strong monoidal functor such that the natural isomorphisms α and β are monoidal. Observe that the monoidality of α means that $\alpha_{\mathbb{1}} = (GF)_0 = G(F_0)G_0$ and, for all $X, Y \in \mathrm{Ob}(\mathcal{C})$,

$$\alpha_{X\otimes Y} = (GF)_2(X,Y)(\alpha_X \otimes \alpha_Y) = G(F_2(X,Y)) G_2(F(X), F(Y))(\alpha_X \otimes \alpha_Y).$$

To prove the uniqueness of the monoidal structure (G_2, G_0) for G such that α is monoidal, deduce from the monoidality of α and the naturality of G_2 that

$$G_2(A,B) = G(f \otimes g) G(F_2(X,Y)^{-1}) \alpha_{X\otimes Y} (\alpha_X^{-1} G(f^{-1}) \otimes \alpha_Y^{-1} G(g^{-1}))$$

for any isomorphisms $f\colon F(X) \to A$ and $g\colon F(Y) \to B$ in $\mathcal{D}$, with $X, Y \in \mathrm{Ob}(\mathcal{C})$ and $A, B \in \mathrm{Ob}(\mathcal{D})$.

Exercise 1.6.5 (Hint)

Assume that $\mathcal{C}$ is left rigid. Let $X \in \mathrm{Ob}(\mathcal{C})$. Pick a left dual $({}^\vee\! X, \mathrm{ev}_X)$ of X. Then $(F({}^\vee\! X), (\mathrm{ev}_X)^F)$ is a left dual of $F(X)$ and $(G({}^\vee\! X), (\mathrm{ev}_X)^G)$ is a left dual of $G(X)$. Define the morphism $\beta_X\colon G(X) \to F(X)$ by

$$\beta_X = (\mathrm{id}_{F(X)} \otimes (\mathrm{ev}_X)^G)(\mathrm{id}_{F(X)} \otimes \alpha_{{}^\vee\! X} \otimes \mathrm{id}_{G(X)})((\mathrm{coev}_X)^F \otimes \mathrm{id}_{G(X)}),$$

where $(\mathrm{coev}_X)^F$ is the inverse of the pairing $(\mathrm{ev}_X)^F$. Verify that $\beta_X \alpha_X = \mathrm{id}_{F(X)}$ and $\alpha_X \beta_X = \mathrm{id}_{G(X)}$.

Exercise 2.4.4 (Solution)

To prove the first equality, insert in a vertical segment of c oriented downward two boxes colored respectively with an isomorphism $X \to Y$ and its inverse $Y \to X$, the arc between the boxes being colored with Y. The diagram $\mathcal{P}'$ is obtained by sliding the lower box around c so that it approaches the upper box from above and removing both resulting boxes and the small X-colored arc between them. The equality $\langle \mathcal{P} \rangle = \langle \mathcal{P}' \rangle$ follows from Theorem 2.6 and the properties of the graphical calculus discussed in Section 2.4.3. The formula $\langle \mathcal{P} \rangle = \langle \mathcal{P}'' \rangle$ is proved similarly inserting in a vertical segment of c oriented upward two boxes colored with id_{X^*}, the arc between the boxes being oriented downward and colored with X^*.

Exercise 3.1.12 (Hint)

Let $X, Y \in \mathrm{Ob}(\mathcal{C})$. If X has a left dual $({}^{\vee}X, \mathrm{ev}_X)$, then the morphism $\tau_{X,Y}$ is invertible with inverse

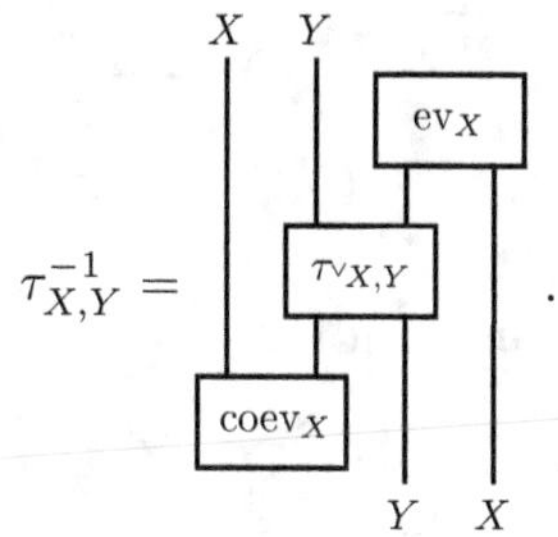

Similarly, if Y has a right dual $(Y^{\vee}, \widetilde{\mathrm{ev}}_Y)$, then $\tau_{X,Y}$ is invertible and

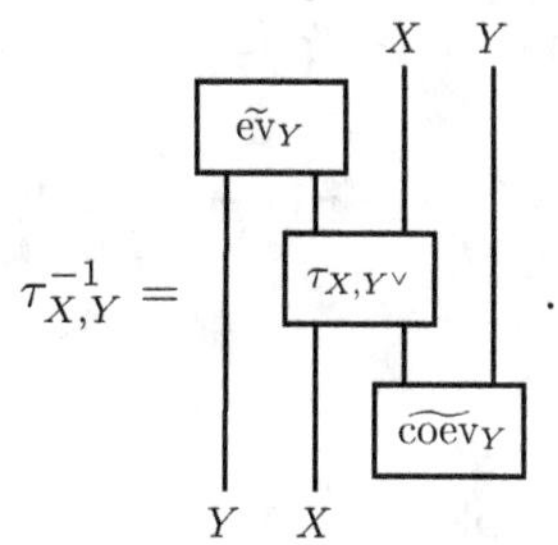

Exercise 3.2.4 (Hint)

Proceed as in Exercise 2.4.4 using the invariance of $\langle\cdot\rangle$ under the Reidemeister moves R_4 and R_5.

Exercise 3.2.5 (Solution)

We have:

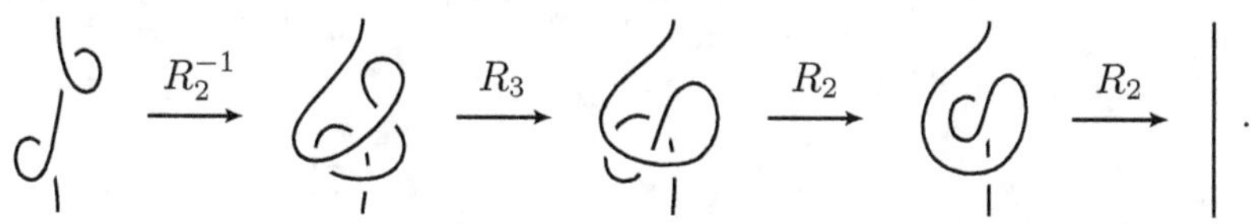

Exercise 3.2.6 (Solution)

We have:

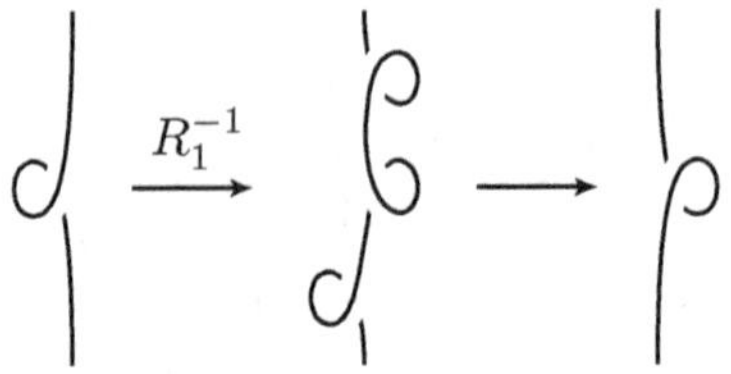

where the second arrow is the move of Exercise 3.2.5.

Exercise 3.2.7 (Solution)

We have:

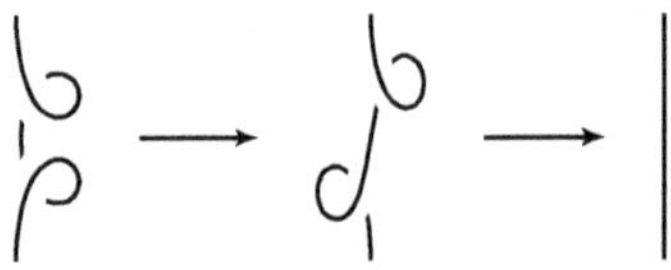

where the first arrow is the inverse of the move of Exercise 3.2.6 and the second arrow is the move of Exercise 3.2.5.

Exercise 4.2.7 (Solution)

The if part follows from Example 1.7.6. Conversely, assume that $\mathcal{C}_1$ and $\mathcal{C}_2$ are equivalent pivotal $\Bbbk$-categories. Pick a $\Bbbk$-linear pivotal equivalence $F\colon \mathcal{C}_1 \to \mathcal{C}_2$. The action of F on objects of $\mathcal{C}_1$ is determined by a map $f\colon G_1 \to G_2$. Since the functor F is monoidal, the map f must be a group homomorphism. Since F is an equivalence, f is an isomorphism. The monoidal constraint F_2 is determined by the map $\chi\colon G_1 \times G_1 \to \Bbbk^*$ defined by $\chi(g,h) = F_2(g,h)$ for all $g,h \in G_1$. The axioms of a monoidal functor (see Section 1.4.1) and the $\Bbbk$-linearity of F imply that $F_0 = \chi(1,1)$ and, for all $g,h,j \in G_1$,

$$\alpha_2(f(g),f(h),f(j)) = \alpha_1(g,h,j)\,\chi(g,h)\,\chi(g,hj)^{-1}\,\chi(gh,j)\,\chi(h,j)^{-1}. \tag{I.1}$$

Thus $[f^*(\alpha_2)] = [\alpha_1]$. Now, for $i \in \{1,2\}$ and any $g \in G_i$, set

$$\lambda_{i,g} = \alpha_i(1,1,g)\,\alpha_i(g,g^{-1},g)\,\alpha_i(g,1,1) \in \Bbbk^*.$$

It follows from the definitions and the $\Bbbk$-linearity of F that the natural isomorphisms F^l and F^r associated with the strong monoidal functor F (see Section 1.6.4) are computed as follows: for all $g \in G_1$,

$$F^l(g) = \chi(1,1)^{-1}\,\chi(g^{-1},g)\,\lambda_{2,f(g)^{-1}}^{-1}\,\lambda_{2,f(g)}^{-1}$$

and

$$F^r(g) = \chi(1,1)^{-1} \lambda_{1,g} \, d_1(g) \, \chi(g, g^{-1}) \, \lambda_{2,f(g)^{-1}} \, d_2(f(g))^{-1}.$$

Since the functor F is pivotal, we have $F^l(g) = F^r(g)$ and so

$$d_2(f(g)) = d_1(g) \, \lambda_{2,f(g)^{-1}} \, \lambda_{2,f(g)} \, \lambda_{2,f(g)^{-1}} \, \lambda_{1,g} \, \chi(g, g^{-1}) \, \chi(g^{-1}, g)^{-1}.$$

Now, since α_2 is a 3-cocycle, we have $\lambda_{2,f(g)^{-1}} = (\lambda_{2,f(g)})^{-1}$. Also, Formula (I.1) implies that

$$\lambda_{2,f(g)} = \lambda_{1,g} \, \chi(g, g^{-1}) \, \chi(g^{-1}, g)^{-1}.$$

Thus $d_2(f(g)) = d_1(g)$ for all $g \in G_1$, that is, $d_2 f = d_1$.

Exercise 4.4.5 (Hint)

For any $X \in \mathrm{Ob}(\mathcal{C})$, we have:

$$\alpha_X = \sum_\alpha G(q_\alpha) \alpha_{i_\alpha} F(p_\alpha) \colon F(X) \to G(X),$$

where $(p_\alpha \colon X \to i_\alpha, q_\alpha \colon i_\alpha \to X)_\alpha$ is an I-partition of X.

Exercise 4.4.6 (Hint)

It follows from Exercise 4.4.5 that any strong monoidal natural automorphism $\varphi = \{\varphi_X \colon X \to X\}_{X \in \mathrm{Ob}(\mathcal{C})}$ of the identity endofunctor $1_\mathcal{C}$ is determined by a family $\{a_i\}_{i \in I}$ of invertible elements of $\Bbbk$ such that $\varphi_i = a_i \, \mathrm{id}_i$ for all $i \in I$. Clearly, $a_\mathbb{1} = 1$. Also, if an object $i \in I$ is a direct summand of the monoidal product of $m \geq 1$ objects $i_1, \ldots, i_m \in I$, then $a_i = \prod_{j=1}^m a_{i_j}$. These properties mean precisely that the formula $[i] \mapsto a_i$ defines a group homomorphism $A_\mathcal{C} \to \Bbbk^*$. For $\mathcal{C} = G_\Bbbk^{\alpha,d}$, the map $G \to A_\mathcal{C}$, $g \mapsto [g]$ is a group isomorphism.

Exercise 5.1.5 (Solution)

We need to prove that for all $X \in \mathrm{Ob}(\mathcal{C})$, the morphism σ_X is invertible. Let $(X^\vee, \widetilde{\mathrm{ev}}_X)$ be a right dual of X and let $\widetilde{\mathrm{coev}}_X$ be the inverse of the pairing $\widetilde{\mathrm{ev}}_X$. Set

$\nu_X =$ 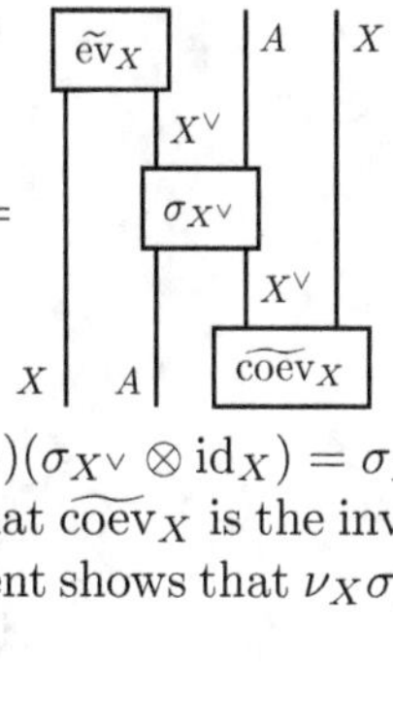 $: X \otimes A \to A \otimes X.$

The formula $(\mathrm{id}_{X^\vee} \otimes \sigma_X)(\sigma_{X^\vee} \otimes \mathrm{id}_X) = \sigma_{X^\vee \otimes X}$, the naturality of σ, the formula $\sigma_\mathbb{1} = \mathrm{id}_A$, and the fact that $\widetilde{\mathrm{coev}}_X$ is the inverse of $\widetilde{\mathrm{ev}}_X$ imply the equality $\sigma_X \nu_X = \mathrm{id}_{X \otimes A}$. A similar argument shows that $\nu_X \sigma_X = \mathrm{id}_{A \otimes X}$. Thus, σ_X is invertible with inverse ν_X.

Exercise 5.2.3 (Hint)

Let $F\colon \mathcal{C} \to \mathcal{D}$ be a monoidal equivalence between monoidal categories $\mathcal{C}, \mathcal{D}$ with quasi-inverse $G\colon \mathcal{D} \to \mathcal{C}$. Pick a natural isomorphism $\alpha\colon 1_{\mathcal{C}} \to GF$. By Exercise 1.4.6, the functor G has a unique structure of a strong monoidal functor so that α is monoidal, and then the natural isomorphism $\beta\colon FG \to 1_{\mathcal{D}}$ associated with α in Exercise 1.1.9 is monoidal. For any $(M,\sigma) \in \mathcal{Z}(\mathcal{C})$, let $\widetilde{F}(M,\sigma) = (F(M), \sigma^\circ)$ be the half-braiding of $\mathcal{D}$ such that for all $Y \in \mathrm{Ob}(\mathcal{D})$, the following diagram commutes:

$$
\begin{array}{ccccc}
F(M)\otimes Y & \xrightarrow{\mathrm{id}_{F(M)}\otimes\beta_Y^{-1}} & F(M)\otimes FG(Y) & \xrightarrow{F_2(M,G(Y))} & F(M\otimes G(Y)) \\
{\scriptstyle \sigma^\circ_Y}\big\downarrow & & & & \big\downarrow{\scriptstyle F(\sigma_{G(Y)})} \\
Y\otimes F(M) & \xleftarrow[\beta_Y\otimes\mathrm{id}_{F(M)}]{} & FG(Y)\otimes F(M) & \xleftarrow[F_2(G(Y),M)^{-1}]{} & F(G(Y)\otimes M).
\end{array}
$$

For any $(N,\gamma) \in \mathcal{Z}(\mathcal{D})$, let $\widetilde{G}(N,\gamma) = (G(N), \gamma^\bullet)$ be the half-braiding of $\mathcal{C}$ such that for all $X \in \mathrm{Ob}(\mathcal{C})$, the following diagram commutes:

$$
\begin{array}{ccccc}
G(N)\otimes X & \xrightarrow{\mathrm{id}_{G(N)}\otimes\alpha_X} & G(N)\otimes GF(X) & \xrightarrow{G_2(N,F(X))} & G(N\otimes F(X)) \\
{\scriptstyle \gamma^\bullet_X}\big\downarrow & & & & \big\downarrow{\scriptstyle G(\gamma_{F(X)})} \\
X\otimes G(N) & \xleftarrow[\alpha_X^{-1}\otimes\mathrm{id}_{G(N)}]{} & GF(X)\otimes G(N) & \xleftarrow[G_2(F(X),N)^{-1}]{} & G(F(X)\otimes N).
\end{array}
$$

For any morphisms f in $\mathcal{Z}(\mathcal{C})$ and g in $\mathcal{Z}(\mathcal{D})$, set $\widetilde{F}(f) = F(f)$ and $\widetilde{G}(g) = G(g)$. This defines functors $\widetilde{F}\colon \mathcal{Z}(\mathcal{C}) \to \mathcal{Z}(\mathcal{D})$ and $\widetilde{G}\colon \mathcal{Z}(\mathcal{D}) \to \mathcal{Z}(\mathcal{C})$ which we endow with monoidal constraints induced by those of F and G. Then $\widetilde{F}$ and $\widetilde{G}$ are quasi-inverse braided strong monoidal functors.

Exercise 6.2.7 (Hint)

If A is a Hopf algebra, then

$H_l^{-1} =$ 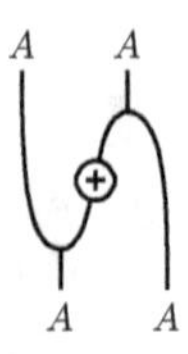

and $H_r^{-1} =$ 

.

Conversely, if H_l and H_r are invertible, then the antipode of A and its inverse are given by

$S =$ 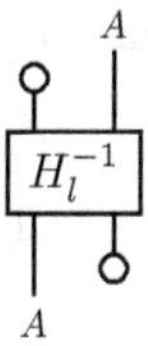 and $S^{-1} =$ 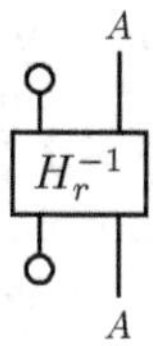 .

Exercise 6.3.3 (Hint)

Consider the convolution monoid $\mathcal{M} = \mathrm{Hom}_{\mathcal{C}}(A \otimes A, \mathbb{1})$, see Exercise 6.1.6. Here $\mathbb{1}$ is endowed with the trivial algebra structure and $A \otimes A$ is endowed with the coalgebra structure (Δ', ε') defined by $\Delta' = (\mathrm{id}_A \otimes \tau_{A \otimes A, A})(\Delta \otimes \Delta)$ and $\varepsilon' = \varepsilon \otimes \varepsilon$, where τ is the braiding of $\mathcal{C}$, Δ is the coproduct of A, and ε is the counit of A. The axioms of a bialgebra pairing and of an antipode imply that both $\omega(S \otimes \mathrm{id}_A)$ and $\omega(\mathrm{id}_A \otimes S)$ are inverse to ω in $\mathcal{M}$. Hence, they are equal.

Exercise 6.4.6 (Hint)

Use that by Section 1.6.3, different choices of a left duality lead to isomorphic left dual functors and apply Lemma B.1.

Exercise 6.4.7 (Solution)

Observe first that the multiplicativity of the coproduct and the anti-multiplicativity of the antipode of A imply that the coadjoint action $\cdot$ is indeed an action of A on A^*. Thus, $C \in \mathrm{Ob}(\mathrm{mod}_A)$. Let us prove that ρ_X is an A-linear homomorphism for any $X \in \mathrm{Ob}(\mathrm{mod}_A)$. Clearly, ρ_X is $\Bbbk$-linear. Pick $g \in {}^{\vee}X = X^*$, $x \in X$, and $h \in A$. For all $a \in A$,

$$\begin{aligned}
&\big(\rho_X(h(g \otimes_{\Bbbk} x))\big)(a) \overset{\text{(i)}}{=} \big(\rho_X(h_{(1)}g \otimes_{\Bbbk} h_{(2)}x)\big)(a) \\
&\overset{\text{(ii)}}{=} (h_{(1)}g)(ah_{(2)}x) \overset{\text{(iii)}}{=} g(S(h_{(1)})ah_{(2)}x) \overset{\text{(iv)}}{=} \big(h \cdot \rho_X(g \otimes_{\Bbbk} x)\big)(a).
\end{aligned}$$

Here (i) follows from the definition of the monoidal product in mod_A, (ii) from the definition of ρ_X, (iii) from the definition of the left duality in mod_A, and (iv) from the definitions of ρ_X and the coadjoint action. Thus, $\rho_X(h \cdot (g \otimes_{\Bbbk} x)) = h \cdot \rho_X(g \otimes_{\Bbbk} x)$. Hence ρ_X is A-linear.

Let us prove that ρ is a dinatural transformation. Let $\phi\colon X \to Y$ be a morphism in the category mod_A. Pick $g \in Y^*$ and $x \in X$. For all $a \in A$,

$$\begin{aligned}
\big(\rho_Y(g \otimes_{\Bbbk} \phi(x))\big)(a) &\overset{\text{(i)}}{=} g(a\phi(x)) \overset{\text{(ii)}}{=} g(\phi(ax)) \\
&\overset{\text{(iii)}}{=} {}^{\vee}\phi(g)(ax) \overset{\text{(iv)}}{=} \big(\rho_X({}^{\vee}\phi(g) \otimes_{\Bbbk} x)\big)(a).
\end{aligned}$$

Here (i) and (iv) follow from the definition of ρ, (ii) from the A-linearity of ϕ, and (iii) from the definition of the left duality in mod_A. Thus, $\rho_Y(g \otimes_{\Bbbk} \phi(x)) = \rho_X({}^{\vee}\phi(g) \otimes_{\Bbbk} x)$. Hence ρ is a dinatural transformation.

We now prove that ρ is universal. To this end, pick a dinatural transformation $d = \{d_X : {}^{\vee}X \otimes_{\Bbbk} X \to D\}_{X \in \mathrm{Ob}(\mathrm{mod}_A)}$ in mod_A. Let us show that there is a unique A-linear homomorphism $\varphi\colon C \to D$ satisfying $d = \varphi\rho$, and that this homomorphism carries any $f \in C$ to

$$\varphi(f) = d_A(f \otimes_{\Bbbk} 1_A), \tag{I.2}$$

where A is endowed with the A-module structure given by left multiplication. Note first that for any $X \in \mathrm{mod}_A$, $g \in X^*$, and $x \in X$, we have

$$d_X(g \otimes_{\Bbbk} x) = d_A(\rho_X(g \otimes_{\Bbbk} x) \otimes_{\Bbbk} 1_A). \tag{I.3}$$

Indeed, consider the A-linear morphism $\phi\colon A \to X$ defined by $\phi(a) = ax$ for all $a \in A$. The dinaturality of d implies that

$$d_X(g \otimes_{\Bbbk} \phi(1_A)) = d_A({}^{\vee}\phi(g) \otimes_{\Bbbk} 1_A).$$

The last equality is exactly (I.3) because $\phi(1_A) = x$ and ${}^{\vee}\phi(g) = \rho_X(g \otimes_{\Bbbk} x)$. Next, consider the $\Bbbk$-linear homomorphism $\varphi\colon C \to D$ defined by (I.2). It is A-linear since, for any $f \in C$ and $h \in A$, we have

$$\begin{aligned} h\varphi(f) &\overset{\text{(i)}}{=} hd_A(f \otimes_{\Bbbk} 1_A) \overset{\text{(ii)}}{=} d_A(h_{(1)}f \otimes_{\Bbbk} h_{(2)}) \\ &\overset{\text{(iii)}}{=} d_A(\rho_A(h_{(1)}f \otimes_{\Bbbk} h_{(2)}) \otimes_{\Bbbk} 1_A) \overset{\text{(iv)}}{=} d_A(h \cdot f \otimes_{\Bbbk} 1_A) \overset{\text{(v)}}{=} \varphi(h \cdot f). \end{aligned}$$

Here (i) and (v) follow from (I.2), (ii) from the A-linearity of d_A, (iii) from (I.3), and (iv) from the definitions of ρ_A and the coadjoint action. The equality (I.3) says exactly that $d_X = \varphi\,\rho_X$ for all $X \in \mathrm{mod}_A$. Also, the uniqueness of φ follows from the equality $\varphi\,\rho_A = d_A$ and the fact that $\rho_A\colon A^* \otimes_{\Bbbk} A \to A$ is surjective (because $d_A(A^* \otimes_{\Bbbk} 1_A) = A^* = C$). Hence ρ is universal.

Finally, let us compute the coproduct Δ and counit ε of the coend C. Pick $f \in C$. First,

$$\varepsilon(f) \overset{\text{(i)}}{=} \mathrm{ev}_A(f \otimes_{\Bbbk} 1_A) \overset{\text{(ii)}}{=} f(1_A)$$

Here (i) follows from (I.2) and the expression of the dinatural transformation defining ε (see Section 6.4.2), and (ii) from the definition of the left duality in mod_A. Second, consider finite families $\{e_i\}_i \subset A$ and $\{\ell_i\}_i \subset A^*$ such that $a = \sum_i \ell_i(a)e_i$ for all $a \in A$ (see the proof of Lemma 1.6). Then

$$\Delta(f) \overset{\text{(i)}}{=} (\rho_A \otimes_{\Bbbk} \rho_A)(f \otimes_{\Bbbk} \mathrm{coev}_A \otimes_{\Bbbk} \mathrm{id}_A) \overset{\text{(ii)}}{=} \sum_i \rho_A(f \otimes_{\Bbbk} e_i) \otimes_{\Bbbk} \rho_A(\ell_i \otimes_{\Bbbk} 1_A).$$

Here (i) follows from (I.2) and the expression of the dinatural transformation defining Δ (see Section 6.4.2), and (ii) from the fact that $\mathrm{coev}_A = \sum_i e_i \otimes_{\Bbbk} \ell_i$. Consequently, for all $a, b \in A$,

$$\begin{aligned}\Delta(f)(a \otimes_{\Bbbk} b) &= \sum_i \rho_A(f \otimes_{\Bbbk} e_i)(a)\, \rho_A(\ell_i \otimes_{\Bbbk} 1_A)(b) \\ &= \sum_i f(ae_i)\ell_i(b) = f\Big(a \sum_i \ell_i(b) e_i\Big) = f(ab).\end{aligned}$$

Exercise 6.6.4 (Solution)

By Lemma 3.2, the family

$$\upsilon = \{\upsilon_X = \theta^r_X (\theta^l_X)^{-1} \colon X \to X\}_{X \in \mathrm{Ob}(\mathcal{C})}$$

is a natural automorphism of the identity functor $1_{\mathcal{C}}$. By Exercise 3.3.3, $\upsilon_{\mathbb{1}} = \mathrm{id}_{\mathbb{1}}$ and $\upsilon_{X \otimes Y} = \upsilon_X \otimes \upsilon_Y$ for all $X, Y \in \mathrm{Ob}(\mathcal{C})$, implying that υ is monoidal. Then $(\upsilon_X)^* = \upsilon_{X^*}^{-1}$ for all $X \in \mathrm{Ob}(\mathcal{C})$. Let $\mathcal{C}_0$ be the full subcategory of $\mathcal{C}$ consisting of all $X \in \mathrm{Ob}(\mathcal{C})$ such that $\upsilon_X = \mathrm{id}_X$. Clearly, $\mathcal{C}_0$ is a braided pivotal subcategory of $\mathcal{C}$ and $\theta^l_X = \theta^r_X$ for all $X \in \mathrm{Ob}(\mathcal{C}_0)$. Thus, $\mathcal{C}_0$ is ribbon, and, in particular, spherical by Corollary 3.4. We need only to prove that $C \in \mathrm{Ob}(\mathcal{C}_0)$. For any $X \in \mathrm{Ob}(\mathcal{C})$,

$$\upsilon_C \rho_X \overset{\text{(i)}}{=} \rho_X \upsilon_{X^* \otimes X} \overset{\text{(ii)}}{=} \rho_X(\upsilon_{X^*} \otimes \upsilon_X) \overset{\text{(iii)}}{=} \rho_X((\upsilon_X)^* \upsilon_{X^*} \otimes \mathrm{id}_X) \overset{\text{(iv)}}{=} \rho_X.$$

Here, (i) follows from the naturality of υ, (ii) from the monoidality of υ, (iii) from the dinaturality of ρ, and (iv) from the relation $(\upsilon_X)^* = \upsilon_{X^*}^{-1}$. Since ρ is universal, we deduce that $\upsilon_C = \mathrm{id}_C$, that is, $C \in \mathrm{Ob}(\mathcal{C}_0)$.

Exercise 7.1.6 (Hint)

If f is a monad morphism, then the assignment $(M, r) \mapsto (M, rf_M)$ gives rise to a functor $f^* \colon \mathcal{C}^{T'} \to \mathcal{C}^T$ acting as the identity on morphisms and such that $U_T f^* = U_{T'}$. Conversely, let $F \colon \mathcal{C}^{T'} \to \mathcal{C}^T$ be a functor such that $U_T F = U_{T'}$. For $X \in \mathrm{Ob}(\mathcal{C})$, consider the action $r_X \colon TT'(X) \to T'(X)$ of the image under F of the T'-module $(T'(X), \mu'_X)$ and set $f_X = r_X T(\eta'_X) \colon T(X) \to T'(X)$. Then $f = \{f_X\}_{X \in \mathrm{Ob}(\mathcal{C})}$ is a monad morphism from T to T' such that $f^* = F$.

Exercise 7.3.4 (Hint)

Construct a functor $F \colon \mathcal{K} \to \mathcal{F}$ as follows. For any $X \in \mathrm{Ob}(\mathcal{K})$, set $F(X) = (T(X), \mu_X)$. For any $f \in \mathrm{Hom}_{\mathcal{K}}(X, Y) = \mathrm{Hom}_{\mathcal{C}}(X, T(Y))$, set $F(f) = \mu_Y T(f)$. Prove that F is essentially surjective and fully faithful. Conclude that F is an equivalence (assuming the axiom of choice, cf. Section 1.1.7).

Exercise 7.3.5 (Hint)

If A is separable, then there is a morphism $\varsigma\colon A \to A \otimes A$ in $\mathcal{C}$ such that

$$m\varsigma = \mathrm{id}_A \quad \text{and} \quad (m \otimes \mathrm{id}_A)(\mathrm{id}_A \otimes \varsigma) = \varsigma m = (\mathrm{id}_A \otimes m)(\varsigma \otimes \mathrm{id}_A). \tag{I.4}$$

Then the natural transformation

$$\gamma = \{\gamma_X = \varsigma u \otimes \mathrm{id}_X : X \to A \otimes A \otimes X\}_{X \in \mathrm{Ob}(\mathcal{C})},$$

where u is the unit of A, satisfies the conditions of Lemma 7.4. By this lemma, the monad $A\otimes ?$ is separable. Conversely, if the monad $A\otimes ?$ is separable and $\gamma = \{\gamma_X\}_{X \in \mathrm{Ob}(\mathcal{C})}$ is a natural transformation as in Lemma 7.4, then the morphism

$$\varsigma = (m \otimes \mathrm{id}_A)(\mathrm{id}_A \otimes \gamma_{\mathbb{1}})\colon A \to A \otimes A$$

satisfies (I.4), and the algebra A is separable.

Exercise 7.3.6 (Solution)

Let $F\colon \mathcal{C} \rightleftarrows \mathcal{D} : U$ be an adjunction with unit η and counit ε. Consider a monomorphism $f\colon X \to Y$ in $\mathcal{D}$ and two morphisms $g, h\colon Z \to U(X)$ in $\mathcal{C}$ such that $U(f)g = U(f)h$. Then

$$\varepsilon_Y F\big(U(f)g\big) = \varepsilon_Y F\big(U(f)h\big).$$

The naturality of ε implies that $f\varepsilon_X F(g) = f\varepsilon_X F(h)$. Since f is a monomorphism, we deduce that $\varepsilon_X F(g) = \varepsilon_X F(h)$. Then

$$U\big(\varepsilon_X F(g)\big)\eta_Z = U\big(\varepsilon_X F(h)\big)\eta_Z.$$

The naturality of η implies that $U(\varepsilon_X)\eta_{U(X)} g = U(\varepsilon_X)\eta_{U(X)} h$. Therefore $g = h$ by (7.1). Hence $U(f)$ is a monomorphism. A similar argument shows that the image of an epimorphism in $\mathcal{C}$ under F is an epimorphism in $\mathcal{D}$.

Exercise 7.4.3 (Solution)

We have:

$$\begin{aligned}
\mathrm{End}_{\mathcal{C}^T}(M, r) &\overset{\text{(i)}}{=} \{f \in \mathrm{End}_{\mathcal{C}}(M) \mid fr = rT(f)\} \\
&\overset{\text{(ii)}}{=} \{k\,\mathrm{id}_M \mid k \in \Bbbk \text{ and } k\,r = rT(k\,\mathrm{id}_M)\} \\
&\overset{\text{(iii)}}{=} \{k\,\mathrm{id}_M \mid k \in \Bbbk\} \overset{\text{(iv)}}{=} \Bbbk\,\mathrm{id}_{(M,r)}.
\end{aligned}$$

Here, (i) follows from the definition of a T-linear morphism, (ii) from the simplicity of M, (iii) from the $\Bbbk$-linearity of T and the $\Bbbk$-bilinearity of the composition in $\mathcal{C}$, and (iv) from the fact that $\mathrm{id}_{(M,r)} = \mathrm{id}_M$. Hence, (M, r) is a simple object of $\mathcal{C}^T$.

Exercise 8.3.7 (Solution)

Pick $X, Y \in \mathrm{Ob}(\mathcal{C})$. Let ε and $\mathbb{H}^l$ be respectively the counit and the left fusion operator of the comonoidal adjunction $F_T \dashv U_T$ associated with T. We endow the left rigid category $\mathcal{C}^T$ with left duality produced by Theorem 8.7(b). Set $e = F_T(X \otimes T(Y)) \in \mathrm{Ob}(\mathcal{C}^T)$ and consider the bijection

$$h_e \colon \mathrm{Hom}_{\mathcal{C}^T}(F_T(X) \otimes F_T(Y), e) \to \mathrm{End}_{\mathcal{C}^T}(e)$$

defined in the proof of Theorem 8.4. We have:

$$\begin{aligned}
(H^l_{X,Y})^{-1} &\overset{\text{(i)}}{=} (\mathbb{H}^l_{X,F_T(Y)})^{-1} \overset{\text{(ii)}}{=} h_e^{-1}(\mathrm{id}_e) \overset{\text{(iii)}}{=} (\mathrm{id}_e \otimes \mathrm{ev}_{F_T(Y)}) \\
&\quad \circ \big(\varepsilon_{e \otimes {}^\vee F_T(Y)} F_T((\eta_{X \otimes T(Y)} \otimes \mathrm{id}_{{}^\vee T(Y)})(\mathrm{id}_X \otimes \mathrm{coev}_{T(Y)})) \otimes \mathrm{id}_{T(Y)}\big) \\
&\overset{\text{(iv)}}{=} \big(\mu_{X \otimes T(Y)} \otimes \mathrm{ev}_{T(Y)}(s^l_{T(Y)} T({}^\vee \mu_Y) \otimes \mathrm{id}_{T(Y)})\big) \\
&\quad \circ \big(T_2(T(X \otimes T(Y)), {}^\vee T(Y)) T((\eta_{X \otimes T(Y)} \otimes \mathrm{id}_{{}^\vee T(Y)})(\mathrm{id}_X \otimes \mathrm{coev}_{T(Y)})) \otimes \mathrm{id}_{T(Y)}\big) \\
&\overset{\text{(v)}}{=} \big(\mathrm{id}_{T(X \otimes T(Y))} \otimes \mathrm{ev}_{T(Y)}(s^l_{T(Y)} T({}^\vee \mu_Y) \otimes \mathrm{id}_{T(Y)})\big) \\
&\quad \circ \big(T_2(X \otimes T(Y), {}^\vee T(Y)) T(\mathrm{id}_X \otimes \mathrm{coev}_{T(Y)}) \otimes \mathrm{id}_{T(Y)}\big).
\end{aligned}$$

Here, (i) follows from the equality (8.2) in the proof of Lemma 8.2 and the fact that the functor U_T is strict monoidal and acts as the identity on morphisms, (ii) from the proof of Theorem 8.4, (iii) from making explicit the bijections in the definition of h_e, (iv) from the definition of ε (see Section 7.2.2) and the computation of duality in $\mathcal{C}^T$ via s^l (see Section 8.3.1), and (v) from the naturality of T_2 and the unitality of μ.

Exercise 8.3.8 (Solution)

Consider an R-matrix R for a left Hopf monad T on a left rigid category $\mathcal{C}$. For any T-modules $(M, r), (N, s)$, set

$$\tau_{(M,r),(N,s)} = (s \otimes r) R_{M,N} \colon (M, r) \otimes (N, s) \to (N, s) \otimes (M, r).$$

It follows from the definition of an R-matrix that this morphism is T-linear and that the family $\tau = \{\tau_{(M,r),(N,s)}\}_{(M,r),(N,s) \in \mathrm{Ob}(\mathcal{C}^T)}$ is a weak braiding in $\mathcal{C}^T$ in the sense of Exercise 3.1.12. Since $\mathcal{C}^T$ is left rigid (by Theorem 8.7), τ is a braiding by Exercise 3.1.12. If η is the unit of T, then the family

$$\{\tau^{-1}_{F_T(Y),F_T(X)}(\eta_X \otimes \eta_Y) \colon X \otimes Y \to T(Y) \otimes T(X)\}_{X,Y \in \mathrm{Ob}(\mathcal{C})},$$

is the $*$-inverse of R (see Section 7.7.1).

Exercise 9.5.4 (Hint)

Let Z be the centralizer of $\mathcal{C}$ from Section 9.4.1. In view of Theorem 9.3, it suffices to prove that the free module functor $F_Z\colon \mathcal{C} \to \mathcal{C}^Z$ is a right adjoint to the forgetful functor $U_Z\colon \mathcal{C}^Z \to \mathcal{C}$. Consider the natural transformation

$$\beta = \{\beta_{(M,r)}\colon (M,r) \to F_Z U_Z(M,r) = F_Z(M)\}_{(M,r)\in \mathrm{Ob}(\mathcal{C}^Z)},$$

where

$$\beta_{(M,r)} = \sum_{i,j\in I} \dim_r(i)$$

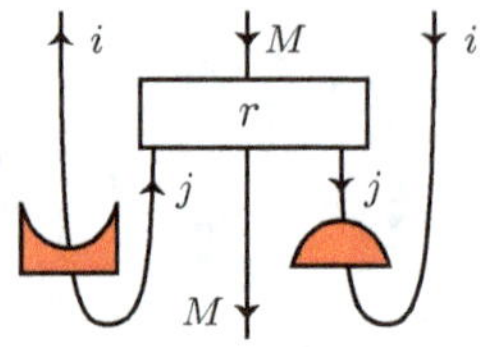

Consider also the natural transformation

$$\pi = \{\pi_X\colon U_Z F_Z(X) = Z(X) \to X\}_{X\in \mathrm{Ob}(\mathcal{C})},$$

where π_X is the projection of $Z(X) = \bigoplus_{i\in I} i^* \otimes X \otimes i$ onto $X = \mathbb{1}^* \otimes X \otimes \mathbb{1}$. Prove that β and π are respectively the unit and counit for an adjunction $U_Z \dashv F_Z$.

Exercise 16.3.4 (Solution)

Any conjugacy class $C \subset G$ gives rise to an object (A,σ) of $\mathcal{Z}(\mathcal{C}^{\mathrm{add}})$ as follows. Set

$$A = \bigoplus_{g\in C} g \in \mathrm{Ob}(\mathcal{C}^{\mathrm{add}}).$$

For $x \in \mathrm{Ob}(\mathcal{C}) = G$, set

$$\sigma_x = \sum_{g,h\in C} (\sigma_x)_g^h\colon A\otimes x = \bigoplus_{g\in C} gx \to x\otimes A = \bigoplus_{h\in C} xh,$$

where

$$(\sigma_x)_g^h = \delta_{gx,xh} \in \mathrm{Hom}_{\mathcal{C}^{\mathrm{add}}}(gx,xh) = \mathrm{Hom}_{\mathcal{C}}(gx,xh).$$

We extend the family $\{\sigma_x\}_{x\in\mathrm{Ob}(\mathcal{C})}$ by additivity (as in the proof of Lemma 16.3) to a half-braiding (A,σ) of $\mathcal{C}^{\mathrm{add}}$. Then (A,σ) is a simple object of $\mathcal{Z}(\mathcal{C}^{\mathrm{add}})$. If (A,σ) is isomorphic to the image under H of an object of $\mathcal{Z}(\mathcal{C})$, then the faithfulness of H implies that this object of $\mathcal{Z}(\mathcal{C})$ is simple. Its underlying object in $\mathcal{C}$ is then a central element of G. Consequently, when $\mathrm{card}(C) \geq 2$, the object (A,σ) is not isomorphic to the image under H of an object of $\mathcal{Z}(\mathcal{C})$. Thus, the functor H is not essentially surjective. The induced functor $\mathcal{Z}(\mathcal{C})^{\mathrm{add}} \to \mathcal{Z}(\mathcal{C}^{\mathrm{add}})$ also is not essentially surjective because all simple objects of $\mathcal{Z}(\mathcal{C})^{\mathrm{add}}$ are isomorphic to simple objects of $\mathcal{Z}(\mathcal{C})$.

Bibliography

[At] Atiyah, M., *Topological quantum field theories*, Inst. Hautes Études Sci. Publ. Math. No. 68 (1988), 175–186 (1989).

[Ba1] Balsam, B., *Turaev–Viro invariants as an extended TQFT II*, arXiv:1010.1222.

[Ba2] ———, *Turaev–Viro invariants as an extended TQFT III*, arXiv:1012.0560.

[BW1] Barrett, J., Westbury, B., *Invariants of piecewise-linear* 3-*manifolds*, Trans. Amer. Math. Soc. 348 (1996), 3997–4022.

[BW2] ———, *The equality of* 3-*manifold invariants*, Math. Proc. Cambridge Philos. Soc. 118 (1995), 503–510.

[Ben] Bénabou, J., *Catégories avec multiplication*, C.R. Acad. Sci. Paris 256 (1963), 1887–1890.

[BLV] Bruguières, A., Lack, S., Virelizier, A., *Hopf monads on monoidal categories*, Adv. in Math. 227 (2011), 745–800.

[BV1] Bruguières, A., Virelizier, A., *Hopf monads*, Advances in Math. 215 (2007), 679–733.

[BV2] ———, *Quantum double of Hopf monads and categorical centers*, Trans. Amer. Math. Soc. 364 (2012), 1225–1279.

[BV3] ———, *On the center of fusion categories*, Pacific J. Math. 264 (2013), 1–30.

[DW] Dijkgraaf, R., Witten, E., *Topological Gauge Theories and Group Cohomology*, Commun. Math. Phys. 129 (1990), 393–429.

[Dri] Drinfeld, V., *Quantum groups*, Proc. ICM 1986, Vol. 1, 2 798–820, AMS 1987.

[EG] Etingof, P., Gelaki, S., *Some properties of finite-dimensional semisimple Hopf algebras*, Math. Res. Lett. 5 (1998), 191–197.

[EGNO] Etingof, P., Gelaki, S., Nikshych, D., Ostrik, V., Tensor categories, Mathematical Surveys and Monographs, 205. American Mathematical Society, Providence, RI, 2015.

[ENO] Etingof, P., Nikshych, D., Ostrik, V., *On fusion categories*, Ann. of Math. (2) 162 (2005), 581–642.

[Fr] Freyd, P., Abelian categories, Harper and Row, New York, 1964. Reprints in Theory and Applications of Categories 3 (2003), 23–164.

[Go] Godement, R., Topologie Algébrique et Théorie des Faisceaux, Hermann, Paris, 1958.

[Ha] Hayashi, T., *A canonical Tannaka duality for finite semisimple tensor categories*, arXiv:math.QA/9904073.

[Hem] Hempel, J., 3-Manifolds. Ann. of Math. Studies, No. 86. Princeton Univ. Press, Princeton, N. J.; University of Tokyo Press, Tokyo, 1976.

[He] Hennings, M., *Invariants of links and 3-manifolds obtained from Hopf algebras*, J. of London Math. Soc. 54 (1996), 594–624.

[JS3] Joyal, A., Street, R., *The geometry of tensor calculus I*, Adv. in Math. 88 (1991), 55–112.

[JS4] ———, *The geometry of tensor calculus II*, unpublished manuscript (1994), available from Ross Street's website.

[Ka] Kassel, C., Quantum groups. Graduate Texts in Math., 155. Springer-Verlag, New York, 1995.

[KR] Kauffman, L., Radford, D., *Invariants of 3-manifolds derived from finite dimensional Hopf algebras*, J. of Knot Theory and its Ramifications 4 (1995), 131–162.

[KSW] Kawahigashi, Y., Sato, N., Wakui, M., $(2+1)$*-dimensional topological quantum field theory from subfactors and Dehn surgery formula for* 3*-manifold invariants*, Adv. Math. 195 (2005), 165–204.

[Ke] Kerler, T., *Genealogy of non-perturbative quantum-invariants of* 3*-manifolds: the surgical family.* Geometry and physics (Aarhus, 1995), 503–547, Lecture Notes in Pure and Appl. Math., 184, Dekker, New York, 1997.

[KL] Kerler, T., Lyubashenko, V., Non-semisimple topological quantum field theories for 3-manifolds with corners. Lecture Notes in Math., 1765. Springer-Verlag, Berlin, 2001.

[KB] Kirillov, A. Jr., Balsam, B., *Turaev-Viro invariants as an extended TQFT*, arXiv:1004.1533.

[Ko] Kock, J., Frobenius algebras and 2D topological quantum field theories. London Mathematical Society Student Texts 59. Cambridge University Press, Cambridge, 2004.

[Ku] Kuperberg, G., *Involutory Hopf algebras and 3-manifold invariants*, Internat. J. Math. 2 (1991), 41–66.

[LR] Larson, R.G., Radford, D. E., *Finite dimensional cosemisimple Hopf algebras in characteristic zero are semisimple*, J. Algebra 117 (1988), 267–289.

[La] Lawrence, R., *Triangulation, categories and extended field theories.* In Quantum Topology, R. Baadhio and L. Kaufman, Editors. World Scientific, Singapore, 1993, 191–208.

[Lu] Lurie, J., *On the classification of topological field theories*, Current developments in mathematics, 2008, Int. Press, Somerville, MA, 2009, 129–280.

[Lyu] Lyubashenko, V., *Invariants of 3-manifolds and projective representations of mapping class groups via quantum groups at roots of unity*, Comm. Math. Phys. 172 (1995), 467–516.

[ML1] Mac Lane, S., *Natural associativity and commutativity*, Rice University Studies 49 (1963), 28–46.

[ML2] ———, Categories for the working mathematician, second ed., Springer-Verlag, New York, 1998.

[Maj] Majid, S., Foundations of quantum group theory, Cambridge: Cambridge Univ. Press. xix, 607 p., 1995.

[Mat] Matveev, S. V. Algorithmic topology and classification of 3-manifolds, Second ed. Algorithms and Computation in Mathematics, 9. Springer, Berlin, 2007.

[Mo] Moerdijk, I. *Monads on tensor categories*, J. Pure Appl. Algebra **168** (2002), 189–208.

[Mü] Müger, M., *From subfactors to categories and topology. II. The quantum double of tensor categories and subfactors*, J. Pure Appl. Algebra 180 (2003), 159–219.

[Pe] Penrose, R., *Applications of negative dimensional tensors.* In: Combinatorial Mathematics and its Applications, 221-244. Academic Press, London 1971.

[RT] Reshetikhin, N., Turaev, V., *Invariants of 3-manifolds via link polynomials and quantum groups*, Invent. Math. 103 (1991), 547–597.

[Sc] Schwarz, A.S., *The partition function of degenerate quadratic functional and Ray–Singer invariants*, Lett. Math. Phys. 2 (1977/78), 247–252.

[Sh] Shum, M.C., *Tortile Tensor Categories*, PhD thesis, Macquarie University, 1989.

[Sp] Spanier, E.H., Algebraic topology, Springer-Verlag, New York-Berlin, 1981.

[Sz] Szlachányi, K., *Finite quantum groupoids and inclusions of finite type*, Mathematical Physics in Mathematics and Physics (Siena, 2000), 393–407, Fields Inst. Commun. 30, A. M. S., Providence, RI (2001).

[Tu1] Turaev, V., Quantum invariants of knots and 3-manifolds, de Gruyter Studies in Mathematics, 18. Walter de Gruyter, Berlin, 1994.

[TV] Turaev, V., Viro, O., *State sum invariants of* 3*-manifolds and quantum 6j-symbols*, Topology 31 (1992), 865–902.

[TVi] Turaev, V., Virelizier, A., *On two approaches to* 3*-dimensional TQFTs*, arXiv:1006.3501.

[Wa] Walker, K., *On Witten's* 3*-manifold invariants*, preprint (1991).

[Wak] Wakui, M., *On Dijkgraaf–Witten invariant for* 3*-manifolds*, Osaka J. Math. 29 (1992), 675–696.

[Wi] Witten, E., *Quantum field theory and the Jones polynomial*, Comm. Math. Phys. 121 (1989), 351–399.

Index

www.ingramcontent.com/pod-product-compliance
Lightning Source LLC
Chambersburg PA
CBHW071606290626
47387CB00006B/1041
* 9 7 8 3 3 1 9 4 9 8 3 3 1 *